IET ENERGY ENGINEERING SERIES 204

Electromagnetic Transients in Large HV Cable Networks

Other volumes in this series:

Volume 1 **Power Circuit Breaker Theory and Design** C.H. Flurscheim (Editor)
Volume 4 **Industrial Microwave Heating** A.C. Metaxas and R.J. Meredith
Volume 7 **Insulators for High Voltages** J.S.T. Looms
Volume 8 **Variable Frequency AC Motor Drive Systems** D. Finney
Volume 10 **SF_6 Switchgear** H.M. Ryan and G.R. Jones
Volume 11 **Conduction and Induction Heating** E.J. Davies
Volume 13 **Statistical Techniques for High Voltage Engineering** W. Hauschild and W. Mosch
Volume 14 **Uninterruptible Power Supplies** J. Platts and J.D. St Aubyn (Editors)
Volume 15 **Digital Protection for Power Systems** A.T. Johns and S.K. Salman
Volume 16 **Electricity Economics and Planning** T.W. Berrie
Volume 18 **Vacuum Switchgear** A. Greenwood
Volume 19 **Electrical Safety: A guide to causes and prevention of hazards** J. Maxwell Adams
Volume 21 **Electricity Distribution Network Design, 2nd Edition** E. Lakervi and E.J. Holmes
Volume 22 **Artificial Intelligence Techniques in Power Systems** K. Warwick, A.O. Ekwue and R. Aggarwal (Editors)
Volume 24 **Power System Commissioning and Maintenance Practice** K. Harker
Volume 25 **Engineers' Handbook of Industrial Microwave Heating** R.J. Meredith
Volume 26 **Small Electric Motors** H. Moczala *et al.*
Volume 27 **AC–DC Power System Analysis** J. Arrillaga and B.C. Smith
Volume 29 **High Voltage Direct Current Transmission, 2nd Edition** J. Arrillaga
Volume 30 **Flexible AC Transmission Systems (FACTS)** Y-H. Song (Editor)
Volume 31 **Embedded generation** N. Jenkins *et al.*
Volume 32 **High Voltage Engineering and Testing, 2nd Edition** H.M. Ryan (Editor)
Volume 33 **Overvoltage Protection of Low-Voltage Systems, Revised Edition** P. Hasse
Volume 36 **Voltage Quality in Electrical Power Systems** J. Schlabbach *et al.*
Volume 37 **Electrical Steels for Rotating Machines** P. Beckley
Volume 38 **The Electric Car: Development and future of battery, hybrid and fuel-cell cars** M. Westbrook
Volume 39 **Power Systems Electromagnetic Transients Simulation** J. Arrillaga and N. Watson
Volume 40 **Advances in High Voltage Engineering** M. Haddad and D. Warne
Volume 41 **Electrical Operation of Electrostatic Precipitators** K. Parker
Volume 43 **Thermal Power Plant Simulation and Control** D. Flynn
Volume 44 **Economic Evaluation of Projects in the Electricity Supply Industry** H. Khatib
Volume 45 **Propulsion Systems for Hybrid Vehicles** J. Miller
Volume 46 **Distribution Switchgear** S. Stewart
Volume 47 **Protection of Electricity Distribution Networks, 2nd Edition** J. Gers and E. Holmes
Volume 48 **Wood Pole Overhead Lines** B. Wareing
Volume 49 **Electric Fuses, 3rd Edition** A. Wright and G. Newbery
Volume 50 **Wind Power Integration: Connection and system operational aspects** B. Fox *et al.*
Volume 51 **Short Circuit Currents** J. Schlabbach
Volume 52 **Nuclear Power** J. Wood
Volume 53 **Condition Assessment of High Voltage Insulation in Power System Equipment** R.E. James and Q. Su
Volume 55 **Local Energy: Distributed generation of heat and power** J. Wood
Volume 56 **Condition Monitoring of Rotating Electrical Machines** P. Tavner, L. Ran, J. Penman and H. Sedding
Volume 57 **The Control Techniques Drives and Controls Handbook, 2nd Edition** B. Drury
Volume 58 **Lightning Protection** V. Cooray (Editor)
Volume 59 **Ultracapacitor Applications** J.M. Miller
Volume 62 **Lightning Electromagnetics** V. Cooray
Volume 63 **Energy Storage for Power Systems, 2nd Edition** A. Ter-Gazarian
Volume 65 **Protection of Electricity Distribution Networks, 3rd Edition** J. Gers
Volume 66 **High Voltage Engineering Testing, 3rd Edition** H. Ryan (Editor)
Volume 67 **Multicore Simulation of Power System Transients** F.M. Uriate
Volume 68 **Distribution System Analysis and Automation** J. Gers
Volume 69 **The Lightening Flash, 2nd Edition** V. Cooray (Editor)
Volume 70 **Economic Evaluation of Projects in the Electricity Supply Industry, 3rd Edition** H. Khatib

Volume 72	**Control Circuits in Power Electronics: Practical issues in design and implementation** M. Castilla (Editor)
Volume 73	**Wide Area Monitoring, Protection and Control Systems: The enabler for Smarter Grids** A. Vaccaro and A. Zobaa (Editors)
Volume 74	**Power Electronic Converters and Systems: Frontiers and applications** A.M. Trzynadlowski (Editor)
Volume 75	**Power Distribution Automation** B. Das (Editor)
Volume 76	**Power System Stability: Modelling, analysis and control** A.A. Sallam and B. Om P. Malik
Volume 78	**Numerical Analysis of Power System Transients and Dynamics** A. Ametani (Editor)
Volume 79	**Vehicle-to-Grid: Linking electric vehicles to the smart grid** J. Lu and J. Hossain (Editors)
Volume 81	**Cyber-Physical-Social Systems and Constructs in Electric Power Engineering** S. Suryanarayanan, R. Roche and T.M. Hansen (Editors)
Volume 82	**Periodic Control of Power Electronic Converters** F. Blaabjerg, K.Zhou, D. Wang and Y. Yang
Volume 86	**Advances in Power System Modelling, Control and Stability Analysis** F. Milano (Editor)
Volume 87	**Cogeneration: Technologies, Optimisation and Implementation** C.A. Frangopoulos (Editor)
Volume 88	**Smarter Energy: From smart metering to the smart grid** H. Sun, N. Hatziargyriou, H.V. Poor, L. Carpanini and M.A. Sánchez Fornié (Editors)
Volume 89	**Hydrogen Production, Separation and Purification for Energy** A. Basile, F. Dalena, J. Tong and T.N.Veziroğlu (Editors)
Volume 90	**Clean Energy Microgrids** S. Obara and J. Morel (Editors)
Volume 91	**Fuzzy Logic Control in Energy Systems with Design Applications in MATLAB®/Simulink®** İ.H. Altaş
Volume 92	**Power Quality in Future Electrical Power Systems** A.F. Zobaa and S.H.E.A. Aleem (Editors)
Volume 93	**Cogeneration and District Energy Systems: Modelling, analysis and optimization** M.A. Rosen and S. Koohi-Fayegh
Volume 94	**Introduction to the Smart Grid: Concepts, technologies and evolution** S.K. Salman
Volume 95	**Communication, Control and Security Challenges for the Smart Grid** S.M. Muyeen and S. Rahman (Editors)
Volume 96	**Industrial Power Systems with Distributed and Embedded Generation** R Belu
Volume 97	**Synchronized Phasor Measurements for Smart Grids** M.J.B. Reddy and D.K. Mohanta (Editors)
Volume 98	**Large Scale Grid Integration of Renewable Energy Sources** A. Moreno-Munoz (Editor)
Volume 100	**Modeling and Dynamic Behaviour of Hydropower Plants** N. Kishor and J. Fraile-Ardanuy (Editors)
Volume 101	**Methane and Hydrogen for Energy Storage** R. Carriveau and D.S-K. Ting
Volume 104	**Power Transformer Condition Monitoring and Diagnosis** A. Abu-Siada (Editor)
Volume 106	**Surface Passivation of Industrial Crystalline Silicon Solar Cells** J. John (Editor)
Volume 107	**Bifacial Photovoltaics: Technology, applications and economics** J. Libal and R. Kopecek (Editors)
Volume 108	**Fault Diagnosis of Induction Motors** J. Faiz, V. Ghorbanian and G. Joksimović
Volume 110	**High Voltage Power Network Construction** K. Harker
Volume 111	**Energy Storage at Different Voltage Levels: Technology, integration, and market aspects** A.F. Zobaa, P.F. Ribeiro, S.H.A. Aleem and S.N. Afifi (Editors)
Volume 112	**Wireless Power Transfer: Theory, technology and application** N. Shinohara
Volume 114	**Lightning-Induced Effects in Electrical and Telecommunication Systems** Y. Baba and V.A. Rakov
Volume 115	**DC Distribution Systems and Microgrids** T. Dragičević, F.Blaabjerg and P. Wheeler
Volume 116	**Modelling and Simulation of HVDC Transmission** M. Han (Editor)
Volume 117	**Structural Control and Fault Detection of Wind Turbine Systems** H.R. Karimi
Volume 119	**Thermal Power Plant Control and Instrumentation: The control of boilers and HRSGs, 2nd Edition** D. Lindsley, J. Grist and D. Parker
Volume 120	**Fault Diagnosis for Robust Inverter Power Drives** A. Ginart (Editor)
Volume 121	**Monitoring and Control Using Synchrophasors in Power Systems with Renewables** I. Kamwa and C. Lu (Editors)

Volume 123 **Power Systems Electromagnetic Transients Simulation, 2nd Edition** N. Watson and J. Arrillaga
Volume 124 **Power Market Transformation** B. Murray
Volume 125 **Wind Energy Modeling and Simulation, Volume 1: Atmosphere and plant** P. Veers (Editor)
Volume 126 **Diagnosis and Fault Tolerance of Electrical Machines, Power Electronics and Drives** A.J.M. Cardoso
Volume 128 **Characterization of Wide Bandgap Power Semiconductor Devices** F. Wang, Z. Zhang and E.A. Jones
Volume 129 **Renewable Energy from the Oceans: From wave, tidal and gradient systems to offshore wind and solar** D. Coiro and T. Sant (Editors)
Volume 130 **Wind and Solar Based Energy Systems for Communities** R. Carriveau and D.S-K. Ting (Editors)
Volume 131 **Metaheuristic Optimization in Power Engineering** J. Radosavljević
Volume 132 **Power Line Communication Systems for Smart Grids** I.R.S. Casella and A. Anpalagan
Volume 134 **Hydrogen Passivation and Laser Doping for Silicon Solar Cells** Brett Hallam and Catherine Chan (Editors)
Volume 139 **Variability, Scalability and Stability of Microgrids** S.M. Muyeen, S.M. Islam and F. Blaabjerg (Editors)
Volume 145 **Condition Monitoring of Rotating Electrical Machines** P. Tavner, L. Ran, C. Crabtree
Volume 146 **Energy Storage for Power Systems, 3rd Edition** A.G. Ter-Gazarian
Volume 147 **Distribution Systems Analysis and Automation 2nd Edition** J. Gers
Volume 151 **SiC Power Module Design: Performance, robustness and reliability** Alberto Castellazzi and Andrea Irace (Editors)
Volume 152 **Power Electronic Devices: Applications, failure mechanisms and reliability** F Iannuzzo (Editor)
Volume 153 **Signal Processing for Fault Detection and Diagnosis in Electric Machines and Systems** Mohamed Benbouzid (Editor)
Volume 155 **Energy Generation and Efficiency Technologies for Green Residential Buildings** D. Ting and R. Carriveau (Editors)
Volume 156 **Lithium-ion Batteries Enabled by Silicon Anodes** Chunmei Ban and Kang Xu (Editors)
Volume 157 **Electrical Steels, 2 Volumes** A. Moses, K. Jenkins, Philip Anderson and H. Stanbury
Volume 158 **Advanced Dielectric Materials for Electrostatic Capacitors** Q Li (Editor)
Volume 159 **Transforming the Grid Towards Fully Renewable Energy** O. Probst, S. Castellanos and R. Palacios (Editors)
Volume 160 **Microgrids for Rural Areas: Research and case studies** R.K. Chauhan, K. Chauhan and S.N. Singh (Editors)
Volume 166 **Advanced Characterization of Thin Film Solar Cells** N. Haegel and M. Al-Jassim (Editors)
Volume 167 **Power Grids with Renewable Energy: Storage, integration and digitalization** A.A. Sallam and B. Om P. Malik
Volume 169 **Small Wind and Hydrokinetic Turbines** Philip Clausen, Jonathan Whale and David Wood (Editors)
Volume 170 **Reliability of Power Electronics Converters for Solar Photovoltaic Applications** Frede Blaabjerg, Ahteshamul Haque, Huai Wang, Zainul Abdin Jaffery and Yongheng Yang (Editors)
Volume 171 **Utility-scale Wind Turbines and Wind Farms** A. Vasel-Be-Hagh and D.S-K. Ting
Volume 172 **Lighting Interaction with Power Systems, 2 Volumes** A. Piantini (Editor)
Volume 174 **Silicon Solar Cell Metallization and Module Technology** T. Dullweber (Editor)
Volume 193 **Overhead Electric Power Lines: Theory and practice** S. Chattopadhyay and A. Das
Volume 194 **Offshore Wind Power: Reliability, availability and maintenance, 2nd Edition** P. Tavner
Volume 198 **Battery Management Systems and Inductive Balancing** A. Van den Bossche and A. Farzan Moghaddam
Volume 199 **Model Predictive Control for Microgrids: From power electronic converters to energy management** J. Hu, J.M. Guerrero and S. Islam
Volume 905 **Power System Protection, 4 Volumes**

Electromagnetic Transients in Large HV Cable Networks

Modeling and calculations

Akihiro Ametani, Haoyan Xue, Teruo Ohno
and Hossein Khalilnezhad

The Institution of Engineering and Technology

Published by The Institution of Engineering and Technology, London, United Kingdom

The Institution of Engineering and Technology is registered as a Charity in England & Wales (no. 211014) and Scotland (no. SC038698).

First published 2021

The Institution of Engineering and Technology
Michael Faraday House
Six Hills Way, Stevenage
Herts, SG1 2AY, United Kingdom

www.theiet.org

British Library Cataloguing in Publication Data
A catalogue record for this product is available from the British Library

ISBN 978-1-83953-431-7 (hardback)
ISBN 978-1-83953-432-4 (PDF)

Typeset in India by MPS Limited
Printed in the UK by CPI Group (UK) Ltd, Croydon

Cover image:

– 500 kV Shin-Toyosu Line
– Route length: 39.8 km
– 2500 mm^2 XLPE cable

Reprinted with permission

Contents

About the authors

Akihiro Ametani is an Adjunct Professor at the University of Manitoba, Canada. He is a CIGRE Distinguished member, and an IEEE Life Fellow. Previously, he was with Bonneville Power Administration, and a Vice-President of the IEE Japan. His research topics include finite difference time-domain analysis, earthing and lightning protection, poles and towers, substations, and transmission lines.

Haoyan Xue is currently with the African Office of Global Energy Interconnection Development and Cooperation Organization, Global Energy Interconnection Group Co. Ltd., and he is also a Research Associate with the Polytechnique Montréal. He received his B.Eng. in Electrical Engineering and its Automation from Hunan University, China in 2010; his M.Sc. in Electrical Engineering from Delft University of Technology, The Netherlands in 2012; and his Ph.D. in Electrical Engineering from Polytechnique Montréal, Canada, in 2018. From 2013 to 2015, he was employed by the State Grid Corporation of China (SGCC) in Yinchuan, Co., working on protections, relays, and SCADA systems of substations.

Teruo Ohno has been with the Tokyo Electric Power Company, Inc, since 1996, where he has been involved in various power system studies including the transient analysis of cable systems in different countries. He received his B.Sc. degree from the University of Tokyo, Tokyo, his M.Sc. degree from the Massachusetts Institute of Technology, Cambridge, and his Ph.D. degree from the Aalborg University, Aalborg, all in Electrical Engineering, in 1996, 2005, and 2012, respectively. He served as a secretary of Cigré WG C4.502, which focused on technical performance issues related to the application of long HVAC cables.

Hossein Khalilnezhad is currently working as Senior Grid Strategist at the Asset Management department of the Dutch-German transmission system operator, TenneT TSO BV. He received his B.Sc. degree in 2010 from the Shiraz University of Technology, Iran, and his M.Sc. degree in 2013 from the Delft University of Technology, the Netherlands, both in Electrical Engineering. From 2013 to 2018, he was working toward his Ph.D. degree in Electrical Power Engineering at the Delft University of Technology, the Netherlands, where he studied the technical performance of (E)HV underground cable systems within the scope of the Dutch 380 kV cable research program. He worked from 2018 to 2021 for the Stedin Group as one of the largest distribution system operators in the Netherlands, where he served as Grid Analyst and Strategist at the Asset Management Department.

Preface

This book shows accurate formulas of the earth-return impedance and admittance of a cable derived based on complete field solutions of Maxwell's equations. Carsons's and Pollaczek's formulas implemented into existing electromagnetic transient (EMT) type simulation tools were derived by assuming a TEM or a quasi-TEM mode of wave propagation, uniform current distribution, and no displacement current. Thus, the formulas are not applicable to high-frequency phenomena as is well-known. The high-frequency phenomena are becoming very important because of electromagnetic compatibility (EMC)-related disturbances in substations and control centers. External electromagnetic fields generated by overhead lines and underground cables becomes also significant from the viewpoints not only of the control circuits but also of human health which is one of the main concerns of replacing existing overhead lines by underground cables in European countries especially in Denmark. Considering the above situation, theoretical formulas of the external electromagnetic fields generated from the overhead lines and underground cables are explained. The impedance and admittance at a very low frequency for HVDC lines are investigated too considering the fact that the conductance of the HVDC lines is estimated with no physical background nor theoretical basis. Modeling methods of the cable prepared in the EMT-type simulation tools are described, and a frequency-dependent soil model is explained.

Wave propagation characteristics of a cable calculated by Carson's and Pollaczek's formulas (classical transmission line (TL) approach) are compared with those by the accurate formulas (extended TL approach) and by numerical electromagnetic code (NEC). A difference is made clear.

Transient and steady-state characteristics of various cable systems are explained, and it is shown that the classical TL approach in the EMT simulation results in numerical instability when simulating a very fast transient (VFT) in a gas-insulated substation (GIS, overhead cable). The instability does not appear when the extended TL approach is adopted. EMT simulation of VFTs is compared with computed results by finite-difference time-domain (FDTD) method.

Finally, transients in large interconnected cable networks in Denmark and in the Netherlands are investigated. The influence of cables on the transients in a network composed of overhead lines and underground cables is made clear. Resonant over-voltages in large networks are explained in detail. Transients due to islanding and ground faults are investigated. Also, the zero-missing phenomenon is investigated.

Aims and reasons for writing this book

1. To show the rigorous formulation of accurate earth-return impedance and admittance for overhead cables (lines) and underground cables, and define the extended and classical TL approaches based on the new formulas.
2. To make clear the applicable limit of the classical TL approach with Carson's and Pollaczek's impedance and space capacitance implemented into existing EMT-type simulation tools by showing various application examples, that is, the extended TL approach to VFTs in GISs and the classical TL approach to large interconnected cable networks.
3. To show the theoretical formulations and calculation examples of the external electromagnetic field components generated by transmission lines and underground cables.
4. To supply the frequency-dependent (FD) soil models as an important consideration in the modeling of cable electromagnetic characteristics.

Chapter 1
Introduction

In this chapter, the summary of each chapter (Chapters 2–7) are given.

1.1 Chapter 2: Series impedance and shunt admittance

The most basic and essential parameters to analyze steady-state, dynamic and transient phenomena in cable networks are the series impedance and shunt admittance of the cable. Since 1928, Carson's and Pollaczek's earth-return impedance formulas with the space admittance (classical transmission line (TL) approach) have been adopted all over the world, and have been implemented into electromagnetic transients program (originally developed in Bonneville Power Administration and called as "EMTP") since the 1970s. The formulas were derived under the assumption of the TEM mode of wave propagation and without consideration of displacement current. Thus, the formulas are not applicable to high-frequency phenomena as is well-known.

In Chapter 2, accurate formulas of the earth-return impedance and admittance (extended TL approach) derived from complete field solutions of Maxwell's equations are explained. Calculated results of the impedance by the classical TL approach are compared with those by the extended TL approach. Also, the impedance and admittance at a very low frequency (in fact, at $f = 0$) for HVDC systems are investigated and a method of calculating the DC conductance is explained.

In Section 2.1, the formulation of the series impedance and shunt admittance of cables, either overhead or underground, are described. In general, the impedance is composed of the cable internal impedance and external (earth-return) impedance. The shunt admittance is formulated with a use of potential coefficients, that is the cable internal and external (earth-return) potential coefficients.

In Section 2.2, the existing formulas of the earth-return impedance, such as Carson's and Pollaczek's formulas, and the earth-return admittance are reviewed and explained.

Sections 2.3 and 2.4 show the accurate and approximate formulas of the earth-return impedance and admittance for overhead cables (lines), respectively.

In Section 2.5, accurate and approximate formulas of the earth-return impedance for underground cables are described and reviewed. Those of the earth-return admittance for underground cables are also explained in Section 2.6. The earth-return admittance has been neglected in most studies of wave propagations and transient characteristics on underground cables.

In Section 2.7, a modified earth-return Green function of overhead cables (lines) for MoM-SO technique is proposed and discussed. The modified earth-return Green function can be used to avoid some possible numerical instabilities at high-frequency regions.

Section 2.8 shows a comparison of series impedance and shunt admittance calculated by extended and classical TL approaches for overhead and underground cables. A single overhead conductor and three typical single-core underground cables are considered in the calculations and investigations.

Section 2.9 has proposed a distribution line concept to evaluate the series impedance and shunt admittance of the HVDC line and investigated the basic electrical parameters of the HVDC line.

In Section 2.10, theoretical formulations of external electromagnetic field components generated from a multiphase overhead line and underground cable are proposed and introduced. Also, the characteristics of terminating conditions and various excitations are included and explained in the formulations.

1.2 Chapter 3: Modeling of cables

In Chapter 3, the TL models prepared in existing electromagnetic transient (EMT) type simulation tools based on the classical TL approach are explained in Section 3.1. Basically, there are two types of line models: one is a lumped-parameter equivalent composed of R, L, and C elements which is widely used in a steady-state analysis and the other is a distributed-parameter line model which can represent the traveling wave nature of the TL. It should be noted that those models are within the TEM mode of wave propagation in a frequency range up to some MHz. If the extended TL approach with the accurate earth-return impedance and admittance formulas are implemented into the EMT-type simulation tools, the frequency-dependent (FD) distributed-line model can deal with very-fast transients (VFTs) in gas-insulated substations (GISs) which involve non-TEM modes of wave propagation. The accurate impedance and admittance formulas have been implemented into PSCAD/EMTDC and EMTP to study the VFTs.

In Section 3.2, various FD soil (earth) models are explained, and calculated results of the models are compared with several independent measured data. If test results of the FD earth resistivity (and permittivity) are available and considered, more accurate studies on wave propagations and transient characteristics can be analyzed.

Section 3.3 explains the modeling methods for a cable installed underneath a steel bridge, a tower-installed cable, a tunnel-installed cable, a cable installed in a high-density polyethylene (HDPE) tube, and a submarine cable. The transient characteristics and performances of different cable systems are investigated and made clear.

Section 3.4 introduces typical bonding methods for underground cables. Various bonding methods are described and differences between the bonding methods are also made clear.

Numerical electromagnetic analysis (NEA) methods such as finite element method (FEM), method of moments (MoM), and finite-difference time-domain (FDTD) are explained and reviewed in Section 3.5. The NEA methods can be used to solve electrical parameters of cables when the parameters are hardly evaluated by the existing EMT-type simulation tools.

1.3 Chapter 4: Wave propagation characteristics of overhead and underground cables

Wave propagation characteristics of overhead cables (lines) and underground cables are investigated by adopting the extended and classical TL approaches, and differences between the propagation characteristics calculated by the two approaches are made clear in this chapter.

In Section 4.1, the basic theory for the evaluation of modal propagation constants on overhead and underground cables is introduced.

Section 4.2 explains the modal attenuation constants and external electromagnetic field components of overhead line and cable calculated by the extended and classical TL approaches.

Section 4.3 investigates the modal propagation constants of underground cables calculated by the extended and classical TL approaches. The effect of the earth-return admittance on the propagation constants is discussed. Furthermore, external electromagnetic fields produced by a three-phase underground cable are discussed. The electromagnetic fields are also calculated by NEC, and the NEC results are compared with those by the TL approaches. The influence of FD soil models on electromagnetic fields is also studied.

In Section 4.4, the input impedance of a cross-bonded cable is investigated in comparison with that of a homogeneous cable. The resonant frequency and attenuation characteristics in the frequency domain are investigated to explain the calculation of the propagation constant and modeling of a cross-bonded cable. Also, the time responses are explained to show the effect of cross-bonding and transposition.

1.4 Chapter 5: Steady-state and transient characteristics on three-phase cables

In Chapter 5, the steady-state and transient characteristics of three-phase cables are explained and discussed.

In Section 5.1, the cable discharging phenomena are investigated based on laboratory test results of cable insulators, and field test results on a 275 kV oil-filled (OF) cable in Japan and a 275 kV pressurized OF (POF) cable in the UK. The test results are theoretically analyzed, and EMT simulations with a simple RC circuit model are performed.

In Section 5.2, the EMT simulations on an RTE 225 kV cable system are thoroughly investigated and studied. Moreover, the simulations are well-validated and compared with the field test results of the cable system.

In Section 5.3, the basic characteristics of transient homogeneous and cross-bonded cables are investigated by adopting the extended and classical TL approaches. Because coaxial mode transients are well-known and have been investigated in many publications, transients due to earth-return and inter-sheath (phase) mode energizations are investigated.

In Section 5.4, EMT simulations of VFTs in GISs are performed, and it is shown that the EMT simulations with the classical TL approach result in numerical instability. When the extended TL approach is adopted, no instability appears, because the extended TL approach can deal with non-TEM modes in a very high-frequency range. The effects of various GIS parameters are investigated based on the simulation results by the extended TL approach.

In Section 5.5, EMT simulation results of various VFTs are compared with test results and FDTD computed results. It is made clear that the constant parameter (CP) line model shows no numerical instability. This is the reason why the CP line model has been widely used in the VFTs studies.

1.5 Chapter 6: Transients in the interconnected EHV cable network in Denmark

As in other countries, the visual and environmental impacts of transmission towers are an important issue in Denmark. In 2008, the Danish Parliament decided that all the existing 132 kV and 150 kV overhead lines are to be replaced by underground cables and new 400 kV lines are to be undergrounded. Based on the decision, the Danish TSO, Energinet.dk, started many projects related to the cable systems for planning and designing extra-high voltage (EHV) cable networks to be constructed.

Chapter 6 deals with transients in existing and planned EHV networks composed of 400 kV cables and overhead lines in Denmark.

Section 6.1 explains the background of the interconnected long EHV cable networks.

Section 6.2 describes the model setup of the EHV networks including the detail of the underground cables, such as physical and electrical parameters and cable layout and route. Similarly, the conductor, tower, and phase configuration of the overhead lines are explained. Modeling methods of various apparatuses, such as a transformer and a shunt reactor, are also explained. The effects of cable models, cross-bonding, span length, and cable armor on EMT simulation results are investigated. Also, simulation results by PSCAD and ATP are compared.

Temporary over-voltages, such as series and parallel resonant over-voltages, in the cable network are investigated in Section 6.3. The dominant frequency in energization over-voltages and the natural frequency of the resonant circuits are explained. Over-voltages caused by system islanding are also discussed.

Ground fault and fault clearing over-voltages are investigated in Section 6.4.

1.6 Chapter 7: Steady-state and transient behavior of hybrid overhead line-underground cable networks in the Netherlands

This chapter deals with the overhead line (OHL) and underground cable hybrid systems in the Netherlands.

Section 7.1 describes the background of the hybrid transmission grids in the Netherlands and the Dutch 380 kV cable research program.

In Section 7.2, the models of the Dutch transmission grid for steady-state and transient studies are elaborated.

Section 7.3 describes the shunt compensation allocation in the hybrid OHL-cable systems. The capacitive behavior of a cable, the negative effects of cable reactive power, and reactive power compensation are described. A study on the Dutch transmission system is described in detail. Finally, cable overloading is explained.

In Section 7.4, resonance behavior of the hybrid systems is explained, where the impact of hybrid OHL-cable systems on the resonance behavior of the Dutch transmission system is investigated by performing the frequency-scan analysis.

Energization overvoltages in hybrid OHL-cable systems are explained in Section 7.5, where they are analyzed based a comprehensive statistical analysis and the EMT simulations.

In Section 7.6, de-energization transients of hybrid OHL-cable systems are investigated. The cable de-energization phenomena are elaborated and the associated voltages, currents, and transient recovery voltages due to de-energization of a hybrid circuit are assessed.

In Section 7.7, the zero-missing phenomenon is investigated in detail. Simulation results involve various countermeasures to avoid or mitigate the zero-missing current, where the requirements, benefits, and deficiencies of each method are discussed.

Chapter 2

Series impedance and shunt admittance

2.1 Formulation of series impedance and shunt admittance

In general, the series impedance matrix **Z** of a cable, either overhead or underground, is given in the following form [1–7]:

$$\mathbf{Z} = \mathbf{Z_e} + \mathbf{Z_i}(\Omega/\text{m}) \tag{2.1}$$

where $\mathbf{Z_e}$ is the earth-return impedance, $\mathbf{Z_i}$ is the cable internal impedance, **Z** is an $n \times n$ series impedance matrix, and n is the number of conductors of a given cable.

For example, $n = 3$ for a three-phase overhead line with no ground wire, and $n = 5$ with two ground wires. In a three-phase ($N = 3$) single-core coaxial cable (SC cable) composed of a core and a metallic sheath ($m = 2$), $n = N \times m = 3 \times 2 = 6$.

The shunt admittance matrix **Y** of a cable is defined in the following form [1–7]:

$$\mathbf{Y} = j\omega\mathbf{C}\ (\text{S/m}) \text{ and } \mathbf{C} = \mathbf{P}^{-1}(\text{F/m}) \tag{2.2}$$

where **C** is the shunt capacitance matrix, **P** is the potential coefficient matrix, and the dimensions of the matrices are the same as the impedance matrix.

The potential coefficient matrix is given in the following forms:

- Overhead line:

$$\mathbf{P} = \mathbf{P_e} \tag{2.3}$$

- Overhead cable:

$$\mathbf{P} = \mathbf{P_e} + \mathbf{P_i} \tag{2.4}$$

- Underground cable:

$$\mathbf{P} = \mathbf{P_e} + \mathbf{P_i} \tag{2.5}$$

where $\mathbf{P_e}$ is the earth-return potential coefficient matrix and $\mathbf{P_i}$ is the cable internal potential coefficient matrix.

The earth-return admittance matrix $\mathbf{Y_e}$ is given in the following form:

$$\mathbf{Y_e} = j\omega\mathbf{C_e} \text{ and } \mathbf{C_e} = \mathbf{P_e}^{-1} \tag{2.6}$$

It should be noted that the earth return admittance $\mathbf{Y_e}$ has not been considered in a conventional transmission line (TL) approach in electromagnetic transients

(EMT/EMTs) program. Instead, a space admittance $\mathbf{Y_0}$ has been adopted into the EMT-type simulation tools since the original software produced in 1970s:

$$\mathbf{Y_0} = j\omega\mathbf{C_0} \text{ and } \mathbf{C_0} = \mathbf{P_0^{-1}} \tag{2.7}$$

where $\mathbf{P_0}$ is the space potential coefficient matrix.

The earth-return impedance and admittance formulas are explained in Sections 2.3–2.6. The internal impedance and potential coefficient (admittance) formulas of a cable are described in Appendix A1.

2.2 Review of existing formulas of earth-return impedance and admittance

2.2.1 Overhead cable (line)

The major formulations of earth-return impedance and admittance for overhead lines and overhead cables are proposed in [8–40]. The following discussions of formulas are based on quasi transverse electromagnetic (TEM) assumption and complete field solution.

2.2.1.1 Formulas based on quasi-TEM assumption

The earth-return impedance of an overhead line is developed independently by Carson in the United States of America [8] and by Pollaczek in Germany [9] in 1926. The electromagnetic characteristics of air and earth permittivity are neglected in their formulas. It is the basic reason that Carson's and Pollaczek's formulas cannot deal with displacement currents of air and earth. Another assumption is the uniform current distribution [41], which means that the unknown propagation constant along a conductor is assumed to be 0 ($\gamma_x = 0$). Furthermore, their formulas are based on the quasi-TEM assumption. Note that "quasi" means the axial component of electric fields is very small compared with the transverse components. The reason is that a true TEM mode is not possible with a lossy earth, because the Ohm's law requires an axial component of the electric field if a current is returning through the earth [42]. Although their approaches involved some inaccurate initial assumptions, the formulas are still extensively used in various software tools. Sunde also derives the earth-return impedance using a different method [10], that is, Hertzian vector and horizontal electric dipole technique. The combined techniques of Hertzian vector and electric dipole are initially investigated by Sommerfeld [43], and further discussed by Wait [44] and Stoyer [45]. The earth permittivity is also included in Sunde's earth-return impedance formula, however, the uniform current distribution is still kept in his formula. Because of some inaccurate problems as discussed in [17], Sunde's formula is not implemented into the Cable Constants (CC) of EMTP-type simulation tools [3].

Since Carson and Pollaczek have proposed the formulas in 1920s, a number of efforts discussing the earth-return impedance of overhead lines above a multilayer earth have been published in [11–13]. Again, the same initial assumption of uniform current distribution has been used in [11–13].

A significant improvement of earth-return parameters is performed by Wise in [14,15]. Although the Wise's formulation is still within quasi-TEM assumption, the uniform current distribution has been removed and replaced by the first-order correction which means that the unknown propagation constant along a conductor is equal to the intrinsic propagation constant of air ($\gamma_x = \gamma_0$). This important modification in the formulation allows to derive the earth-return admittance formula together with full expressions of electromagnetic properties of air and earth from Maxwell's equations. Nakagawa [16], Ametani [20,21], and Papadopoulos [22,23] further adopt the Wise's method to investigate and derive the formulas of earth-return impedance and admittance of overhead lines above a multilayer earth.

Because a numerical instability (divergence) is occasionally observed in Pollaczek's formula [24], a closed-form formula of the earth-return impedance for an overhead line is developed by Gary in 1976 [25], and the detailed derivation of the formula is explained in [26]. The formula shows no instability.

Recently, a new line/cable parameter calculation code with a Method of Moments and Surface Admittance Operator (MoM-SO) method has been developed in [27–29]. The code can deal with the arbitrary number of circular conductors, either overhead and/or underground. However, it should be noted that the MoM-SO code is within TEM wave propagation, and thus cannot be used for very-fast transient (VFT) simulation in a gas-insulated substation (GIS).

2.2.1.2 Formulas based on complete field solution

In 1955, Kikuchi made clear that Carson's and Pollaczek's formulas are based on TEM wave propagation, and thus could not be applied to study high-frequency phenomena, because the wave propagation at high frequencies is not TEM, but TM and/or TE (non-TEM) [30,31]. Kikuchi proposed a more accurate formula for both earth-return impedance and admittance of an overhead conductor based on a complete field solution. However, the Kikuchi's formula requires a complicated solution of Maxwell's equations, and it causes numerical difficulties in 1950s.

In 1976, Wedepohl and Efthymiadis proposed a different formulation of earth-return impedance and admittance which is based on the vertical electric field, characteristics impedance, unknown propagation constant, voltage, and current propagation equations of an overhead conductor [32,33]. In addition to earth-return parameters, the external electromagnetic fields generated by the overhead conductor in the air and earth can be also calculated by adopting the complete field solutions in [32,33], and further studied in [46]. The other similar complete field solutions are proposed in [34–40]. It should be noted that the zero potential references used in the formulation of voltage and current propagation equations are thoroughly studied, investigated, and documented in [32,38,47,48].

2.2.2 Underground cable

The major formulations of earth-return impedance and admittance for underground cables are derived and presented in [49–59]. The following discussions of formulas are also based on quasi-TEM assumption and complete field solution.

2.2.2.1 Formulas based on quasi-TEM assumption

Similar to overhead lines, the earth-return impedance formula of underground cables is initially derived by Pollaczek under the quasi-TEM assumption and uniform current distribution [9]. Sunde and Wedepohl independently produce the earth-return impedance formula to consider the influence of earth permittivity [10,49]. Considering the same approach of Sunde, a formula of earth-return impedance of underground cables buried in a multilayer earth is proposed by Tsiamitros [50].

A significant formulation of the earth-return impedance and admittance of underground cables is presented by Papadopoulos in [51]. The method [51] involves the first-order correction of an unknown propagation constant along the cable conductors. Also, the method in [51] is extended to a scenario with a multilayer earth [52,53]. Moreover, the application limit of the earth-return impedance and admittance on underground cables with the quasi-TEM approach is studied and investigated in [54,55].

The closed-form expression of earth-return impedance and admittance shows several advantages, such as high computational efficiency, numerical stability and flexible implementation, etc. Vance derives an approximate formula considering the earth as a lossy cylindrical dielectric around the cable [56]. By adopting a logarithmic approximation of Pollaczek's and Sunde's integral in the earth-return impedance, Wedepohl, Saad, Petrache and Theethayi derive a series of closed-form formulas of the earth-return impedance and admittance for the underground cables [49,57–59]. Also, Salvador independently produces a closed-form formula of the earth-return parameters based on the method in [55].

2.2.2.2 Formulas based on complete field solution

Generalized formulas of the earth-return impedance and admittance for a single underground cable have been proposed and derived by Wait and Bridges in [60–63]. The method is based on exact expressions of an electromagnetic theory.

Recently, Xue has derived a generalized formulation for earth-return impedance and admittance based on the complete field solution and telegrapher's equations for a multiphase underground cable [64]. Three different line voltage calculations related to different earth-return impedance and admittance formulas are presented in [64]. Also, the quasi-TEM-based earth-return impedance and admittance formulas are deduced from the formulas of complete field solution [64].

2.3 Accurate and approximate earth-return impedance formulas for overhead cable (line)

The configuration of overhead conductors above a three-layer earth is illustrated in Figure 2.1, where the radii of conductors are r_i and r_j, respectively. The depths of earth layers are given by d_1 and d_2, and the depth of the lowest layer extends to infinity. The permeability and permittivity of air are μ_0 and ε_0. The permeability, resistivity, and permittivity of the stratified earth are characterized by μ_k, ρ_k (conductivity $\sigma_k = 1/\rho_k$), and ε_k, where $k = 1$ to 3. The conductor heights are h_i and h_j, and the horizontal distance between the two conductors is y.

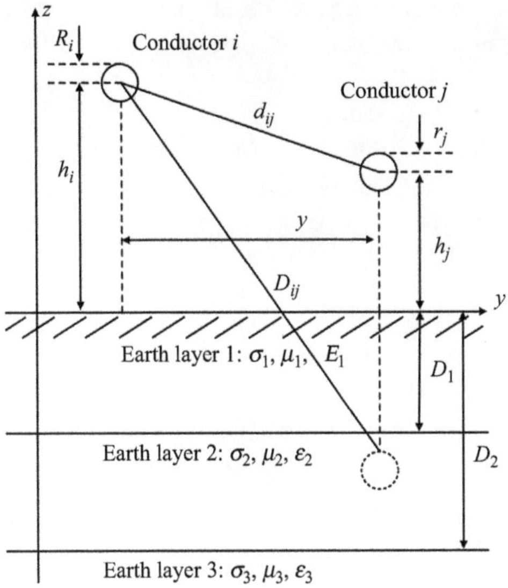

Figure 2.1 An overhead multiconductor system above a multilayer earth

In general, the earth-return impedance formula of the overhead line is given in the following form:

$$Z_{ij} = \frac{j\omega\mu_0}{2\pi}[P_0 + (Q - jR)] \tag{2.8}$$

where

$$P_0 = \ln\left(\frac{D_{ij}}{d_{ij}}\right) \tag{2.9}$$

$$Q - jR = 2\int_0^{+\infty} AF(s)ds \tag{2.10}$$

with

$$F(s) = \exp\left[-(h_i + h_j)s\right]\cos(ys) \tag{2.11}$$

$$d_{ij}^2 = (h_i - h_j)^2 + y^2 \tag{2.12}$$

$$D_{ij}^2 = (h_i + h_j)^2 + y^2 \tag{2.13}$$

The irrational A is a function of earth electromagnetic properties and differs in accurate and approximate formulas. Also, the physical meaning of each term in (2.8) is clear. The term of P_0 represents the mutual inductance due to perfect conducting earth, and the term of $Q - jR$ is the correction of lossy earth for mutual impedance.

If the self-impedance of (2.8) is calculated, $h_i = h_j$ and $y = r_i/y = r_j$ are adopted. Also, the mutual impedance can be calculated at the center of two conductors (cables).

2.3.1 *Accurate impedance formulas for stratified earth*

Nakagawa derived a generalized formula of the earth-return impedance of a multiphase overhead line above a three-layer stratified earth shown in Figure 2.1, and it has the following form [16]:

$$A = \frac{c_1 + c_2}{(s + \mu_0 b_1)c_1 + (s - \mu_0 b_1)c_2} \tag{2.14}$$

where

$$c_1 = (b_1 + b_2)(b_2 + b_3) + (b_1 - b_2)(b_2 - b_3)\exp[2a_2(d_1 - d_2)] \tag{2.15}$$

$$c_2 = \{(b_1 - b_2)(b_2 + b_3) + (b_1 + b_2)(b_2 - b_3)\exp[2a_2(d_1 - d_2)]\}\exp(-2a_1 d_1) \tag{2.16}$$

with

$$a_k = \sqrt{s^2 + \gamma_k^2 - \gamma_0^2} \text{ and } b_k = a_k/\mu_k \tag{2.17}$$

The intrinsic propagation constant of air and each layer of earth are given below:

$$\gamma_0^2 = j\omega\mu_0 j\omega\varepsilon_0 \text{ and } \gamma_k^2 = j\omega\mu_k(\sigma_k + j\omega\varepsilon_k) \tag{2.18}$$

For a two-layer earth, it has the following expression:

$$A = \frac{b_1 + b_2 + (b_1 - b_2)\exp(-2a_1 d_1)}{(s + \mu_0 b_1)(b_1 + b_2) + (s - \mu_0 b_1)(b_1 - b_2)\exp(-2a_1 d_1)} \tag{2.19}$$

For a homogeneous earth, the irrational A has been further simplified to the following formula:

$$A = \frac{1}{s + \mu_0 b_1} = \frac{1}{s + \frac{\mu_0}{\mu_1}\sqrt{s^2 + \gamma_1^2 - \gamma_0^2}} \tag{2.20}$$

When the earth permeability μ_1 is taken to be the same as that of air in (2.20), then the following formula which is the same as Wise's earth-return impedance formula [14] is deduced:

$$A = \frac{1}{s + \sqrt{s^2 + \gamma_1^2 - \gamma_0^2}} \tag{2.21}$$

where

$$\gamma_1^2 = j\omega\mu_0(\sigma_1 + j\omega\varepsilon_1) \tag{2.22}$$

2.3.2 Approximate impedance formulas

2.3.2.1 Pollaczek

Pollaczek derives the following irrational A:

$$A = \frac{1}{s + \sqrt{s^2 + \gamma_1^2}} \tag{2.23}$$

where

$$\gamma_1^2 = j\omega\mu_0\sigma_1 \tag{2.24}$$

It is clear that Pollaczek assumes the earth permeability is the same as that of a free space. Further, the earth permittivity has been also neglected.

2.3.2.2 Carson

It is well-known that Carson's formula in an infinite integral form is the same as Pollaczek's formula (2.23). In the 1920s, the infinite integral shown in (2.10) and (2.23) causes numerical difficulties. Thus, a formula of asymptotic expansion which has been widely used since then is developed by Carson [8].

2.3.2.3 Sunde

Sunde also improves the formula of earth-return impedance by giving the following expression:

$$A = \frac{1}{s + \sqrt{s^2 + \gamma_1^2}} \tag{2.25}$$

where

$$\gamma_1^2 = j\omega\mu_0(\sigma_1 + j\omega\varepsilon_1) \tag{2.26}$$

2.3.3 Further approximation of formulas

There are a number of approximate formulas of the earth-return impedance. Most widely used one is that developed originally by Gary [25] and later confirmed theoretically by Deri [26]:

$$Z_{ij} = \frac{j\omega\mu_0}{2\pi} \ln\left(\frac{S_{ij}}{d_{ij}}\right) \tag{2.27}$$

where d_{ij} is given in (2.12) and

$$S_{ij} = \sqrt{\left(h_i + h_j + 2h_e\right)^2 + y^2} \text{ and } h_e = \frac{1}{\sqrt{j\omega\mu_0\sigma_1}} \tag{2.28}$$

2.3.4 Discussion

It is said quite often that Pollaczek's and Carson's formulas cannot deal with displacement currents in earth and thus the formulas are not applicable to high frequencies. This is true when the relative permittivity of earth is other than unity, that is, $\varepsilon_1 \neq \varepsilon_0$. When $\varepsilon_1 = \varepsilon_0$ is assumed in (2.21), it gives

$$A = \frac{1}{s + \sqrt{s^2 + j\omega\mu_0\sigma_1}} \tag{2.29}$$

This result means that the displacement currents in earth are not excluded. The permittivity term in the air and earth are canceled out because of $\varepsilon_1 = \varepsilon_0$. On the contrary, if the displacement currents in the earth is neglected, that is, $\varepsilon_1 = 0$, in Sunde's formula (2.25), then (2.29) is also obtained. Thus, it is clear that Pollaczek's and Carson's formulas become (2.29) if we consider the displacement currents with $\varepsilon_1 = \varepsilon_0$, while if we neglect the currents, that is, $\varepsilon_1 = 0$, Sunde's formula becomes (2.29). As discussed in [17], if we consider the displacement currents, the formula should be either (2.29) for $\varepsilon_1 = \varepsilon_0$ and $\varepsilon_1 = 0$, or (2.20) for arbitrary ε_1.

Considering the above facts, Sunde's formula in (2.25) is not correct although it is said that Sunde's formula considers the displacement currents in earth, Pollaczek's and Carson's formulas neglect the displacement currents. Also, Sunde stated in his book that "Formulas and curves for $W(Q - jR$ in (2.8)) have been given by Carson" (p. 112, Section 4.5 in [10]). Pollaczek starts the impedance formula derivation with the intrinsic propagation constants γ_0 for air and γ_1 for earth in (2.8). However, further simplifications are made by Pollaczek in order to simplify the formulation procedure, and it results in (2.23). Therefore, the following approximate formula with consideration of displacement currents is easily derived:

$$Z_{ij} = \frac{j\omega\mu_0}{2\pi} \ln\left(\frac{S'_{ij}}{d_{ij}}\right) \tag{2.30}$$

where d_{ij} is given in (2.12) and

$$S'_{ij} = \sqrt{\left(h_i + h_j + 2h'_e\right)^2 + y^2} \text{ and } h'_e = \frac{1}{\sqrt{\gamma_1^2 - \gamma_0^2}}$$

$$= \frac{1}{\sqrt{j\omega\mu_0[\sigma_1 + j\omega\varepsilon_0(\varepsilon_{r1} - 1)]}} \tag{2.31}$$

It is noteworthy that Kikuchi's formula [30,31] can deal with not only TEM mode of propagation but also TM/TE mode propagation.

2.4 Accurate and approximate earth-return admittance formulas for overhead cable (line)

2.4.1 Accurate admittance formulas

The following formula of earth-return potential coefficient of overhead cable above a homogeneous earth is proposed by Wise in [15], and it is the same as that derived by Nakagawa [18,19]:

$$P_{ij} = \frac{P_0 + M + jN}{2\pi\varepsilon_0} \tag{2.32}$$

where

$$M + jN = 2\int_0^{+\infty} \frac{F(s)}{\frac{\gamma_1^2}{\gamma_0^2}s + \sqrt{s^2 + \gamma_1^2 - \gamma_0^2}} ds \tag{2.33}$$

with P_0 and $F(s)$ are given in (2.9) and (2.11).

2.4.2 Approximate admittance formulas

It should be noted that only the space potential coefficient is implemented into the Line/Cable Constants of EMT-type simulation tools [3–7], that means $M + jN = 0$ in (2.32). Thus, it has the following expressions:

$$P_{ij} = \frac{P_0}{2\pi\varepsilon_0} \tag{2.34}$$

2.5 Accurate and approximate earth-return impedance formulas for underground cable

As shown in Figure 2.2, the air is the medium in the upper region, and characterized by permeability μ_0 and permittivity ε_0. The lower region is the lossy earth, and it is characterized by permeability μ_0, resistivity ρ_1 (conductivity $\sigma_1 = 1/\rho_1$) and permittivity ε_1. It should be noted that the buried depths of cables are set to positive values in the following formulas.

2.5.1 Accurate impedance formulas

The earth-return impedance and admittance formulas of underground cables based on the complete electromagnetic field solutions are proposed by Xue in [64], and explained in Appendix A2. Because of numerical difficulties for complete field solutions as discussed in [64], the assumption of quasi-TEM propagation is adopted to derive a quite simple form of the earth-return impedance and admittance so that

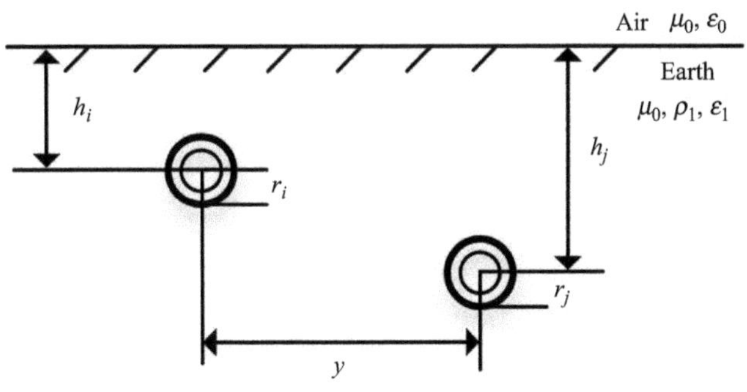

Figure 2.2 An underground cable system buried in homogeneous earth

anyone can use the formulas in a study of wave propagation and power system transients.

For the earth-return impedance between cables *i* and *j* shown in Figure 2.2, it has the following expressions:

$$Z_{ij} = \frac{j\omega\mu_0}{2\pi}\left[K_0(\gamma_1 d_{ij}) - K_0(\gamma_1 D_{ij}) + 2\underset{=11}{S^c} + 2\gamma_1^2 \underset{=13}{S^c}\right] \tag{2.35}$$

where K_0 is modified Bessel function of second kind with zero order, and the irrationals have

$$d_{ij}^2 = (h_i - h_j)^2 + y^2 \tag{2.36}$$

$$D_{ij}^2 = (h_i + h_j)^2 + y^2 \tag{2.37}$$

$$\underset{=11}{S^c} = \int_0^{+\infty} \frac{e^{-(h_i+h_j)\sqrt{\lambda^2+\gamma_1^2}}}{(\lambda^2+\gamma_1^2)\left(\sqrt{\lambda^2+\gamma_0^2}+\sqrt{\lambda^2+\gamma_1^2}\right)}\lambda^2 \cos(y\lambda)d\lambda \tag{2.38}$$

$$\underset{=13}{S^c} = \int_0^{+\infty} \frac{e^{-(h_i+h_j)\sqrt{\lambda^2+\gamma_1^2}}}{(\lambda^2+\gamma_1^2)\left(\sqrt{\lambda^2+\gamma_0^2}+\sqrt{\lambda^2+\gamma_1^2}\right)} \cos(y\lambda)d\lambda \tag{2.39}$$

with λ is integral variable and

$$\gamma_0^2 = j\omega\mu_0 j\omega\varepsilon_0 \text{ and } \gamma_1^2 = j\omega\mu_0(\sigma_1 + j\omega\varepsilon_1) \tag{2.40}$$

If the self-impedance of (2.35) is calculated, $h_i = h_j$ and $y = r_i/y = r_j$ are adopted. Also, the mutual impedance can be calculated at the center of two cables.

2.5.2 Approximate impedance formulas

2.5.2.1 Pollaczek

Pollaczek develops the earth-return impedance of underground cables for a frequency below several hundreds of kHz in [9], and it has the below formula:

$$Z_{ij} = \frac{j\omega\mu_0}{2\pi}\left[K_0(\gamma_1 d_{ij}) - K_0(\gamma_1 D_{ij}) + 2\underset{=}{S}\right] \tag{2.41}$$

where

$$\underset{=}{S} = \int_0^{+\infty} \frac{e^{-(h_i+h_j)\sqrt{\lambda^2+\gamma_1^2}}}{\lambda+\sqrt{\lambda^2+\gamma_1^2}} \cos(y\lambda)d\lambda \tag{2.42}$$

with

$$\gamma_1^2 = j\omega\mu_0\sigma_1 \tag{2.43}$$

2.5.2.2 Sunde

Sunde's earth-return impedance formula can be derived when (2.43) is replaced by (2.44):

$$\gamma_1^2 = j\omega\mu_0(\sigma_1 + j\omega\varepsilon_1) \tag{2.44}$$

It has been well-known that Pollaczek's formula cannot deal with displacement currents because of neglecting ε_1 in (2.43). Although Sunde's formula includes ε_1, it neglects the intrinsic propagation constant of air.

2.5.2.3 Papadopoulos

A modified earth-return impedance formula has been proposed by Papadopoulos in [51], and it is further extended to a two-layer earth case as discussed in [52,53]. The earth-return impedance formula for a homogeneous earth by adopting $\mu_1 = \mu_0$ in [51] is given below:

$$Z_{ij} = \frac{j\omega\mu_0}{2\pi}\left(\underset{=1}{\Delta} + 2\underset{=}{S}\right) \tag{2.45}$$

where

$$\underset{=1}{\Delta} = \int_0^{+\infty} \frac{\left[e^{-|h_i-h_j|\sqrt{\lambda^2+\gamma_1^2+k_x^2}} - e^{-(h_i+h_j)\sqrt{\lambda^2+\gamma_1^2+k_x^2}}\right]}{\sqrt{\lambda^2+\gamma_1^2+k_x^2}}\cos(y\lambda)d\lambda \tag{2.46}$$

$$\underset{=}{S} = \int_0^{+\infty} \frac{e^{-(h_i+h_j)\sqrt{\lambda^2+\gamma_1^2+k_x^2}}}{\sqrt{\lambda^2+\gamma_0^2+k_x^2} + \sqrt{\lambda^2+\gamma_1^2+k_x^2}}\cos(y\lambda)d\lambda \tag{2.47}$$

with $k_x = \omega\sqrt{\mu_0\varepsilon_1}$ and γ_1^2 is given in (2.40).

In fact, the issue of Papadopoulos' formulas comes from the less generalized formulation because only the electric Hertzian vector is adopted in the derivation of external electromagnetic fields generated by underground cables [51].

2.5.3 Further approximation of formulas

There are several closed-form expressions of earth-return impedance formulas are proposed in references [49,56–59].

2.5.3.1 Wedepohl

A similar earth-return impedance formula to Sunde's one was derived by Wedepohl in [49], and the formula is further simplified by adopting a series expansion method as discussed in [49]. The approximate formula for the mutual earth-return impedance is given in the following form [49]:

$$Z_{ij} = \frac{j\omega\mu_0}{2\pi}\left[-\ln\left(\frac{e_c\gamma_1 y}{2}\right) + \frac{1}{2} - \frac{2}{3}\gamma_1(h_i + h_j)\right] \tag{2.48}$$

where Euler's constant has $e_c = 1.7811$ and γ_1 is given in (2.43).

For the self-impedance of the *i*th cable becomes:

$$Z_{ij} = \frac{j\omega\mu_0}{2\pi}\left[-\ln\left(\frac{e_c\gamma_1 r_i}{2}\right) + \frac{1}{2} - \frac{4}{3}\gamma_1 h_i\right] \tag{2.49}$$

2.5.3.2 Vance

The following formula of a single phase underground cable for earth-return impedance considering the earth as a lossy cylindrical dielectric surrounding the cable has been proposed by Vance [56]:

$$Z_{ij} = \frac{\omega\mu_0}{2\pi\gamma_1 r_i}\frac{H_0^{(1)}(j\gamma_1 r_i)}{H_1^{(1)}(j\gamma_1 r_i)} \tag{2.50}$$

where γ_1 is given in (2.43), and r_i is the outer radius of cable i and can be replaced by the outer radius r_j of cable j.

2.5.3.3 Bridges

Adopting from an exact scattering solution for a single phase underground cable, a frequency domain integral equation for the cable induced current and a general formula of earth-return impedance are proposed and presented by Bridges in [62,63]. Using TL approach and neglecting displacement currents in earth, the following approximate formula of earth-return impedance applied to a single phase underground cable has been set up:

$$Z_{ij} = -\frac{j\omega\mu_0}{2\pi}\ln\left(\frac{e_c\gamma_1 r_i}{2}\right) \tag{2.51}$$

where Euler's constant has $e_c = 1.7811$ and γ_1 is given in (2.43).

The same manner of outer radius of cable of Vance's formula can be adopted in (2.51). It should be noted that the Bridge's formula (2.51) could be regarded as that the Wedepohl's formula (2.49) neglects the buried depth of cable.

2.5.3.4 Saad

Saad derived an approximate formula of Pollaczek's earth-return impedance [57]. It is shown in [59] that the formula is identical or better compared with the other low-frequency approximations proposed by Wedepohl (2.48) and (2.49):

$$Z_{ij} = \frac{j\omega\mu_0}{2\pi}\left[K_0(\gamma_1 R_{ab}) + \frac{2e^{-(h_i+h_j)\gamma_1}}{4+\gamma_1^2 R_{ab}^2}\right] \tag{2.52}$$

where γ_1^2 is given in (2.43).

Note that in (2.52) and the formulas below, for self-impedance, R_{ab} is the same as the outer insulator radius of a cable. For the mutual impedance between two parallel cables, R_{ab} needs to be replaced by the horizontal distance between the buried cables.

2.5.3.5 Petrache

Petrache has proposed a simple logarithmic approximation for the earth-return impedance, and it is given by the following expression [58]:

$$Z_{ij} = \frac{j\omega\mu_0}{2\pi} \ln\left(\frac{1 + \gamma_1 R_{ab}}{\gamma_1 R_{ab}}\right) \tag{2.53}$$

where γ_1^2 is given in (2.44).

However, neglecting the buried depth of cable in (2.53) may result in inaccuracies, especially for the cases in low frequencies.

2.5.3.6 Theethayi

A modified empirical logarithmic-exponential formula has been developed by Theethayi [59], in which he combines (2.52) and (2.53) to account for the cable buried depth:

$$Z_{ij} = \frac{j\omega\mu_0}{2\pi} \left[\ln\left(\frac{1 + \gamma_1 R_{ab}}{\gamma_1 R_{ab}}\right) + \frac{2e^{-(h_i + h_j)|\gamma_1|}}{4 + \gamma_1^2 R_{ab}^2}\right] \tag{2.54}$$

where γ_1^2 is given in (2.44).

Theethayi's formula (2.54) can be applied to a multiphase underground cable, and in the meantime, the formula considers the buried depth of cable and full expression of electromagnetic properties of earth.

2.6 Accurate and approximate earth-return admittance formulas for underground cable

2.6.1 Accurate admittance formulas

As shown in Figure 2.2, the earth-return admittance between cables i and j can be represented by the earth-return potential coefficient. The concept of the earth-return impedance together with the admittance of a single-phase underground cable has been originated by Bridges [62,63]. Recently, starting from the rigorous formulation of complete field solution for underground cables, a set of earth-return potential coefficient formulas based on quasi-TEM approach are derived and presented by Xue in [64]. The formulas proposed in [64] are based on different voltage integral references in the formulation of current and voltage propagation equations. In general, three earth-return potential coefficient formulas are found, and all the details are given in Appendix A2. In the following parts, three significant earth-return potential coefficient formulas using quasi-TEM approach are shown and discussed, and they could be applied to major transient studies of underground cables in power system.

2.6.1.1 Line voltage integral equation referred to infinite depth in the earth

The earth-return potential coefficient formula between cables i and j based on the integral path to infinite depth in the earth ($h_r = -\infty$) has the following expression:

$$P_{ij} = \frac{j\omega}{2\pi(\sigma_1 + j\omega\varepsilon_1)}\left[K_0(\gamma_1 d_{ij}) - K_0(\gamma_1 D_{ij}) + 2\underset{=12}{S^c} + 2\gamma_1^2 \underset{=13}{S^c}\right] \tag{2.55}$$

where $\underset{=13}{S^c}$ is given in (2.39) and

$$\underset{=12}{S^c} = \int_0^{+\infty} \frac{e^{-(h_i+h_j)\sqrt{\lambda^2+\gamma_1^2}}}{(\lambda^2+\gamma_1^2)\left(\sqrt{\lambda^2+\gamma_0^2} + \frac{\gamma_0^2}{\gamma_1^2}\sqrt{\lambda^2+\gamma_1^2}\right)} \lambda^2 \cos(y\lambda) d\lambda \tag{2.56}$$

with γ_0^2 and γ_1^2 are given in (2.40).

2.6.1.2 Line voltage integral equation referred to the earth surface

The earth-return potential coefficient formula between cables i and j based on the integral path to the earth surface ($h_r = 0$) is

$$P_{ij} = \frac{j\omega}{2\pi(\sigma_1 + j\omega\varepsilon_1)}\left[K_0(\gamma_1 d_{ij}) - K_0(\gamma_1 D_{ij}) + 2\underset{=12}{S^c} + 2\gamma_1^2 \underset{=13}{S^c}\right] - P_{ecij} \tag{2.57}$$

where $\underset{=12}{S^c}$ and $\underset{=13}{S^c}$ are given in (2.56) and (2.39), and the earth surface correction term P_{ecij} has

$$P_{ecij} = \frac{j\omega}{\pi(\sigma_1 + j\omega\varepsilon_1)}\left(\underset{=14}{S^c} + \gamma_1^2 \underset{=15}{S^c}\right) \tag{2.58}$$

with

$$\underset{=14}{S^c} = \int_0^{+\infty} \frac{e^{-\frac{(h_i+h_j)}{2}\sqrt{\lambda^2+\gamma_1^2}}}{(\lambda^2+\gamma_1^2)\left(\sqrt{\lambda^2+\gamma_0^2} + \frac{\gamma_0^2}{\gamma_1^2}\sqrt{\lambda^2+\gamma_1^2}\right)} \lambda^2 \cos(y\lambda) d\lambda \tag{2.59}$$

$$\underset{=15}{S^c} = \int_0^{+\infty} \frac{e^{-\frac{(h_i+h_j)}{2}\sqrt{\lambda^2+\gamma_1^2}}}{(\lambda^2+\gamma_1^2)\left(\sqrt{\lambda^2+\gamma_0^2} + \sqrt{\lambda^2+\gamma_1^2}\right)} \cos(y\lambda) d\lambda \tag{2.60}$$

2.6.1.3 Line voltage integral equation referred to penetration depth in the earth

The earth-return potential coefficient formula between cables i and j based on penetration depth in the earth $(h_r = -\delta_e)$ is

$$P_{ij} = \frac{j\omega}{2\pi(\sigma_1 + j\omega\varepsilon_1)}\left[K_0(\gamma_1 d_{ij}) - K_0(\gamma_1 D_{ij}) + 2\underset{=12}{S^c} + 2\gamma_1^2 \underset{=13}{S^c}\right] - P_{\delta_e ij} \quad (2.61)$$

where the penetration depth correction term $P_{\delta_e ij}$ has

$$P_{\delta_e ij} = \frac{j\omega}{2\pi(\sigma_1 + j\omega\varepsilon_1)}\left[K_0(\gamma_1 d_{\delta_e ij}) - K_0(\gamma_1 D_{\delta_e ij}) + 2\underset{=16}{S^c} + 2\gamma_1^2 \underset{=17}{S^c}\right] \quad (2.62)$$

with

$$d_{\delta_e ij} = \sqrt{y^2 + \left(h_r - \frac{h_i + h_j}{2}\right)^2} \quad (2.63)$$

$$D_{\delta_e ij} = \sqrt{y^2 + \left(h_r + \frac{h_i + h_j}{2}\right)^2} \quad (2.64)$$

$$\underset{=16}{S^c} = \int_0^{+\infty} \frac{e^{\left[h_r - \frac{(h_i + h_j)}{2}\right]\sqrt{\lambda^2 + \gamma_1^2}}}{(\lambda^2 + \gamma_1^2)\left(\sqrt{\lambda^2 + \gamma_0^2} + \frac{\gamma_0^2}{\gamma_1^2}\sqrt{\lambda^2 + \gamma_1^2}\right)} \lambda^2 \cos(y\lambda) d\lambda \quad (2.65)$$

$$\underset{=17}{S^c} = \int_0^{+\infty} \frac{e^{\left[h_r - \frac{(h_i + h_j)}{2}\right]\sqrt{\lambda^2 + \gamma_1^2}}}{(\lambda^2 + \gamma_1^2)\left(\sqrt{\lambda^2 + \gamma_0^2} + \sqrt{\lambda^2 + \gamma_1^2}\right)} \cos(y\lambda) d\lambda \quad (2.66)$$

and h_r has

$$h_r = -\delta_e = -\left\{\frac{\omega^2 \varepsilon_1 \mu_0}{2}\left[\sqrt{1 + \left(\frac{\sigma_1}{\omega\varepsilon_1}\right)^2} - 1\right]\right\}^{-\frac{1}{2}} \quad (2.67)$$

It should be noted that $h_i = h_j$ and $y = r_i$ or $y = r_j$ are used for the self-potential coefficient. The mutual element can be calculated at the center of two cables. By substituting $h_r = 0$ into (2.61), (2.57) is deduced. If $h_r = -\infty$ is assumed in (2.61), $P_{\delta_e ij} \to 0$, (2.55) is derived.

It is very important to discuss the influence of h_r on the earth-return impedance and potential coefficient. By adopting quasi-TEM assumption, the earth-return impedance becomes independent on h_r.

However, h_r shows a significant influence on the earth-return potential coefficient. In fact, correction term $P_{\delta_e ij}$ of (2.61) comes from the influence of vertical electric field E_{1z} generated by underground cables in the earth at the effective zero potential level "h_r", as shown in Appendix A2. In theory, E_{1z} should have decayed sufficiently at h_r, otherwise nonzero E_{1z} contributes to the voltage and results in the correction term $P_{\delta_e ij}$ of (2.61). Thus, (2.57) should not be used considering the fact that a cable is often buried close to the earth surface and E_{1z} cannot decay sufficiently to permit the earth surface as an effective zero potential level. Instead of (2.57), (2.55) is adopted to keep $E_{1z} = 0$ for the entire region.

2.6.2 Approximate admittance formulas

2.6.2.1 Papadopoulos

A modified earth-return potential coefficient equation of underground cables has been proposed by Papadopoulos [51] together with the earth-return impedance formula shown in (2.45). Also, the formula is extended to a two-layer earth model [52]. The formula of earth-return potential coefficient for a homogeneous earth $\mu_1 = \mu_0$ is given by [51]

$$P_{ij} = \frac{j\omega}{2\pi(\sigma_1 + j\omega\varepsilon_1)}\left(\underset{=1}{\Delta} + 2\underset{=}{S}\right) \tag{2.68}$$

where $\underset{=1}{\Delta}$ is given in (2.46) and

$$\underset{=}{S} = \int_0^{+\infty} \frac{e^{-(h_i+h_j)\sqrt{\lambda^2+\gamma_1^2+k_x^2}}}{\sqrt{\lambda^2+\gamma_0^2+k_x^2} + \frac{\gamma_0^2}{\gamma_1^2}\sqrt{\lambda^2+\gamma_1^2+k_x^2}} \cos(y\lambda)d\lambda \tag{2.69}$$

and γ_0^2 and γ_1^2 are given in (2.40), and $k_x = \omega\sqrt{\mu_0\varepsilon_1}$.

2.6.2.2 Magalhães

The following formula of the earth-return potential coefficient based on quasi-TEM assumption is presented by Magalhães [55], in which the numerical results are the same as (2.57):

$$P_{ij} = \frac{j\omega}{2\pi(\sigma_1 + j\omega\varepsilon_1)}\left[K_0(\gamma_1 d_{ij}) - K_0(\gamma_1 D_{ij}) - 2\underset{=}{S}\right] \tag{2.70}$$

where

$$\underset{=}{S} = \int_0^{+\infty} \frac{\sqrt{\lambda^2+\gamma_0^2}}{\sqrt{\lambda^2+\gamma_1^2}}\left[\frac{e^{-\frac{(h_i+h_j)}{2}\sqrt{\lambda^2+\gamma_1^2}} - e^{-(h_i+h_j)\sqrt{\lambda^2+\gamma_1^2}}}{\sqrt{\lambda^2+\gamma_0^2} + \frac{\gamma_0^2}{\gamma_1^2}\sqrt{\lambda^2+\gamma_1^2}}\right]\cos(y\lambda)d\lambda \tag{2.71}$$

where γ_0^2 and γ_1^2 are given in (2.40).

The problem of the above formula by Magalhães is that the effective zero potential level h_r is assumed to be the earth surface ($h_r = 0$). As already explained in Section 2.6.1.3, this assumption gives inaccurate results.

2.6.3 Further approximation of formulas

The following approximate expression [58,59] shows a straightforward relation between the earth-return impedance and admittance, thus it can be used as a further approximation of earth-return admittance formulas discussed above. A significant advantage is that the numerical integrals can be avoided in the calculations:

$$Y_{ij} = \frac{\gamma_1^2}{Z_{ij}} \tag{2.72}$$

where γ_1^2 can be (2.43) or (2.44), and it depends on the approximate formula of Z_{ij}.

Equation (2.72) gives an expression of a single phase underground cable. If the multiphase underground cable is needed, it can be modified to a matrix expression. Also, (2.48)–(2.54) can be used in (2.72) with simplification of both earth-return impedance and admittance parts.

In addition to (2.72), the following closed-form formula of the earth-return potential coefficient is proposed by Xue based on (2.52) and (2.53) in this book:

$$P_{ij} = \frac{j\omega}{2\pi(\sigma_1 + j\omega\varepsilon_1)}\left[\ln\left(\frac{1+\gamma_1 R_{ab}}{\gamma_1 R_{ab}}\right) + \frac{2e^{-(h_i+h_j)\gamma_1}}{4+\gamma_1^2 R_{ab}^2}\right] \tag{2.73}$$

where γ_1^2 and R_{ab} are given in (2.44) and (2.52).

2.6.4 Applicable limit of quasi-TEM-based formulas

It should be noted that the formulas of earth-return parameters (2.35) and (2.55) are limited to the highest frequency of about 10 MHz, because only the first region of quasi-TEM assumption is adopted [54]. The configuration of a cable should be much smaller than the wavelength in the earth, and the wavelength is greater than the penetration depth (2.67) in the soil. According to the critical frequency analysis [53], the quasi-TEM assumption or a TL based approach can be applicable up to 10 MHz by assuming that the earth resistivity does not exceed 1,000 Ωm. If necessary, a much higher frequency can be handled by adopting [54] in the complete field solutions (3.55) and (3.58).

2.6.5 Definition of classical and extended TL approaches

The series impedance and shunt admittance of a cable system, such as an overhead line, overhead cable and underground cable, are defined in the two matrix equations based on TL approach [1]:

$$-\frac{d\mathbf{V}}{dx} = \mathbf{ZI} \tag{2.74}$$

$$-\frac{d\mathbf{I}}{dx} = \mathbf{YV} \tag{2.75}$$

where **V** and **I** are an $N \times 1$ vector of the voltages and currents at distance x along the cable. **Z** and **Y** are an $N \times N$ matrix of the series impedance and shunt admittance of the cable.

Backtracking the reviews in Section 2.1, the series impedance and shunt admittance matrices of a cable can be expressed in the following forms:

$$\mathbf{Z} = \mathbf{Z}_i + \mathbf{Z}_e \tag{2.76}$$

$$\mathbf{Y} = j\omega \mathbf{P}^{-1} \tag{2.77}$$

where $\mathbf{Z}_i$ and $\mathbf{Z}_e$ are the internal and earth-return impedances of the cable. The potential coefficient matrix **P** is defined as

$$\mathbf{P} = \mathbf{P}_i + \mathbf{P}_e \tag{2.78}$$

where $\mathbf{P}_i$ and $\mathbf{P}_e$ are the internal and earth-return potential coefficients of the cable.

Hereafter, (2.76) and (2.77) are designated as "extended TL approach." Carson's or Pollaczek's earth-return impedance with constant space or insulator potential coefficient is called "classical TL approach." The formulas adopted into extended and classical TL approaches are summarized in Table 2.1. The internal impedance and potential coefficients of the cable is based on the reviews in Section 2.1 and Appendix A1.

2.7 Derivation of the modified earth-return Green function for MoM-SO technique

Recently, the Method of Moments and Surface Admittance Operator (MoM-SO) [27–29] technique has been proposed to calculate the frequency-dependent series impedance of overhead lines and underground cables. It is based on the surface admittance operator combined with the method of moments [27–29]. The MoM-SO has several advantages for calculating the series impedance of the overhead lines and underground cables. For example, the skin and proximity effects of conductors can be considered. A stratified earth with arbitrary earth parameters (permeability, permittivity, and conductivity) can be properly modeled. Also, arbitrary cable configurations, such as tunnel installed and submarine cables, can be handled.

Table 2.1 Summary of formulas adopted into extended and classical TL approaches

Item		Overhead cable (line)	Underground cable
Extended TL approach	$\mathbf{Z}_e$	(2.8) and (2.14)	(2.35)
	$\mathbf{P}_e$	(2.32)	(2.55)
Classical TL approach	$\mathbf{Z}_e$	(2.8) and (2.23)	(2.41)
	$\mathbf{P}_e$	(2.34)	–

Furthermore, the MoM-SO provides much higher computational efficiency than a software based on the finite element method (FEM).

A numerical instability for overhead lines has been observed when using MoM-SO as described in [28]. Considering the above facts, a modified earth-return Green function for a multiphase overhead line with N conductors is derived in this section. The numerical instability inherent in the existing MoM-SO [28] can be eliminated by adopting the newly derived Green function.

The configuration of a multiphase overhead line with N conductors illustrated in Appendix A3 is used. Green function considering a lossy earth can be derived based on the electromagnetic field components in the air and earth [65]:

$$\overline{\mathbf{E}} = -j\omega\overline{\mathbf{A}}_{\mathbf{M}} - j\frac{\nabla\nabla\cdot\overline{\mathbf{A}}_{\mathbf{M}}}{\omega\mu\varepsilon} - \frac{\nabla\times\overline{\mathbf{A}}_{\mathbf{E}}}{\varepsilon} \tag{2.79}$$

$$\overline{\mathbf{H}} = -j\omega\overline{\mathbf{A}}_{\mathbf{E}} - j\frac{\nabla\nabla\cdot\overline{\mathbf{A}}_{\mathbf{E}}}{\omega\mu\varepsilon} + \frac{\nabla\times\overline{\mathbf{A}}_{\mathbf{M}}}{\mu} \tag{2.80}$$

where $\overline{\mathbf{A}}_{\mathbf{M}}$ and $\overline{\mathbf{A}}_{\mathbf{E}}$ are vector magnetic and electric potentials and given by [65]

$$\overline{\mathbf{A}}_{\mathbf{M}} = j\omega\mu\varepsilon\overline{\mathbf{\Pi}}_{\mathbf{E}}, \overline{\mathbf{A}}_{\mathbf{E}} = j\omega\mu\varepsilon\overline{\mathbf{\Pi}}_{\mathbf{M}} \tag{2.81}$$

where $\overline{\mathbf{\Pi}}_{\mathbf{E}}$ and $\overline{\mathbf{\Pi}}_{\mathbf{M}}$ are given in (A3.3) and (A3.4).

By adopting the same modal analysis in Appendix A2, the x-component of Green function in the air can be derived:

$$G_{0x}(x, y, z) = \mathbf{\Phi}^{G}_{0x}\mathbf{B}^{-1}\mathbf{I} \tag{2.82}$$

where $\mathbf{B}$ is given in (A3.1), $\mathbf{I}$ is an $N \times 1$ phase domain current vector and each element is defined in (A3.1), and the $1v$-th element of $1 \times N$ vector $\mathbf{\Phi}^{G}_{0x}$ is given in (A3.47).

Equation (2.82) requires to find k_v and $\mathbf{B}$ as the first step. However, the quasi-TEM assumption $k_v \approx k_a$ and a uniform current assumption $k_v \approx 0$ can be used to simplify (2.82). It should be noted that MoM-SO calculates only the series impedance matrix but not the admittance, and the second term of (A3.25) (kernel function of (A3.47)) can be neglected because it represents the earth-return admittance. Then, the Green function based on quasi-TEM assumption has

$$\underline{G}_{0x}(x, y, z) = \underline{\mathbf{\Phi}}^{G}_{0x}\mathbf{I} \tag{2.83}$$

where $\mathbf{I}$ is given in (2.82) and the $1m$-th element of $1 \times N$ vector $\underline{\mathbf{\Phi}}^{G}_{0x}$ is

$$\underline{\Phi}^{G}_{0x(1m)}(y, z, d_m, h_m) = -\frac{j\omega\mu_0}{2\pi}\left\{\ln\left[\frac{D_{nm}(y, z, d_m, h_m)}{d_{nm}(y, z, d_m, h_m)}\right] + 2\underset{=1}{S}(y, z, d_m, h_m)\right\} \tag{2.84}$$

where $\underset{=1}{S}(y, z, d_m, h_m)$, $d_{nm}(y, z, d_m, h_m)$ and $D_{nm}(y, z, d_m, h_m)$ are given in

$$\underset{=1}{S}(y, z, d_m, h_m) = \int_0^{+\infty} \frac{e^{-(z+h_m)\lambda}}{\lambda + \sqrt{\lambda^2 + k_a^2 - k_e^2}} \cos[(y - d_m)\lambda] d\lambda \tag{2.85}$$

$$d_{nm}(y, z, d_m, h_m) = \sqrt{(y - d_m)^2 + (z - h_m)^2} \tag{2.86}$$

$$D_{nm}(y, z, d_m, h_m) = \sqrt{(y - d_m)^2 + (z + h_m)^2} \tag{2.87}$$

The self-elements of (2.84) are calculated by adopting $h_n = h_m$ and $d_n - d_m = r_n / d_n - d_m = r_m$. The mutual elements are calculated at the center of the two conductors.

Green function based on the uniform current assumption gives the same expression as (2.83), but with different elements of $\underline{\mathbf{\Phi}}_{0x}^{G}$:

$$\begin{aligned} \underline{\Phi}_{0x(1m)}^{G}(y, z, d_m, h_m) &= -\frac{j\omega\mu_0}{2\pi}\{K_0[jk_a d_{nm}(y, z, d_m, h_m)] \\ &\quad -K_0[jk_a D_{nm}(y, z, d_m, h_m)] + 2\underset{=1}{S^{\mathrm{u}}}(y, z, d_m, h_m)\} \end{aligned} \tag{2.88}$$

where K_0 is the modified Bessel function of the second kind with zero order and $\underset{=1}{S^{\mathrm{u}}}(y, z, d_m, h_m)$ is Sommerfeld integral (A3.35) based on the uniform current assumption:

$$\underset{=1}{S^{\mathrm{u}}}(y, z, d_m, h_m) = \int_0^{+\infty} \frac{e^{-(z+z_m)\sqrt{\lambda^2 - k_a^2}}}{\sqrt{\lambda^2 - k_a^2} + \sqrt{\lambda^2 - k_e^2}} \cos[(y - y_m)\lambda] d\lambda \tag{2.89}$$

Also, K_0 can be represented by the Bessel function of the third kind (Hankel function) [65]:

$$K_0(jw) = \frac{j\pi}{2} H_0^{(2)}(w) \tag{2.90}$$

where w is the argument.

2.8 Comparison of calculated impedance and admittance by different methods

2.8.1 Overhead lines

Figure 2.3 illustrates a single overhead conductor investigated in this section. The conductor resistivity is $\rho_c = 1.68 \times 10^{-8}$ Ωm. The earth resistivity and permittivity are ρ_e and ε_e.

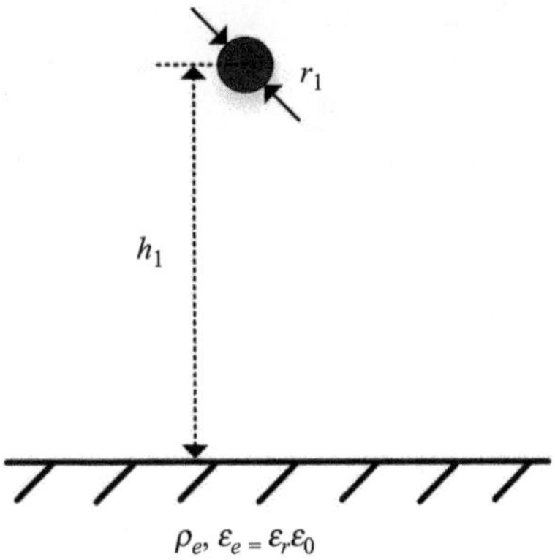

Figure 2.3 A single overhead conductor: $h_1 = 10$ m and $r_1 = 1$ cm

2.8.1.1 Series impedance

Figure 2.4 shows calculated results of the series impedance of a single overhead conductor by the extended TL approach. The influence of earth parameters is clearly observed. The resistance increases monotonously as frequency increases. A saturated behavior of the resistance is observed in high frequencies if a large earth permittivity is used such as $\varepsilon_r = 20$. This is caused by displacement currents in the earth. Also, a large earth resistivity increases the resistance in high frequencies.

Moreover, the earth permittivity has a negligible effect on the inductance in the whole frequency range. Similar to the resistance, the earth resistivity shows a significant influence on the inductance.

Figure 2.5 shows a comparison of the impedance between the extended and classical approaches. No difference is observed between the extended and classical TL approaches for $\varepsilon_r = 1$. It is well-known that the earth-return impedance of the classical approach is the same as the extended TL approach for $\varepsilon_r = 1$, as discussed and reviewed in Section 2.1.

2.8.1.2 Numerical instability of MoM-SO and solution

Figure 2.6 shows the series impedance of the conductor illustrated in Figure 2.3 calculated by original MoM-SO corresponding to (2.88), improved MoM-SO with (2.84) and the extended TL approach. It is apparent that a numerical instability is observed in the results calculated by the original MoM-SO above several hundreds of kHz. The reason for this is due to the uniform current assumption adopted in the Green function of the earth-return path. The two Bessel functions in (2.88) cause the unstable numerical values at high frequencies.

Further as shown in Figure 2.6, the results of the improved MoM-SO with (2.84) agree very well with those by the extended TL approach. Therefore, the quasi-TEM-based Green function (2.84) should be adopted to model an overhead line or an overhead cable to avoid the instability in the original MoM-SO.

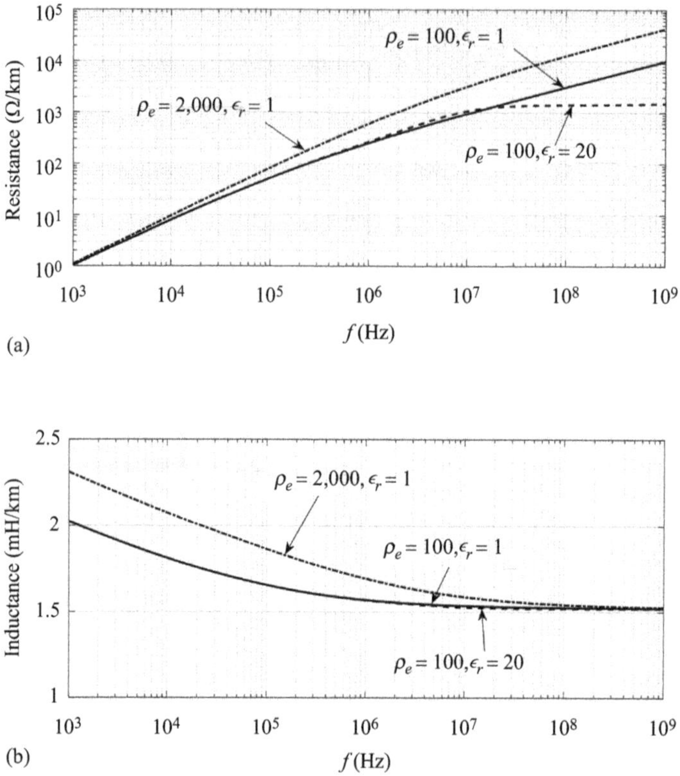

Figure 2.4 Series impedance of the conductor illustrated in Figure 2.3 by the extended TL approach. (a) Resistance and (b) inductance

2.8.1.3 Shunt admittance

The calculated conductance and capacitance of the shunt admittance in Figure 2.3 are shown in Figure 2.7 for various earth resistivities and permittivities. The earth permittivity shows less influence on the admittance in the entire frequency region. The admittance is more sensitive to the earth resistivity.

Figure 2.8 shows a comparison of the shunt admittance between the extended and classical TL approaches. It is recalled that the potential coefficient in the classical TL approach is a real constant, that is, the admittance is purely capacitive. Thus, no frequency-dependent characteristic of the shunt admittance appears in the classical TL approach. The influence of a lossy earth correction term of (2.33) in a high-frequency region on the propagation function calculated by the extended TL approach is significantly different from that calculated by the classical TL approach.

2.8.2 Underground cables

Three important configurations and parameters of SC cables are summarized firstly [7]. The series impedance and shunt admittance of a cable system are investigated

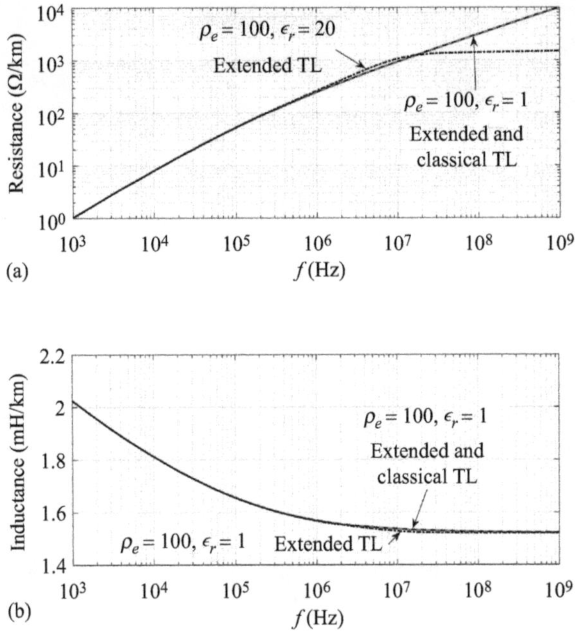

Figure 2.5 Series impedance of the conductor illustrated in Figure 2.3 by the extended and classical TL approaches. (a) Resistance and (b) inductance

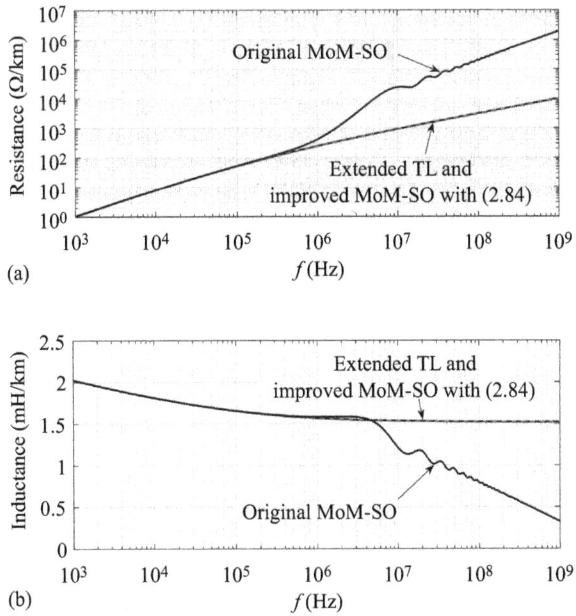

Figure 2.6 Series impedance of the conductor illustrated in Figure 2.3 by the extended TL approach and MoM-SO, $\rho_e = 100$ Ωm, and $\varepsilon_r = 1$. (a) Resistance and (b) inductance

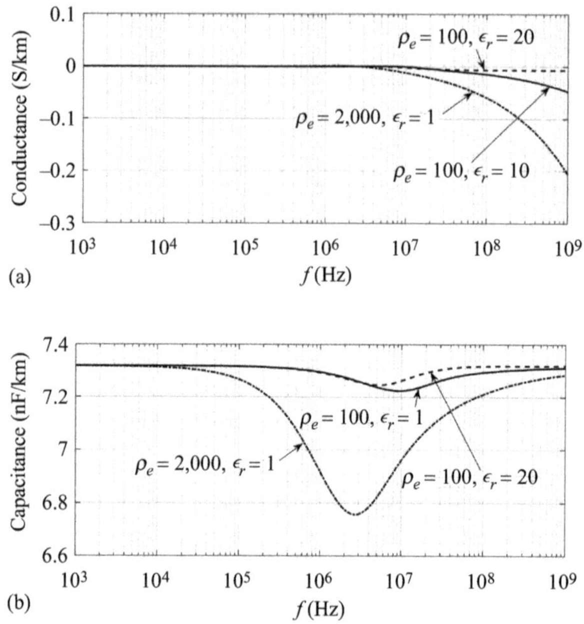

Figure 2.7 Shunt admittance of the conductor illustrated in Figure 2.3 by the extended TL approaches. (a) Conductance and (b) capacitance

based on different approaches and configuration of Cable 1 in this section. The Cables 2 and 3 are further used in the discussions of wave propagation characteristics and transient simulations in next several chapters.

The cross-section of a single core cable is shown in Figure 2.9. Three different cable arrangements are illustrated in Figure 2.10 [7].

The parameters adopted into three cables are summarized in Table 2.2.

2.8.2.1 Series impedance

Cable 1 shown in Figures 2.9 and 2.10(a) is used to calculate the series impedance (2.76).

Self-impedance of phase-a sheath

Figure 2.11 shows the calculated series impedance of phase-a sheath by the extended TL approach. The earth resistivity and permittivity have a minor influence on the resistance in high frequencies. However, a clear difference in the inductance is observed between $\rho_e = 100$ and $\rho_e = 500$ Ωm. Also, the earth permittivity shows a minor influence on the inductance in high frequencies.

The calculated series impedance of phase-a sheath is compared in Figure 2.12 by the extended TL approach, equation (2.45), MoM-SO, and classical TL approach. It is clear that the results by MoM-SO agree very well with the results by

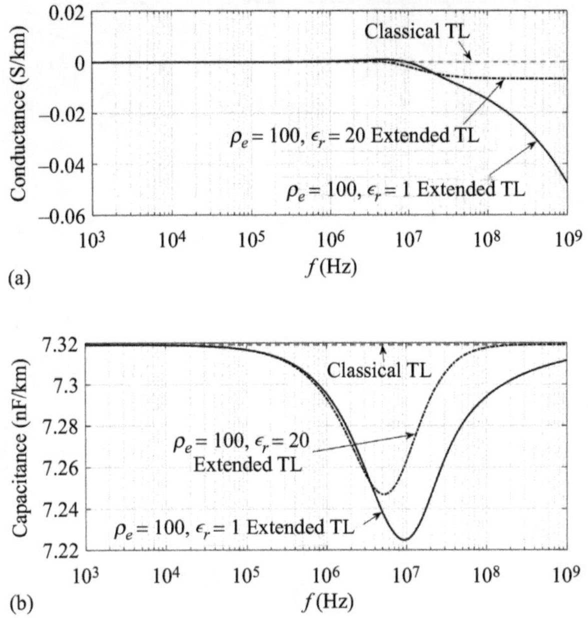

Figure 2.8 Shunt admittance of the conductor illustrated in Figure 2.3 by the extended and classical TL approaches. (a) Conductance and (b) capacitance

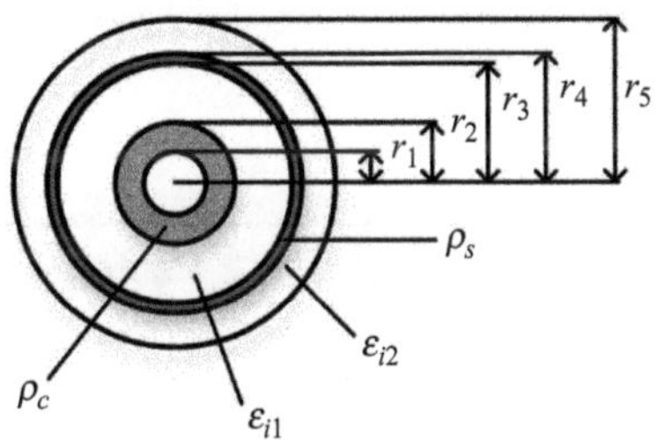

Figure 2.9 Cross-section of a single core cable

the extended TL approach. No significant difference is observed at low frequency between all the methods, that is, $f \leq 1$ MHz.

Mutual impedance between phase-a and phase-b sheaths
Similarly to self-impedance, the earth permittivity causes a minor effect on the mutual impedance in high frequencies as shown in Figure 2.13. The earth resistivity shows a significant influence on the inductance.

As shown in Figure 2.14, only a minor difference is observed between different formulas in high frequencies. MoM-SO also agrees very well with the extended TL approach.

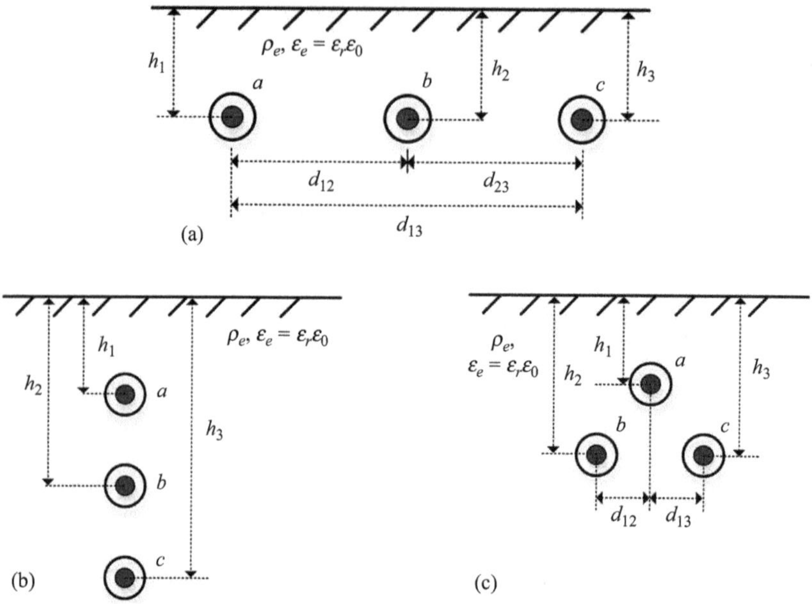

Figure 2.10 Arrangement of three-phase single core cable. (a) Flat arrangement of Cable 1, (b) vertical arrangement of Cable 2, and (c) trefoil arrangement of Cable 3

Table 2.2 Parameters of cables

	Cable 1	**Cable 2**	**Cable 3**
Voltage (kV)	132	110	150
Arrangement	Flat	Vertical	Trefoil
h_1 (m)	1	1	1.213
h_2	1	1.35	1.3
h_3	1	1.7	1.3
d_{12} (m)	0.35	0	0.05
d_{13}	0.7	0	0.05
d_{23}	0.35	0	0.1
r_1 (cm)	1.03	1.049	0
r_2	1.9	1.78	2.075
r_3	3.45	3.075	4.025
r_4	3.85	3.255	4.276
r_5	4.25	4.075	4.794
ρ_c (Ωm)	1.7×10^{-8}	1.7×10^{-8}	3.19×10^{-8}
ρ_s	2.1×10^{-7}	2.1×10^{-7}	2.228×10^{-7}
ε_{i1}	3.5	3.5	2.68
ε_{i2}	4	4	2.3

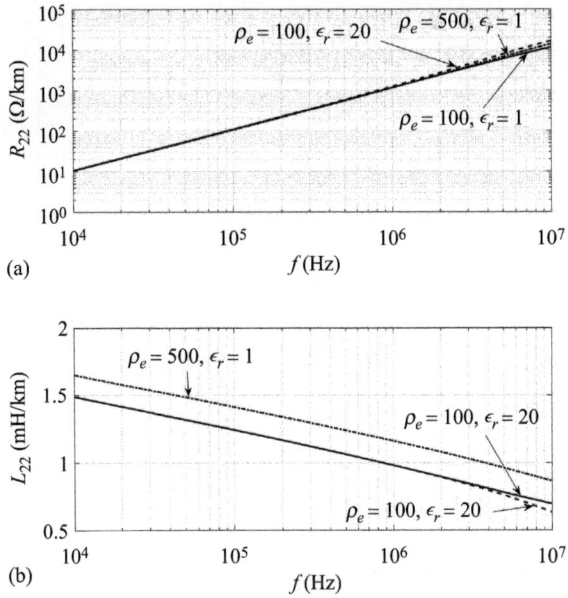

Figure 2.11 Self-impedance of phase-a sheath by the extended TL approach. (a) Self-resistance and (b) self-inductance

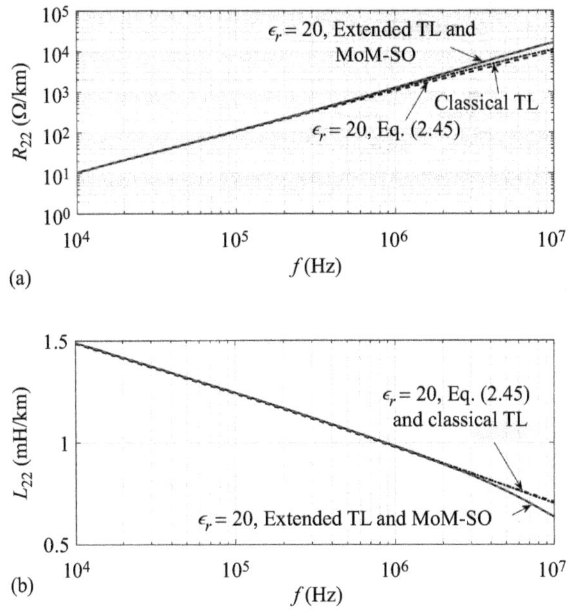

Figure 2.12 Self-impedance of phase-a sheath, $\rho_e = 100\ \Omega m$. (a) Self-resistance and (b) self-inductance

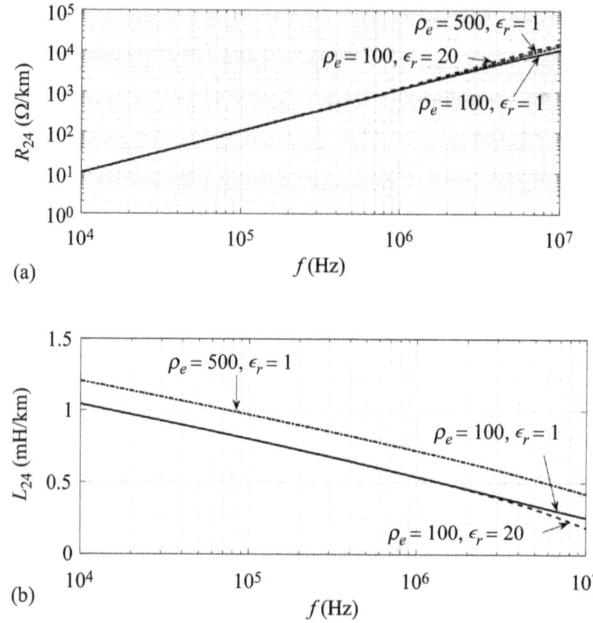

Figure 2.13 Mutual impedance between phase-a and phase-b sheaths by the extended TL approach. (a) Mutual resistance and (b) mutual inductance

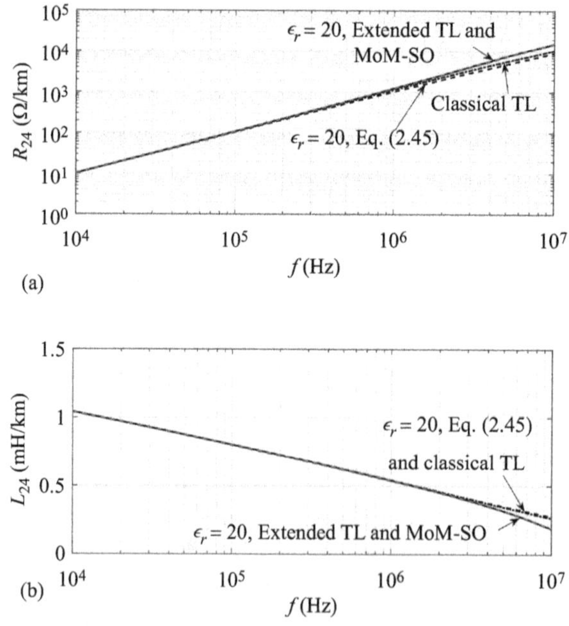

Figure 2.14 Mutual impedance between phase-a and phase-b sheaths, $\rho_e = 100$ Ωm. (a) Mutual resistance and (b) mutual inductance

2.8.2.2 Shunt admittance

The shunt admittance matrix (2.77) of Cable 1 is calculated and the configuration is shown in Figures 2.9 and 2.10(a).

Self-admittance of phase-a sheath

The self-admittance of phase-*a* sheath is calculated and shown in Figures 2.15 and 2.16. The earth resistivity has a more significant influence on shunt admittance than earth permittivity. By adopting the earth-return admittance, the conductance increases monotonously at high frequency. Capacitance is nearly the same as the insulator capacitance in a low frequency and it decreases as frequency increases. The classical TL approach shows no frequency-dependent characteristics of admittance, and it is because only the insulator admittance has been adopted in (2.77). The difference between extended TL approach and (2.68) comes from the different methods of formulations.

Mutual admittance between phase-a and phase-b sheaths

The mutual admittances between phase-*a* and phase-*b* sheaths are calculated and shown in Figures 2.17 and 2.18. Similar to the self-admittance of phase-*a* sheath, the mutual admittance is more sensitive to earth resistivity than earth permittivity. Again, the frequency-dependent characteristics of mutual admittance have been observed in high frequencies if the earth-return admittance is considered. A more noticeable difference is observed between the extended TL approach and (2.68). The classical TL approach deals with no mutual admittance as is well-known.

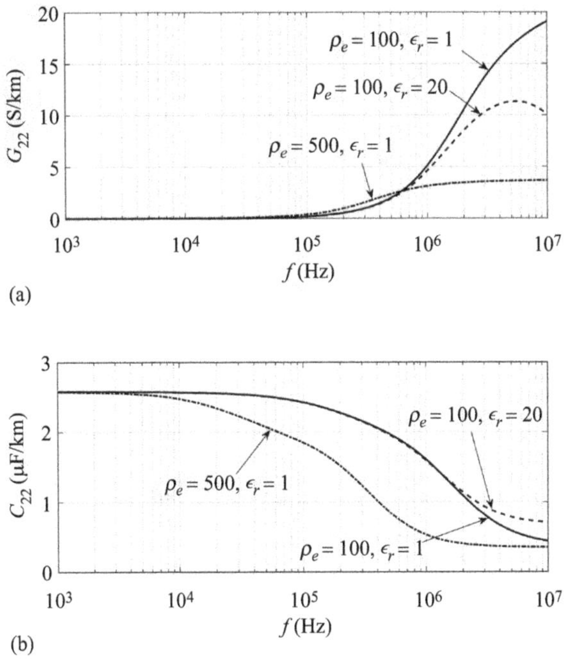

Figure 2.15 Self-admittance of phase-a sheath by the extended TL approach. (a) Self-conductance and (b) self-capacitance

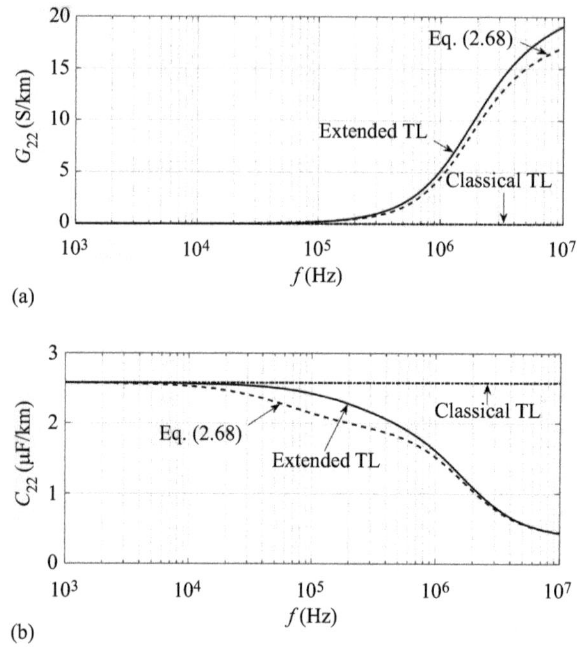

Figure 2.16 Self-admittance of phase-a sheath, $\rho_e = 100\ \Omega m$ and $\varepsilon_r = 1$. (a) Self-conductance and (b) self-capacitance

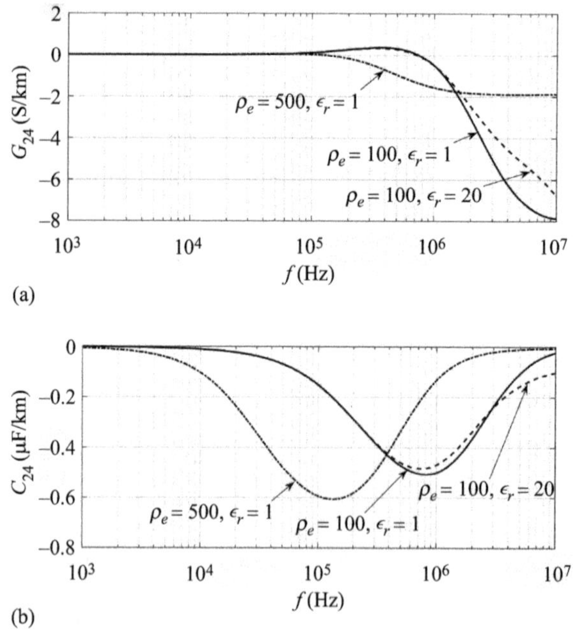

Figure 2.17 Mutual admittance between phase-a and phase-b sheaths by the extended TL approach. (a) Mutual conductance and (b) Mutual capacitance

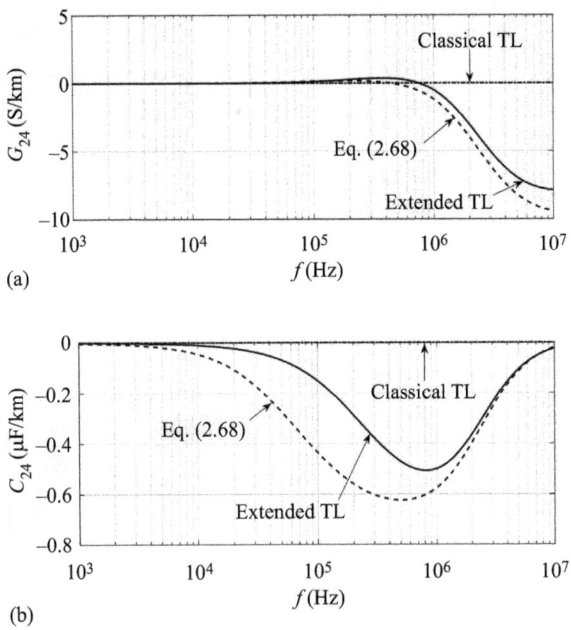

Figure 2.18 Mutual admittance between phase-a and phase-b sheaths, $\rho_e = 100$ Ωm and $\varepsilon_r = 1$. (a) Mutual conductance and (b) Mutual capacitance

Influence of integral reference h_r

As discussed in Section 2.6.1, the quasi-TEM-based earth-return potential coefficient is dependent on the integral reference h_r. Figures 2.19 and 2.20 show the calculated self and mutual admittances by adopting different earth-return potential coefficient formulas in (2.78). Formulas (2.55), (2.57), and (2.61) are derived under the condition of $h_r = -\infty$, $h_r = 0$, and $h_r = -\delta_e$, respectively. In addition, the cases of $h_r = -0.2$, -2, and -20 m, are calculated in comparison with (2.55), (2.57), and (2.61). It should be noted that the results by (2.70) are the same as those by (2.57).

As shown in Figures 2.19 and 2.20, only minor differences of admittance are observed between the proposed formulas with (2.55) and (2.61). A significant difference of the results is observed between $h_r = 0$ and $h_r = -\infty$. This difference is explained by the penetration depth of earth in Figure 2.21. The penetration depth is ranging from 100 m to 10 m for frequencies 10 kHz $< f <$ 1 MHz. The depth is much larger than the distance between the cable and the earth surface (1 m in this case). Thus, the vertical electric field component cannot decay sufficiently at the earth surface and results in the non-negligible effects in (A2.74). Moreover, it is clear that admittance converges to the results calculated by $h_r = -\infty$ if h_r is changed from -0.2 m to -20 m.

Validation of approximate earth-return potential coefficient formula

In this section, the proposed approximate earth-return potential coefficient formula (2.73) is validated in comparison to the results evaluated by extended TL approach.

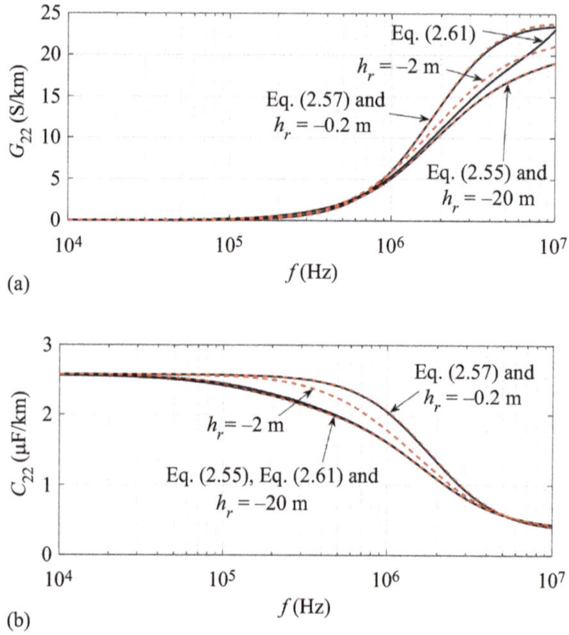

Figure 2.19 Self-admittance of phase-a sheath, $\rho_e = 100$ Ωm and $\varepsilon_r = 1$. (a) Self-conductance and (b) self-capacitance

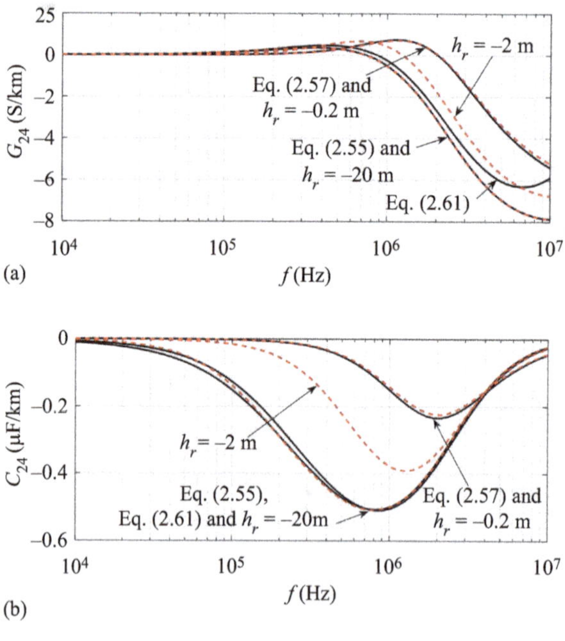

Figure 2.20 Mutual admittance between phase-a and phase-b sheaths, $\rho_e = 100$ Ωm and $\varepsilon_r = 1$. (a) Mutual conductance and (b) Mutual capacitance

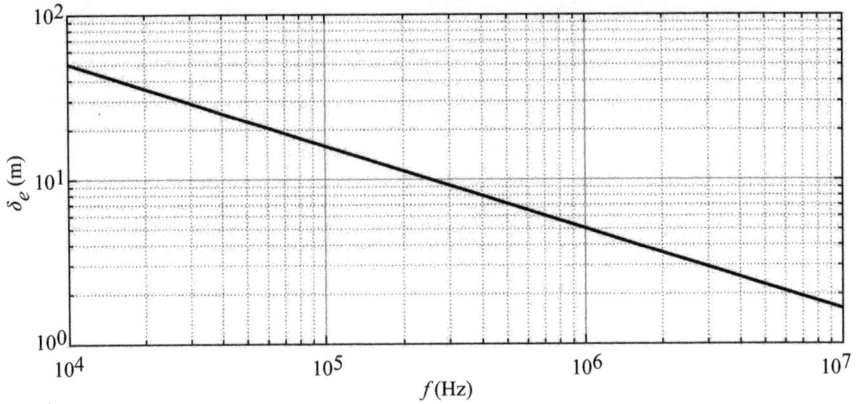

Figure 2.21 Penetration depth of earth calculated by (2.67), $\rho_e = 100\ \Omega m$ and $\varepsilon_r = 1$

As shown in Figures 2.22 and 2.23, the results calculated by (2.73) agree well with the results evaluated by extended TL approach. Therefore, (2.73) can be used as a reasonable approximation in cable transient calculations, and it avoids solving any types of Sommerfeld integrals shown in accurate formulas.

2.8.3 Concluding remarks

This section shows a comparison of the earth-return impedance and admittance calculated by different formulas. Based on the comparison, the following remarks are made clear.

1. Overhead line
    1. The earth resistivity has impacts on the resistance of series impedance and conductance of shunt admittance at high-frequency regions, while it influences the inductance of series impedance at low-frequency regions. The capacitance of shunt admittance noticeably depends on the earth resistivity at middle-frequency regions. Note that the classical TL approach cannot consider the frequency-dependent conductance and capacitance.
    2. The effect of earth permittivity is noticeable at high-frequency regions.
    3. The original MoM-SO shows a numerical instability at high-frequency regions because of assumption of uniform current distribution in earth-return Green function.

2. Underground cable
    1. The influence of earth resistivity and permittivity on the resistance of series impedance is less noticeable in underground cables than that in overhead lines.
    2. The earth resistivity influences the inductance of series impedance in a wide range of frequencies.

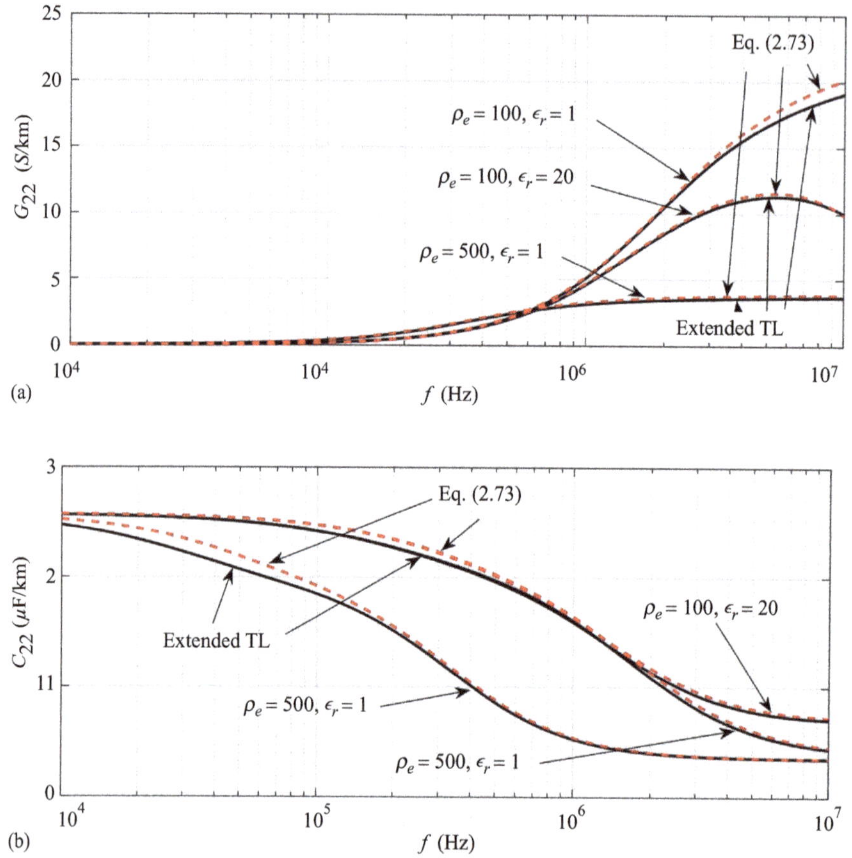

Figure 2.22 Self-admittance of phase-a sheath by the extended TL approach and (2.73). (a) Self-conductance and (b) self-capacitance

3. The earth resistivity and permittivity influence the conductance of shunt admittance at high-frequency regions, while these influence the capacitance of shunt admittance in a wide range of frequencies.

3. The influences of integral reference of earth-return admittance and various formulas are made clear.

2.9 Impedance and admittance at $f = 0$ for HVDC line [66]

Nowadays, a large number of high voltage direct current (HVDC) transmission lines are under operation and more lines are being constructed. In China alone, over 40,000 km of HVDC lines have been constructed. In 2019, the world highest voltage (1,100 kV) and longest (3,300 km) HVDC transmission line was completed and commissioned. On the American Continent, HVDC lines cover more than

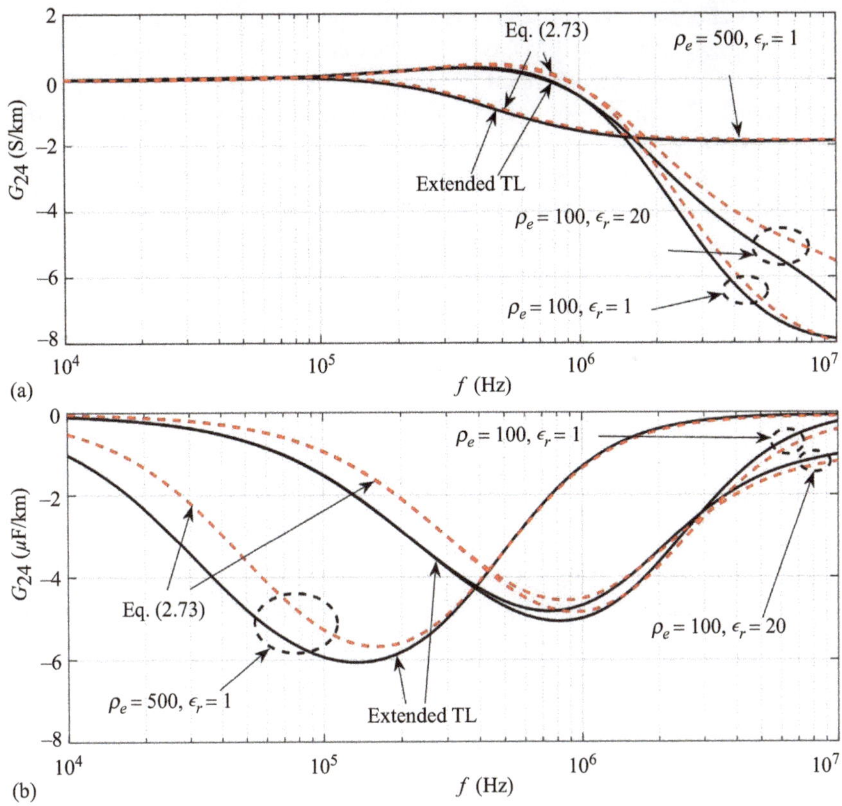

Figure 2.23 Mutual admittance between phase-a and phase-b sheaths by the extended TL approach and (2.73). (a) Mutual conductance and (b) Mutual capacitance

20,000 km, and in Africa and Europe, over 10,000 km, which makes the total length of HVDC lines greater than the earth's circumference!

Planning and designing new HVDC systems, and to protect and control existing HVDC systems, the steady state, dynamics, and transient operation of these lines have to be accurately predicted. For this purpose, the EMT-type simulation tools have been widely used [4,6]. Accurate determination of the line impedance and admittance as functions of frequency is very important for accurate simulations. The steady-state operating frequency of HVDC lines is 0 Hz; however, the transients on the line, say due to a fault or lightning strike are at much higher frequencies. Hence, a very wide-band model of the line is required. Most previous studies try to fit the line parameters accurately in the 0.001 Hz to 1 Hz frequency range [66–71]. However, because the nominal operating frequency of the HVDC system is $f = 0$, we also require the electric parameters at 0 Hz [66].

For calculating the resistances, inductances, and capacitances of overhead lines and underground cables, formulas by Carson's and Pollaczek's (for the earth-return impedance) and Schelukunoff (for conductor internal impedance) [1,2,8,9,25,26] have been implemented in EMT-type simulation tools [4,6]. Although Carson's formula agrees well with measured impedance values in the 0.1 Hz to 10 Hz frequency range [72], there is no investigation into the accuracy of Carson's formula at extremely low (near-zero) frequency. The EMT-type line models also require the propagation velocity and the characteristic impedance, and again, no previous research has been performed to discuss their behavior at extremely low frequencies. Carson's formula shows that the propagation velocity converges to zero, and characteristic impedance becomes infinite as $f \to 0$.

This section investigates the impedance and admittance at very low frequencies using extended and classical TL approaches defined in Section 2.6.5 [15–18,26,73–75] based on a newly proposed approach in [66].

2.9.1 *Impedance in a low-frequency condition*

2.9.1.1 Classical TL approach

Carson's (Pollaczek's) formulas of earth-return impedance (2.8) and (2.23) have been implemented into classical TL approach as reviewed and defined in Sections 2.3 and 2.6.5. Using a low-frequency assumption in (2.8) and (2.23), the following closed-form formula [76] for low frequency is easily deduced:

$$Z_{ij} = f + jf\left[8.253 + 0.628\ln\left(\frac{\rho_1}{fr^2}\right)\right] \times 10^{-6} \tag{2.91}$$

where ρ_1 is the earth resistivity and r is the radius of a single conductor.

If the mutual impedance of a multiphase conductor system is considered in (2.91), r should be replaced by d_{ij} shown in (2.12).

Also, Gary/Deri derived another approximate formula [25,26] as explained in Section 2.3.3.

2.9.1.2 Impedance calculated by classical TL approach in a low-frequency condition

Figure 2.24 shows a comparison between measured results and impedances calculated by classical TL approach. The voltage level of tested overhead line is 750 kV, and earth conductivity is specified as $\sigma_e = 0.0143$ (S/m), that is, the resistivity is $\rho_e = 69.9$ (Ωm). It should be noted that the measured and calculated results are zero-sequence impedance, but not the self-impedance of a phase conductor. It is stated in [75] that a difference observed between the measured and calculated results is caused by unknown parameters such as resistivity of stratified earth.

Figure 2.25 shows very low-frequency characteristics of earth-return impedance (resistance and inductance) of a single overhead conductor with height $h = 20$ m, radius $r = 2.0$ cm, conductor resistivity $\rho_c = 3.78\times10^{-8}$ Ωm and earth resistivity $\rho_e = 100$ Ωm. Also, the earth permittivity ε_e is set to ε_0. In the following calculations, the same geometrical and physical parameters are used. In Figure 2.25, the red solid line shows the results evaluated by extended TL approach, and the blue dashed line shows the results calculated by Carson's

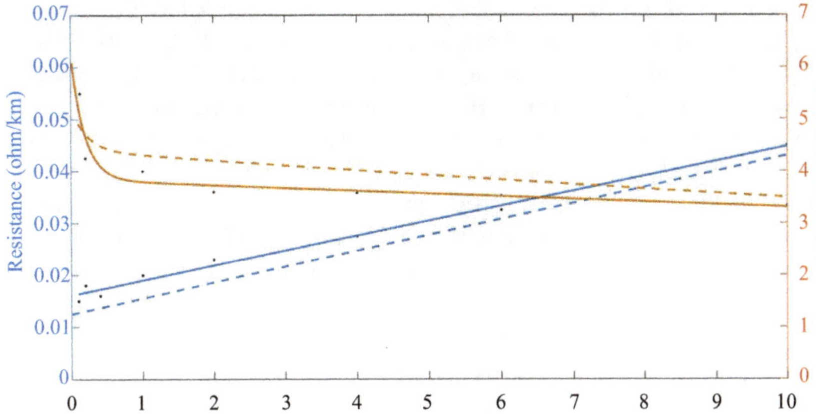

Figure 2.24 Measured results [75] of zero-sequence impedance at low frequencies, solid line: measured results and dashed line: classical TL approach

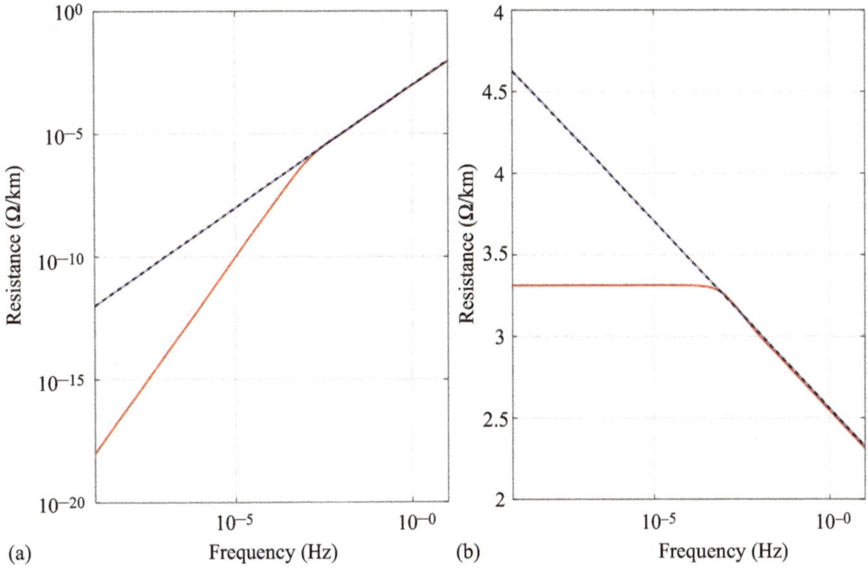

Figure 2.25 Earth-return impedance calculated by low-frequency condition, red solid line: extended TL, blue dashed line: (2.91) and black dotted line: classical TL with $h = 20$ m, $r = 2$ cm, $\rho_e = 100$ Ωm, $\varepsilon_e = \varepsilon_0$, and $\rho_c = 3.78 \times 10^{-8}$ Ωm. (a) Resistance R and (b) inductance L

low-frequency formula (2.91), and the results produced by classical TL approach are marked by black dotted line. No significant difference is observed in the case of $\rho_e = 2000$ Ωm, although the result is not shown.

It is observed that the impedances calculated by Carson's low-frequency formula (2.91) are identical to the results evaluated by classical TL approach, and the results are nearly the same as those calculated by the extended TL approach in frequency region higher than 0.01 Hz. For the frequency lower than 0.01 Hz, the results obtained by classical TL and (2.91) deviate significantly from the results produced by extended TL approach, and the deviation tends to increase monotonously as frequency decreases. The reason for the deviation is explained in Section 2.9.4. Remind that the impedance produced by (2.91) and classical TL approach agree well with the results calculated by extended TL in a high-frequency region (lower than 2 MHz) [15,26].

2.9.2 Admittance in a low-frequency condition

Figure 2.26 shows the very low-frequency characteristics of earth-return admittance, that is, conductance and capacitance, which are calculated by extended TL approach (2.32) and (2.34) as explained in Section 2.4. The same overhead conductor shown in Figure 2.25 is adopted in the calculations illustrated in Figure 2.26. Also, the space capacitance used in EMT-type simulation tools as the classical TL approach is calculated and drawn by blue solid line in Figure 2.26(b). Only a minor difference of capacitance is observed between the results evaluated by classical and extended TL approaches. The conductance produced by extended TL approach in Figure 2.26(a) decreases monotonously as frequency decreases, and can be regarded as zero (10^{-37} S/m) at $f = 10^{-10}$ Hz.

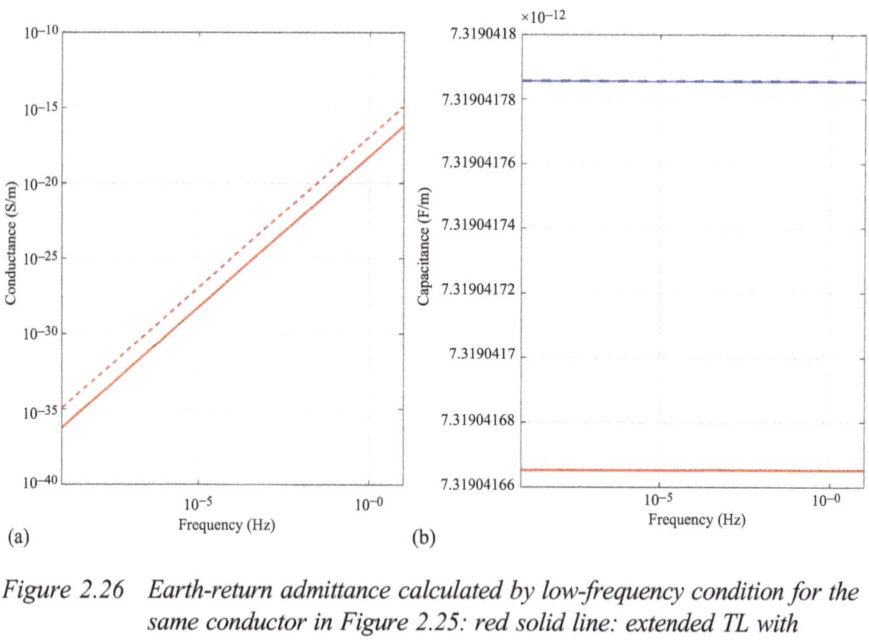

Figure 2.26 Earth-return admittance calculated by low-frequency condition for the same conductor in Figure 2.25: red solid line: extended TL with $\rho_e = 100\ \Omega m$, red dashed line: extended TL with $\rho_e = 2{,}000\ \Omega m$, and blue solid line: classical TL. (a) Conductance G and (b) Capacitance C

2.9.3 *Wave propagation characteristics in a low-frequency condition*

2.9.3.1 Propagation constant

The propagation constant γ of the single overhead conductor discussed in Figure 2.25 can be calculated by the following formula:

$$\gamma = \sqrt{ZY} \tag{2.92}$$

where Z and Y are series impedance and shunt admittance of conductor, and can be evaluated by classical and extended TL approaches.

Figure 2.27 shows the calculated attenuation constant and phase velocity in a low-frequency region. The attenuation constant and propagation velocity decrease as frequency decreases, and tend to converge to zero. No significant difference of propagation constant is observed for extended and classical TL approaches. Above 1 Hz, the phase velocity evaluated by $\rho_e = 100\ \Omega\text{m}$ becomes greater than the results calculated by $\rho_e = 2{,}000\ \Omega\text{m}$.

2.9.3.2 Characteristic impedance

Figure 2.28 shows the characteristic impedance calculated by extended and classical TL approaches in a low-frequency condition. The same overhead conductor used in Figure 2.25 is adopted in the following calculations. It is observed that the characteristic impedance monotonously increases as frequency decreases, and tends to become infinite at $f = 0$. No noticeable difference is observed between the results calculated by extended and classical TL approaches.

2.9.4 *Investigation of parameters at a frequency approaching zero*

2.9.4.1 Impedance and capacitance

Having confirmed that (2.91) shows a reasonable accuracy in a low-frequency region higher than 0.001 Hz; therefore, the low-frequency approximate formula (2.91) is used to find an analytical solution of earth-return impedance. At a low frequency, the following expression is derived under the condition of "h_e" being far greater than "h" in (2.91) with a very small "f":

$$Z = R_{dc} + R_e + j\omega(L_c + L_e) \tag{2.93}$$

where

$$R_{dc} = \frac{\rho_c}{S}\,(\Omega/\text{m}) \tag{2.94}$$

$$R_e = f = \frac{\omega\mu_0}{8}\ (\mu\Omega/\text{m}) \tag{2.95}$$

$$L_c = \frac{\mu_0}{8\pi}\ (\text{H}/m) = 0.05\ (\mu\text{H/m}) \tag{2.96}$$

$$L_e = \left[1.3135 + 0.1\left\{\ln\left(\frac{\rho_e}{r^2}\right) - \ln(f)\right\}\right]\ (\mu\text{H/m}) \tag{2.97}$$

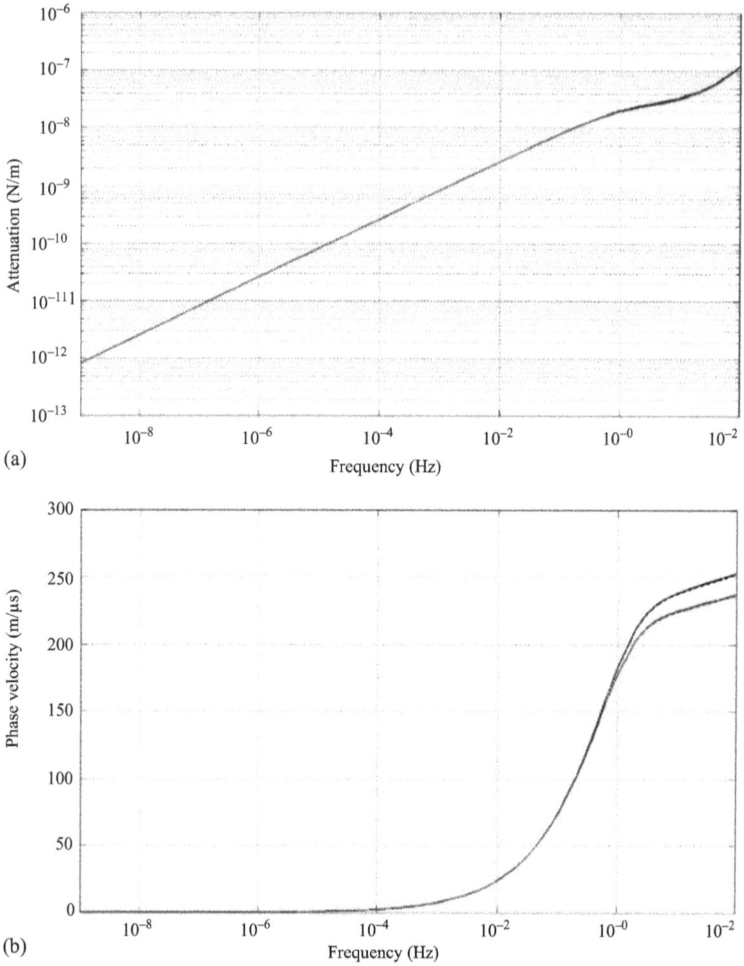

Figure 2.27 Propagation constant calculated by low-frequency condition for the same conductor in Figure 2.25: solid line: extended TL, dashed line: classical TL, black: $\rho_e = 100\ \Omega\text{m}$, and red: $\rho_e = 2{,}000\ \Omega\text{m}$. (a) Attenuation and (b) phase velocity

with

$$S = \pi r^2 \tag{2.98}$$

The above formula can be also deduced from (2.27).

It is clear from the impedance formula (2.93) that the resistance converges to the DC resistance R_{dc} of a conductor at $f = 0$. The inductance increase proportionally to $-\ln(f)$, and tends to be infinite at $f = 0$. This analytical investigation explains the monotonous increase of the inductance as frequency decreases

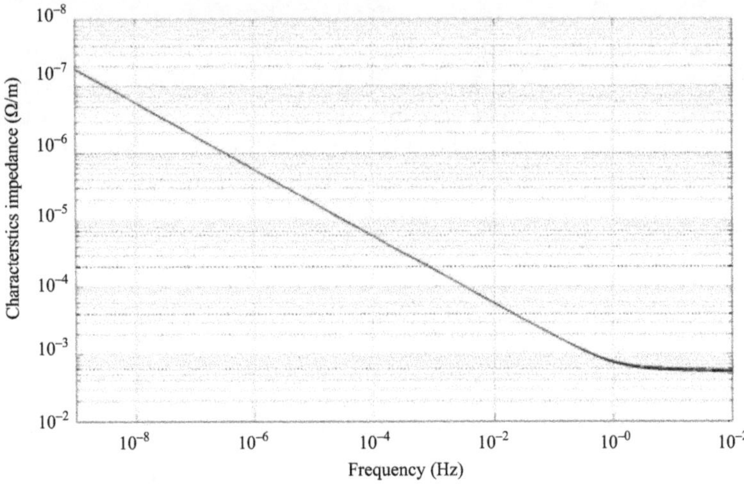

Figure 2.28 Real part of characteristic impedance calculated by low-frequency condition for the same conductor in Figure 2.25: solid line: extended TL, dashed line: classical TL, black: ρ_e = 100 Ωm, and red: ρ_e = 2,000 Ωm

observed in the result calculated by classical TL approach in Figure 2.25(b). On the contrary, the extended TL approach tends to converge a constant inductance in the frequency region lower than 10^{-3} Hz. It is hard to accept physically an infinite inductance obtained by (2.91) and (2.27). Therefore, it can be concluded that classical TL approach is incorrect at a low-frequency region.

The converged value L_{e0} of the inductance estimated by earth-return impedance of extended TL approach is a function of the conductor height, radius, and the earth resistivity.

From the above investigation, the series impedance for a frequency at $f = 0$ can be given by

$$Z_0 = R_{dc} + j\omega L_0 \tag{2.99}$$

where

$$L_0 = L_c + L_{e0} \tag{2.100}$$

It is observed in Figure 2.26 that there is no significant difference between classical space capacitance and capacitance calculated by the accurate extended TL approach. Therefore, we can adopt classical TL approach (2.34) for the calculation of capacitance at a very low frequency, that is

$$C_0 = \frac{2\pi\epsilon_0}{P_0}, \quad P_0 = \ln\left(\frac{D_{ij}}{dij}\right) (2.34)$$

2.9.4.2 Phase velocity and characteristic impedance

As observed in Figure 2.27(b), the propagation velocity decreases as frequency decreases and tends to converge zero at $f = 0$. The characteristic impedance monotonously increases as frequency decreases as observed in Figure 2.28. The observation can be explained analytically from the following equations. The propagation constant and characteristic impedance are obtained by

$$\gamma = \sqrt{ZY} \text{ and } \quad Z_c = \sqrt{\frac{Z}{Y}} \tag{2.101}$$

where

$$Z = R + j\omega L \text{ and } Y = G + j\omega C \tag{2.102}$$

When f reaches zero, $Z = Z_0 = R_{dc} + j\omega L_0$, and $Y = j\omega C$ as G is nearly zero as observed in Figure 2.26(a).

Then, γ is given in the following form:

$$\gamma = \sqrt{j\omega C(R_{dc} + j\omega L_0)} = \sqrt{R_{dc} j\omega C} \tag{2.103}$$

which can be further simplified into

$$\gamma = e^{\frac{j\pi}{4}} \sqrt{\omega C R_{dc}} = \alpha + j\beta \tag{2.104}$$

The phase velocity v is defined as

$$v = \frac{\omega}{\beta} = \sqrt{\frac{\omega}{R_{dc} C}} \tag{2.105}$$

The characteristic impedance Z_c is given by

$$Z_c = \sqrt{\frac{R_{dc} + j\omega L_0}{j\omega C}} = \sqrt{\frac{R_{dc}}{j\omega C}} = \mathrm{e}^{-\frac{j\pi}{4}} \sqrt{\frac{R_{dc}}{\omega C}} \tag{2.106}$$

It is clear from (2.105) and (2.106) that the phase velocity converges to zero and characteristic impedance tends to be infinite at $f = 0$. The investigation corresponds to conventional approaches to deal with a steady state in HVDC transmission systems. However, it is hard to accept physically the phase velocity reaches zero and the characteristic impedance becomes infinite at $f = 0$.

2.9.5 *New approach to determine the parameters at $f = 0$*

2.9.5.1 Distortion-less line model based on circuit theory

In a steady state of an HVDC transmission line, it can be assumed that no wave distortion occurs during current propagation from the sending end to receiving end.

Only a voltage drop $V = R_{dc}\, I_{dc}$ might be observed between the sending and receiving ends due to DC resistance R_{dc}. The phenomenon can be expressed by a distortion-less distributed line which has the following characteristic [76]:

$$\frac{R}{L} = \frac{G}{C} \tag{2.107}$$

The circuit parameters R, L, and C are given in (2.93) and (2.34) for a very low-frequency condition, the conductance G_0 is estimated by

$$G_0 = \frac{R_{dc}C_0}{L_0} \tag{2.108}$$

The characteristic impedance in a very low frequency is defined by

$$Z_c = Z_{c0} = \sqrt{\frac{L_0}{C_0}} \tag{2.109}$$

2.9.5.2 Method-1: Theoretical estimation of G_0

R_{dc}, and C_0 are calculated from (2.93) and (2.34). L_0 is estimated from the converged inductance in Figure 2.25(b) with $\rho_e = 100\ \Omega\text{m}$ based on extended TL approach. Therefore, it has

$$R_{dc} = 30.08 \times 10^{-6}\ (\Omega/\text{m}) \tag{2.110}$$

$$L_0 = 3.23 \times 10^{-6}\ (\text{H/m}) \tag{2.111}$$

$$C_0 = 7.31 \times 10^{-12}\ (\text{F/m}) \tag{2.112}$$

Then, the conductance and characteristic impedance are evaluated and given by

$$G_0 = 68.08 \times 10^{-12}\ (\text{S/m}) \tag{2.113}$$

$$Z_{c0} = 664.7\ (\Omega) \text{ and } Z_{cf} = 456.1(\Omega)\ (\text{in high} - \text{frequency region}) \tag{2.114}$$

From the above values, the attenuation constant α and phase velocity v in low-frequency region are calculated as

$$\alpha = \sqrt{R_{dc}G_0} = 45.25 \times 10^{-9}\ (\text{Np/m}) \tag{2.115}$$

$$v = \frac{1}{\sqrt{L_0C_0}} = 205.8\ (\text{m}/\mu\text{s}) \tag{2.116}$$

Also, the phase constant β is calculated as

$$\beta = \omega\sqrt{L_0C_0} = \omega 4.859 \times 10^{-9}\ (\text{rad/m}) \tag{2.117}$$

2.9.5.3 Method-2: Estimating G_0 at any location x along HVDC line from monitored voltage V_x

The line voltage V_x is monitored in a substation along an HVDC transmission line. Assume $V_x = 460$ kV at $x = 900$ km on a ± 500 kV ($=V_0$) HVDC line. The line geometry and physical parameters: conductor height $h = 30$ m, bundled conductor radius $r = 0.02$ m, bundle separation $s = 0.457$ m, equivalent radius for earth-return $r_e = 0.0956$ m and earth resistivity $\rho_e = 100\ \Omega$m.

Calculated results of DC resistance is $R_{dc} = 14.26 \times 10^{-6}\ (\Omega/\mathrm{m})$ and space capacitance $C_0 = 8.64 \times 10^{-12}$ (F/m).

The following relation is given in a circuit theory [76]:

$$\alpha = \ln\left(\frac{v_0}{v_x}\right) = 9.265 \times 10^{-8}\ (\mathrm{Np/m}) \tag{2.118}$$

For $\alpha = \sqrt{R_{dc} G_0}$, G_0 is determined as

$$G_0 = \frac{\alpha^2}{R_{dc}} = 6.02 \times 10^{-10}\ (\mathrm{S/m}) \tag{2.119}$$

Then, L_0 is determined as

$$L_0 = \frac{R_{dc} C_0}{G_0} = 2.045 \times 10^{-7}\ (\mathrm{H/m}) \tag{2.120}$$

It should be noted that no estimation of L_0 using the accurate earth-return impedance of extended TL approach is required in the above method, that is, G_0 and L_0 are evaluated by the DC resistance, space capacitance, and a monitored voltage V_x.

2.9.5.4 Low-frequency responses of wave propagation related parameters

By adopting the conductance G_0 determined in Section 2.9.5, the shunt admittance Y of an HVDC transmission line is re-defined as

$$Y = G_0 + Y_e \tag{2.121}$$

Then, the propagation-related parameters in a low-frequency region are calculated by adopting extended TL approach and (2.121). Calculated results corresponding to Figures 2.27 and 2.28 are shown in Figures 2.29 and 2.30. It is clear that the attenuation constant converges to results given in (2.115), phase velocity to results in (2.116), and characteristic impedance to results in (2.114). Moreover, the results shown in Figures 2.29 and 2.30 are physically more reasonable than those illustrated in Figures 2.27 and 2.28, as theoretically explained in Sections 2.9.3 and 2.9.4.

Calculated examples of G_0, L_0, attenuation constant α, phase velocity v, and the characteristic impedance Z_c are shown in Table 2.3. The conductance and converged inductance are dependent on the conductor geometry but not on the earth resistivity.

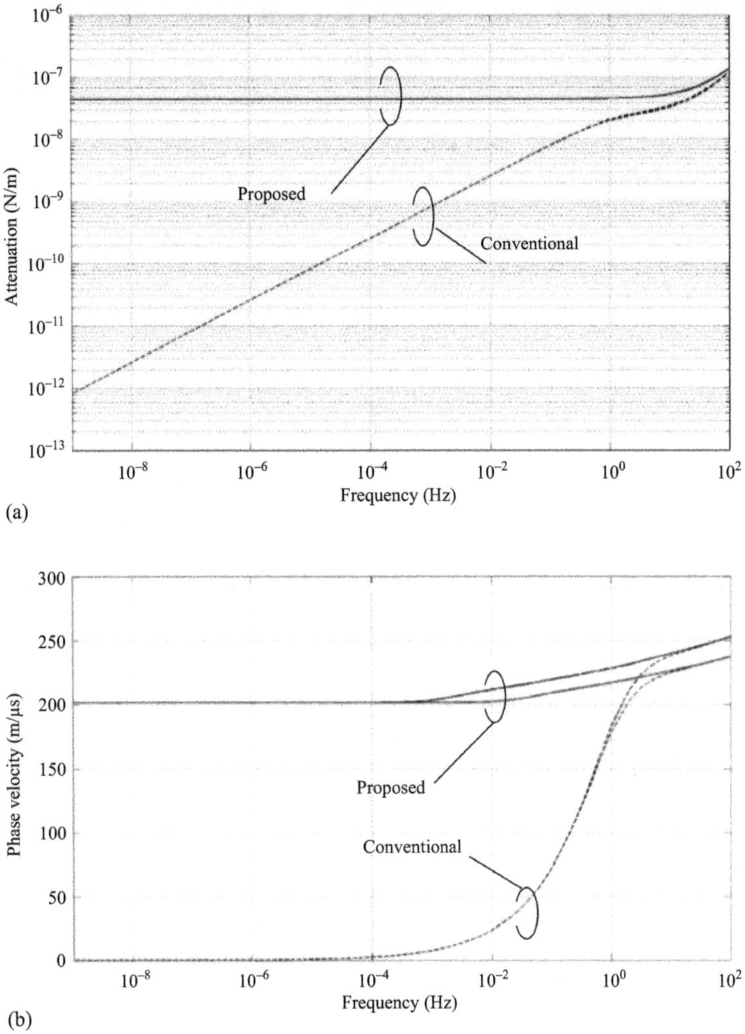

Figure 2.29 Propagation constant calculated by low-frequency condition for the same conductor in Figure 2.25: solid line: (2.121), dashed line: extended TL, black: ρ_e = 100 Ωm, and red: ρ_e = 2,000 Ωm. (a) Attenuation constant and (b) phase velocity

2.9.6 Concluding remarks

This section has proposed a distribution line concept to evaluate the impedance and admittance of an HVDC transmission line, and investigated the basic electrical parameters of the line. The following remarks are made clear.

1. Carson's earth-return impedance cannot be used to study extremely low-frequency phenomena in an HVDC transmission line as well as very high-frequency phenomena.

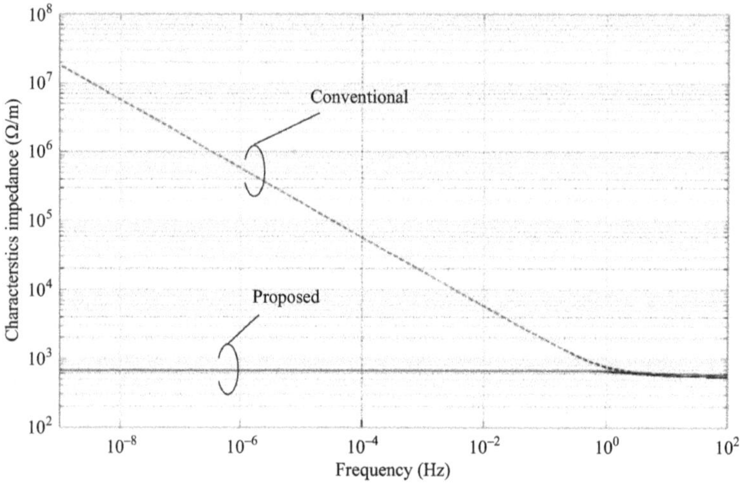

Figure 2.30 Real part of characteristic impedance calculated by low-frequency condition for the same conductor in Figure 2.25: solid line: (2.121), dashed line: extended TL, black: $\rho_e = 100\ \Omega m$, and red: $\rho_e = 2{,}000\ \Omega m$

Table 2.3 Calculated low-frequency parameters

h (m)	r (cm)	G_0 (pS/m)	R_{dc} (μΩ/m)	C_0 (pF/m)	L_0 (μH/m)	α (nNp/m)	v (m/μs)	Z_c (Ω)
10	1	261.8	120.3	7.319	3.363	177.5	201.6	677.9
	2	75.13	30.08	8.053	3.224	47.54	196.2	632.8
	5	14.69	4.813	9.285	3.041	8.410	188.2	572.3
20	1	230.47	120.3	6.707	3.502	166.5	206.3	722.5
	2	65.46	30.08	7.319	3.363	44.38	201.6	677.9
	5	12.60	4.813	8.322	3.180	7.786	194.4	618.1
30	1	214.8	120.3	6.395	3.582	160.7	208.9	748.5
	2	60.69	30.08	6.948	3.444	42.73	204.4	704.0
	5	11.58	4.813	7.846	3.261	7.466	197.7	644.7
50	1	197.2	120.3	6.040	3.685	154.0	212.0	781.1
	2	55.40	30.08	6.532	3.546	40.82	207.8	736.9
	5	10.47	4.813	7.319	3.363	7.100	201.6	677.9

2. A conventional approach even with accurate earth-return impedance and admittance formulae shows that the propagation velocity converges to zero and the characteristic impedance tends infinity.
3. A distortion-less line model is proposed to evaluate the conductance from known values of the DC resistance, space capacitance, and inductance which is given as a converged value of the accurate earth-return inductance.

4. From a monitored voltage in an HVDC transmission line, the attenuation constant of the line is calculated. Then, the conductance and inductance of the line are easily estimated at various locations along the line.
5. The frequency responses of the propagation-related parameters in a low-frequency region are calculated using the proposed approach. The results are physically more reasonable than those calculated by the conventional approach.

2.10 Theoretical formulation of external electromagnetic fields generated by overhead lines and underground cables

2.10.1 Overhead lines

As discussed, and shown in Section 2.7 and Appendix A3, the vectors of external electromagnetic field components in air and earth generated from a multiphase overhead lines are deduced and presented. A similar formulation has been performed by Kikuchi [30,31], Wedepohl [32], Wait [34], and Olsen [41] for a scenario with a single overhead conductor. Furthermore, a series of investigations of external electromagnetic fields produced by overhead lines with considerations of unidirectional/bidirectional propagations, conductor length, different loads, and earth characteristics in frequency and time domains are conducted and investigated by D'Amore and Sarto in [38,39,77–79].

2.10.2 Underground cables

Unlike the overhead lines, only a few studies of external electromagnetic field components have been conducted for an underground cable with single conductor in [60–63]. The accurate formulas of external electromagnetic fields generated from the single-phase underground cable are derived in [60–63]. The electromagnetic field components of the cable are calculated based on approximate formulas of TL approach, and the results are validated by exact analytical solutions [62,63]. Similarly to an overhead line, there exists a full spectrum of the wave propagation for the underground cable [63]. However, those additional modes except the TL mode are expected to be small and could be neglected as explained in [63].

Recently, an exact formulation of vectors of external electromagnetic field components produced from a multiphase underground cable has been developed based on the rigorous modal theory by Xue [80,81]. The formulation proposed in [80,81] sets up a correspondence between the extended and classical TL approaches discussed in Section 2.6.5 based on different current excitations and vectors of external electromagnetic fields generated by a multiphase underground cables. The studies of calculations and analysis of external electromagnetic field components in [80] of cables adopt the unidirectional propagation of currents which carried by the cables, and in the meantime, the terminating conditions, that is, the load/source impedances applied at the sending end and receiving end of cables, and effects of length of cables are neglected in the analysis of external electromagnetic field components. The detail of the formulation of external electromagnetic field components based on unidirectional propagation condition is given in (A2.61) of Appendix A2.

Based on the rigorous modal theory shown in [82], a formulation of excitation current using a bidirectional propagation together with various terminating conditions is proposed in [81]. Also, a way to set up a connection between the proposed excitation current and vectors of external electromagnetic field components using an N-phase underground cable is developed. The above newly proposed method [81] can consider the influences of terminating conditions and length of cable on characteristics of external electromagnetic fields. The details of the formulation are given and discussed in this section.

2.10.2.1 Formulation of an excitation current based on a bidirectional propagation with terminating conditions

Figure 2.31 shows the terminating boundary conditions for a formulation of an excitation current on a multiphase underground cable using a bi-directional propagation. The configuration and buried condition of multiphase underground cable is shown in Figure A2.1. The phase number is set to N. The cable length is set by the variable x. The terminating conditions used for the cable, that is, the sending end at the $x = 0$ and the receiving end at the $x = l$, are represented by the equivalent circuits of current sources and admittance matrices.

As illustrated in Figure 2.31, the general solutions in the phase domain of an $N \times 1$ voltage and current vectors of the underground cables at the length x have the following expressions:

$$\mathbf{V}(x) = \mathbf{Z_c B}\left(\mathbf{e}^{-\hat{\gamma}x}\mathbf{I}_\mathbf{m}^+ + \mathbf{e}^{\hat{\gamma}x}\mathbf{I}_\mathbf{m}^-\right) \tag{2.122}$$

$$\mathbf{I}(x) = \mathbf{B}\left(\mathbf{e}^{-\hat{\gamma}x}\mathbf{I}_\mathbf{m}^+ - \mathbf{e}^{\hat{\gamma}x}\mathbf{I}_\mathbf{m}^-\right) \tag{2.123}$$

where $\mathbf{Z_c}$ is the $N \times N$ phase domain characteristic impedance matrix of the cable, $\mathbf{B}$ represents the current transformation matrix of the cable, $\mathbf{I}_\mathbf{m}^+$ and $\mathbf{I}_\mathbf{m}^-$ represent respectively the $N \times 1$ forward and backward traveling modal current vectors of the cable, and the $N \times N$exponential term $\mathbf{e}^{\pm\hat{\gamma}x}$ are given by the following expression:

$$\mathbf{e}^{\pm\hat{\gamma}x} = \begin{bmatrix} e^{\pm\hat{\gamma}_1 x} & & \\ & \ddots & \\ & & e^{\pm\hat{\gamma}_N x} \end{bmatrix} \tag{2.124}$$

with $\hat{\gamma}_N$ is the modal propagation constant of the Nth mode.

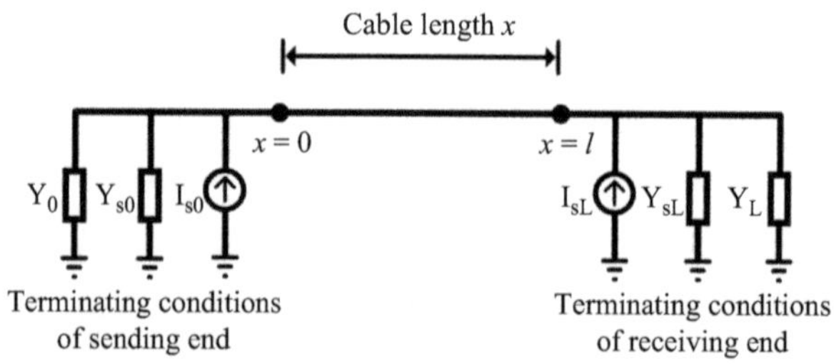

Figure 2.31 Boundary conditions applied at the two ends of a multiphase underground cable

Moreover, based on the boundary conditions shown in [82], it has the following expressions of the current vectors at $x = 0$ and $x = l$:

$$\mathbf{I}(0) = \mathbf{I}_{s0} - (\mathbf{Y}_0 + \mathbf{Y}_{s0})\mathbf{V}(0) \tag{2.125}$$

$$\mathbf{I}(l) = -\mathbf{I}_{sL} + (\mathbf{Y}_L + \mathbf{Y}_{sL})\mathbf{V}(l) \tag{2.126}$$

where $\mathbf{I}_{s0}$, $\mathbf{Y}_{s0}$, $\mathbf{I}_{sL}$, and $\mathbf{Y}_{sL}$ represent the external source circuits applied at the each end of the cable, the dimensions of the current vectors $\mathbf{I}_{s0}$ and $\mathbf{I}_{sL}$, and the source admittance matrices $\mathbf{Y}_{s0}$ and $\mathbf{Y}_{sL}$ are $N \times 1$ and $N \times N$, respectively, and $\mathbf{Y}_0$ and $\mathbf{Y}_L$ with $N \times N$ dimensions are the load admittance matrices at the both ends.

By adopting the general solutions (2.122) and (2.123) into the boundary conditions (2.125) and (2.126), the vectors $\mathbf{I}_\mathbf{m}^+$ and $\mathbf{I}_\mathbf{m}^-$ can be solved and have the following expressions:

$$\mathbf{I}_\mathbf{m}^+ = \mathbf{A}_1^{-1}\mathbf{I}_{s0} + \mathbf{A}_1^{-1}\mathbf{A}_2\left(\mathbf{A}_4 - \mathbf{A}_3\mathbf{A}_1^{-1}\mathbf{A}_2\right)^{-1}\left(\mathbf{I}_{sL} + \mathbf{A}_3\mathbf{A}_1^{-1}\mathbf{I}_{s0}\right) \tag{2.127}$$

$$\mathbf{I}_\mathbf{m}^- = \left(\mathbf{A}_4 - \mathbf{A}_3\mathbf{A}_1^{-1}\mathbf{A}_2\right)^{-1}\left(\mathbf{I}_{sL} + \mathbf{A}_3\mathbf{A}_1^{-1}\mathbf{I}_{s0}\right) \tag{2.128}$$

where

$$\mathbf{A}_1 = [\mathbf{U} + (\mathbf{Y}_0 + \mathbf{Y}_{s0})\mathbf{Z}_c]\mathbf{B} \tag{2.129}$$

$$\mathbf{A}_2 = [\mathbf{U} - (\mathbf{Y}_0 + \mathbf{Y}_{s0})\mathbf{Z}_c]\mathbf{B} \tag{2.130}$$

$$\mathbf{A}_3 = [\mathbf{U} - (\mathbf{Y}_L + \mathbf{Y}_{sL})\mathbf{Z}_c]\mathbf{B}\mathbf{e}^{-\hat{\gamma}l} \tag{2.131}$$

$$\mathbf{A}_4 = [\mathbf{U} + (\mathbf{Y}_L + \mathbf{Y}_{sL})\mathbf{Z}_c]\mathbf{B}\mathbf{e}^{\hat{\gamma}l} \tag{2.132}$$

with $\mathbf{U}$ is an $N \times N$ unit matrix.

Furthermore, the following alternative method using the chain-parameter submatrices [7] can be also adopted into the formulation of the vectors $\mathbf{I}_\mathbf{m}^+$ and $\mathbf{I}_\mathbf{m}^-$.

$$\mathbf{V}(l) = \mathbf{C}_{11}(l)\mathbf{V}(0) + \mathbf{C}_{12}(l)\mathbf{I}(0) \tag{2.133}$$

$$\mathbf{I}(l) = \mathbf{C}_{21}(l)\mathbf{V}(0) + \mathbf{C}_{22}(l)\mathbf{I}(0) \tag{2.134}$$

where

$$\mathbf{C}_{11}(l) = \frac{1}{2}\mathbf{Z}_c\mathbf{B}\left(\mathbf{e}^{\hat{\gamma}l} + \mathbf{e}^{-\hat{\gamma}l}\right)\mathbf{B}^{-1}\mathbf{Y}_c \tag{2.135}$$

$$\mathbf{C}_{12}(l) = -\frac{1}{2}\mathbf{Z}_c\mathbf{B}\left(\mathbf{e}^{\hat{\gamma}l} - \mathbf{e}^{-\hat{\gamma}l}\right)\mathbf{B}^{-1} \tag{2.136}$$

$$\mathbf{C}_{21}(l) = -\frac{1}{2}\mathbf{B}\left(\mathbf{e}^{\hat{\gamma}l} - \mathbf{e}^{-\hat{\gamma}l}\right)\mathbf{B}^{-1}\mathbf{Y}_c \tag{2.137}$$

$$\mathbf{C}_{22}(l) = \frac{1}{2}\mathbf{B}\left(\mathbf{e}^{\hat{\gamma}l} + \mathbf{e}^{-\hat{\gamma}l}\right)\mathbf{B}^{-1} \tag{2.138}$$

with the characteristic admittance matrix $\mathbf{Y}_c = \mathbf{Z}_c^{-1}$.

Next, solving the boundary conditions (2.125) and (2.126) together with (2.122), (2.123), (2.133), and (2.134), it gives

$$\mathbf{I}_\mathbf{m}^+ = -\mathbf{A}_5^{-1}\mathbf{I}_{sL} + \mathbf{A}_5^{-1}\mathbf{A}_6\left(\mathbf{A}_1\mathbf{A}_5^{-1}\mathbf{A}_6 - \mathbf{A}_2\right)^{-1}\left(\mathbf{I}_{s0} + \mathbf{A}_1\mathbf{A}_5^{-1}\mathbf{I}_{sL}\right) \tag{2.139}$$

$$\mathbf{I}_\mathbf{m}^- = \left(\mathbf{A}_1\mathbf{A}_5^{-1}\mathbf{A}_6 - \mathbf{A}_2\right)^{-1}\left(\mathbf{I}_{s0} + \mathbf{A}_1\mathbf{A}_5^{-1}\mathbf{I}_{sL}\right) \tag{2.140}$$

where

$$\begin{aligned}\mathbf{A}_5 = \mathbf{C}_{21}(l)\mathbf{Z}_c\mathbf{B} + \mathbf{C}_{22}(l)\mathbf{B} \quad &- (\mathbf{Y}_L + \mathbf{Y}_{sL})\mathbf{C}_{11}(l)\mathbf{Z}_c\mathbf{B} \\ -(\mathbf{Y}_L + \mathbf{Y}_{sL})\mathbf{C}_{12}(l)\mathbf{B}&\end{aligned} \tag{2.141}$$

$$\begin{aligned}\mathbf{A}_6 = (\mathbf{Y}_L + \mathbf{Y}_{sL})\mathbf{C}_{11}(l)\mathbf{Z}_c\mathbf{B} \quad &- \mathbf{C}_{21}(l)\mathbf{Z}_c\mathbf{B} + \mathbf{C}_{22}(l)\mathbf{B} \\ -(\mathbf{Y}_L + \mathbf{Y}_{sL})\mathbf{C}_{12}(l)\mathbf{B}&\end{aligned} \tag{2.142}$$

Therefore, the current expression based on a bidirectional propagation with different terminating conditions at both ends can be solved by adopting (2.127) and (2.128) or (2.139) and (2.140) into (2.123). The matrices of $\mathbf{Z_c}$, $\mathbf{B}$, and $\mathbf{e}^{\pm\hat{\gamma}x}$ can be calculated by the eigen-analysis of the underground cables using the extended TL approach shown in Section 2.6.5. Also, the expressions of (2.135)–(2.138) can be further manipulated to the hyperbolic functions shown in [7].

Moreover, the calculated currents of (2.123) using (2.127) and (2.128) are validated and compared with the alternative method (2.139) and (2.140); in general, no difference is observed between the two methods. Therefore, only (2.127) and (2.128) are considered in the calculations of Section 2.10.2.2.

2.10.2.2 Formulation of vectors of external electromagnetic field components

The underground cables have the following expressions of vectors of external electromagnetic fields in a lossy earth considering (A2.57) of Appendix A2 and (2.123), the vectors of external electromagnetic fields are given in the rectangular coordinate with the following formulas:

$$\left.\begin{aligned}\mathbf{E}_\mathbf{e}(x,y,z) &= \mathbf{\Phi}_\mathbf{e}(y,z)\mathbf{e}^{-\hat{\gamma}x}\mathbf{I}_\mathbf{m}^+ - \mathbf{\Phi}_\mathbf{e}(y,z)\mathbf{e}^{\hat{\gamma}x}\mathbf{I}_\mathbf{m}^- \\ \mathbf{H}_\mathbf{e}(x,y,z) &= \mathbf{\Upsilon}_\mathbf{e}(y,z)\mathbf{e}^{-\hat{\gamma}x}\mathbf{I}_\mathbf{m}^+ - \mathbf{\Upsilon}_\mathbf{e}(y,z)\mathbf{e}^{\hat{\gamma}x}\mathbf{I}_\mathbf{m}^-\end{aligned}\right\} \tag{2.143}$$

where $\mathbf{\Phi}_\mathbf{e}(y,z)$ and $\mathbf{\Upsilon}_\mathbf{e}(y,z)$ are $3 \times N$ operator's vectors of external electromagnetic field components, and the full expressions are given in (A2.51)–(A2.56) of Appendix A2, $\mathbf{I}_\mathbf{m}^+$ and $\mathbf{I}_\mathbf{m}^-$ are given in (2.127) and (2.128), $\mathbf{E}_\mathbf{e}(x,y,z)$ and $\mathbf{H}_\mathbf{e}(x,y,z)$ are 3×1 vectors of external electromagnetic field components in the phase domain, and

$$\mathbf{E}_\mathbf{e}(x,y,z) = \begin{bmatrix} E_{ex}(x,y,z) \\ E_{ey}(x,y,z) \\ E_{ez}(x,y,z) \end{bmatrix}, \mathbf{H}_\mathbf{e}(x,y,z) = \begin{bmatrix} H_{ex}(x,y,z) \\ H_{ey}(x,y,z) \\ H_{ez}(x,y,z) \end{bmatrix} \tag{2.144}$$

In comparison to the current excitation adopted into (A2.61) of Appendix A2, the calculated electromagnetic field components of (2.143) using (2.123) can consider the influence of various terminating conditions adopted at the each end of the cable, and also the effect of the cable length. Moreover, (2.143) represents that the external electromagnetic fields at any field observation point are generated by the superposition of the forward and backward traveling current vectors of the cable. Also, it has no limitations of arrangement, such as flat, vertical, or trefoil of cable for (2.143).

The general steps for evaluation of (2.143) are suggested below. The bidirectional modal current vectors (2.127) and (2.128) should be calculated with consideration of extended TL approach, cable terminating conditions, energizations at the both ends, and earth parameters as the first step. Then, the operator's vectors of external electromagnetic field components in (2.143) can be calculated at the selected field observation points. Finally, the external electromagnetic field components in phase domain (2.143) and (23) can be obtained.

2.11 Conclusions

In this chapter, Section 2.1 describes and summarizes a general formulation of series impedance and shunt admittance of overhead and underground cables. The calculations of series impedance and shunt admittance matrices of cables are performed by the Line and Cable Constants in the existing EMT-type simulation tools.

Section 2.2 reviews the existing formulas of earth-return impedance and admittance of overhead and underground cables which are derived under different assumptions of wave propagations.

Sections 2.3 and 2.4 describe accurate and approximate formulas of earth-return impedance and admittance for overhead cables (lines). The characteristics of different formulas are made clear.

Sections 2.5 and 2.6 explain and summarize the accurate and approximate formulations of earth-return impedance and admittance for underground cables. The newly proposed earth-return admittance formulas are based on different integral references of voltage propagation equation. Also, the concept of classical and extended TL approaches is defined.

Section 2.7 discusses the derivation of the modified earth-return Green function of overhead cables (lines) for MoM-SO technique. The modified earth-return Green function can be used to avoid possible numerical instability.

Section 2.8 shows a comparison of series impedance and shunt admittance calculated by different formulas for overhead and underground cables. A single overhead conductor and three typical single-core underground cables are adopted in the calculations and investigations.

Section 2.9 has proposed a distribution line concept to evaluate the series impedance and shunt admittance of the HVDC transmission line, and investigated the basic electrical parameters of the line.

Section 2.10 explains and proposes theoretical formulations of external electromagnetic field components generated from a multiphase overhead and underground cables. Also, the characteristics of terminating conditions are included in the formulations, and different excitations of current are discussed.

Appendix A1 Cable internal impedance and admittance including semiconducting layer [1]

A1.1 Formulation of internal impedance matrix

A1.1.1 Single-core coaxial (SC) cable [1]

When the total number of conductors in a given cable is "n", then the internal impedance matrix Z_i is defined as

$$\mathbf{Z}_i = \text{SC cable internal impedance matrix}$$

$$= \begin{bmatrix} \mathbf{Z}_{i1} & 0 & \ldots & 0 \\ 0 & \mathbf{Z}_{i2} & \ldots & 0 \\ \ldots & \ldots & \ldots & \ldots \\ 0 & 0 & \ldots & \mathbf{Z}_{in} \end{bmatrix} \tag{A1.1}$$

A1.1.1.1 SC cable composed of core, sheath, and armor

Element matrix $\boldsymbol{Z}_{ij}$ in (A1.1) is given in the following form for the SC cable illustrated in Figure A1.1

$$\mathbf{Z}_{ij} = \begin{bmatrix} Z_{ccj} & Z_{csj} & Z_{caj} \\ Z_{csj} & Z_{ssj} & Z_{saj} \\ Z_{caj} & Z_{saj} & Z_{aaj} \end{bmatrix} \tag{A1.2}$$

where

$$Z_{ccj} = \text{core self} - \text{impedance} = z_{cs} + z_{za} + z_{a4} - 2z_{2m} - 2z_{3m} \tag{A1.3}$$

$$Z_{ssj} = \text{sheath self} - \text{impedance} = z_{sa} + z_{a4} - 2z_{3m} \tag{A1.4}$$

$$Z_{aaj} = \text{armour self} - \text{impedance} = z_{a4} \tag{A1.5}$$

$$Z_{csj} = \text{mutual impedance between the core and the sheath}$$

$$= z_{sa} + z_{a4} - z_{2m} - 2z_{3m} \tag{A1.6}$$

$$Z_{caj} = \text{mutual impedance between the core and the armour} = z_{a4} - z_{3m} \tag{A1.7}$$

$$Z_{saj} = \text{mutual impedance between the sheath and the armour} = Z_{caj} \tag{A1.8}$$

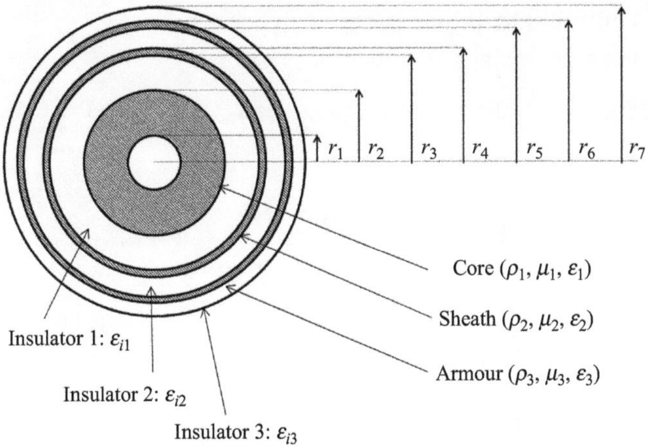

Figure A1.1 Cross-section of SC cable composed of core, metallic sheath, and armor

Impedances z_{cs}, z_{sa}, and z_{a4} in (A1.3) to (A1.7) are defined by

$$\left.\begin{aligned} z_{cs} &= z_{11} + z_{12} + z_{2j} \\ z_{sa} &= z_{20} + z_{23} + z_{3j} \\ z_{a4} &= z_{30} + z_{34} \end{aligned}\right\} \tag{A1.9}$$

A1.1.1.2 SC cable composed of core and sheath

When an SC cable is composed of a core and a sheath, $\boldsymbol{Z}_{ij}$ becomes a 2 × 2 matrix in the following form:

$$\mathbf{Z}_{ij} = \begin{bmatrix} Z_{ccj} & Z_{csj} \\ Z_{csj} & Z_{ssj} \end{bmatrix} \tag{A1.10}$$

where

$$\left.\begin{aligned} Z_{ccj} &= z_{cs} + z_{s3} - 2z_{2m} \\ Z_{ssj} &= z_{s3} \\ Z_{csj} &= z_{s3} - z_{2m} \\ z_{s3} &= z_{20} + z_{23} \end{aligned}\right\} \tag{A1.11}$$

A1.1.1.3 SC cable composed only of core

$$\mathbf{Z}_{ij} = Z_{ccj} = z_{11} + z_{12} \tag{A1.12}$$

A1.1.1.4 Component impedances of SC cable

1. z_{11} : internal impedance of core outer surface

$$z_{11} = (s\mu_0\mu_1/2\pi) \cdot (1/x_2D_1) \cdot \{I_0(x_2) \cdot K_1(x_1) + K_0(x_2) \cdot I_1(x_1)\} \quad \text{(A1.13)}$$

2. z_{12} : core outer insulation impedance

$$z_{12} = (s\mu_0\mu_{i1}/2\pi) \cdot \ln(r_3/r_2) \quad \text{(A1.14)}$$

3. z_{2i} : internal impedance of the sheath inner surface

$$z_{2i} = (s\mu_0\mu_2/2\pi) \cdot (1/x_3D_2) \cdot \{I_0(x_3) \cdot K_1(x_4) + K_0(x_3) \cdot I_1(x_4)\} \quad \text{(A1.15)}$$

4. z_{2m} : sheath mutual impedance

$$z_{2m} = \rho_2/2\pi r_3r_4D_2 \quad \text{(A1.16)}$$

5. z_{20} : internal impedance of the sheath outer surface

$$z_{20} = (s\mu_0\mu_2/2\pi) \cdot (1/x_4D_2) \cdot \{I_0(x_4) \cdot K_1(x_3) + K_0(x_4) \cdot I_1(x_3)\} \quad \text{(A1.17)}$$

6. z_{23} : sheath outer insulation impedance

$$z_{23} = (s\mu_0\mu_{i2}/2\pi) \cdot \ln(r_5/r_4) \quad \text{(A1.18)}$$

7. z_{3i} : internal impedance of the arm our inner surface

$$z_{3i} = (s\mu_0\mu_3/2\pi) \cdot (1/x_5D_3) \cdot \{I_0(x_5) \cdot K_1(x_6) + K_0(x_5) \cdot I_1(x_6)\} \quad \text{(A1.19)}$$

8. z_{3m} : arm our mutual impedance

$$z_{3m} = \rho_3/2\pi r_5r_6D_3 \quad \text{(A1.20)}$$

9. z_{30} : internal impedance of the arm our outer surface

$$z_{30} = (s\mu_0\mu_3/2\pi) \cdot (1/x_6D_3) \cdot \{I_0(x_6) \cdot K_1(x_5) + K_0(x_6) \cdot I_1(x_5)\} \quad \text{(A1.21)}$$

10. z_{34} : arm our outer insulator impedance

$$z_{34} = (s\mu_0\mu_{i3}/2\pi) \cdot \ln(r_7/r_6) \quad \text{(A1.22)}$$

where

$$\begin{aligned}
&D_1 = I_1(x_2) \cdot K_1(x_1) - I_1(x_1) \cdot K_1(x_2) \\
&D_1 = I_1(x_4) \cdot K_1(x_3) - I_1(x_3) \cdot K_1(x_4) \\
&D_1 = I_1(x_6) \cdot K_1(x_5) - I_1(x_5) \cdot K_1(x_6) \\
&x_k = \beta_k\sqrt{s},\ \beta_2 = r_2\sqrt{\mu_0\mu_1/\rho_1} \\
&\beta_3 = r_3\sqrt{\mu_0\mu_2/\rho_2},\ \beta_4 = r_4\sqrt{\mu_0\mu_2/\rho_2} \\
&\beta_5 = r_5\sqrt{\mu_0\mu_3/\rho_3},\ \beta_6 = r_6\sqrt{\mu_0\mu_3/\rho_3}
\end{aligned} \quad \text{(A1.23)}$$

A1.1.2 Pipe-enclosed type (PT) cable [1,83,84]

A1.1.2.1 Formulation of PT cable impedance matrix

The total impedance of a PT cable composed of "n" SC cables within the pipe illustrated in Figure A1.2 is defined in the following form:

$$\boldsymbol{Z} = \boldsymbol{Z}_i + \boldsymbol{Z}_p + \boldsymbol{Z}_c + \boldsymbol{Z}_e \tag{A1.24}$$

In the above equation, $\mathbf{Z}_i$ = SC cable internal impedance matrix defined in (A1.1)

$\mathbf{Z}_p$ = pipe internal impedance matrix

$$= \begin{bmatrix} \mathbf{Z}_{p11} & \mathbf{Z}_{p12} & \cdots & \mathbf{Z}_{p1n} & 0 \\ \mathbf{Z}_{p12} & \mathbf{Z}_{p22} & \cdots & \mathbf{Z}_{p2n} & 0 \\ \cdots & \cdots & \cdots & \cdots & \cdots \\ \mathbf{Z}_{p1n} & \mathbf{Z}_{p12} & \cdots & \mathbf{Z}_{pnn} & 0 \\ 0 & 0 & \cdots & 0 & 0 \end{bmatrix} \tag{A1.25}$$

$\mathbf{Z}_c$ = connection impedance matrix between pipe inner and outer surfaces

$$= \begin{bmatrix} \mathbf{Z}_{c1} & \mathbf{Z}_{c1} & \cdots & \mathbf{Z}_{c1} & Z_{c2} \\ \mathbf{Z}_{c1} & \mathbf{Z}_{c1} & \cdots & \mathbf{Z}_{c1} & Z_{c2} \\ \cdots & \cdots & \cdots & \cdots & \cdots \\ \mathbf{Z}_{c1} & \mathbf{Z}_{c1} & \cdots & \mathbf{Z}_{c1} & Z_{c2} \\ Z_{c2} & Z_{c2} & \cdots & Z_{c2} & Z_{c3} \end{bmatrix} \tag{A1.26}$$

$\boldsymbol{Z}_e$ = earth-return impedance.

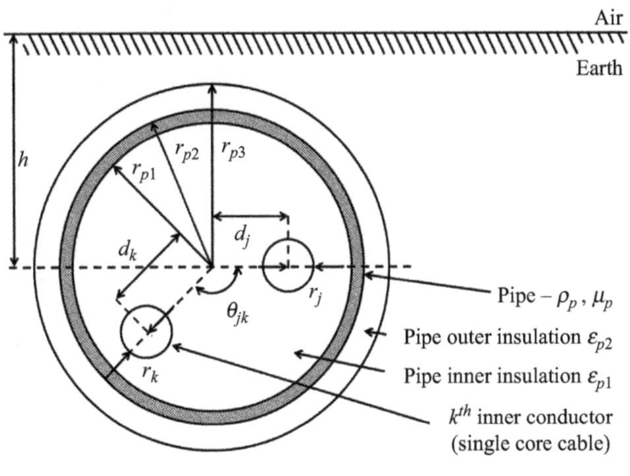

Figure A1.2 Cross-section of underground PT cable

When the SC cable in the PT cable is composed of three conductors as in Figure A1.1, element $\mathbf{Z}_{pjk}$ in (A1.25) is given by a 3 × 3 matrix in the following form.

$$\mathbf{Z}_{pjk} = \begin{bmatrix} Z_{pjk} & Z_{pjk} & Z_{pjk} \\ Z_{pjk} & Z_{pjk} & Z_{pjk} \\ Z_{pjk} & Z_{pjk} & Z_{pjk} \end{bmatrix} \tag{A1.27}$$

A1.1.3 Component impedance of PT cable

$$Z_{pjk} = (s\mu_0/2\pi) \cdot \left[\mu_p K_0(x_1)/\mu_1 K_1(x_1) + Q_{jk} + 2\mu_p \sum_{n=1}^{\infty} C_n / \left\{n\left(1+\mu_p\right) + x_1 K_{n-1}(x_1)/K_n(x_1)\right\}\right] \tag{A1.28}$$

$$\left.\begin{aligned} Q_{jj} &= \ln\left[\left(r_{p1}/r_j\right) \cdot \left\{1 - \left(d_j/r_{p1}\right)^2\right\}\right] \\ Q_{jk} &= \ln\left[r_{p1}/\sqrt{d_j^2 + d_k^2 - 2d_j d_k \cos\theta_{jk}}\right] - \sum_{n=1}^{\infty} C_n/n \\ &C_n\left(d_j d_k/r_{p1}^2\right)^n \cdot \cos\left(n\theta_{jk}\right) \end{aligned}\right\} \tag{A1.29}$$

where $x_1 = \beta_1\sqrt{s},\ \beta_1 = r_{p1}\sqrt{\mu_0\mu_p/\rho_p}$

A1.2 Formulation of potential coefficient matrix

A1.2.1 Single-core coaxial (SC) cable

A1.2.1.1 SC cable composed of core, sheath, and armor

The potential coefficient of an SC cable is determined by its insulators: (1) core outer insulator between core and sheath, (2) sheath outer insulator between sheath and armor, and (3) armor outer insulator:

$$P_{ij} = \begin{bmatrix} P_{cj} + P_{sj} + P_{aj} & P_{sj} + P_{aj} & P_{aj} \\ P_{sj} + P_{aj} & P_{sj} + P_{aj} & P_{aj} \\ P_{aj} & P_{aj} & P_{aj} \end{bmatrix} \tag{A1.30}$$

$$\left.\begin{aligned} P_{cj} &= (1/2\pi\varepsilon_0\varepsilon_{i1}) \cdot \ln(r_3/r_2) \\ P_{sj} &= (1/2\pi\varepsilon_0\varepsilon_{i2}) \cdot \ln(r_5/r_4) \\ P_{aj} &= (1/2\pi\varepsilon_0\varepsilon_{i3}) \cdot \ln(r_7/r_6) \end{aligned}\right\} \tag{A1.31}$$

A1.2.1.2 SC cable composed of core and sheath

The potential coefficient of an SC cable is determined by its insulators: (1) core outer insulator between core and sheath and (2) sheath outer insulator:

$$\mathbf{P}_{ij} = \begin{bmatrix} P_{cj} + P_{sj} & P_{sj} \\ P_{sj} & P_{sj} \end{bmatrix} \tag{A1.32}$$

A1.2.1.3 SC cable composed only of core

$$\mathbf{P}_{\mathrm{ij}} = \boldsymbol{P}_{cj} \tag{A1.33}$$

A1.2.2 Pipe-enclosed type (PT) cable

(a) Underground cable

$$\mathbf{P} = \mathbf{P}_{\mathrm{i}} + \mathbf{P}_{\mathrm{p}} + \mathbf{P}_{\mathrm{c}} + \mathbf{P}_{\mathrm{e}} \tag{A1.34}$$

(b) Overhead cable

$$\mathbf{P} = \mathbf{P}_{i} + \mathbf{P}_{p} + \mathbf{P}_{c} + \mathbf{P}_{0} \tag{A1.35}$$

where

$\mathbf{P}_i$ = SC cable internal potential coefficient matrix

$$= \begin{bmatrix} \mathbf{P}_{i1} & \mathbf{0} & \cdots & \mathbf{0} & 0 \\ \mathbf{0} & \mathbf{P}_{i2} & \cdots & \mathbf{0} & 0 \\ \cdots & \cdots & \cdots & \cdots & \cdots \\ \mathbf{0} & \mathbf{0} & \cdots & \mathbf{P}_{in} & 0 \\ 0 & 0 & \cdots & 0 & 0 \end{bmatrix} \tag{A1.36}$$

Submatrix $\mathbf{P}_{ij}$ is defined in (A1.30) for an SC cable composed of three conductors, and in (A1.32) for an SC cable composed of core and sheath conductors.

$\mathbf{P}_p$ = pipe internal potential coefficient matrix

$$= \begin{bmatrix} \mathbf{P}_{p11} & \mathbf{P}_{p12} & \cdots & \mathbf{P}_{p1n} & 0 \\ \mathbf{P}_{p12} & \mathbf{P}_{p22} & \cdots & \mathbf{P}_{p2n} & 0 \\ \cdots & \cdots & \cdots & \cdots & \cdots \\ \mathbf{P}_{p1n} & \mathbf{P}_{p12} & \cdots & \mathbf{P}_{pnn} & 0 \\ 0 & 0 & \cdots & 0 & 0 \end{bmatrix} \tag{A1.37}$$

$\boldsymbol{P}_{pij}$ is defined by

$$P_{p_{jj}} = Q_{jj}/2\pi\varepsilon_0\varepsilon_{p1}, \quad P_{p_{jk}} = Q_{jk}/2\pi\varepsilon_0\varepsilon_{p1}$$

Q_{ij} is given in (A1.29).

$\mathbf{P}c$ = potential coefficient matrix between pipe inner and outer surfaces

$$= \begin{bmatrix} \mathbf{P}_c & \mathbf{P}_c & \cdots & \mathbf{P}_c & P_c \\ \mathbf{P}_c & \mathbf{P}_c & \cdots & \mathbf{P}_c & P_c \\ \cdots & \cdots & \cdots & \cdots & \cdots \\ \mathbf{P}_c & \mathbf{P}_c & \cdots & \mathbf{P}_c & P_c \\ P_c & P_c & \cdots & P_c & P_c \end{bmatrix} \tag{A1.38}$$

$\mathbf{P}c = (1/2\pi\varepsilon_0\varepsilon_{p2})\ln(r_{p3}/r_{p2})$

$\mathbf{P}_0$ = potential coefficient matrix of the pipe in air = space potential coefficient

$$= \begin{bmatrix} \mathbf{P}_0 & \mathbf{P}_0 & \cdots & \mathbf{P}_0 & P_0 \\ \mathbf{P}_0 & \mathbf{P}_0 & \cdots & \mathbf{P}_0 & P_0 \\ \cdots & \cdots & \cdots & \cdots & \cdots \\ \mathbf{P}_0 & \mathbf{P}_0 & \cdots & \mathbf{P}_0 & P_0 \\ P_0 & P_0 & \cdots & P_0 & P_0 \end{bmatrix} \tag{A1.39}$$

$\mathbf{P}_0$ is given in (2.6') and (2.8).

A1.3 Semiconducting layer (two-layered conductor) impedance [85]

A1.3.1 Derivation of impedance

Figure A1.3 illustrates the cross-section of a two-layered coaxial cylindrical conductor composed of medium 1 (resistivity ρ_1, permeability μ_1, inner and outer radii a and b) and medium 2 (ρ_2, μ_2, b', c). Medium 1 corresponds to a conductor and medium 2 to a semiconducting layer in a cable of which the conductor has a semiconducting layer on its outer surface. The same is applicable to an outer

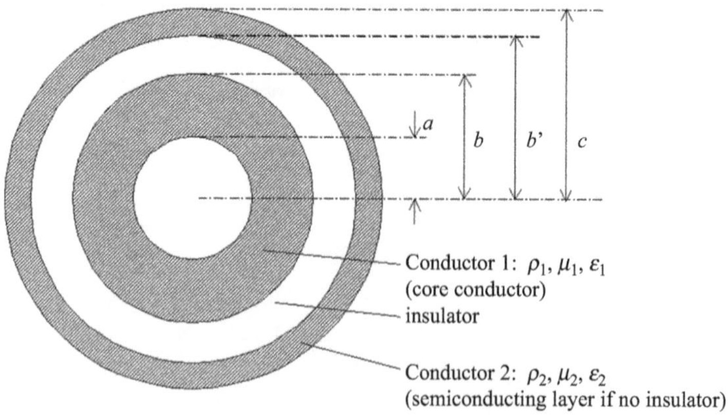

Figure A1.3 Cross-section of a two-layered conductor

semiconducting layer on the insulation. In the case of a semiconducting layer on the inner surface of a conductor shield, medium 1 is the semiconducting layer and medium 2 is the conductor shield.

The derivation of an accurate impedance of the two-layered conductor in Figure A1.3 by solving Maxwell's equation is explained in [86] and [87]. Since the resistivity of a cable conductor (ρ_1 : 10^{-8} Ωm) is far smaller than that of a semiconducting layer (ρ_2 : 0.01–10 Ωm [88,89]), no current can be assumed to flow between the media 1 and 2 across the boundary considering that the penetration depth d becomes greater than the thickness (c-b' : 1–5 mm) of the medium 2, that is

$$d = \sqrt{\rho^2/\omega\mu^2} = 35.6/\sqrt{f} \text{ to } 356/\sqrt{f}[\text{m}] < c - b' \quad \text{(A1.40)}$$

where $\mu_2 = \mu_0$: permeability of medium 2 and f: frequency of the current.

The above condition is almost always satisfied in a frequency range less than 1 GHz which is enough to deal with a wave propagation characteristic and a transient on a cable. This means that the impedance of the semiconducting layer is evaluated as an impedance of an independent (isolated) coaxial cylindrical conductor, which was derived by Schelkunoff [2], and the total impedance of the two-layered (conductor/semiconducting layer) is obtained as the parallel impedance of the conductor and the semiconducting layer impedances based on a circuit theory.

The impedance of a cable consisting of inner conductor 1, outer conductor 2, and an insulator in Figure A1.3 is given in the following form [1]:

$$\begin{aligned} Z_{11} &= Z_{10} + Z_{12} + Z_{2i} - Z_{2m} + Z_{12} \\ Z_{12} &= Z_{22} - Z_{2m}, Z_{22} = Z_{20} \end{aligned} \quad \text{(A1.41)}$$

where $Z_{10} = (m_1\rho_1/2\pi b)A/B$, $Z_{2i} = (m_2\rho_2/2\pi b')E/F$, $Z_{2m} = \rho_2/2\pi b' cF$, $Z_{20} = (m_2\rho_2/2\pi c)R/F$, $Z_{12} = j\omega(\mu_0/2\pi)\ln(b'/b)$

$$\begin{aligned} A &= K_1(x_1)\cdot I_0(x_2) + I_1(x_1)\cdot K_0(x_2) \\ B &= K_1(x_1)\cdot I_1(x_2) - I_1(x_1)\cdot K_1(x_2) \\ E &= I_0(x_3)\cdot K_1(x_4) + K_0(x_3)\cdot I_1(x_4) \\ F &= K_1(x_3)\cdot I_1(x_4) - I_1(x_3)\cdot K_1(x_4) \\ R &= K_1(x_3)\cdot I_0(x_4) + I_1(x_3)\cdot K_0(x_4) \end{aligned} \quad \text{(A1.42)}$$

and $I_n(x), K_n(x)$: modified Bessel function of order n

$$\begin{aligned} &x1 = m_1 a, \ x_2 = m_1 b, \ x_3 = m_2 b', \ x_4 = m_2 c \\ &m_1 = \sqrt{j\omega\mu_1/\rho_1}, \ m_2 = \sqrt{j\omega\mu_2/\rho 2} \end{aligned} \quad \text{(A1.43)}$$

When there is no insulator, that is, $b = b'$ in Figure A1.3, the following result is obtained for $Z_{12} = 0$:

$$Z_{11} = z_{10} + z_{2i} + z_{20} - 2z_{2m}, Z_{12} = z_{20} - z_{2m}, \text{and } Z_{22} = z_{20} \quad \text{(A1.44)}$$

A1.3.2 Impedance of two-layered conductor

When the conductors 1 and 2 are short-circuited at the both ends of a cable with the receiving-end grounded, and voltage V is applied to the sending end, then the following relation is obtained:

$$\begin{bmatrix} V \\ V \end{bmatrix} = \begin{bmatrix} Z_{11} & Z_{12} \\ Z_{12} & Z_{22} \end{bmatrix} \begin{bmatrix} I_1 \\ I_2 \end{bmatrix} \tag{A1.45}$$

where I_1, I_2: current of the conductors 1 and 2.

The total current I of the conductors are given by

$$I = I_1 + I_2 \tag{A1.46}$$

The total (outer surface) impedance of the two-layered conductor seen from the sending end is obtained from (A1.45) and (A1.46):

$$Z_{out} = V/I = \left(Z_{11}Z_{22} - Z_{12}^2\right)/(Z_{11} + Z_{22} - 2Z_{12}) \tag{A1.47}$$

Substitution of (A1.44) into (A1.47) yields the impedance of the two-layered conductor in the following form:

$$Z_{out} = Z_{20} - Z_{2m}^2/(Z_{10} + Z_{2i}) \tag{A1.48}$$

Similarly to the above, the inner surface impedance and the mutual impedance between the inner and the outer surfaces of the two-layered conductor are given by

$$\begin{aligned} Z_{in} &= Z_{1i} - Z_{1m}/(Z_{10} + Z_{2i}) \\ Z_m &= Z_{1m} \cdot Z_{2m}/(Z_{10} + Z_{2i}) \end{aligned} \tag{A1.49}$$

The above equation is a general formula of the impedance of a two-layered conductor, and it is applicable to a semiconducting layer on the outer surface of a core, on the inner surface of a sheath and on the outer surface of the sheath of a cable. The formula is identical to the formula derived by solving Maxwell's equations [86,87].

A1.4 Investigation of cable internal impedance [90]

A1.4.1 Accurate formula by Schelkunoff [2]

The accurate formula of the internal impedance for a cylindrical conductor in Figure A1.4 was derived by Schelkunoff in 1934.

Inner surface impedance Z_{is}:

$$Z_{i_s} = \frac{\rho \cdot m}{2 \cdot \pi \cdot a \cdot D} \cdot (I_0(m \cdot a) \cdot K_1(m \cdot b) + K_0(m \cdot a) \cdot I_1(m \cdot b)) \tag{A1.50}$$

Mutual impedance between the inner and the outer surfaces Z_{ms}:

$$Z_{m_s} = \frac{\rho \cdot m}{2 \cdot \pi \cdot a \cdot b \cdot D} \tag{A1.51}$$

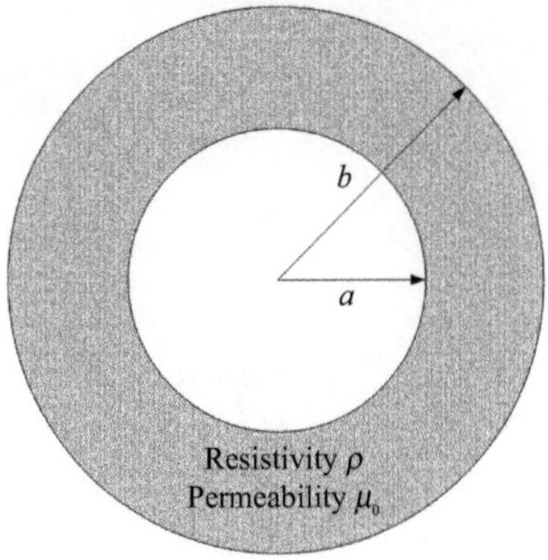

Figure A1.4 Cross-section of a cylindrical conductor

Outer surface impedance Z_{0s}:

$$Z_{0_s} = \frac{\rho \cdot m}{2 \cdot \pi \cdot b \cdot D} \cdot (I_0(m \cdot b) \cdot K_1(m \cdot a) + K_0(m \cdot b) \cdot I_1(m \cdot a)) \quad \text{(A1.52)}$$

$$D = I_1(m \cdot b) \cdot K_1(m \cdot a) - K_1(m \cdot b) \cdot I_1(m \cdot a) \quad \text{(A1.53)}$$

$$\text{and } m = \sqrt{j\omega\mu_0/\rho} \quad \text{(A1.54)}$$

where $I_n(x), K_n(x)$ is the modified Bessel's function of kind 1 and 2, respectively, with order n.

As is clear from the above equation, there exist three-component impedances for a cylindrical conductor. In the case of a circular solid conductor, $z_i = z_m = 0$ for $a = 0$, and z_0 becomes:

$$z_0 = j(\omega\mu)I_1(mb)/I_2(mb) \quad \text{(A1.55)}$$

When we say a conductor internal impedance (z_i), it means z_0, the outer surface impedance given in (A1.50) or in (A1.52) as far as an overhead line is concerned. However, in the case of a cable, z_i is composed of a number of component impedances and also of an insulator impedance between metallic conductors as explained in Section A1.1.1, because the cable is, in general, composed of a core conductor carrying a current and a metallic sheath (shield or screen) for a current return path.

A1.4.2 Approximate formula by Wedepohl/Wilcox [49]

$$Z_{i_w} = \frac{\rho \cdot m}{2 \cdot \pi \cdot a} \cdot \coth(m \cdot (b - a)) - \frac{\rho}{2 \cdot \pi \cdot b \cdot (a + b)} \quad \text{(A1.56)}$$

$$Z_{m_w} = \frac{\rho \cdot m}{\pi \cdot (a+b)} \cdot \operatorname{cosech}(m \cdot (b-a)) \tag{A1.57}$$

$$Z_{i_w} = \frac{\rho \cdot m}{2 \cdot \pi \cdot b} \cdot \coth(m \cdot (b-a)) + \frac{\rho}{2 \cdot \pi \cdot b \cdot (a+b)} \tag{A1.58}$$

A1.4.3 Conductor outer surface impedance Zo with arbitrary cross-section [91]

$$Z_i = R_{dc}\sqrt{\{1 + j\omega\mu_c S/(R_{dc} \cdot \ell^2)\}} \tag{A1.59}$$

where S is the cross-section area of the conductor (m^2) and ℓ is the circumferential length of the conductor outer surface (m).

It is clear that the above equation becomes identical to the DC resistance in a low-frequency region by assuming a small ω, and to an approximate formula at a high frequency by assuming a large ω:

$$Z_{\mathrm{ohf}} = \rho m / 2\pi b \tag{A1.60}$$

A1.4.4 Discussion on accurate and approximate impedances [90]

A1.4.4.1 Numerical instability

Figure A1.5 is an example of numerical instability observed in D defined in (A1.53) calculated by a commercial software. Figure A1.5(a) shows the real part of D, and Figure A1.5(b) the imaginary part. The software could not calculate the real part in Figure A1.5(a) above the frequency 800 kHz. The same for the imaginary part in Figure A1.5(b).

A1.4.4.2 Countermeasure of the numerical instability

It is a common practice to adopt an asymptotic expansion when a numerical instability is expected as in EMTP Cable Constants [3]. By adopting the asymptotic expansion in a high-frequency region, the numerical instability observed in Figure A1.5 is solved.

When variable "x" is very large corresponding to a high frequency, Bessel functions are approximated by

$$I_n(x) \approx \frac{e^x}{\sqrt{2\pi \cdot x}} \tag{A1.61}$$

$$K_n(x) \approx \frac{e^{-x}}{\sqrt{\frac{2}{\pi} \cdot x}} \tag{A1.62}$$

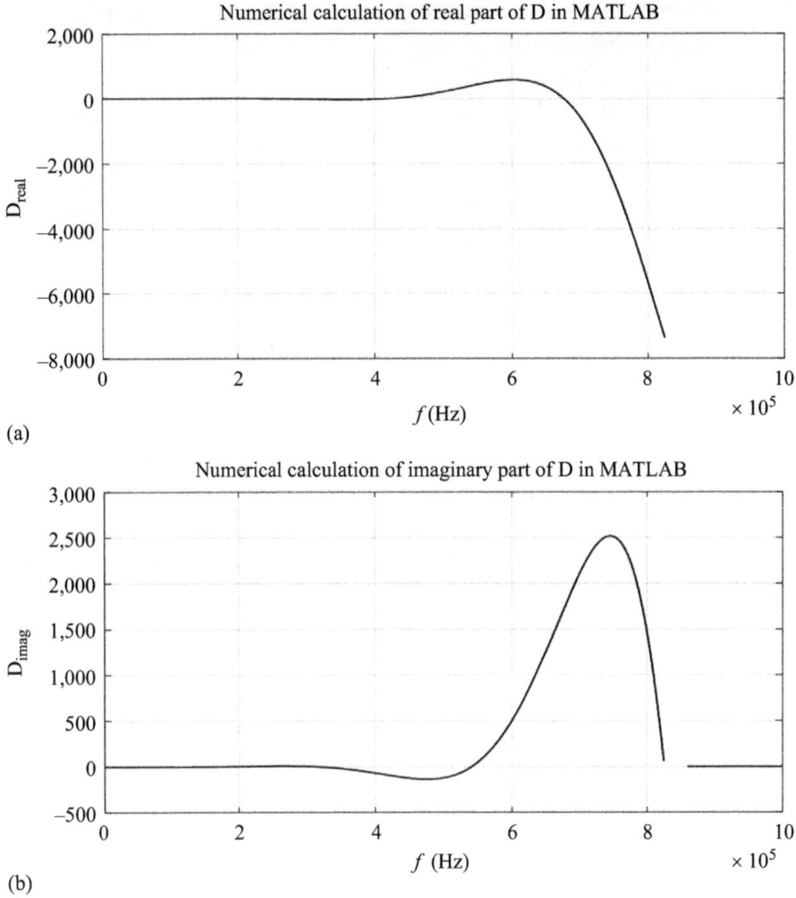

Figure A1.5 Examples of numerical instability of D in (4) when a commercial software is used

Substituting the above approximation of the Bessel functions into Schelkunoff's formulas in (A1.50) to (A1.52), the following approximate formulas derived at once:

$$Z_{i_a} = \frac{\rho \cdot m}{2 \cdot \pi \cdot a} \cdot \coth(m \cdot (b - a)) \tag{A1.63}$$

$$Z_{m_a} = \frac{\rho \cdot m}{2\pi \cdot \sqrt{a \cdot b}} \cdot \operatorname{cosech}(m \cdot (b - a)) \tag{A1.64}$$

$$Z_{o_a} = \frac{\rho \cdot m}{2 \cdot \pi \cdot b} \cdot \coth(m \cdot (b - a)) \tag{A1.65}$$

The above formulas are the same as those derived by Wedepohl/Wilcox [49] except an additional term in (A1.56) and (A1.58), and "$a + b$" in (A1.57).

A1.4.4.3 Comparison of calculated results by accurate and approximate formulas

1. Inner surface impedance Z_i

   Figure A1.6 shows the inner surface impedance Z_i calculated by (1) Schelkunoff, (2) Wedepohl/Wilcox, and (3) new formula in (A1.63)–(A1.64) explained in Section A1.4.4.2. It is observed in the figure that the three formulae give essentially identical results, except that Schellkunoff's formula cannot be calculated above about 800 kHz due to the numerical divergence.
2. Mutual impedance Z_m

   Figure A1.7 shows the mutual impedance Z_m calculated by Schelkunoff, Wedepohl/Wilcox and new approximate formulas explained in Section A1.4.4.2. No significant difference is observed between the calculated results in Figure A1.7.
3. Outer surface impedance Z_o

   Figure A1.8 shows the outer surface impedance Z_i. Calculated results by the three formulas are again, closely overlapping, with Schelkunoff not giving a result beyond approximately 800 kHz due to numerical problems.

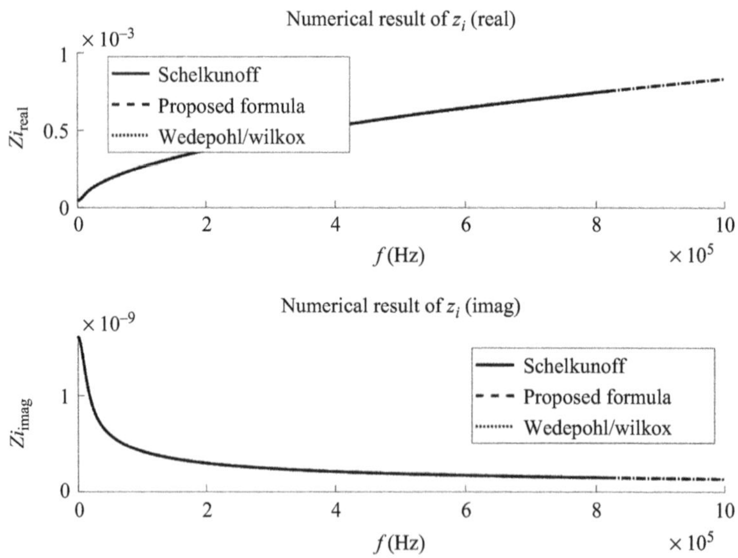

Figure A1.6 Inner surface impedance Z_i: (1) solid red line: Schelkunoff's formula, (2) broken line: Wedepohl/Wilcox formula, and (3) dashed line: new approximate formula

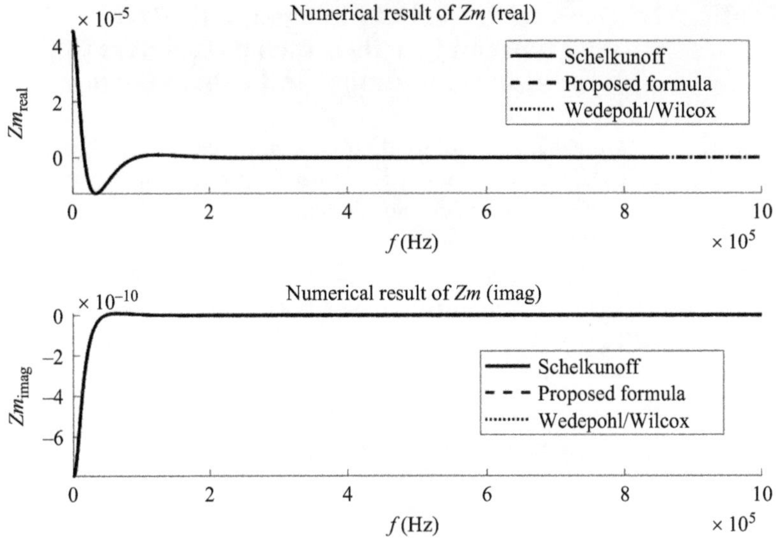

Figure A1.7 *Mutual impedance Z_m: (1) solid line: Schelkunoff, (2) broken line: Wedepohl/Wilcox, and (3) dashed line: new formula*

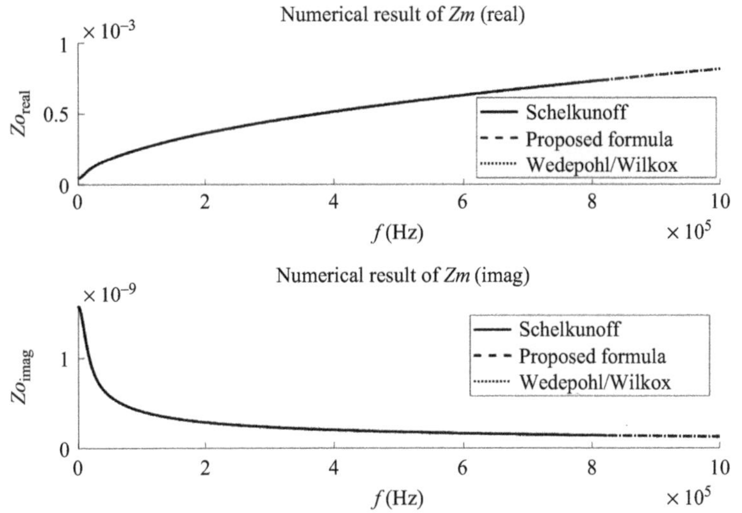

Figure A1.8 *Outer surface impedance Z_o, (1) solid red line: Schelkunoff, (2) broken line: Wedepohl/Wilcox, and (3) dashed line: new formula*

Appendix A2 Derivation of electromagnetic field equations and earth-return parameters for a multiphase underground cable system

The complete electromagnetic field solution of a multiphase underground cable system with N cables is developed based on rigorous Maxwell equations. Then, the generalized earth-return impedance and admittance formulas are derived by adopting the complete field solution with exact voltage and current propagation equations for a multiphase cable. Three different approaches of a cable voltage calculation related to different earth-return impedance and admittance formulas are discussed. Moreover, quasi-TEM-based earth-return impedance and admittance formulas are deduced from the formulas of the complete field solution.

A2.1 Derivation of electromagnetic field formulas

The configuration of an underground cable system with N cables is illustrated in Figure A2.1, where the cable is composed of an ideal conductor of radius r_i and an insulator (μ_{ins} and ε_{ins}) of radius r_o. The air is the medium in the region $z > 0$, and characterized by permeability μ_0 and permittivity ε_0. The region of $z < 0$ is the lossy earth, and it is characterized by permeability μ_1, permittivity ε_1 and conductivity σ_1. It is assumed that the radius of each conductor is small compared with the height and separation. Thus, a filament current approximation can be adopted where all the conductor currents are only expressed by the axial component [60].

It should be noted that the cables are located at the negative side of z-axis, that is, the buried depth of a cable is negative.

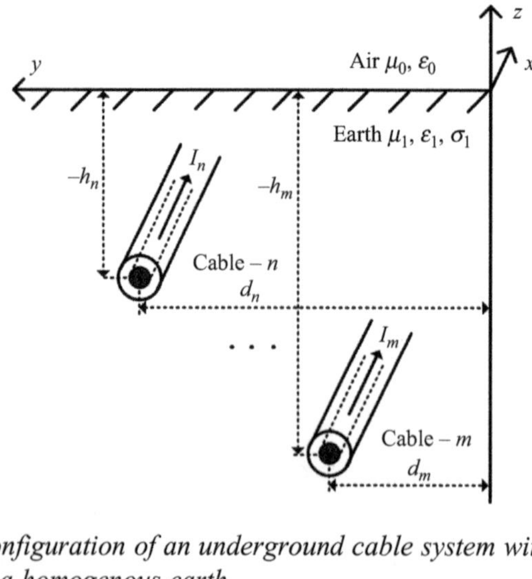

Figure A2.1 Configuration of an underground cable system with N cables buried in a homogenous earth

Moreover, there exist as many propagation modes as the number of insulated conductors in the system of Figure A2.1. The Cable-m with radius r_{om} carries the following current in phase domain [39]:

$$I_m(x) = \sum_{\nu=1}^{N} B_{m\nu} i_\nu(x) \tag{A2.1}$$

where $B_{m\nu}$ is the $m\nu$th element of the $N \times N$ unknown current transformation matrix **B** (to be determined by impedance boundary condition) and the νth modal current $i_\nu(x)$ is defined by

$$i_\nu(x) = i_{\nu p} e^{-jk_\nu x} \tag{A2.2}$$

where $i_{\nu p}$ is the amplitude of the modal current and k_ν is the unknown wave number of the νth mode.

The electromagnetic vector fields produced by underground cables in the earth can be defined by using Hertzian vectors as given below [65]:

$$\overline{\mathbf{E}} = \nabla\nabla \cdot \overline{\mathbf{\Pi}}_{\mathbf{E}} + \omega^2 \mu\varepsilon \overline{\mathbf{\Pi}}_{\mathbf{E}} - j\omega\mu \nabla \times \overline{\mathbf{\Pi}}_{\mathbf{M}} \tag{A2.3}$$

$$\overline{\mathbf{H}} = \nabla\nabla \cdot \overline{\mathbf{\Pi}}_{\mathbf{M}} + \omega^2 \mu\varepsilon \overline{\mathbf{\Pi}}_{\mathbf{M}} + j\omega\varepsilon \nabla \times \overline{\mathbf{\Pi}}_{\mathbf{E}} \tag{A2.4}$$

where ω is the angular frequency.

$\overline{\mathbf{\Pi}}_{\mathbf{E}}$ and $\overline{\mathbf{\Pi}}_{\mathbf{M}}$ are electric and magnetic Hertzian vectors. It should be noted that any continuously differentiable vector field is defined by [65]

$$\overline{\mathbf{F}} = \mathbf{e_x} F_x + \mathbf{e_y} F_y + \mathbf{e_z} F_z \tag{A2.5}$$

where $\mathbf{e_x}$, $\mathbf{e_y}$, and $\mathbf{e_z}$ be the corresponding basis of the unit vector **e**.

For the present problem, it is convenient to adopt only x-component of Hertzian vectors because $\overline{\mathbf{\Pi}}_{\mathbf{E}}$ and $\overline{\mathbf{\Pi}}_{\mathbf{M}}$ are generated by $I_m(x)$ and will be directed parallel to the x-axis [32].

Thus, the expressions of the x-component of the electric and magnetic Hertzian vectors at any point in the air ("0"), and earth ("1") can be derived as a sum of contributions by the current in each conductor. By adopting the superposition theorem, it gives

$$\Pi_{0E}(x, y, z) = \sum_{m=1}^{N} \Pi_{0Em}(x, y, z) \tag{A2.6}$$

$$\Pi_{0M}(x, y, z) = \sum_{m=1}^{N} \Pi_{0Mm}(x, y, z) \tag{A2.7}$$

$$\Pi_{1E}(x, y, z) = \sum_{m=1}^{N} \Pi_{1Em}(x, y, z) \tag{A2.8}$$

$$\Pi_{1M}(x, y, z) = \sum_{m=1}^{N} \Pi_{1Mm}(x, y, z) \tag{A2.9}$$

where $\Pi_{0Em}(x,y,z)$, $\Pi_{0Mm}(x,y,z)$, $\Pi_{1Em}(x,y,z)$, and $\Pi_{1Mm}(x,y,z)$ are the x-components of electric and magnetic Hertzian vectors in the air and earth, generated by the current $I_m(x)$ of the mth cable.

By adopting the rigorous modal analysis [39] in the formulations, the x-component of the electric and magnetic Hertz vectors at any point in the air and earth can be given by

$$\Pi_{0E}(x,y,z) = -\frac{j\omega\mu_1}{4\pi k_e^2}\sum_{m=1}^{N}\sum_{v=1}^{N} B_{mv}i_v(x)\int_{-\infty}^{+\infty}\frac{R_{Ev}e^{h_m u_{1v}-zu_{0v}}}{u_{1v}}e^{-j(y-d_m)\lambda}d\lambda \tag{A2.10}$$

$$\Pi_{0M}(x,y,z) = -\frac{j\omega\mu_1}{4\pi k_e^2}\sum_{m=1}^{N}\sum_{v=1}^{N} B_{mv}i_v(x)\int_{-\infty}^{+\infty}\frac{R_{Mv}e^{h_m u_{1v}-zu_{0v}}}{u_{1v}}e^{-j(y-d_m)\lambda}d\lambda \tag{A2.11}$$

$$\Pi_{1E}(x,y,z) = -\frac{j\omega\mu_1}{4\pi k_e^2}\sum_{m=1}^{N}\sum_{v=1}^{N} B_{mv}i_v(x)\int_{-\infty}^{+\infty}\frac{e^{-|z-h_m|u_{1v}} + T_{Ev}e^{(z+h_m)u_{1v}}}{u_{1v}}e^{-j(y-d_m)\lambda}d\lambda \tag{A2.12}$$

$$\Pi_{1M}(x,y,z) = -\frac{j\omega\mu_1}{4\pi k_e^2}\sum_{m=1}^{N}\sum_{v=1}^{N} B_{mv}i_v(x)\int_{-\infty}^{+\infty}\frac{T_{Mv}e^{(z+h_m)u_{1v}}}{u_{1v}}e^{-j(y-d_m)\lambda}d\lambda \tag{A2.13}$$

where λ is the integral variable and

$$u_{0v} = \sqrt{\lambda^2 + k_v^2 - k_a^2}, u_{1v} = \sqrt{\lambda^2 + k_v^2 - k_e^2} \tag{A2.14}$$

with

$$k_a^2 = -j\omega\mu_0 j\omega\varepsilon_0, k_e^2 = -j\omega\mu_1(\sigma_1 + j\omega\varepsilon_1) \tag{A2.15}$$

The irrationals R_{Ev}, R_{Mv}, T_{Ev}, and T_{Mv} can be derived by adopting the following boundary conditions at the air/earth interface ($z = 0$), referred to the vth mode:

$$\left.\begin{aligned} E_{1x} &= E_{0x} \\ E_{1y} &= E_{0y} \\ H_{1x} &= H_{0x} \\ H_{1y} &= H_{0y} \end{aligned}\right\} \tag{A2.16}$$

where E_x, E_y, H_x and H_y are axial and horizontal electric and magnetic field components.

Solving (A2.16) with $\mu_1 = \mu_0$, T_{Ev} and T_{Mv} can be deduced as

$$T_{Ev} = -1 + 2\frac{k_e^2 u_{1v}}{k_e^2 - k_v^2}\left(\frac{1}{u_{0v} + u_{1v}} - \frac{k_v^2}{k_e^2 u_{0v} + k_a^2 u_{1v}}\right) \tag{A2.17}$$

$$T_{Mv} = -2\frac{k_e^2 k_v \lambda}{j\omega\mu_0\left(k_e^2 - k_v^2\right)}\left(\frac{1}{u_{0v} + u_{1v}} - \frac{k_e^2}{k_e^2 u_{0v} + k_a^2 u_{1v}}\right) \tag{A2.18}$$

By substituting (A2.17) and (A2.18) into (A2.12) and (A2.13), and solving (A2.3) and (A2.4), the formulas of electromagnetic field components at an arbitrary point in the earth are derived:

$$E_{1x}(x,y,z) = \sum_{m=1}^{N}\sum_{v=1}^{N}\Lambda_{1x}(y,z,d_m,h_m,v)B_{mv}i_v(x) \tag{A2.19}$$

$$E_{1y}(x,y,z) = \sum_{m=1}^{N}\sum_{v=1}^{N}\Lambda_{1y}(y,z,d_m,h_m,v)B_{mv}i_v(x) \tag{A2.20}$$

$$E_{1z}(x,y,z) = \sum_{m=1}^{N}\sum_{v=1}^{N}\Lambda_{1z}(y,z,d_m,h_m,v)B_{mv}i_v(x) \tag{A2.21}$$

$$H_{1x}(x,y,z) = \sum_{m=1}^{N}\sum_{v=1}^{N}\Psi_{1x}(y,z,d_m,h_m,v)B_{mv}i_v(x) \tag{A2.22}$$

$$H_{1y}(x,y,z) = \sum_{m=1}^{N}\sum_{v=1}^{N}\Psi_{1y}(y,z,d_m,h_m,v)B_{mv}i_v(x) \tag{A2.23}$$

$$H_{1z}(x,y,z) = \sum_{m=1}^{N}\sum_{v=1}^{N}\Psi_{1z}(y,z,d_m,h_m,v)B_{mv}i_v(x) \tag{A2.24}$$

where the kernel functions $\Lambda_{1x}(y,z,d_m,h_m,v)$, $\Lambda_{1y}(y,z,d_m,h_m,v)$, $\Lambda_{1z}(y,z,d_m,h_m,v)$, $\Psi_{1x}(y,z,d_m,h_m,v)$, $\Psi_{1y}(y,z,d_m,h_m,v)$, and $\Psi_{1z}(y,z,d_m,h_m,v)$ are given below:

$$\begin{aligned}\Lambda_{1x}(y,z,d_m,h_m,v) = &-\frac{j\omega\mu_0}{4\pi}\left[\underline{\Delta}_1^c(y,z,d_m,h_m,v) + 2\underline{S}_1^c(y,z,d_m,h_m,v)\right]\\ &+\frac{j\omega\mu_0 k_v^2}{4\pi k_e^2}\left[\underline{\Delta}_1^c(y,z,d_m,h_m,v) + 2\underline{S}_2^c(y,z,d_m,h_m,v)\right]\end{aligned} \tag{A2.25}$$

$$\Lambda_{1y}(y,z,d_m,h_m,v) = \frac{j\omega\mu_0 k_v}{4\pi k_e^2}\left[\underline{\Delta}_2^c(y,z,d_m,h_m,v) + 2\underline{S}_4^c(y,z,d_m,h_m,v)\right] \tag{A2.26}$$

$$\begin{aligned}\Lambda_{1z}(y,z,d_m,h_m,v) \quad = \quad &-\frac{\omega\mu_0 k_v}{4\pi k_e^2}[\underline{\Delta}_3^c(y,z,d_m,h_m,v) - 2k_e^2\underline{S}_7^c(y,z,d_m,h_m,v)\\ &+2k_v^2\underline{S}_8^c(y,z,d_m,h_m,v) + 2\underline{S}_{10}^c(y,z,d_m,h_m,v)]\end{aligned} \tag{A2.27}$$

$$\Psi_{1x}(y,z,d_m,h_m,v) = \frac{k_v}{2\pi}\left[\underline{S}_5^c(y,z,d_m,h_m,v) - \underline{S}_6^c(y,z,d_m,h_m,v)\right] \tag{A2.28}$$

$$\begin{aligned}\Psi_{1y}(y,z,d_m,h_m,v) &= \frac{1}{4\pi}\left[\underline{\Delta}_3^c(y,z,d_m,h_m,v) - 2k_e^2\underline{S}_7^c(y,z,d_m,h_m,v)\right.\\ &\quad \left.+2k_v^2\underline{S}_8^c(y,z,d_m,h_m,v) + 2\underline{S}_9^c(y,z,d_m,h_m,v)\right]\end{aligned} \tag{A2.29}$$

$$\Psi_{1z}(y,z,d_m,h_m,v) = \frac{j}{4\pi}\left[\underline{\Delta}_2^c(y,z,d_m,h_m,v) + 2\underline{S}_3^c(y,z,d_m,h_m,v)\right] \tag{A2.30}$$

with

$$\underline{\Delta}_1^c(y,z,d_m,h_m,v) = \int_{-\infty}^{+\infty} \frac{e^{-|z-h_m|u_{1v}} - e^{(z+h_m)u_{1v}}}{u_{1v}} e^{-j(y-d_m)\lambda} d\lambda \tag{A2.31}$$

$$\underline{\Delta}_2^c(y,z,d_m,h_m,v) = \int_{-\infty}^{+\infty} \frac{e^{-|z-h_m|u_{1v}} - e^{(z+h_m)u_{1v}}}{u_{1v}} e^{-j(y-d_m)\lambda} \lambda d\lambda \tag{A2.32}$$

$$\underline{\Delta}_3^c(y,z,d_m,h_m,v) = \int_{-\infty}^{+\infty} \left[e^{(z-h_m)u_{1v}} - e^{(z+h_m)u_{1v}}\right] e^{-j(y-d_m)\lambda} d\lambda \text{ for } z < h_m \tag{A2.33}$$

$$\underline{\Delta}_3^c(y,z,d_m,h_m,v) = \int_{-\infty}^{+\infty} \left[-e^{(-z+h_m)u_{1v}} - e^{(z+h_m)u_{1v}}\right] e^{-j(y-d_m)\lambda} d\lambda \text{ for } h_m < z < 0 \tag{A2.34}$$

$$\underline{S}_1^c(y,z,d_m,h_m,v) = \int_{-\infty}^{+\infty} \frac{e^{(z+h_m)u_{1v}}}{u_{0v} + u_{1v}} e^{-j(y-d_m)\lambda} d\lambda \tag{A2.35}$$

$$\underline{S}_2^c(y,z,d_m,h_m,v) = \int_{-\infty}^{+\infty} \frac{e^{(z+h_m)u_{1v}}}{u_{0v} + k_a^2 k_e^{-2} u_{1v}} e^{-j(y-d_m)\lambda} d\lambda \tag{A2.36}$$

$$\underline{S}_3^c(y,z,d_m,h_m,v) = \int_{-\infty}^{+\infty} \frac{e^{(z+h_m)u_{1v}}}{u_{0v} + u_{1v}} e^{-j(y-d_m)\lambda} \lambda d\lambda \tag{A2.37}$$

$$\underline{S}_4^c(y,z,d_m,h_m,v) = \int_{-\infty}^{+\infty} \frac{e^{(z+h_m)u_{1v}}}{u_{0v} + k_a^2 k_e^{-2} u_{1v}} e^{-j(y-d_m)\lambda} \lambda d\lambda \tag{A2.38}$$

$$\underline{S}_5^c(y,z,d_m,h_m,v) = \int_{-\infty}^{+\infty} \frac{e^{(z+h_m)u_{1v}}}{u_{1v}(u_{0v} + u_{1v})} e^{-j(y-d_m)\lambda} \lambda d\lambda \tag{A2.39}$$

$$\underline{S}_6^c(y,z,d_m,h_m,v) = \int_{-\infty}^{+\infty} \frac{e^{(z+h_m)u_{1v}}}{u_{1v}\left(u_{0v} + k_a^2 k_e^{-2} u_{1v}\right)} e^{-j(y-d_m)\lambda} \lambda d\lambda \tag{A2.40}$$

$$\underline{S}_7^c(y,z,d_m,h_m,v) = \int_{-\infty}^{+\infty} \frac{e^{(z+h_m)u_{1v}}}{u_{1v}(u_{0v} + u_{1v})} e^{-j(y-d_m)\lambda} d\lambda \tag{A2.41}$$

$$\underline{S}_8^c(y,z,d_m,h_m,v) = \int_{-\infty}^{+\infty} \frac{e^{(z+h_m)u_{1v}}}{u_{1v}\left(u_{0v} + k_a^2 k_e^{-2} u_{1v}\right)} e^{-j(y-d_m)\lambda} d\lambda \tag{A2.42}$$

$$\underline{S}_9^c(y,z,d_m,h_m,v) = \int_{-\infty}^{+\infty} \frac{e^{(z+h_m)u_{1v}}}{u_{1v}(u_{0v}+u_{1v})} e^{-j(y-d_m)\lambda} \lambda^2 d\lambda \tag{A2.43}$$

$$\underline{S}_{10}^c(y,z,d_m,h_m,v) = \int_{-\infty}^{+\infty} \frac{e^{(z+h_m)u_{1v}}}{u_{1v}\left(u_{0v}+k_a^2 k_e^{-2} u_{1v}\right)} e^{-j(y-d_m)\lambda} \lambda^2 d\lambda \tag{A2.44}$$

$$\underline{S}_{11}^c(y,z,d_m,h_m,v) = \int_{-\infty}^{+\infty} \frac{e^{(z+h_m)u_{1v}}}{u_{1v}^2(u_{0v}+u_{1v})} e^{-j(y-d_m)\lambda} \lambda^2 d\lambda \tag{A2.45}$$

$$\underline{S}_{12}^c(y,z,d_m,h_m,v) = \int_{-\infty}^{+\infty} \frac{e^{(z+h_m)u_{1v}}}{u_{1v}^2\left(u_{0v}+k_a^2 k_e^{-2} u_{1v}\right)} e^{-j(y-d_m)\lambda} \lambda^2 d\lambda \tag{A2.46}$$

$$\underline{S}_{13}^c(y,z,d_m,h_m,v) = \int_{-\infty}^{+\infty} \frac{e^{(z+h_m)u_{1v}}}{u_{1v}^2(u_{0v}+u_{1v})} e^{-j(y-d_m)\lambda} d\lambda \tag{A2.47}$$

$$\underline{S}_{14}^c(y,z,d_m,h_m,v) = \int_{-\infty}^{+\infty} \frac{e^{(z+h_m)u_{1v}}}{u_{1v}^2\left(u_{0v}+k_a^2 k_e^{-2} u_{1v}\right)} e^{-j(y-d_m)\lambda} d\lambda \tag{A2.48}$$

Equations (A2.19)–(A2.24) can be rewritten into matrix expression as a function of the exciting modal currents:

$$\begin{bmatrix} E_{1x}(x,y,z) \\ E_{1y}(x,y,z) \\ E_{1z}(x,y,z) \end{bmatrix} = \begin{bmatrix} \Phi_{1x}(y,z,d_m,h_m,1) & \cdots & \Phi_{1x}(y,z,d_m,h_m,N) \\ \Phi_{1y}(y,z,d_m,h_m,1) & \cdots & \Phi_{1y}(y,z,d_m,h_m,N) \\ \Phi_{1z}(y,z,d_m,h_m,1) & \cdots & \Phi_{1z}(y,z,d_m,h_m,N) \end{bmatrix} \begin{bmatrix} i_1(x) \\ \vdots \\ i_N(x) \end{bmatrix} \tag{A2.49}$$

$$\begin{bmatrix} H_{1x}(x,y,z) \\ H_{1y}(x,y,z) \\ H_{1z}(x,y,z) \end{bmatrix} = \begin{bmatrix} \Upsilon_{1x}(y,z,d_m,h_m,1) & \cdots & \Upsilon_{1x}(y,z,d_m,h_m,N) \\ \Upsilon_{1y}(y,z,d_m,h_m,1) & \cdots & \Upsilon_{1y}(y,z,d_m,h_m,N) \\ \Upsilon_{1z}(y,z,d_m,h_m,1) & \cdots & \Upsilon_{1z}(y,z,d_m,h_m,N) \end{bmatrix} \begin{bmatrix} i_1(x) \\ \vdots \\ i_N(x) \end{bmatrix} \tag{A2.50}$$

where

$$\Phi_{1x}(y,z,d_m,h_m,v) = \sum_{m=1}^{N} \Lambda_{1x}(y,z,d_m,h_m,v) B_{mv} \tag{A2.51}$$

$$\Phi_{1y}(y,z,d_m,h_m,v) = \sum_{m=1}^{N} \Lambda_{1y}(y,z,d_m,h_m,v) B_{mv} \tag{A2.52}$$

$$\Phi_{1z}(y,z,d_m,h_m,v) = \sum_{m=1}^{N} \Lambda_{1z}(y,z,d_m,h_m,v) B_{mv} \tag{A2.53}$$

$$\Upsilon_{1x}(y,z,d_m,h_m,v) = \sum_{m=1}^{N} \Psi_{1x}(y,z,d_m,h_m,v) B_{mv} \tag{A2.54}$$

$$\Upsilon_{1y}(y, z, d_m, h_m, v) = \sum_{m=1}^{N} \Psi_{1y}(y, z, d_m, h_m, v) B_{mv} \tag{A2.55}$$

$$\Upsilon_{1z}(y, z, d_m, h_m, v) = \sum_{m=1}^{N} \Psi_{1z}(y, z, d_m, h_m, v) B_{mv} \tag{A2.56}$$

The matrix expression (A2.49) and (A2.50) is rewritten in the following simple form:

$$\left. \begin{aligned} \mathbf{E}_1 &= \mathbf{\Phi}_1 \mathbf{i} \\ \mathbf{H}_1 &= \mathbf{\Upsilon}_1 \mathbf{i} \end{aligned} \right\} \tag{A2.57}$$

where $\mathbf{E}_1$ and $\mathbf{H}_1$ are an $N \times 1$ electromagnetic field vector, $\mathbf{\Phi}_1$ and $\mathbf{\Upsilon}_1$ are a $3 \times N$ modal current to field operator's matrix, and $\mathbf{i}$ is an $N \times 1$ modal current vector.

It should be noted that (A2.57) derived in the above is based on the assumption of unidirectional propagation of voltage and current waves. The responses of electromagnetic fields should be related to the exciting sources which are impressed by the transmitting device. Thus, the transfer function from voltage to field can be defined by using $\mathbf{\Phi}_1$ and $\mathbf{\Upsilon}_1$ matrices together with the following modal current vector:

$$\mathbf{i} = \mathbf{B}^{-1} \mathbf{Y}_0 \mathbf{A} \zeta \mathbf{A}^{-1} \mathbf{V_f} \tag{A2.58}$$

where $\mathbf{Y}_0$ is an $N \times N$ characteristic admittance matrix of underground cables in phase domain, $\mathbf{A}$ is an $N \times N$ voltage transformation matrix, ζ is an $N \times N$ reflection free transition diagonal matrix, $\mathbf{V_f}$ is an $N \times 1$ impressed voltage source vector, and

$$\mathbf{A} = \left(\mathbf{B}^T\right)^{-1} \tag{A2.59}$$

$$\zeta = \operatorname{diag}\left(e^{-jk_v x}\right) \tag{A2.60}$$

Equation (A2.57) can be rewritten by replacing $\mathbf{i}$ by (A2.58)

$$\left. \begin{aligned} \mathbf{E}_1 &= \mathbf{\Phi}_1 \mathbf{B}^{-1} \mathbf{Y}_0 \mathbf{A} \zeta \mathbf{A}^{-1} \mathbf{V_f} \\ \mathbf{H}_1 &= \mathbf{\Upsilon}_1 \mathbf{B}^{-1} \mathbf{Y}_0 \mathbf{A} \zeta \mathbf{A}^{-1} \mathbf{V_f} \end{aligned} \right\} \tag{A2.61}$$

Therefore, the generated electromagnetic fields of a multiphase underground cable can be calculated by applying the impressed voltage source $\mathbf{V_f}$ in (A2.61).

A2.2 Derivation of modal equation

The modal equation of underground cables is used to solve unknown modal wave number k_v for (A2.61). The derivation of the modal equation is based on the continuity of the axial electric field on the nth cable:

$$\sum_{v=1}^{N} Z_{sn}^{c}(r_{on}, r_{in}, v) B_{nv} i_v(x) = \sum_{m=1}^{N} \sum_{v=1}^{N} \Lambda_{1x}(d_n, h_n, d_m, h_m, v) B_{mv} i_v(x) \tag{A2.62}$$

where $Z_{sn}^{c}(r_{on}, r_{in}, \nu)$ is the surface impedance of nth cable for νth mode and it is given by

$$Z_{sn}^{c}(r_{on}, r_{in}, \nu) = \frac{k_{\nu}^{2} - k_{ins}^{2}}{j\omega 2\pi\varepsilon_{ins}} \ln\left(\frac{r_{on}}{r_{in}}\right) \tag{A2.63}$$

where r_{on} and r_{in} are outer and inner radii of nth cable, and

$$k_{ins}^{2} = \omega^{2}\mu_{ins}\varepsilon_{ins} \tag{A2.64}$$

In the evaluation of the right side of (A2.62), the self-component, $n = m$, is calculated at the surface of the cable. The mutual element, $n \neq m$, can be calculated at the center of two cables. If $N = 1$, (A2.62) is the same as the exact approach of references [60–63]. Taking further mathematical manipulation to (A2.62) and considering the same eigenvalue analysis in [39], the matrix expression of (A2.62) can be derived:

$$\mathbf{P^{c}B_{\nu}} = -k_{\nu}^{2}\mathbf{B_{\nu}} \tag{A2.65}$$

where the full $N \times N$ matrix $\mathbf{P^{c}}$ is given by

$$\mathbf{P^{c}} = \left(\mathbf{P_{2}^{c}}\right)^{-1}\mathbf{P_{1}^{c}} \tag{A2.66}$$

with

$$\mathbf{P_{1}^{c}} = \mathbf{Z_{ins}} + \frac{j\omega\mu_{0}}{4\pi}\left(\underline{\mathbf{\Delta}}_{1}^{\mathbf{c}} + 2\underline{\mathbf{S}}_{1}^{\mathbf{c}}\right), \mathbf{P_{2}^{c}} = \mathbf{Y_{ins}^{-1}} + \frac{1}{4\pi(\sigma_{1} + j\omega\varepsilon_{1})}\left(\underline{\mathbf{\Delta}}_{1}^{\mathbf{c}} + 2\underline{\mathbf{S}}_{2}^{\mathbf{c}}\right) \tag{A2.67}$$

$\mathbf{Z_{ins}}$ and $\mathbf{Y_{ins}}$ are an $N \times N$ diagonal matrix and each element is related to the insulator impedance and admittance of the nth cable [1]. Also, each element of $N \times N$ matrices $\underline{\mathbf{\Delta}}_{1}^{\mathbf{c}}$, $\underline{\mathbf{S}}_{1}^{\mathbf{c}}$ and $\underline{\mathbf{S}}_{2}^{\mathbf{c}}$ are given in (A2.62) together with (A2.31), (A2.35), and (A2.36).

A2.3 Derivation of generalized earth-return impedance and admittance formulas

In this section, the generalized earth-return impedance and admittance formulas of a single underground cable are derived based on the exact voltage and current propagation equations. The same method is also extended to a multiphase underground cable. Then, the earth-return impedance and admittance formulas based on a quasi-TEM assumption which avoids the numerical difficulties due to k_{ν} in the complete field solution are proposed.

A2.3.1 Complete field solution

The generalized earth-return impedance and admittance formulas are derived based on the complete field solution with exact voltage and current propagation equations for a single and multiphase underground cable. Curl Maxwell equations of electromagnetic field and Kelvin–Stokes theorem are given by [65]

$$\nabla \times \overline{\mathbf{E}}_{1} = -j\omega\mu_{0}\overline{\mathbf{H}}_{1} \tag{A2.68}$$

$$\nabla \times \overline{\mathbf{H}}_1 = (j\omega\varepsilon_1 + \sigma_1)\overline{\mathbf{E}}_1 \tag{A2.69}$$

$$\int_c \overline{\mathbf{F}} \cdot d\mathbf{l} = \iint_s \nabla \times \overline{\mathbf{F}} \cdot d\mathbf{s} \tag{A2.70}$$

where $\overline{\mathbf{F}}$ is given in (A2.5), ∇ is Laplacian operator, $\overline{\mathbf{E}}_1$ and $\overline{\mathbf{H}}_1$ are

$$\overline{\mathbf{E}}_1 = (\mathbf{E}_1)^T\mathbf{e}, \overline{\mathbf{H}}_1 = (\mathbf{H}_1)^T\mathbf{e} \tag{A2.71}$$

with $\mathbf{E}_1$, $\mathbf{H}_1$ and $\mathbf{e}$ are given in (A2.57) and (A2.5).

By substituting (A2.68) into (A2.70), it gives:

$$\int_c \overline{\mathbf{E}}_1 \cdot d\mathbf{l} = -j\omega\mu_0 \iint_s \overline{\mathbf{H}}_1 \cdot d\mathbf{s} \tag{A2.72}$$

Equation (A2.72) involves a line integral to the electric vector field and a surface integral to the magnetic vector field. Then, taking z-component of (A2.69), it becomes:

$$\frac{\partial H_{1y}(x,y,z)}{\partial x} - \frac{\partial H_{1x}(x,y,z)}{\partial y} = j\omega\left(\varepsilon_1 - j\frac{\sigma_1}{\omega}\right)E_{1z}(x,y,z) \tag{A2.73}$$

where $E_{1z}(x,y,z)$, $H_{1x}(x,y,z)$ and $H_{1y}(x,y,z)$ are given in (A2.21), (A2.22), and (A2.23).

The exact voltage and current propagation equations can be derived by (A2.72) and (A2.73) together with the following line voltage integral equation on the nth cable:

$$V_n(x) = -\int_{h_r}^{h_n - r_{on}} E_{1z}(x, d_n, z)dz \tag{A2.74}$$

where h_r indicates a position (depth) in the earth, at which E_{1z} has decayed sufficiently to permit this position to be taken as an effective zero potential level [32,47,48].

The accurate way to evaluate (A2.74) requires to find the unknown modal wave number and current transformation matrix in modal equation (A2.65) firstly. Then, the vertical electric field component is calculated to depth h_r which is adopted by considering the contributions of E_{1z} at different depths until these can be neglected. Next, this depth is adopted in (A2.74). However, the accurate method requires the numerical calculations of (A2.65) and E_{1z} at each frequency step, and thus it causes complexities. Instead, h_r can be assumed to be the earth surface $h_r = 0$ [47] and the infinite depth in the earth $h_r = -\infty$ [47]. Also, the following penetration depth of soil can be used [47]:

$$h_r = -\delta_e = -\left\{\frac{\omega^2\varepsilon_1\mu_0}{2}\left[\sqrt{1 + \left(\frac{\sigma_1}{\omega\varepsilon_1}\right)^2} - 1\right]\right\}^{-\frac{1}{2}} \tag{A2.75}$$

Therefore, the generalized telegrapher's equations are proposed based on different h_r.

A2.3.1.1 A single underground cable

Line voltage integral equation referred to infinite depth in the earth

Figure A2.2 illustrates the integral path of the infinite depth in the earth. It should be noted that a single phase underground cable is considered and assumes $d_1 = 0$ in Figure A2.1. The positive direction of the integral path: $C_1 - C_2 - C_3 - C_4$ adopts the anti-clock direction. Also, the direction of the normal vector $\mathbf{n}$ of area S is $-\mathbf{e_y}$.

The following expression is deduced by solving (A2.72):

$$\int_{-\infty}^{h_1-r_{o1}} E_{1z}(x+\Delta x, 0, z)dz - \int_{-\infty}^{h_1-r_{o1}} E_{1z}(x, 0, z)dz + \int_{x}^{x+\Delta x} E_{1x}(x, 0, -\infty)dx$$
$$- \int_{x}^{x+\Delta x} E_{1x}(x, 0, h_1 - r_{o1})dx = j\omega\mu_0 \int_{-\infty}^{h_1-r_{o1}} \int_{x}^{x+\Delta x} H_{1y}(x, 0, z)dxdz \tag{A2.76}$$

Based on the radiation condition, the electromagnetic fields are zero at the infinite position, and it has

$$\int_{x}^{x+\Delta x} E_{1x}(x, 0, -\infty)dx = 0 \tag{A2.77}$$

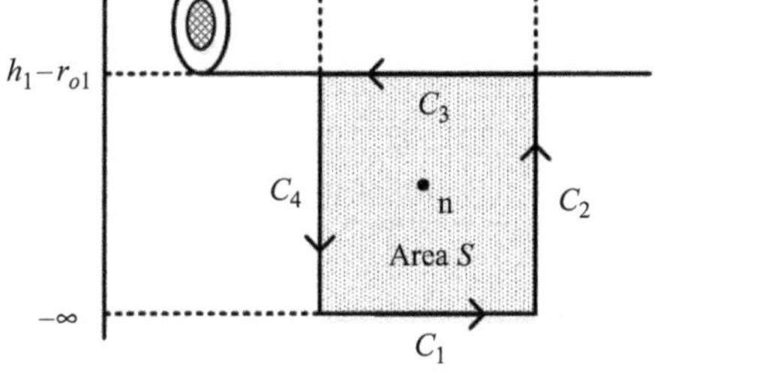

Figure A2.2 The integral path with the reference at the infinite depth

By substituting (A2.77) into (A2.76) and taking the limit $\Delta x \to 0$, the following relation is obtained:

$$\lim_{\Delta x\to 0}\frac{\int_{-\infty}^{h_1-r_{o1}} E_{1z}(x+\Delta x,0,z)dz - \int_{-\infty}^{h_1-r_{o1}} E_{1z}(x,0,z)dz}{\Delta x}$$

$$-\lim_{\Delta x\to 0}\frac{\int_{x}^{x+\Delta x} E_{1x}(x,0,h_1-r_{o1})dx}{\Delta x}$$

$$=\lim_{\Delta x\to 0}\frac{j\omega\mu_0\int_{-\infty}^{h_1-r_{o1}}\int_{x}^{x+\Delta x} H_{1y}(x,0,z)dxdz}{\Delta x} \tag{A2.78}$$

The first term on the left side of (A2.78) represents the voltage propagation characteristic. The voltage propagation equation of the cable is derived by substituting (A2.74) into (A2.78):

$$-\frac{dV_1(x)}{dx} = E_{1x}(x,0,h_1-r_{o1}) + j\omega\mu_0\int_{-\infty}^{h_1-r_{o1}} H_{1y}(x,0,z)dz \tag{A2.79}$$

where $E_{1x}(x,0,h_1-r_{o1})$ represents the axial electric field at the cable surface, and it can be replaced by the surface impedance of the cable in (A2.63). $H_{1y}(x,y,z)$ is given in (A2.23).

The second term of the right side in (A2.79) represents the external magnetic flux between the cable and earth. Then, (A2.79) is rewritten in the following form:

$$-\frac{dV_1(x)}{dx} = Z_{s1}^c(r_{o1},r_{i1},1)I_1(x) + Z_e(0,h_1-r_{o1},d_1,h_1,1)I_1(x) \tag{A2.80}$$

where $Z_e(0,h_1-r_{o1},d_1,h_1,1)$ is the earth-return impedance of the cable given by

$$\begin{aligned} Z_e(y,z,d_1,h_1,v) &= \frac{j\omega\mu_0}{4\pi}\underline{\Delta}_1^c(y,z,d_1,h_1,v) \\ &\quad + \frac{j\omega\mu_0}{2\pi}\left[\underline{S}_{11}^c(y,z,d_1,h_1,v) - k_e^2\underline{S}_{13}^c(y,z,d_1,h_1,v) + k_v^2\underline{S}_{14}^c(y,z,d_1,h_1,v)\right] \end{aligned} \tag{A2.81}$$

where $\underline{\Delta}_1^c(y,z,d_1,h_1,v)$, $\underline{S}_{11}^c(y,z,d_1,h_1,v)$, $\underline{S}_{13}^c(y,z,d_1,h_1,v)$, and $\underline{S}_{14}^c(y,z,d_1,h_1,v)$ are given in (A2.31), (A2.45), (A2.47), and (A2.48).

Integrating (A2.73) with the same integral path:

$$\int_{-\infty}^{h_1-r_{o1}}\left[\frac{\partial H_{1y}(x,0,z)}{\partial x} - \frac{\partial H_{1x}(x,y,z)}{\partial y}\Big|_{y=0}\right]dz = (j\omega\varepsilon_1+\sigma_1)\int_{-\infty}^{h_1-r_{o1}} E_{1z}(x,0,z)dz \tag{A2.82}$$

where $E_{1z}(x,y,z)$ and $H_{1x}(x,y,z)$ are given in (A2.21) and (A2.22).

The current propagation equation of the cable is deduced by solving (A2.82) and (A2.74).

$$-\frac{dI_1(x)}{dx} = Y_e(0, h_1 - r_{o1}, d_1, h_1, 1)V_1(x) \tag{A2.83}$$

where $Y_e(0, h_1 - r_{o1}, d_1, h_1, 1)$ is the earth-return admittance of the cable given by

$$Y_e(y,z,d_1,h_1,v) = \frac{4\pi(j\omega\varepsilon_1 + \sigma_1)}{\left[\underline{\Delta}_1^c(y,z,d_1,h_1,v) + 2\underline{S}_{12}^c(y,z,d_1,h_1,v) - 2k_e^2\underline{S}_{13}^c(y,z,d_1,h_1,v) + 2k_v^2\underline{S}_{14}^c(y,z,d_1,h_1,v)\right]} \tag{A2.84}$$

where $\underline{S}_{12}^c(y,z,d_1,h_1,v)$ is given in (A2.46).

Line voltage integral equation referred to the earth surface

In this section, the earth-return impedance and admittance formulas are developed by adopting the earth surface as the integral reference in (A2.74). Figure A2.3 shows the same direction of the integral path and the normal vector of area S as those in Figure A2.2.

In the same manner as (A2.80), the voltage propagation equation of the cable is given by

$$-\frac{dV_1(x)}{dx} = E_{1x}(x,0,h_1 + r_{o1}) - E_{1x}(x,0,0) + j\omega\mu_0 \int_0^{h_1+r_{o1}} H_{1y}(x,0,z)dz \tag{A2.85}$$

Equation (A2.85) is rewritten in the following form similarly to (A2.80)

$$-\frac{dV_1(x)}{dx} = Z_{s1}^c(r_{o1}, r_{i1}, 1)I_1(x) + Z_e(0, h_1 - r_{o1}, d_1, h_1, 1)I_1(x) \tag{A2.86}$$

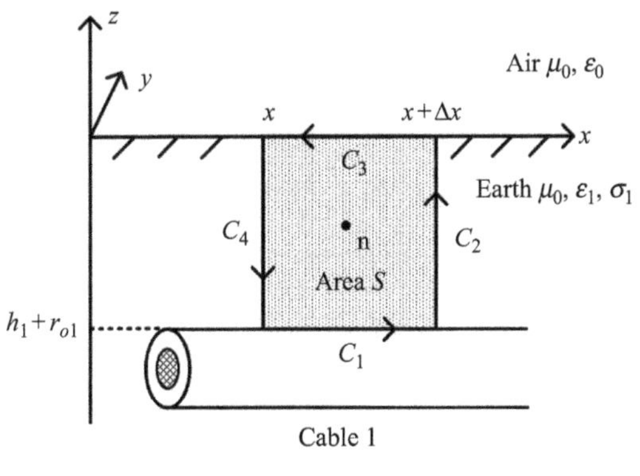

Figure A2.3 The integral path with the reference at the earth surface

where the earth-return impedance $Z_e(0, h_1 - r_{o1}, d_1, h_1, 1)$ becomes

$$Z_e(0, h_1 - r_{o1}, d_1, h_1, 1) = Z_{es}(0, 0, d_1, h_1, 1) + Z_{e\varphi}(0, h_1 - r_{o1}, d_1, h_1, 1) \tag{A2.87}$$

where $Z_{es}(0, 0, d_1, h_1, 1)$ and $Z_{e\varphi}(0, h_1 - r_{o1}, d_1, h_1, 1)$ are the earth surface impedance and external magnetic flux impedance defined as

$$Z_{es}(y, z, d_1, h_1, v) = \frac{j\omega\mu_0}{2\pi}\left[\underline{S}_1^c(y, z, d_1, h_1, v) - \frac{k_v^2}{k_e^2}\underline{S}_2^c(y, z, d_1, h_1, v)\right] \tag{A2.88}$$

$$\begin{aligned} Z_{e\varphi}(y, z, d_1, h_1, v) = & \frac{j\omega\mu_0}{4\pi}\underline{\Delta}_1^c(y, z, d_1, h_1, v) \\ & + \frac{j\omega\mu_0}{2\pi}\left[\underline{S}_{11}^c(y, z, d_1, h_1, v) - k_e^2\underline{S}_{13}^c(y, z, d_1, h_1, v) + k_v^2\underline{S}_{14}^c(y, z, d_1, h_1, v)\right] \\ & - \frac{j\omega\mu_0}{2\pi}\left[\underline{S}_{11}^c(y, 0, d_1, h_1, v) - k_e^2\underline{S}_{13}^c(y, 0, d_1, h_1, v) + k_v^2\underline{S}_{14}^c(y, 0, d_1, h_1, v)\right] \end{aligned} \tag{A2.89}$$

In (A2.86) and (A2.87), $Z_{es}(0, 0, d_1, h_1, 1)I_1(x)$ represents the per-unit length voltage drop on the earth surface due to the current $I_1(x)$. $Z_{e\varphi}(0, h_1 - r_{o1}, d_1, h_1, 1)$ is the external magnetic flux impedance due to the field scattered from the earth surface.

The current propagation equation is developed in the same manner as (A2.83) but referring to the earth surface as the integral reference as

$$-\frac{dI_1(x)}{dx} = Y_e(0, h_1 - r_{o1}, d_1, h_1, 1)V_1(x) \tag{A2.90}$$

where the earth-return admittance $Y_e(0, h_1 - r_{o1}, d_1, h_1, 1)$ is

$$\begin{aligned} Y_e(y, z, d_1, h_1, v) = & 4\pi(j\omega\varepsilon_1 + \sigma_1)[\underline{\Delta}_1^c(y, z, d_1, h_1, v) + 2\underline{S}_{12}^c(y, z, d_1, h_1, v) \\ & -2k_e^2\underline{S}_{13}^c(y, z, d_1, h_1, v) + 2k_v^2 S_{14}^c(y, z, d_1, h_1, v) \\ & -2\underline{S}_{12}^c(y, 0, d_1, h_1, v) \\ & +2k_e^2\underline{S}_{13}^c(y, 0, d_1, h_1, v) - 2k_v^2 S_{14}^c(y, 0, d_1, h_1, v)]^{-1} \end{aligned} \tag{A2.91}$$

A2.3.1.2 Multiphase underground cable

Line voltage integral equation referred to infinite depth in the earth

The earth-return impedance and admittance formulas of a multiphase cable are derived based on the modal analysis and the formulation of voltage and current propagation equations as explained in Section A2.3.1.1. The voltage propagation equation of the nth cable to the reference of the infinite depth is

$$-\frac{dV_n(x)}{dx} = E_{1x}(x, d_n, h_n - r_{on}) + j\omega\mu_0\int_{-\infty}^{h_n - r_{on}} H_{1y}(x, d_n, z)dz \tag{A2.92}$$

Similarly to (A2.79), the first term on the right side of (A2.92) represents the axial electric field at the nth cable surface. The second term represents the external magnetic flux between the nth cable and earth. Substituting (A2.19) and (A2.23) into (A2.92), rewriting in a matrix expression and transferring back to the phase domain, the following earth-return impedance formula is obtained:

$$\mathbf{Z_e} = \mathbf{Z_f B}^{-1} \tag{A2.93}$$

where $\mathbf{Z_f}$ is the earth flux impedance matrix in modal domain and the $n\nu$th element of matrix $\mathbf{Z_f}$ is given by

$$Z_{f(n\nu)}(d_n, h_n - r_{on}, d_m, h_m, \nu) = \sum_{m=1}^{N} Z_{fn}(d_n, h_n - r_{on}, d_m, h_m, \nu) B_{m\nu} \tag{A2.94}$$

where

$$Z_{fn}(y, z, d_m, h_m, \nu) = \frac{j\omega\mu_0}{4\pi} \underline{\Delta}_1^c(y, z, d_m, h_m, \nu)$$
$$+\frac{j\omega\mu_0}{2\pi}\left[\underline{S}_{11}^c(y, z, d_m, h_m, \nu) - k_e^2 \underline{S}_{13}^c(y, z, d_m, h_m, \nu) + k_\nu^2 \underline{S}_{14}^c(y, z, d_m, h_m, \nu)\right] \tag{A2.95}$$

In the same manner, the following earth-return admittance formula in phase domain is obtained from the current propagation equation of the nth cable:

$$\mathbf{Y_e} = \left(\mathbf{Y_f B}^{-1}\right)^{-1} \tag{A2.96}$$

where $\mathbf{Y_f}$ is a coefficient matrix in modal domain and the $n\nu$th element of $\mathbf{Y_f}$ is given by

$$Y_{f(n\nu)}(d_n, h_n - r_{on}, d_m, h_m, \nu) = \sum_{m=1}^{N} Y_{fn}(d_n, h_n - r_{on}, d_m, h_m, \nu) B_{m\nu} \tag{A2.97}$$

where

$$Y_{fn}(y, z, d_m, h_m, \nu) = \frac{1}{4\pi(\sigma_1 + j\omega\varepsilon_1)}\left[\underline{\Delta}_1^c(y, z, d_m, h_m, \nu)\right.$$
$$\left.+2\underline{S}_{12}^c(y, z, d_m, h_m, \nu) - 2k_e^2 \underline{S}_{13}^c(y, z, d_m, h_m, \nu) + 2k_\nu^2 \underline{S}_{14}^c(y, z, d_m, h_m, \nu)\right] \tag{A2.98}$$

Line voltage integral equation referred to the earth surface

If the integral reference is changed to the earth surface, the earth-return impedance and admittance formulas have the same expressions as (A2.93) and (A2.96) with different $\mathbf{Z_f}$ and $\mathbf{Y_f}$. For the earth-return impedance formula, the $n\nu$th element of matrix $\mathbf{Z_f}$ is given by

$$Z_{f(n\nu)}(d_n, h_n - r_{on}, d_m, h_m, \nu) = \sum_{m=1}^{N} Z_{fn}(d_n, h_n - r_{on}, d_m, h_m, \nu) B_{m\nu} \tag{A2.99}$$

where

$$Z_{fn}(y,z,d_m,h_m,v) = Z_{es}(y,0,d_m,h_m,v) + Z_{e\varphi}(y,z,d_m,h_m,v) \tag{A2.100}$$

and

$$Z_{es}(y,0,d_m,h_m,v) = \frac{j\omega\mu_0}{2\pi}\left[\underline{S}_1^c(y,0,d_m,h_m,v) - \frac{k_v^2}{k_e^2}\underline{S}_2^c(y,0,d_m,h_m,v)\right] \tag{A2.101}$$

$$\begin{aligned} Z_{e\varphi}(y,z,d_m,h_m,v) \;=\;& \frac{j\omega\mu_0}{4\pi}\underline{\Delta}_1^c(y,z,d_m,h_m,v) \\ &+\frac{j\omega\mu_0}{2\pi}\left[\underline{S}_{11}^c(y,z,d_m,h_m,v) - k_e^2\underline{S}_{13}^c(y,z,d_m,h_m,v) + k_v^2\underline{S}_{14}^c(y,z,d_m,h_m,v)\right] \\ &-\frac{j\omega\mu_0}{2\pi}\left[\underline{S}_{11}^c(y,0,d_m,h_m,v) - k_e^2\underline{S}_{13}^c(y,0,d_m,h_m,v) + k_v^2\underline{S}_{14}^c(y,0,d_m,h_m,v)\right] \end{aligned} \tag{A2.102}$$

Similarly, the earth-return admittance formula is given by

$$Y_{f(nv)}(d_n,h_n-r_{on},d_m,h_m,v) = \sum_{m=1}^{N} Y_{fn}(d_n,h_n-r_{on},d_m,h_m,v)B_{mv} \tag{A2.103}$$

where

$$\begin{aligned} Y_{fn}(y,z,d_m,h_m,v) \;=\;& \frac{1}{4\pi(\sigma_1+j\omega\varepsilon_1)}\left[\underline{\Delta}_1^c(y,z,d_m,h_m,v)\right. \\ &+2\underline{S}_{12}^c(y,z,d_m,h_m,v) - 2k_e^2\underline{S}_{13}^c(y,z,d_m,h_m,v) + 2k_v^2\underline{S}_{14}^c(y,z,d_m,h_m,v) \\ &\left.-2\underline{S}_{12}^c(y,0,d_m,h_m,v) + 2k_e^2\underline{S}_{13}^c(y,0,d_m,h_m,v) - 2k_v^2\underline{S}_{14}^c(y,0,d_m,h_m,v)\right]) \end{aligned} \tag{A2.104}$$

A2.3.2 Quasi-TEM solution

The earth-return impedance and admittance formulas based on the complete electromagnetic field solutions are proposed in Sections A2.3.1 and A2.3.2. In this section, the assumption of quasi-TEM propagation is adopted to derive a quite simple form of the earth-return impedance and admittance so that anyone can use the formulas in a study of wave propagation and power system transients.

A2.3.2.1 Numerical difficulties for the formulas based on complete electromagnetic field solutions

The modal equation and earth-return impedance and admittance formulas of a multiphase underground cable are summarized as

$$\mathbf{P^c B_v} = -k_v^2\mathbf{B_v} \tag{A2.105}$$

$$\left.\begin{aligned}\mathbf{Z_e}&=\mathbf{Z_f}\mathbf{B}^{-1}\\ \mathbf{Y_e}&=\left(\mathbf{Y_f}\mathbf{B}^{-1}\right)^{-1}\end{aligned}\right\}\tag{A2.106}$$

Equation (A2.105) is the modal equation of a multiphase underground cable, which is obtained by enforcing the boundary condition for the electric field at the earth/cable interface. Equation (A2.106) shows the matrices of the earth-return parameters of the multiphase cable based on $h_r = -\infty$ or $h_r = 0$. The details can be found in Sections A2.3.1 and A2.3.2. Before solving (A2.106), unknown modal wave number k_v and current transformation matrix **B** should be calculated by solving a nonlinear integral eigenvalue equation in (A2.105). Then, by substituting k_v, **B**, geometrical data of the cable and the earth parameters into (A2.106), $\mathbf{Z_e}$ and $\mathbf{Y_e}$ can be calculated. However, the numerical calculation of (A2.105) is very complicated in any case because each element of target matrix $\mathbf{P^c}$ is a nonlinear integral function of k_v. The quasi-TEM assumption, $k_v \approx 0$ [54], can be adopted and it covers the majority of surge phenomena in a cable system [54].

A2.3.2.2 Earth-return impedance and admittance formulas based on quasi-TEM assumption

By substituting $k_v \approx 0$ into the formulation of (A2.106), the earth-return impedance and admittance formulas can be significantly simplified. For the earth-return impedance between cables n and m:

$$\begin{aligned}\underline{Z}_{e(nm)}(d_n,h_n,d_m,h_m) \;&=\; \frac{j\omega\mu_0}{2\pi}\{K_0[jk_e d_{nm}(d_n,h_n,d_m,h_m)] - K_0[jk_e D_{nm}(d_n,h_n,d_m,h_m)]\\ &+ 2\underset{=11}{S^c}(d_n,h_n,d_m,h_m) - 2k_e^2\underset{=13}{S^c}(d_n,h_n,d_m,h_m)\}\end{aligned}\tag{A2.107}$$

where K_0 is the modified Bessel function of the second kind with zero order and

$$d_{nm}(y,z,d_m,h_m) = \sqrt{(y-d_m)^2 + (z-h_m)^2}\tag{A2.108}$$

$$D_{nm}(y,z,d_m,h_m) = \sqrt{(y-d_m)^2 + (z+h_m)^2}\tag{A2.109}$$

with $\underset{=11}{S^c}(y,z,d_m,h_m)$ and $\underset{=13}{S^c}(y,z,d_m,h_m)$ are Sommerfeld integrals of $\underline{S}_{11}^c(y,z,d_m,h_m,v)$ and $\underline{S}_{13}^c(y,z,d_m,h_m,v)$ based on quasi-TEM assumption:

$$\underset{=11}{S^c}(y,z,d_m,h_m) = \int_0^{+\infty} \frac{e^{(z+h_m)\sqrt{\lambda^2-k_e^2}}}{\left(\lambda^2-k_e^2\right)\left(\sqrt{\lambda^2-k_a^2}+\sqrt{\lambda^2-k_e^2}\right)}\lambda^2\cos[(y-d_m)\lambda]d\lambda\tag{A2.110}$$

$$\underset{=13}{S^c}(y,z,d_m,h_m) = \int_0^{+\infty} \frac{e^{(z+h_m)\sqrt{\lambda^2-k_e^2}}}{\left(\lambda^2-k_e^2\right)\left(\sqrt{\lambda^2-k_a^2}+\sqrt{\lambda^2-k_e^2}\right)}\cos[(y-d_m)\lambda]d\lambda\tag{A2.111}$$

The earth-return admittance between cables n and m can be represented by the earth-return potential coefficient. The different formulas are derived based on different voltage integral references.

A2.3.2.3 Line voltage integral equation referred to infinite depth in the earth

The earth-return potential coefficient formula between mth and nth cables based on the integral path to infinite depth in the earth ($h_r = -\infty$) is

$$\begin{aligned} \underline{P}_{e(nm)}(d_n, h_n, d_m, h_m) &= \frac{j\omega}{2\pi(\sigma_1 + j\omega\varepsilon_1)} \{K_0[jk_e d_{nm}(d_n, h_n, d_m, h_m)] \\ &\quad - K_0[jk_e D_{nm}(d_n, h_n, d_m, h_m)] + 2\underset{=12}{S^c}(d_n, h_n, d_m, h_m) \\ &\quad - 2k_e^2 \underset{=13}{S^c}(d_n, h_n, d_m, h_m)\Big\} \end{aligned} \tag{A2.112}$$

where $\underset{=12}{S^c}(y, z, d_m, h_m)$ is Sommerfeld integral $\underline{S}_{12}^c(y, z, d_m, h_m, v)$ based on quasi-TEM assumption:

$$\underset{=12}{S^c}(y, z, d_m, h_m) = \int_0^{+\infty} \frac{e^{(z+h_m)\sqrt{\lambda^2 - k_e^2}}}{(\lambda^2 - k_e^2)\left(\sqrt{\lambda^2 - k_a^2} + k_a^2 k_e^{-2}\sqrt{\lambda^2 - k_e^2}\right)} \lambda^2 \cos[(y - d_m)\lambda] d\lambda \tag{A2.113}$$

A2.3.2.4 Line voltage integral equation referred to the earth surface

The earth-return potential coefficient formula between cables n and m based on the integral path to the earth surface ($h_r = 0$) is

$$\begin{aligned} \underline{P}_{e(nm)}(d_n, h_n, d_m, h_m) &= \frac{j\omega}{2\pi(\sigma_1 + j\omega\varepsilon_1)} \{K_0[jk_e d_{nm}(d_n, h_n, d_m, h_m)] \\ &\quad - K_0[jk_e D_{nm}(d_n, h_n, d_m, h_m)] + 2\underset{=12}{S^c}(d_n, h_n, d_m, h_m) \\ &\quad - 2k_e^2 \underset{=13}{S^c}(d_n, h_n, d_m, h_m)\Big\} - \underline{P}_{ec(nm)}(d_n, h_n, d_m, h_m) \end{aligned} \tag{A2.114}$$

where the earth surface correction term $\underline{P}_{ec(nm)}(y, z, d_m, h_m)$ has

$$\underline{P}_{ec(nm)}(y, z, d_m, h_m) = \frac{j\omega}{\pi(\sigma_1 + j\omega\varepsilon_1)} \left[\underset{=14}{S^c}(y, z, d_m, h_m) - k_e^2 \underset{=15}{S^c}(y, z, d_m, h_m)\right] \tag{A2.115}$$

where

$$\underset{=14}{S^c}(y, z, d_m, h_m) = \int_0^{+\infty} \frac{e^{\frac{z+h_m}{2}\sqrt{\lambda^2 - k_e^2}}}{(\lambda^2 - k_e^2)\left(\sqrt{\lambda^2 - k_a^2} + k_a^2 k_e^{-2}\sqrt{\lambda^2 - k_e^2}\right)} \lambda^2 \cos[(y - d_m)\lambda] d\lambda \tag{A2.116}$$

$$\underset{=15}{S^c}(y,z,d_m,h_m) = \int_0^{+\infty} \frac{e^{\frac{z+h_m}{2}\sqrt{\lambda^2-k_e^2}}}{(\lambda^2-k_e^2)\left(\sqrt{\lambda^2-k_a^2}+\sqrt{\lambda^2-k_e^2}\right)} \cos[(y-d_m)\lambda]d\lambda \tag{A2.117}$$

A2.3.2.5 Line voltage integral equation referred to penetration depth in the earth

The earth-return potential coefficient formula between cable n and m based on penetration depth in the earth ($h_r = -\delta_e$) is

$$\begin{aligned} \underline{P}_{e(nm)}(d_n,h_n,d_m,h_m) &= \frac{j\omega}{2\pi(\sigma_1+j\omega\varepsilon_1)}\{K_0[jk_e d_{nm}(d_n,h_n,d_m,h_m)] \\ &\quad -K_0[jk_e D_{nm}(d_n,h_n,d_m,h_m)] + 2\underset{=12}{S^c}(d_n,h_n,d_m,h_m) \\ &\quad -2k_e^2\underset{=13}{S^c}(d_n,h_n,d_m,h_m)\Big\} - \underline{P}_{ec(nm)}(d_n,h_n,d_m,h_m,h_r) \end{aligned} \tag{A2.118}$$

where the penetration depth correction term $\underline{P}_{ec(nm)}(y,z,d_m,h_m,h_r)$ has

$$\begin{aligned} \underline{P}_{ec(nm)}(y,z,d_m,h_m,h_r) &= \frac{j\omega}{2\pi(\sigma_1+j\omega\varepsilon_1)}\{K_0[jk_e d_{nmc}(y,z,d_m,h_m,h_r)] \\ &\quad -K_0[jk_e D_{nmc}(y,z,d_m,h_m,h_r)] + 2\underset{=16}{S^c}(y,z,d_m,h_m,h_r) \\ &\quad -2k_e^2\underset{=17}{S^c}(y,z,d_m,h_m,h_r)\Big\} \end{aligned} \tag{A2.119}$$

where h_r is given in (A2.75) and

$$d_{nmc}(y,z,d_m,h_m,h_r) = \sqrt{(y-d_m)^2 + \left(h_r - \frac{z+h_m}{2}\right)^2} \tag{A2.120}$$

$$D_{nmc}(y,z,d_m,h_m,h_r) = \sqrt{(y-d_m)^2 + \left(h_r + \frac{z+h_m}{2}\right)^2} \tag{A2.121}$$

$$\underset{=16}{S^c}(y,z,d_m,h_m,h_r) = \int_0^{+\infty} \frac{e^{\left(h_r+\frac{z+h_m}{2}\right)\sqrt{\lambda^2-k_e^2}}}{(\lambda^2-k_e^2)\left(\sqrt{\lambda^2-k_a^2}+k_a^2k_e^{-2}\sqrt{\lambda^2-k_e^2}\right)}\lambda^2 \cos[(y-d_m)\lambda]d\lambda \tag{A2.122}$$

$$\underset{=17}{S^c}(y,z,d_m,h_m,h_r) = \int_0^{+\infty} \frac{e^{\left(h_r+\frac{z+h_m}{2}\right)\sqrt{\lambda^2-k_e^2}}}{(\lambda^2-k_e^2)\left(\sqrt{\lambda^2-k_a^2}+\sqrt{\lambda^2-k_e^2}\right)} \cos[(y-d_m)\lambda]d\lambda \tag{A2.123}$$

It should be noted that $h_n = h_m$ and $d_n - d_m = r_{on}$ or $d_n - d_m = r_{om}$ are used for the self-impedance and potential coefficient. The mutual element can be calculated at the center of two cables. By substituting $h_r = 0$ into (A2.118), (A2.114) is deduced. When $h_r = -\infty$ rather than $-\delta_e$ in (A2.118), $\underline{P}_{ec(nm)}(d_n, h_n, d_m, h_m, h_r) \rightarrow 0$, (A2.112) is derived. Furthermore, the formulas shown and discussed in Chapter 2 are further modified to avoid complicated expressions in this appendix.

Appendix A3 Derivation of electromagnetic field equations for a multiphase overhead line system

In this section, formulas of the complete electromagnetic field solutions on a multiphase overhead line with N conductors are derived based on rigorous Maxwell equations. The derived formulas of electromagnetic fields can be applied to formulate a modified earth-return Green function and can be used for the recently developed software called MoM-SO [27–29]. By adopting the newly derived Green function, MoM-SO becomes able to generate accurate parameters of overhead lines within quasi-TEM mode of wave propagation, and the numerical instability observed in the original MoM-SO is eliminated [29].

A3.1 Derivation of electromagnetic field formulas

A multiphase overhead line with N conductors is located above a homogenous earth, as illustrated in Figure A3.1.

The air is the medium in the region $z > 0$, and characterized by permeability μ_0 and permittivity ε_0. The region of $z < 0$ is the lossy earth, and it is characterized by permeability μ_1, permittivity ε_1 and conductivity σ_1. It is assumed that the radius of each conductor is small when compared with the height and separation. Thus, a

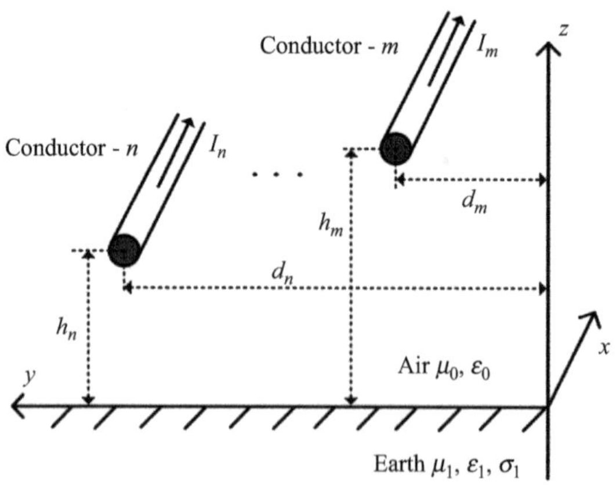

Figure A3.1 Configuration of an overhead line with N conductors above a homogenous earth

filament current approximation can be adopted where all the conductor currents are only expressed by the axial component [39].

There exists as many propagation modes as the number of conductors in the system. The conductor-m with radius r_m carries the following current in phase domain [39]:

$$I_m(x) = \sum_{v=1}^{N} B_{mv} i_v(x) \tag{A3.1}$$

where B_{mv} is the mvth element of the $N \times N$ unknown current transformation matrix $\mathbf{B}$ (to be determined by impedance boundary condition) and the vth modal current $i_v(x)$ is defined by

$$i_v(x) = i_{vp} e^{-jk_v x} \tag{A3.2}$$

where i_{vp} is the amplitude of the modal current and k_v is the unknown wave number of the vth mode.

The electromagnetic vector fields can be defined by using Hertzian vectors as [65]

$$\overline{\mathbf{E}} = \nabla\nabla \cdot \overline{\mathbf{\Pi}}_{\mathbf{E}} + \omega^2 \mu\varepsilon \overline{\mathbf{\Pi}}_{\mathbf{E}} - j\omega\mu \nabla \times \overline{\mathbf{\Pi}}_{\mathbf{M}} \tag{A3.3}$$

$$\overline{\mathbf{H}} = \nabla\nabla \cdot \overline{\mathbf{\Pi}}_{\mathbf{M}} + \omega^2 \mu\varepsilon \overline{\mathbf{\Pi}}_{\mathbf{M}} + j\omega\varepsilon \nabla \times \overline{\mathbf{\Pi}}_{\mathbf{E}} \tag{A3.4}$$

where ω is the angular frequency.

$\overline{\mathbf{\Pi}}_{\mathbf{E}}$ and $\overline{\mathbf{\Pi}}_{\mathbf{M}}$ are electric and magnetic Hertzian vectors. It should be noted that any continuously differentiable vector field is defined by

$$\overline{\mathbf{F}} = \mathbf{e_x} F_x + \mathbf{e_y} F_y + \mathbf{e_z} F_z \tag{A3.5}$$

where $\mathbf{e_x}$, $\mathbf{e_y}$, and $\mathbf{e_z}$ be the corresponding basis of the unit vector $\mathbf{e}$.

For the present problem, it is convenient to adopt only x-component of Hertzian vectors because $\overline{\mathbf{\Pi}}_{\mathbf{E}}$ and $\overline{\mathbf{\Pi}}_{\mathbf{M}}$ are generated by $I_m(x)$ and will be directed parallel to the x-axis.

Thus, the expressions of the x-component of the electric and magnetic Hertzian vectors at any point in the air ("0") and earth ("1") can be derived as a sum of contributions by the current in each conductor. By adopting the superposition theorem, it gives

$$\Pi_{0E}(x,y,z) = \sum_{m=1}^{N} \Pi_{0Em}(x,y,z) \tag{A3.6}$$

$$\Pi_{0M}(x,y,z) = \sum_{m=1}^{N} \Pi_{0Mm}(x,y,z) \tag{A3.7}$$

$$\Pi_{1E}(x,y,z) = \sum_{m=1}^{N} \Pi_{1Em}(x,y,z) \tag{A3.8}$$

$$\Pi_{1M}(x,y,z) = \sum_{m=1}^{N} \Pi_{1Mm}(x,y,z) \tag{A3.9}$$

where $\Pi_{0Em}(x, y, z)$, $\Pi_{0Mm}(x, y, z)$, $\Pi_{1Em}(x, y, z)$, and $\Pi_{1Mm}(x, y, z)$ are the x-components of electric and magnetic Hertzian vectors in the air and earth, generated by the current $I_m(x)$ of the mth cable.

By adopting the formulas of Hertzian vectors [39] into (A3.6)–(A3.9), the following formulas are deduced:

$$\Pi_{0E}(x, y, z) = -\frac{j\omega\mu_0}{4\pi k_a^2}\sum_{m=1}^{N}\sum_{v=1}^{N} B_{mv} i_v(x) \int_{-\infty}^{+\infty} \frac{e^{-|z-h_m|u_{0v}} + R_{Ev} e^{-(z+h_m)u_{0v}}}{u_{0v}} e^{-j(y-d_m)\lambda} d\lambda \tag{A3.10}$$

$$\Pi_{0M}(x, y, z) = -\frac{j\omega\mu_0}{4\pi k_a^2}\sum_{m=1}^{N}\sum_{v=1}^{N} B_{mv} i_v(x) \int_{-\infty}^{+\infty} \frac{R_{Mv} e^{-(z+h_m)u_{0v}}}{u_{0v}} e^{-j(y-d_m)\lambda} d\lambda \tag{A3.11}$$

$$\Pi_{1E}(x, y, z) = -\frac{j\omega\mu_0}{4\pi k_a^2}\sum_{m=1}^{N}\sum_{v=1}^{N} B_{mv} i_v(x) \int_{-\infty}^{+\infty} \frac{T_{Ev} e^{zu_{1v}-h_m u_{0v}}}{u_{0v}} e^{-j(y-d_m)\lambda} d\lambda \tag{A3.12}$$

$$\Pi_{1M}(x, y, z) = -\frac{j\omega\mu_0}{4\pi k_a^2}\sum_{m=1}^{N}\sum_{v=1}^{N} B_{mv} i_v(x) \int_{-\infty}^{+\infty} \frac{T_{Mv} e^{zu_{1v}-h_m u_{0v}}}{u_{0v}} e^{-j(y-d_m)\lambda} d\lambda \tag{A3.13}$$

where λ is the integral variable and

$$u_{0v} = \sqrt{\lambda^2 + k_v^2 - k_a^2}, u_{1v} = \sqrt{\lambda^2 + k_v^2 - k_e^2} \tag{A3.14}$$

with

$$k_a^2 = -j\omega\mu_0 j\omega\varepsilon_0, k_e^2 = -j\omega\mu_1(\sigma_1 + j\omega\varepsilon_1) \tag{A3.15}$$

The irrationals R_{Ev}, R_{Mv}, T_{Ev}, and T_{Mv} are derived by adopting the following boundary conditions at the air/earth interface ($z = 0$), referred to the vth mode:

$$\left.\begin{aligned} E_{0x} &= E_{1x} \\ E_{0y} &= E_{1y} \\ H_{0x} &= H_{1x} \\ H_{0y} &= H_{1y} \end{aligned}\right\} \tag{A3.16}$$

where E_x, E_y, H_x, and H_y are axial and horizontal electric and magnetic field components.

Substituting (A3.10)–(A3.13) into (A3.16) and assuming $\mu_1 = \mu_0$, R_{Ev} and R_{Mv} can be deduced:

$$R_{Ev} = -1 + 2\frac{k_a^2 u_{0v}}{k_a^2 - k_v^2}\left(\frac{1}{u_{0v} + u_{1v}} - \frac{k_v^2}{k_e^2 u_{0v} + k_a^2 u_{1v}}\right) \tag{A3.17}$$

$$R_{Mv} = -2\frac{k_a^2 k_v \lambda}{j\omega\mu_0 \left(k_a^2 - k_v^2\right)}\left(\frac{1}{u_{0v} + u_{1v}} - \frac{k_a^2}{k_e^2 u_{0v} + k_a^2 u_{1v}}\right) \tag{A3.18}$$

By adopting (A3.17) and (A3.18) into (A3.10) and (A3.11), and solving (A3.3) and (A3.4), the formulas of electromagnetic field components at an arbitrary point in the air are derived:

$$E_{0x}(x,y,z) = \sum_{m=1}^{N}\sum_{v=1}^{N} \Lambda_{0x}(y,z,d_m,h_m,v) B_{mv} i_v(x) \tag{A3.19}$$

$$E_{0y}(x,y,z) = \sum_{m=1}^{N}\sum_{v=1}^{N} \Lambda_{0y}(y,z,d_m,h_m,v) B_{mv} i_v(x) \tag{A3.20}$$

$$E_{0z}(x,y,z) = \sum_{m=1}^{N}\sum_{v=1}^{N} \Lambda_{0z}(y,z,d_m,h_m,v) B_{mv} i_v(x) \tag{A3.21}$$

$$H_{0x}(x,y,z) = \sum_{m=1}^{N}\sum_{v=1}^{N} \Psi_{0x}(y,z,d_m,h_m,v) B_{mv} i_v(x) \tag{A3.22}$$

$$H_{0y}(x,y,z) = \sum_{m=1}^{N}\sum_{v=1}^{N} \Psi_{0y}(y,z,d_m,h_m,v) B_{mv} i_v(x) \tag{A3.23}$$

$$H_{0z}(x,y,z) = \sum_{m=1}^{N}\sum_{v=1}^{N} \Psi_{0z}(y,z,d_m,h_m,v) B_{mv} i_v(x) \tag{A3.24}$$

where $E_z(x,y,z)$ and $H_z(x,y,z)$ are vertical electric and magnetic field components. The kernel functions have the following expressions:

$$\begin{aligned}\Lambda_{0x}(y,z,d_m,h_m,v) &= -\frac{j\omega\mu_0}{4\pi}\left[\underline{\Delta}_1(y,z,d_m,h_m,v) + 2\underline{S}_1(y,z,d_m,h_m,v)\right] \\ &\quad + \frac{j\omega\mu_0 k_v^2}{4\pi k_a^2}\left[\underline{\Delta}_1(y,z,d_m,h_m,v) + 2\underline{S}_2(y,z,d_m,h_m,v)\right]\end{aligned} \tag{A3.25}$$

$$\Lambda_{0y}(y,z,d_m,h_m,v) = \frac{j\omega\mu_0 k_v}{4\pi k_a^2}\left[\underline{\Delta}_2(y,z,d_m,h_m,v) + 2\underline{S}_4(y,z,d_m,h_m,v)\right] \tag{A3.26}$$

$$\begin{aligned}\Lambda_{0z}(y,z,d_m,h_m,v) = -\frac{\omega\mu_0 k_v}{4\pi k_a^2}\Big[&\underline{\Delta}_3(y,z,d_m,h_m,v) + 2k_a^2\underline{S}_7(y,z,d_m,h_m,v) \\ &-2k_v^2\underline{S}_8(y,z,d_m,h_m,v) - 2\underline{S}_{10}(y,z,d_m,h_m,v)\Big]\end{aligned} \tag{A3.27}$$

$$\Psi_{0x}(y,z,d_m,h_m,v) = \frac{k_v}{2\pi}\left[\underline{S}_5(y,z,d_m,h_m,v) - \underline{S}_6(y,z,d_m,h_m,v)\right] \tag{A3.28}$$

$$\Psi_{0y}(y,z,d_m,h_m,v) = -\frac{1}{4\pi}\left[\underline{\Delta}_3(y,z,d_m,h_m,v) + 2k_a^2\underline{S}_7(y,z,d_m,h_m,v) \right. \\ \left. -2k_v^2\underline{S}_8(y,z,d_m,h_m,v) - 2\underline{S}_9(y,z,d_m,h_m,v)\right] \tag{A3.29}$$

$$\Psi_{0z}(y,z,d_m,h_m,v) = -\frac{j}{4\pi}\left[\underline{\Delta}_2(y,z,d_m,h_m,v) + 2\underline{S}_3(y,z,d_m,h_m,v)\right] \tag{A3.30}$$

where

$$\underline{\Delta}_1(y,z,d_m,h_m,v) = \int_{-\infty}^{+\infty} \frac{e^{-|z-h_m|u_{0v}} - e^{-(z+h_m)u_{0v}}}{u_{0v}} e^{-j(y-d_m)\lambda} d\lambda \tag{A3.31}$$

$$\underline{\Delta}_2(y,z,d_m,h_m,v) = \int_{-\infty}^{+\infty} \frac{e^{-|z-h_m|u_{0v}} - e^{-(z+h_m)u_{0v}}}{u_{0v}} e^{-j(y-d_m)\lambda} \lambda d\lambda \tag{A3.32}$$

$$\underline{\Delta}_3(y,z,d_m,h_m,v) = \int_{-\infty}^{+\infty} \left[e^{(z-h_m)u_{0v}} + e^{-(z+h_m)u_{0v}}\right] e^{-j(y-d_m)\lambda} d\lambda \text{ for } 0 < z < h_m \tag{A3.33}$$

$$\underline{\Delta}_3(y,z,d_m,h_m,v) = \int_{-\infty}^{+\infty} \left[e^{-(z+h_m)u_{0v}} - e^{-(z-h_m)u_{0v}}\right] e^{-j(y-d_m)\lambda} d\lambda \text{ for } z > h_m \tag{A3.34}$$

$$\underline{S}_1(y,z,d_m,h_m,v) = \int_{-\infty}^{+\infty} \frac{e^{-(z+h_m)u_{0v}}}{u_{0v} + u_{1v}} e^{-j(y-d_m)\lambda} d\lambda \tag{A3.35}$$

$$\underline{S}_2(y,z,d_m,h_m,v) = \int_{-\infty}^{+\infty} \frac{e^{-(z+h_m)u_{0v}}}{k_e^2 k_a^{-2} u_{0v} + u_{1v}} e^{-j(y-d_m)\lambda} d\lambda \tag{A3.36}$$

$$\underline{S}_3(y,z,d_m,h_m,v) = \int_{-\infty}^{+\infty} \frac{e^{-(z+h_m)u_{0v}}}{u_{0v} + u_{1v}} e^{-j(y-d_m)\lambda} \lambda d\lambda \tag{A3.37}$$

$$\underline{S}_4(y,z,d_m,h_m,v) = \int_{-\infty}^{+\infty} \frac{e^{-(z+h_m)u_{0v}}}{k_e^2 k_a^{-2} u_{0v} + u_{1v}} e^{-j(y-d_m)\lambda} \lambda d\lambda \tag{A3.38}$$

$$\underline{S}_5(y,z,d_m,h_m,v) = \int_{-\infty}^{+\infty} \frac{e^{-(z+h_m)u_{0v}}}{u_{0v}(u_{0v} + u_{1v})} e^{-j(y-d_m)\lambda} \lambda d\lambda \tag{A3.39}$$

$$\underline{S}_6(y,z,d_m,h_m,v) = \int_{-\infty}^{+\infty} \frac{e^{-(z+h_m)u_{0v}}}{u_{0v}\left(k_e^2 k_a^{-2} u_{0v} + u_{1v}\right)} e^{-j(y-d_m)\lambda} \lambda d\lambda \tag{A3.40}$$

$$\underline{S}_7(y,z,d_m,h_m,v) = \int_{-\infty}^{+\infty} \frac{e^{-(z+h_m)u_{0v}}}{u_{0v}(u_{0v} + u_{1v})} e^{-j(y-d_m)\lambda} d\lambda \tag{A3.41}$$

$$\underline{S}_8(y,z,d_m,h_m,v) = \int_{-\infty}^{+\infty} \frac{e^{-(z+h_m)u_{0v}}}{u_{0v}\left(k_e^2 k_a^{-2} u_{0v} + u_{1v}\right)} e^{-j(y-d_m)\lambda} d\lambda \tag{A3.42}$$

$$\underline{S}_9(y,z,d_m,h_m,v) = \int_{-\infty}^{+\infty} \frac{e^{-(z+h_m)u_{0v}}}{u_{0v}(u_{0v}+u_{1v})} e^{-j(y-d_m)\lambda}\lambda^2 d\lambda \tag{A3.43}$$

$$\underline{S}_{10}(y,z,d_m,h_m,v) = \int_{-\infty}^{+\infty} \frac{e^{-(z+h_m)u_{0v}}}{u_{0v}\left(k_e^2 k_a^{-2} u_{0v}+u_{1v}\right)} e^{-j(y-d_m)\lambda}\lambda^2 d\lambda \tag{A3.44}$$

Equations (A3.19)–(A3.24) can be rewritten into matrix expressions as a function of the exciting modal currents:

$$\begin{bmatrix} E_{0x}(x,y,z) \\ E_{0y}(x,y,z) \\ E_{0z}(x,y,z) \end{bmatrix} = \begin{bmatrix} \Phi_{0x}(y,z,d_m,h_m,1) & \cdots & \Phi_{0x}(y,z,d_m,h_m,N) \\ \Phi_{0y}(y,z,d_m,h_m,1) & \cdots & \Phi_{0y}(y,z,d_m,h_m,N) \\ \Phi_{0z}(y,z,d_m,h_m,1) & \cdots & \Phi_{0z}(y,z,d_m,h_m,N) \end{bmatrix} \begin{bmatrix} i_1(x) \\ \vdots \\ i_N(x) \end{bmatrix} \tag{A3.45}$$

$$\begin{bmatrix} H_{0x}(x,y,z) \\ H_{0y}(x,y,z) \\ H_{0z}(x,y,z) \end{bmatrix} = \begin{bmatrix} \Upsilon_{0x}(y,z,d_m,h_m,1) & \cdots & \Upsilon_{0x}(y,z,d_m,h_m,N) \\ \Upsilon_{0y}(y,z,d_m,h_m,1) & \cdots & \Upsilon_{0y}(y,z,d_m,h_m,N) \\ \Upsilon_{0z}(y,z,d_m,h_m,1) & \cdots & \Upsilon_{0z}(y,z,d_m,h_m,N) \end{bmatrix} \begin{bmatrix} i_1(x) \\ \vdots \\ i_N(x) \end{bmatrix} \tag{A3.46}$$

where

$$\Phi_{0x}(y,z,d_m,h_m,v) = \sum_{m=1}^{N} \Lambda_{0x}(y,z,d_m,h_m,v)B_{mv} \tag{A3.47}$$

$$\Phi_{0y}(y,z,d_m,h_m,v) = \sum_{m=1}^{N} \Lambda_{0y}(y,z,d_m,h_m,v)B_{mv} \tag{A3.48}$$

$$\Phi_{0z}(y,z,d_m,h_m,v) = \sum_{m=1}^{N} \Lambda_{0z}(y,z,d_m,h_m,v)B_{mv} \tag{A3.49}$$

$$\Upsilon_{0x}(y,z,d_m,h_m,v) = \sum_{m=1}^{N} \Psi_{0x}(y,z,d_m,h_m,v)B_{mv} \tag{A3.50}$$

$$\Upsilon_{0y}(y,z,d_m,h_m,v) = \sum_{m=1}^{N} \Psi_{0y}(y,z,d_m,h_m,v)B_{mv} \tag{A3.51}$$

$$\Upsilon_{0z}(y,z,d_m,h_m,v) = \sum_{m=1}^{N} \Psi_{0z}(y,z,d_m,h_m,v)B_{mv} \tag{A3.52}$$

The matrices expressions (A3.45) and (A3.46) describing the electromagnetic field components can be also represented in the following compact forms:

$$\left.\begin{aligned} \mathbf{E}_0 &= \mathbf{\Phi}_0\mathbf{i} \\ \mathbf{H}_0 &= \mathbf{\Upsilon}_0\mathbf{i} \end{aligned}\right\} \tag{A3.53}$$

where $\mathbf{E}_0$ and $\mathbf{H}_0$ are $N \times 1$ electromagnetic field vectors, $\boldsymbol{\Phi}_0$ and $\boldsymbol{\Upsilon}_0$ are $3 \times N$ modal current to field operator's matrices, and $\mathbf{i}$ is an $N \times 1$ modal current vector.

It should be noted that (A3.53) derived in the above is based on the assumption of unidirectional propagation of voltage and current waves. The responses of electromagnetic fields should be related to the exciting sources which are impressed by the transmitting device. Thus, the transfer function from voltage to field can be defined by using $\boldsymbol{\Phi}_0$ and $\boldsymbol{\Upsilon}_0$ matrices together with the following modal current vector:

$$\mathbf{i} = \mathbf{B}^{-1}\mathbf{Y}_0\mathbf{A}\zeta\mathbf{A}^{-1}\mathbf{V_f} \tag{A3.54}$$

where $\mathbf{Y}_0$ is an $N \times N$ characteristic admittance matrix of an overhead line in phase domain, $\mathbf{A}$ is an $N \times N$ voltage transformation matrix, ζ is an $N \times N$ reflection free transition diagonal matrix, $\mathbf{V_f}$ is an $N \times 1$ impressed voltage source vector, and

$$\mathbf{A} = \left(\mathbf{B}^T\right)^{-1} \tag{A3.55}$$

$$\zeta = \mathrm{diag}\left(e^{-jk_v x}\right) \tag{A3.56}$$

Equation (A3.53) can be rewritten by replacing $\mathbf{i}$ by (A3.54):

$$\left.\begin{aligned} \mathbf{E}_0 &= \boldsymbol{\Phi}_0\mathbf{B}^{-1}\mathbf{Y}_0\mathbf{A}\zeta\mathbf{A}^{-1}\mathbf{V_f} \\ \mathbf{H}_0 &= \boldsymbol{\Upsilon}_0\mathbf{B}^{-1}\mathbf{Y}_0\mathbf{A}\zeta\mathbf{A}^{-1}\mathbf{V_f} \end{aligned}\right\} \tag{A3.57}$$

Therefore, the radiated electromagnetic fields of a multiphase overhead line can be calculated by applying the impressed voltage source $\mathbf{V_f}$ in (A3.57).

References

[1] Ametani, A. "A general formulation of impedance and admittance of cables." *IEEE Transactions on Power Apparatus and Systems*. 1980; 99(3): 902–10.

[2] Schelkunoff, S. A. "The electromagnetic theory of coaxial transmission line and cylindrical shields." *Bell System Technical Journal*. 1934; 13(4): 532–79.

[3] Ametani, A. *Cable constants rule book*. Portland: Bonneville Power Administration; 1982.

[4] Dommel, H. W. *EMTP theory book*. Portland: Bonneville Power Administration; 1992.

[5] Ametani, A., Nagaoka, N., Baba Y., Ohno T., and Yamabuki K. *Power system transient: Theory and applications*. 2nd edn. Boca Raton, FL: CRC Press; 2016.

[6] Ametani, A. (ed). *Numerical analysis of power system transients and dynamics*. London: IET; 2015.

[7] Ametani, A., Ohno, T., and Nagaoka N. *Cable system transients: Theory, modeling and simulation*. Singapore: Wiley; 2015.

[8] Carson, J. R. "Wave propagation in overhead wires with ground return." *Bell System Technical Journal*. 1926; 5(4): 539–54.

[9] Pollaczek, F. "Uber das feld einer unendlich langen wechselstrom durchlossenen einfachleitung." *Elektrische Nac hrichtentech.* 1926; 3(4): 339–59.

[10] Sunde, E. D. *Earth conduction effects in transmission Systems.* New York, NY: Dover; 1968.

[11] Wedepohl, L. M., and Wasley, R. G. "Wave propagation in multiconductor overhead lines-calculation of series impedance for multilayer earth." *Proceedings of the Institution of Electrical Engineers.* 1966; 113(4): 627–32.

[12] Moghram, I. S. "Effects of earth stratification on impedances of power transmission lines." *European Transactions on Electrical Power.* 1998; 8(6): 445–9.

[13] Tsiamitros, D., Papagiannis, G., and Dokopoulos, P. "Earth return impedances of conductor arrangements in multilayer soils-Part I: Theoretical model." *IEEE Transactions on Power Delivery.* 2008; 23(4): 2392–00.

[14] Wise, W. H. "Propagation of high-frequency currents in ground return circuits." *Proceedings of the Institute of Radio Engineers.* 1934; 22(4): 522–27.

[15] Wise, W.H. "Potential coefficients for ground return circuits." *Bell System Technical Journal.* 1948; 27(2): 365–71.

[16] Nakagawa, M., Ametani, A., and Iwamoto, K. "Further studies on wave propagation in overhead lines with earth return: Impedance of stratified earth." *Proceedings of the Institution of Electrical Engineers.* 1973; 120(12): 1521–28.

[17] Ametani, A., Miyamoto, Y., and Mahseredjian, J. "Derivation of an earth-return impedance of an overhead multi-conductor considering displacement currents." *IEE Japan Transactions on Power and Energy.* 2014; 134(11): 936–40.

[18] Nakagawa, M. "Admittance correction effects of a single overhead line." *IEEE Transactions on Power Apparatus and Systems.* 1981; 100(3): 1154–61.

[19] Nakagawa, M. "Further studies on wave propagation along overhead transmission lines: Effects of admittance correction." *IEEE Transactions on Power Apparatus and Systems.* 1981; 100(7): 3626–33.

[20] Ametani, A., Nagaoka, N., Koide, R. "Wave propagation characteristics on an on overhead conductor above snow." *Electrical Engineering in Japan.* 2001; 134(3): 26–33.

[21] Ametani, A. "Stratified earth effects on wave propagation: Frequency-dependent parameters." *IEEE Transactions on Power Apparatus and Systems.* 1974; 93(5): 1233–39.

[22] Papadopoulos, T., Papagiannis, G., and Labridis, D. P. "Wave propagation characteristics of overhead conductors above imperfect stratified earth for a wide frequency range." *IEEE Transactions on Magnetics.* 2009; 45(3): 1064–67.

[23] Papadopoulos, T. A., Papagiannis, G. K., and Labridis, D. P. "A generalized model for the calculation of the impedances and admittances of overhead power lines above stratified earth." *Electric Power Systems Research.* 2010; 80(9): 1160–70.

[24] Ametani, A., Yoned, T., Baba, Y., and Nagaoka, N. "An investigation of earth-return impedance between overhead and underground conductors and its approximation." *IEEE Transactions on Electromagnetic Compatibility*. 2009 ; 51(3): 860–67.

[25] Gary, C. "Approche complete de la propagation multifilaire en haute fréquence par utilization des matrices complexes." *EDF Bulletin de la Direction des Etudes et Recherches*. 1976 ; Série B(3/4) : 5–20.

[26] Deri, A., Tevan, G., Semlyen, A., and Castanheira, A. "The complex ground return plane: A simplified model for homogeneous and multi-layer earth return." *IEEE Transactions on Power Apparatus and Systems*. 1981; 100(8): 3686–93.

[27] Patel, U. R., Gustavsen, B., and Triverio, P. "An equivalent surface current approach for the computation of the series impedance of power cables with inclusion of skin and proximity effects." *IEEE Transactions on Power Delivery*. 2013; 28(4): 2474–82.

[28] Patel, U. R., and Triverio, P. "MoM-SO: A complete method for computing the impedance of cable systems including skin, proximity, and ground return effects." *IEEE Transactions on Power Delivery*. 2014; 30(5): 2110–18.

[29] Xue, H., Ametani, A., Mahseredjian, J., and Kocar, I. "Computation of overhead line/underground cable parameters with improved MoM-SO method." *Power Systems Computation Conference* (PSCC); Dublin, Ireland, June 2018.

[30] Kikuchi, H. "Wave propagation on the ground return circuit in high frequency regions." *Electrical Engineering in Japan*. 1955; 75: 1176–87.

[31] Kikuchi, H. "Electro-magnetic field on infinite wire at high frequencies above plane-earth." *Electrical Engineering in Japan*. 1957; 77(825): 721–33.

[32] Wedepohl, L. M., and Efthymiadis, A. E. "Wave propagation in transmission lines over lossy ground: A new, complete field solution." *Proceedings of the Institution of Electrical Engineers*. 1978; 125(6): 505–10.

[33] Efthymiadis, A. E., and Wedepohl, L. M. "Propagation characteristics of infinitely-long single conductor lines by the complete field solution method." *Proceedings of the Institution of Electrical Engineers*. 1978; 125(6): 511–17.

[34] Wait, J. R. "Theory of wave propagation along a thin wire parallel to an interface." *Radio Science*. 1972; 7(6): 675–79.

[35] Wait, J. R. "Tutorial note on the general transmission line theory for a thin wire above the ground." *IEEE Transactions on Electromagnetic Compatibility*. 1991; 33(1): 65–67.

[36] Bridges, G., Atta, O. A., and Shafai, L. "Solution of discrete modes for wave propagation along multiple conductors structures above a dissipative earth." *Canadian Journal of Physics*. 1988; 66(5): 428–38.

[37] Bridges, G., and Shafai, L. "Plane wave coupling to multiple conductor transmission lines above a lossy earth." *IEEE Transactions on Electromagnetic Compatibility*. 1989; 31(1): 21–33.

[38] D'Amore, M., and Sarto, M. "Simulation models of a dissipative transmission line above a lossy ground for a wide-frequency range-Part I: Single

conductor configuration." *IEEE Transactions on Electromagnetic Compatibility*. 1996; 38(2): 127–38.

[39] D'Amore, M., and Sarto, M. "Simulation models of a dissipative transmission line above a lossy ground for a wide-frequency range-Part II: Multiconductor configuration." *IEEE Transactions on Electromagnetic Compatibility*. 1996; 38(2): 139–49.

[40] Pettersson, P. "Propagation of waves on a wire above a lossy ground-different formulations with approximations." *IEEE Transactions on Power Delivery*. 1999; 14(3): 1173–79.

[41] Olsen, R. G., and Pankaskie, T. A. "On the exact, Carson and image theories for wires at or above the earth's interface." *IEEE Transactions on Power Apparatus and Systems*. 1983; 102(4): 769–78.

[42] Olsen, R.G., Young, J.L., and Chang, D.C. "Electromagnetic wave propagation on a thin wire above earth." *IEEE Transactions on Antennas and Propagation*. 2000; 48(9): 1413–19.

[43] Sommerfeld, A. *Partial differential equations in physics*. New York, NY: AP; 1949.

[44] Wait, J. R. "The electromagnetic fields of a horizontal dipole in the presence of a conducting half-space." *Canadian Journal of Physics*. 1961; 39(7): 1017–28.

[45] Stoyer, C. H. "Electromagnetic fields of dipoles in stratified media." *IEEE Transactions on Antennas and Propagation*. 1977; 25(4): 547–52.

[46] Xue, H., Ametani, A., Mahseredjian, J., Baba, Y., and Rachidi, F. "Frequency response of electric and magnetic fields of overhead conductors with particular reference to axial electric field." *IEEE Transactions on Electromagnetic Compatibility*. 2018; 60(6): 2029–32.

[47] Tesche, F.M., Ianoz, M.V., and Karlsson, T. *EMC analysis methods and computational Models*. New York, NY: John Wiley and Sons Inc.; 1997.

[48] Rachidi, F., and Tkachenko, S. *Electromagnetic field interaction with transmission lines from classical theory to HF radiation effects*. Southampton: WIT Press; 2008.

[49] Wedepohl, L. M., and Wilcox, D. J. "Transient analysis of underground power transmission systems." *Proceedings of the Institution of Electrical Engineers*. 1973; 120(2): 253–60.

[50] Tsiamitros, D., Papagiannis, G., Labridis, D., and Dokopoulos, P. "Earth return path impedances of underground cables for the two-layer earth case." *IEEE Transactions on Power Delivery*. 2005; 20(3): 2174–81.

[51] Papadopoulos, T., Tsiamitros, D., and Papagiannis, G. "Impedances and admittance of underground cables for the homogeneous earth case." *IEEE Transactions on Power Delivery*. 2010; 25(2): 961–69.

[52] Papadopoulos, T., Tsiamitros D., and Papagiannis G. "Earth return admittances and impedances of underground cables in non-homogeneous earth." *IET Generation, Transmission and Distribution*. 2011; 5(2): 161–71.

[53] Papadopoulos, T., Tsiamitros, D., and Papagiannis G. "Analytical study of the frequency-dependent earth conduction effects on underground power

cables." *IET Generation, Transmission and Distribution*. 2013; 7(3): 276–87.

[54] Magalhães, A. P. C., Silva, J. C. L. V., Lima, A. C. S., and de Barros, M. T. C. "Validation limits of quasi-TEM approximation for buried bare and insulated cables." *IEEE Transactions on Electromagnetic Compatibility*. 2015; 57(6): 1690–97.

[55] Magalhães, A. P. C., de Barros, M. T. C., and Lima, A. C. S. "Earth return admittance effect on underground cable system modeling." *IEEE Transactions on Power Delivery*. 2018; 33(2): 662–70.

[56] Vance, E. F. *Coupling with shielded cables*. New York, NY: Wiley; 1978.

[57] Saad O., Gaba G., Giroux M. "A closed-form approximation for ground return impedance of underground cables." *IEEE Transactions on Power Delivery*. 1996; 11(3): 1536–45.

[58] Petrache, E., Rachidi, F., Paolone, M., Nucci, C. A., Rakov, V. A., and Uman, M. A. "Lightning induced disturbances in buried cables-Part I: Theory." *IEEE Transactions on Electromagnetic Compatibility*. 2005; 47(3): 498–08.

[59] Theethayi, N., Thottappillil, R., Paolone, M., Nucci, C. A., and Labridis, D. "External impedance and admittance of buried horizontal wires of transient studies using transmission line analysis." *IEEE Transactions on Dielectrics and Electrical Insulation*. 2007; 14(3): 751–61.

[60] Wait, J. R. "Electromagnetic wave propagation along a buried insulated wire." *Canadian Journal of Physics*. 1972; 50(20): 2402–09.

[61] Wait, J. R. "Excitation of currents on a buried insulated cable." *Journal of Applied Physics*. 1978; 49(2): 876–80.

[62] Bridges, G. "Fields generated by bare and insulated cables buried in a lossy half-space." *IEEE Transactions on Geoscience and Remote Sensing*. 1992; 30(1): 140–46.

[63] Bridges, G. "Transient plane wave coupling to bare and insulated cables buried in a lossy half-space." *IEEE Transactions on Electromagnetic Compatibility*. 1995; 37(1): 62–70.

[64] Xue, H., Ametani, A., Mahsersdjian, J., and Kocar, I. "Generalized formulation of earth-return impedance/admittance and surge analysis on underground cables." *IEEE Transactions on Power Delivery*. 2018; 33(6): 2654–63.

[65] Stratton, J. A. *Electromagnetic theory*. New York, NY: McGraw-Hill; 1941.

[66] Ametani, A. Liu, J., Gole, A., De Silva, J., and Yamamoto, K. "Impedance and admittance of HVDC overhead transmission lines." *IEE Japan, High-Voltag*e Eng. Conference, Paper HV-21-045, January 2021.

[67] De Silva, J., Gole, A. M., and Wedepohl, L. M. "Accurate electromagnetic transient simulations of HVDC cables and overhead transmission lines." *International Conference on Power Systems Transients*; Lyon, France, June 2007.

[68] Suonan, J., Zhang, J., Yang, L., and Song, G. "Distance protection for HVDC transmission lines considering frequency-dependent parameters." *IEEE Transactions on Power Delivery*. 2013; 28(2): 723–32.

[69] Ramirez, A., and Iravani, R. "Enhanced fitting to obtain an accurate dc response of transmission lines in the analysis of electromagnetic transients." *IEEE Transactions on Power Delivery*. 29(6): 2614–21.

[70] Kocar, I., and Mahseredjian, J. "Accurate frequency dependent cable model for electromagnetic transients." *IEEE Transactions on Power Delivery*. 2016; 31(3): 1281–88.

[71] Cervantes, M., Kocar, I., Mahseredjian, J., and Ramirez, A. "Partitioned fitting and dc correction for the simulation of electromagnetic transients in transmission lines/cables." *IEEE Transactions on Power Delivery*. 2018; 33 (6): 3264–3248.

[72] Xue, H., Ametani, A., Mahseredjian, J., Baba, Y., Rachidi, F., and Kocar, I. "Transient responses of overhead cables due to mode transition in high frequencies." *IEEE Transactions on Electromagnetic Compatibility*. 2018; 60 (3): 785–94.

[73] Ametani, A., Miyamoto, Y., Baba, Y., and Nagaoka, N. "Wave propagation on an overhead multiconductor in a high-frequency region." *IEEE Transactions on Electromagnetic Compatibility*. 2014; 56(6): 1638–48.

[74] Xue, H., Ametani, A., Mahseredjian, J., Kocar, I. "Review of earth-return-impedance/admittance and the application to surge propagation on overhead lines." *Institute of Electrical Engineers of Japan High Voltage Engineering Conference*; Montreal, Canada, August 2018.

[75] L'Vov, Y. N., and Vhernova, Y. P. "Experimental determination of the zero-sequence parameters of a proto-type 750 kV transmission." *Electrical Technology USSR*. 1973; 1: 54–8.

[76] Ametani, A. *Distributed-parameter circuit theory*. Tokyo: Corona Co; 1990 (in Japanese).

[77] D'Amore, M., and Sarto, M. S. "Electromagnetic field radiated from broadband signal transmission on power line carrier channels." *IEEE Transactions on Power Delivery*. 1997; 12(2): 624–31.

[78] Sarto, M. S. "Electromagnetic interference from carrier channels on finite-length power lines above a lossy ground in a wide frequency range." *IEEE Transactions on Power Delivery*. 1998; 13(2): 336–43.

[79] Sarto, M. S. "Time-domain analysis of electromagnetic interference from broadband digital signal transmission on finite-length power networks." *IEEE Transactions on Power Delivery*. 1998; 13(4): 972–78.

[80] Xue, H., Ametani, A., and Yamamoto, K. "Theoretical and NEC calculations of electromagnetic fields generated from a multi-phase underground cable." *IEEE Transactions on Power Delivery*. 2020; doi: 10.1109/TPWRD.2020.3005521.

[81] Xue, H., Ametani, A., and Yamamoto, K. "A study on external electromagnetic characteristics of underground cables with considerations of terminations." *IEEE Transactions on Power Delivery*. 2020; doi: 10.1109/TPWRD.2020.3037335.

[82] Wedepohl, L. M. "Application of matrix methods to the solution of travelling-wave phenomena in polyphase systems." *Proceedings of the Institution of Electrical Engineers*. 1963; 110(12): 2200–12.

[83] Brown, G. W., and Rocamora, R.G. "Surge propagation in three-phase pipe-type cables part I-unsaturated pipe." *IEEE Transactions on Power Apparatus and Systems*. 1976; 95(1): 89–95.

[84] Amekawa, N., Nagaoka, N., and Ametani, A. "Impedance derivation and wave propagation characteristics of pipe-enclosed and tunnel-installed cables." *IEEE Transactions on Power Delivery*. 2004; 19(1): 380–86.

[85] Ametani, A., Miyamoto, Y., and Nagaoka, N. "Semiconducting layer impedance and its effect on cable wave-propagation and transient characteristics." *IEEE Transactions on Power Delivery*. 2004; 19(4): 1523–31.

[86] Amekawa, N. Derivation of a semiconducting layer impedance and its effect on a wave propagation characteristic of a cable. Doshisha University. M. Eng. Thesis; 2001.

[87] Amekawa, A., Nagaoka, N., Baba, Y., and Ametani, A. "Derivation of a semiconducting layer impedance and its effect on wave propagation characteristics on a cable." *IEE Proceedings – Generation, Transmission and Distribution*. 2003; 150(4): 434–40.

[88] Yasui, T., and Miyauchi, H. "Investigation of a semiconducting layer for coating an EHV CV cable." *Journal of Sumitomo Electric*. 1966; 91: 43–8.

[89] Institute of Electrical Engineers of Japan. High Temperature Capability of and CV Cables and Factors Controlling the Capability. Technical Report 858; 2001.

[90] Zhao, H., Ametani, A., Gole, A., and De Silva, J. "A study on the accuracy and stability of cable internal impedances." *Institute of Electrical Engineers of Japan High Voltage Engineering Conference*; Tsushima, Japan, January 2020.

[91] Ametani, A., and Fuse, I. "Approximate method for calculating impedance of multi-conductor with arbitrary cross-Section." *Electrical Engineering in Japan*. 1992; 112(2): 117–23.

Chapter 3
Modeling of cables

This chapter describes the modeling of cables by the well-known electromagnetic transient (EMT) simulation techniques which are widely used and implemented into various EMT-type simulation tools [1–15]. The modeling methods based on numerical electromagnetic analysis (NEA) approaches such as finite-difference time-domain (FDTD) and method of moments (MoM) [16–24] are also discussed. The EMT-type simulation tools are very powerful to analyze phenomena in a large and complex network provided that the phenomena are associated with transverse electromagnetic (TEM) mode of propagation. On the contrary, the NEA approaches can deal with TEM and non-TEM mode phenomena, due to that it solves Maxwell's equation directly. However, it is computationally heavy, and infeasible to deal with a large network.

3.1 Transmission line models in EMT-type simulation tools

The EMT-type simulation software has been originally developed by the Bonneville Power Administration (BPA) of the US Department of Energy from 1966 to 1984 [1–8]. Currently, there exist three well-known EMT-type simulation tools: ATP Program, PSCAD/EMTDC and EMTP [9–14]. Also, MicroTran [14] and eXpandable Transient Analysis Program (XTAP) [15] are independently developed by University of British Columbia and Central Research Institute of Electric Power Industry.

The EMT-type simulation tools are fundamentally based on electric circuit theory which assumes TEM mode of wave propagation. Thus, those cannot solve phenomena associated with full electromagnetic wave propagations. This is not only from the viewpoint of circuit theories, but also from viewpoint of the parameters used in a circuit analysis. For example, if the impedance and admittance of an overhead line are derived under the assumption of TEM mode, those are not applicable to analyze and calculate phenomena involving non-TEM characteristics.

3.1.1 Equivalent π-circuit model [4]

It is well-known that transmission lines and cables can be represented by a lumped π-circuit equivalent model as illustrated in Figure 3.1(a). In Figure 3.1(a), the per

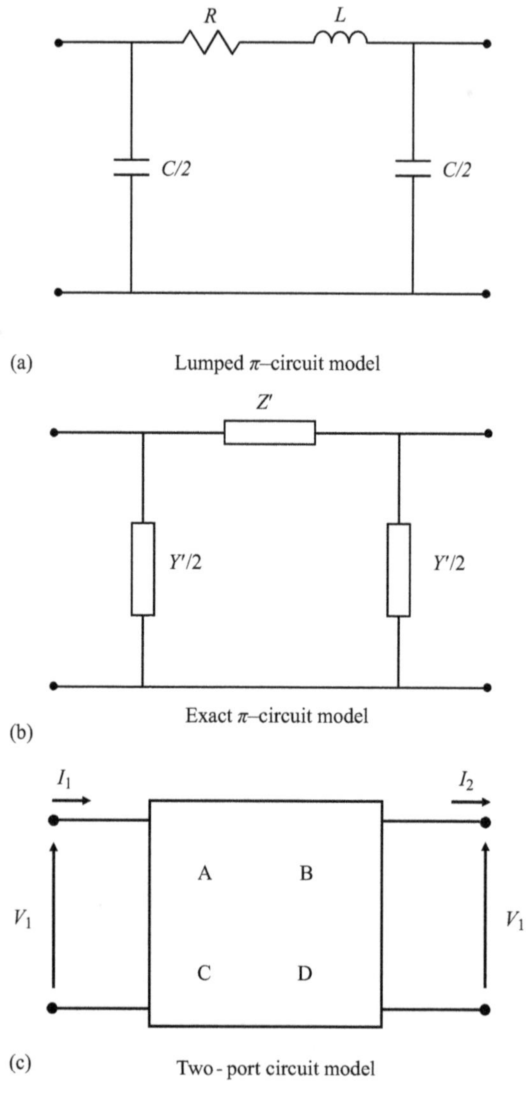

Figure 3.1 Equivalent π-circuit models

unit length parameters of resistance *R*, inductance *L*, and capacitance *C* are calculated by an internal routine of EMT-type simulation tools which is named Line/Cable Constants. In the Line/Cable Constants, the series impedance and shunt admittance of lines and cables at a defined frequency can be evaluated with consideration of different mutual couplings of conductors and physical parameters of the earth, and the details of theory are explained in Chapter 2. A lumped π-circuit model is easily produced by a user using *R*, *L*, and *C* elements prepared by the internal functions of EMT-type simulation tools. The accuracy of π-circuit model is reasonable at the given frequency in a steady-state analysis, but it is not satisfactory for transient studies.

Figure 3.1(b) shows an exact π-circuit model based on a single conductor, and the parameters are given in the following expressions:

$$Z' = Z\frac{\sinh(\gamma l)}{\gamma} \tag{3.1}$$

$$Y' = Y\frac{\tanh\left(\frac{\gamma l}{2}\right)}{\frac{\gamma}{2}} \tag{3.2}$$

where the conductor length is l and propagation constant is

$$\gamma = \sqrt{ZY} \tag{3.3}$$

The series impedance Z and shunt admittance Y of the conductor are given in Chapter 2.

Figure 3.1(c) illustrates a single conductor transmission line represented by a two-port circuit. The voltages (V_1 and V_2) and currents (I_1 and I_2) at the sending and receiving ends are related by adopting a four-terminal parameters in the following expression:

$$\begin{bmatrix} V_2 \\ I_2 \end{bmatrix} = \begin{bmatrix} A & B \\ C & D \end{bmatrix} \begin{bmatrix} V_1 \\ I_2 \end{bmatrix} \tag{3.4}$$

$$\begin{aligned} A &= \cosh(\gamma l) \\ &= \frac{e^{\gamma l} + e^{-\gamma l}}{2} \end{aligned} \tag{3.5}$$

$$\begin{aligned} B &= -Z_0 \sinh(\gamma l) \\ &= -Z_0\frac{e^{\gamma l} - e^{-\gamma l}}{2} \end{aligned} \tag{3.6}$$

$$\begin{aligned} C &= -Y_0 \sinh(\gamma l) \\ &= -Y_0\frac{e^{\gamma l} - e^{-\gamma l}}{2} \end{aligned} \tag{3.7}$$

$$D = A \tag{3.8}$$

where the characteristic impedance Z_0 and admittance Y_0 of a single conductor are defined by

$$Z_0 = \frac{1}{Y_0} = \sqrt{\frac{Z}{Y}} \tag{3.9}$$

3.1.2 Distributed parameters transmission line model

3.1.2.1 Basic concept of model

Figure 3.2 shows a single conductor system and a distributed parameters transmission line model. In Figure 3.2, the voltage and the current are related by adopting Schnyder–Bergeron method to solve a partial differential equation (wave equation) using the following expressions [1,25–28]:

$$v(x,t) + Z_0 \cdot i(x,t) = 2F_1\left(t - \frac{x}{v}\right) \tag{3.10}$$

$$v(x,t) - Z_0 \cdot i(x,t) = 2F_2\left(t + \frac{x}{v}\right) \tag{3.11}$$

where v is propagation velocity.

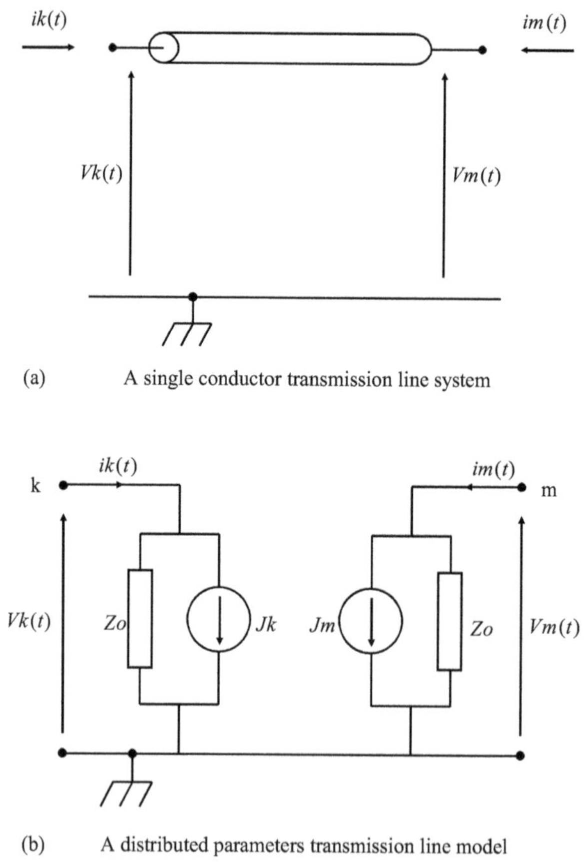

Figure 3.2 Illustration of a single conductor system and a distributed parameters transmission line model

The above equations can be rewritten at node 1 ($k = 1$) and node 2 ($m = 2$) in Figure 3.2.

$$v_1(t-\tau) + Z_0 i_1(t-\tau) = v_2(t) - Z_0 i_2(t) \tag{3.12}$$

$$v_1(t) - Z_0 i_1(t) = v_2(t-\tau) + Z_0 i_2(t-\tau) \tag{3.13}$$

where the traveling time τ from node 1 to node 2 has

$$\tau = \frac{l}{v} \tag{3.14}$$

It is observed in (3.12) that voltage $v_1(t)$ and current $i_1(t)$ at node 1, the sending end of the line, influence $v_2(t)$ and $i_2(t)$ at the receiving end for $t \geq \tau$. Similarly, $v_2(t)$ and $i_2(t)$ influence $v_1(t)$ and $i_1(t)$ with time delay τ. In a lumped-parameter model, the time delay becomes Δt, and it is a time step for time discretization to solve numerically a differential equation describing the relation between the voltage and the current of the lumped element.

From the above equations, the following relations are obtained:

$$i_1(t) = \frac{v_1(t)}{Z_0} + J_1(t-\tau) \tag{3.15}$$

$$i_2(t) = \frac{v_2(t)}{Z_0} + J_2(t-\tau) \tag{3.16}$$

where

$$J_1(t-\tau) = -\frac{v_2(t-\tau)}{Z_0} - i_2(t-\tau) \tag{3.17}$$

$$J_2(t-\tau) = -\frac{v_1(t-\tau)}{Z_0} - i_1(t-\tau) \tag{3.18}$$

The above results give the representations of a distributed parameters transmission line model shown in Figure 3.2(b).

In (3.12) and (3.13), the time delay τ and characteristic impedance Z_0 are frequency-dependent. Also, the current sources J_1 and J_2 representing the past history of a distributed-parameter line are frequency-dependent corresponding to traveling wave attenuation between node 1 and node 2. When the frequency dependence of a distributed-parameter line is to be considered, a frequency-dependent line model such as JMarti line model [29] is prepared as a subroutine in the EMT-type simulation tools.

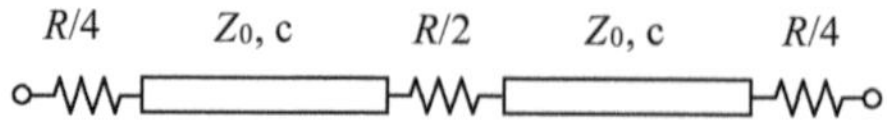

Figure 3.3 A constant-parameter line model

3.1.2.2 Constant-parameter (CP) line model

When the traveling time τ and characteristic impedance Z_0 are fixed as constants at a given frequency, the distributed parameters transmission line model shown in Figure 3.2 becomes a constant-parameter (CP) line model. The CP line model illustrated in Figure 3.3 is called as "Dommel model (or K.C. Lee model)" in ATP Program [4], and is also called as "Bergeron model" in PSCAD/EMTDC [10].

3.1.3 Frequency-dependent line model in modal domain

3.1.3.1 Frequency-dependent effect

Fig. 3. 3 A constant-parameter line model

It is well-known that the impedances of overhead lines and underground cables show the frequency-dependence due to skin effects of conductor and earth-return phenomena. The frequency-dependent characteristics of propagation constant γ and characteristic impedance Z_0 are due to frequency-dependent series impedance and shunt admittance, and are given in (3.3) and (3.9).

When a step function voltage is applied to the sending end of a line, the receiving-end voltage is deformed due to the frequency-dependent effect of the line. The receiving-end voltage is defined by the following form in a frequency domain:

$$S(\omega) = \exp(-\gamma \cdot x) \tag{3.19}$$

The time-domain response $s(t)$ is obtained by using numerical Fourier or Laplace transform.

$$s(t) = F^{-1}\left[\frac{S(\omega)}{j\omega}\right] \tag{3.20}$$

where F^{-1} is Fourier inverse transform.

Figure 3.4 shows the typical step responses s(t) of overhead lines and underground cables in time domain. Figure 3.4(a) illustrates a step response of a 500 kV vertical twin-circuit line with length of 101.1 km, and Figure 3.4(b) is for a single phase of underground single core cables with length of 30 km. It is clear that the original step function voltage at the sending end is distorted due to the frequency-dependent propagation parameters (attenuation and time delay) at the receiving end of the conductor.

Table 3.1(a) shows the frequency response of characteristic impedance $Z_0(\omega)$, and Table 3.1(b) is its time response $z_0(t)$ for the upper phase of a 500 kV vertical twin-circuit line [30]. The frequency dependence and time dependence of the characteristic impedance are clearly observed in the table.

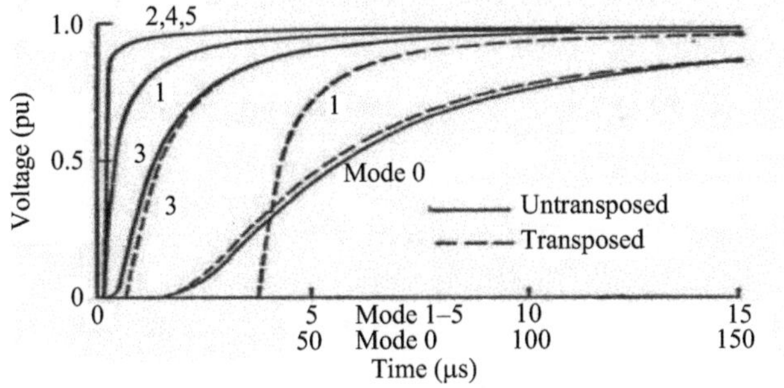

(a) A step responses of a 500 kV vertical overhead line with the length of 101.1 km

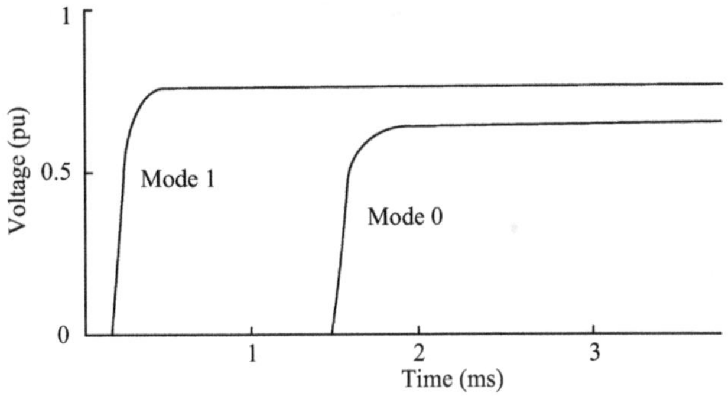

(b) A step responses of a single phase of underground single core cables with the length of 30 km

Figure 3.4 An example of modal step responses $s(t)$ of wave deformation due to the frequency-dependent effect

Table 3.1 (a) Frequency response $Z_0(\omega)$

f **(kHz)**	$Z_0(\omega)$ **(Ω)**
0.01	350
0.1	318
1	312
10	300
100	295
1,000	294

(b) Time response $z_0(t)$

t (μs)	$z_0(t)$ (Ω)
2	298
5	300
10	302
50	310
100	312
500	320

3.1.3.2 Frequency-dependent line model in modal domain

Assuming a traveling wave $E_0(\omega)$ propagates from the sending end to the receiving end on a transmission line, then at the receiving end, the traveling wave is given as

$$E(\omega) = S(\omega) \cdot E_0(\omega) \tag{3.21}$$

In the time domain, the above equation is expressed by a real-time convolution:

$$e(t) = s(t) * e_0(t) = s(0) \cdot e_0(t) + \int_0^t s'(\tau) \cdot e_0(t - \tau) \cdot d\tau \tag{3.22}$$

where

$$e(t) = F^{-1} \quad [E(\omega)] \tag{3.23}$$

$$e_0(t) = F^{-1}[E_0(\omega)] \tag{3.24}$$

It should be noted that (3.21) is evaluated by single multiplication of two functions $S(\omega)$ and $E_0(\omega)$, while (3.22) in the time domain requires a number of multiplication, that is, multiplication of two matrices. It is not easy for a time-domain transient simulation tool to deal with the frequency-dependent effect. The direct inclusion of the frequency-dependent effect needs a large memory storage and computational time. In fact, it required three full days to calculate switching surges on a three-phase line with the real-time convolution to include the frequency-dependent effect of propagation parameters in 1960s.

To avoid this heavy computational burden, a recursive convolution in time domain was developed using exponential functions by Semlyen [31] for approximating $s(t)$ in (3.22), and using linear approximation of $s(t)$ by Ametani [32]. Later, a more sophisticated approach approximating $S(\omega)$ in (3.21) by lumped CR ladder circuits in frequency domain has been developed by Marti [29]:

$$S(s) = \sum_{i=1}^{m} (a_i/s + b_i) \tag{3.25}$$

where Laplace operator has

$$s = j\omega \tag{3.26}$$

The above equations are analytically transformed into time domain by Laplace transform:

$$s(t) = \sum_{i=1}^{m} a_i \cdot \exp(-b_i t) \tag{3.27}$$

In the approaches explained above, the transformation matrices between phase and modal domains are assumed to be constant. However, the transformation matrices are heavily frequency-dependent especially in un-transposed vertical overhead lines and underground cables, and the frequency-dependence transformation matrices influence transient simulations significantly [30,33,34]. An approach to consider the frequency-dependent transformation matrices was proposed in [33–35]. However, a numerical instability and large computer resource became significant disadvantages.

3.1.4 Frequency-dependent line model in the phase domain

To avoid the issues associated with the constant transformation matrices mentioned above and perform more accurate transient simulations, the phase domain approach has been proposed in [36–50]. The methods discussed in [36] and [37] have been implemented into the original BPA-EMTP and ATP Program. However, fitting the line functions directly in the phase domain [36–50] shows complexities because the elements of propagation function matrix **H** involve modal contributions with multiple time delays. Some models suggest to extract a single time delay from each entry of matrix **H** [39,40]. However, this method may stiffen the fitting procedure, needing a high-order fitting technique. This problem has been addressed in other models by containing modal time delays in the phase domain formulations [37,42].

One of the most well-known and accepted phase domain models is the universal line model (ULM) [42], which has been implemented into the existing EMT-type simulation tools. The ULM has a two-step fitting approach that means the poles of matrix **H** are identified in the modal domain together with time delays, and residues in the phase domain. Repetitive or close modal delays are also grouped. The condition for grouping is the difference in angle and magnitude at a high frequency [43,44]. The ULM has received several improvements, for example, the fitting accuracy [45], out-of-band passivity violations [46,47], matrix symmetry issues [48], and real-time implementation [49].

Although the ULM is considered as an accurate method for overhead lines and underground cables, it has several problems of numerical stability [50–53]. These problems potentially occur when the propagation function matrix **H** is fitted with poles and delays coming from different but close delay groups. This is considered as unbalanced modal contributions and it causes errors of integration in time-domain simulations.

3.1.4.1 Curve fitting techniques of ULM

The ULM [42] considers frequency-dependent parameters and a wideband model. It resolves multiconductor propagation relations in the phase domain. Thus, in comparison to the results produced by the models in modal domain, the ULM generates accurate results for the simulations of coupling effects between parallel conductors.

3.1.4.1.1 Fitting of the characteristic admittance

The fitting of characteristic admittance $\mathbf{Y_c}$ has the following form [42]:

$$\mathbf{Y_c} \cong \mathbf{G}_0 + \sum_{i=1}^{N_y} \frac{\mathbf{G_i}}{s - q_i} \tag{3.28}$$

where N_y is the order of approximation, q_i represents the ith fitting pole, $\mathbf{G_i}$ is the matrix of residues, and $\mathbf{G_0}$ is a constant matrix representing the limit of $\mathbf{Y_c}$ when s $\rightarrow \infty$.

3.1.4.1.2 Fitting of the propagation function

In the ULM, the propagation function matrix **H** is firstly decoupled into single-delay terms through modal decomposition:

$$\mathbf{H} = \mathbf{T}\mathbf{H^m}\mathbf{T}^{-1} \tag{3.29}$$

where **T** is the matrix of eigenvectors of the product **YZ** (is also equal to current transformation matrix explained in Chapter 2), and $\mathbf{H^m}$ is a matrix of the form:

$$\mathbf{H^m} = diag\left(e^{\lambda_1}, e^{\lambda_2}, \ldots, e^{\lambda_N}\right) \tag{3.30}$$

where the terms e^{λ_i} are the eigenvalues of **H**, representing the propagation modes $\mathbf{H_i}$. To compensate excessive phase lag and receive a low-order rational approximation, a constant time delay is removed prior to fitting. Thus, the poles and delays are identified by fitting each modal propagation function with

$$H_i \cong \sum_{j=1}^{M_i} \frac{c_j}{s - p_{i,j}} e^{(-s\tau_i)} \tag{3.31}$$

where M_i is the order of the approximation for the ith mode, $p_{i,j}$ represents the jth pole, c_j represents its jth residue, and τ_i is the time associated to the ith mode. Repetitive or close modal delays are grouped based on the conditions in [42]. The matrix of residues can be determined by solving the following equation:

$$\mathbf{H} \cong \sum_{i=1}^{N_{gr}} \left(\sum_{j=1}^{M_i} \frac{\mathbf{R}_{i,j}}{s - p_{i,j}} e^{(-s\tau_i)} \right) \tag{3.32}$$

where N_{gr} is the number of modal groups, and $\mathbf{R}_{i,j}$, corresponds to the matrix of residues.

The basic idea is to use rational approximations for matrices $\mathbf{Y_c}$ and **H**. The vector fitting (VF) method has been widely adopted in the fitting of calculated frequency domain functions with rational function approximations [54]. Furthermore, the VF method has been also modified to improve the numerical performances [55,56]. The orthonormal vector fitting and weighted vector fitting have been developed in [57] and [58] in order to resolve some ill-conditioned problems due to wide frequency ranges.

3.2 Modeling of frequency-dependent soil parameters

3.2.1 *Review of models of frequency-dependent soil parameters*

The formulas of frequency-dependent (FD) soil models [59–62] are briefly reviewed and summarized in this section. Based on widely measured results of soil, all the formulas are specifically derived for the representations of FD soil electrical parameters which are conductivity and permittivity.

3.2.1.1 Longmire/Smith (LS) model

By examining the experimental soil data of Scott and Wilkenfeld, Longmire and Smith have derived a universal FD soil model by adopting a curve-matching method [59]. Moreover, the universal soil model is further scaled by adopting the DC conductivity instead of the water content [63], and it gives the following formulas:

$$\sigma(f) = \sigma_{DC} + 2\pi\varepsilon_0 \sum_{i=1}^{13} a_n F_n \frac{\left(\frac{f}{F_n}\right)^2}{1 + \left(\frac{f}{F_n}\right)^2} \text{(S/m)} \tag{3.33}$$

$$\varepsilon_r(f) = \varepsilon_\infty + \sum_{n=1}^{13} \frac{a_n}{1 + \left(\frac{f}{F_n}\right)^2} \tag{3.34}$$

where f is the frequency in Hz and ranges from 100 Hz to 200 MHz, $\sigma(f)$ is the FD soil conductivity, $\varepsilon_r(f)$ is the FD soil relative permittivity, σ_{DC} is the DC soil conductivity in S/m, a_n is the empirical coefficient and is given in Table 3.2, ε_∞ is the high-frequency permittivity and is set to 5 [63], ε_0 is the vacuum permittivity, and the scaled F_n is given by

$$F_n = F(\sigma_{DC})10^{n-1} \tag{3.35}$$

with

$$F(\sigma_{DC}) = (125\sigma_{DC})^{0.8312} \tag{3.36}$$

Table 3.2 Values of empirical coefficients for Longmire/Smith model. Adapted from [63]

n	1	2	3	4	5
a_n	3.4×10^6	2.74×10^5	2.58×10^4	3.38×10^3	5.26×10^2
n	6	7	8	9	10
a_n	1.33×10^2	2.72×10^1	1.25×10^1	4.8	2.17
n	11	12	13	-	-
a_n	9.8×10^{-1}	3.92×10^{-1}	1.73×10^{-1}	-	-

3.2.1.2 Alipio/Visacro (AV) model

Based on a large number of field measurements, Alipio and Visacro proposed the formulation to take into account the dispersion of soil frequency dependence by adopting solid physical assumptions [60]. The expressions of the soil frequency dependence are given by [60]

$$\sigma(f) = \left[\sigma_0 + 1.26\sigma_0^{0.27}\left(\frac{f}{10^6}\right)^{\gamma}\right]\cdot 10^{-3}(\mathrm{mS/m}) \tag{3.37}$$

$$\varepsilon_r(f) = \varepsilon_{r\infty} + \frac{1.26\cdot 10^{-3}\tan\left(\frac{\pi\gamma_{av}}{2}\right)\sigma_0^{0.27}f^{\gamma_{av}-1}}{2\pi\varepsilon_0\cdot 10^{6\gamma_{av}}} \tag{3.38}$$

where f ranges from 100 Hz to 4 MHz, γ_{av} is the empirical coefficient and is set to 0.54 pu [60], and $\varepsilon_{r\infty}$ is the high-frequency relative permittivity and is set to 12 [60].

3.2.1.3 Scott (SC) model

Field and laboratory measurements of various soil samples with different water contents have been performed by Scott, and the following curve-matching expressions of the FD soil model are proposed in [61,62]:

$$\sigma(f) = 10^K\ (\mathrm{mS/m}) \tag{3.39}$$

$$\varepsilon_r(f) = 10^D \tag{3.40}$$

where f ranges from 100 Hz to 1 MHz, and the coefficients K and D are given by

$$\begin{aligned}K &= 0.028 + 1.098\log_{10}(\sigma_0) - 0.068\log_{10}(f) + 0.036\log_{10}^2(\sigma_0)\\ &\quad -0.046\log_{10}(f)\log_{10}(\sigma_0) + 0.018\log_{10}^2(f)\end{aligned} \tag{3.41}$$

$$\begin{aligned}D &= 5.491 + 0.946\log_{10}(\sigma_0) - 1.097\log_{10}(f) + 0.069\log_{10}^2(\sigma_0)\\ &\quad -0.114\log_{10}(f)\log_{10}(\sigma_0) + 0.067\log_{10}^2(f)\end{aligned} \tag{3.42}$$

with σ_0 is the soil conductivity at 100 Hz in mS/m.

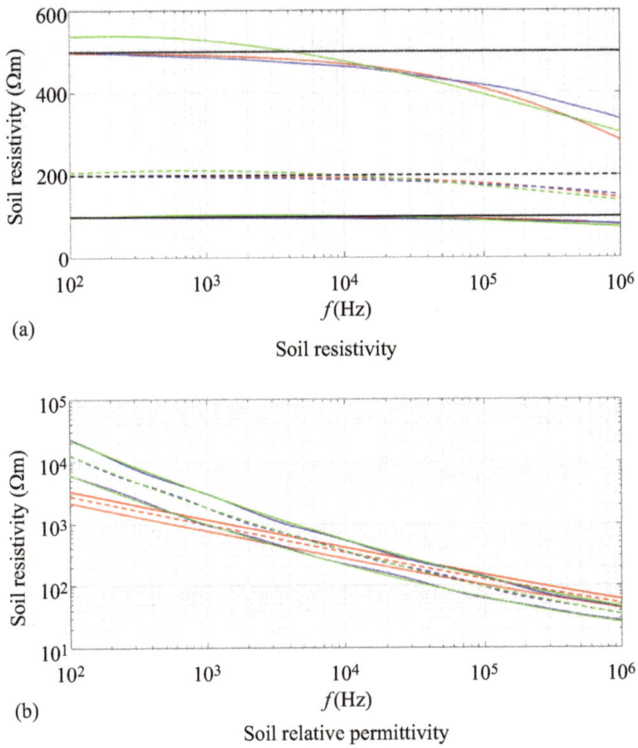

Figure 3.5 Calculated results of FD soil parameters: solid line: $\rho_0 = \rho_{cs} = 100$ Ωm, dashed line: $\rho_0 = \rho_{cs} = 200$ Ωm, dotted line: $\rho_0 = \rho_{cs} = 500$ Ωm, black: CS model, blue: LS model, red: AV model, and green: SC model

3.2.2 Calculated results of FD soil models

Figures 3.5 and 3.6 show the calculated FD soil resistivity ($1/\sigma(f)$) and FD soil relative permittivity respectively using previously discussed FD soil models.

Based on a logarithmic sampling method, low and high values of the soil resistivity at 100 Hz ($\rho_0 = 1/\sigma_0$) are adopted in the following calculations. The low values of ρ_0 with 100 Ωm, 200 Ωm, and 500 Ωm cover the major surge simulations performed in power systems. The high values of ρ_0 with 1,000 Ωm, 2,000 Ωm, and 5,000 Ωm can be used to investigate the transients of overhead lines and underground cables for the cases with very large soil resistivity. The CS resistivity ρ_{cs} are also compared with the calculated FD soil resistivity.

The general characteristics are summarized below:

1. The values of FD soil resistivity and FD soil relative permittivity decrease significantly as frequency increases.
2. The FD soil relative permittivity shows much larger values at a low frequency, that is, $f < 10$ kHz than that at a high frequency.

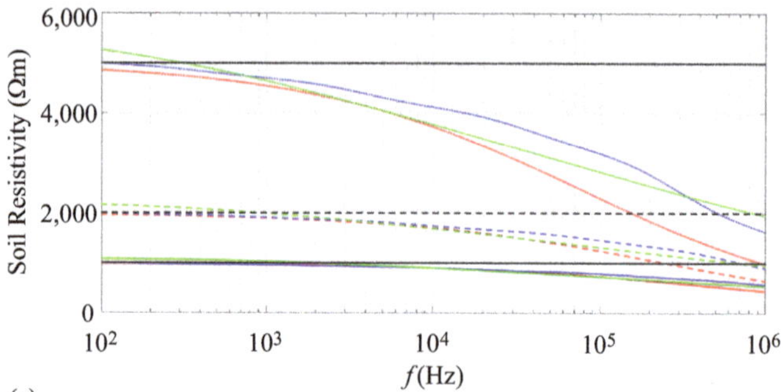

(a) Soil resistivity

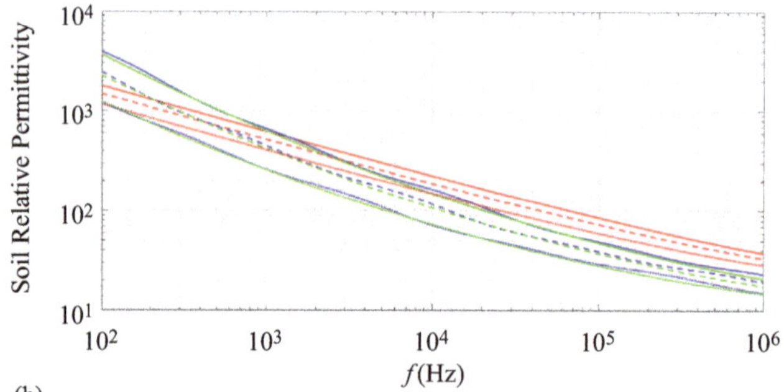

(b) Soil relative permittivity

Figure 3.6 Calculated results of FD soil parameters: solid line: $\rho_0 = \rho_{cs} = 1{,}000$ Ωm, dashed line: $\rho_0 = \rho_{cs} = 2{,}000$ Ωm, dotted line: $\rho_0 = \rho_{cs} = 5{,}000$ Ωm, black: CS model, blue: LS model, red: AV model, and green: SC model

3. As ρ_0 increases, the FD soil relative permittivity decrease.
4. The FD soil relative permittivity calculated by the LS and SC models have minor differences.
5. As frequency increases, the differences between the CS resistivity and the FD soil resistivity increase.
6. The differences between the CS resistivity and the FD soil resistivity are more significant for the large values of ρ_0 and ρ_{cs}.

3.2.3 Comparison with measured data

The FD soil models discussed in Section 3.2.1, the Portela (PR) and Visacro/Portela (VP) FD soil models [64,65] are used to calculate the FD resistivity and relative permittivity. The results are compared with the independent measured data [66,67] are shown in Tables 3.3–3.12.

The FD electrical parameters calculated by the LS, AV, and SC models match qualitatively with the independent wide-band measured data of soil in [66,67], and the results of LS model show a better agreement than the other two models.

Table 3.3 Calculated resistivity by FD soil models with $\rho_0 = 93.2\ \Omega m$ and measured data

Frequency (Hz)	10^2	10^3	10^4	10^5	10^6	10^7
Measured data (Ωm) [66]	93.2	89.6	87.1	84.7	82.2	74.8
LS (Ωm)	93.2	91.5	89.18	84.77	76.21	56.19
AV (Ωm)	93.06	92.7	91.5	87.57	76.21	52.57
SC (Ωm)	91.41	96.92	94.58	84.96	70.25	53.46
PR (Ωm)	93.12	92.82	91.28	84.2	60.39	24.79
VP (Ωm)	93.2	78.96	66.9	56.68	48.02	40.68

Table 3.4 Calculated relative permittivity by FD soil models with $\rho_0 = 93.2\ \Omega m$ and measured data

Frequency (Hz)	10^2	10^3	10^4	10^5	10^6	10^7
Measured data [66]	42,500	3,820	384	112	20.4	16.0
LS	24,160	3,171	556.1	141.4	44.35	21.42
AV	3,385	1,182	417.5	152.6	60.75	28.91
SC	23,890	3,153	566.6	138.6	46.16	20.93
PR	3,157	1,604	815.1	414.2	210.5	107
VP	13,230	3,346	846.4	214.1	54.15	13.7

Table 3.5 Calculated resistivity by FD soil models with $\rho_0 = 198\ \Omega m$ and measured data

Frequency (Hz)	10^2	10^3	10^4	10^5	10^6	10^7
Measured data (Ωm) [66]	198	191	185	180	171	139
LS (Ωm)	198	193.6	187.2	173.9	150	95.74
AV (Ωm)	197.5	196.2	191.8	178.2	142.8	84.63
SC (Ωm)	204.5	209.4	197.4	171.3	136.8	100.6
PR (Ωm)	197.7	196.3	189.5	161.4	91.92	28.85
VP (Ωm)	198	167.8	142.1	120.4	102	86.43

Table 3.6 Calculated relative permittivity by FD soil models with $\rho_0 = 198\ \Omega m$ and measured data

Frequency (Hz)	10^2	10^3	10^4	10^5	10^6	10^7
Measured data [66]	22,900	2,030	194	41.5	18.1	15.4
LS	12,960	1,824	354.6	94.41	34.38	17.23
AV	2,764	966.2	342.9	126.7	51.78	25.79
SC	12,710	1,828	357.9	95.41	34.63	17.11
PR	3,157	1,604	815.1	414.2	210.5	107
VP	8,841	2,236	565.6	143.1	36.18	9.152

Table 3.7 Calculated resistivity by FD soil models with $\rho_0 = 296\ \Omega m$ and measured data

Frequency (Hz)	10^2	10^3	10^4	10^5	10^6	10^7
Measured data (Ωm) [66]	296	287	280	275	254	200
LS (Ωm)	296	288.7	277.8	254.9	213.9	127
AV (Ωm)	294.9	292.4	283.8	257.5	195	105.8
SC (Ωm)	311.9	313.6	290.2	247.2	193.8	139.8
PR (Ωm)	295.2	292.2	277.5	221	108.6	30.31
VP (Ωm)	296	250.8	212.5	180	152.5	129.2

Table 3.8 Calculated relative permittivity by FD soil models with $\rho_0 = 296\ \Omega m$ and measured data

Frequency (Hz)	10^2	10^3	10^4	10^5	10^6	10^7
Measured data [66]	14,100	1,210	127	34.4	18.3	15.5
LS	9,141	1,345	280.7	76.61	30.54	15.56
AV	2,481	868.1	308.8	114.9	47.69	24.37
SC	9,202	1,386	284	79.27	30.12	15.58
PR	3,157	1,604	815.1	414.2	210.5	107
VP	7,130	1,803	456.1	115.4	29.18	7.38

Table 3.9 Calculated resistivity by FD soil models with $\rho_0 = 343\ \Omega m$ and measured data

Frequency (Hz)	10^2	10^3	10^4	10^5	10^6	10^7
Measured data (Ωm) [66]	343	330	317	306	282	219
LS (Ωm)	343	334.2	320.9	293.1	243.1	140.5
AV (Ωm)	341.6	338.3	327.2	294.1	217.5	114.3
SC (Ωm)	363.6	363.1	333.7	282.3	219.9	157.6
PR (Ωm)	341.9	337.8	318.3	246.1	114.3	30.7
VP (Ωm)	343	290.5	246.2	208.5	176.7	149.7

Table 3.10 Calculated relative permittivity by FD soil models with $\rho_0 = 343\ \Omega m$ and measured data

Frequency (Hz)	10^2	10^3	10^4	10^5	10^6	10^7
Measured data [66]	14,600	1,430	161	42.1	19	15.9
LS	8,096	1,213	259.9	71.59	29.42	15.07
AV	2,385	834.7	297.3	110.9	46.3	23.89
SC	8,195	1,255	261.6	74.24	28.68	15.09
PR	3,157	1,604	815.1	414.2	210.5	107
VP	6,589	1,667	421.5	106.6	26.97	6.82

Table 3.11 Calculated resistivity by FD soil models with $\rho_0 = 1{,}800\ \Omega m$ and measured data

Frequency (Hz)	10^2	10^3	10^4	10^5	10^6	10^7
Measured data (Ωm) [67]	1,800	1,600	1,500	1,400	1,000	360
LS (Ωm)	1,800	1,724	1,582	1,334	839	337.4
AV (Ωm)	1,776	1,720	1,550	1,155	613.3	233.5
SC (Ωm)	1,952	1,806	1,538	1,206	869.9	577.8
PR (Ωm)	1,772	1,667	1,280	587.4	156.6	33.14
VP (Ωm)	1,800	1,525	1,292	1,095	927.4	785.7

Table 3.12 Calculated relative permittivity by FD soil models with $\rho_0 = 1{,}800\ \Omega m$ and measured data

Frequency (Hz)	10^2	10^3	10^4	10^5	10^6	10^7
Measured data [67]	19,500	1,000	105	55	45	40
LS	2,670	482.3	124.6	40.77	19.96	11.2
AV	1,528	537.8	194.3	75.22	33.92	19.6
SC	2,433	450.1	113.3	38.86	18.14	11.53
PR	3,157	1,604	815.1	414.2	210.5	107
VP	2,714	686.5	173.6	43.92	11.11	2.81

3.2.4 Summary

In this section, the existing FD soil models are reviewed and discussed. Several typical values of soil resistivity at low frequencies are adopted in the calculations and studies. Also, the calculated FD soil resistivity and permittivity are compared with independent measured data of soil in respect to different regions and locations. In general, the differences between CS model and FD soil model are more significant at high-frequency region with large soil resistivity.

3.3 Various cable installation

3.3.1 Cable installed underneath a bridge

This section briefly explains the modeling of a cross-bonded cable installed underneath an iron bridge, which is connected with overhead lines at the both ends as illustrated in Figure 3.7(a) [68,69]. Figure 3.7(b) shows a model circuit for EMT simulation. The cable underneath the bridge is overhead. The both sides of the cable are underground. The underground part of the cable is horizontally arranged, and the cable from the bridge to the underground part is with vertical configuration.

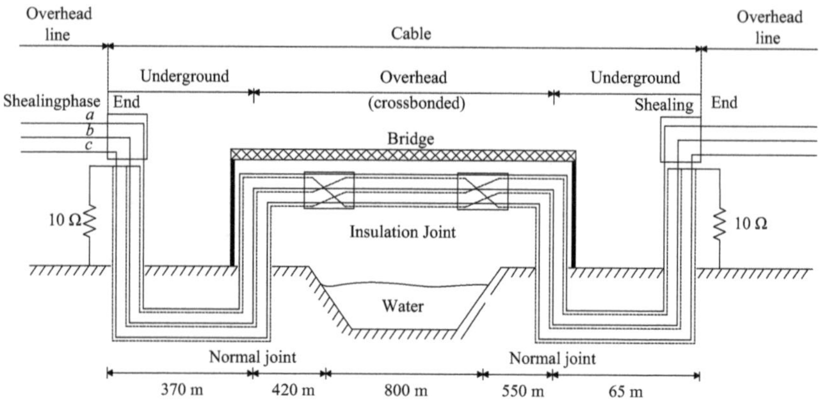

(a) Cable installed underneath a bridge

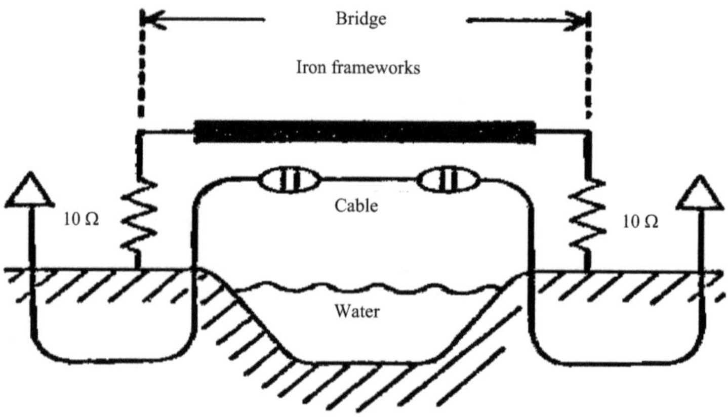

(b) Model circuit of the cable installed underneath a bridge

Figure 3.7 Cable installed underneath a bridge and its model circuit for EMT simulation

The iron framework of the bridge is represented as an overhead line parallel to the cable with mutual coupling.

The maximum over-voltages are reduced by the iron frameworks due to the mutual coupling between the cable and the conductors simulating the iron frameworks of a bridge. The rate of reduction reaches to more than 50 % on the maximum voltages and the oscillation of the voltage waveforms are dying out much faster than the case with no mutual coupling. But, it should be noticed that the maximum sheath over-voltages even in the case of considering the mutual coupling between the cable and nearby iron frameworks of a bridge are higher than those in the underground cable case. More details can be found in [68,69].

3.3.2 Tower-installed (vertical) cable

It is rather common to install a power cable directly at the arms of a transmission tower or pole and to connect it with an overhead line so as to reduce space required for a substation. This configuration, however, may produce a high-lightning overvoltage on the cable when lightning strikes the tower or a nearby tower, and the insulation and the protection of the cable has to be carefully examined in conjunction with grounding of the cable metallic sheath, that is, either the metallic sheath being grounded to the arm or grounded through an independent grounding lead [70–72]. To investigate a lightning surge on a cable installed along a tower, a field measurement was carried out on a 77 kV transmission tower of Kansai Electric Power Co. [73]. The measured results are summarized in this section. Also, modeling of the tower installed cable is explained. EMT simulation is carried out to estimate the lightning surge characteristics on the 77 kV transmission tower by adopting the earth-return impedance formula of a vertical conductor developed in [74]. The simulation results are compared with the measured results. A theoretical analysis is also performed.

3.3.2.1 Field measurement

3.3.2.1.1 Experimental setup

Figure 3.8(a) illustrates an experimental setup in a system composed of a 77 kV transmission tower and a power cable along the tower. The tower height is 20.8 m. The surge impedance Z_0 of the tower is measured to be about 119 Ω, and the tower footing resistance R_t is 8.4 Ω. The grounding resistance R_g of a sheath grounding lead is 5.0 Ω. The grounding lead is a covered conductor (conductor radius 0.65 mm). The earth resistivity in this area is measured to be 111 Ωm.

The cross-section of a power cable is given in Figure 3.8(b). The power cable is connected to a capacitance C_s of 2,000 pF representing an open-circuited transformer in a substation (S/S).

R_s= 7.1 Ω represents the grounding mesh resistance in the substation, and the cable sheath is grounded to the mesh. A pulse generator (P.G.) with a 5.0 kΩ resistance in series is used to represent a lightning current.

An applied current to the tower top is shown in Figure 3.8(c). The rise time T_f at the wave front is about 70 ns and the peak value is about 0.55A. To measure a transient voltage, a grounding wire is used as a voltage reference wire with the length of 350 m of which the other end is grounded at the next tower. Thus, the observation time of the field measurement is up to 2.3 μs corresponding to the two travel time of a traveling wave along the reference wire.

Experimental conditions and results are summarized in Table 3.13. Cases 1 and 2 are a case of the cable metallic sheath at the tower being connected to the tower arm, and cases 3 and 4 are a case of the sheath connected to a grounding lead.

3.3.2.1.2 Measured results

Measured voltage waveforms are shown in Figures 3.9 and 3.10 for cases 1–4, and the maximum voltage is summarized in Table 3.13. It should be noted that a spike-

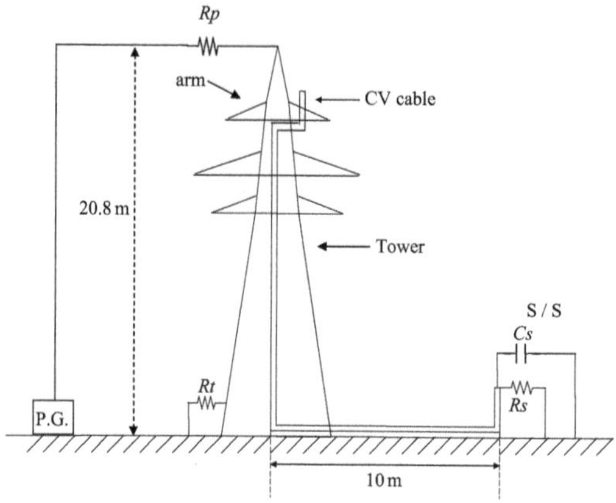

(a) Field test setup

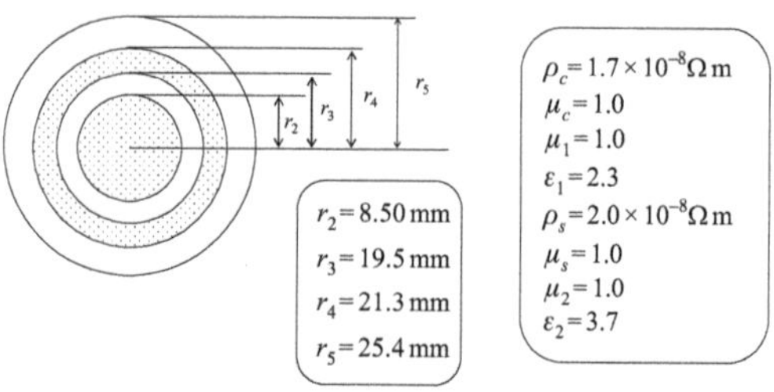

(b) Cross-section of a power cable

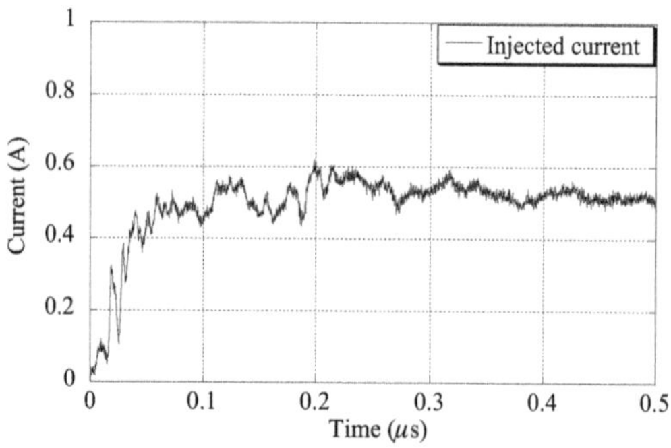

(c) Applied current waveform

Figure 3.8 Experimental setup

like voltage with the peak value of about 70 V at the beginning (ns) of the tower top voltage may be a noise during the measurement.

It is observed that the tower top voltage is about 35(V) while the cable core voltage at the top is about 24(V) in the case of the cable sheath being grounded to the

Table 3.13 Test conditions and measured maximum voltages

Case	Sending-end sheath connected to	Flashover	Tower top	Tower bottom	Core at arm	Sheath at arm
1	Tower arm	Yes	35.2	11.23	25.2	25.5
2	Tower arm	no	35.5	10.82	25.8	25.8
3	Grounding lead	yes	34.7	11.35	26.6	24.4
4	Grounding lead	no	35.1	11.03	26.4	23.6

grounding lead (cases 3 and 4). The cable sheath and core voltages are determined by the mutual impedance between the tower and the cable in the grounding lead case. The mutual impedance Z_{ij} is given approximately in the following form [74,75]:

$$Z_{ij} = 60\ \ln\left(\frac{D_{ij}}{2ed_{ij}}\right) \tag{3.43}$$

where

$$D_{ij} = \sqrt{(h_i + h_j)^2 + y_{ij}{}^2} \tag{3.44}$$

$$d_{ij} = \sqrt{(h_i - h_j)^2 + y_{ij}{}^2} \tag{3.45}$$

where h_i and h_j are heights of conductors i and j, y_{ij} is horizontal separation between conductors i and j, and Euler constant $e = 2.714$.

A further approximation of $h >> y$ gives the following simple formulas:

$$Z_m = Z_{ij} = 60\ \ln\left(\frac{h}{ed}\right) \tag{3.46}$$

$$Z_s = Z_{ii} = 60\ \ln\left(\frac{h}{er}\right) \tag{3.47}$$

where r *is* the conductor radius.

With $h = 17.5$ m and $d = 1.37$ m , the mutual impedance becomes $Z_m = 113\ \Omega$. The tower self-impedance with equivalent radius $r = 1.22$ m is $Z_s = 119\ \Omega$.

The ratio of Z_m/Z_s somehow corresponds to the ratio of the measured tower voltage and the core voltage. The sheath voltage is nearly the same as the core voltage because the cable is coaxial cylindrical cable. The fact corresponds to the theory of coaxial mode wave propagation [76].

1. Cable sheath grounded at the tower top

It is observed in cases 1–4 in Table 3.13 that the cable voltage is higher in the case of the sheath being grounded to the arm. The result is quite reasonable because

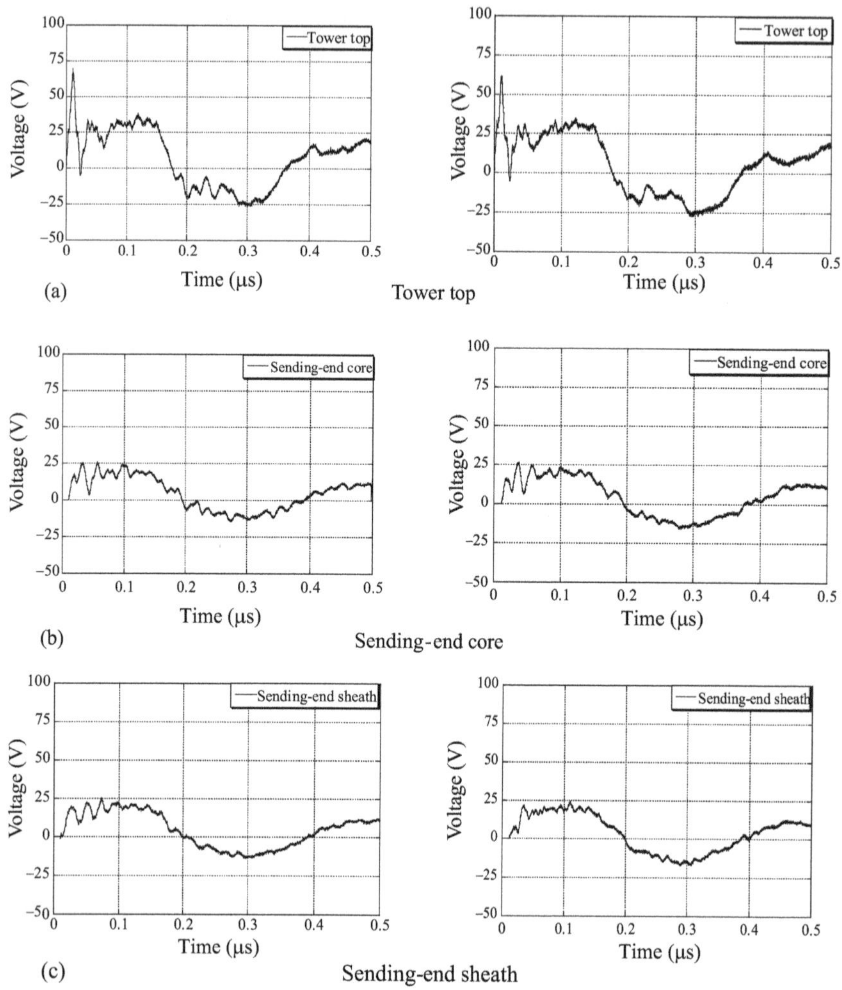

(1) Case 1: grounded to arm

(2) Case 3: grounding lead

Figure 3.9 Measured results of transient voltages for Cases 1 and 3 (F.O.)

the tower arm voltage, which is the same as the sheath voltage when the sheath is grounded to the tower arm, is higher than the sheath voltage produced by mutual coupling between the tower and the sheath in the case of the sheath grounded to the grounding lead.

However, the sheath and tower voltages are dependent on the grounding resistances of the tower (8.4 Ω in this measurement) and the grounding lead (5 Ω), and thus it is not possible to draw a general conclusion from the measured result.

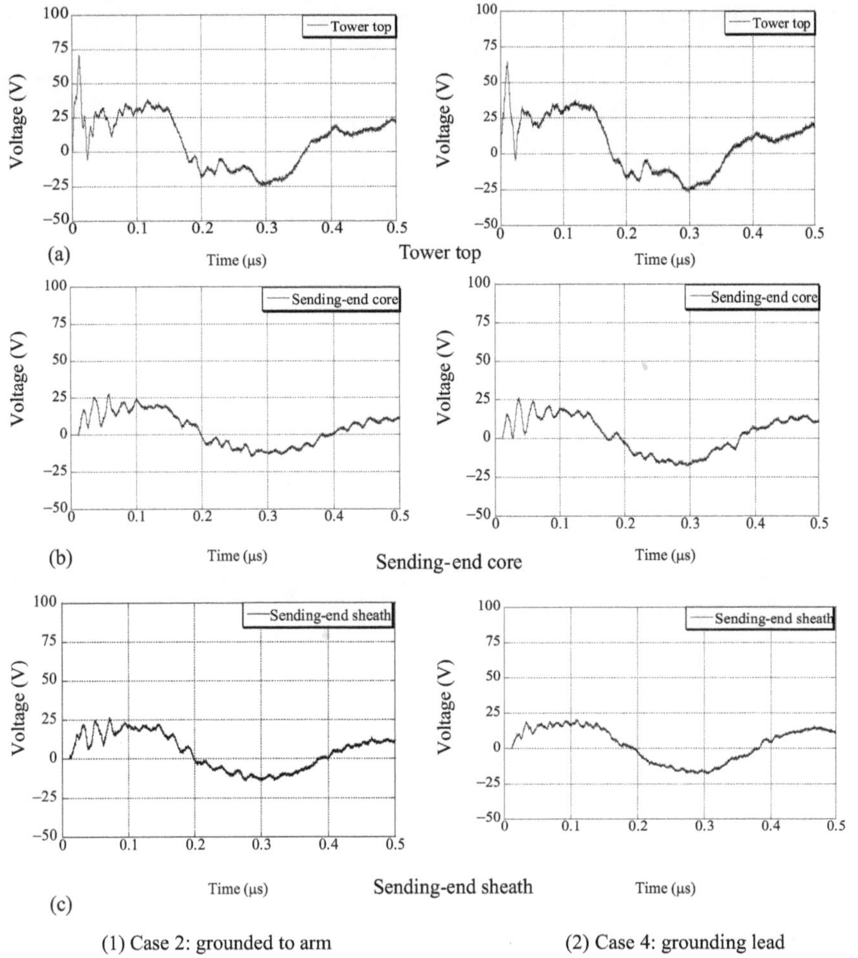

Figure 3.10 Measured results of transient voltages for Cases 2 and 4 (no F.O.)

2. Flashover between a tower arm and a cable head (core)

Traveling waves due to a flashover (FO) between the tower arm and the cable core flows into the cable core. Thus, it is expected that the flashover generates a higher voltage at the cable terminals. The measured results in Table 3.13 show that a flashover causes a higher voltage at the cable core, by about 10% in the case of the sheath connected to the arm (Cases 1/2) and by about 1% in the grounding lead case (cases 3/4). The FO is modeled by short-circuiting the arm and the core with a covered conductor, and thus, it may not be possible to draw a conclusion from the measured result to a real flashover case.

3.3.2.2 Theoretical analysis and modeling for EMTP simulation

3.3.2.2.1 Modeling of tower part (vertical conductor)

Figure 3.11 illustrates a model circuit for a theoretical analysis.

The parameters of the circuit are given as: $E_o = 3.5$ kV step function, $R_o = 5.0$ kΩ, $Z_1 = Z_t Z_G/(Z_t + Z_G) = 89\ \Omega$

where $Z_G = 517\ \Omega$: impedance of the ground wire.

A surge impedance of a vertical cylindrical conductor is evaluated approximately by (3.47). The internal impedance of a conductor is calculated by Cable Constants explained in Chapter 2. When the cable core is neglected, the surge impedance matrix in this approach is given by

$$[Z_2] = \begin{bmatrix} 89 & 113 & 112 \\ 113 & 362 & 338 \\ 112 & 338 & 507 \end{bmatrix} \begin{matrix} \textit{tower "t"} \\ \textit{cable sheath "s"} \\ \textit{grounding lead "g"} \end{matrix} \tag{3.48}$$

The surge impedance matrix at Node 3 is given in the following equation:

$$[Z_3] = \begin{bmatrix} R_f & 0 & 0 \\ 0 & Z_s' & 0 \\ 0 & 0 & R_g \end{bmatrix} = \begin{bmatrix} 7.1 & 0 & 0 \\ 0 & 137 & 0 \\ 0 & 0 & 5 \end{bmatrix} \tag{3.49}$$

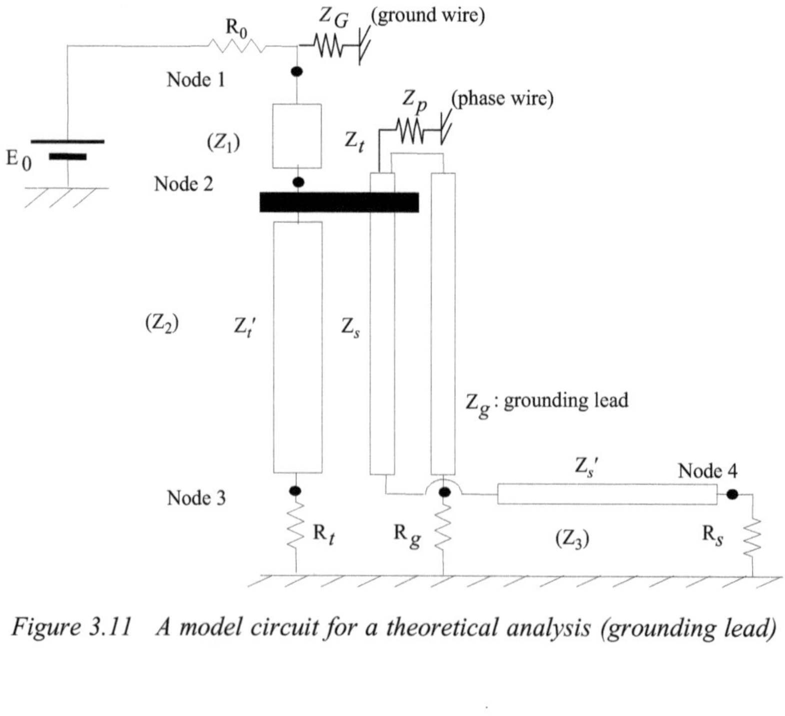

Figure 3.11 A model circuit for a theoretical analysis (grounding lead)

3.3.2.2.2 Theoretical analysis

When the sheath is grounded to a grounding lead, the refraction coefficient at each node is given in the following form:

$$\lambda_{01} = \frac{2Z_1}{R_0 + Z_1} = 0.034 \tag{3.50}$$

$$\lambda_{34} = 0.098 \tag{3.51}$$

$$\lambda_{10} = 1.96 \tag{3.52}$$

$$[\lambda_{12}] = \begin{bmatrix} 0.95 & 0 & 0 \\ 0.86 & 0 & 0 \\ 0.85 & 0 & 0 \end{bmatrix} \tag{3.53}$$

$$[\lambda_{23}] = \begin{bmatrix} 0.212 & -0.03 & -0.03 \\ -0.59 & 1.05 & -0.55 \\ -0.02 & -0.03 & 0.04 \end{bmatrix} \tag{3.54}$$

$$[\lambda_{21}] = \begin{bmatrix} 1.02 & 0 & 0 \\ -0.86 & 2 & 0 \\ -0.84 & 0 & 2 \end{bmatrix} \tag{3.55}$$

Then, the following analytical solution is obtained:

1. $t = 0 : V_1 = 34.6$ V
2. $t = \tau_1 = 3.3/300 = 10$ ns :

$$(V_2) = \begin{pmatrix} V_{2t} \\ V_{2s} \\ V_{2g} \end{pmatrix} = \begin{pmatrix} 33.2 \\ 22.9 \\ 8.58 \end{pmatrix} \tag{3.56}$$

3. $t = \tau_2 = 69$ ns :

$$(V_3) = \begin{pmatrix} V_{3t} \\ V_{3s} \\ V_{3g} \end{pmatrix} = \begin{pmatrix} 6.17 \\ 1.26 \\ 0.80 \end{pmatrix} \tag{3.57}$$

4. $t = \tau_2 + \tau_3 = 102$ ns : $V_4 = 0.96$ V
5. $t = 2\tau_2 - \tau_1 = 128$ ns

$$(V_2) = \begin{pmatrix} -4.96 \\ -4.05 \\ -1.17 \end{pmatrix} \tag{3.58}$$

6. $t = 2\tau_2 = 138$ ns: $V_1 = -31.9$V

The above result is shown in Figure 3.12.

3.3.2.3 EMT simulation results

Figure 3.13 and Table 3.14 show simulation results by the above-explained approach. It is observed that the simulation results agree satisfactorily with the experimental results in Figure 3.9 and Table 3.13 except the spike-like voltage with the peak value of about 70 (V) observed at the beginning (at around $t = 10$ ns) of the tower top voltage in Figures 3.9 and 3.10.

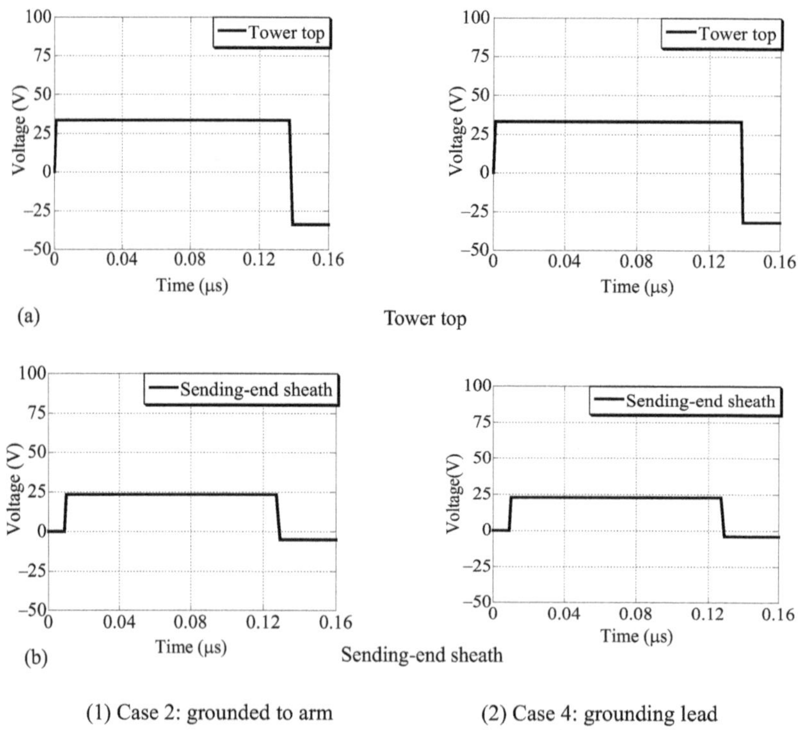

Figure 3.12 Analytical results of transient voltages for Cases 2 and 4 (no F.O.)

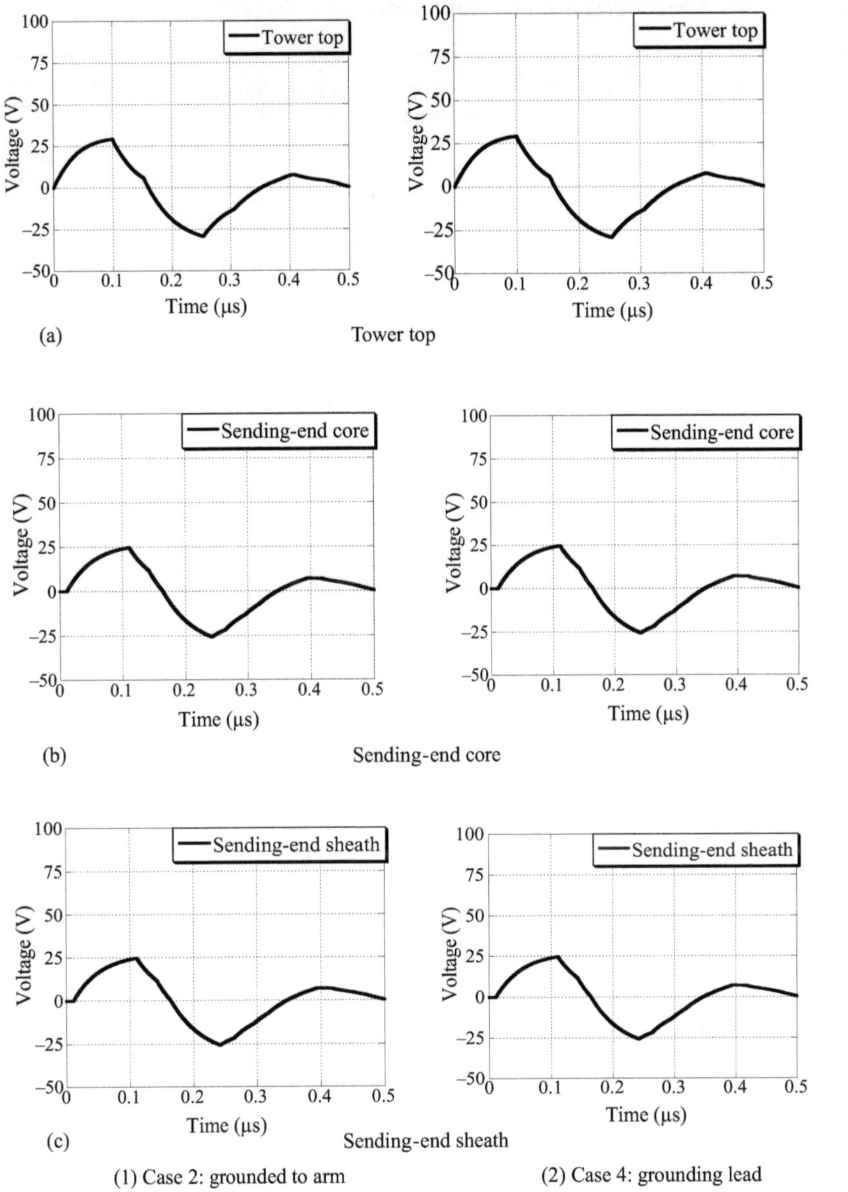

Figure 3.13 Simulation results of transient voltages for Cases 2 and 4 (no F.O.)

It is observed that the maximum sheath voltage is 29.8 (V) in Figure 3.13 (2), while it is 28.8 (V) in Figure 3.12 (1).

The analytical result in Figure 3.12 corresponds qualitatively to the experimental and EMT simulation results. Thus, the surge characteristics discussed in Section 3.3.2.2 are theoretically confirmed to be appropriate.

Table 3.14 Simulation conditions and maximum voltages (V)

Case	Sending-end sheath connected to	F. O.	Tower top	Tower bottom	Core at arm	Sheath at arm	Voltage difference	
							Tower/ sheath at bottom	Core/ sheath at S/S
1	Tower arm	Yes	34.7	9.09	30.4	30.4	5.18	0.54
2	Tower arm	No	34.3	9.01	29.8	29.8	5.13	0.53
3	Grounding lead	Yes	34.7	8.96	29.6	29.6	6.30	0.34
4	Grounding lead	No	34.3	8.92	28.8	28.8	6.25	0.34

If the cable voltage at the tower top is of dominant importance, the theoretical analysis neglecting a cable core is useful to catch the physical insight of the overvoltage on the cable.

3.3.2.4 Voltage difference between tower and cable

In Sections 3.3.2.2 and 3.3.2.3, only a voltage to ground is discussed. However, a voltage difference between a tower, cable sheath and core is necessary to discuss a possibility of breakdown between the tower and the cable. For this purpose, an EMT simulation is carried out in a model circuit in Figure 3.11.

Figure 3.14 shows a simulation result. It is clear in the figure that a significant voltage difference appears at the tower bottom independently of the system condition. Thus, it is concluded that a flashover between the cable sheath and the tower is quite probable at the tower bottom. The case of the sheath grounded to a grounding lead produces a higher voltage difference than that in the ease of the sheath grounded to the tower arm.

3.3.2.5 Concluding remarks

Based on experimental and theoretical observations together with EMT simulations, the following remarks have been obtained when lightning strikes a tower where a power cable is installed.

(a) Because of short distance between a tower structure and a power cable, the self-surge impedance of the tower is nearly the same as the mutual impedance between the tower and the cable and thus no significant voltage difference appears between the tower and the cable. Also, the cable core voltage is nearly the same as the cable sheath voltage due the electrostatic shield effect of the cable metallic sheath.

(b) The cable voltage is determined by electromagnetic coupling between a tower and the cable when the cable sheath at the tower top is grounded through a grounding lead. When the sheath is grounded to a tower arm at the top, the cable voltage is determined by conductive coupling, that is, the sheath voltage becomes nearly the same as the arm voltage. In general, the conductive

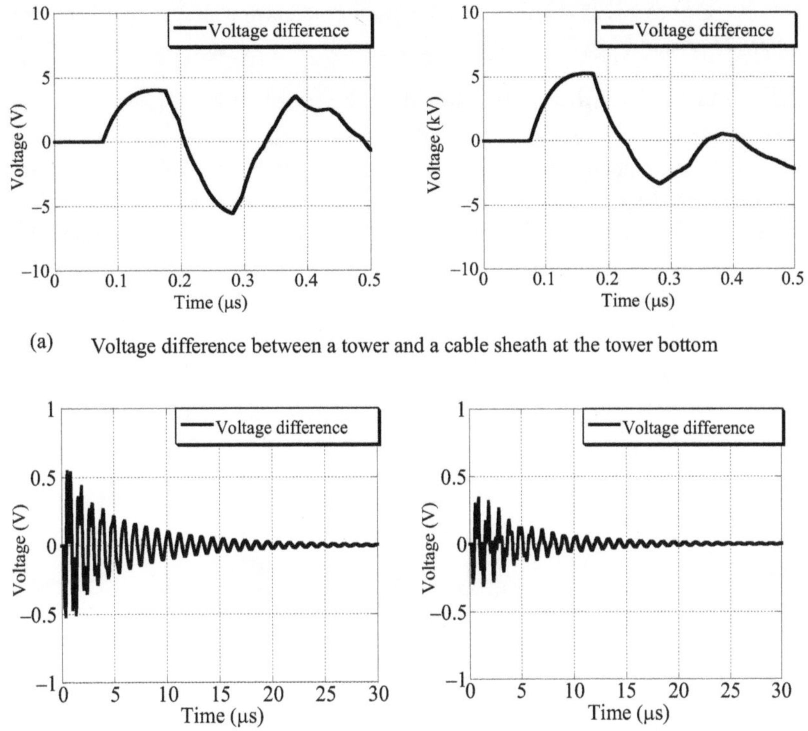

(a) Voltage difference between a tower and a cable sheath at the tower bottom

(b) Voltage difference between a cable core and sheath at S/S

(1) Case 2: grounded to arm (2) Case 4: grounding lead

Figure 3.14 Simulation results of transient voltages for Cases 2 and 4 (no F.O.)

coupling being greater than the electromagnetic coupling, the cable voltage to earth becomes higher in the case of the sheath is grounded to the arm.

(c) The voltage difference between the tower and the cable sheath becomes the largest at the tower bottom, and is found to be greater in the case of the sheath being grounded through a grounding lead. The phenomenon may be a result of a lower voltage of the cable at the tower top.

(d) It is suggested to adopt an impedance of a vertical conductor to simulate a lightning surge on a cable installed along a tower, although a conventional approach of adopting a horizontal conductor equivalence is very convenient.

3.3.3 Cable installed in a tunnel

A number of power transmission cables are installed in an underground tunnel together with water pipelines and telecommunication lines. For example, in Singapore, Transmission Cable Tunnel Project was planned with two cross-island cable tunnels buried with a depth of 60 m underground. North-South and East-West tunnels are of a total length of 35 km. The tunnel is composed of steel frameworks and concrete and is buried underground.

3.3.3.1 Existing model of tunnel-installed cable [69,77]

(a) The earth-return impedance of a cable installed in a tunnel becomes similar to that of a pipe-enclosed type (PT) cable. Therefore, the PT cable impedance implemented into Cable Constants of EMT-type simulation tools [69,77] has been widely used to study wave propagation and transient phenomena [78].
(b) The simplest approach to deal with the tunnel-installed cable is to represent the cable as an overhead cable. The height of the overhead cable is taken as the shortest distance between the cable and the tunnel surface [78].

It should be noted that the resistivity of the pipe conductor (steel or aluminum, 10^{-7} to 10^{-8} Ωm) in approach (a) and that of the earth in approach (b) is assumed to be the same as that of the pipe conductor or soil (10–1,000 Ωm). In the case of the tunnel, the resistivity of the inner conductor (steel framework) differs from that of the outer conductor (soil). Thus, the wave propagation characteristic of the tunnel-installed cable significantly different from those in the cable in approach-(a) and approach-(b).

3.3.3.2 Accurate tunnel model [79]

The earth-return impedance of the tunnel has been derived in [79]. Because the tunnel is composed of steel frameworks, concrete, and soil, the tunnel is similar to the two-layer conductor explained in Appendix A1.3 in Chapter 2. The accurate impedance formula in [79] is quite complicated similar to the impedance of the semiconducting layer in Appendix A1.3.

The impedance evaluated by the accurate impedance shows a characteristic that is nearly the same as the impedance of an infinitely thick outer medium calculated by the existing PT cable impedance formula [69,77] in a low-frequency region and reaches that of the inner medium of the finite thickness in a high-frequency region. The wave propagation and transient voltage characteristics on the PT and tunnel-installed cables calculated by using the accurate impedance are quite different from those calculated by the existing impedance formula based on an assumption of an infinitely thick pipe or tunnel. The propagation velocity shows a rather flat characteristic as a function of a frequency, and the transient voltage waveform is far less distorted in the case of a steel PT cable. The above observations are physically interpreted as a result of the penetration depth of a two-layered conductor and cannot be explained by the existing impedance formula of the infinitely thick pipe.

3.3.4 Cable installed in an HDPE tube

The underground cables are conventionally buried directly into soils. Some of the cables are installed in a tunnel, where a three-phase cable becomes eccentric to the center of the tunnel. Occasionally, each phase cable is installed within a high-density poly-ethylene (HDPE) tube which is buried in soils or concrete [80–83].

Transients on a directly buried cable and a tunnel-installed cable are easily simulated by the EMT in which all the parameters of the cables, that is, the impedance, admittance, propagation constants, characteristic impedance, and transformation matrices, are evaluated by the subroutine Cable Constants provided in the EMT-type simulation tools.

The parameters of a cable in an HDPE tube cannot be evaluated by the Cable Constants Program, because the outer insulator of the metallic sheath is composed of three layers, that is, the cable outer insulator, the air gap, and the HDPE tube, and those are not concentric. The impedance and the admittance of such an irregular structure of a cable have not been discussed in detail and there exists no analytical formula.

In this section, a finite element method COMSOL [84] is adopted to evaluate the impedance and admittance of a cable installed in an HDPE tube. Considering the COMSOL calculated results, which show the equipotential surface is not much different from the boundaries of the HDPE tube and the air gap, an equivalent cylindrical insulator is proposed, similar to the concept of the geometrical mean distance. Then, all the parameters are calculated by the Cable Constants. Transient simulations by the EMT-type tool are carried out to confirm the accuracy of the equivalent insulator of the HDPE tube and air gap in comparison with a field test result. Finally, the effect of the HDPE tube with an air gap on a cable transient is investigated based on EMTP simulations.

3.3.4.1 COMSOL and cable constants calculations of impedance and capacitance

When a cable is installed in the poly-ethylene (HDPE) tube as illustrated in Figure 3.15 or in a small hole, there exist an air gap and the poly-ethylene of which the permittivities are different from that of the cable outer insulator. Furthermore, the tube or the hole is not concentric to the cable outer insulator, although each is of circular shape almost always.

It is necessary to find an equivalent concentric and homogeneous (one layer) insulator model to deal with a cable installed in an HDPE tube or in a hole.

3.3.4.1.1 Model cable

An underground cable with a cross-section of 2,000 mm^2 [82] is taken as a model cable in this chapter. The cross-section of the cable is illustrated in Figure 3.15.

3.3.4.1.2 Impedance

Figure 3.16 shows the calculated results of (a) core self-impedance Z_{11}, (b) core to sheath mutual impedance Z_{12}, and (c) sheath self-impedance Z_{22} as a function of frequency. Dotted lines in the figures are calculated results by EMTP cable constants assuming that the cable is directly buried underground, that is, no air gap and no HDPE tube. It is observed in the figures that there exists about 3% difference at frequency $f = 1$ MHz between the COMSOL and EMT simulation results. The mutual impedances show a similar trend to that of the self-impedance in Figure 3.16. No significant difference is observed in the mutual impedance case too. The maximum difference is about 2.7%.

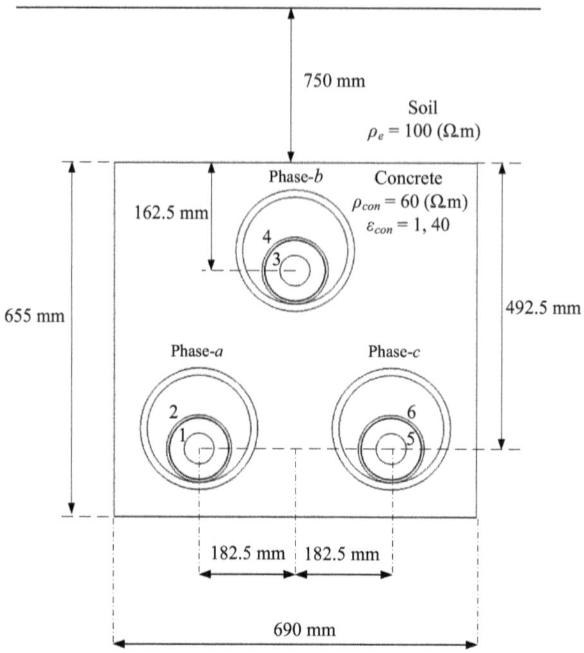

(a) Cable configuration

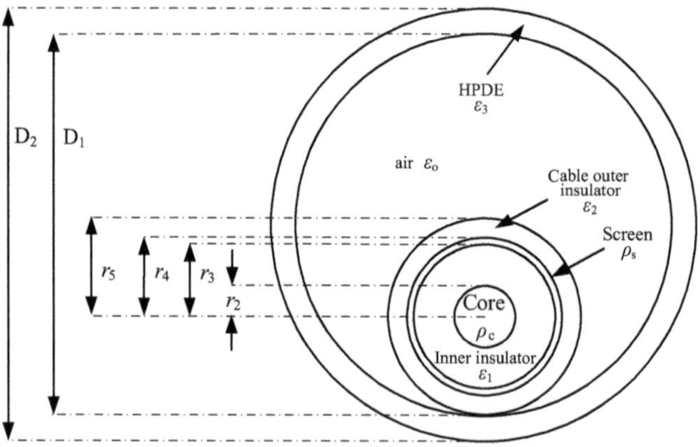

(b) Cross-section of a cable (2,000 mm²)

Figure 3.15 An underground cable installed in HDPE tube. ε_1, ε_2, ε_3: relative permittivity and ρ_c, ρ_s, ρ_e: resistivity. Adapted from [82].

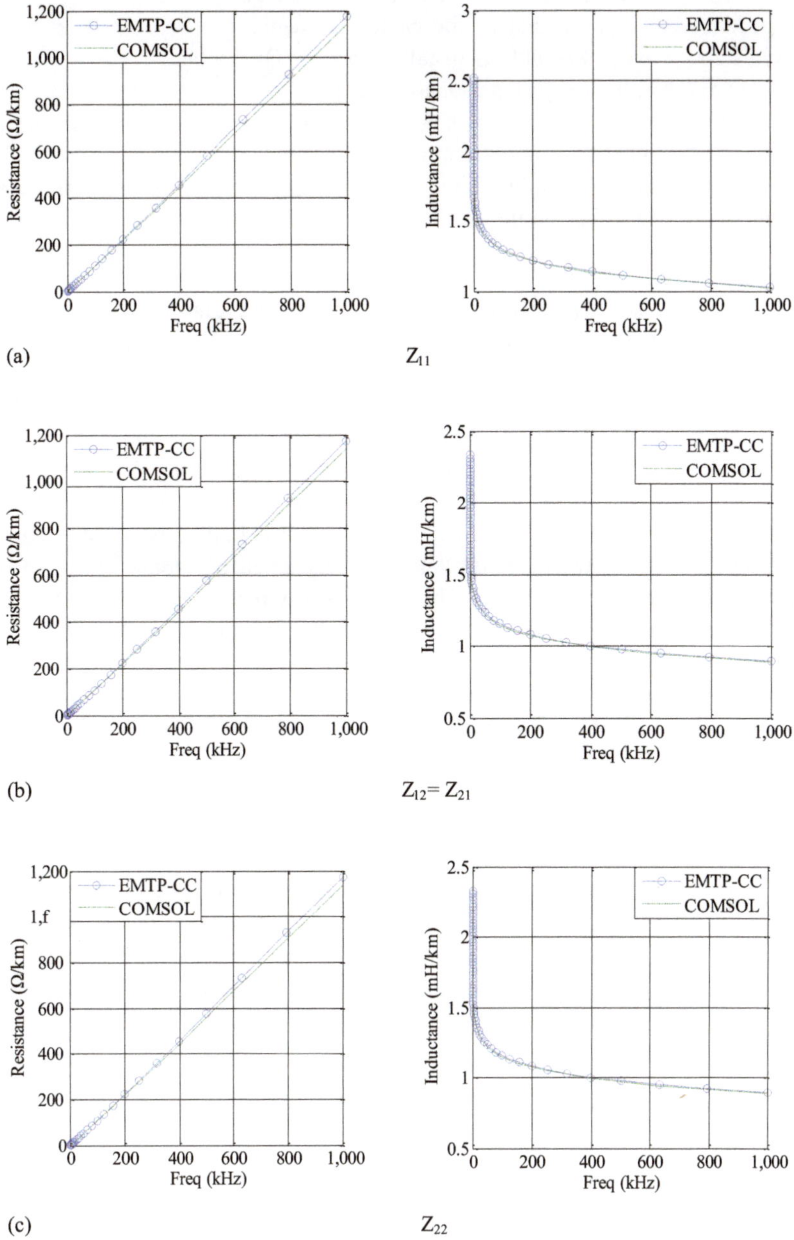

Figure 3.16 Impedances of a single-phase cable

Thus, it is concluded that the HDPE tube (and the air gap) can be neglected in the impedance calculation, that is, the cable impedance can be calculated by the cable constants even if the cable is installed in the HDPE tube.

The above results are just reasonable because the insulator impedance Z_i is given by $j\omega(\mu_o/2\pi)\ln(r_5/r_4)$ which is far smaller than the conductor impedance Z_c. The earth-return impedance is determined by the conductor's outer surface radius and the separation of two cables. Thus, the outer insulator radius and permittivity cause no significant effect on the earth-return impedance.

3.3.4.1.3 Capacitances

Figure 3.17 shows surface electric potentials in insulators. Figure 3.17(a) is for the cable inner insulator in Figure 3.15(b), when a source voltage V_o is applied to the cable core to ground. Figure 3.17(b) is the potential when V_o is applied to the sheath to ground, and Figure 3.17(c) is the case when V_o is applied between the core and the sheath. It is observed in Figure 3.17(a) that the inner insulator potential is not influenced by the HDPE tube. The HDPE tube significantly affects the potential when the source voltage is applied to the sheath as observed in Figure 3.17(b). It is clear that the capacitance calculated by COMSOL which considers the HDPE tube, the air gap, and the cable (metallic sheath) outer insulation is very small in comparison with that neglecting the tube and the air gap. This is physically reasonable because the total capacitance of the insulator composed of the sheath outer insulator, the air gap, and the HDPE tube is determined by the series connected capacitances and the capacitance of the air gap becomes dominant to determine the total capacitance. The calculated values are given in Table 3.15. It should be noted in Figure 3.17 that the equipotential surface follows the boundaries of the earth, the HDPE tube, the air gap, and the cable. It should be noted in Figure 3.17 that the equipotential surface nearly follows the boundaries of the earth, the HDPE tube, the air gap, and the cable.

3.3.4.2 Equivalent model for eccentric cable insulators

EMT-type simulations tools can handle cables of which the outer insulator is homogeneous (one layer only) and concentric with the core and sheath of the same phase. The modeling of a cable like that in Figure 3.15, which is laid on the bottom of an HDPE air-filled tube requires some modifications to the data used as input to the cable constants.

3.3.4.2.1 Equivalent radius

As explained in Figure 3.17, the equipotential surface is similar to the boundaries of the earth, the HDPE tube, the air gap, and the cable, and thus the concept of geometrical mean distance (GMD) can be adopted. It is known that the geometrical mean distance R between the point P (center of the cable) and the cross-section area of a circle with radius a illustrated in Figure 3.18(a) is defined by the following equation [85]:

$$\ln R = \ln a + (b/a)^2/2 - 1/2 = n, \; R = \exp n \tag{3.59}$$

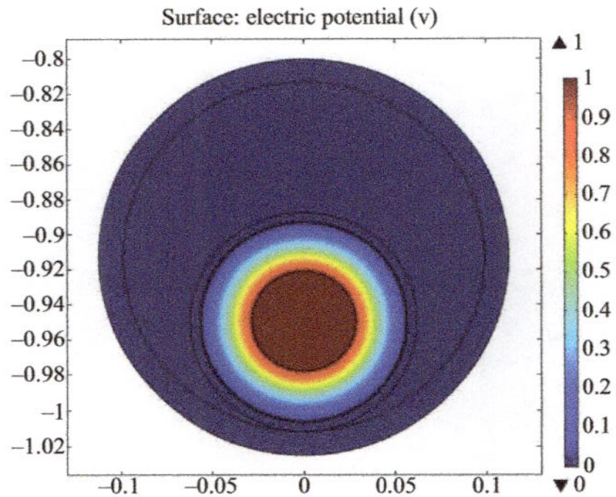

(a) Source applied to the core

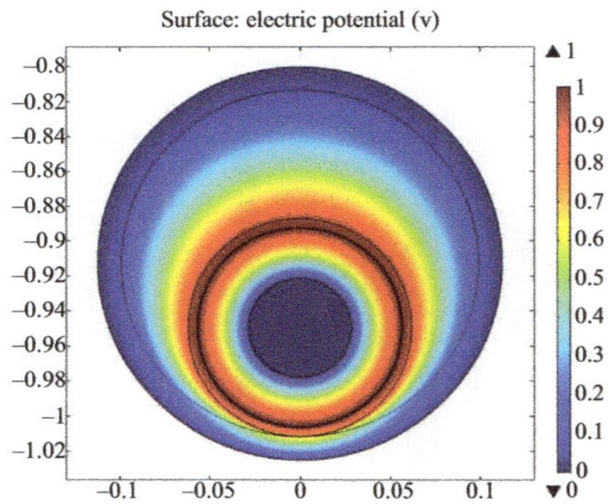

(b) Source applied to the sheath

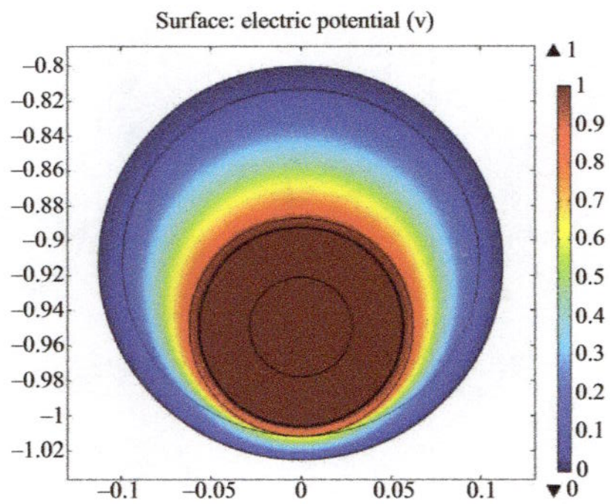

(c) Source applied between the core and sheath

Figure 3.17 Electric potential

Table 3.15 Capacitances calculated by COMSOL and cable constants

Cases		ε_a	r_0 [mm]	$C_{22}\left(\frac{\mu F}{km}\right)$
COMSOL	**Case 0**	**2.3/1.0**	**62.2**	**0.3625**
Cable constants	Case 1	2.3	$r_5 = 62.2$	1.7134
	Case 2	1.47	$D_2/2 = 112.5$	0.3074
	Case 31	0.329	62.2	0.4056
	Case 32	0.972	73.26	0.4050
	Case 33	1.293	79.3	0.4045
	Case 4	0.7	62.2	0.6512
	Case 5	1	62.2	0.8504
	Case 6	1	68.3	0.5002

$C_{11} = 0.1865$ (μF/km).

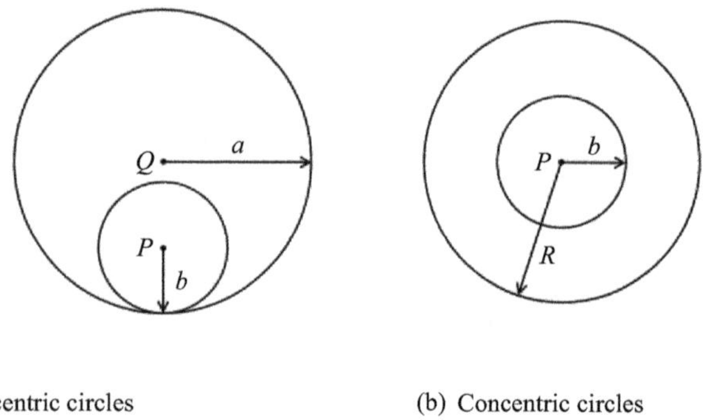

Figure 3.18 Equivalent radius R of a concentric circle to an eccentric circle with radius a

where b is the outer radius of the cable = nearest distance between P and the surface of the circle with radius a.

The outer radius of the air gap r_6 is calculated as follows:

$$r_6 = R_1 = \exp n_1 \tag{3.60}$$

$$n_1 = \ln(D_1/2) + (2r_5/D_1)^2/2 - 1/2 \tag{3.61}$$

where r_5 and D_1 are defined in Figure 3.15(b).

Now, the insulation layer composed of the air gap is a cylinder with inner and outer radii r_5 and r_6 concentric to the cable as illustrated in Figure 3.18(b). In the same manner, the outer radius r_7 of the HDPE tube is obtained by

$$r_7 = R_2 = \exp n_2 \tag{3.62}$$

$$n_2 = \ln(D_2/2) + (2r_5/D_2)^2/2 - 1/2 \tag{3.63}$$

In this case, the permittivity of the HDPE tube has to be modified so as to give the same capacitance as the original HDPE tube. That is

$$\frac{2\pi\varepsilon_0\varepsilon_3}{\ln(D_2/D_1)} = \frac{2\pi\varepsilon_0\varepsilon_3^{'}}{\ln(r_7/r_6)} \tag{3.64}$$

$$\therefore \varepsilon_3^{'} = [\ln(r_7/r_6)/\ln(D_2/D_1)] \cdot \varepsilon_3 \tag{3.65}$$

Now, we obtain concentric insulators composed of an HDPE tube, an air gap, and a cable (metallic sheath) outer insulator. Define the capacitances of the three insulators as C_H, C_a and C_s, respectively. Then, the total capacitance C_0 is calculated as the series connection of C_H, C_a, and C_s:

$$C_0 = (1/C_s + 1/C_a + 1/C_H)^{-1} \tag{3.66}$$

where $C_s = 2\pi\varepsilon_0\varepsilon_2/\ln\left(\frac{r_5}{r_4}\right)$, $C_a = 2\pi\varepsilon_0/\ln\left(\frac{r_6}{r_5}\right)$ and $C_H = 2\pi\varepsilon_0\varepsilon_3^{'}/\ln\left(\frac{r_7}{r_6}\right)$.

To keep the same impedance of the cable, it is better to keep the outer radius of the cable to be the same as the original one "r_5". Then, the following equation is to be satisfied:

$$C_0 = 2\pi\varepsilon_0\varepsilon_a/\ln\left(\frac{r_5}{r_4}\right) \tag{3.67}$$

Thus, we have to adopt the following equivalent permittivity ε_a:

$$\therefore \varepsilon_a = C_0 \ln\left(\frac{r_5}{r_4}\right)2\pi\varepsilon_0 \tag{3.68}$$

3.3.4.2.2 *Calculated results of capacitance*

Calculated results of the capacitance are given in Table 3.15. It is observed that neglecting the HDPE tube and the air gap, that is, directly buried cable (Case 1) gives a too large capacitance, while assuming the cable outer radius to be that of the HDPE tube (Case 2) gives a too small capacitance. Cases 31–33 are based on the theory in Section 3.3.4.2.1 and show a similar value to the capacitance calculated by COMSOL. The outer radius r_0 of the equivalent insulator in Case 31 is taken as the cable outer radius r_5 in Figure 3.15(b). Then, to keep the same capacitance C_0 in (3.67), the equivalent relative permittivity ε_a in (3.68) becomes smaller than unit because $\ln(r_5/r_4)$ is small. Such a permittivity is just artificial and is not realistic from the physical viewpoint. The same is often observed in physical parameters used for a simulation by a numerical electromagnetic analysis method [85]. If this is to be avoided, then we need to take a larger radius. Case 32 adopts the radius r_6 as r_0, and ε_a becomes 0.972. If r_7 is adopted as r_0, then ε_a is given as $\varepsilon_a = 1.293$. It should be noted that all the three cases give the same capacitance.

Cases 4 and 5 adopt the same radius $r_0 = r_5$, but the relative permittivity ε_a is changed to 0.7 in Case 4 and 1.0 in Case 5 which is larger than that in Case 31. Case 6 adopts $\varepsilon_a = 1$, but the radius r_0 is taken a little greater than r_5.

3.3.4.3 Transient simulation in comparison with the field test result

Figure 3.19 illustrates a test circuit of a 220 kV cable with a length 1,252 m. An impulse voltage of 1.2 μs/50 μs with an amplitude of about 2 kV is applied between phases *b* and *c* of the metallic sheath. All the other conductors are open-circuited.

3.3.4.3.1 Comparison of measured and simulation results

Figure 3.20 shows the measured results of the transient voltage on the phase-b core at the receiving end [82]. Figure 3.21 shows the EMT simulation results of the receiving-end core voltage by adopting six insulator models given in Table 3.15:

Case 0: COMSOL
Case 1: Neglecting the HDPE tube
Case 2: $\varepsilon_a = 1.47,\ r_0 = D_2/2$
Case 31: $\varepsilon_a = 0.329,\ r_0 = r_5$
Case 32: $\varepsilon_a = 0.972,\ r_0 = r_6$
Case 33: $\varepsilon_a = 1.293\ ,\ r_0 = r_7$
Case 4: $\varepsilon_a = 0.7,\ r_0 = r_5$
Case 5: $\varepsilon_a = 1,\ r_0 = r_5$
Case 6: $\varepsilon_a = 1,\ r_0 = 68.3$ mm.

It is observed in Figure 3.21 that the directly buried model (Case 1) shows a too small transient voltage, and thus it should not be used to represent a cable installed in an HDPE tube. Case 2 with the cable outer radius given as the HDPE tube outer radius shows a too high voltage in comparison with the measured result. Case 0 with COMSOL calculated capacitance shows not satisfactory compared with the measured result. Cases 31–33 show an identical result as it should be expected from the theory explained in Section 3.3.4.2.1. The results are better than that in Case 0, but still not satisfactory enough in comparison with the measured result which is in between the results given by Cases 3–6. If only the maximum voltage is concerned, Case 6 shows a satisfactory result.

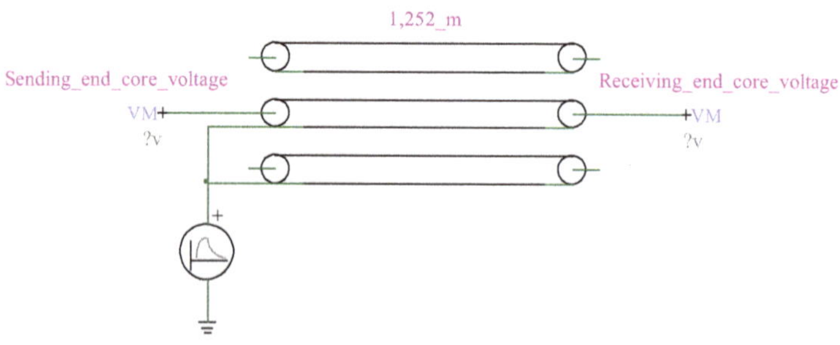

Figure 3.19 A test circuit of an underground cable with the cross-section of 2,000 mm²

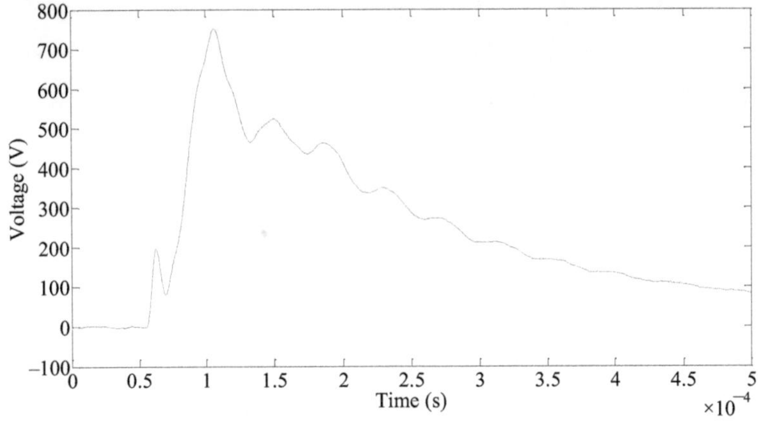

Figure 3.20 Measured results of phase-b core voltage at the receiving end

It should be noted that a test result in Figure 3.20 involves a number of unknown factors. For example, earth resistivity, grounding, etc., which is required as an input data to an EMT simulation.

3.3.3.4 Effect of HDPE tube

It has been shown that an equivalent insulator model to represent an HDPE tube, an air gap and a cable outer insulator gives a reasonable accuracy of a transient simulation on a cable installed in the HDPE tube in comparison with a measured result.

In this chapter, the effect of an HDPE tube, the outer and inner radii, on a transient waveform is investigated in comparison with that based on the capacitance evaluated by COMSOL.

3.3.3.4.1 Simulation conditions

Figure 3.22 illustrates a model circuit investigated. Because the mutual coupling between the phases is not influenced by the HDPE tube and the air gap, a single-phase cable is considered.

A step function voltage with the amplitude $E_0 = 1$ kV is applied to the metallic sheath at the sending end as in Figure 3.22. All the other conductors are open-circuited. The parameters of the HDPE tube and the cable investigated in this chapter are given in Table 3.16 together with the equivalent permittivity ε_a, equivalent radius r_0 and the self-capacitance of the metallic sheath C_{22} calculated from ε_a, r_4, and r_0 by the EMT-type tool and by COMSOL.

3.3.3.4.2 Simulation results

Figure 3.23 shows EMT simulation results of the receiving-end core voltage V_c. It is observed in Figure 3.23 that Case 12 shows the longest time delay (or traveling time) τ and the lowest oscillating frequency f which is given by $f = 1/4\tau$ except Case 10 which is directly buried. The reason for this is that the smallest HDPE tube

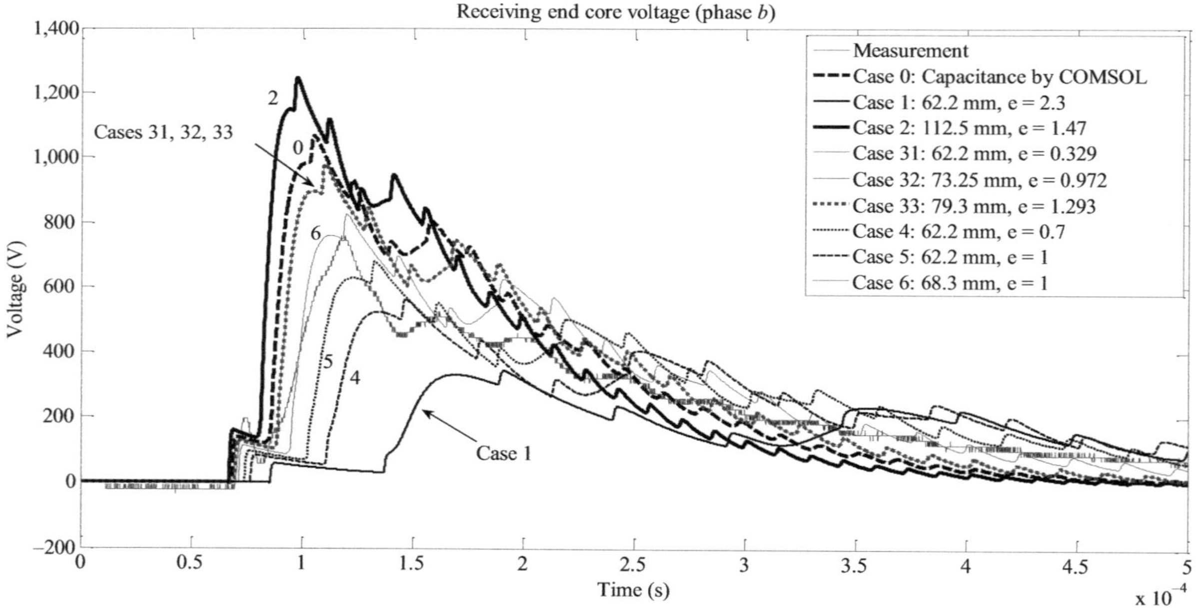

Figure 3.21 EMT simulation results of the receiving-end core voltage for the cases in Table 3.15

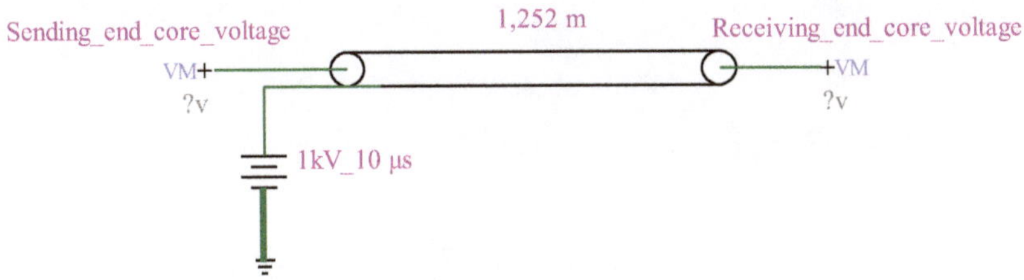

Figure 3.22 A model circuit of a cable with length x = 1,252 m

Table 3.16 Conditions of HDPE installed cable

Case	Cable dimensions				HDPE tube radius (mm)		EMT			COMSOL
	Size (mm^2)	Radius (mm)		ε_2	$D_1/2$	$D_2/2$	ε_a	r_0	C_{22}	C_{22}
		r_4	r_5							
10	2,000	57.2	62.2	2.3	–	–	2.30	62.2	1.7134	1.7134
11	2,000	57.2	62.2	2.3	99.25	112.5	1.29	79.5	0.4056	0.3625
							0.97	73.3		
							0.33	62.2		
12	2,000	57.2	62.2	2.3	65	75	1.14	64.2	0.7375	0.5959
							0.85	62.3		
							0.83	62.2		
13	2,000	57.2	62.2	2.3	135	150	1.19	99.1	0.3060	0.3292
							1.00	91.1		
							0.18	62.2		
14	2,000	57.2	62.2	3.5	99.25	112.5	1.36	79.5	0.4189	0.3737
							1.02	73.3		
							0.35	62.2		
21	2,500	60.7	65.47	2.3	99.25	112.5	1.30	80.83	0.4531	0.3929
							0.95	74.83		
							0.34	65.47		

results in the smallest air gap and thus, the effect of the air gap which reduces the capacitance becomes the smallest.

In fact, we can neglect the air gap in this case, and make the sheath outer insulator and the HDPE tube as one equivalent insulator because the permittivity of the sheath outer insulator is the same as that of the HDPE tube. Thus, the capacitance C_0 of the equivalent insulator with the outer radius r_0 is defined in the following equation:

$$C_0 = 2\pi\varepsilon_0\varepsilon_2/\ln\left(\frac{r_0}{r_4}\right) \tag{3.69}$$

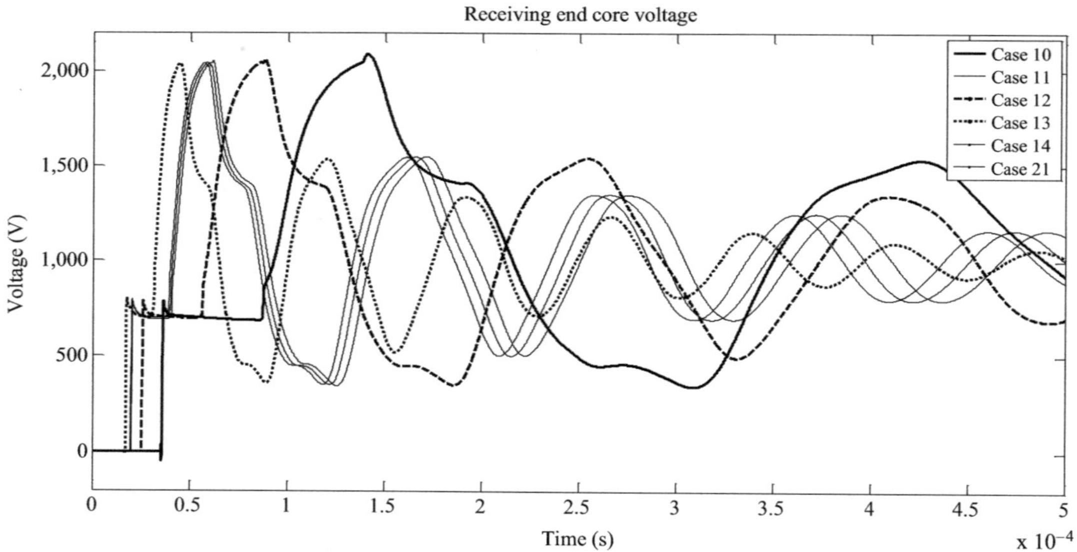

Figure 3.23 EMT simulation results of the receiving-end core voltage for the cases in Table 3.2

The capacitance C_s of the sheath outer insulator and C_H of the HDPE tube are calculated by

$$C_s = 2\pi\varepsilon_0\varepsilon_2/\ln\left(\frac{r_5}{r_4}\right), \; C_H = 2\pi\varepsilon_0\varepsilon_2/\ln\left(\frac{D_2}{D_1}\right) \tag{3.70}$$

Since C_0 is the total capacitance of the series-connected capacitances C_s and C_H, the following relation is obtained:

$$C_0 = (1/C_s + 1/C_H)^{-1} \tag{3.71}$$

Also, the unknown radius r_0 can be calculated.

$$\ln(r_0/r_4) = \left\{\ln\left(\frac{r_5}{/r_4}\right) + \ln\left(\frac{D_2}{D_1}\right)\right\}\left\{\ln\left(\frac{r_5}{r_4}\right) \cdot \ln\left(\frac{D_2}{D_1}\right)\right\} = m \tag{3.72}$$

$$r_0 = r_4 \cdot \exp(m) \tag{3.73}$$

For Case 12, the following values are obtained:

$$C_s = 1.525\ (\mu\text{F/km}),\ C_H = 0.892\ (\mu\text{F/km}),\ C_0 = 0.5628\ (\mu\text{F/km}) \tag{3.74}$$

$$m = 0.22706,\ r_0 = 57.2 \times \exp(0.22706) = 57.2 \times 1.255 = 71.8\ (\text{mm}) \tag{3.75}$$

Remind that the capacitance C_{22} is given as

$$C_{22} = C_0 + C_{11} = 0.5628 + 0.1865 = 0.7493\ (\mu\text{F/km}) \tag{3.76}$$

Contrary to Case 12, Case 13 with the largest HDPE tube shows the smallest time delay and correspondingly the highest oscillating frequency. The other three cases, Cases 11, 14, and 21 (same HDPE tube, different cable size, and permittivity ε_2), show no significant difference between each other. The greater permittivity of the sheath outer insulator (Case 14: $\varepsilon_2 = 3.5$) decreases the propagation velocity and results in the lower oscillating frequency in comparison with Case 11 ($\varepsilon_2 = 2.3$). The larger cable size (Case 21: 2,500 mm^2) shows a lower oscillating frequency than that in Case 11 (2,000 mm^2). The reason for this is estimated due to the impedance of the 2,500 mm^2 cable being different from that of the 2,000 mm^2 cable.

It should be noted that the peak voltages show no significant difference from each other, although the oscillating frequency is different. The peak voltage is heavily dependent on the attenuation constant which is determined dominantly by the resistance of the cable impedance.

3.3.3.4 Concluding remarks

This section has dealt with a transient simulation on a cable installed in a high-density poly-ethylene (HDPE) tube which is eccentric to the cable with an air gap. The impedance and the capacitance of the HDPE installed cable have been evaluated by COMSOL considering the real geometry of the system and by EMT-type simulation tools neglecting the HDPE tube and the air gap. No significant difference of the impedance is found between the calculated results by COMSOL and the EMT-type tools. On the contrary, the capacitance of the cable is found to be far smaller when the air gap and the tube are taken into account.

Then, an equivalent insulator to the HDPE tube, the air gap, and the cable outer insulator has been presented. The outer radius of the equivalent insulator is determined in a similar manner to the above, and the equivalent insulator becomes concentric to the cable. The equivalent permittivity is calculated from the total capacitance of the HDPE tube, the air gap and the cable sheath insulator. By adopting this equivalent insulator, the cable capacitance calculated by the EMT-type tools becomes nearly the same as that calculated by the COMSOL. A transient on an underground cable installed in the HDPE tube is simulated by the EMT-type tools with this model and the results show reasonable agreement with a measured result on the cable. A further improvement is necessary, but it requires an information on the unknown parameters such as the earth resistivity, grounding resistances of the impulse generator, and so on.

Finally, the effect of the HDPE tube and the air gap on a cable transient is examined and it has been made clear that the air gap affects the transient significantly because its permittivity is far smaller than that of the HDPE tube and the cable outer insulator.

3.3.5 Submarine cable

Many cables are installed in sea, either on the sea bed or under the surface of the sea bed, and are called "submarine cable." A basic difference between the submarine cable and a cable buried underground in soil is the resistivity. The soil resistivity ranges from 10 to 10,000 Ωm, while that of sea water is 0.1 to 0.2 Ωm. However, the depth of a submarine cable from the sea water surface, that is, 50 m to more than some hundreds m, is much greater than that of an underground cable from the earth surface, that is, 1–10 m in most cases. Thus, it is possible to assume that a submarine cable is laid down in an infinite uniform space of sea water. This makes the earth-return impedance much simpler.

3.4 Cable bonding

3.4.1 Bonding methods

Cross-bonding and grounding of sheaths are used to improve the performance of long HVAC cable systems by reducing sheath currents, thus improving cable ampacity, and by reducing sheath voltages for preventing breakdown of cable insulation. Transposition of cores may also be used to further reduce sheath currents [86].

Studies concerning cross-bonded cables were first published in the 1960s motivated by new cable projects and continued steadily in the 1980s [87–101]. Recently, interest toward transmission cables and particularly cross-bonded systems has been renewed [83,102–112]. This is related to new projects for the integration of renewable energies in the grid and for the replacement of overhead lines by underground cables [103–106].

The steady-state and time-domain responses of cross-bonded cables are more complex compared to no-bonding cables because the sheaths change phase throughout the cable route and because bonding points create complicated reflection phenomena. The first paper concerning cross-bonded cables used complicated equations and approximate lumped parameter models to calculate power frequency and transient responses [87–90]. Due to the complexity of calculations and limited computer resources at that time, studies were mainly limited to one major section and to the first incident wave [87–90].

To overcome the limitation in computer resources for the simulation of long cross-bonded systems, a homogeneous equivalent was derived for modeling of cross-bonded cables [91–93] (a comprehensive derivation of the equivalent is given in [99]). The homogeneous equivalent option is available in various EMT-type software and it is generally used for evaluating the propagation constants (i.e., attenuation and velocity) of cross-bonded cables. Despite its computational performance, the homogeneous equivalent model has several limitations for transient studies as it does not represent the reflections from bonding points nor the impedance of bonding wires and it cannot calculate sheath over-voltages within major sections. Recently the homogeneous equivalent was still used in frequency domain analysis [107], but accurate transient analysis studies opted for the discrete representation of each minor section [83,108–111].

This section presents a comprehensive analysis of the most common cable bonding methods illustrated in Figure 3.24 normal-bonding (NB) or no-bonding at all (i.e., submarine cables); cross-bonding (XB), which includes normal cross-bonding, that is, transposition of sheaths (i.e., in France, Denmark, and Japan); and transposed cable (TC), that is, transposition of cores (also called Kirke–Searing bonding, for example, in United Kingdom and Japan).

The impact of cross-bonding and sheath grounding has been studied in [83,108–111] in time-domain, but not in frequency domain. A frequency response analysis of cross-bonded cables are used in this section.

The exact-π circuit model can be used to accurately calculate the propagation constants (attenuation and velocity) of a homogeneous structure, that is, a minor section. It is demonstrated that the exact-pi circuit representation can be also used for non-uniform cross-bonded cables. Such calculations are useful for analysis of

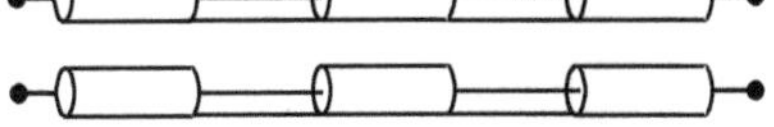

(a) Normal-bonded cable (NB)

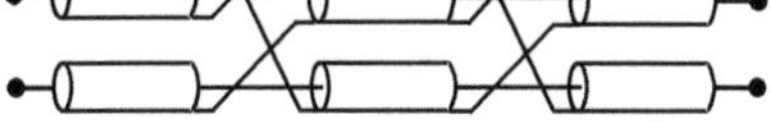

(b) Normal cross-bonded cable (XB)

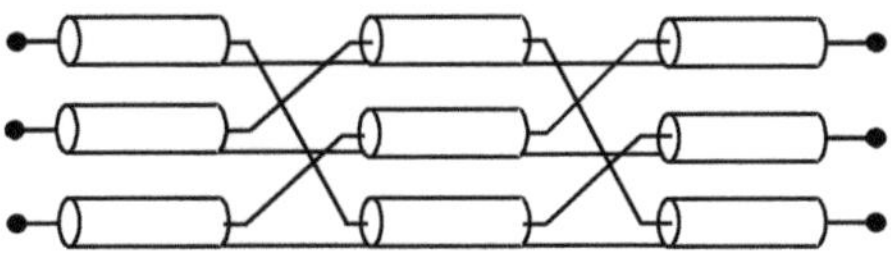

(c) Transposed cable (TC)

Figure 3.24 Cable bonding methods

cross-bonded cables and eliminate the need for finding a homogeneous equivalent for the studied system.

3.4.2 Various bonding methods

3.4.2.1 Definition of modal components

The modal components can be identified using modal analysis [68,69,76,107]. The relation between modal and phase currents is given by

$$\mathbf{I_{ph}} = \mathbf{T_i}\ \mathbf{I_m} \tag{3.77}$$

where $\mathbf{I_{ph}}$ is the vector of phase currents, $\mathbf{I_m}$ is the vector of modal currents, and $\mathbf{T_i}$ is the current transformation matrix. It is found from the calculation of eigenvalues $\mathbf{\Lambda}$ in

$$\mathbf{\Lambda} = \mathbf{T_i^{-1}}\ \mathbf{Y'}\ \mathbf{Z'}\ \mathbf{T_i} \tag{3.78}$$

where $\mathbf{Y'}$ is the per-unit-length shunt admittance matrix and $\mathbf{Z'}$ is the per-unit-length series impedance matrix. These matrices are calculated using the approach and assumptions presented in [98].

Equations (3.79)–(3.81) give the current transformation matrices ($\mathbf{T_i}$) at 1 MHz for the 225 kV cable given in Figure 3.25 and Table 3.17 [83] for the different bonding methods shown in Figure 3.24.

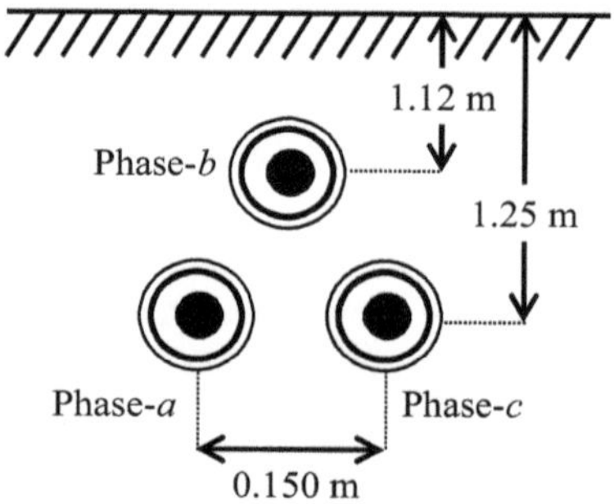

Figure 3.25 Layout of the 225 kV cable. Adapted from [83]

Table 3.17 Cable data. Adapted from [83]

Parameter	Value
Core conductor	$r_1 = 0$, $r_2 = 28.4$ mm, $\rho = 2.6 \times 10^{-8}\ \Omega$m
Core insulation	$\varepsilon_{r1} = 3.23$, $\tan\delta = 0.0008$
Metallic sheath	$r_3 = 56.4$ mm, $r_4 = 57.2$ mm, $\rho = 2.84 \times 10^{-8}\ \Omega$m
Outer insulation	$r_5 = 62.2$ mm, $\varepsilon_{r2} = 0.321$, $\tan\delta = 0.001$
Earth resistivity $\rho = 100\ \Omega - $m, sheath grounding $R_g = 5\ \Omega$ (by RTE)	

$$\mathbf{T}_{NB} = \begin{array}{l} \text{Core-a} \\ \text{Core-b} \\ \text{Core-c} \\ \text{Sheath-a} \\ \text{Sheath-b} \\ \text{Sheath-c} \end{array} \left[\begin{array}{c:c:c:c:c:c} 0 & 0.57 & -0.707 & -0.41 & 0 & 0 \\ 0 & 0.59 & 0 & 0.81 & 0 & 0 \\ 0 & 0.57 & 0.707 & -0.41 & 0 & 0 \\ 0.81 & -0.57 & 0.707 & 0.41 & 1 & -0.59 \\ 0.83 & -0.59 & 0 & -0.81 & 0 & 1.15 \\ 0.81 & -0.57 & -0.707 & 0.41 & -1 & -0.59 \end{array}\right] \tag{3.79}$$

$$\mathbf{T}_{XB} = \begin{array}{l} \text{Core-a} \\ \text{Core-b} \\ \text{Core-c} \\ \text{Sheath-a} \\ \text{Sheath-b} \\ \text{Sheath-c} \end{array} \left[\begin{array}{c:c:c:c:c:c} 0 & 0.58 & 0.71 & -0.41 & 0 & 0 \\ 0 & 0.58 & 0 & 0.82 & 0 & 0 \\ 0 & 0.58 & -0.71 & -0.41 & 0 & 0 \\ 0.82 & -0.58 & 0 & 0 & 0.72 & -0.25 \\ 0.82 & -0.58 & 0 & 0 & 0.03 & 0.83 \\ 0.82 & -0.58 & 0 & 0 & -0.75 & -0.58 \end{array}\right] \tag{3.80}$$

$$\mathbf{T}_{TC} = \begin{array}{l} \text{Core-a} \\ \text{Core-b} \\ \text{Core-c} \\ \text{Sheath-a} \\ \text{Sheath-b} \\ \text{Sheath-c} \end{array} \left[\begin{array}{c:c:c:c:c:c} 0 & 0.58 & -0.07 & 0.82 & 0 & 0 \\ 0 & 0.58 & -0.68 & -0.41 & 0 & 0 \\ 0 & 0.58 & 0.75 & -0.41 & 0 & 0 \\ 0.81 & -0.58 & 0 & 0 & -0.71 & -0.41 \\ 0.83 & -0.58 & 0 & 0 & 0 & 0.82 \\ 0.81 & -0.58 & 0 & 0 & 0.71 & -0.41 \end{array}\right] \tag{3.81}$$

The subscripts *NB*, *XB*, and *TC* are used for normal-bonded, cross-bonded, and transposed cores, respectively. These matrices are calculated using 18 major sections of 3 km (with three minor sections of 1 km). The homogeneous model [69,101] is used for (3.79) to (3.81). The bonding wire impedances and grounding resistances are zero. The frequency of 1 MHz is used because the modes are nearly constant at this frequency and thus they are more easily identified. Each column of $\mathbf{T}_i$ represents an eigenvector: column 1 is a zero-sequence sheath mode and column 2 is a zero-sequence coaxial mode. Columns 3 and 4 are inter-core modes (except for $\mathbf{T}_{NB}$) and columns 5 and 6 correspond to inter-sheath modes.

It is clear from columns 3 and 4 in (3.79) that the inter-core modes in the NB cable are in fact a set of single-phase coaxial modes because the value for a phase core appears with an opposite sign for the same phase sheath. This is not observed

Table 3.18 Propagation constant for various bonding at $f = 1$ MHz

Attenuation (dB/km)						
Mode	**1**	**2**	**3**	**4**	**5**	**6**
NB	222	0.65	0.65	0.65	0.89	0.78
XB	222	0.65	0.88	0.83	0.95	0.95
TC	222	0.65	0.83	0.83	1.07	0.95
Velocity (m/μs)						
Mode	**1**	**2**	**3**	**4**	**5**	**6**
NB	28.7	167	167	167	101	120
XB	28.7	167	108	119	89.1	89.1
TC	28.7	167	113	113	98.0	82.2

in (3.80) and (3.81). The reason is that in XB cables, the sheaths change phase at every minor section and it becomes impossible to have a single-phase coaxial mode propagating in the cable.

Table 3.18 gives the attenuations and velocities calculated at 1 MHz for the given bonding methods. Numbers 1 to 6 describe the different modes and correspond to the columns of the matrices in (3.79)–(3.81).

It is observed that zero-sequence sheath and core modes (modes 1 and 2) do not depend on the bonding method.

The NB cable has three pure coaxial modes (modes 2, 3, and 4) characterized by the same attenuation and same velocity found $c/\sqrt{\varepsilon_{r1}} = 300/\sqrt{3.23} = 167$ m/μs with c being the speed of light in vacuum and ε_{r1} the insulation relative permittivity given in Table 3.17. For XB and TC cases, modes 3 and 4 are inter-core modes, slower, and more attenuated than the zero-sequence coaxial mode 2. The TC has two equal inter-core modes because of core transposition. The XB has instead two equal inter-sheath modes due to transposition of sheaths.

3.5 Numerical electromagnetic analysis [113,114]

It is hard to handle a transient associated with non-TEM mode propagation by conventional circuit-theory-based approaches such as the EMT-type simulation tools because the tools are based on TEM mode propagation. To overcome the problem, a numerical electromagnetic analysis (NEA) method looks most promised among existing transient analysis approaches as it solves Maxwell's equation directly without assumptions often made for the circuit-theory based approach.

The NEA method has been widely used to solve surge phenomena in power systems such as an induced lightning surge, a surge response of a grounding electrode, and corona pulse propagation on a gas-insulated bus which are very hard to solve by an existing circuit-theory based tool, because those surge phenomena involve TM (transverse magnetic) and TE (transverse electric) modes of wave propagation rather than TEM mode propagation. Also, the circuit parameters, which are necessary to analyze a phenomenon by the circuit-theory-based tool, are quite often not clear or not available. Because the numerical electromagnetic analysis method solves Maxwell's equation directly based on the boundary conditions and the physical and geometrical parameters of a surge phenomenon to be solved, it requires no circuit parameter. Also, the method can solve the TE, TM, and TEM mode propagation without any difficulties.

This section describes the basic theory of two representative methods, that is, method of moment (MoM) and finite-difference time-domain (FDTD) method, of the numerical electromagnetic analysis. Also, a software based on a finite element method (FEM) is briefly explained.

Table 3.19 categorizes various methods of numerical electromagnetic analysis (NEA) which are presently used. Appendix A3.5 summarizes Maxwell's equations related to the NEA.

The method of moments (MoM) in the frequency and time domains and the finite-difference time-domain (FDTD) method, both for solving Maxwell's equations numerically, have frequently been used in calculating surges on power systems. Applications of the transmission matrix method (TLM) to surge calculations are rare at present.

3.5.1 Method of moments (MoM) in time and frequency domains

3.5.1.1 MoM in the time domain

The MoM in the time domain [115,116] is widely used in analyzing responses of thin-wire metallic structures to external time-varying electromagnetic fields. The entire conducting structure representing the lightning channel is modeled by a combination of cylindrical wire segments whose radii are much smaller than the wavelengths of interest. The so-called electric-field integral equation for a perfectly conducting thin wire in air as in .

Assuming that current and charge q are confined to the wire axis (thin-wire approximation) and that the boundary condition on the tangential electric field on the surface of the wire (this field must be equal to zero) is fulfilled, is given by

$$\hat{s} \cdot E_{inc}(r,t) = \frac{\mu_0}{4\pi} \int_C \left[\frac{\hat{s} \cdot \hat{s}'}{R} \frac{\partial I(s',t')}{\partial t'} + c \frac{\hat{s} \cdot R}{R^2} \frac{\partial I(s',t')}{\partial s'} - c^2 \frac{\hat{s} \cdot R}{R^3} q(s',t') \right] ds' \quad (3.82)$$

where

$$q(s',t') = -\int_{-\infty}^{t'} \frac{\partial I(s',\tau)}{\partial s'} d\tau \quad (3.83)$$

C is an integration path along the wire axis, $\boldsymbol{E}_{\mathbf{inc}}$ denotes the incident electric field that induces current I, $\boldsymbol{R}=\boldsymbol{r}-\boldsymbol{r}'$, $\boldsymbol{r}$ and t denote the observation location (a point on the wire surface) and time, respectively, $\boldsymbol{r}'$ and t' denote the source location (a point on the wire axis) and time, respectively, s and s' denote the distance along the wire surface at $\boldsymbol{r}$ and that along the wire axis at $\boldsymbol{r}'$, $\hat{s}$ and $\hat{s}'$ denote unit vectors

Table 3.19 Various methods of numerical electromagnetic analysis

Partition	**Space**					**Boundary**
Discretization/ domain	**Finite difference**		**Boundary length**		**Finite element**	
Time-domain	FDTD	TD-FI	3D circuit	TLM	TD-FEM	MOM (TWTDA)
Frequency	–	FI	–	–	FEM	MOM
Base equation	Maxwell differential	Maxwell integral	Maxwell characteristic	D'Alembert solution		Field integral
Feature	Easy programming	Multi media	Circuit theory extension Hard program	Easy program	Wide application	Small CPU nonlinear in Time domain

tangent to path $\boldsymbol{C}$ in Maxwell's time-domain equation at $\boldsymbol{r}$ and $\boldsymbol{r'}$, μ_0 is the permeability of vacuum, and c is the speed of light. Through numerically solving the Maxwell's equation, the time-dependent current distribution along with the wire structure (lightning channel), excited by a lumped source, is obtained.

The thin-wire time-domain (TWTD) code [115] (available from the Lawrence Livermore National Laboratory) is based on the MoM in the time domain. One of the advantages of the use of the time-domain MoM is that it can incorporate nonlinear effects such as the lightning attachment process [117], although it does not allow lossy ground and wires buried in the lossy ground to be incorporated.

3.5.1.2 MoM in the frequency domain

The MoM in the frequency domain [118] is widely used in analyzing the electromagnetic scattering by antennas and other metallic structures. In order to obtain the time-varying responses, Fourier and inverse Fourier transforms are employed. The electric-field integral equation derived for a perfectly conducting thin wire in the air as in Figure 3.26 in the frequency domain is given by

$$-\hat{s} \cdot E_{inc}(r) = \frac{j\eta}{4\pi k} \int_C I(s') \left(k^2 \hat{s} \cdot \hat{s}' - \frac{\partial^2}{\partial s \partial s'} \right) g(r, r') ds' \tag{3.84}$$

where

$$g(r, r') = \exp\left(\frac{-jk|r - r'|}{|r - r'|} \right), \quad k = \omega\sqrt{\mu_0 \varepsilon_0}, \quad \eta = \sqrt{\frac{\mu_0}{\varepsilon_0}} \tag{3.85}$$

and ω is the angular frequency, μ_0 is the permeability of vacuum, and ε_0 is the permittivity of vacuum. Other quantities in the above equation are the same as those in (3.82). Current distribution along the lightning channel can be obtained numerically solving (3.84).

This method allows lossy ground and wires in lossy ground (e.g., grounding of a tall strike object) to be incorporated into the model. The commercially available numerical electromagnetic codes [119,120] are based on the MoM in the frequency domain.

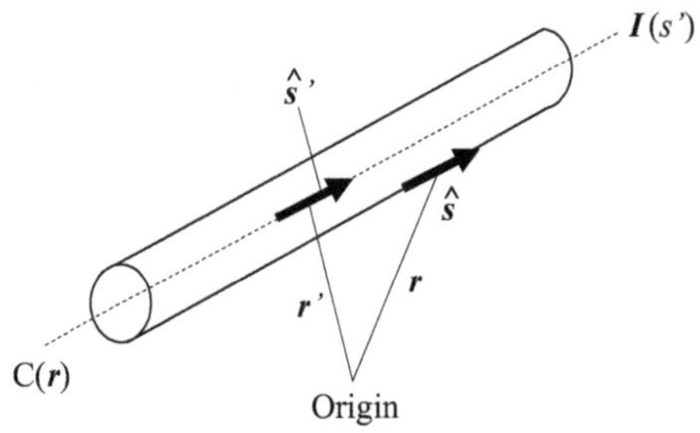

Figure 3.26 Thin-wire segment for MoM-based calculations. Current is confined to the wire axis, and the tangential electric field on the surface of the wire is set to zero

3.5.2 Finite-difference time-domain (FDTD) method

The FDTD method [121] employs a simple way to discretize Maxwell's equations in differential form. In the Cartesian coordinate system, it requires discretization of the entire space of interest into small cubic or rectangular-parallelepiped cells. Cells for specifying or computing electric field (electric field cells) and magnetic field cells are placed relative to each other as shown in Figure 3.27. Electric and magnetic fields of the cells are calculated using the discretized Maxwell's equations given below:

$$E_z^n\left(i,j,k+\frac{1}{2}\right)=\frac{1-\sigma(i,j,k+1/2)\Delta t/[2\varepsilon(i,j,k+1/2)]}{1+\sigma(i,j,k+1/2)\Delta t/[2\varepsilon(i,j,k+1/2)]}\times E_z^{n-1}\left(i,j,k+\frac{1}{2}\right)$$
$$+\frac{\Delta t/\varepsilon(i,j,k+1/2)}{1+\sigma(i,j,k+1/2)\Delta t/[2\varepsilon(i,j,k+1/2)]}\frac{1}{\Delta x\Delta y}$$
$$\times\begin{bmatrix} H_y^{n-\frac{1}{2}}(i+1/2,j,k+1/2)\Delta y-H_y^{n-\frac{1}{2}}(i-1/2,j,k+1/2)\Delta y \\ -H_x^{n-\frac{1}{2}}(i,j+1/2,k+1/2)\Delta x+H_x^{n-\frac{1}{2}}(i,j-1/2,k+1/2)\Delta x \end{bmatrix} \tag{3.86}$$

$$H_x^{n+\frac{1}{2}}\left(i,j-\frac{1}{2},k+\frac{1}{2}\right)=H_x^{n-\frac{1}{2}}\left(i,j-\frac{1}{2},k+\frac{1}{2}\right)+\frac{\Delta t}{\mu(i,j-1/2,k+1/2)}\frac{1}{\Delta y\Delta z}$$
$$\times\begin{bmatrix} -E_z^n(i,j,k+1/2)\Delta z+E_z^n(i,j-1,k+1/2)\Delta z \\ +E_y^n(i,j-1/2,k+1)\Delta y-E_y^n(i,j-1/2,k)\Delta y \end{bmatrix} \tag{3.87}$$

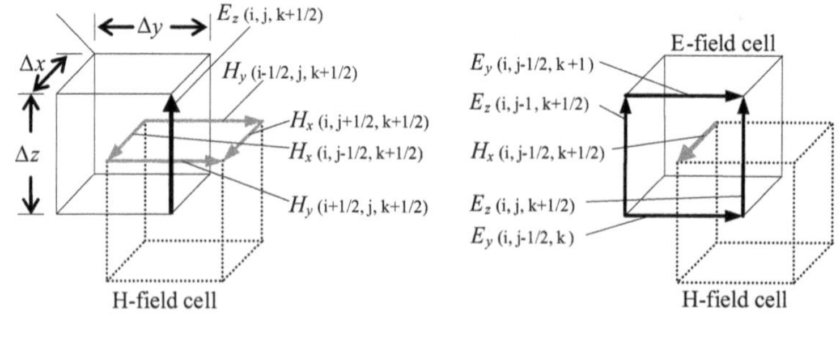

Figure 3.27 Placement of electric-field and magnetic-field cells for solving discretized Maxwell's equations using the FDTD method

Equation (3.86), which is based on Ampere's law, is an equation updating z component of electric field, $E_z(i, j, k+1/2)$, at point $x = i\Delta x$, $y = j\Delta y$, and $z = (k+1/2)\ \Delta z$, and at time $t = n\Delta t$. Equation (3.87) which is based on Faraday's law, is an equation updating x component of magnetic field, $H_x(i, j-1/2, k+1/2)$, at point $x = i\Delta x$, $y = (j-1/2)\ \Delta y$, and $z = (k+1/2)\ \Delta z$, and at time $t = (n+1/2)\ \Delta t$. Equations updating x and y components of electric field, and y and z components of magnetic field can be written in a similar manner. Note that $\sigma(i, j, k+1/2)$ and $\varepsilon(i, j, k+1/2)$ are the conductivity and permittivity at point $x = i\Delta x$, $y = j\Delta y$, and $z = (k+1/2)\ \Delta z$, respectively, $\sigma\ (i, j-1/2, k+1/2)$ is the permeability at point $x = i\Delta x$, $y = (j-1/2)\ \Delta y$, and $z = (k+1/2)\ \Delta z$. By updating electric and magnetic fields at every point using (3.86) and (3.87), transient fields throughout the computational domain are obtained. Since the material constants of each cell can be specified individually, a complex inhomogeneous medium can be analyzed easily.

In order to analyze fields in unbounded space, an absorbing boundary condition has to be set on each plane which limits the space to be analyzed, so as to avoid reflections there. The FDTD method allows one to incorporate wires buried in the lossy ground, such as strike-object grounding electrodes, and nonlinear effects [122].

3.5.3 Finite element method (FEM)

3.5.3.1 Basic theory of FEM

The finite element method (FEM) is the most widely used technique for solving complex problems of engineering and mathematical models [84]. Typical domains of problems include the traditional fields of structural analysis, heat transfer, fluid flow, mass transport, and electromagnetic characteristics. The FEM is a particular numerical method for solving partial differential equations in two or three space variables (i.e., some boundary value problems) [84]. In order to solve the above-mentioned problems, a large system is subdivided into smaller parts by the FEM, which are called finite elements. A particular space discretization is performed by the construction of a mesh of the object [84]. The FEM formulation of a boundary value problem results in a set of algebraic equations. The simple equations represented by finite elements are assembled and used to model the entire problem with a larger equation system.

3.5.3.2 COMSOL multiphysics

A commercial software package COMSOL has the capability of dealing with three-dimensional (3D) problems. COMSOL offers a variety of modules as graphical interface environments to solve the problem according to its physical nature. Design and simulation of electromagnetic models in COMSOL can be performed using the "AC/DC module." Using this module, static, quasi-static, and transient simulations can be carried out. The cable impedance matrix $\mathbf{Z}$ is calculated by [83]

$$Z_{ij} = \frac{Js_i}{I_j\sigma_i}(i\ =\ 1,2,\ \ldots\ N) \tag{3.88}$$

where Js is the current density, I is the current, σ is the conductor conductivity, and N is the the number of conductors.

J and I in a given conductor are solved in the magnetic and electric fields (MEF) platform.

The admittance matrix **Y** can be found by simulating the cable in the Electrostatics (ES) platform. For the parameters related to the conductance of a cable are not clear, for example, "tanδ (loss factor)" in a high-frequency corresponding to a transient, the conductance is neglected. Then, the admittance is simplified to the capacitance matrix **C**. The energy method is employed to calculate the matrix **C**, which is based on the evaluation of electric energy stored in insulators. The elements of matrix **C** are independent of frequency and of the cable depth under the ground [83]. The self-capacitance C_{ii} is evaluated in the following equation, when a potential V_o is applied to the surface of the conductor while the other conductors are grounded:

$$C_{ii} = 2W_{ii}/V_o^2 \tag{3.89}$$

where W_{ii} is the stored electric energy.

The mutual capacitance between two conductors is calculated by applying a potential V_o to the surfaces of both conductors, while the other conductors are grounded. The following equation is used:

$$W_{ij} = \frac{1}{2}\left(C_{ii} + C_{jj} + 2C_{ij}\right)V_o^2 \tag{3.90}$$

In the case of a cable composed of a core and a metallic sheath (screen or shield), (3.89) and (3.90) can be combined in a matrix form as follow:

$$\begin{bmatrix} 1 & 0 & 0 \\ 0 & 1 & 0 \\ 1 & 1 & 2 \end{bmatrix} \begin{bmatrix} C_{11} \\ C_{22} \\ C_{12} \end{bmatrix} = \frac{2}{V_o^2} \begin{bmatrix} W_{11} \\ W_{22} \\ W_{12} \end{bmatrix} \tag{3.91}$$

3.6 Conclusions

In this chapter, Section 3.1 describes different models of overhead lines and underground cables prepared in the existing EMT-type simulation tools. The equivalent π-circuit model, constant parameters model, frequency-dependent line models in modal, and phase domains are also reviewed and introduced.

Section 3.2 reviews the existing frequency-dependent soil models. Also, the calculated results of FD soil resistivity and permittivity are investigated and compared with independently measured data.

Section 3.3 explains how to model a cable installed underneath a steel bridge, a tower-installed cable, a tunnel-installed cable, a cable installed in an HDPE tube, and a submarine cable. The transient characteristics and performances of different cable systems are investigated and made clear.

Section 3.4 introduces typical bonding methods for underground cables. Various bonding methods are described and differences between the bonding methods are also made clear.

Numerical electromagnetic analysis (NEA) methods such as FEM, MoM, and FDTD are explained and reviewed in Section 3.5. The NEA methods are powerful and useful when the impedance and admittance of a cable are hardly calculated by the existing EMT-type simulation tools, for example, a phenomenon contains a full-wave mode of electromagnetic propagation and also a three-dimensional structure.

Appendix A3.1 Maxwell's equations

There are two types of equations: one is in the time domain and the other in the frequency domain.

In the time domain, Maxwell's equations are expressed as

$$\nabla \times \boldsymbol{E}(\boldsymbol{r}, t) = -\frac{\partial \boldsymbol{B}(\boldsymbol{r}, t)}{\partial t} \tag{A3.1}$$

$$\nabla \times \boldsymbol{H}(\boldsymbol{r}, t) = \frac{\partial \boldsymbol{D}(\boldsymbol{r}, t)}{\partial t} + \boldsymbol{J}(\boldsymbol{r}, t) \tag{A3.2}$$

$$\nabla \cdot \boldsymbol{D}(\boldsymbol{r}, t) = \rho(\boldsymbol{r}, t) \tag{A3.3}$$

$$\nabla \cdot \boldsymbol{B}(\boldsymbol{r}, t) = 0 \tag{A3.4}$$

where $\boldsymbol{E}(\boldsymbol{r},t)$ is the electric field and $\boldsymbol{H}(\boldsymbol{r},t)$ is the magnetic field, $\boldsymbol{D}(\boldsymbol{r},t)$ is the electric flux density, $\boldsymbol{B}(\boldsymbol{r},t)$ is the magnetic flux density, $\boldsymbol{J}(\boldsymbol{r},t)$ is the conduction current density, and $\rho(\boldsymbol{r},t)$ is the volume charge density, each at the space point $\boldsymbol{r}$ and at the time t.

Equation (A3.1) represents Faraday's law, and Equation (A3.2) represents Ampere's law. Equations (A3.3) and (A3.4) represent Gauss's law for electric and magnetic fields, respectively.

In the frequency domain, Maxwell's equations are expressed as

$$\nabla \times \boldsymbol{E}(\boldsymbol{r}) = -j\omega B\,(\boldsymbol{r}) \tag{A3.5}$$

$$\nabla \times \boldsymbol{H}(\boldsymbol{r}) = \boldsymbol{J}(\boldsymbol{r}) + \mathrm{j}\omega D(\boldsymbol{r}) \tag{A3.6}$$

$$\nabla \cdot \boldsymbol{D}(\boldsymbol{r}) = \rho(\boldsymbol{r}) \tag{A3.7}$$

$$\nabla \cdot \boldsymbol{B}(\boldsymbol{r}) = 0 \tag{A3.8}$$

where j is the imaginary unit and ω is the angular frequency.

The relations of the electric and magnetic flux densities to the electric and magnetic fields are given as

$$D(r) = \varepsilon E(r) = \varepsilon_r \varepsilon_0 E(r) \tag{A3.9}$$

$$B(r) = \mu H(r) = \mu_r \mu_0 H(r) \tag{A3.10}$$

where ε_0 is the permittivity of the vacuum $= 8.854 \times 10^{-12}$ F/m, μ_0 is the permeability of the vacuum $= 4\pi \times 10^{-7}$ H/m, ε_r is the relative permittivity of the medium, μ_r is the relative permeability of the medium, ε is the permittivity of the medium, and μ is the permeability of the medium.

In a conductive medium, the following relation of the electric field to the conduction current density, Ohm's law, is fulfilled:

$$J(r) = \sigma E(r) \tag{A3.11}$$

where σ is the conductivity of the medium.

References

[1] Dommel, H. W. "Digital computer solution of electromagnetic transients in a single- and multiphase networks." *IEEE Transactions on Power Apparatus and Systems*. 1969; 88(4): 388–98

[2] Dommel, H. W., and Scott-Meyer, W. S. "Computation of electromagnetic transients." *Proceedings of the Institute of Electrical and Electronics Engineers*. 1974; 62(7): 983–93

[3] Scott-Meyer, W. *EMTP rule book*. Portland: Bonneville Power Administration; 1982.

[4] Dommel, H. W. *EMTP theory book*. Portland: Bonneville Power Administration; 1986.

[5] Tinney, W. F., and Walker, J. W. "Direct solution of sparse network equations by optimally ordered triangular factorization." *Proceedings of the Institute of Electrical and Electronics Engineers*. 1967; 55(11): 1801–09

[6] Tinney, W. F., and Meyer, W. S. "Solution of large sparse systems by ordered triangular factorization." *IEEE Transactions on Automatic Control*. 1973; 18(4): 333–46

[7] Dube, L., and Dommel, H. W. "Simulation of control systems in an electromagnetic transients program with TACS." *IEEE Power and Energy Society Conference Record*. 1977; 10: 266–71

[8] Ametani, A., and Nagaoka, N. *EMTP test cases*. Japan: Japanese EMTP Committee; 1987.

[9] European EMTP User Group. *ATP rule book*. EEUG; 2007.

[10] Manitoba HVDC Research Centre. *User's guide on the use of PSCAD*. Winnipeg: Manitoba HVDC Research Centre; 2004.

[11] EPRI/DCG. "EMTP Ver. 02 Revised source code documentation." *EMTP Review*. 1990; 4.

[12] Mahseredjian, J., Dennetière, S., Dubé, L., Khodabakhchian, B., and Gérin-Lajoie, L. "On a new approach for the simulation of transients in power systems." *Electric Power Systems Research*. 2007; 77(11): 1514–20

[13] Ametani, A. (eds). *Numerical analysis of power system transients and dynamics*. London: IET; 2015.

[14] Ametani, A. "Electromagnetic transients program: History and future." *IEEJ Transactions on Electrical and Electronic Engineering*. 2020; doi:10.1002/tee.23192

[15] Central Research Institute of Electric Power Industry. *Introduction of eXpandable Transient Analysis Program*. https://www.xtap.jp/.

[16] Working Group of Institute of Electrical Engineers of Japan (chair Ametani A.). *Numerical electromagnetic analysis method for surge phenomenon*. Tokyo: Institute of Electrical Engineers of Japan; 2007

[17] CIGRE Working Group C4.501 (convenor Ametani A.). *Guideline for Numerical Electromagnetic Analysis Method and its Application to Surge Phenomena*. Paris: CIGRE Technical Brochure 543; 2013.

[18] Van Baricum, M., and Miller, E. K. *TWTD–A computer program for time-domain analysis of thin-wire structures*. California: Lawrence Livermore Laboratory UCRL-51-277; 1972.

[19] Miller, E. K., Poggio, A. J., and Burke, G. J. "An integro-differential equation technique for the time-domain analysis of thin wire structures." *Journal of Computational Physics*. 1973; 12(1): 24–48

[20] Podgorski, A. S., and Landt, J. A. "Three dimensional time domain modelling of lightning." *IEEE Transactions on Power Delivery*. 1987; 2(3): 931–38

[21] Harrington, R. F. *Field computation by moment methods*. New York, NY: Macmillan Co.; 1968.

[22] Burke, G. J., and Poggio, A. J. "Numerical Electromagnetic Code (NEC)–Method of Moments." San Diego: Naval Ocean Systems Center Technical Document 116; 1980.

[23] Burke, G. J. *Numerical Electromagnetic Code (NEC-4) - Method of Moments*. California: Lawrence Livermore National Laboratory UCRL-MA-109338; 1992.

[24] Yee, K. S. "Numerical solution of initial boundary value problems involving Maxwell's equations in isotropic media." *IEEE Transactions on Antennas and Propagation*. 1966; 14(3): 302–07

[25] Schnyder, O. "Druckstosse in pumpensteigleitungen." *Schweizerische Bauzeitung*. 1929; 94 (22): 271–86

[26] Bergeron, L. "Etude des variations de regime dans les conduites d'eau: Solution graphique generale." *Revue Générale de l'Hydraulique*. 1935; 1 (1): 12–25

[27] Parmakian, J. Water hammer analysis. Dover; 1963.

[28] Frey, W., and Althammer, P. "The calculation of electromagnetic transients on lines by means of a digital computer." *Brown Boveri Review*. 1961; 48(5/6): 344–55

[29] Marti, J. R. "Accurate modelling of frequency-dependent transmission lines in electromagnetic transient simulations." *IEEE Transactions on Power Apparatus and Systems*. 1982; 101(1): 147–55

[30] Ametani, A., and Tanaka, A. "Wave propagation characteristics on an un-transposed vertical twin-circuit line." *IEEJ Transactions on Power and Energy*. 1981; 101-B(11): 675–682

[31] Semlyen, A., and Dabuleanu, A. "Fast and accurate switching transient calculations on transmission lines with ground return using recursive convolution." *IEEE Transactions on Power Apparatus and Systems*. 1975; 94 (2): 561–71

[32] Ametani, A. "A highly efficient method for calculating transmission line transients." *IEEE Transactions on Power Apparatus and Systems*. 1976; 95 (5): 1545–51

[33] Ametani, A. *Refraction Coefficient Theory and Surge Phenomena in Power Systems*. University of Manchester. Ph.D. Thesis; 1973.

[34] Ametani, A. "Refraction coefficient method for switching surge calculations on un-transposed transmission line." Proceedings of the IEEE Summer Meeting. 1973; C73: 444–47

[35] Marti, L. "Simulation of transients in underground cables with frequency dependent modal transformation matrices." *IEEE Transactions on Power Delivery*. 1988; 3(3): 1099–1110

[36] Nakanishi, N., and Ametani, A. "Transient calculation of a transmission line using superposition law." *IEE Proceedings C–Generation, Transmission and Distribution. Part-C*. 1986; 133(5): 263–69

[37] Noda, T., Nagaoka, N., and Ametani, A. "Phase domain modeling of frequency-dependent transmission lines by means of an ARMA model." *IEEE Transactions on Power Delivery*. 1996; 11(1): 401–11

[38] Gustavsen, B., Sletbak, J., and Henriksen, T. "Calculation of electromagnetic transients in transmission cables and lines taking frequency dependent effects accurately into account." *IEEE Transactions on Power Delivery*. 1995; 10(2): 1076–84

[39] Nguyen, H. V., Dommel, H. W., and Marti, J. R. "Direct phase-domain modelling of frequency-dependent overhead transmission lines." *IEEE Transactions on Power Delivery*. 1997; 12(3): 1335–42

[40] Gustavsen, B., and Semlyen, A. "Combined phase and modal domain calculation of transmission line transients based on vector fitting." *IEEE Transactions on Power Delivery*. 1998; 13(2): 596–04

[41] Angelidis, G., and Semlyen, A. "Direct phase-domain calculation of transmission line transients using two-sided recursions." *IEEE Transactions on Power Delivery*. 1995; 10(2): 941–49

[42] Morched, A., Gustavsen, B., and Tartibi, M. "A universal model for accurate calculation of electromagnetic transients on overhead lines and underground cables." *IEEE Transactions on Power Delivery*. 1999; 14(3): 1032–38

[43] Gustavsen, B. "Time delay identification for transmission line modeling." *Signal Propagation on Interconnects, Proceedings*. 2004; 103–06

[44] Kocar, I., and Mahseredjian, J. "New procedure for computation of time delays in propagation function fitting for transient modeling of cables." *IEEE Transactions on Power Delivery*. 2016; 31(2): 613–21

[45] Gustavsen, B., and Nordstrom, J. "Pole identification for the universal line model based on trace fitting." *IEEE Transactions on Power Delivery*. 2008; 23(1): 472–79

[46] Gustavsen, B. "Passivity enforcement for transmission line models based on the method of characteristics." *IEEE Transactions on Power Delivery*. 2008; 23(4): 2286–93

[47] Silva, H. M. J. D., Gole, A. M., Nordstrom, J. E., and Wedepohl, L. M. "Robust passivity enforcement scheme for time-domain simulation of multi-conductor transmission lines and cables." *IEEE Transactions on Power Delivery*. 2010; 25(2): 930–38

[48] Semlyen, A., and Gustavsen, B. "Phase-domain transmission-line modeling with enforcement of symmetry via the propagated characteristic admittance matrix." *IEEE Transactions on Power Delivery*. 2012; 27(2): 626–31

[49] Ramos-Leanos, O., Naredo, J. L., Mahseredjian, J., Dufour, C., Gutierrez-Robles, J. A., and Kocar, I. "A wideband line/cable model for real-time simulations of power system transients." *IEEE Transactions on Power Delivery*. 2012; 27(4): 2211–18

[50] Koçar, I., Mahseredjian, J., and Olivier, G. "Improvement of numerical stability for the computation of transients in lines and cables." *IEEE Transactions on Power Delivery*. 2010; 25(2): 1104–11

[51] Gustavsen, B. "Avoiding numerical instabilities in the universal line model by a two-segment interpolation scheme." *IEEE Transactions on Power Delivery*. 2013; 28(3): 1643–51

[52] Ramos-Leanos, O., Mahseredjian, J., Naredo, J. L., Kocar, I., Gutierrez-Robles, J. A., and Martinez, J. A. "Phase-domain line/cable model through second-order blocks." *IEEE Transactions on Power Delivery*. 2015; 30(6): 2460–67

[53] Kocar, I., and Mahseredjian, J. "Accurate frequency dependent cable model for electromagnetic transients." *IEEE Transactions on Power Delivery*. 2016; 31(3): 1281–88

[54] Gustavsen, B., and Semlyen, A. "Rational approximations of frequency domain responses by vector fitting." *IEEE Transactions on Power Delivery*. 1999; 14(3): 1052–61

[55] Gustavsen, B. "Computer code for rational approximation of frequency dependent admittance matrices." *IEEE Transactions on Power Delivery*. 2002; 17(4): 1093–98

[56] Gustavsen, B. "Improving the pole relocating properties of vector fitting." *IEEE Transactions on Power Delivery*. 2006; 21(3): 1587–92

[57] Deschrijver, D., and Dhaene, T. "Broadband macromodelling of passive components using orthonormal vector fitting." *Electronics Letters*. 2005; 41(21): 1160–61

[58] Kocar, I., Mahseredjian, J., and Olivier, G. "Weighting method for transient analysis of underground cables." *IEEE Transactions on Power Delivery*. 2008; 23(3): 1629–35

[59] Longmire, C. L., and Smith, K. S. A Universal Impedance for Soils. Santa Barbara: Mission Research Corporation DNA3788T; 1975.

[60] Alipio, R., and Visacro, S. "Modeling the frequency dependence of electrical parameters of soil." *IEEE Transactions on Electromagnetic Compatibility*. 2014; 56(5): 1163–71

[61] Scott, J. H. *Electrical and Magnetic Properties of Rock and Soil*. U.S. Geological Survey: Technical Report; 1983.

[62] Scott, J. H., Carroll, R. D., and Cunningham, D. R. "Dielectric constant and electrical conductivity measurements of moist rock: A new laboratory method." Journal of Geophysical Research. 1967; 72(20): 5101–15

[63] Cavka, D., Mora, N., and Rachidi, F. "A comparison of frequency-dependent soil models: Application to the analysis of grounding systems." *IEEE Transactions on Electromagnetic Compatibility*. 2014; 56(1): 177–87

[64] Portela, C. "Measurement and modeling of soil electromagnetic behavior." *Proceedings of International Symposium on EMC*. 1999; 2: 1004–09

[65] Visacro, F. S., and Portela, C. M. "Soil permittivity and conductivity behavior on frequency range of transient phenomena in electric power systems." Proceedings of the Symposium on High Voltage Engineering; 1987

[66] Bigelow, R. C., and Eberle, W. R. *Empirical Predictive Curves for Resistivity and Dielectric Constant of Earth Materials: 100 Hz to 100 MHz*. U.S. Geological Survey: Open-File Report 83-911; 1972.

[67] He, J., Zeng, R., and Zhang, B. *Methodology and Technology for Power System Grounding*. Singapore: Wiley-IEEE Press; 2013.

[68] Itoh, Y., Nagaoka, N., and Ametani, A. "Transient analysis of a cross-bonded cable system underneath a bridge." *IEEE Transactions on Power Delivery*. 1990; 5(2): 527–32

[69] Ametani, A., Ohno, T., and Nagaoka, N. *Cable system transients: theory, modeling and simulation*. Singapore: Wiley-IEEE Press; 2015

[70] Mori, N., Inoue, T., Yamashita, Y., Yoshida, T., Maruyama, Y., and Tanabe, T. "Surge propagation characteristics of cables fixed to a tower." *IEE Japan Transactions on Power and Energy*. 1991; 111(1): 99–06

[71] Ohashi, Y., Hirata, N., Nagata, T., Shinagawa, J., and Fujiwara, Y. "Measurement of the surge current in the EHV underground cable system and the cable lines connected directly to the overhead lines." *IEE Japan Transactions on Power and Energy*. 1994; 114(1): 53–59

[72] Oka, K., Hirai, T., Murata, K., Hashimoto, E., and Ariga, Y. "A calculation of transient overvoltage for rationalization of insulation design of 22kV cable system." *IEE Japan Transactions on Power and Energy*. 2002; 122(6): 732–37

[73] Tanemura, K., Ametani, A., and Omura, H. "Surge characteristics on a cable vertically installed along a transmission tower." *IEE Japan International Workshop on High Voltage Engineering*. Kyoto, Japan, 2008.

[74] Ametani, A., Kasai, Y., Sawada, J., Mochizuki, A., and Yamada, T. "Frequency-dependent impedance of vertical conductors and a multiconductor tower model." *IEE Proceedings – Generation, Transmission and Distribution*. 1994; 141(4): 339–44

[75] Ametani, A., and Kawamura, T. "A method of a lightning surges analysis recommended in Japan using EMTP." *IEEE Transactions on Power Delivery*. 2005; 20(2): 867–75

[76] Ametani, A. *Distributed-circuit theory*. Tokyo: Corona Pub. Co.; 1990.

[77] Ametani, A. "A general formulation of impedance and admittance of cables." *IEEE Transactions on Power Apparatus and Systems*. 1980; 99(3): 902–10

[78] Nagaoka, N., Yoshida, Y., Ametani, A., and Karaski, T. "Transient sheath characteristic of a cross-bonded cable installed in a tunnel." *IEEJ Transactions on Power and Energy*. 1991; 111(7): 784–90

[79] Amekawa, N., Nagaoka, N., and Ametani, A. "Impedance derivation and wave propagation characteristics of pipe-enclosed and tunnel-installed cables." *IEEE Transactions on Power Delivery*. 2004; 19(1): 380–86

[80] Jinno, M., Yokota, K., Tanabe, T., and Yoshida, T. "Basic investigation of the sheath surge on a cable installed in a tube." *IEE Japan Annual Conference*. Japan; 1989.

[81] IEE Japan Working Group. *Surge Phenomena and Countermeasures on Cable Systems*. IEE Japan Technical Report No. 399; 1991.

[82] Lafaia, I., Alatawneh, N., Ametani, A., Mahseredjian, J., Naud, A., and Correia de Barros, M. T. "Model for an underground cable installed in an HDPE tube." International Symposium of EMC/Transients and Student Session (ISET/ISS-2014), Danang, Vietnam; 2014.

[83] Lafaia, I., Ametani, A., Mahseredjian, J., Naud, A., Correia de Barros, M. T., and Kocar, I. "Field test and simulation of transients on the RTE 225 kV cable." *IEEE Transactions on Power Delivery*. 2017; 32(2): 628–37

[84] *COMSOL Multiphysics 5.5 Reference Manual*. COMSOL; 2019.

[85] Takeyama, S. *Theory of electro-magnetic phenomena*. Tokyo: Maruzen Pub. Co; 1943.

[86] IEEE Guide for Bonding Shields and Sheaths of Single-Conductor Power Cables Rated 5 kV through 500 kV. IEEE Standard 575; 2014.

[87] Ball, E. H., Occhini, E., and Luoni, G. "Sheath overvoltages in high-voltage cables resulting from special sheath-bonding connections." *IEEE Transactions on Power Apparatus and System*. 1965; 84(10): 974–88

[88] Rhodes, D. J., and Wright, A. "Induced voltages in the sheaths of cross-bonded ac cables." *Proceedings of the Institution of Electrical Engineers*. 1966; 113(1): 99–10

[89] Adamson, C., Taha, E. A.-Z. H., and Wedepohl, L. M. "Comparative steady-state performance of cross-bonded cable systems." *Proceedings of the Institution of Electrical Engineers*. 1968; 115(8): 1147–55

[90] Heaton, A. G. "Transient response of cross-bonded cable systems." *Proceedings of the Institution of Electrical Engineers*. 1970; 117(3): 578–86

[91] Dang, N. D. "Transient performance of cross-bonded cable systems." University of Manchester, Ph.D. Thesis; 1972.

[92] Wedepohl, L. M., and Indulkar, C. S. "Wave propagation in non-homogeneous systems. Properties of the chain matrix." *Proceedings of the Institution of Electrical Engineers*. 1974; 121(9): 997–00

[93] Wedepohl, L. M., and Indulkar, C. S. "Switching overvoltages in short cross-bonded cable systems using the Fourier transform." *Proceedings of the Institution of Electrical Engineers*. 1975; 122(11): 1217–21

[94] Wilcox, D. J., and Lawler, K. J. "Implementation of nonhomogeneous theory in the transient analysis of cross-bonded cable systems." *Proceedings of the Institution of Electrical Engineers*. 1978; 125(10): 993–98

[95] Wilcox, D. J., and Lawler, K. J. "Transient phenomena in cross-bonded cable systems: Analytical results." *Proceedings of the Institution of Electrical Engineers*. 1978; 125(10): 999–05

[96] Kersten, W. F. J. "Surge arresters for sheath protection in cross-bonded cable system." *Proceedings of the Institution of Electrical Engineers*. 1979; 126(12): 1255–62

[97] Wedepohl, L. M., and Indulkar, C. S. "Switching overvoltages in long cross-bonded cable systems using the Fourier transform." *IEEE Transactions on Power Apparatus and System*. 1979; 98(4): 1476–80

[98] Jackson, R. J., Wilson, A., and Giesner, D. B. "Partial discharges in power-cable joints: Their propagation along a cross-bonded circuit and methods for their detection." *IEE Proceedings C (Generation, Transmission and Distribution)*. 1980; 127(6): 420–29

[99] Humpage, W. D., Wong, K. P., and Nguyen, T. T. "Z-transform electromagnetic transient analysis of cross-bonded cable transmission systems." *IEE Proceedings C (Generation, Transmission and Distribution)*. 1981; 128(2): 55–62

[100] Kuffel, E., and Poltz, J. "AC losses in cross-bonded and bonded at both ends high voltage cables." *IEEE Transactions on Power Apparatus and System*. 1981; 100(1): 369–74

[101] Nagaoka, N., and Ametani, A. "Transient calculations on cross-bonded cables." *IEEE Transactions on Power Apparatus and System*. 1983; 102(4): 779–87

[102] CIGRE WG C4.502. *Power System Technical Performance Issues Related to the Application of Long HVAC Cables*. CIGRE Technical Brochure No. 556; 2013.

[103] Colla, L., Rebolini, M., and Iliceto, F. "400 kV AC new submarine cable links new between Sicily and the Italian mainland." *CIGRE 2008 General Session Paper C4-116*, Paris, France; 2008.

[104] Energinet.dk. *Cable Action Plan*. www.energinet.dk; 2009.

[105] Lauria, S., and Palone F. "Optimal operation of long inhomogeneous AC cable lines: The Malta–Sicily interconnector." *IEEE Transactions on Power Delivery*. 2014 ; 29(3): 1036–44

[106] *Réseaux de Transport d'Électricité. Filet de Sécurité PACA: Pour Une Sécurisation Electrique Durable de la Région*. http://www.rte-france.com/fr/projet/filet-de-securite-paca-pour-une-securisation-electrique-durable-de-la-region; 2015.

[107] Ohno, T., Bak, C. L., Ametani, A., Wiechowski, W., and Sorensen, T. K. "Derivation of theoretical formulas of the frequency component contained

in the overvoltage related to long EHV cables." IEEE Transactions on Power Delivery. 2012; 27(2): 866–76

[108] Gudmundsdottir, U. S., Gustavsen, B., Bak, C. L., and Wiechowski, W. "Field test and simulation of a 400-kV cross-bonded cable system." IEEE Transactions on Power Delivery. 2011; 26(3): 1403–10

[109] Bak, C. L., and Jensen, C. F. "Distance protection of cross-bonded transmission cable-systems." *IET International Conference on Developments in Power System Protection*. Copenhagen, Denmark; 2014.

[110] Jensen, C. F., Gundmundsdottir, U. S., Bak, C. L., and Abur, A. "Field test and theoretical analysis of electromagnetic pulse propagation velocity on cross-bonded cable systems." *IEEE Transactions on Power Delivery*. 2014; 29(3): 1028–35

[111] Jensen, C. F. *Online location of faults on AC cables in underground transmission systems*. Switzerland: Springer; 2014.

[112] Wedepohl, L. M. "Application of matrix methods to the solution of traveling wave phenomena in polyphase systems." *Proceedings of the Institution of Electrical Engineers*. 1963; 110(12): 2200–12

[113] IEE Japan. Application of Numerical Transient Electromagnetic Analysis Methods to Surge Phenomena. ISBN 978-4-88686-263-1; 2008.

[114] CIGRE WG C4.501. *Guideline for Numerical Electromagnetic Analysis Methods and the Application to Surge Phenomena*. CIGRE TB-543; 2013.

[115] Van Baricum, M., and Miller, E. K. *TWTD–A Computer Program for Time-Domain Analysis of Thin-Wire Structures*. Lawrence Livermore Laboratory UCRL-51-277; 1972.

[116] Miller, E. K., Poggio, A. J., and Burke, G. J. "An integro-differential equation technique for the time-domain analysis of thin wire structures." *Journal of Computational Physics*. 1973; 12(2): 24–48

[117] Podgorski, A. S., and Landt, J. A. "Three dimensional time domain modelling of lightning." *IEEE Transactions on Power Delivery*. 1987; 2 (3): 931–38

[118] Harrington, R. F. *Field computation by moment methods*. New York, NY: Macmillan Co.; 1968.

[119] Burke, G. J., and Poggio, A. J. *Numerical Electromagnetic Code (NEC)–Method of Moments*. Naval Ocean Systems Center Technical Document 116; 1980.

[120] Burke, G. J. *Numerical Electromagnetic Code (NEC-4) - Method of Moments*. Lawrence Livermore National Laboratory UCRL-MA-109338; 1992

[121] Yee, K. S. "Numerical solution of initial boundary value problems involving Maxwell's equations in isotropic media." *IEEE Transactions on Antennas and Propagation*. 1966; 14 (3): 302–07

[122] Tanabe, K. "Novel method for analyzing dynamic behavior of grounding systems based on the finite-difference time-domain method." *IEEE Power Engineering Society Winter Meeting*. Columbus, USA, 2001

Chapter 4

Wave propagation characteristics of overhead and underground cables

4.1 Evaluation of propagation constant for overhead and underground cables

The frequency-dependent wave propagation characteristics are investigated for an overhead line, an overhead cable, and a multiphase underground cable based on the extended and classical transmission line (TL) approaches.

It has the following matrix operation by adopting [1]:

$$\mathbf{A}^{-1}(\mathbf{ZY})\mathbf{A} = \mathbf{Q} \tag{4.1}$$

where **Z** is the series impedance matrix, **Y** is the shunt admittance matrix, **A** is the voltage transformation matrix, and **Q** is an $N \times N$ diagonal matrix.

The modal propagation constant related to v-th mode is given by

$$\gamma_v = \alpha_v + j\beta_v = \sqrt{Q_v} : \quad v = 1, \ldots, N \tag{4.2}$$

where α_v is the modal attenuation constant of the v-th mode, β_v is the modal phase constant of the v-th mode, and Q_v is the v-th diagonal element of **Q**.

The modal phase velocity related to v-th mode is defined:

$$c_v = \frac{\omega}{\beta_v} \tag{4.3}$$

The modal attenuation constant and phase velocity are calculated and analyzed using (4.2) and (4.3) in Sections 4.2 and 4.3. Also, the extended and classical TL approaches discussed in Chapter 2 are adopted respectively in the following calculations.

4.2 Overhead cables

4.2.1 Overhead line

Figure 4.1 illustrates a single overhead conductor investigated in this section. The conductor resistivity is $\rho_c = 1.68 \times 10^{-8}$ Ωm. The earth resistivity and permittivity are ρ_e and ε_e.

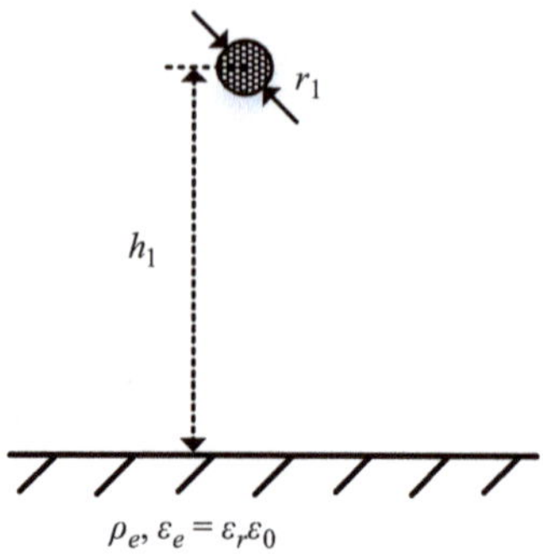

Figure 4.1 A single overhead conductor: $h_1 = 10$ m and $r_1 = 1$ cm

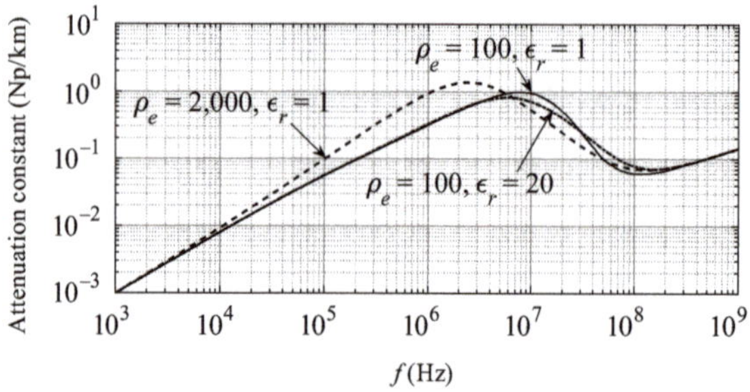

Figure 4.2 Attenuation constant of the conductor illustrated in Figure 4.1 by the extended TL approach

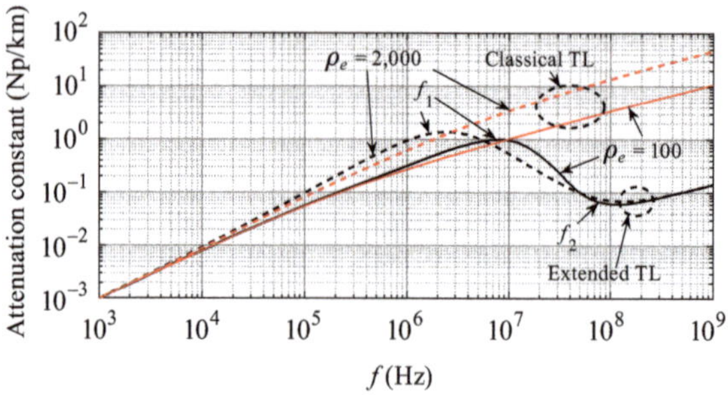

Figure 4.3 Attenuation constant of the conductor illustrated in Figure 4.1 by the extended and classical TL approaches, $\varepsilon_r = 1$

4.2.1.1 Attenuation constant

Figures 4.2 and 4.3 show the frequency responses of the attenuation constant for the overhead conductor illustrated in Figure 4.1. In Figure 4.2, the attenuation constant is calculated by extended TL approach, considering three different earth parameters. The earth permittivity has a minor influence on the attenuation constant in the entire frequency range. Also, a significant influence due to a large earth resistivity is observed.

The attenuation calculated by the classical TL approach in Figure 4.3 monotonously increases as a function of frequency. On the contrary, the extended approach shows that the attenuation starts to decrease at the frequency f_1, and increases again monotonously after the frequency f_2. This phenomenon is referred to as "mode transition" by Kikuchi [2,3]. Later, Chiba [4] presented an experimental validation of the Kikuchi's theory. The attenuation in frequencies lower than f_1 is related to the low-frequency quasi-TEM (earth-return wave) propagation. In the frequency region up to f_1, the classical TL approach agrees with the extended TL approach. For frequencies higher than f_1, the propagation mode changes from the quasi-TEM propagation, and reaches the Sommerfeld–Goubau (surface wave) propagation at the frequency f_2 [5]. The surface wave is a result of large displacement currents which represent the return current in air nearby the conductor. The displacement currents are generated by high electric field strength at the conductor surface [5]. It is also interpreted physically that the earth shows a capacitive nature at high frequencies, while it is inductive at low frequencies [5]. This phenomenon cannot be explained by the classical TL approach. In Figures 4.2 and 4.3, only the TL mode is calculated by the iteration of the modal equation as shown in [2,3]. In [6,7], it is shown that the frequency spectra of the TL and FW or SA modes are much closed to each other until a break-point where two different branches appear. Above the frequency of this break-point, the attenuation constants of the TL and FW modes are respectively decreasing and increasing as frequency increases. The TL mode approaches to zero if the conductor is perfectly conducting. If the conductor is lossy, the TL mode decreases to a critical frequency f_2 and then increases again, and eventually it reaches the Sommerfeld–Goubau propagation wave. The attenuation constant of the FW mode approaches infinite. Therefore, the TL mode can be regarded as the dominant one for the wave propagation with finite attenuation constant even in the high-frequency region [8]. It should be noted that the attenuation at frequencies higher than f_2 is significantly smaller than that at frequencies lower than f_1. The reason for the lower attenuation is that the dominant cause of the attenuation in this region is the conductor internal impedance which is much smaller than the earth-return impedance [5].

4.2.1.2 Calculated results of electromagnetic field components

The formulas of $\mathbf{E}_0$ and $\mathbf{H}_0$ have been derived and given in Chapter 2 based on the complete field solutions.

The basic characteristics of axial and vertical electric field components of the conductor illustrated in Figure 4.1 have been already investigated in [9] for an earth resistivity $\rho_e = 200\ \Omega\text{m}$ and relative permittivity $\varepsilon_r = 1$. Considering the same parameters, the calculated field components are identical to those in [9]. This section focuses on the high-frequency characteristics of both $\mathbf{E}_0$ and $\mathbf{H}_0$ with various earth resistivities and permittivities to observe the mode transition described in Section 4.2.1.1.

High-frequency characteristics of E_{0x}, E_{0z}, *and* H_{0y}

Figure 4.1 illustrates a single overhead conductor investigated in this section and the rectangular coordinate.

Figure 4.4 shows (a) axial E_{0x}, (b) vertical E_{0z}, and (c) horizontal H_{0y} field components in the air as a function of frequency. In each figure, one group of curves corresponds to a vertical distance z from the earth surface being 9.99 m, that is, at the conductor surface. The other group corresponds to $z = 0.01$ m, that is, 1 cm from the earth surface. Also, included in the figure are the cases of (1) $\rho_e = 100\ \Omega$m, $\varepsilon_r = 1$, (2) $\rho_e = 100\ \Omega$m, $\varepsilon_r = 20$, and (3) $\rho_e = 2{,}000\ \Omega$m, $\varepsilon_r = 1$.

It is clear in Figure 4.4(b) and (c) that E_{0z} and H_{0y} at $z = 9.99$ m, that is, at the conductor surface, are almost independent of the frequency, that is, $E_{0z} \approx 7{,}000$ V/m and $H_{0y} = 16$ A/m. At $z = 0.01$ m, E_{0z} and H_{0y} are nearly constant or little increasing up to $f = 4$ MHz in the case of $\rho_e = 100\ \Omega$m, and up to $f = 1$ MHz in the case of $\rho_e = 2{,}000\ \Omega$m. Then, those field components start to decrease as

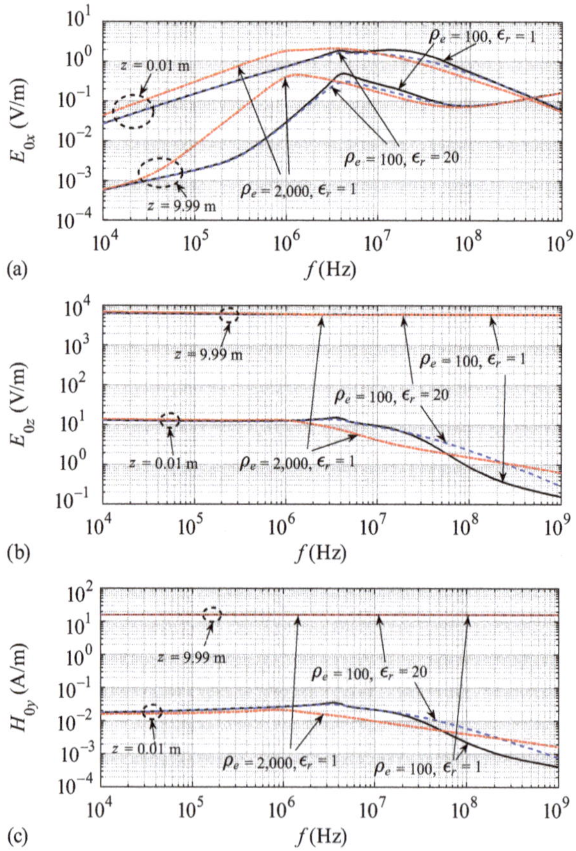

Figure 4.4 Frequency responses of electromagnetic field components in the air, $y = 0$ m. (a) Axial electric field, (b) vertical electric field and (c) horizontal magnetic field

frequency increases, because the earth becomes less influential at high frequencies where the surface wave becomes dominant, as it will be explained later for E_{0x}.

No noticeable difference is observed between $\varepsilon_r = 1$ and 20 with $\rho_e = 100\ \Omega$m up to 20 MHz. E_{0z} and H_{0y} in the case of $\varepsilon_r = 20$ become greater than those in the case of $\varepsilon_r = 1$. For $\rho_e = 2{,}000\ \Omega$m, E_{0z} is slightly greater than that for $\rho_e = 100\ \Omega$m up to about 1 MHz. E_{0z} becomes smaller up to 50 MHz. Above this frequency, E_{0z} for $\rho_e = 2{,}000\ \Omega$m is greater than that for $\rho_e = 100\ \Omega$m. H_{0y} for $\rho_e = 2{,}000\ \Omega$m is smaller up to 50 MHz, and becomes greater than that for $\rho_e = 100\ \Omega$m.

The axial electric field E_{0x} shows a more complicated behavior than those of E_{0z} and H_{0y}. It should be noted that E_{0x} is far smaller than E_{0z} especially at the conductor surface ($z = 9.99$ m). E_{0x} for $\rho_e = 2{,}000\ \Omega$m is greater than that for $\rho_e = 100\ \Omega$m up to frequencies of some MHz. Then, it becomes smaller. E_{0x} with $\varepsilon_r = 20$ is always smaller than that with $\varepsilon_r = 1$, even though the difference is quite small. However, when frequency becomes very high, above 100 MHz, no significant difference is observed between various ρ_e and ε_r. This fact indicates that the earth causes no significant effect on the axial electric field, and corresponds to the surface wave propagation explained in Section 4.2.1.1. In fact, E_{0x} near the earth surface becomes smaller than that at the conductor surface above 300 MHz.

Moreover, E_{0x} increases as frequency increases up to some MHz. Then, E_{0x} at the earth surface decreases as frequency increases. E_{0x} at the conductor surface decreases up to a frequency of about 80 MHz. Then, it increases again. This characteristic is very similar to that of attenuation constant as shown in Figure 4.5, attenuation constant, and E_{0x} are shown in the same plot for an easy observation of the correlation for the same conductor geometry and earth electrical parameters. It should be noted that E_{0x} on the conductor surface at various positions, for example, ($z = 9.99$ m, $y = 0$ m), ($z = 10.01$ m, $y = 0$ m), and ($z = 9.9929$ m, $y = 0.00707$ m) is the same.

Comparison with classical TL approach

Based on the classical TL approach with well-known Pollaczek's earth-return impedance and the space admittance, the attenuation is calculated and the result is

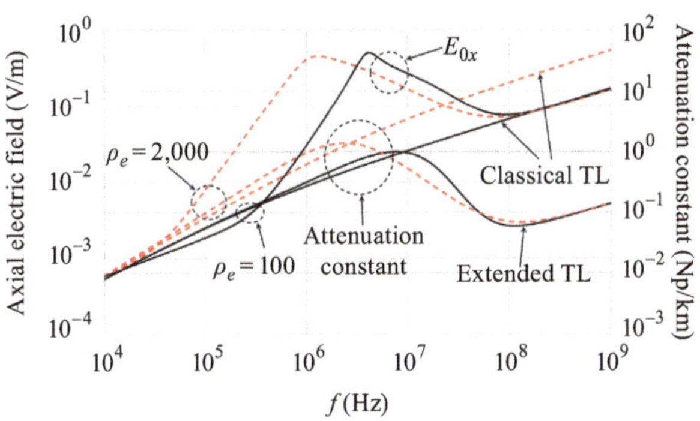

Figure 4.5 Frequency responses of attenuation constant and axial electric field component by the extended and classical TL approaches

shown in Figure 4.5. The attenuation constant by the classical TL approach is a monotonously increasing function of frequency and shows no mode transition.

Figures 4.6 and 4.7 show a comparison of electromagnetic field components evaluated by the extended and classical TL approaches. At low frequencies, the wave propagation on an overhead conductor above a lossy earth is a quasi-TEM mode and no noticeable difference between the classical TL approach and the complete field solution can be observed. However, above a frequency of about 5 MHz, a noticeable difference for E_{0x} is observed. E_{0x} calculated by the classical TL approach shows a monotonous increase as a function of frequency in Figure 4.6 (a), similarly to the monotonously increasing attenuation constant in Figure 4.5, while E_{0x} evaluated by the complete field solution shows mode transition as already explained in Section 4.2.1.1.

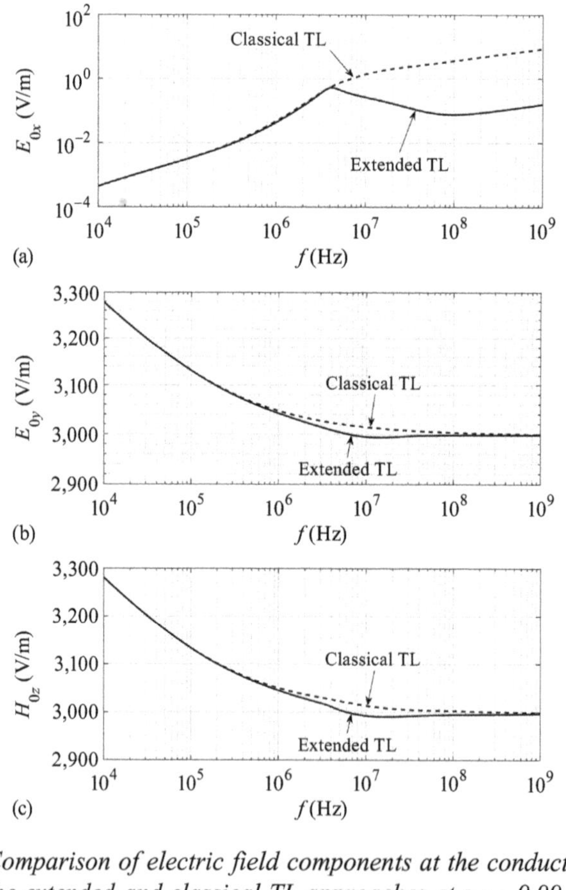

Figure 4.6 Comparison of electric field components at the conductor surface by the extended and classical TL approaches at $z = 9.99$ m and $y = 0.01$ m, $\rho_e = 100$ Ωm, and $\varepsilon_r = 1$. (a) Axial electric field, (b) vertical electric field and (c) horizontal magnetic field

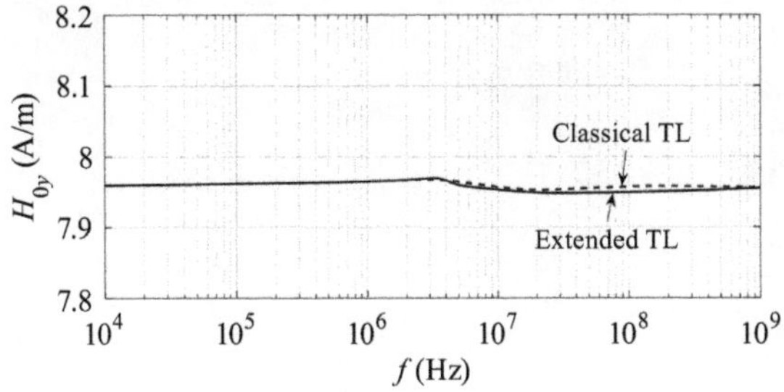

Figure 4.7 Comparison of horizontal magnetic field component by the extended and classical TL approaches at $z = 9.99$ m and $y = 0.01$ m, $\rho_e = 100$ Ωm, and $\varepsilon_r = 1$

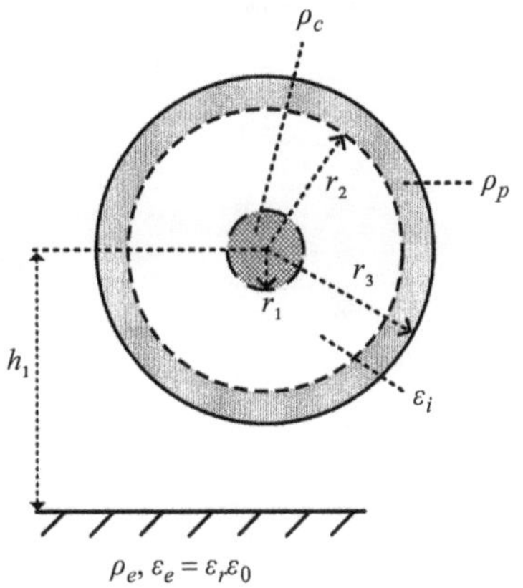

Figure 4.8 Cross-section of a 500 kV GIB, $h_1 = 2.45$ m, $r_1 = 12.5$ cm, $r_2 = 46$ cm, $r_3 = 48$ cm, $\rho_c = 1.68\times10^{-8}$ Ωm, $\rho_p = 2.82\times10^{-8}$ Ωm, and SF6 gas $\varepsilon_i = 1$

4.2.2 Overhead cable

This section focuses on mode transition from a low-frequency TEM mode to high-frequency Sommerfeld–Goubau propagation and associated propagation functions for a 500 kV GIB. The GIB is a typical overhead cable and the one illustrated in Figure 4.8 is used as an example in this section [1,8].

4.2.2.1 Attenuation constant

Figure 4.9 shows the frequency response of the modal attenuation constants for the GIB illustrated in Figure 4.8. No difference of the coaxial mode between the extended and classical TL approaches is observed, as expected. The earth-return mode calculated by the extended TL approach shows the mode transition as explained in Section 4.2.1.1. However, the result by the classical TL approach increases monotonously.

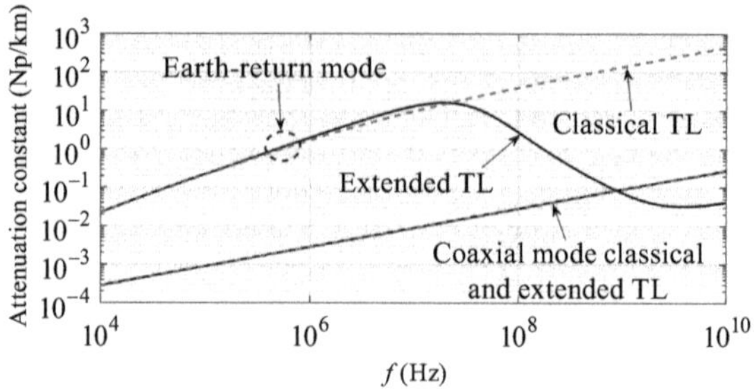

Figure 4.9 Modal attenuation constants of the GIB illustrated in Figure 4.8 by the extended and classical TL approaches, $\rho_e = 100\ \Omega m$ and $\varepsilon_r = 1$

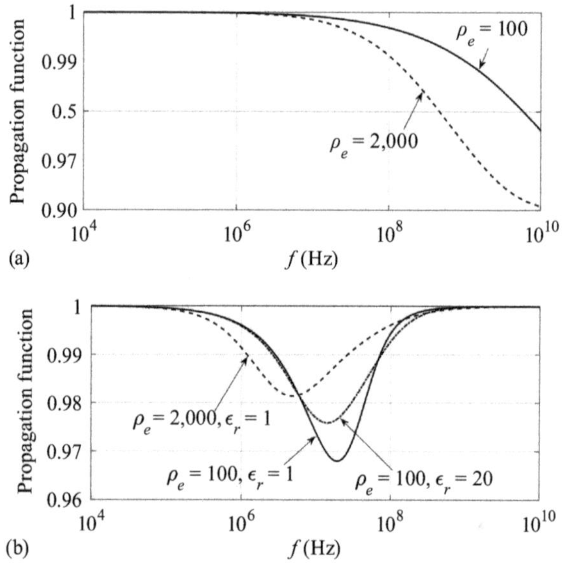

Figure 4.10 Earth-return mode propagation function on the GIB illustrated in Figure 4.8 at distance $x = 2$ m from the sending end by the extended and classical TL approaches. (a) Classical TL and (b) extended TL

It should be noted that the applicable range of the extended approach is about 100 MHz, because of the relation between the wavelength and the pipe radius. However, in order to show the mode transition clearly, the attenuation is given up to 10 GHz in Figures 4.9 and 4.10.

4.2.2.2 Propagation function

This section shows the earth-return mode propagation function H_w which is defined as

$$H_w = e^{-\alpha_1 x} \tag{4.4}$$

where x is the distance along the conductor and α_1 is the earth-return mode attenuation constant.

It is clear from Figure 4.10(a) that H_w monotonously decreases as the frequency increases in the classical TL approach, following the characteristic of the earth-return mode propagation constant in Figure 4.9. With the more accurate extended TL approach, H_w decreases until the frequency up to f_1. Then it increases until the frequency reaches f_2, f_1, and f_2 are explained in Section 4.2.1.1. Because of lower attenuation in the frequency region is higher than f_1 as in Figure 4.9, H_w in the extended TL approach is much greater than that in the classical TL approach.

4.2.2.3 Transition frequency f_1

The mode transition explained in Section 4.2.1.1 starts at the frequency f_1. The frequency can be used as an indicator of low attenuation in high-frequency switching surges in a GIS. The frequency is related to conductor height h_1, earth resistivity ρ_e, and relative earth permittivity ε_r as shown in Figure 4.11. For example, f_1 in Figure 4.11 decreases as the conductor height or earth resistivity increases.

The reason for this behavior is that the mode transition is influenced dominantly by the earth-return path. When the conductor height is small, the influence of the earth-return path is stronger and the earth-return wave stays dominant up to higher frequencies and thus the transition frequency becomes higher.

4.3 Underground cables

4.3.1 Propagation constant

Cable 1 illustrated in Figure 2.10 is used to calculate modal propagation characteristics. It is clear in Figure 4.12 that the attenuation constant and phase velocity

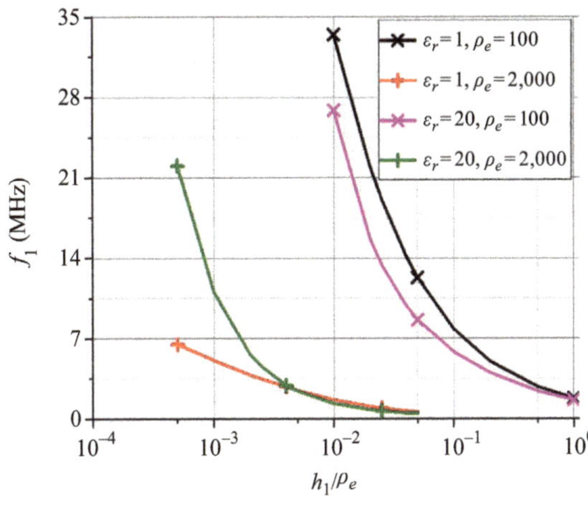

Figure 4.11 Transition frequency f_1 for the GIB illustrated in Figure 4.8

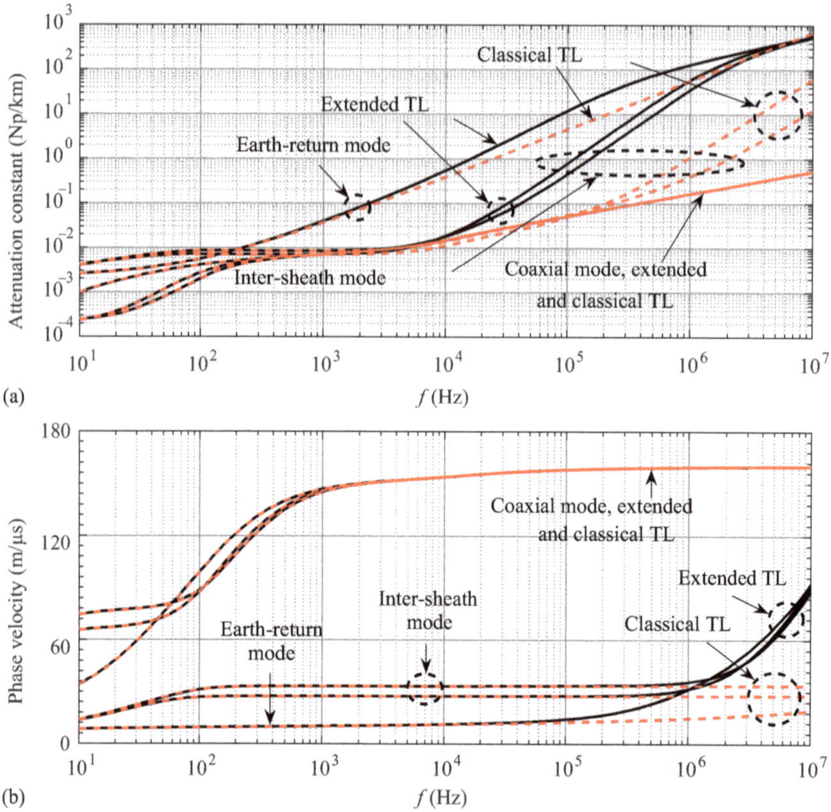

Figure 4.12 Modal propagation constants on Cable 1 in Figure 2.10 by the extended and classical TL approaches, $\rho_e = 100\ \Omega m$ and $\varepsilon_r = 1$. (a) Attenuation constant and (b) phase velocity

of the earth-return mode and inter-sheath mode show a significant difference between the extended and classical TL approaches above 10 kHz.

The phase velocity of the extended TL approach increases faster as frequency increases at high frequencies, whereas the velocity evaluated by the classical TL approach is nearly constant. The higher attenuation and higher velocity at high frequencies in the extended approach are caused mostly by the earth-return admittance. Also, the effect of earth-return impedance is minor.

Before discussing the effect of earth-return admittance, it should be noted that the capacitance C is the total capacitance of the series circuit of the cable outer insulator capacitance C_i and earth-return capacitance C_e when we neglect a conductance for simplicity, that is,

$$C = \frac{C_i C_e}{C_i + C_e} \tag{4.5}$$

where $C_i = 2.57$ μF/km for Cable 1 as shown in Figure 2.10.

As frequency approaches to zero, C_e tends to be far greater than C_i:

$$C_e \gg C_i: \quad f \to 0 \tag{4.6}$$

Then, the total capacitance becomes C_i at low frequencies as observed in (4.6).

When the frequency starts to increase, C_e starts to decrease. At high frequencies, C_e is far smaller than C_i:

$$C_e \ll C_i: \quad f \to 10\ \text{MHz} \tag{4.7}$$

Therefore, the total capacitance decreases as frequency increases. It should be clear that the propagation velocity involved in the cable outer insulator, that is, the earth-return mode and inter-sheath modes, becomes higher as C becomes smaller because the phase constant becomes smaller. On the contrary, the capacitance C_i is independent of frequency, that is, constant in the classical TL approach, and thus the velocity is kept nearly constant even at high frequencies. The higher attenuation in the extended one in Figure 4.12(a) is explained similarly to the velocity considering the decreasing C and the existence of the conductance.

No significant differences are observed in Figure 4.12 between the extended and classical TL approaches at low frequencies. When the earth relative permittivity is taken to be 20, no significant differences from those with $\varepsilon_r = 1$ are observed, although the results are not shown in Figure 4.12. The coaxial mode shows no differences for all the cases, as expected.

Figure 4.13 shows the modal propagation constants calculated by Cable 2 in Figure 2.10. A significant difference is observed between extended and classical TL approaches for earth-return and inter-sheath modes above some tens of kHz, similarly to that explained for Cable 1. Only minor differences are observed between Figures 4.12 and 4.13, although the cable arrangement is different.

The modal propagation constants calculated using Cable 3 are shown in Figure 4.14. A difference of the coaxial and inter-sheath modes in comparison with those calculated by Cables 1 and 2 becomes more noticeable. This is caused by the configuration of Cable 3 being more symmetrical than Cables 1 and 2, and it makes two coaxial modes and two inter-sheath modes overlapped.

4.3.2 External electromagnetic field components

4.3.2.1 Uni-directional wave propagation

A three-phase single core underground cable using a horizontal arrangement is shown in Figure 4.15 [10,11]. The external electromagnetic field components, shielding effect, and phase current produced by the cable are calculated and analyzed using extended and classical TL approaches. The calculated results are also verified with numerical electromagnetic code (NEC) [12,13]. The generated electromagnetic fields are evaluated by the uni-directional current propagation which is given in (A2.58).

Energization of electromagnetic field components

Considering the Phases *a–b* energization in (A2.58), the voltage vector $\mathbf{V_f}$ is defined by the following expression:

$$V_{fa} = 10\ \text{kV}, \quad V_{fb} = -10\ \text{kV} \quad \text{and} \quad V_{fc} = 0\ \text{kV} \tag{4.8}$$

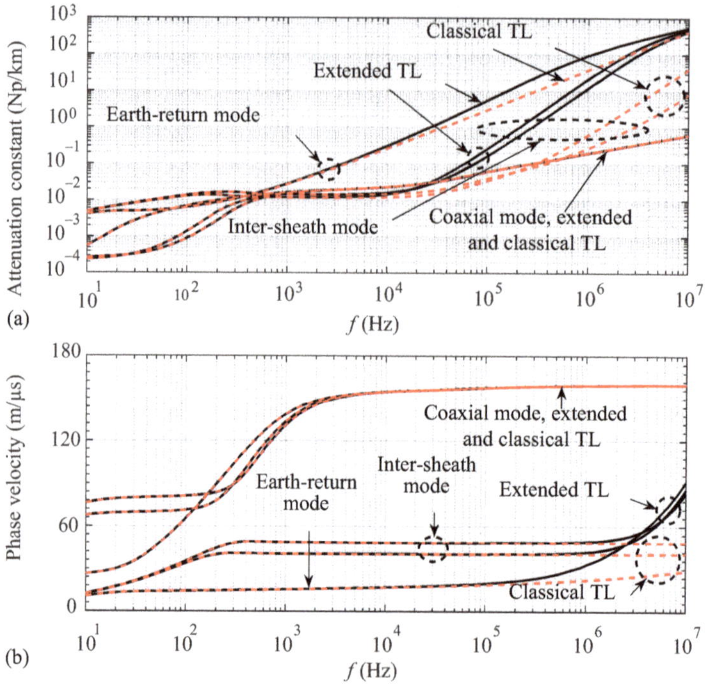

Figure 4.13 Modal propagation constants on Cable 2 in Figure 2.10 by the extended and classical TL approaches, ρ_e = 100 Ωm and ε_r = 1. (a) Attenuation constant and (b) phase velocity

where V_{fa}, V_{fb}, and V_{fc} are voltages applied at the different phases of the cable, and refer to the zero potential level.

A mixed propagation mode of earth-return and inter-phase effects on electromagnetic fields is excited by the Phases *a–b* energization.

The voltage vector $\mathbf{V_f}$ has the following expressions based on the Phase-*b* energization:

$$V_{fa} = 0\,\text{kV}, \quad V_{fb} = 10\,\text{kV} \quad \text{and} \quad V_{fc} = 0\,\text{kV} \tag{4.9}$$

The Phase-*b* energization mainly produces the earth-return effect on electromagnetic field components.

Characteristics of NEC

The calculated results of electromagnetic field components are validated and verified by adopting NEC 4.2. NEC is a numerical program for analyzing three-dimensional boundary problems using the electric-field and magnetic-field integral equations based on method of moment (MoM) [12,13].

A major restriction and condition in using NEC is the electrical dimension. For example, the length and radius of cable segment should correspond to the wavelength in the earth. The length of the segment is used to be 1/30 of the wavelength. Also, the condition of thin wire approximation should be fulfilled [12,13]. A similar

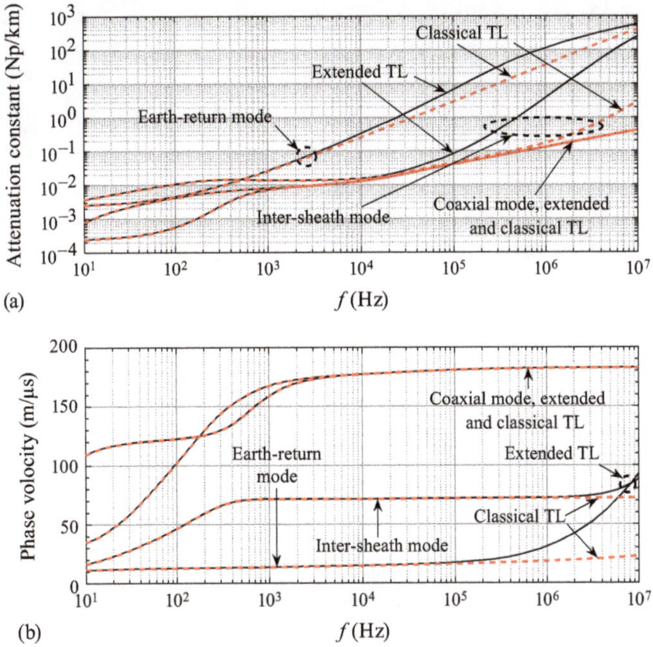

Figure 4.14 Modal propagation constants on Cable 3 in Figure 2.10 by the extended and classical TL approaches, $\rho_e = 100\ \Omega m$ and $\varepsilon_r = 1$. (a) Attenuation constant and (b) phase velocity

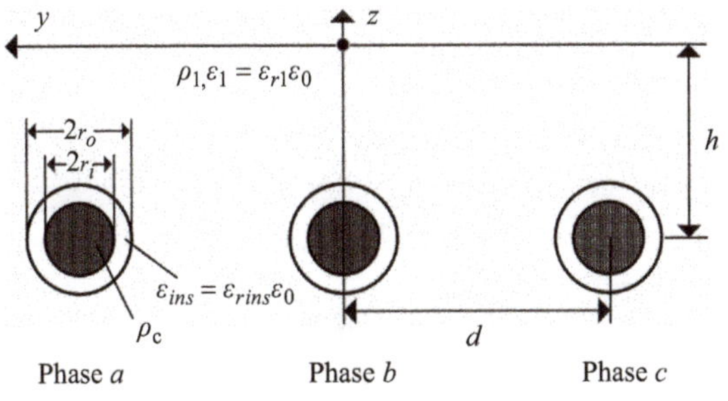

Figure 4.15 A three phase underground cable with a horizontal arrangement, $h = -1$ m, $d = 0.2$ m, $r_i = 1$ cm, $r_o = 1.2$ cm, $\rho_c = 1.68\times10^{-8}$ Ωm, and $\varepsilon_{rins} = 3$

procedure of NEC in [12,13] is used to model the cable structure shown in Figure 4.15.

Propagation constant

The modal propagation constants of the cable in Figure 4.15 are calculated using the extended and classical TL approaches. The earth parameters are set to $\rho_1 = 1{,}000\ \Omega\text{m}$ and $\varepsilon_{r1} = 1$.

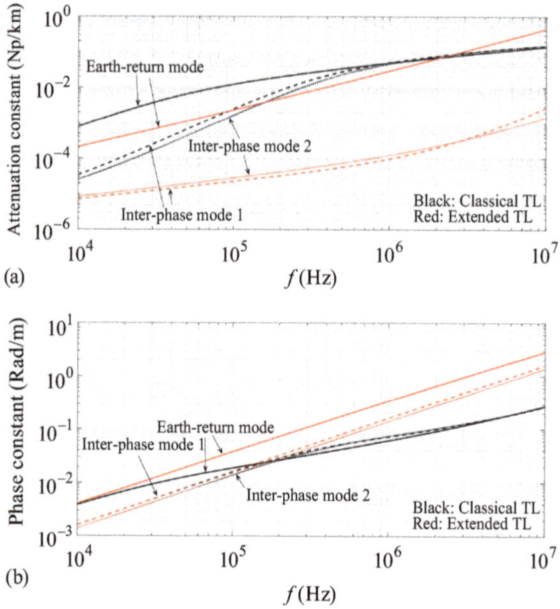

Figure 4.16 Modal propagation constants evaluated by the extended and classical TL approaches. (a) Modal attenuation constants and (b) modal phase constants

It is clear in Figure 4.16 that the modal attenuation constants evaluated by the extended TL approach increase significantly as frequency increases, especially the inter-phase modes. Also, a significant difference is observed for the modal phase constants evaluated by the two approaches above $f = 100$ kHz. The visible differences of the modal propagation constants calculated by the two TL approaches are mainly caused by the influence of earth-return admittance.

Calculations of external electromagnetic field components and comparisons with NEC

External electromagnetic field components for Phases a–b energization: The near-field amplitudes of lateral external electromagnetic field components are calculated at 100 kHz, 1 MHz, and 10 MHz using Phases *a–b* energization, and the results are shown in Figures 4.17–4.19. The observation point of electromagnetic field is referred to the coordinate system shown in Figure 4.15, and set to $z = -1.5$ m which is below the cable.

It is clear that the dominant electromagnetic field components E_y, E_z, H_y, and H_z decrease smoothly as the lateral distance increases. At the frequency of 100 kHz, the electromagnetic field components calculated using the classical TL approach show a minor difference to the results calculated by the extended TL approach and NEC. It is because the earth-return admittance gives less influence to the wave propagation at this frequency point. Also, if the frequency increases, such as 1 MHz and 10 MHz, the electromagnetic field components calculated by the classical TL approach shows

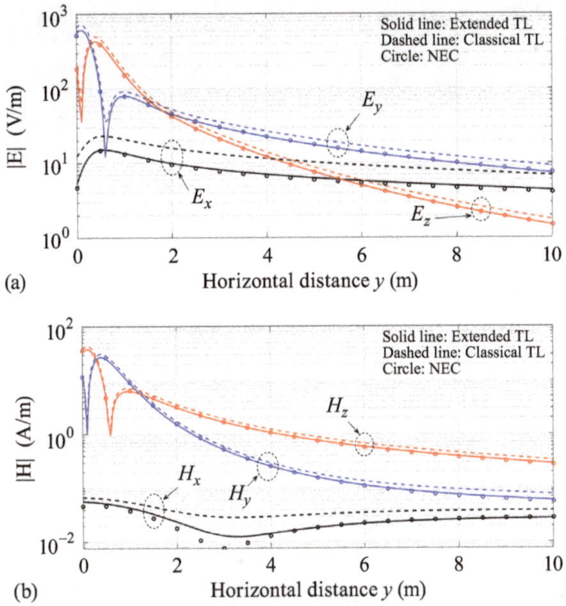

Figure 4.17 Lateral electromagnetic field components calculated at 100 kHz for Phases a–b energization, $x = 50$ m, $z = -1.5$ m, $\rho_1 = 1{,}000$ Ωm, and $\varepsilon_{r1} = 1$. (a) Electric field components and (b) magnetic field components

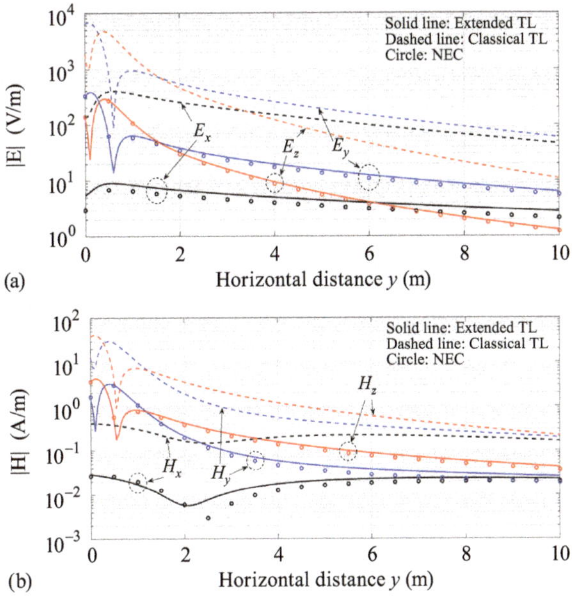

Figure 4.18 Lateral electromagnetic field components calculated at 1 MHz for Phases a–b energization, $x = 35$ m, $z = -1.5$ m, $\rho_1 = 1{,}000$ Ωm, and $\varepsilon_{r1} = 1$. (a) Electric field components and (b) magnetic field components

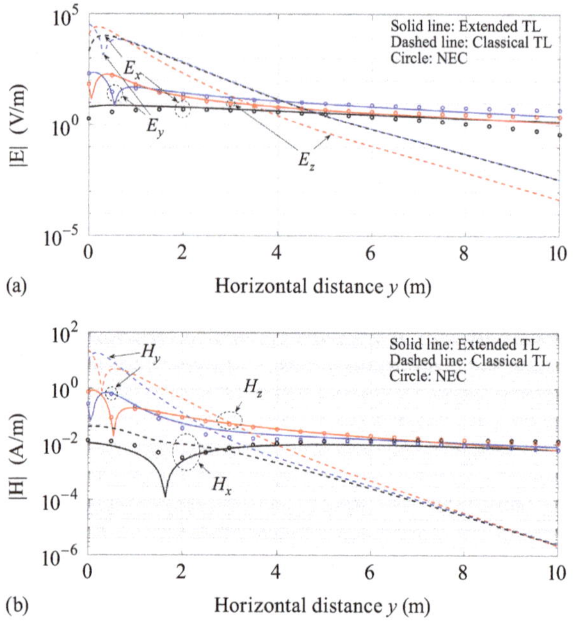

Figure 4.19 Lateral electromagnetic field components calculated at 10 MHz for Phases a–b energization, $x = 15$ *m,* $z = -1.5$ *m,* $\rho_1 = 1{,}000\ \Omega m$*, and* $\varepsilon_{r1} = 1$

significant differences to the results obtained by the other two methods. This phenomenon is caused by the transition of earth characteristic which is inductive initially in the low-frequency region and then becomes capacitive in the high-frequency region due to the impact of earth-return admittance. It is further proved the importance of earth-return admittance in the high-frequency wave propagation of underground cables. Moreover, the results calculated by the extended TL approach are verified by the results of NEC, and a good agreement is achieved.

The lateral profiles of external electromagnetic field components nearby the earth surface with $z = -0.1$ are calculated using the Phases *a–b* energization, and the results are illustrated in Figures 4.20–4.22.

In comparison with the electromagnetic fields calculated below the cable ($z = -1.5$ m), similar damping effects of electromagnetic field components along lateral distances are observed nearby the earth surface. The dominant electromagnetic field components observed at $z = -0.1$ m are slightly more attenuated than the results calculated below the cable, and also the electromagnetic field components have not been effectively attenuated at the earth surface as theoretically predicted in [14]. The major reason is the buried depth which is not deep enough to damp the electromagnetic field. Again, the results calculated by the extended TL approach show a good agreement with the results of NEC, particularly for the dominant electromagnetic field components.

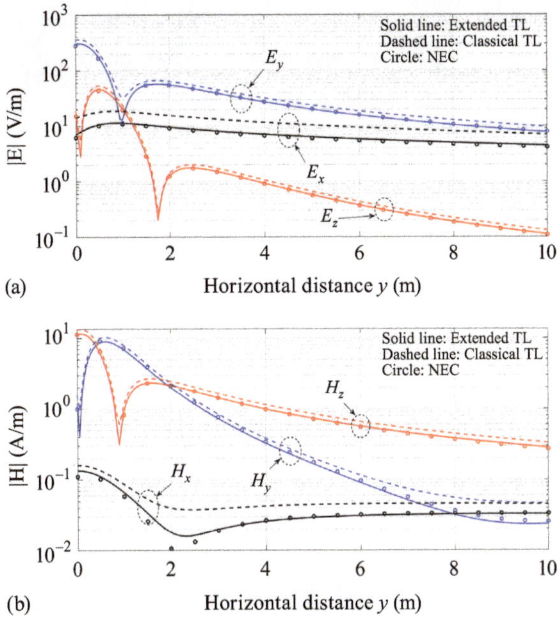

Figure 4.20 Lateral electromagnetic field components calculated at 100 kHz for Phases a–b energization, $x = 50$ m, $z = -0.1$ m, $\rho_1 = 1{,}000\ \Omega$m, and $\varepsilon_{r1} = 1$. (a) Electric field components and (b) magnetic field components

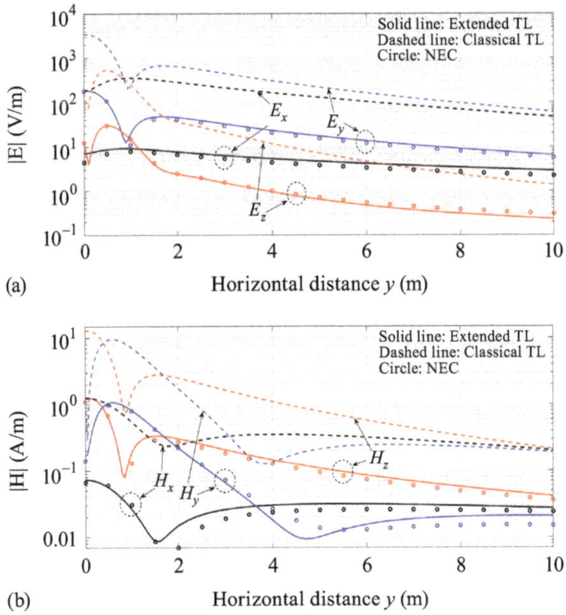

Figure 4.21 Lateral electromagnetic field components calculated at 1 MHz for Phases a–b energization, $x = 35$ m, $z = -0.1$ m, $\rho_1 = 1{,}000\ \Omega$m, and $\varepsilon_{r1} = 1$. (a) Electric field components and (b) magnetic field components

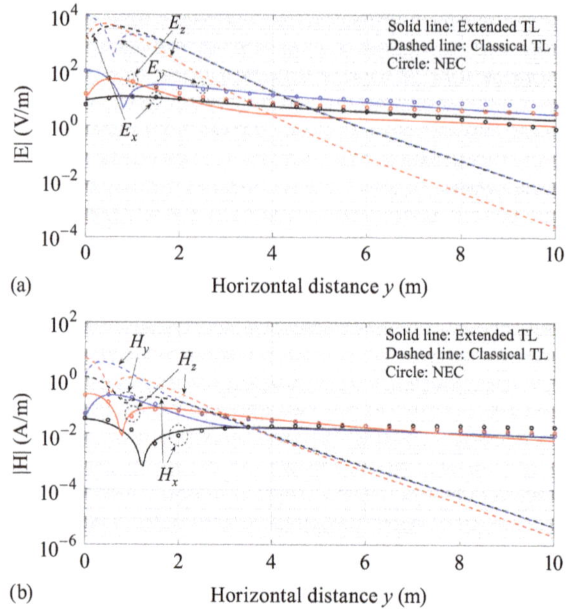

Figure 4.22 Lateral electromagnetic field components calculated at 10 MHz for Phases a–b energization, $x = 15$ m, $z = -0.1$ m, $\rho_1 = 1{,}000$ Ωm, and $\varepsilon_{r1} = 1$. (a) Electric field components and (b) magnetic field components

In general, the run time of NEC for each figure (Figures 4.17–4.22) is around 2,000 s depending on the segment and cable lengths in the modeling. For the same figures, the analytical solutions using extended TL approach need around 90 s for each figure.

Some minor differences in electromagnetic field components can be observed between the results calculated by the extended TL approach and NEC for the field region $y > 5$ m with $f = 10$ MHz. Because the TL approach may reach the application limit in such a high-frequency region.

Figure 4.23 shows the calculated electromagnetic field components as a function of frequency. It is clear that no significant difference is observed between the results calculated by the extended and classical TL approaches below 100 kHz. It matches with the general conclusion that the classical TL approach is applicable to the underground cables up to a few hundreds of kHz [36]. If the frequency is higher than 300 kHz, the electromagnetic field components evaluated by the extended TL approach start to gradually decrease and the results agree well with the NEC results. The results obtained by the classical TL approach show instable characteristics.

External electromagnetic field components for Phase-b energization: The calculation of electromagnetic field components under the Phase-*b* energization is performed and shown in Figure 4.24. In comparison with the results produced by the

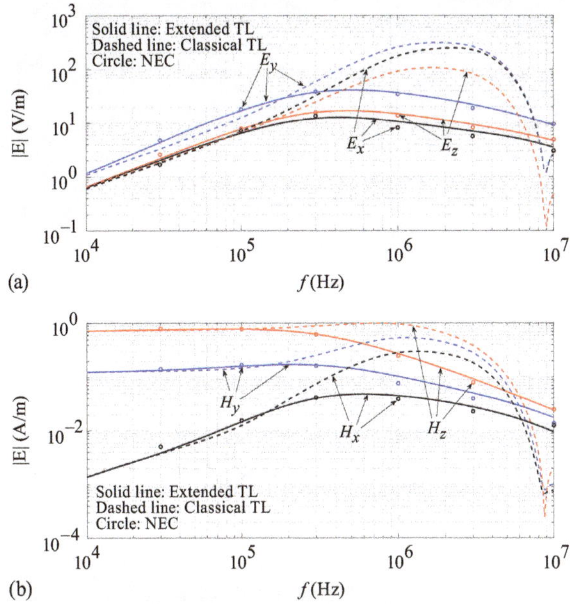

Figure 4.23 Electromagnetic field components calculated at $x = 15$ m, $y = 5$ m, and $z = -1.5$ m for Phases a–b energization, $\rho_1 = 1{,}000\ \Omega$m, and $\varepsilon_{r1} = 1$. (a) Electric field components and (b) magnetic field components

Phases *a–b* energization, it is clear that the electromagnetic field components excited by the Phase-*b* energization give a smoother distribution nearby the cable. Moreover, unlike the results shown in Figure 4.17, the electromagnetic field components calculated by the classical TL approach have a significant difference compared to the results calculated by the extended TL approach and NEC in Figure 4.24. In fact, the earth-return effect is dominant in the Phase-*b* energization, which is importantly influenced by the earth-return admittance.

Figure 4.25 shows the influence of earth resistivity and earth permittivity on the calculated electromagnetic field components using the extended TL approach. Three different earth parameters $\rho_1 = 100\ \Omega$m, $\varepsilon_{r1} = 1$, $\rho_1 = 100\ \Omega$m, $\varepsilon_{r1} = 20$, and $\rho_1 = 1{,}000\ \Omega$m, $\varepsilon_{r1} = 1$are applied to the calculations. The earth permittivity has no visible impact on electromagnetic field components. However, all the electromagnetic field components are significantly affected by the earth resistivity since the earth-return admittance is sensitive to the earth resistivity. The magnetic field components evaluated by $\rho_1 = 1{,}000\ \Omega$m are smaller than the results calculated by $\rho_1 = 100\ \Omega$m. The dominant electric field components E_y and E_z calculated by $\rho_1 = 1{,}000\ \Omega$m are larger than the results obtained by $\rho_1 = 100\ \Omega$m. Again, the NEC results validated well with the electromagnetic field components calculated by the extended TL approach.

Shielding effects of cable: Using the same conditions of energization shown in Figure 4.24, of that the nonenergized phases *a* and *c* are removed, and then the electromagnetic field components are calculated by the extended TL approach and

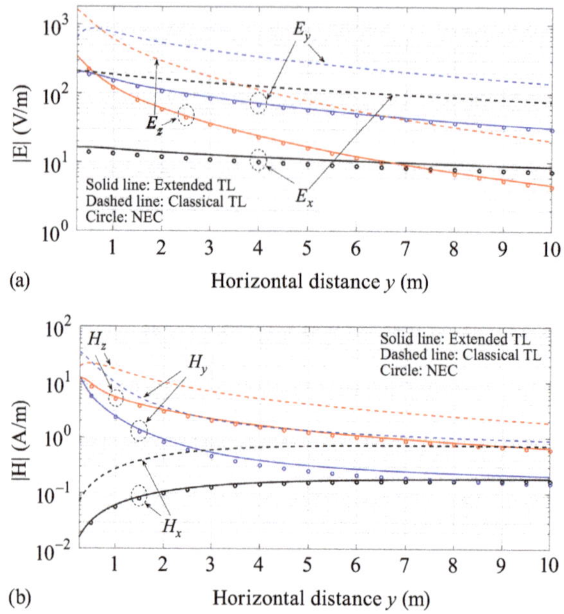

Figure 4.24 *Lateral electromagnetic field components calculated at 100 kHz for Phase-b energization, $x = 50$ m, $z = -1.5$ m, $\rho_1 = 1{,}000\ \Omega$m, and $\varepsilon_{r1} = 1$. (a) Electric field components and (b) magnetic field components*

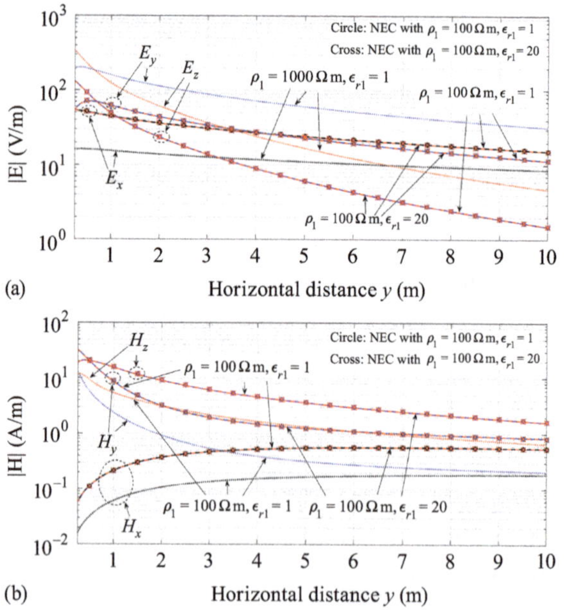

Figure 4.25 *Lateral electromagnetic field components calculated at 100 kHz for Phase-b energization by the extended TL approach and NEC, $x = 50$ m and $z = -1.5$ m. (a) Electric field components and (b) magnetic field components*

NEC. As shown in Figure 4.26, the shielding effects of cable are clearly illustrated. The nonenergized Phases *a* and *c* give a clear reduction of the amplitude of electromagnetic field components.

In general, at the field point of $y = 1$ m, the amplitudes of E_z and H_y with Phases *a* and *c* decrease more than 60% than the results with only Phase *b*, which means E_z from 334.3 V/m to 138.2 V/m and H_y from 8.9 A/m to 2.6 A/m, respectively. Also, the results are well-validated by the calculations of NEC.

Phase current in high frequencies: Figure 4.27 illustrates the current amplitudes of energized Phase *a* and induced Phase *c* as a function of axial direction in the high-frequency regions, such as 1 and 10 MHz.

The currents calculated by the extended TL approach and NEC show a clear characteristic of convergence as the axial distance increases. Also, the currents calculated by the extended TL approach are well-validated with those evaluated by NEC.

The currents calculated by the classical TL approach which has been implemented into the existing EMT-type simulation tools [36] show numerical oscillations. A major reason is the lack of consideration of earth-return admittance.

4.3.2.2 Bidirectional wave propagation

The evaluation and calculation of external electromagnetic field components in Section 4.3.2.1 are based on the uni-directional wave propagation, thus the influence

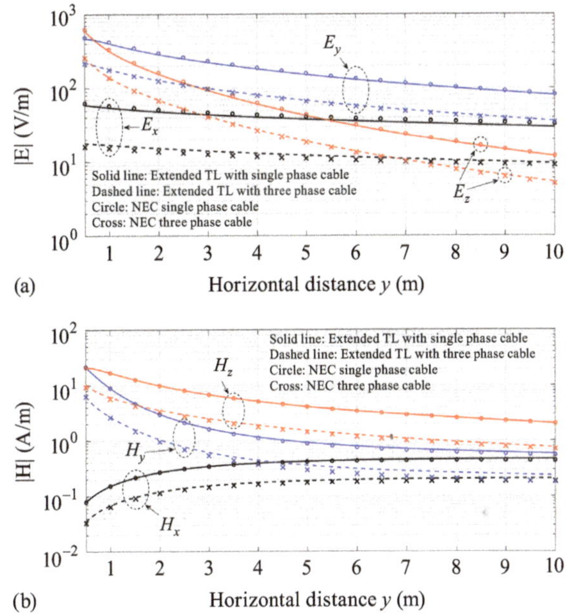

Figure 4.26 Lateral EM field components calculated at 100 kHz for shielding effect of cables with Phase - b energization, $x = 50$ m, $z = -1.5$ m, $\rho_1 = 1{,}000$ Ωm, and $\varepsilon_{r1} = 1$. (a) Electric field components and (b) magnetic field components

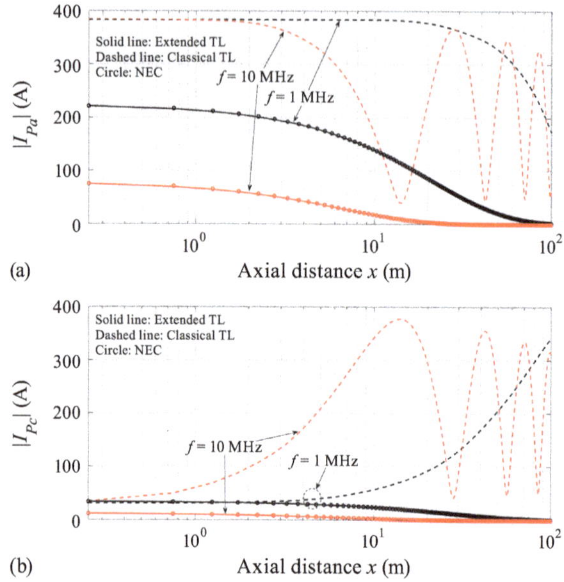

Figure 4.27 Amplitude of phase currents for Phases a–b energization, $\rho_1 = 1{,}000\ \Omega m$, and $\varepsilon_{r1} = 1$. (a) Phase a and (b) phase c

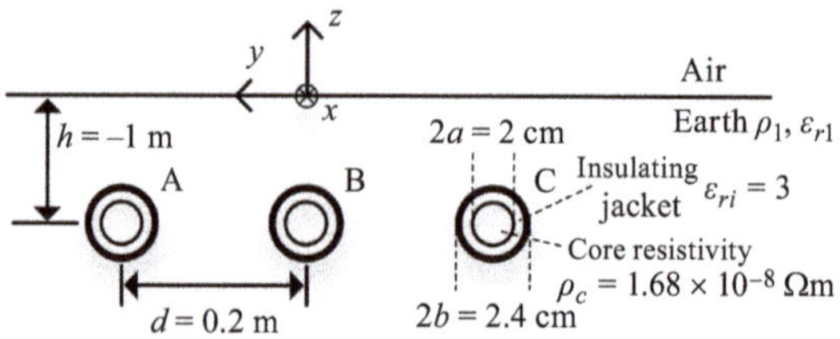

Figure 4.28 A three-phase cable with horizontal arrangement

of cable terminations and lengths has been neglected. As discussed in Chapter 2, (2.143) further improves the assumption of uni-directional propagation by adopting the concept of bi-directional propagation with cable terminations.

Equation (2.143) is evaluated using the extended TL approach which includes the generalized earth-return impedance and admittance formulas derived in Chapter 2. A three-phase underground cable shown in Figure 4.28 has been used in the following calculations of external electromagnetic field components [15].

Energization of external electromagnetic field components

The conditions of energization are similar to the cases used in the calculations of uni-directional wave propagation. However, since the terminations of cables are considered, the expressions of energizations should be re-defined.

The underground cable adopted into the calculations has been energized at the sending end by two phases and single-phase voltage sources. The energization of the two phases means that Phase *a* and Phase *b* of the cable are energized by $V_a = 10$ kV and $V_b = -10$ kV constant voltage sources, respectively, and it gives the following expressions for $\mathbf{I}_{s0}$ in (2.127) and (2.128):

$$\mathbf{I}_{s0} = [\mathbf{Y}_{s0}(1,1)V_a \quad \mathbf{Y}_{s0}(2,2)V_b \quad 0]^T \tag{4.10}$$

where $\mathbf{Y}_{s0}(1,1)$ and $\mathbf{Y}_{s0}(2,2)$ are the first and second diagonal elements of $\mathbf{Y}_{s0}$, also mean the self-admittances of Phases *a* and *b* sources.

The single-phase energization only involves that the Phase *b* is energized by $V_b = 10$ kV constant voltage source, and $\mathbf{I}_{s0}$ has

$$\mathbf{I}_{s0} = [0 \quad \mathbf{Y}_{s0}(2,2)V_b \quad 0]^T \tag{4.11}$$

It should be noted that the source admittance matrix $\mathbf{Y}_{s0}$ is not necessary to be the only diagonal matrix as used in (4.10) and (4.11), the nondiagonal elements in $\mathbf{Y}_{s0}$ can be further adapted to represent the mutual couplings of source circuits, for example, the Phase to Phase coupling of two cables.

Using the energizations of (4.10) and (4.11), the vectors of external electromagnetic field components in the lossy earth are computed as the lateral profiles and functions of the frequency with different conditions. The field observation points used in the following calculations are respected to the Cartesian coordinate system shown in Figure 4.27. The influences of energizations, terminating conditions, length of cable, and also frequency-dependent (FD) soil parameters on external electromagnetic field components are investigated and studied.

Influence of different energizations

The influences of two phases and single-phase energizations on the vectors of external electromagnetic field components are investigated and analyzed. The resistance R_{s0} and inductance L_{s0} of the source at the sending end of the cable are assumed to be 1 Ω and 0.5 μH for each phase [1]. The load on the sending end of the cable is assumed to be open-circuited, that is, $Z_0 = \infty$. Also, the source at the receiving end is neglected. The load impedance Z_L of receiving end is assumed to be resistive and the value is 200 Ω for each phase. The length l of the cable is assumed to be 500 m.

The amplitudes of total electromagnetic field components in the phase and modal domains are calculated by the following formulas:

$$\left.\begin{aligned} E_t &= |E_{ex}(x,y,z)| + |E_{ey}(x,y,z)| + |E_{ez}(x,y,z)| \\ H_t &= |H_{ex}(x,y,z)| + |H_{ey}(x,y,z)| + |H_{ez}(x,y,z)| \end{aligned}\right\} \tag{4.12}$$

where E_t and H_t are the amplitudes of total external electromagnetic field components in phase domain:

$$\left.\begin{aligned} E_{t-\text{mode}} &= |E_{ex-\text{mode}}| + |E_{ey-\text{mode}}| + |E_{ez-\text{mode}}| \\ H_{t-\text{mode}} &= |H_{ex-\text{mode}}| + |H_{ey-\text{mode}}| + |H_{ez-\text{mode}}| \end{aligned}\right\} \tag{4.13}$$

where $E_{t-\text{mode}}$ and $H_{t-\text{mode}}$ are the amplitudes of total external electromagnetic field components in modal domain and

$$\begin{cases} E_{ex-\text{mode}} = \sum_{m=1}^{N} E_{exmv} \\ E_{ey-\text{mode}} = \sum_{m=1}^{N} E_{eymv} \\ E_{ez-\text{mode}} = \sum_{m=1}^{N} E_{ezmv} \end{cases} \tag{4.14}$$

$$\begin{cases} H_{ex-\text{mode}} = \sum_{m=1}^{N} H_{exmv} \\ H_{ey-\text{mode}} = \sum_{m=1}^{N} H_{eymv} \\ H_{ez-\text{mode}} = \sum_{m=1}^{N} H_{ezmv} \end{cases} \tag{4.15}$$

with $v = 1$ to N and represents the v-th mode.

The axial "x," horizontal "y," and vertical "z" components in (4.12) and (4.13) are calculated from the vectors of external electromagnetic field components in (2.143). The modal external electromagnetic field components generated by the v-th mode are calculated considering the contributions of all cables. Also, (4.14) and (4.15) allow the calculations of electromagnetic fields produced by the separated mode by each cable, such as the v-th modal current on the m-th cable.

Considering the characteristics of different methods, the NEC which is based on MoM has been used as a validation for the calculated electromagnetic field components based on (2.143).

Phases a–b *energization:* The lateral profiles of external electromagnetic field components are calculated using (2.143) and (4.12)–(4.15), and the results are shown in Figure 4.29. The field observation point is set to $x = 30$ and $z = -1.5$ m. It is clear that E_{ex}, E_{ey}, H_{ey}, and H_{ez} are dominant field components at the calculated field observation points. The vertical electric field component E_{ez} is dominant nearby the cable; however, it decreases significantly as the horizontal distance increases. Also, because of the Phases *a–b* energization, the field components shown in Figure 4.29(a) and (c) have nonsymmetrical distributions in respect of the center of lateral cable arrangement. The electromagnetic field components calculated using (2.143) are well-validated and verified by NEC.

Figure 4.29(b) and (d) shows the calculated total external electromagnetic field components in the phase and modal domains. The electromagnetic fields produced by the earth-return mode and inter-phase mode 1 majorly contribute to the total field components E_t and H_t for the Phases *a–b* energization. It is important to notice that modal fields $E_{t-\text{mode}}$ and $H_{t-\text{mode}}$ show symmetrical distributions although the

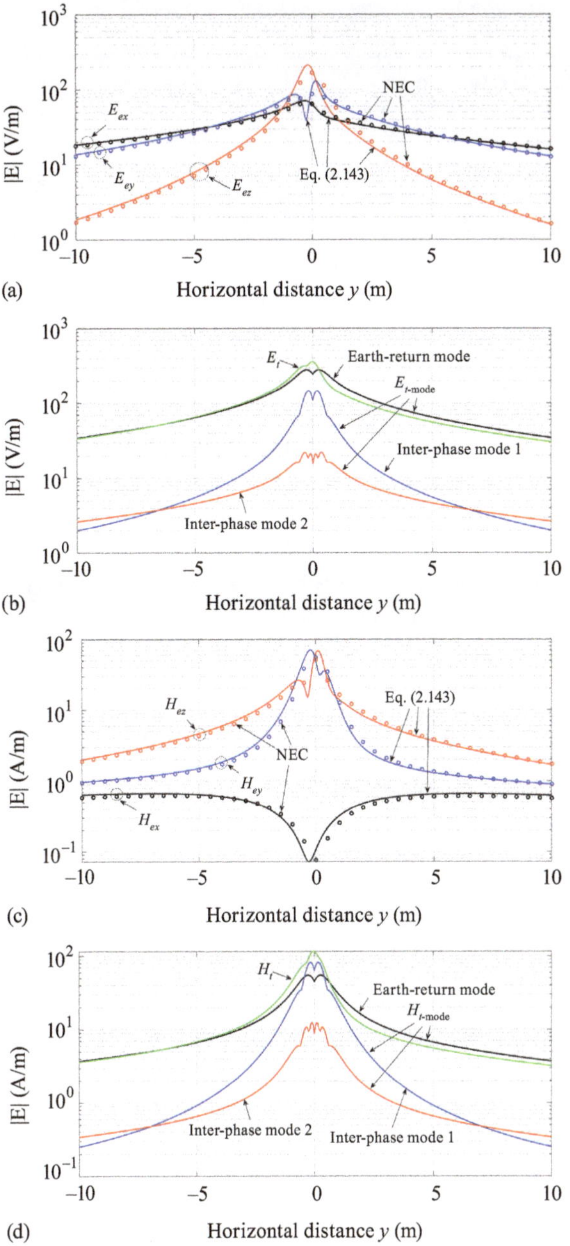

Figure 4.29 Lateral profiles of external electromagnetic field components calculated by Phases a–b energization with $f = 100$ kHz, $\rho_1 = 100$ Ωm, $\varepsilon_{r1} = 1$, $R_{s0} = 1$ Ω, $L_{s0} = 0.5$ μH, $Z_L = 200$ Ω, and $l = 500$ m at $x = 30$ m and $z = -1.5$ m. (a) Electric field components, (b) total and modal domain electric field components, (c) magnetic field components, and (d) total and modal domain magnetic field components

Table 4.1 Distribution of modal currents in the cable

Phase	**Earth-return mode**	**Distribution of earth-return mode currents**
A	**B**(1,1)**i**(1,1) = 49.5∠−99.3°	
B	**B**(2,1)**i** (1,1) = 52.1∠−98.6°	
C	**B**(3,1)**i** (1,1) = 49.5∠−99.3°	
Phase	**Inter-phase mode 1**	**Distribution of inter-phase mode 1 currents**
A	**B**(1,2)**i**(2,1) = 291.5∠31.9°	
B	**B**(2,2)**i**(2,1) = 0	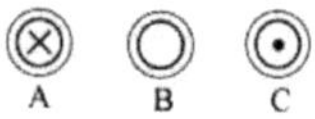
C	**B**(2,3)**i**(2,1) = 291.5∠−148.1°	
Phase	**Inter-phase mode 2**	**Distribution of inter-phase mode 2 currents**
A	**B**(1,3)**i**(3,1) = 157.3∠34.9°	
B	**B**(2,3)**i**(3,1) = 299.5∠−145.7°	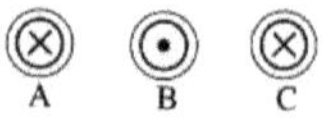
C	**B**(3,3)**i**(3,1) = 157.3∠34.9°	

energization is nonsymmetrical in the phase domain. The reason is the distributions of modal currents in the three-phase cable, as shown in Table 4.1.

Based on the results shown in Table 4.1, the following characteristics of distributions of modal currents are summarized:

- All three phases of underground cable carry the currents of earth-return mode, the current on Phase *a* is the same as the one on Phase *c*, but both are slightly different from the current on Phase *b*.
- Only phases *a* and *c* carry the currents of inter-phase mode 1. The current flows through Phase *a* and returns from Phase *c* due to that the phase shift between two currents is 180°.
- The currents of inter-phase mode 2 flow through Phases *a* and *c*, and return from Phase *b*, and also a 180° phase shift is observed between the current of Phase *a*/*c* and Phase *b*.
- Because of the above facts, it is clear that the distributions of modal currents in three modes are symmetrical in respect of Phase *b* of the cable. Thus, it results in the symmetrical distributions of electromagnetic fields in the modal domain although the energization is nonsymmetrical in the phase domain.

The calculated external electromagnetic field components using the classical and extended TL approaches at $f = 1$ MHz are illustrated in Figure 4.30. The circuit and cable conditions are the same as in Figure 4.29. It is clear that the higher frequency gives more nonsymmetrical distributions to external electromagnetic field components evaluated by extended TL in respect of cable's arrangement. Also, the amplitudes of all calculated external electromagnetic field components using extended TL decrease as the frequency increases.

The influence of earth-return admittance is also illustrated in Figure 4.30. A significant difference is observed for the results evaluated by the classical and

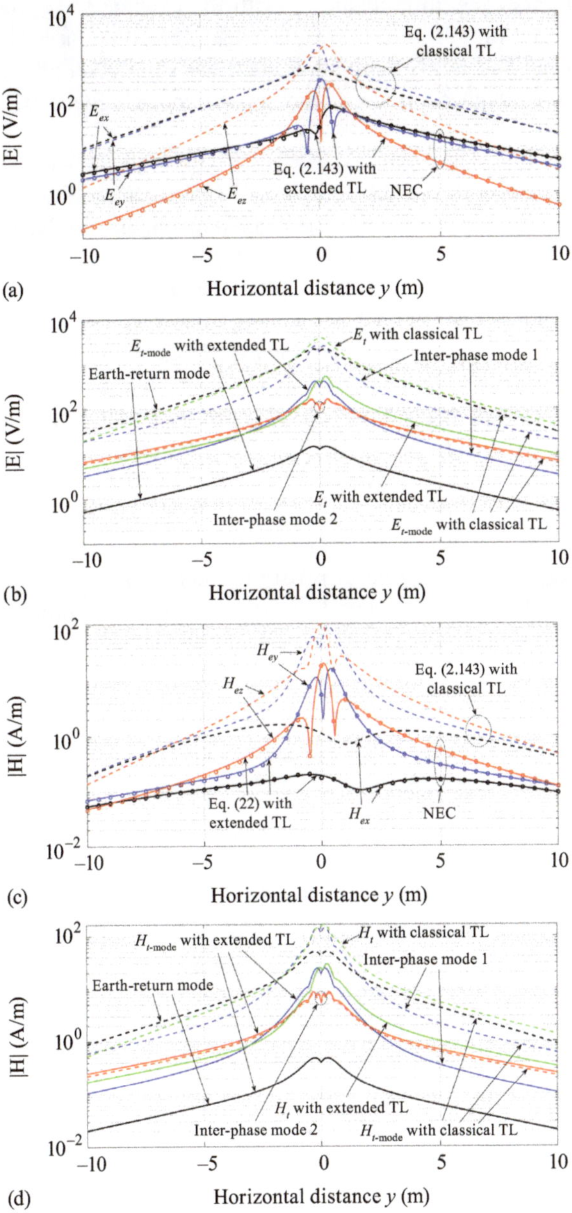

Figure 4.30 Lateral profiles of external electromagnetic field components calculated by Phases a–b energization with $f = 1$ MHz, $\rho_1 = 100$ Ωm, $\varepsilon_{r1} = 1$, $R_{s0} = 1$ Ω, $L_{s0} = 0.5$ μH, $Z_L = 200$ Ω, and $l = 500$ m at $x = 30$ m and $z = -1.5$ m. (a) Electric field components, (b) total and modal domain electric field components, (c) magnetic field components, and (d) total and modal domain magnetic field components

extended TL approaches. In comparison with the extended TL, the earth permittivity and earth-return admittance are excluded from the classical TL approach. Also, it is clear that the earth-return admittance has significant effects on the earth-return mode and inter-phase mode 1 of external electromagnetic field components.

Phase-b energization: The calculated external electromagnetic field components using the Phase-*b* energization in both phase and modal domains are shown in Figure 4.31. Due to the symmetrical energization in the phase domain, the distributions of evaluated external electromagnetic field components are also symmetrical. Moreover, the earth-return mode shows a significant impact in the calculations of total field components. The modal field components produced by the inter-phase modes 1 and 2 give negligible effects on the results of total fields; in fact, the modal field components calculated by the inter-phase mode 1 are nearly 0, and thus are not shown in the following figures.

Figure 4.32 shows the influences of earth resistivity on external electromagnetic field components in comparison with the cases shown in Figure 4.31. In Figure 4.32, the earth resistivity ρ_1 is set to 1,000 Ωm. The amplitudes of transverse components of the electric field E_{ey} and E_{ez} increase due to that the earth resistivity increases. However, the amplitudes of axial electric field E_{ex} and all magnetic fields decrease as the earth resistivity increases.

Furthermore, the results of external electromagnetic field components evaluated by the analytical solution (2.143) agree well with the results produced by NEC.

Influence of terminating conditions

In this section, the impact of load impedance at receiving end and source impedance at sending end of the cable on the amplitudes of total external electromagnetic field components are analyzed and investigated.

Load impedance at receiving end: The load impedance of the cable is adopted by different resistive values, that is, 10 (nearly short-circuited), 200, and 1,000 Ω. Also, the open-ended condition and loaded condition by the cable characteristic impedance Z_c are used in the calculations.

The amplitudes of external electromagnetic field components produced by Phases *a–b* energization with different load conditions and earth resistivities are calculated and shown in Figures 4.33 and 4.34. The field observation point is set to $x = 30$, $y = 5$, and $z = -1.5$ m. No resonance of total electromagnetic field components is observed if the characteristic impedance of the cable is used as the load impedance. The resonant phenomena of electromagnetic field components are obvious for the rest of the cases due to mismatch at the termination of receiving end of the cable. The resonance is highly frequency-dependent; however, in the low and high-frequency regions, that is, $f < 10$ kHz and $f > 400$ kHz, no resonance is observed.

Also, the load impedances with 200, 1,000 Ω, and open-ended condition have influences on the amplitude of resonance, but not on the resonant frequency. The load impedance of 10 Ω has a visible influence on the amplitude and resonant frequency due to that it is an extreme case because of nearly short-circuited condition. Moreover, the total external electromagnetic field components are essentially influenced by the earth resistivity, as shown in Figures 4.33 and 4.34.

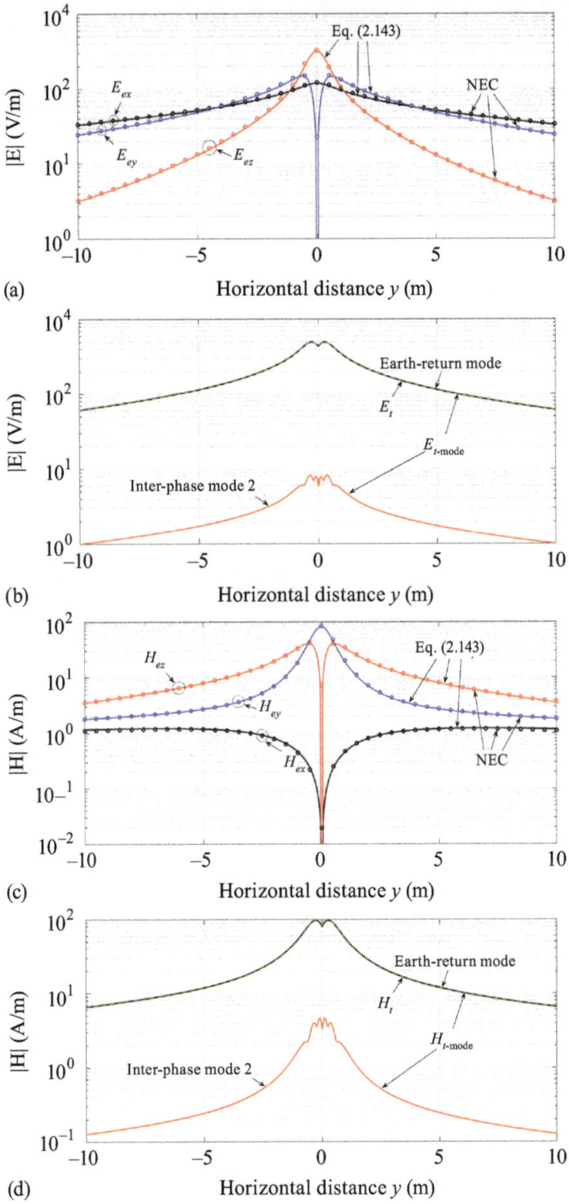

Figure 4.31 Lateral profiles of external electromagnetic field components calculated by Phase b energization with $f = 100$ kHz, $\rho_1 = 100\ \Omega$m, $\varepsilon_{r1} = 1$, $R_{s0} = 1\ \Omega$, $L_{s0} = 0.5\ \mu$H, $Z_L = 200\ \Omega$, and $l = 500$ m at $x = 30$ m and $z = -1.5$ m. (a) Electric field components, (b) total and modal domain electric field components, (c) magnetic field components, and (d) total and modal domain magnetic field components

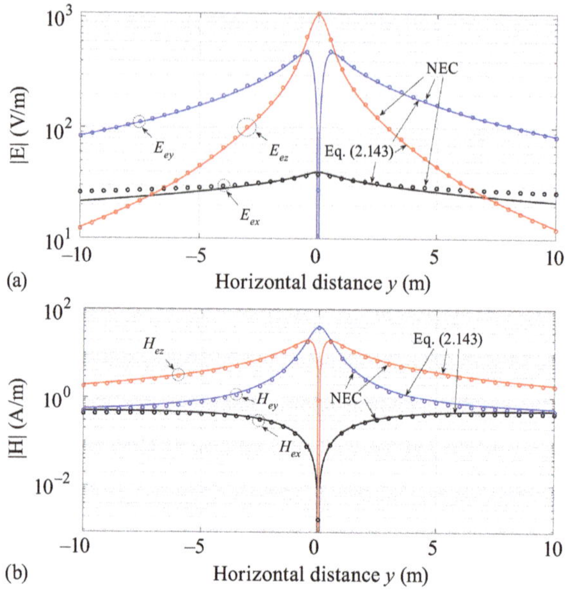

Figure 4.32 Lateral profiles of external electromagnetic field components calculated by Phase b energization with f = 100 kHz, ρ_1 = 1,000 Ωm, ε_{r1} = 1, R_{s0} = 1 Ω, L_{s0} = 0.5 μH, Z_L = 200 Ω, and l = 500 m at x = 30 m and z = −1.5 m. (a) Electric field components and (b) magnetic field components

The external electromagnetic field components close to the surface of the earth are shown and observed in Figure 4.35. The single-phase energization is adopted to the calculations. The similar resonant characteristics of external electromagnetic field components which are calculated at the field point below the cable are observed. It should be noted that the first peaks of resonance shift to a lower frequency in Figure 4.35 than the frequency observed in Figure 4.33.

Source impedance at the sending end: Figure 4.36 presents the influences of source inductance on the total external electromagnetic field components. The load resistance of 200 Ω is implemented into the calculations. No visible influences are found for the field components with $L_{s0} = 0$ and 0.5 μH. The inductance of $L_{s0} = 5$ μH gives minor influence on the field components at the high frequencies, that is, $f > 400$ kHz.

Influence of the cable length

As shown in Figures 4.37 and 4.38, the external electromagnetic field components evaluated by different lengths of the cable with the various earth resistivities are evaluated and compared. The lengths of the cable are varied by the values of 100, 500, 1,000, and 2,000 m. The receiving end of the cable is terminated by a resistive load of 200 Ω.

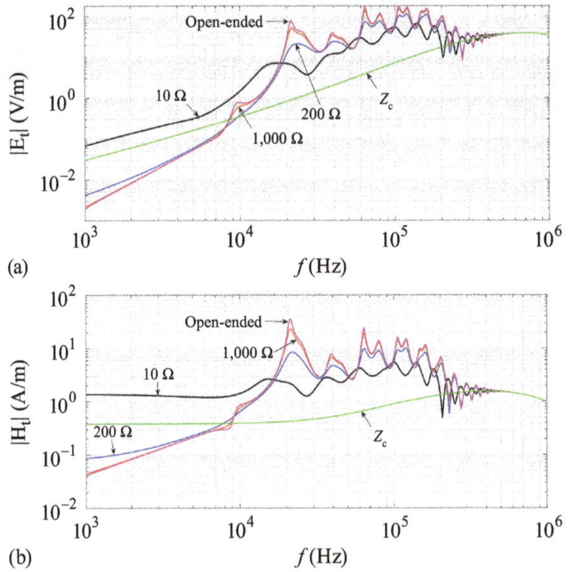

Figure 4.33 Total external electromagnetic field components calculated by Phases a–b energization with $\rho_1 = 100\ \Omega m$, $\varepsilon_{r1} = 1$, $R_{s0} = 1\ \Omega$, $L_{s0} = 0.5\ \mu H$, and $l = 500$ m at $x = 30$ m, $y = 5$ m, and $z = -1.5$ m. (a) Electric field components and (b) magnetic field components

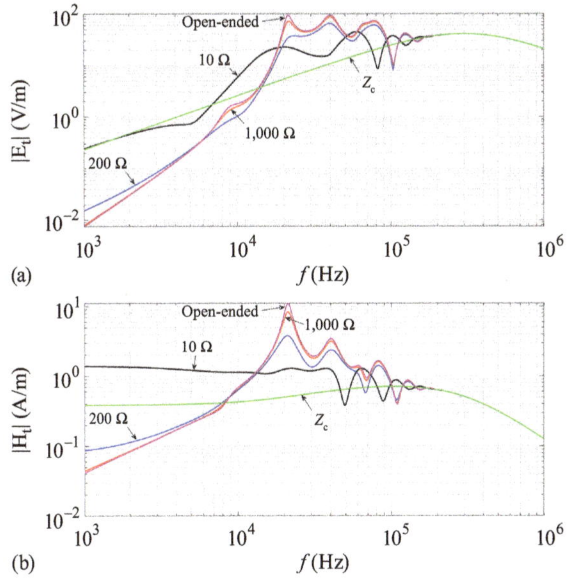

Figure 4.34 Total external electromagnetic field components calculated by Phases a–b energization with $\rho_1 = 1{,}000\ \Omega m$, $\varepsilon_{r1} = 1$, $R_{s0} = 1\ \Omega$, $L_{s0} = 0.5\ \mu H$, $l = 500$ m at $x = 30$ m, $y = 5$ m, and $z = -1.5$ m. (a) Electric field components and (b) magnetic field components

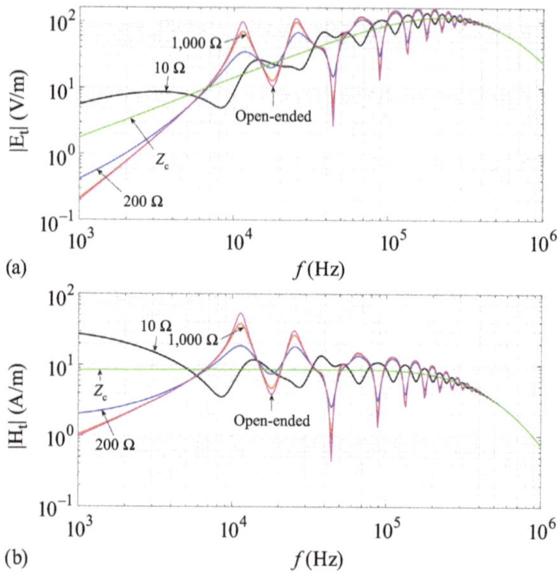

Figure 4.35 Total external electromagnetic field components calculated by Phase b energization with $\rho_1 = 100\ \Omega m$, $\varepsilon_{r1} = 1$, $R_{s0} = 1\ \Omega$, $L_{s0} = 0.5$ μH, and $l = 500$ m at $x = 30$ m, $y = 5$ m, and $z = -0.1$ m. (a) Electric field components and (b) magnetic field components

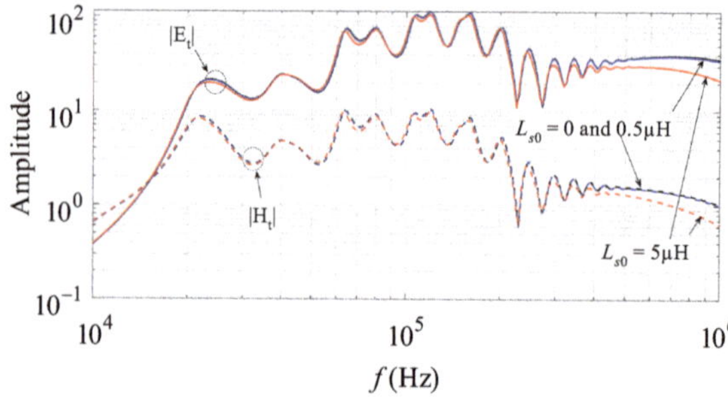

Figure 4.36 Total external electromagnetic field components calculated by Phases a–b energization with $\rho_1 = 100\ \Omega m$, $\varepsilon_{r1} = 1$, $R_{s0} = 1\ \Omega$, $Z_L = 200\ \Omega$, and $l = 500$ m at $x = 30$ m, $y = 5$ m, and $z = -1.5$ m

It is clear that the resonance of electromagnetic field components is highly frequency-dependent. The resonant frequency of the first peak decreases as the length of the cable increases, as illustrated in Figure 4.37. For example, the resonant frequency of the first peak with the cable length 100 m is around 107.2 kHz which corresponds to the cable delay of 9.3 μs; however, the first resonance appears at the frequency of 5.6 kHz with the cable of 2,000 m and has the cable delay of 178.6 μs. Also, the longer cable gives more serious resonant characteristics.

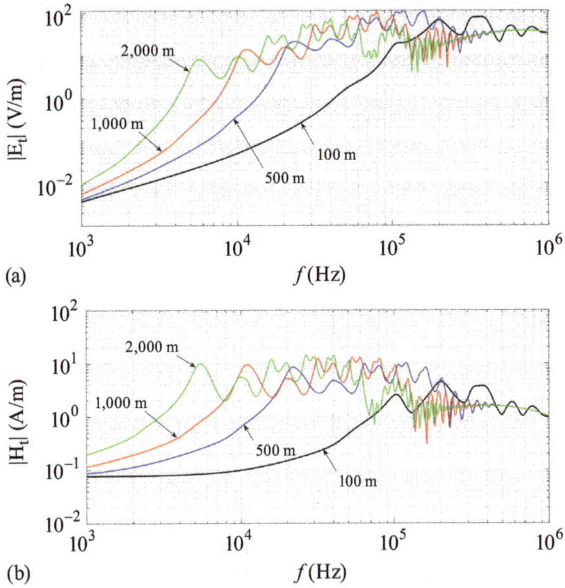

Figure 4.37 Total external electromagnetic field components calculated by Phases a–b energization with $\rho_1 = 100\ \Omega m$, $\varepsilon_{r1} = 1$, $R_{s0} = 1\ \Omega$, $L_{s0} = 0.5\ \mu H$, and $Z_L = 200\ \Omega$ at $x = 30$ m, $y = 5$ m, and $z = -1.5$ m. (a) Electric field components and (b) magnetic field components

The calculated total external electromagnetic field components with earth resistivity $\rho_1 = 1{,}000\ \Omega$m are given in Figure 4.38. In general, the earth resistivity has no visible influence on the frequency of the first peak of resonance with various lengths of the cable. However, the external electromagnetic field components evaluated by the large earth resistivity, that is, $\rho_1 = 1{,}000\ \Omega$m, show smoother and less resonances than the cases produced by $\rho_1 = 100\ \Omega$m.

Influence of frequency-dependent soil models

In this section, the frequency-dependent (FD) soil models which investigated in Section 3.2 of Chapter 3 are adopted into the following calculations of total external electromagnetic field components. The calculated results are shown in Figures 4.39–4.41. In fact, the electrical properties of soil (resistivity and permittivity) are naturally FD; however, the investigation and calculation discussed in the above sections neglect the impact of FD soil resistivity and permittivity on external electromagnetic field components.

The influences of Longmire/Smith (LS), Alipio/Visacro (AV), and Scott (SC) FD soil models which reviewed in Section 3.2 of Chapter 3 are applied into the calculations of external electromagnetic field components, and the results are compared to the field components computed by the constant soil (CS) model.

The earth resistivity at 100 Hz $\rho_{1(100\text{Hz})}$ is set to 100 and 1,000 Ωm for the FD soil models and the same constant values for the CS model. The FD earth relative

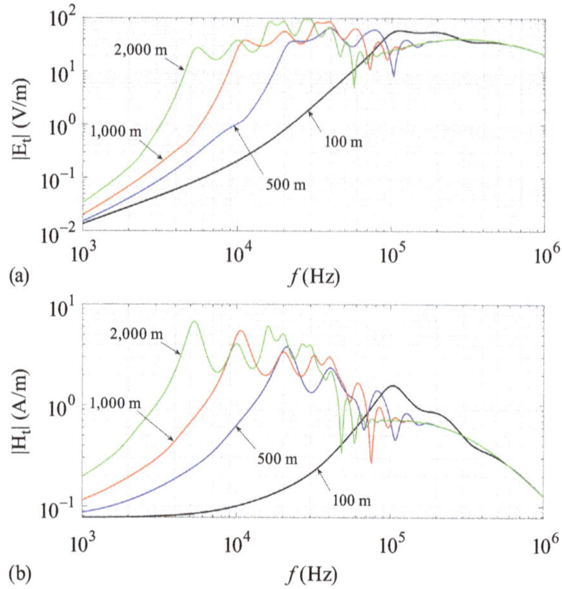

Figure 4.38 Total external electromagnetic field components calculated by Phases a–b energization with $\rho_1 = 1{,}000\ \Omega$m, $\varepsilon_{r1} = 1$, $R_{s0} = 1\ \Omega$, $L_{s0} = 0.5\ \mu$H, and $Z_L = 200\ \Omega$ at $x = 30$ m, $y = 5$ m, and $z = -1.5$ m. (a) Electric field components and (b) magnetic field components

permittivity is also included in the calculations. The upper limit of frequency is set to 1 MHz which generally fulfills the application limit of the three FD earth models.

As shown in Figure 4.39(a), the total external electromagnetic field components evaluated using both $\rho_{1(100\text{ Hz})} = 100$ and 1,000 Ωm are less sensitive to the FD soil parameters in comparison to the field components calculated by the CS model. Also, only minor influences of FD soil models with $\rho_{1(100\text{ Hz})} = 100$ Ωm on the total magnetic fields are observed in Figure 4.39(b) at the high frequency, that is, $f > 300$ kHz. Moreover, it is clear that the FD soil models with $\rho_{1(100\text{ Hz})} = 1{,}000$ Ωm have more visible influences on the magnetic field components above several tens of kHz. It is worth mentioning that the external electromagnetic field components computed by the LS and SC models have similar results, and external electromagnetic field components evaluated by adopting the AV model show a minor difference compared to the field components produced by the other two FD earth models.

Figure 4.40 illustrates the influence of FD soil models on the total external electromagnetic field components using single-phase energization. The similar characteristics shown in Figure 4.39 are observed.

Figure 4.41 illustrates the external electromagnetic field components calculated by AV soil model with $\rho_{1(100\text{ Hz})} = 1{,}000$ Ωm. The energization and load conditions are the same as those in Figure 4.34. It should be noted that Figure 4.34 is plotted by the CS model. Thus, the influence of FD soil model together with different load conditions can be further investigated and examined. It is clear that the FD soil

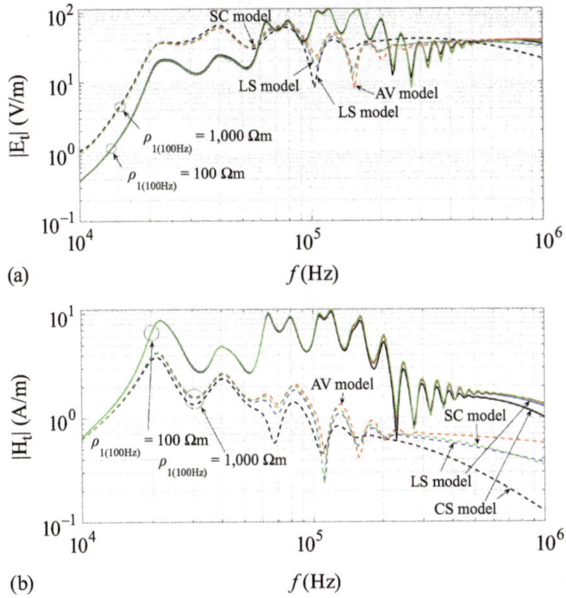

Figure 4.39 *Total external electromagnetic field components calculated by Phases a–b energization with $R_{s0} = 1\ \Omega$, $L_{s0} = 0.5\ \mu H$, $Z_L = 200\ \Omega$, and $l = 500$ m at $x = 30$ m, $y = 5$ m, and $z = -1.5$ m. (a) Electric field components and (b) magnetic field components*

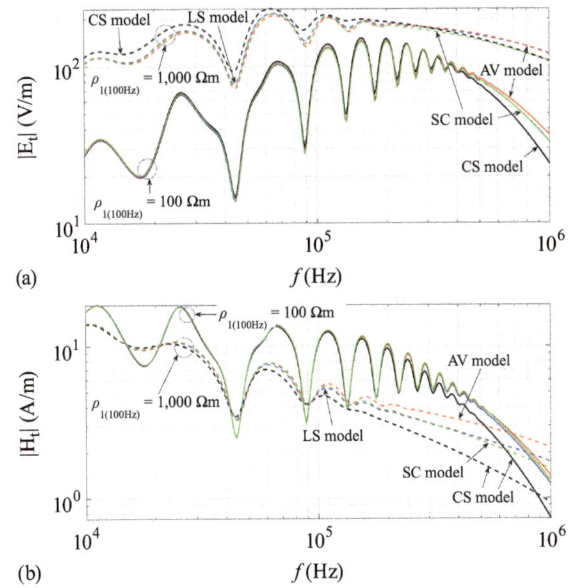

Figure 4.40 *Total external electromagnetic field components calculated by Phase b energization with $R_{s0} = 1\ \Omega$, $L_{s0} = 0.5\ \mu H$, $Z_L = 200\ \Omega$, and $l = 500$ m at $x = 30$ m, $y = 5$ m and $z = -0.1$ m. (a) Electric field components and (b) magnetic field components*

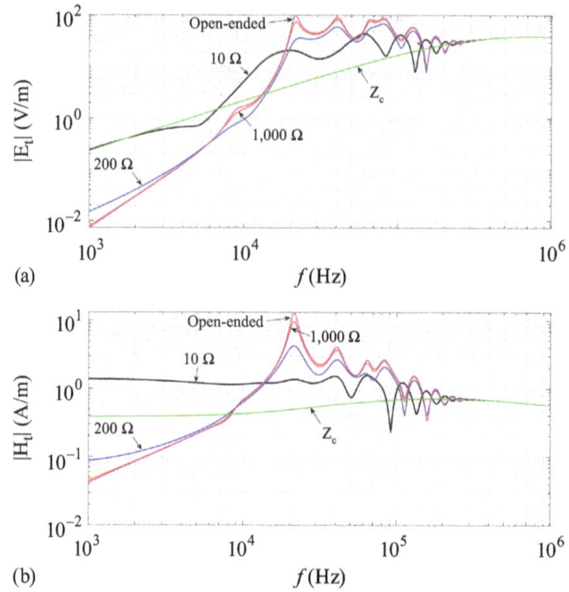

Figure 4.41 Total external electromagnetic field components calculated by Phases a–b energization with AV FD earth model, $\rho_{1(100Hz)} = 1{,}000$ *Ωm,* $R_{s0} = 1\ \Omega$*,* $L_{s0} = 0.5\ \mu H$*, and* $l = 500$ *m at* $x = 30$ *m,* $y = 5$ *m, and* $z = -1.5$ *m. (a) Electric field components and (b) magnetic field components*

model shows important influences on the external electromagnetic field components using different loads above several tens of kHz. For example, more resonances on electromagnetic field components are observed by adopting the FD soil model than the results calculated using the CS model between 50 kHz and 300 kHz.

4.4 Input impedance of cross-bonded cable

4.4.1 Bonding methods for cross-bonded cable

The objective of this section is to study the frequency domain responses of NB, XB, and TC cables illustrated in Figure 4.42. Resonance effects due to nonuniformities in overhead lines (OHLs) have been studied in many publications and are the cause of abrupt changes in line propagation characteristics [16–21]. There are no such studies for underground cables which also constitute nonuniform structures and the literature is limited to time-domain effects of nonuniformities [22–29].

4.4.2 Model circuit

The circuit shown in Figure 4.43, which corresponds to inter-core mode excitation in Figure 4.42, is used to compare the frequency responses of the cable given in Figure 3.25 and Table 3. 18. Each major section of 3 km is divided into three equal

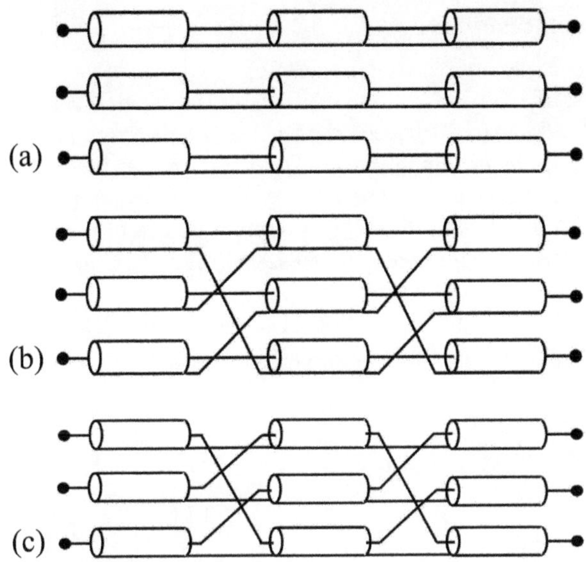

Figure 4.42 *Cable bonding methods. (a) Normal-bonded cable (NB), (b) normal cross-bonded cable (XB), and (c) transposed cable (TC)*

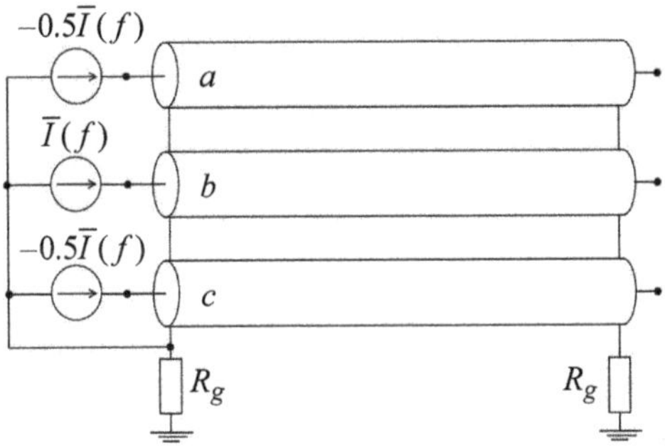

Figure 4.43 *Circuit for calculating frequency responses*

minor sections. The sheaths are bonded and grounded at each major section through a resistance $R_g = 5\ \Omega$.

The current source illustrated in Figure 4.43 is given in phasor notation by

$$\bar{I}(f) = e^{j2\pi f} \tag{4.16}$$

with frequency f scanned from 1 Hz to 1 MHz. The input impedance is calculated from the amplitude $|V_{src}|$ and phase θ of the source voltage on phase b as follows:

$$\bar{Z} = \frac{\bar{V}_{src}(f)}{\bar{I}}(f) = \frac{|V_{src}|e^{j2\pi f + j\theta}}{e^{j2\pi f}} = |V_{src}|e^{j\theta} \tag{4.17}$$

4.4.3 Input impedance

Figure 4.44 shows the input impedances for one major section. Contrary to the XB and TC cables, the response of the NB cable is regular with resonance dips occurring at equally spaced frequencies. The irregular responses of XB and TC are caused by the sheaths changing phase within each major section and by the different excitations of each phase. The inter-core excitation applied to the first minor section gives rise to inter-sheath mode currents which in turn also impact the core voltages of all phases. The irregular responses of XB and TC in Figure 4.44 are thus caused by the superposition of modes with different propagation constants.

Due to the bonding and grounding of sheaths, inter-sheath modes (modes 5 and 6, velocity $c_5 = c_6$ in Table 3.18) cannot propagate between major sections.

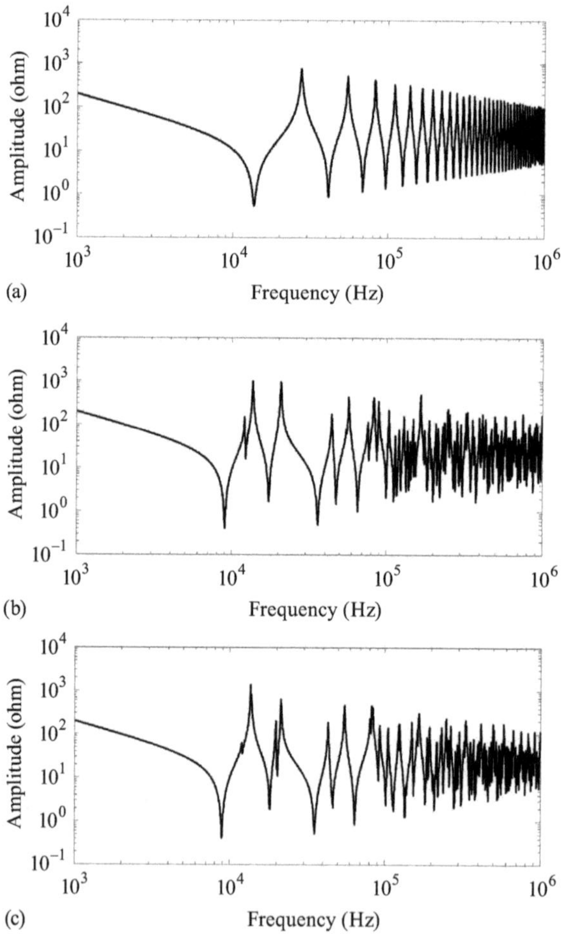

Figure 4.44 Input impedance magnitudes for three-phase inter-core mode excitation, one major section (3 km). (a) Normal-bonded (NB), (b) cross-bonded (XB) and (c) transposed cores (TC)

The first resonance associated to the inter-sheath modes in XB cable is calculated from the velocity in Table 3.18 at 1 MHz with one major section length $l = 3$ km: $f_0 = c_5/(4l) = 89.1\times10^6/12{,}000 = 7.425$ kHz. The impact of the inter-sheath modes is thus more noticeable above this frequency. It should be noted that $f_0 = 7.425$ kHz is evaluated by the propagation velocity at $f = 1$ MHz. Because the velocity is frequency-dependent and increases as frequency becomes higher, the accurate value of f_0 is estimated to be lower than 7.425 kHz.

Figure 4.45 shows the three-phase excitation of 18 major sections with cross-bonded sheaths for reference. The irregular response is again due to mode superposition.

4.4.4 *Resonant frequency*

Table 4.2 gives the first resonant frequencies (impedance scan dips/minima) in the response of NB and XB cables for a different number of major sections and for the excitation in Figure 4.43. The TC cable response is very similar to that of the XB cable and it is not shown.

In the NB cable response, resonant frequencies occur at f_0, $3f_0$, $5f_0, \ldots$ where $f_0 = v/(4l)$ and the inter-core mode velocity $c_2 = c_3 = c_4 = 167$ m/s in Table 3.18 (at $f = 1$ MHz). For one major section, it is found that $f_0 = 13.9$ kHz, $3f_0 = 41.9$ kHz,

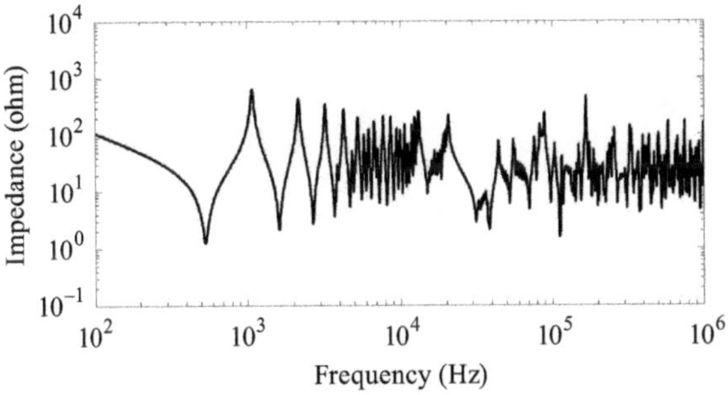

Figure 4.45 Input impedance magnitude for three-phase inter-core mode excitation of 18 major sections (54 km) with cross-bonded (XB) sheaths

Table 4.2 Resonant frequencies in open-circuited cable (kHz)

Major sections	NB cable			XB cable		
	f_0	f_1	f_2	f_0	f_1	f_2
1	13.8	41.4	68.9	9.02	17.4	36.0
3	4.56	13.8	23.0	3.24	8.93	11.7
6	2.28	6.84	11.4	1.62	4.79	7.64
18	0.73	2.26	3.79	0.54	1.63	2.70

and $5f_0 = 69.5$ kHz agree with Table 4.2. The same regularity is not observed for the XB cable in the table and the cause is the superposition of induced inter-sheath modes over the excited inter-core modes, as explained in the previous paragraph.

4.4.5 Attenuation characteristics

Figure 4.46 shows the zero-sequence and positive-sequence attenuations of the NB cable. The attenuation characteristics of an NB cable are perfectly smooth functions of frequency and no signs of resonance are observed. This demonstrates that NB cables are not affected by sheath grounding as far as core conductor propagation is concerned. The reason is that the first frequency at which resonance could be observed is high enough that the sheaths become perfect shields for the cores. Thus any phenomenon outside the sheath has no impact on the core propagation, including sheath grounding.

For the XB case, the zero-sequence attenuation plot is the same as in Figure 4.46. Figure 4.47 shows the irregular positive-sequence attenuation of the XB cable. This irregularity is due to the resonance of superimposed inter-core and inter-sheath modes which have very different characteristics.

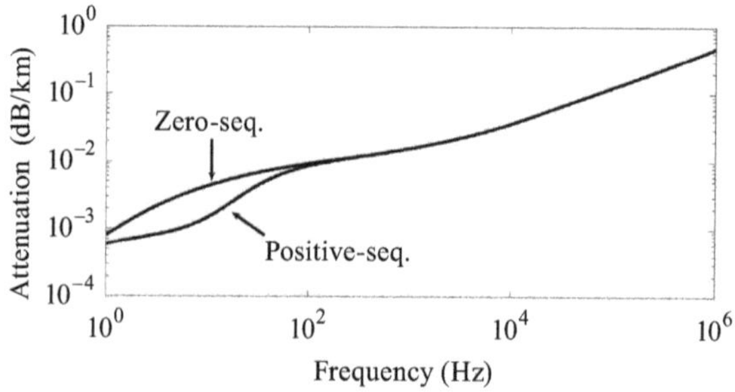

Figure 4.46 Zero-sequence and positive-sequence attenuation of a NB cable with one major section of 3 km

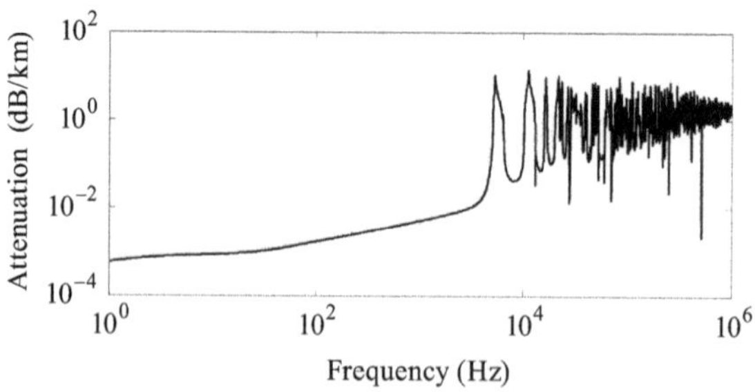

Figure 4.47 Positive-sequence attenuation of an XB cable with 1 major section of 3 km

If the grounding of sheaths uses a high resistance (e.g. $R_g = 30\ \Omega$), the positive-sequence attenuation remains exactly the same as in Figure 4.47. This means that what affects the cable response is not the grounding of sheaths but rather the bonding them together at the end of the major section. On the other hand, the irregularities in Figure 4.47 would not exist if the sheaths were not cross-bonded, that is, it is the cross-bonding that causes the superimposition of different modal components.

4.4.6 Time responses

The cable illustrated in Figure 3.25 with its 18 major sections (of 3 km) for a total length of 54 km is simulated in the time-domain for the bonding methods of Figure 4.42. Each major section is divided into three equal minor sections. The sheaths are bonded and grounded at each major section through 5 Ω resistances. A 225 kV balanced source (phase-*a* is at 0 degree) is applied to the cable at $t = 0$ s (all phases close at the same time).

Figure 4.48 shows the voltage of phase-*b* at the receiving end. The transients start at different times for the three-bonding methods in Figure 4.42, because of different propagation velocities given in Table 3.18. The XB and TC cables present initially small-amplitude high-frequency oscillations due to multiple reflections between the receiving end and the closest cable junction.

The above illustrates how XB and TC cables differ from NB cables in the time-domain. The slower propagation speed and lower attenuation result in transients with longer damping times.

4.4.7 Application of frequency responses for calculating the propagation constants of cross-bonded cables

To pursue the analysis of Sections 4.4.1–4.4.6, this section demonstrates that it is possible to calculate the propagation characteristics of nonuniform cross-bonded cable structures using directly the exact-pi circuit equivalent. This approach is useful for accounting for nonuniformities, such as grounding resistances and inductances of bonding wires.

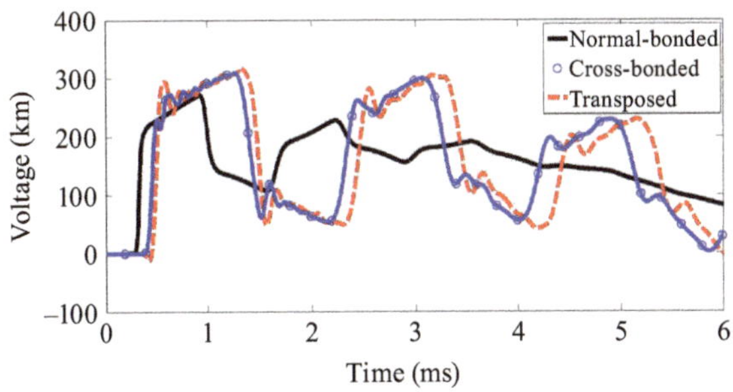

Figure 4.48 Phase-b receiving end voltage, energization of 18 major sections (54 km) with a 225 kV balanced infinite source

Since a modal theory was first applied to the study of power transmission systems [1,16,28,29], the usual approach for calculating the propagation constant $\gamma = \alpha + j\beta$ for a multiphase system is to calculate the eigenvalues of the product of per-unit-length matrices **ZY**. For nonuniform systems such as XB cables, this usually requires deriving the per-unit-length parameters of a homogeneous equivalent. The derivations of the homogeneous equivalent of a cross-bonded cable were first published in [30–32] (various derivation methods), but the most detailed and complete approach can be found in [33]. The method in [33] provides only the means (input data) for calculating **ZY** for the method presented below.

For a perfectly uniform structure, that is, a cable minor section, γ is accurately calculated from the equivalent exact-pi circuit representation of a mode, given by

$$\begin{pmatrix} I_1 \\ I_2 \end{pmatrix} = \begin{bmatrix} Y_{11} & -Y_{12} \\ -Y_{12} & Y_{11} \end{bmatrix} \begin{pmatrix} V_1 \\ V_2 \end{pmatrix} \tag{4.18}$$

$$\mathrm{Y}_{11} = \mathrm{Y}_0 \coth(\gamma l), \mathrm{Y}_{12} = \mathrm{Y}_0 \,\mathrm{cosech}\, (\gamma l) \tag{4.19}$$

where subscripts 1 and 2 are used for sending end and receiving end quantities and l is the total length of the cable. In the open-circuit case, $I_1 = 0$ and with the applied voltage $\mathrm{V}_1 = \mathrm{V}_{\mathrm{src}}$ in (4.18) and (4.19) are reduced to the following equation [28]:

$$\bar{V}_2 = \frac{1}{\cosh(\bar{\gamma} l)} \bar{V}_{src} \overset{\bar{V}_{src}=1\text{ V}}{\longrightarrow} \bar{\gamma} = \frac{1}{l}\cosh^{-1}\left(\frac{1}{\bar{V}_2}\right) \tag{4.20}$$

From (4.20), it is possible to evaluate γ simply from the open-circuit receiving end voltage and cable length.

If one assumes that (4.18) and (4.19) are also applicable to a nonuniform structure, that is, an XB cable, then γ can be calculated from the receiving end voltage response of the cable by applying (4.20) as proposed in [20].

Equation (4.20) can be used to calculate γ for the cable described in Figure 3.25 and Table 3.17 for a 3 km major section (with three minor sections of 1 km). Figure 4.49 shows propagation velocities using $c = \omega/\beta$ with β being the imaginary part of γ. The zero-sequence velocity is the same for NB and for XB sheaths. The

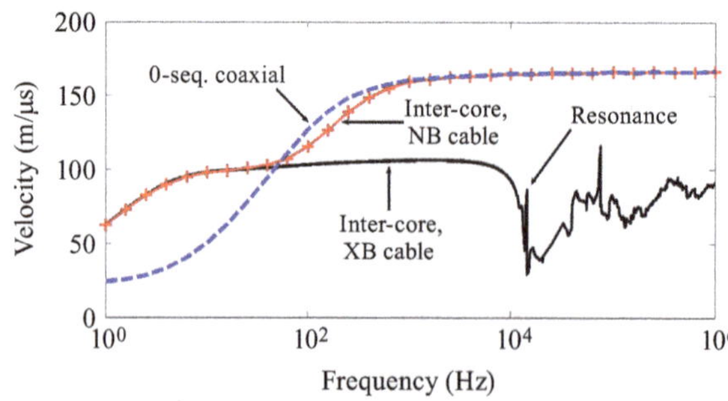

Figure 4.49 Propagation velocities calculated from the frequency scan of the open-circuit response of normal-bonded and cross-bonded cables

inter-core velocity of the XB cable increases up to about 7 kHz, after which it becomes irregular (resonant inter-sheath modes are superposed on the inter-core modes). The inter-core mode velocity in an NB cable goes from purely inter-core to purely coaxial as the sheath skin depth becomes less than the sheath thickness.

The velocity for zero-sequence coaxial and inter-core excitation in the NB cable converges to 167 m/μs. The velocity for inter-core excitation in the XB cable approaches 108 m/μs as in Table 4.2, but then deviates due to mode superimposition. The values agree with Table 3.18 before mode superposition starts to have an impact.

Another useful application of (4.20) is to calculate γ for an XB cable taking into account the grounding resistance R_g and the inductances of bonding wires L_w, as in the circuit of Figure 4.50, which excites an earth-return mode [34–36]. By applying (4.20) to the voltage $\bar{V}_2$ in Figure 4.50, the earth-return mode velocity of Figure 4.51 is obtained and shows the impact of R_g.

The calculation of $\bar{\gamma}$ using the per-unit-length matrices **ZY** (as provided in EMT-type simulation tools for each minor section) cannot account for $R_g \neq 0$, $L_w \neq 0$ which results in inaccurate estimation of modal constants as observed in Figure 4.51.

The results in Figures 4.49 and 4.51 show that the theory in (4.18) to (4.20), developed for uniform structures, can still be used to accurately calculate the

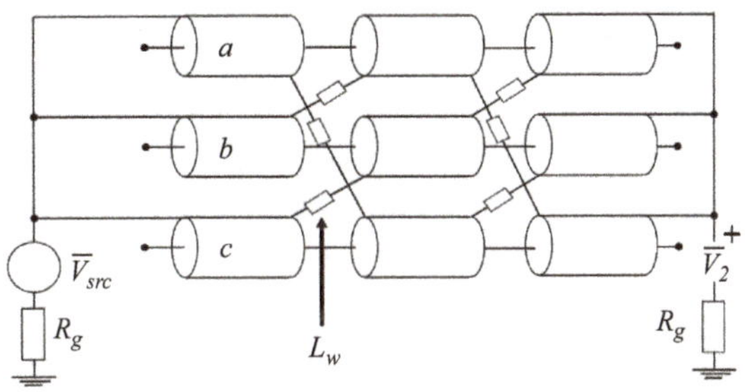

Figure 4.50 Excitation circuit of earth-return mode on an XB cable, one major section of 3 km (three minor sections of 1 km)

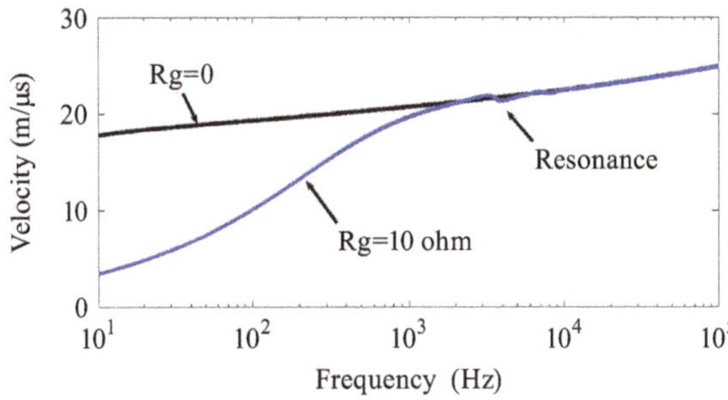

Figure 4.51 Earth-return mode velocity on an XB cable with sheath grounding resistance, one major section of 3 km (three minor sections of 1 km)

propagation characteristics of non-uniform systems such as XB cables [34–36]. The calculation is accurate as long as only one mode is excited. In the case of the XB cable, the method is applicable below modal resonant frequencies.

4.4.8 Modeling methods of complete cross-bonded cable systems

This section explains different modeling methods for XB cables within accuracy and computing time constraints. Validation from measurements is also presented.

The most straightforward way for modeling XB cables is by modeling each minor section separately (discrete model) [1,23–29,34–36]. This approach can be cumbersome and result in excessive computing times if the number of cable sections to model is very large. It also imposes a small numerical integration time-step for the surrounding network. As an alternative, a major section can be modeled by a single homogeneous equivalent with the three-bonded sheaths represented as a single conductor [30–36]. Despite clear gains in computing times, the homogeneous equivalent does not allow calculating sheath over-voltages within major sections and has limitations which affect the accuracy of time and frequency domain simulations. Previous studies demonstrated the importance of modeling bonding wire impedances, the nonuniform lengths of minor sections [37], and reflections caused by discontinuities at bonding points during a transient [25,27,34]. As explained below, even if bonding wire impedances are not important, the homogenous equivalent is not accurate in the frequency domain (wide range of frequencies).

Three alternative models for XB cables are presented in Figure 4.52 [27,34–36]. They are compared below based on the frequency and time-domain responses for a cable of 18 major sections in Figure 3.25 of 3 km (1 km minor sections). The sheaths are bonded and grounded at each major section through 5 Ω resistances and the cores are open-circuited at the receiving end.

As shown in Figure 4.52, Model 1 is the most accurate and models each minor section separately. Model 2 represents each major section by a homogeneous equivalent. Model 3 uses homogeneous equivalents for all, except the first and last major sections.

Figure 4.53 presents the responses for the above models when submitted to the excitation scheme in Figures 4.42 and 4.43. The response of Model 1 is regular for the first three impedance dips, but becomes irregular afterward due to the superposition of resonant inter-sheath modes, as explained in Sections 4.4.3–4.4.5. Model 2 cannot account for this phenomenon and presents a regular impedance response in the entire frequency spectrum. Model 3 can account for the resonance in the first and last major sections and therefore it has an impedance response closer to that of Model 1.

4.5 Conclusions

In this chapter, the wave propagation characteristics and external electromagnetic field components calculated based on extended and classical TL approaches on overhead and underground cables are investigated and analyzed in Sections 4.1–4.3.

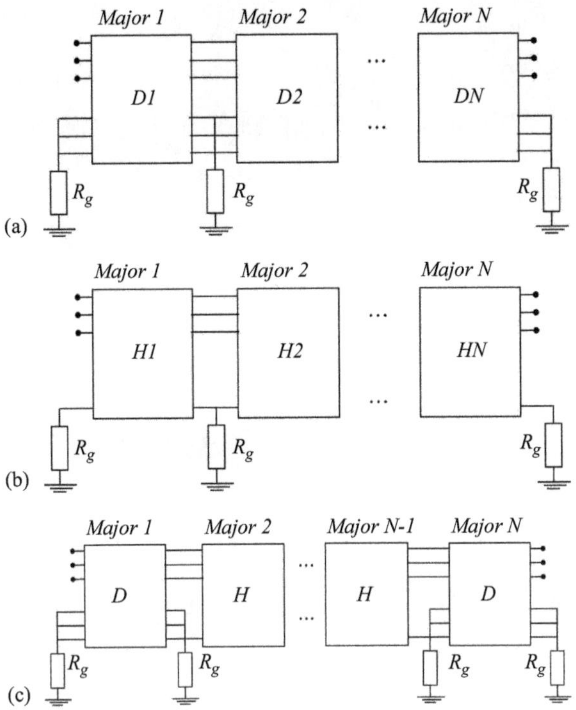

Figure 4.52 Alternate models for a cable with N major sections. (a) Model 1–discrete modeling of each major section, (b) model 2–each major section modeled with a homogeneous equivalent, and (c) model 3–discrete modeling of terminal major sections only. Adapted from [27,34–36]

Because of the earth-return admittance adopted into the extended TL approach, the following characteristics of propagation constants and electromagnetic field are observed.

- Sommerfeld–Goubau wave propagation can be observed at high frequencies for an overhead line and an overhead cable. This characteristic shows mode transition from a low-frequency earth-return wave to a high-frequency surface wave.
- A similar mode transition of the axial electric field component can be observed at high frequencies on an overhead conductor.
- In the surface wave propagation, the earth resistivity and permittivity have no influence on attenuation constants and axial electric field component.
- A significant difference in modal propagation constants on underground cables is observed between the extended and classical TL approaches. The attenuation constant and phase velocity of the earth-return and inter-sheath modes increase as frequency increases. Such phenomena cannot be discussed in the classical TL approach.

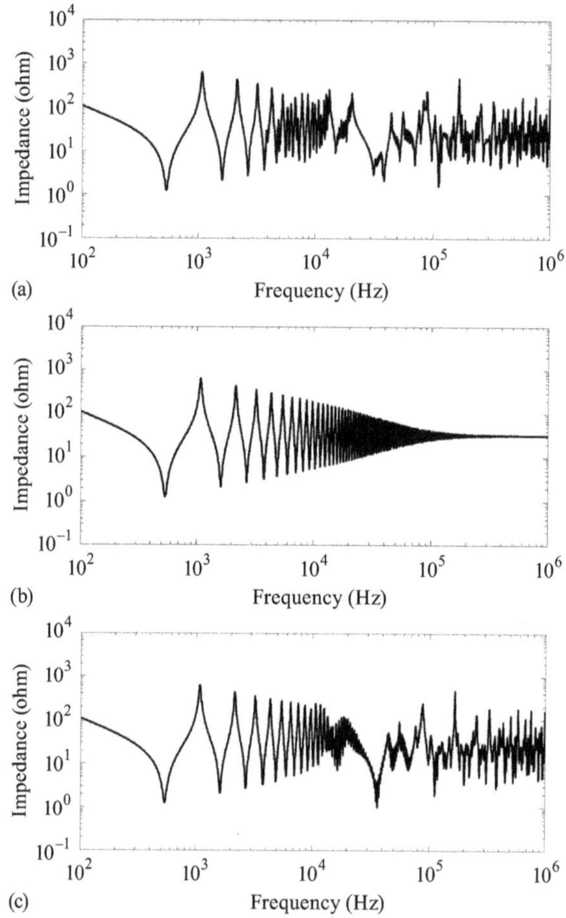

Figure 4.53 Input impedance for three-phase inter-core mode excitation in 18 major sections (54 km) of a cross-bonded cable using the models in Figure 4.52. (a) Model 1, (b) Model 2, and (c) Model 3

- The cable configuration, such as flat, vertical, or trefoil arrangement, shows a minor influence on the modal propagation constants of underground cables.
- The propagation constant of the coaxial mode on overhead and underground cables is independent from earth-return parameters.

Moreover, the external electromagnetic fields components generated by a multiphase underground cable are investigated and studied based on excitation currents of uni- and bi-directional propagation. The formulation using bi-directional current propagation allows the considerations of terminating conditions and length of underground cables on the calculations of external electromagnetic fields components. Also, the proposed formulation can be applied to arbitrary cable arrangements, such as flat, vertical, and trefoil configurations.

Considering a typical three-phase underground cable, the lateral and total external electromagnetic fields components at the various field observation points using the two phases and single-phase energizations are investigated and analyzed. The following general conclusions are summarized.

- The Phases *a–b* and Phase *b* energizations give nonsymmetrical and symmetrical distributions of external electromagnetic field components in respect of the center of the flat cable arrangement. The external electromagnetic field components evaluated by NEC agree well with the results calculated by the analytical solutions.
- If the receiving end of the cable is loaded by the noncharacteristic impedance of the cable, the resonant phenomena are observed in external electromagnetic field components due to the mismatch of termination. The load resistance at the receiving end of the cable has major influences on the amplitude of the resonance. The source inductance generally shows negligible influences on the components of external electromagnetic field components.
- The cable length has significant influences on the resonance and the resonant frequency of the external electromagnetic field components. The first resonant peak shifts to a lower frequency as the cable length increases. Also, the earth resistivity shows important influences on external electromagnetic field components.
- In general, the influences of FD soil models on external electromagnetic field components mainly occur at the high-frequency region together with the large earth resistivity. Also, the FD soil model produces more resonances on external electromagnetic field components with different load conditions in a high-frequency region than the results evaluated by the CS model.

Furthermore, a comprehensive study of frequency domain responses for HVAC cross-bonded cables has been performed in Section 4.4. The results explained in this section suggest the reader for making appropriate choices for modeling cables within computing time and accuracy constraints.

It is shown that cable inter-sheath modes induced because of cross-bonding, affect the propagation characteristics of inter-core modes.

Three alternatives for modeling cross-bonded cables with major sections have been described and analyzed. Modeling cross-bonded systems through homogeneous equivalents becomes inaccurate in the frequency domain when the frequency increases. This is because it cannot represent the inter-sheath modes. In time-domain studies, the models based on homogeneous equivalents are inaccurate in sheath voltage results. Detailed modeling of the first and last major sections does not improve accuracy substantially and limits the numerical integration time step. Accuracy becomes very important when designing cable protections (i.e., sheath voltages limiters) and in such cases, a full discrete modeling of each major section should be used.

The full discrete model is the most accurate, but computing times can become an issue, due to large surrounding networks and/or small numerical integration time-step requirements for short section lengths of such a model. In that case, it is acceptable to use the homogenous equivalents when considering only the core voltage results.

References

[1] Ametani, A., Ohno, T., and Nagaoka N. *Cable system transients: Theory, modeling and simulation*. Singapore: Wiley; 2015

[2] Kikuchi, A. H. "Wave propagation on the ground return circuit in high frequency regions." *IEE Japan Journal.* 1955; 75(805): 1176–87

[3] Kikuchi, A. H. "Electromagnetic fields on infinite wire at high frequencies above plane earth." *IEE Japan Journal*. 1957; 77(825): 721–33

[4] Chiba, J. "Studies in overhead wire – Goubau line above ground." *IEEE Transactions on Microwave Theory and Techniques*. 1977; 25(2): 83–93

[5] Ametani, A., Miyamoto, Y., Baba, Y., and Nagaoka, N. "Wave propagation on an overhead multiconductor in a high-frequency region." *IEEE Transactions on Electromagnetic Compatibility*. 2014; 56(6): 1638–48

[6] D'Amore, M., and Sarto, M. "Simulation models of a dissipative transmission line above a lossy ground for a wide-frequency range–Part I: Single conductor configuration." *IEEE Transactions on Electromagnetic Compatibility*. 1996; 38(2): 127–38

[7] D'Amore M., and Sarto, M. "Simulation models of a dissipative transmission line above a lossy ground for a wide - frequency range–Part II: Multiconductor configuration." *IEEE Transactions on Electromagnetic Compatibility*. 1996; 38(2): 139–49

[8] Xue, H., Ametani, A., Mahseredjian, J., Baba, Y., Rachidi, F., and Kocar, I. "Transient responses of overhead cables due to mode transition in high frequencies." *IEEE Transactions on Electromagnetic Compatibility*. 2018; 60 (3): 785–94

[9] Efthymiadis, A. E., and Wedepohl, L. M. "Propagation characteristics of infinitely-long single-conductor lines by the complete field solution method." *Proceedings of the IEE*. 1978; 125(6): 511–17

[10] Petrache, E., Rachidi, F., Paolone, M., Nucci, C., Rakov, V. A., and Uman, M. A. "Lightning induced disturbances in buried Cables–Part I: Theory." *IEEE Transactions on Electromagnetic Compatibility*. 2005; 47(3): 498–08

[11] Xue, H., Ametani, A., and Yamamoto, K. "Theoretical and NEC calculations of electromagnetic fields generated from a multi-phase underground cable." *IEEE Transactions on Power Delivery*. 2020; 10.1109/TPWRD.2020.3005521

[12] Burke, G. J. *Numerical Electromagnetics Code–NEC–4.2 Method of Moments Part I: User's Manual*. LLNL-SM-490875; 2011.

[13] Burke, G. J. *Numerical Electromagnetics Code–NEC–4 Method of Moments Part II: Program Description–Theory*. UCRL-MA-109338 Pt. II; 1992.

[14] Xue, H., Ametani, A., Mahseredjian, J., and Kocar, I. "Generalized formulation of earth-return impedance/admittance and surge analysis on underground cables." *IEEE Transactions on Power Delivery*. 2018; 33(6): 2654–63

[15] Xue, H., Ametani, A., and Yamamoto, K. "A study on external electromagnetic characteristics of underground cables with considerations of terminations." *IEEE Transactions on Power Delivery*. 2020; 10.1109/TPWRD.2020.3037335

[16] Wedepohl, L. M. "Application of matrix methods to the solution of traveling wave phenomena in polyphase systems." *Proceedings of the IEE*. 1963; 110 (2): 2200–12

[17] Wedepohl, L. M., and Wasley, R. G. "Wave propagation in polyphase transmission systems. Resonant effects due to discretely bonded wires." *Proceedings of the IEE*. 1965; 112(11): 2113–19

[18] Semlyen, A. "Some frequency domain aspects of wave propagation on nonuniform lines." *IEEE Transactions on Power Delivery*. 2003; 18(1): 315–22

[19] Faria, J. A. B., and Borges da Silva, J. F. "The effect of randomly earthed ground wires on PLC transmission–a simulation experiment." *IEEE Transactions on Power Delivery*. 1990; 5(4): 1669–77

[20] Faria, J. A. B. "On the resonance effects due to ground wires in transmission lines with non-uniform soil conductivity and non-uniform tower resistance." *IEEE Transactions on Power Delivery*. 1992; 7(1): 29–38

[21] Faria, J. A. B., and Neves, M. V. G. "Resonance effects due to conductor transposition in three-phase power lines." *Power Systems Computation Conference* (PSCC), Seville, Spain 2002.

[22] CIGRE WG C4.502. *Power System Technical Performance Issues Related to the Application of Long HVAC Cables*. CIGRE Technical Brochure 556; 2013

[23] Scott-Meyer, W. *EMTP rule book*. Portland: Bonneville Power Administration; 1981

[24] Dommel, H. W. *EMTP theory book*. Portland: Bonneville Power Administration; 1984

[25] Gudmundsdottir, U. S., Gustavsen, B., Bak, C. L., and Wiechowski, W. "Field test and simulation of a 400-kV cross-bonded cable system." *IEEE Transactions on Power Delivery*. 2011; 26(3): 1403–10

[26] Jensen, C. F., Gundmundsdottir, U. S., Bak, C. L., and Abur, A. "Field test and theoretical analysis of electromagnetic pulse propagation velocity on cross-bonded cable systems." *IEEE Transactions on Power Delivery*. 2014; 29(3): 1028–35

[27] Lafaia, I., Ametani, A., Mahseredjian, J., Naud, A., Correia de Barros, M. T., and Koçar, I. "Field test and simulation of transients on the RTE 225 kV cable." *IEEE Transactions on Power Delivery*. 2017; 32(2): 628–37

[28] Ametani, A. *Distributed-parameter circuit theory*. Tokyo: Corona Pub. Co.; 1990

[29] Ametani, A., Nagaoka, N., Baba, Y., Ohno, T., and Yamabuki, K. *Power system transient: Theory and applications*. 2nd edn. Boca Raton, FL: CRC Press; 2016

[30] Dang, N. D. *Transient Performance of Cross-Bonded Cable Systems*. University of Manchester, Ph.D. Thesis; 1972

[31] Wedepohl, L. M., and Indulkar, C. S. "Wave propagation in non-homogeneous systems: Properties of the chain matrix." *Proceedings of the IEE*. 1974; 121(9): 997–00

[32] Wedepohl, L. M., and Indulkar, C. S. "Switching overvoltages in short cross-bonded cable systems using the Fourier transform." *Proceedings of the IEE*. 1975; 122(11): 1217–21

[33] Nagaoka, N., and Ametani, A. "Transient calculations on cross-bonded cables." *IEEE Transactions on Power Apparatus and Systems*. 1983; 102(4): 779–87

[34] Ametani, A., and Kasai, Y. "Application of an equivalent homogeneous line model to a transient analysis on a distribution system." *IEE Japan Transactions on Power and Energy*. 1994; 114(10):1059–65

[35] Ametani, A. "An efficient method to deal with boundary conditions in an electromagnetic transient analysis." *International Conference on Power Systems Transients*, Lisbon, Portugal, 1995.

[36] Ametani, A. *Cable parameters rule book*. Portland: Bonneville Power Administration, 1994.

[37] Adamson, C., Taha, E. A.-Z. H., and Wedepohl, L. M. "Comparative steady-state performance of cross-bonded cable systems." *Proceedings of the IEE*. 1968; 115(8): 1147–55

Chapter 5

Steady-state and transient characteristics on three-phase cables

5.1 Cable discharge

5.1.1 Introduction

The significance of cable charging and discharging phenomena has been known by cable engineers especially when performing AC and DC tests of cables and also during maintenance work [1–5]. Recently, the phenomena are focused on adopting a high-voltage (HV) cable to regulate a grid voltage, the same as a shunt capacitor [6]. However, only a few test results of cable charging and discharging characteristics are found [7–10]. Considering this fact, the UK Grid performed a series of cable discharge tests on a 275 kV cable [11–13].

In this section, the charging and discharging characteristics of cables are investigated based on laboratory and field test results [7–13], theoretical analysis, and EMT simulation. Section 5.1.2 explains the basic formulas and parameters necessary for studying the cable charging and discharging. The derivation of the formulas is described in Appendix A5.1. In Section 5.1.3, field test results measured on a 275 kV POF cable with a length of 20 km in Japan and the UK are described. Also, laboratory test results of the charging and discharging characteristics of a thin insulator film are shown under various conditions of temperature and electric field strength. In Section 5.1.4, approximate but simple theoretical formulas are applied to estimate the charging and discharging times measured in the tests. The theoretical results are compared with the test results and the effectiveness of the analytical but simple formulas is discussed. In Section 5.1.5, the leakage resistance of cables and the time constant of cable discharging are surveyed. In Section 5.1.6, EMT simulations are performed, and simulation results are compared with the measured and theoretical results.

When a long cable is compensated by a shunt reactor, the discharge time is some 10 s [3,4]. If not compensated, it becomes much longer, more than an hour. A shunt reactor compensated cable is out of the scope of this chapter.

5.1.2 Basic formulas and parameters for charging and discharging phenomena

Figure 5.1(a) illustrates a typical circuit for cable charging and discharging tests. Figure 5.1(b) is a simplified circuit of Figure 5.1(a) for charging a cable, and

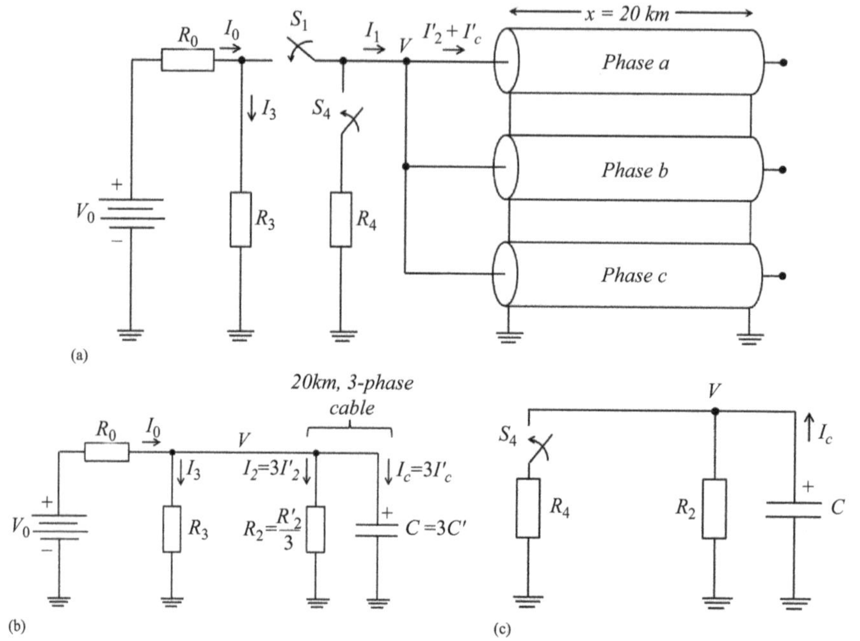

Figure 5.1 Field test circuit of a Japanese 275 kV POF cable with length 20 km estimated from [7] and equivalent circuits for theoretical analysis

Figure 5.1(c) for discharging. The derivation of the formulas in this section is explained in Appendix A5.1.

5.1.2.1 Charging

Cable sending-end voltage:

$$V(t) = V_0[1 - \exp(-t/\tau)] \tag{5.1}$$

where $V_0 = R_0\, I_0$ in Figure 5.1(a) and $V_0 = E$ in Figure A5.1.

The time constant "τ" for the charging is defined by

$$\tau = CR = C/G \tag{5.2}$$

Assuming that the cable is charged until the time T_1 when the voltage reaches V_1, T_1 is defined as

$$T_1 = \tau \ln[V_0/(V_0 - V_1)] \tag{5.3}$$

5.1.2.2 Discharging

Cable voltage:

$$V(t) = V_1 \exp(-t/\tau) \tag{5.4}$$

where V_1 is the voltage when a discharge starts at $t = T_1$, that is, $V_1 = V(T_1)$ in (5.1).

Assume that the cable voltage becomes V_2 at $t = T_2$ by the discharge. Then

$$T_2 - T_1 = \tau \ln\{V(T_1)/V(T_2)\} \tag{5.5}$$

5.1.2.3 Parameters related to charging and discharging

5.1.2.3.1 Cable capacitance C

$$C = \Delta C(nl)(\mathrm{F}), \quad \Delta C = 2\pi\varepsilon_0\varepsilon i/\ln(r_{out}/r_{in})(\mathrm{F/m}) \tag{5.6}$$

where l is the cable length, n is the number of phases, r_{out} and r_{in} are the outer and inner radii of the cable insulator, and εi is the relative permittivity of the insulator.

5.1.2.3.2 Leakage resistance R for discharging

$$R = 1/G = \Delta R/(nl)(\Omega), \quad \Delta R = 1/\Delta G = \rho ln(r_{out}/r_{in})/2\pi\,(\Omega/\mathrm{m}) \tag{5.7}$$

where G is the insulator conductance (S).

It is possible to calculate R using the cable capacitance C and permittivity parameters commonly found in cable datasheets:

$$R = \varepsilon_0\varepsilon_r\rho/C \tag{5.8}$$

where ρ is the resistivity of the cable insulator.

In Figure 5.1, the details of each element are explained below.

I_0 is the rated charging current, about 3 mA and $I_1 = I_2 + I_c$. I_c is the cable charging (discharging) current. I_2 is the cable leakage current, about 1.2 mA/three-phase. I_3 represents the leakage current of equipment used in the test, about 0.6 mA. The voltage $V_0 = 500$ kV/η is the DC generator's internal voltage with a working factor of DC generator $\eta = 0.9$. Cable voltage has $V = 414$ kV. C represents the cable capacitance/three-phase for 20 km. R_0 is the charging resistance. R_2 shows the leakage resistance of the cable, about 22 GΩ-km/phase. R_3 is the leakage resistance of equipment (bushing, insulator strings, etc.). Water resistance used for forced discharge after $t = T_2 = 144$ ms is $R_4 = 30$ MΩ.

5.1.3 Test results

5.1.3.1 Field test on a 275 kV POF cable in Japan (Test-A)

A field measurement on a 275 kV POF cable with a length of 20 km was performed as a commissioning test in the presence of inspectors from the Japanese Ministry of International Trade and Industries, because the 275 kV POF cable installation in 1971 was the world first experience, and there existed no testing standard. It was planned that the test results would be the basis of revising the existing testing standard for POF cables. Thus, the field measurement involved a series of tests for the characteristics of oil behavior, DC withstand voltage, surge propagation, thermal/mechanical behavior of steel pipes, etc. [7]. As a part of the tests, cable charging and discharging characteristics were measured on the 20 km long cable.

5.1.3.1.1 Test circuit

The circuit configuration of the field test is not shown in [7]. From the test conditions and theoretical analysis described in the reference, the test circuit is estimated as in Figure 5.1. The test was somehow similar to an AC voltage withstand test, that is, 1-h application of specified voltage E, 10 min/1.8E, h/E. For the measurement, a DC 500 kV generator (Cock-Croft type composed of an AC source, a transformer, diodes, and capacitors) is connected to the short-circuited three-phase cores at the sending end and measuring equipment is installed. At the receiving end, the three-phase cores are open-circuited.

5.1.3.1.2 Tested cable specifications and parameters [7]

The most significant contribution of this reference is that the leakage current of equipment used in the test and the leakage resistance of the cable are described, although the given parameters are not clear enough. Without the parameters, it is impossible to perform theoretical calculations and EMT simulations of cable charging and discharging characteristics. Although the cable capacitance is easily calculated from the cross-section geometry and the insulator permittivity, the leakage currents of insulator strings, bushings, and the cable itself are not available in general, and very difficult to measure or theoretically to predict. It is noteworthy that the authors of [7], Test-A, are in the same manufacturer as those of [8], Test-B in Section 5.1.2.2. Table 5.1 gives the tested cable specification.

The rated charging current of the DC generator is said to be $I_o = 3$ mA in [7]. Leakage current I_3 of the DC source and the external circuit is stated to be about 0.6 mA at the voltage of 414 kV, and the cable leakage current is said to be 1.2 mA/three phases estimated from the leakage resistance of the cable $R_2 = 22\times10^9$ Ω-km, or 1,100 MΩ for 20 km per phase, which corresponds to tan (δ) of $21.57\times10^{-6} f$, or 0.432×10^{-6} for $f = 50$ Hz. Note that the phenomena discussed in this section is in DC, and thus it is impossible, in theory, to define tan (δ) which is a function of frequency (frequency-dependent).

5.1.3.1.3 Test result

A measured charging and discharging characteristic of the cable is shown in Figure 5.2. In the test, (1) at $t = 0$, a DC voltage is applied to the three-phase cores to observe a

Table 5.1 Specification of tested 275 kV POF cable (cross-section 1,400 mm^2)

Core conductor inner radius	0 mm	Core conductor resistivity	$1.728\cdot10^{-8}$ Ω-m
Core conductor outer radius	23.65 mm	Screen resistivity	$73\cdot10^{-8}$ Ω-m
Screen inner radius	43.65 mm	Pipe resistivity	$10\cdot10^{-8}$ Ω-m
Screen outer radius	44.65 mm	Pipe permeability	300 μ_0
Cable outer radius	46.65 mm	Core/screen/pipe insulation permittivity	3.7 ε_0
Pipe inner radius	127.3 mm	Semiconducting layer thickness	3 mm
Pipe outer radius	133.7 mm	Buried depth	2.5 m

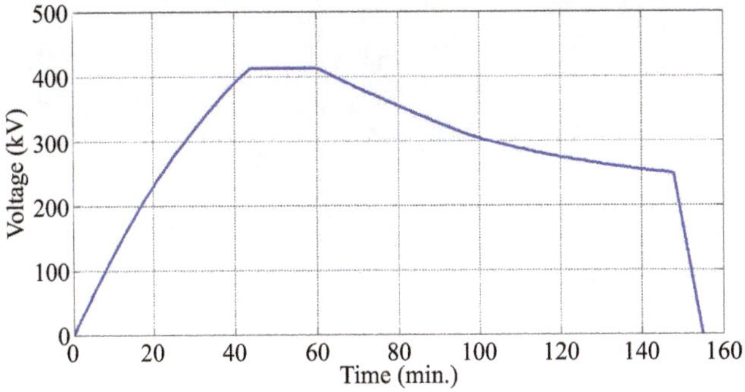

Figure 5.2 Measured result of charging and discharging characteristics of a 275 kV POF cable. Reproduced from [7]

cable charging characteristic. (2) At about 44 min, when the cable voltage reaches 414 kV, the DC source and the measuring equipment are switched off, thus, the voltage is kept to be about 414 kV. (3) At about $t = 60$ min, voltage measurement starts to observe the cable discharge characteristic. (4) At about $t = 148$ min, when the cable voltage is reduced to about 250 kV, a forced discharge is made by using a water resistor.

5.1.3.2 Laboratory test (Test-B)

Figure 5.3 shows measured curves of discharge currents as a function of time [8]. Figure 5.3(a) is for a temperature of 20 °C and (b) for 80°C (near to the cable operating temperature). Note that both the horizontal axis (time) and the vertical axis (discharge current) are on a logarithmic scale. Three curves in each figure correspond to the applied electric field strength, that is, 30 kV/mm, 20 kV/mm, and 10 kV/mm. The measurement was carried out in a laboratory by adopting a specially designed test chamber with the inner chamber composed of a Cu electrode of diameter 70 mm as a cable core and a Pb tape as a metallic sheath. The Cu electrode is lapped over by an OF cable insulator (oil immersed paper with thickness 125 μm, relative permittivity 3.6 and length 1 m) up to the total thickness of 0.5–5 mm. Then, the inner chamber is set into the main chamber and the spaces between the main and inner chambers are filled up by oil with a pressure of 1 kg/cm^2. The oil temperature is controlled by circulating the oil.

A test insulator is charged for 20 h. Then a discharge test is performed to make the stored charge in the thin insulator film flowing out. To measure the discharge current, a switch of the test circuit is made repeatedly on and off by every certain time period. Thus, the discharge current is measured as a function of time.

The figure clearly shows the influence of the insulator temperature, that is, the leakage currents for 80°C in Figure 5.3(b) is higher by more than 10 times those for 20°C in Figure 5.3(a). A similar observation can be found in [14].

It is observed that the leakage current decreases as electric field strength E decrease, but the three curves for different E are somehow parallel in Figure 5.3(a), which indicates that the leakage resistance is not much dependent on the field strength.

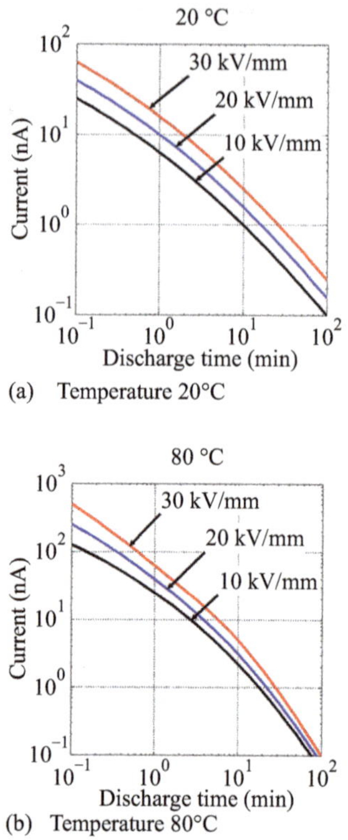

Figure 5.3 Measured results of discharging time as a function of electric field strength. Adapted from [8]. (a) Temperature 20 °C and (b) temperature 80 °C

If we assume that the electric field strength in the insulator of the POF cable described in Section 5.1.3.1 is uniform, then electric field strength E in the cable is given as 20.7 kV/mm for $V = 414$ kV with the insulator thickness of 20 mm, and 10 kV/mm for $V = 200$ kV. Then, it becomes possible to estimate the leakage resistance of the cable from the test results in Figure 5.3 considering the leakage current is in the range of some mA. As the first step, the leakage current, vertical axis, is to be converted from nA/m to mA/20km to consider a longitudinal difference from Test-A. Then, the leakage resistance can be estimated from the ratio of capacitances in Test-A and Test-B and that of voltages which take into account a radial difference between the tests. The estimated resistance "R_2" ranges from 700 to 2500 MΩ per phase for 20 km. This result is also confirmed by test results in a UK 275 kV POF cable described in Section 5.1.3.3.

The test results in Figure 5.3 can be used to estimate the voltage of the model cables during discharge as

$$v(t) = \frac{E_0 d}{I_0} i(t) \tag{5.9}$$

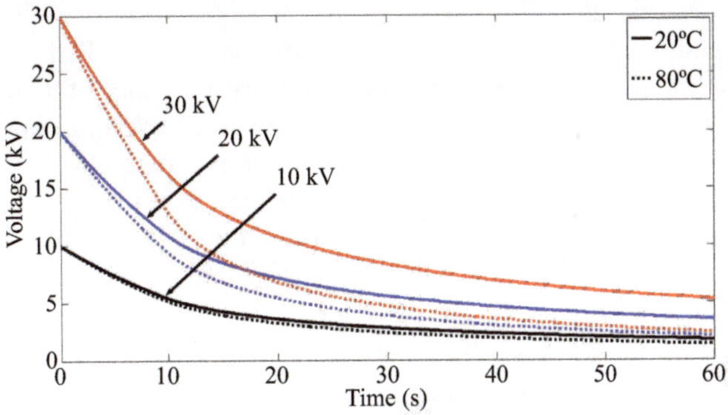

Figure 5.4 Voltage of model cable of 1 mm thick insulation during discharge test for different values of initial applied voltage and temperature

where E_0 is the initial electric field applied to the model cable, $i(t)$ is the leakage current in Figure 5.3, and I_0 is its initial value and d is the insulation thickness of the model cable. The voltages for the case of 1 mm thick insulation are shown in Figure 5.4, where the impact of temperature on the rate of decay is clearly observed.

The leakage currents in Figure 5.3 have a characteristic given by

$$i(t) = I_0 e^{-t/\tau} \tag{5.10}$$

where I_0 is the current at $t = 0$.

The capacitance for the model cable with 1 mm insulation thickness is calculated from (5.6) with $\varepsilon_i = 3.6$, $r_{\text{in}} = 35$ mm, $r_{\text{out}} = 36$ mm, $l = 1$ m, and $n = 1$, resulting into $C = 7.11$ nF. References [8] and [9] do not give details on the leakage resistance. The leakage resistance can be estimated from the cable capacitance and the time constant of the current in Figure 5.3(a) by using (5.2) and (5.5).

The time constant τ is evaluated from Figure 5.3(a) as

$$\tau = \frac{60(100 - 1)}{\log_e(10/0.1585)} = 1433.2 \text{ s} \tag{5.11}$$

Using $C = 7.11$ nF and $\tau = 1433.2$ s, the leakage resistance is found to be $R = 201.6$ GΩ. This value includes the leakage of the model cable and the external circuit. Such a high leakage resistance may be explained by very low losses in paper insulated cables and by the fact that a laboratory test has better conditions (low humidity, low levels of contamination) than a field test.

5.1.3.3 Field test on a 275 kV POF Cable in UK (Test-C)

National Grid (UK), in collaboration with Cardiff University, performed field test measurements of a POF cable discharge. The cable voltage, temperature, pressure, and the relative humidity of air were measured simultaneously allowing us to see the impact of these factors on discharge time [6,11–13].

5.1.3.3.1 Tested cable configuration and specification

Figure 5.5 and Table 5.2 show the details of the 275 kV POF cable with a length of 21 km. The cable is used in a trench with a depth of 1 m. The cable sealing end at ground level is connected to a piece of overhead horizontal bus bar with a length of about 13.5 m. The bus bar is supported by two vertical insulator poles, each with an axial length (not creep-age) of 2.4 m. The capacitance of each cable phase is calculated using the data in Table 5.2, that is, $\varepsilon_r = 3.8$, $r_{\text{in}} = 29.75$ mm, $r_{\text{out}} = 40$ mm, $l = 21 \times 10^3$ m, and $n = 1$ into (5.6), resulting in $C = 15.0$ μF.

The three-phase cables are fed by a balanced (positive sequence) AC voltage source. When the source is switched off, the cable has 2 pu $\left(2 \times 275\sqrt{2/3} \approx 450 \text{ kV}\right)$ on phase-c, −1.3 pu in phase-b, and −1.6 pu in phase-a. The source is switched off and the voltage on the cable is measured for 10 h.

5.1.3.3.2 Measured result

Figure 5.6 shows the measured cable voltage during discharge. It took about 8 h to fully discharge the cable. The voltage measured in each phase in Figure 5.6 follows a characteristic given by (5.4).

The time constant for discharging each phase can be obtained using any two points t_1 and t_2 of a curve in Figure 5.6 and (5.5), resulting $\tau_a = 3.04$ h, $\tau_b = 3.20$ h, and $\tau_c = 3.50$ h. The leakage resistance can be estimated using the time constants and the capacitance $C = 15.0$ μF in (5.2) giving $R_{\text{total}-a} = 729.8$ MΩ, $R_{\text{total}-b} = 768.2$ MΩ, and $R_{\text{total}-c} = 840.2$ MΩ. These values are the total resistance of each phase including the leakage in the cable itself and in the external circuit (test circuit).

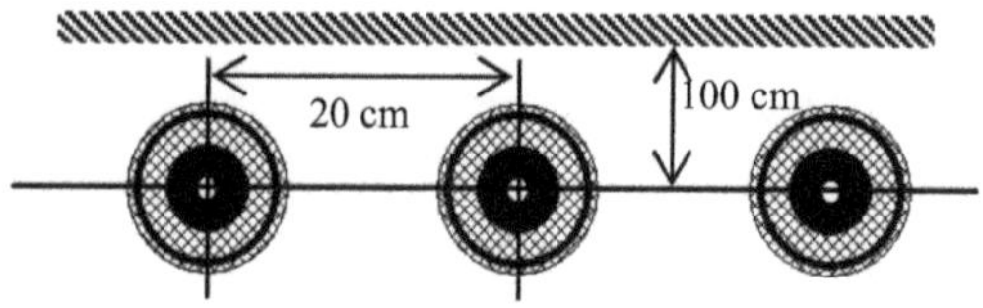

Figure 5.5 Configuration of a 275 kV POF cable in the UK. Adapted from [6]

Table 5.2 Specification of a 275 kV POF cable in the UK. Adapted from [6]

Core conductor inner radius	6.8 mm	Core conductor resistivity	$1.724 \cdot 10^{-8}$ Ωm
Core conductor outer radius	26.75 mm	Screen resistivity	$21.3 \cdot 10^{-8}$ Ωm
Screen inner radius	43.00 mm	Core insulation permittivity	3.8 ε_0
Screen outer radius	47.20 mm		
Semiconducting layers thickness	3 mm	Screen insulation permittivity	8 ε_0
Cable outer radius	51.56 mm	Depth	1 m

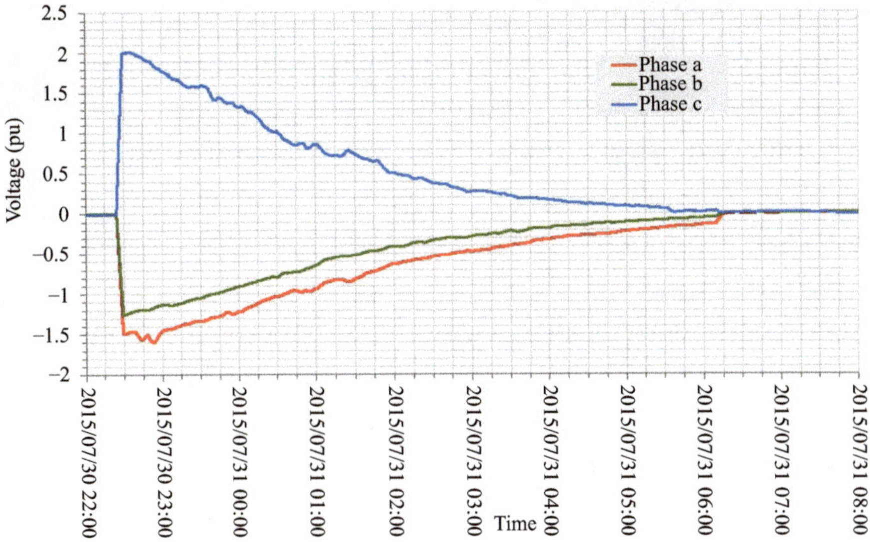

Figure 5.6 Test results of discharge characteristic, 275 kV POF cable. Courtesy of National Grid, UK

5.1.4 Theoretical analysis

5.1.4.1 Test-A

5.1.4.1.1 Parameters of a cable

From the specifications of a 275 kV POF cable in Table 5.1: cross-section 1400 mm^2, core inner radius $r_i = 10$ mm, core outer radius (= insulator inner radius) $r_{in} = 23.65$ mm, insulator outer radius $r_{out} = 43.65$ mm, and relative permittivity $\varepsilon_i = 3.7$.

(a) Cable capacitance:

$$C = 20.124\ \mu\text{F} \tag{5.12}$$

(b) Leakage resistances:

(b1) Charging resistance:

$$R_0 = 500\ \text{kV}/3\ \text{mA} = 166.7\ \text{M}\Omega \tag{5.13}$$

(b2) Cable:

$$\Delta R_2 = 22 \times 10^9\ \Omega\ \text{km/phase},\ R_2 = 22 \times 10^9/20 = 1{,}100\ \text{M}\Omega \text{ for 20 km} \tag{5.14}$$

(b3) Measuring circuit:

$$R_3 = V_1/I_3 = 414 \times 10^3/0.6 \times 10^{-3} = 690\ \text{M}\Omega \tag{5.15}$$

(c) Total leakage resistance:

(c1) Charging:

$$R_c = R_1//R_2//R3 = 112.4\ \text{M}\Omega \tag{5.16}$$

(c2) Discharging:

$$R_d = R_2//R_3 = 424\ \text{M}\Omega \tag{5.17}$$

(d) Time constant

(d1) Charging:

$$\tau_c = CR_c = 2{,}262\ \text{s} = 37.7\ \text{min} \tag{5.18}$$

(d2) Discharging:

$$\tau_d = CR_d = 8904\ \text{s} = 148.4\ \text{min} \tag{5.19}$$

5.1.4.1.2 Calculated results

(a) Time T_1 to charge until $V_1 = 414$ kV, $V_0 = 500/0.9 = 555.5$ kV, see Figure 5.1:

$$T_1 = \tau \ln\{V_0/(V_0 - V_1)\} = 37.7 \ln(555.5/141.55) = 37.7 \times 1.367 = 51.5\ \text{min} \tag{5.20}$$

The above result is reasonable in comparison with the measured result of 44 min in Figure 5.2.

(b) Time T_2 to discharge until V_2, $T_2 = \tau \ln(V_1/V_2)$

Substituting $V_1 = 414$ kV, $V_2 = 250$ kV, and $\tau = 148.4$ min into the above equation,

$$T_2 = 74.85\ \text{min} \tag{5.21}$$

The above theoretical result agrees reasonably well with the measured result of 85 min in Figure 5.2.

If we adopt an approximate formula to estimate the leakage resistance explained in (A5.20) in Appendix A5.1, the resistance is evaluated from the measured result in Figure 5.2 as $R = 531$ MΩ. Then, $\tau = 178.1$ min and $T_1 = 89.76$ min which agrees very well with the measured result as expected.

5.1.4.2 Test-B

Figure 5.3 in Section 5.1.3.2 shows a discharge current characteristic. Differentiating (A5.4) by time

$$dq/dt = I = -(Q/\tau)\exp(-\mathrm{t}/\tau) = -(E/R)\exp(-t/\tau) = -I_0\exp(-t/\tau) \tag{5.22}$$

Considering the current direction is opposite to the charging current

$$I = I_0 \exp(-\mathrm{t}/\tau) \tag{5.23}$$

Time T_2 for $I = I_2$ is given in the following form:

$$T_2 = \tau \ln(I_0/I_1) \tag{5.24}$$

where $I_0 = 75$ nA.

In [8], it is stated that $\varepsilon_i = 3.6$, but no detail of the leakage resistance and insulator capacitance is given. Assuming the insulator thickness of 0.5 mm, $r_{in} = 35$ mm, and $r_{out} = 35.5$ mm. From (5.6),

$$C = 0.14\ \mu\text{F/m} \tag{5.25}$$

Also, assuming the total leakage resistance R is nearly the same as that in Test-A in Section 5.1.3.1:

$$R = 400\ \text{M}\Omega \tag{5.26}$$

Thus

$$\tau = CR = 56\ \text{s} \tag{5.27}$$

T_2 for $I_2 = 15$ nA is now given by

$$T_2 = \tau \ln(75/15) = 56 \times 1.609 = 90.1\ \text{s} = 1.5\ \text{min} \tag{5.28}$$

The above theoretical result shows a reasonable agreement with the test result in Figure 5.3 for 20°C.

5.1.4.3 Test-C

In reality, the leakage resistances estimated in Section 5.1.3.3.2 are average values based on the initial voltage characteristics in Figure 5.6. The time constant can be estimated from (5.5), rewritten as

$$-t/\tau = \ln\{v(\text{t})/v(0)\}, t = T_1 - T_2 \tag{5.29}$$

The above equation shows that if the vertical axis in Figure 5.6 is transformed to logarithmic scale, the value of τ can be observed simply from the slope of the voltage characteristic. This is illustrated in Figure 5.7, and it is easily observed that the time constant changed during the test. This is because the humidity in the air increased during the test, which caused an increase in the leakage current and a reduction of the time constant.

Figure 5.8 shows temperature, pressure, and relative humidity measured during the field test. The humidity increased from 80% to 100%. The impact of this change is that the time constant is reduced from 3.5 h (initial) to 1.5 h and the leakage resistance is reduced from $R_{\text{total}-c} = 840.2$ MΩ to 360.1 MΩ.

The impact of air humidity is caused by the bus-bar, insulator poles connected to the cable, as well as the cable sealing ends which are exposed to air.

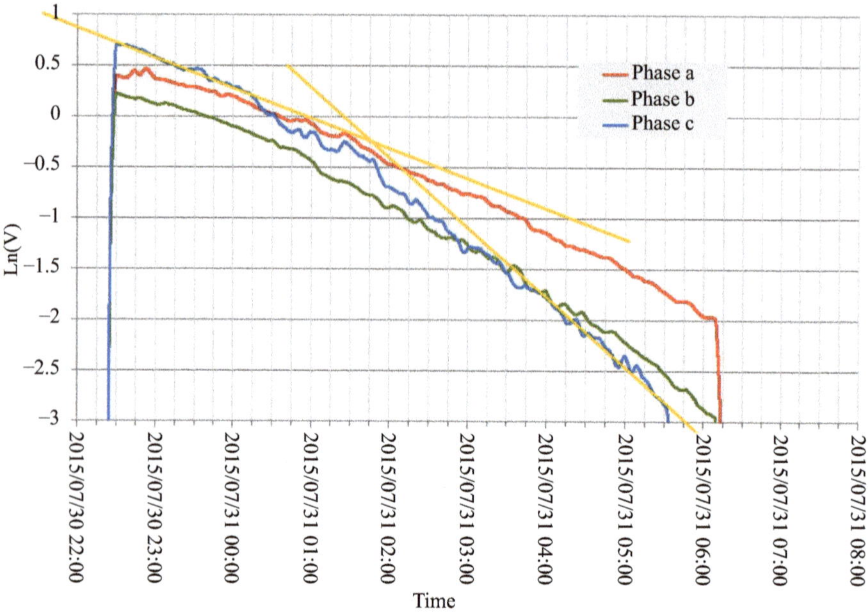

Figure 5.7 Test results of Figure 5.6, logarithmic scale. Courtesy of National Grid, UK

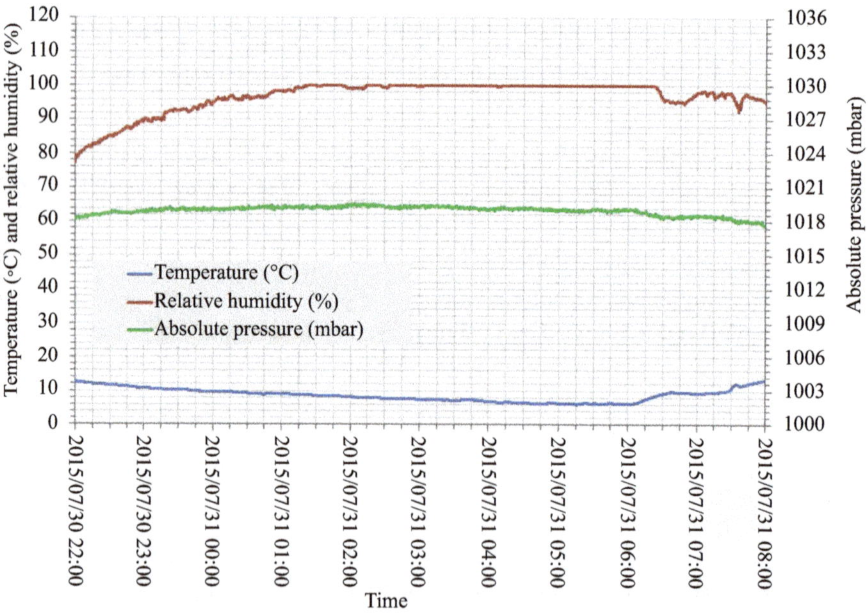

Figure 5.8 Temperature, humidity, and absolute pressure measured during the field tests by National Grid, UK. Courtesy of National Grid

5.1.5 *Survey of leakage resistance and time constant for cable discharge*

As observed in Sections 5.1.3 and 5.1.4, the time required to discharge a cable depends not only on the leakage of cable insulation but also on the leakage of components connected to the cable system. The term “insulator” is henceforward used for all components providing a path for leakage currents, for example, cable sealing ends, insulator poles, and bushing.

The discharge of distributed-parameter systems, for example, cables and lines, is in general different from the discharge of lumped-parameter systems, for example, reactive power compensation capacitor. However, since the phenomenon of cable discharge is very slow, the wavelength will be much larger than the cable length. At 60 Hz, assuming a typical propagation velocity of 150 m/μs, it is $\lambda = \left(150 \times 10^6\right)/60 = 2{,}500$ km, being even larger for lower frequencies. In such a case, the cable is considered as “short” and the response of the cable and insulators connected to it follows the equations of a simple parallel *RC* circuit with a time constant given by (5.2).

It must be noted that discharge through a grounding switch, which represents a short circuit across a charged cable, or equipment such as wound voltage transformers would be different to discharge considered in this section as in those cases the cable series parameters influence the discharge.

The value of *R* in (5.2) depends both on the leakage through the cable insulation and through the insulators connected to the cable. The cable insulation is practically not affected by external factors, like weather and pollution. On the other hand, insulators (cable sealing ends, insulator poles, and bushing) are generally exposed to air and thus are directly affected by humidity and contamination.

This section presents a method for calculating a model of a cable system during discharge. Typical values of cable and insulator parameters are given and allow estimating the discharge time constants based only on the voltage level, cable insulation material (EPR, XLPE, and paper insulation), and insulator type (ceramic or composite).

5.1.5.1 Cable parameters

A cable response during discharge is influenced by its per-unit shunt parameters, that is, the shunt capacitance in (5.6) and shunt conductance in (5.7), or more commonly used, the shunt resistance of its insulation.

The total cable capacitance *C* and shunt resistance *R* are calculated using (5.6) and (5.7) with *l* being the length of the cable and *n* being the number of phases.

It is possible to calculate *R* using the cable capacitance *C* and permittivity parameters commonly found in cable datasheets as defined in (5.8).

Table 5.3 gives the resistivity and relative permittivity for typical cable insulation [15–17]. The capacitance for 275 kV cables is generally within 0.12–0.45 μF/km [18,19].

Based on Table 5.3 and typical capacitance values, it is possible to calculate the leakage resistance and leakage current for 275 kV cable systems in Table 5.4. Paper insulated cables have by far the least leakage current.

Table 5.3 Physical parameters of cable insulation materials. Adapted from [15–17]

Insulation	EPR	XLPE	Paper (Kapton, Polyimide)
Resistivity	10 GΩ-km	100 GΩ-km	1,000–1,500 GΩ-km
Relative permittivity	2.5–2.6	2.3–2.5	3.4–3.8

Table 5.4 Leakage resistance and current in 275 kV cables

	EPR	XLPE	Paper
Leakage resistance	0.49–1.92 GΩ-km	4.53–18.4 GΩ-km	66.9–280 GΩ-km
Typical values	$C=0.35$ μF/km, $R=0.63$ GΩ-km	$C=0.25$ μF/km, $R=8.1$ GΩ-km	$C=0.45$ μF/km, $R=74.8$ GΩ-km
Leakage current	0.35 mA/km	0.028 mA/km	0.003 mA/km

Table 5.5 Surface resistance per mm of leakage distance (kΩ/mm) for typical insulator materials. Adapted from [20]

	EPDM	Silicone	Porcelain
Uncontaminated	80–210	>5,300	3.1–10
Contaminated	0.27–5.3	2.7–8	Not measured

5.1.5.2 Insulator parameters and leakage current

The performance of insulators (cable sealing ends, insulator poles, and bushing) is generally evaluated through specific tests under wet and contaminated conditions and it is quantified in terms of equivalent salt deposit density (ESDD) which is used to calculate the dielectric strength of the insulator. These tests are good enough for porcelain and glass insulators, but they are insufficient for composite insulators which, in addition to contamination and moisture, are also affected by aging of the housing material. Furthermore, they do not inform on the leakage current or leakage resistance of the insulator.

The IEEE Working Group on Insulator Contamination proposed an alternative method for evaluating the performance of insulators using measures of surface resistance and published results for three types of insulators: porcelain, ethylene propylene diene monomer (EPDM), and silicone rubber [20]. Table 5.5 shows values of surface resistance measured by the IEEE working group [20]. The surface resistance is expressed in Ω/mm of leakage (or creep-age) distance. Table 5.5 shows that composite insulators (EPDM and silicone) have substantially better

Table 5.6 Typical resistance and leakage current at a nominal voltage of 275 kV insulators

	EPDM	Silicone	Porcelain
Resistance (MΩ)	400–1,638	>26,500	25–100
Leakage current (mA)	0.1–0.6	<0.008	2–9

performance (higher surface resistance) than porcelain. It is also clear from Table 5.5 that the presence of contaminants in the walls of the insulator greatly reduces its performance.

The leakage resistance of an insulator can be obtained as

$$R_{\text{insulator}} = R_{\text{surf}}\, d_{\text{leakage}} \tag{5.30}$$

where R_{surf} is the surface resistance given in Table 5.5 and d_{leakage} is the leakage distance of the insulator. Values of the leakage distance are easily found in manufacturer catalogs. For 275 kV insulators, the usual values are 5,000–7,800 mm for composite and 8,000–10,000 mm for ceramic [21]. Using these values and those in Table 5.5, allows calculating typical values of resistance and leakage current at a nominal voltage for 275 kV insulators as given in Table 5.6.

5.1.5.3 Formula and typical values discharge time constant

The time constant associated with the discharge of a cable and surrounding insulators is calculated as

$$\tau = \left(\frac{R_{\text{cable}} R_{\text{ext}}}{R_{\text{cable}} + R_{\text{ext}}}\right) C_{\text{cable}} \tag{5.31}$$

where R_{cable} and C_{cable} are given in (5.6) and (5.7) and R_{ext} is the leakage resistance of external components calculated from (5.30) as

$$R_{\text{ext}} = \mathrm{R}_{\text{insulator}}/m \tag{5.32}$$

where m is the number of insulator components connected to the cable.

Using the formula for cable discharge voltage in (5.4) with $V_{\max} = V_n\sqrt{2/3}$, the time required to discharge the cable voltage to a safe level (25 V according to IEC 60479-1) is 10τ for 275 kV cables.

Table 5.7 shows time constants τ and time required to discharge the cable to the safe voltage level for different combinations of cables and insulators. These values are based on the typical values of cable insulation and insulator properties given in Tables 5.4 and 5.6, respectively. It is worth noting that the most severe cases, that is, the longest discharge times (lowest leakage currents), are found when the insulators and cable insulation are at their best condition. This is observed in Table 5.7 for paper insulated cables with silicone insulators.

Table 5.7 Typical discharge time constants for 275 kV/20 km cable with cable sealing end

(a) Discharge time constants

		Insulator		
		EPDM	**Silicone**	**Porcelain**
Cable insulation	EPR	3.4–3.6 min	3.7 min	1.6–2.8 min
	XLPE	17–27 min	33 min	2.0–6.7 min
	Paper	54–170 min	8–9 h	3.7–14.6 min

(b) Time to discharge the cable to safe voltage level

		Insulator		
		EPDM	**Silicone**	**Porcelain**
Cable insulation	EPR	34–36 min	37 min	16–28 min
	XLPE	2.8–4.5 h	5.5 h	20 min–1 h
	Paper	9–28 h	3.3–3.75 days	0.6–2.4 h

Table 5.8 Data for resistance R(t) in the circuit of Figure 5.11(a)

Time	**0 h**	**0 h 30 min**	**3 h 15 min**	**10 h 00**
Resistance (MΩ)	1,359.3	1,359.3	430.6	430.6

5.1.5.4 Correction factor for humidity

The field tests results provided by National Grid represent a very important contribution to this chapter. Not only they show the evolution of voltage during the discharge of a 275 kV POF cable, a result rarely found in the literature, but they also show the evolution of relative humidity, temperature, and atmospheric pressure during those measurements, carried out during a total of 10 h.

These results, presented in Section 5.1.3.3, allow the definition of a correction factor to apply to the leakage resistance of external insulators to account for a modification in the relative humidity of the air. From Figure 5.8, it is observed that the relative humidity increased from 80% to 100%. At the same time, the leakage resistance of external insulators, excluding leakage of cable insulation, measured from the test results, was reduced from 1,359.3 MΩ to 430.6 MΩ (see Table 5.8). Let's define h_1 and $R_{\text{ext}-h_1}$, the initial values of relative humidity and insulator resistance, and h_2 and $R_{\text{ext}-h_2}$, the values after a change in the relative humidity. In the case of the field test by National Grid, $h_1 = 80\%$, $R_{\text{ext}-h_1} = 1,359.3$ MΩ,

$h_2 = 100\%$, and $R_{ext-h_2} = 430.6$ MΩ. The correction factor to apply to insulator resistance to account for a modification in the relative humidity is

$$R_{ext-h_2} = \frac{h_1}{2.5\, h_2} R_{ext-h_1} \tag{5.33}$$

5.1.6 EMT simulation

The field tests in Section 5.1.3 were simulated in EMT-type simulation tools. The cables are modeled using a π-equivalent. Opting for a lumped-parameter circuit rather than a distributed-parameter model is based on the following:

- Due to the very slow phenomena under analysis, it is a good approximation to model the cable by its DC parameters, that is, the DC resistance and the capacitance. The inductance can be modeled as zero as it has no impact on the phenomena.
- Due to the very long simulation time (160 min in Test-A and 10 h in Test-C), it becomes unpractical to use time steps smaller than some 100 ms. Using a distributed-parameter model would require a time step smaller than the longest propagation delay which is of order 0.4 ms for this specific cable system. This would require long processing times and large memory resources.

5.1.6.1 POF cable in Japan

5.1.6.1.1 Parameters in Section 5.1.4.1

In the field test circuit of Figure 5.1, the DC source charges the cable from $t = 0$ until $t = 44$ min as observed in Figure 5.2. Resistance $R_2 = 22$ GΩ connected to the cable represents the leakage of the cable insulation. At $t = 60$ min, a measuring circuit (represented by $R_3 = 690$ MΩ resistance) is connected to the cable, and at $t = 148$ min, the cable is forced to discharge using a water resistor modeled as $R_4 = 30$ MΩ.

Figure 5.9 shows EMT simulation results which agree satisfactorily with the field test result of Figure 5.2. The figure also shows the effect of cable leakage

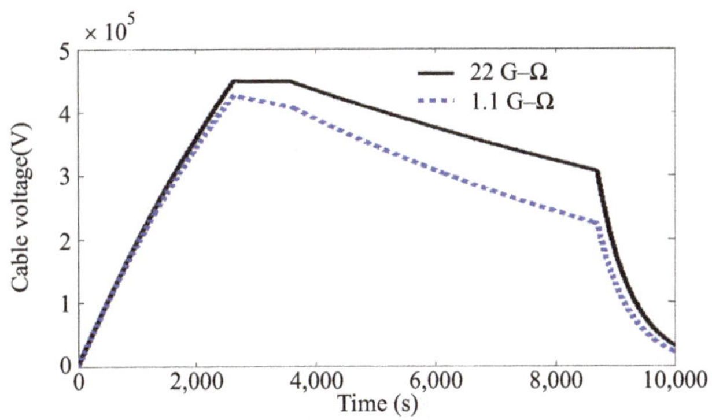

Figure 5.9 Simulated voltage during charging and discharging of a cable for different leakage resistance

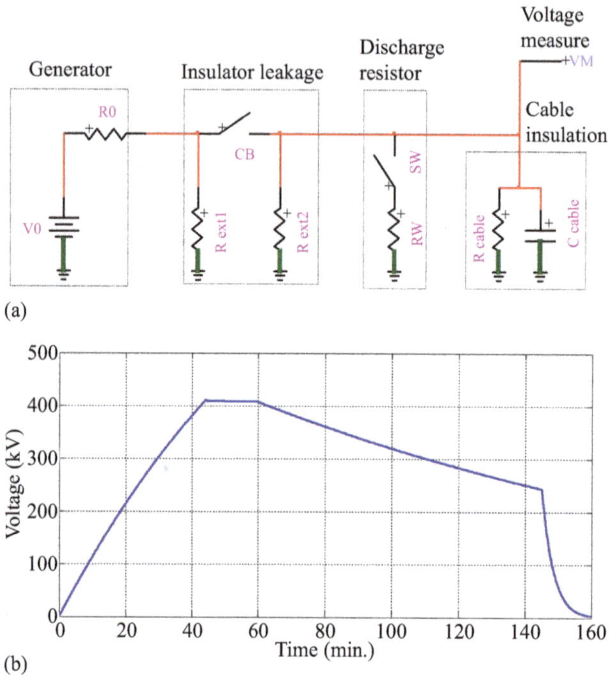

Figure 5.10 Simulation of charging/discharging voltage. (a) Circuit for simulation of field test in Figure 5.2 and (b) simulated result

resistance on charging and discharging phenomena. The cable leakage is often expressed in terms of the loss angle tan δ defined at power frequency, 50 Hz in this case. The two leakage resistances in Figure 5.10 correspond to the following values of tan δ:

$$\tan\delta = G/\omega C = 1/\omega CR, \quad C = 20.1\ \mu\text{F}$$
$$\xrightarrow{R=22\ \text{G}\Omega} \tan\delta = 7.2\times 10^{-9}$$
$$\xrightarrow{R=1.1\ \text{G}\Omega} \tan\delta = 1.4\times 10^{-7} \tag{5.34}$$

In the above equation, the capacitance and resistance are those for the three cable phases in parallel. As observed in Figure 5.9, if the cable insulation has more losses, that is, higher tan δ, then the discharge will be faster. The factor "tan δ" is actually a measure of the quality of the cable insulation. Thus, high-quality cables are of more concern for they need longer times to discharge if no external circuit is connected.

5.1.6.1.2 Modified parameters

The paper in which the field test of Figure 5.2 was first published [7] gives estimated values of leakage resistance which result in a discharge time nearly half (49 min) of that observed in the field test (88 min). This means that the leakage in the system was not as severe as estimated and the components presented a resistance higher than expected.

The cable leakage resistance is calculated using half of the resistivity for paper insulation in Table 5.3, that is, $\rho = 500 \times 10^{12}$ Ω-m, together with $n = 3$, $l = 20 \times 10^3$m, $r_{in} = 26.65$ mm, and $r_{out} = 40.65$ mm considering the thickness, 3 mm, of the semiconducting layer in Table 5.1 into (5.4) and (5.26) resulting $R_{cable} = 560$ MΩ.

The lower value of resistivity is based on typical insulation properties in POF cables of 1970s, which have been improved since then [10].

The leakage resistance of external components is calculated using the surface resistance of EPDM $\rho = 210 \times 10^3$ Ω/mm in Table 5.5 and a leakage distance of $d_{leak} = 5{,}000$ mm into (5.30) resulting into $R_{ext} = R_{insulator} = 1{,}050$ MΩ.

The total leakage resistance is calculated as $R_{total} = R_{cable} \| R_{ext} = 365.2$ MΩ. The time constant of the system is obtained using this value and the cable capacitance $C_{cable} = 29.3$ μF into (5.5) resulting $\tau = 178.3$ min.

The time required to discharge the cable from $V(T_1) = 414$ kV to $V(T_2) = 250$ kV is calculated using $\tau = 178.3$ min into (5.5) as $T_2 - T_2 = 89.9$ min. This value is only 2% above the one observed in the field tests (88 min).

The values of the leakage resistance for cable insulation and external circuit calculated above were used to simulate the field test in Figure 5.2. The circuit used to simulate the cable system is shown in Figure 5.10(a). The 20 km cable is modeled using a parallel RC circuit using the parameters of cable insulation for three phases in parallel, that is, $C_{cable} = 29.3$ μF and $R_{cable} = 560.0$ MΩ. The leakage of the external circuit is $R_{ext1} = R_{ext2} = 1{,}050$ MΩ. Other circuit parameters are

- $V_0 = 414$ kV, the voltage of DC generator,
- $R_0 = 138$ MΩ, charging resistance, and
- $R_W = 30$ MΩ, resistor used for full discharge.

The switch S_0 is closed from $t = 0$ to $t = 60$ min and S_W closes at $t = 148$ min (S_0 being open).

The simulation result in Figure 5.10(b) agrees well with the field test result of Figure 5.2, which validates the modeling approach.

It is observed from a comparison of Figure 5.10(b) with Figure 5.9 that the parameters except for the cable leakage resistance for EMT simulation cause no significant effect on the simulation result.

5.1.6.2 275 kV POF cable in the UK

From the field test result in Figure 5.6 and from the value of cable capacitance, the total leakage resistance of the system has been estimated as 840.2 MΩ for low humidity and 360.1 MΩ for increased humidity.

A cable leakage resistance is calculated using (5.4) and (5.26) with the resistivity for paper insulation $\rho = 1{,}000 \times 10^{12}\ \Omega/\text{m}$ in Table 5.3, $l = 21 \times 10^3$ m, $r_{in} = 29.75$ mm, and $r_{out} = 40$ mm in Table 5.2, resulting into $R_{cable} - 2.2\ \text{G}\Omega$.

The leakage resistance for three insulators in parallel (two insulator poles and cable sealing end) is calculated using 840.2, the surface resistance of silicone $\rho = 5{,}300 \times 10^3 \Omega/\text{mm}$ in Table 5.5 and a leakage distance of $d_{leak} = 5{,}334$ mm (supplied by National Grid) into (5.12) and (5.14) resulting into $R_{ext} = 9.4\ \text{G}\Omega$. With previously found $R_{cable} = 2.2\ \text{G}\Omega$, the total leakage resistance is calculated as $R_{total} = R_{cable} \| R_{ext} = 1811.4\ \text{M}\Omega$. This value is about two times the value observed in the field tests (840.2 MΩ). However, it must be observed that the calculated insulator resistance considers uncontaminated material. With a total leakage resistance of $R_{total} = 840.2\ \text{M}\Omega$ calculated from the field test in Section 5.1.2.3, the leakage resistance of external components can be reevaluated from

$$\frac{2200 \times R_{ext}}{2200 + R_{ext}} = 840.2\ \text{M}\Omega, \quad R_{ext} = 1{,}359.3\ \text{M}\Omega \tag{5.35}$$

Using $R_{total} = 360.1\ \text{M}\Omega$ instead of 840.2 MΩ results into $R_{ext} = 430.6\ \text{M}\Omega$ when humidity rises to 100%.

The field test in Figure 5.6 was simulated using the circuit in Figure 5.11(a). The cable is modeled with a parallel *RC* circuit using the parameters of its insulation, that is, $R_{cable} = 2.2\ \text{G}\Omega$ and $C_{cable} = 15.0\ \mu\text{F}$. The leakage in the external circuit (two insulator poles and cable bushing) is modeled with a time-varying resistance to account for the rise in ambient humidity during the field test. The values for this leakage resistance are given in Table 5.8. The charging resistance is $R_0 = 100\ \text{M}\Omega$ (calculated from the rise time of voltage in Figure 5.6). Figure 5.11 (b) shows the simulated discharge voltage of the cable and the agreement with the field test result in Figure 5.6 is evident.

5.1.7 Summary

Cable charging and discharging phenomena are received new consideration from utilities which are using HV cables for voltage control during periods of light load. This section has investigated discharge characteristics of 275 kV POF cables based on laboratory and field tests, theoretical analysis, and EMT simulations. The following remarks are made clear:

1. The natural discharge of cable systems can be explained using simple formulas of parallel *RC* circuits. This has been verified from the simulation of field test results for two cable systems.
2. The factors affecting the discharge time are the cable capacitance, which is easily found in manufacturer catalogs, the leakage resistance of cable insulation, and external components, such as insulator poles, bushing, and cable sealing ends. Leakage in insulators exposed to air is highly affected by temperature, humidity, and contamination.
3. Leakage in the cable insulation and external components can be estimated based only on the voltage level, types of cable insulation (EPR, XLPE, paper),

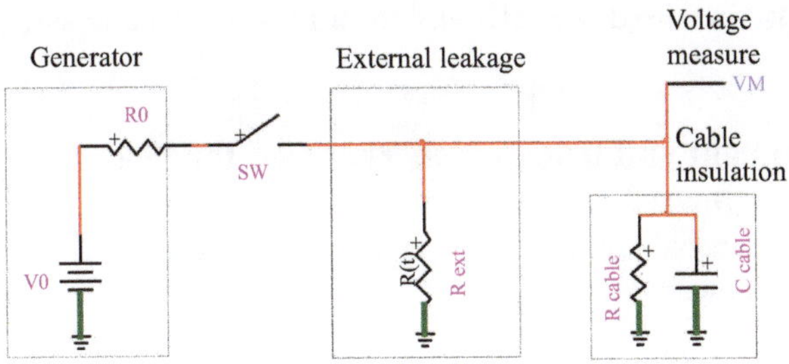

(a)

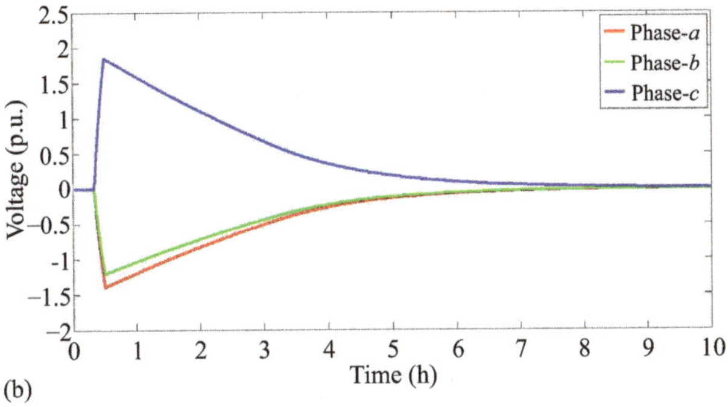

(b)

Figure 5.11 Simulation of discharge of a POF-cable in the UK. (a) Circuit for simulation of discharging of a 275 kV POF-cable and (b) simulation result

insulator material (composite, ceramic), and the number of parallel insulator poles connected to the cable terminations. Estimation of leakage and knowledge of cable capacitance, directly provide a sufficiently accurate estimate of the discharge time.

4. The discharge time constants observed in the two field tests on 275 kV POF cables of 20 km, range from 1.5 h to 3.5 h. The time to discharge the cable to a safe voltage level (25 V according to IEC 60479-1), is in theory ten times the discharge time constant.
5. The leakage resistance of POF cables is about 74.5 GΩ-km (3.7 GΩ for 20 km). For XLPE insulated cables, the value is 1/10 of that. The leakage resistance of 275 kV insulators is about 100 (porcelain) to 1,600 MΩ (EPDM) per element. For silicone insulators, this value can rise to 26.5 GΩ.

5.2 Field measurement of cable transients and EMT simulations

5.2.1 Field measurement

It is not well understood that a field measurement of a transient in a power system is not straightforward at all. A measured result is not necessarily reliable, although the measured result is often used as a reference to validate a computer simulation result of the transient.

5.2.1.1 Equipment and instruments required for field measurement

In a field measurement, the following equipment is necessary to perform the measurement of a power system transient:

1. AC or DC power source to supply power to all the equipments and instruments used for the measurement.
2. Voltage source (impulse generator) for a transient, see the details in Appendix A5.2.1.
3. Measuring instruments: oscilloscope, voltage probe, voltage transformer (VT) for high-voltage measurement, current sensor, and current transformer (CT). The sampling time Δt/frequency Δf and the frequency band are to be carefully chosen.
4. Lead wires for source circuits, measuring instruments, voltage reference, and grounding.
5. Grounding electrodes.

It should be noted that there exists no zero-potential reference (grounding electrode) in a field. Many grounding electrodes have to be prepared for the power source, impulse generator, measuring instruments. Figure 5.12 shows a basic configuration of the equipment and instruments with grounding electrodes. Note that the power source requires its own grounding. It happened that construction workers died in an EHV line construction site because of poor grounding of the power source.

5.2.1.2 Conditions of test site

The voltage reference wire is grounded through a grounding electrode at the remote end, and the other end is connected to the grounding electrode of an oscilloscope in Figure 5.12. Because the voltage at the grounding electrode cannot be zero especially

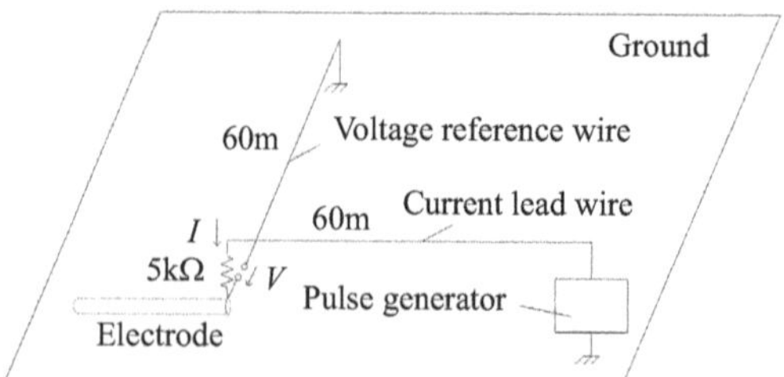

Figure 5.12 Basic configuration of a field measurement (electrode in the figure: testing object)

during a transient, we assume the voltage along the reference wire is zero until the reflection of a traveling wave from the remote end comes back to the oscilloscope, that is, the sending end of the reference wire. Therefore, a possible observation time using the reference wire is $T_{max} = 2\tau$, where τ is one travel time of a traveling wave along the reference wire. For example, when the length of the reference wire is $l = 60$ m in Figure 5.12, the observation time becomes: $T_{max} = 2\tau = 2l / c_0 = 120/300 = 0.4$ μs assuming the propagation velocity $c_0 = 300$ m/μs. For a very fast transient (VFT) such as a lightning surge containing very high-frequency components, the above T_{max} gives the limit of an accurate field measurement.

For a switching surge, the observation time is much longer. When the length of a conductor is l, the oscillating period of the open-circuited condition is defined as $T = 4\tau = 4l/c$, where c is the propagation velocity. The observation time has to cover about 10–20 times of the oscillating period to observe the switching surge characteristics. When the length of an underground coaxial cable is $l = 2{,}000$ m and the relative permittivity of the insulator between the core and metallic sheath is $\varepsilon_i = 2.5$, then the propagation velocity of the coaxial mode becomes $c_c = c_0 / \sqrt{\varepsilon_i} = 300/1.5811 = 189.7$ m/μs. Therefore, $T = 4\tau = 4l/c_c = 42.16$ μs, and $T_{max} = 422$–845 μs. To perform a field measurement under the above condition, we have to assume the voltages at the grounding electrode and of the reference wire are nearly zero. This is one of the critical conditions of a field measurement of a switching surge and is a cause of an in-accurate measured result.

Also, it should be noted that surrounding electrical/power facilities, mostly power transmission lines, near to the test site results in an induced voltage to measuring instruments including voltage reference and current lead wires and grounding electrodes. If there is an EHV transmission line, the induced voltage reaches more than some 10 V which makes it very hard to measure a small voltage response. This induced voltage is well-known as telephone line and pipeline interferences, that is, electromagnetic interferences or disturbances (EMI or EMD). Summarizing the conditions required for a test site:

1. A field test site should be far from any electrical/power facilities.
2. The earth resistivity of the site is not high (not rocky for grounding electrode installation) to reduce the transient impedance of a grounding electrode.
3. In the case of a cable transient, the test site should be within a cable junction and/or insulating joint along the cable route under construction. Otherwise, it is impossible to measure the cable voltages. Note that most field measurements on a transmission line/cable are carried out when the line/cable is under construction or before the commissioning inspection. Figure 5.13(a) shows a cable drum of a 225 kV XLPE cable for installation, and Figure 5.13(b) is a photo of the test site selected in this field measurement.
4. Accommodation and restaurant during a field measurement which takes more than a few days.
5. The climate condition is also important. Note that the earth resistivity changes significantly during dry days and after rain.

(a)

(b)

Figure 5.13 A 225 XLPE underground cable. (a) Cable drum for installing a 225 kV XLPE cable and (b) field test site

5.2.1.3 Time schedule [22,23]

The following note shows the time schedule of the RTE (French TSO: Réseau de Transport d'Electricité) 225 kV underground cable constructed in 2014.

1. July 2013: visited the RTE Boutre-Trans cable construction site to find the place for a field measurement site satisfying the conditions explained in Section 5.2.1.2.

2. July 2013 to May 2014: communication of all the necessary things related to the field measurements explained in Sections 5.2.1.1 and 5.2.1.3.
3. July 2014 for 8 days: field test performed.

5.2.2 Measured results of switching surges on a homogenous cable system (one minor section)

In June of 2014, the French TSO (Réseau de Transport d'Electricité – RTE) performed field tests on a 64 km 225 kV XLPE cable (Boutre-Trans) installed in the south-east of France [24]. The main objective was to improve knowledge on the implemented system and validate cable modeling for electromagnetic transient (EMT) studies. The tests were based on a CIGRÉ working group report for cable model validation [3]. The different propagation modes were excited separately to allow the evaluation of the specific propagation characteristics. The tests were performed first on a minor section. Besides the typical estimation of attenuation, velocity, and characteristic impedance, the PUL (R', L', C') parameters for each mode have been evaluated. The acronym *RLC* is used hereinafter to refer to PUL parameters. Later on a major section with cross-bonding of sheaths.

This section explains the cable system configuration and measured results which are compared with theoretical results of the cable impedance and admittance [25] and with EMT simulation [26–28]. The model uses manufacturer data with necessary corrections to account for wired conductors and semiconducting screens.

5.2.2.1 RTE 225 kV cable and field measurement

5.2.2.1.1 Cable system configuration

In 2015, the Boutre-Trans underground link started operating in the transmission grid of the so-called PACA region (Provence-Alpes-Côte d'Azur) in the south-east of France [24]. This XLPE cross-bonded 225 kV cable uses stranded enamel copper cores and metallic sheaths in aluminum foil. The cable outer insulation and the tube enclosing each phase are made of HDPE (high-density polyethylene). The system is embedded in concrete and buried on earth as shown in Figure 5.14. The details of the cable configuration have been explained in Section 3.3.4 to find an equivalent coaxial cable.

The first field measurement was performed on one minor section, that is, a three-phase homogeneous cable with the length 1,080 m and the cross-section area 2,000 mm^2 of the RTE 225 kV XLPE cable system. The test site is shown in Figure 5.13(b).

5.2.2.1.2 Cable parameter calculation

(a) Core and metallic sheath

Existing cable parameter calculation routines such as Cable Constants (CC) of EMT-type simulation tools allow only for solid or hollow cylindrical conductors [4,29–31]. The metallic sheath of the Boutre-Trans cable is made of aluminum foil with a resistivity of 2.84×10^{-8} Ωm and can be simply modeled as a hollow cylindrical conductor. The core conductor is made of stranded enameled copper $\left(\rho_{\text{copper}} = 1.76 \times 10^{-8}\ \Omega\text{m}\right)$ and requires a

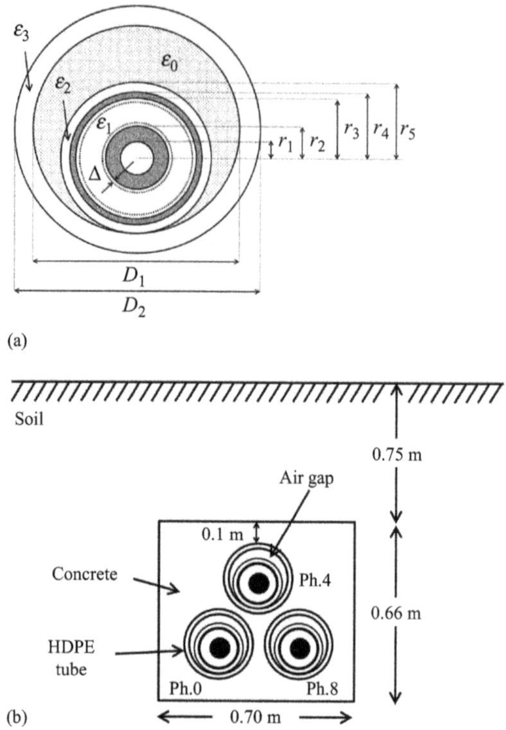

Figure 5.14 Cable configuration. (a) Cross-section of a single-phase cable in an HDPE tube and (b) cross-section of a three-phase cable installed in an HDPE tube

modification of the resistivity to account for the gaps between the copper wires which reduce the effective section of the core conductor. The modified resistivity ρ' is calculated from the core nominal area $A_n = \pi r_{2n}^2$ and the actual area of the copper strands $A = \pi r_2^2$ as follows [32,33]:

$$\rho' = \rho \frac{r_{2n}^2}{r_2^2} \tag{5.36}$$

where ρ, r_{2n}, and r_2 are found in Table 5.9. The ρ' values become 2.6×10^{-8} Ωm and 2.83×10^{-8} Ωm for the 2,000 mm² and 2,500 mm² core cables, respectively.

(b) Core insulator and semi-conducting layer

The cable semiconducting layers cannot be handled directly in EMT-CC. However, these layers may have an important impact on the propagation of coaxial waves [33,34]. The impedance of the semiconducting (SC) layers has a minor impact on the cable series impedance, for which the conductor impedance is predominant, but it has an important effect on the shunt admittance. To account for SC layers, it is generally accurate enough to simply modify the permittivity of the insulation to account for the new

Table 5.9 Specification of the RTE cable

Parameter	2,000 mm² cable	2,500 mm² cable
Core conductor ($r_i = 0$)	$r_{2n} = 34.5$ mm* $r_2 = 28.4$ mm ** $\rho = 1.76 \times 10^{-8}\ \Omega$m	$r_{2n} = 37.1$ mm* $r_2 = 31.9$ mm ** $\rho = 1.76 \times 10^{-8}\ \Omega$m
Semiconductor screen 1	$\Delta = 3$ mm (thickness)	
Core insulation	$\varepsilon_{r1} = 2.5$, $\tan\delta = 0.0008$	
Semiconductor screen 2	$\Delta = 3$ mm (thickness)	
Metallic sheath	$r_3 = 56.4$ mm $r_4 = 57.2$ mm $\rho = 2.84 \times 10^{-8}\ \Omega$m	$r_3 = 59.9$ mm $r_4 = 60.7$ mm $\rho = 2.84 \times 10^{-8}\ \Omega$m
Outer insulation	$r_5 = 62.2$ mm $\varepsilon_{r2} = 2.5$ $\tan\delta = 0.001$	$r_5 = 65.7$ mm $\varepsilon_{r2} = 2.5$ $\tan\delta = 0.001$
HDPE tubes	$D_1 = 198.5$ mm, $D_2 = 225$ mm $\varepsilon_{r3} = 2.5$, $\tan\delta = 0.001$	

Note: *Nominal radius of core conductor and **corrected core radius accounting for effect section of wires in the strand.

thickness as follows [32–34]:

$$\varepsilon'_{r1} = \varepsilon_{r1} \frac{\ln(r_3/r_2)}{\ln[(r_3 - \Delta)/(r_2 + \Delta)]} \tag{5.37}$$

where the values of $\varepsilon_{r1}, r_2, r_3$, and Δ are found in Table 5.9. The modified values of ε'_{r1} become 3.23 and 3.22 for the 2,000 mm² and 2,500 mm², respectively.

(c) Outer insulator and HDPE tube

A further complication arises from the presence of the HDPE tubes as observed in Figure 5.14. No analytical formula of impedance or admittance has been derived for such a system as explained in Section 3.3.4, and the equivalent sheath insulation which is concentric and has an outer radius r_5 in Table 5.9. The relative permittivity of the equivalent insulation was calculated from the capacitance obtained from the field test results of the inter-sheath mode capacitance given in Table 5.10, and becomes $\varepsilon_{r2_{eq}} = 0.321$. A permittivity lower than 1 comes from the representation of a dielectric (composed of sheath insulation and HDPE tube) with relative permittivity of 2.5 and air gap with relative permittivity 1 by an equivalent insulation of smaller dimensions as explained in Section 3.3.4 [38].

5.2.2.1.3 Measuring circuit and grounding

A Haefely Axos 5 impulse generator (IG) with an output impedance of 2 Ω was used in the field to apply a 2 kV 1.2/50 μs impulse. The single oscilloscope used in

Table 5.10 Calculated and measured RLC parameters

Mode	Source	R' (Ω/km)	L' (mH/km)	C' (μF/km)
Coaxialf = 37 kHz*	Field	2.84	0.160	0.249
	CC	0.352	0.139	0.264
Inter-sheathf = 27 kHz*	Field	6.70	0.212	0.343
	CC	0.208	0.203	0.343
Earth-returnf = 13 kHz*	Field	69.9	0.413	0.609
	CC	13.4	1.16	1.65

Note: *Dominant frequency of the transient waveform.

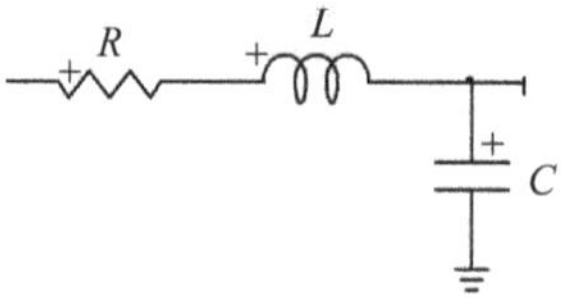

Figure 5.15 L-equivalent circuit (RLC circuit) model for cable

the tests was a WaveJet 324A. Voltage was measured using PHV1000 passive probes and current was obtained using Pearson 110 probes.

The connection wires used for the minor section testing had 5 m and those for the major section had 20 m. The wire inductance is estimated as 5 μH and 20 μH (considering 1 μH/m [25]), respectively.

The two terminals of the minor section and the receiving end of the major section are located in remote sites and required the (manual) installation of grounding rods to protect the equipment and operators. The major section sending end is located in a substation where a grounding network is available. The grounding impedance is estimated as 30 Ω for the remote areas and 5 Ω for the substation.

5.2.2.2 Validation of cable parameters

5.2.2.2.1 Theoretical formulas of voltages and currents in an equivalent circuit of a field test

The accuracy of a cable model depends on impedance and admittance parameters calculated by CC [29–31]. The CC calculated results of R, L, and C of the cable are to be compared with measured results to verify the accuracy. It is possible to evaluate the *RLC* parameters of an equivalent circuit from measured cable transient responses. The cable is represented by an L-equivalent lumped-parameter circuit presented in Figure 5.15. A unit step voltage can be applied to approximate the field test impulse voltage and simplify the derivation of analytical responses in the Laplace domain [25,35]. PI-equivalent and T-equivalent lumped circuits are well-known. However, those circuits create double frequency oscillations, the determination of *RLC* parameters becomes complicated and requires two times more information from measured results. Furthermore, the sending end is in parallel with the test voltage source and quite often the receiving end is either open-circuited or short-circuited. This results in either the shunt admittance in the PI-equivalent or the series impedance in the T-equivalent being missed [25,35].

The Laplace transform of the current in an L-equivalent circuit in Figure 5.15 and its time-domain form are given in the following equation:

$$I(s) = \frac{E}{s\left(R + sL + \frac{1}{sC}\right)} \tag{5.38}$$

$$i(t) = \frac{E}{L\,\omega_d} e^{-at} \sinh(\omega_d t) \tag{5.39}$$

where E is the amplitude of the step function, $\alpha = \frac{R}{2L}$, $\omega_d = \sqrt{|\alpha^2 - \omega^2|}$, and $\omega = 1/\sqrt{LC}$.

The initial conditions are zero. Equation (5.39) is valid for the over-damped condition with $\omega^2 < \alpha^2$. For the oscillatory condition:

$$i(t) = \frac{E}{L\,\omega_d} e^{-at} \sin(\omega_d t), \quad \omega^2 > \alpha^2 \tag{5.40}$$

and for the critically damped condition:

$$i(t) = \frac{E}{L} te^{-at}, \quad \omega^2 = \alpha^2 \tag{5.41}$$

The characteristic impedance is obtained from $Z_c = \sqrt{L/C}$. For the oscillatory condition, Z_c is found from the ratio of peak voltage and current. The attenuation constant is calculated from two current peaks (I_{n-1} and I_n) and the time t_p between them, giving $\alpha = \ln(I_{n-1}/I_n)/t_p$. It is then possible to extract the following relation [25,35]:

$$L = Z_c/\omega, C = 1/(\omega Z_c) \quad \text{and} \quad R = 2L\alpha. \tag{5.42}$$

For the critically damped condition, the current maximum $I_{\max} = 2E/(eR)$ is reached when $t_{\max} = 1/\alpha$. Thus it is possible to extract:

$$\alpha = 1/t_{\max}, R = 2E/(eI_{\max}), \quad L = R/(2\alpha) \text{ and } C = 4L/R^2 \tag{5.43}$$

5.2.2.2.2 *R, L, and C evaluated from measured results by using the formulas explained in the above*

(a) Coaxial mode with the oscillatory condition, see Figures 5.16 and 5.18.

The RLC values for the coaxial mode test are calculated below. The parameter T is the period of oscillations, the subscript *avg* stands for average value, and the samples of current use the index "*n*":

$$Z_c = V_{\max}/I_{\max} = 1752/69.07 = 25.37\ \Omega \tag{5.44}$$

$$(I_{n-1}/I_n)_{avg} = 1.273, T_{avg} = 27.18\ \mu\text{s}, \tau_{avg} = 6.80\ \mu\text{s} \tag{5.45}$$

$$v_{\text{avg}} = 158.92\ \text{m}/\mu\text{s}, f_{\text{avg}} = 36.79\ \text{kHz} \tag{5.46}$$

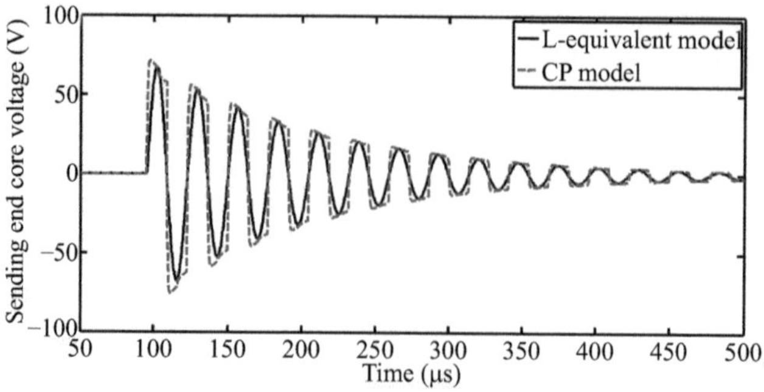

Figure 5.16 Sending end currents, L-equivalent and CP model, coaxial mode, and a minor section

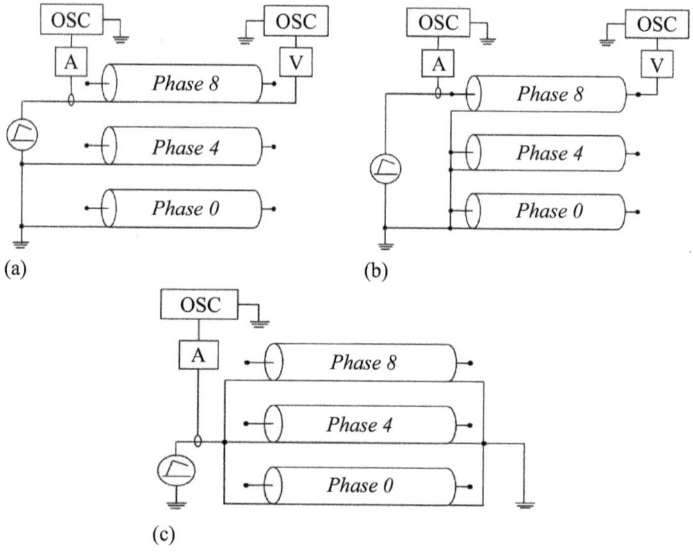

Figure 5.17 Test circuit for a minor section. (a) Coaxial mode, (b) inter-sheath mode and (c) earth-return mode

$$\alpha = \ln\left(1.273/27.18 \times 10^{-6}\right) = 8{,}888.71, L = Z_c/(2\pi f)$$
$$= 0.1098\ \text{mH}/1{,}080\ \text{m} \quad (5.47)$$

$$C = 1/(Z_c 2\pi f) = 0.171\ \mu\text{F}/1{,}080\ \text{m}, \quad R = 2L\alpha = 1.952\ \Omega/1{,}080\ \text{m} \quad (5.48)$$

Calculated results by *L*-equivalent and CP model using the above parameters in (5.45)–(5.48) are shown in Figure 5.17. Calculated results by the frequency-dependent WB (wide-band [84]) model and *L*-equivalent are compared with field test results in Figure 5.18.

(b) Inter-sheath mode with the oscillatory condition, see Figure 5.19.

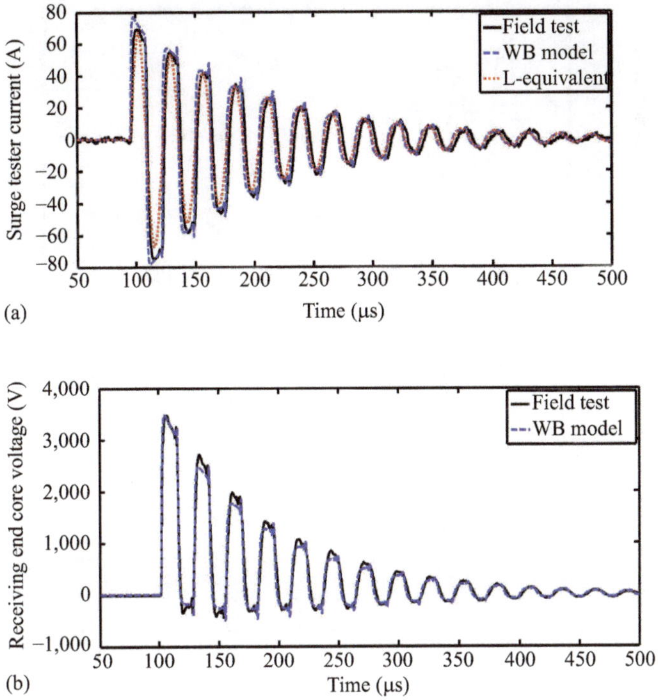

Figure 5.18 Field test and simulation results, coaxial mode, and a minor section. (a) Core current (IG) at the sending end and (b) core voltage at the receiving end

The *RLC* values for the inter-sheath mode are calculated as

$$Z_c = 1221/49.01 = 24.91\ \Omega, (I_{n-1}/I_n)_{\text{avg}} = 1.791 \tag{5.49}$$

$$T_{\text{avg}} = 36.9\ \mu\text{s}, \tau_{\text{avg}} = 9.23\ \mu\text{s}, c_{\text{avg}} = 117.07\ \text{m}/\mu\text{s} \tag{5.50}$$

$$\begin{aligned} f_{\text{avg}} &= 27.10\ \text{kHz}, \alpha = \ \ln 1.791/36.9 \times 10^{-6} \\ &= 15.793 \times 10^3 L = Z_c/(2\pi f) = 0.146\ \text{mH}/1{,}080\ \text{m} \end{aligned} \tag{5.51}$$

$$C = 1/(Z_c 2\pi f) = 0.236\ \mu\text{F}/1{,}080\ \text{m}, R = 2L\alpha = 4.61\ \Omega/1{,}080\ \text{m} \tag{5.52}$$

(c) Earth-return mode with critical damping condition, see Figure 5.20.

As observed in Figure 5.20, the earth-return mode has a damped response due to the high resistance of the grounding system. The *RLC* values for the case of a zero-sequence sheath energization are estimated as

$$Z_c = 1{,}767/27.04 = 65\ \Omega \tag{5.53}$$

$$t_{\max} = 106.7 - 94.9 = 11.8\ \mu\text{s} \rightarrow \alpha = 1/t_{\max} = 84.7 \times 10^3 \tag{5.54}$$

$$I_{\max} = 27.04\ \text{A}, E = 1{,}767\ \text{V} \rightarrow R = 2E/(\text{e}I_{\max}) = 48.1\ \Omega/1{,}080\ \text{m} \tag{5.55}$$

$$L = R/(2\alpha) = 0.284\ \text{mH}/1{,}080\ \text{m}, C = 4L/R^2 = 0.491\ \mu\text{F}/1{,}080\ \text{m} \tag{5.56}$$

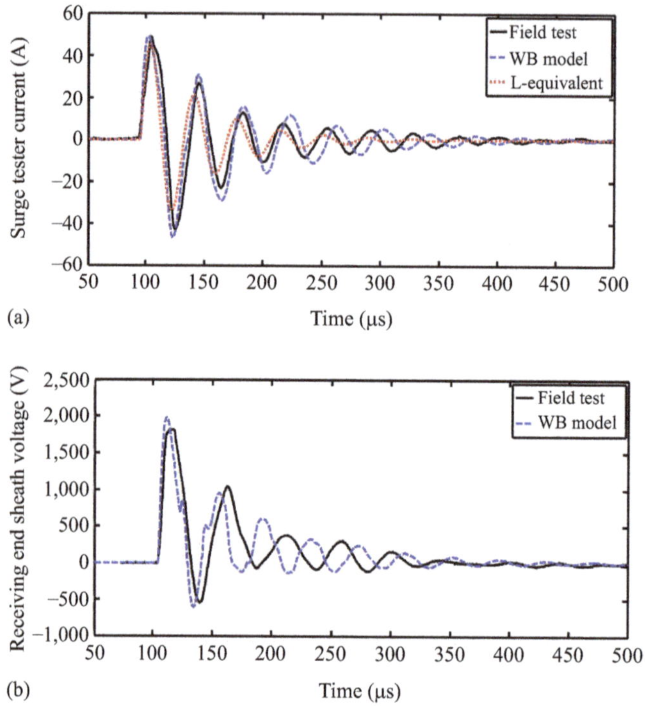

Figure 5.19 Field test and simulation results, inter-sheath mode, and a minor section. (a) IG current and (b) receiving end sheath voltage

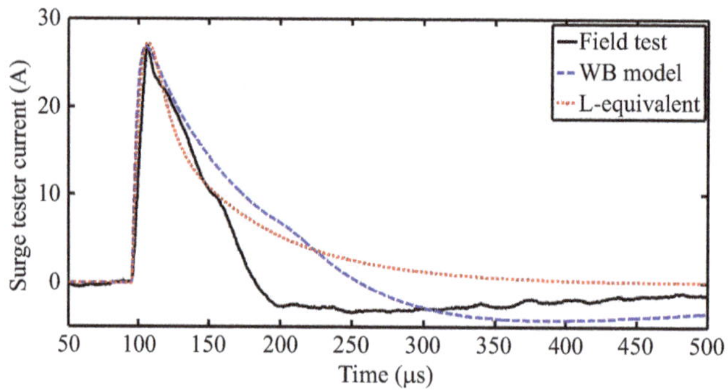

Figure 5.20 Field test and simulation results, earth-return mode, and a minor section

5.2.2.2.3 Modification of R, L, and C

The advantage of the L-equivalent is the simplicity in identifying its parameters from the transient response. On the other hand, an L-equivalent is a lumped-parameter circuit and thus it has no means to represent the propagation delay of a traveling wave.

The parameters of a lumped circuit are not exactly the same as those of a distributed circuit, as can be observed by their corresponding natural frequencies [25]. For the L-equivalent, the natural frequency is found from $f = 1/\left(2\pi\sqrt{LC}\right)$ whereas in the actual cable, it is found from $f' = 1/\left(4\ell\sqrt{L'C'}\right)$ (quarter wavelength), with ℓ being the cable length and when the receiving end is open-circuited. In order to obtain the parameters of a distributed-circuit which has the same transient response as the L-equivalent (except for the propagation delay, which does not exist for the L-equivalent), the two frequencies f and f' are set to be equal and allow calculating:

$$L' = \eta L, \quad C' = \eta C \tag{5.57}$$

where L' and C' are now the per-unit-length parameters with $\eta = \pi/(2\ell)$. To ensure that attenuation remains the same for lumped and distributed circuits, the resistance value is also corrected by the factor η.

The above can be verified by comparing the response of the L-equivalent circuit to that of a distributed circuit modeled using the single-phase constant-parameter (CP) line model available in the EMT-type simulation tool. For the coaxial mode, the parameters of the CP model are:

$$R' = (\pi/2)1.81 = 2.84\ \Omega/\text{km} \tag{5.58}$$

$$L' = (\pi/2)0.102 = 0.160\ \text{mH/km} \tag{5.59}$$

$$C' = (\pi/2)0.158 = 0.249\ \mu\text{F/km} \tag{5.60}$$

5.2.2.2.4 *Validation of equivalent circuit*

To validate the accuracy of *RLC* parameters, the above calculations are compared with the test result in Figure 5.16 and Figure 5.18(a) for the coaxial mode. The agreement between the field test and calculated results by the L-equivalent circuit is sufficiently good for coaxial mode. Also, a satisfactory agreement is observed in Figure 5.19(a) for the inter-sheath mode. For the earth-return mode, the L-equivalent circuit gives a slower decaying current as observed in Figure 5.20. This is caused by the impedances of the earth-return path and grounding system.

The current responses for the L-equivalent circuit and the CP-model corresponding to the coaxial propagation mode are given in Figure 5.16. As observed in this figure, the frequency and attenuation of the two models are exactly the same. The square-shaped wave of the CP model is due to the fact that it does not account for the frequency dependencies of R' and L'.

Table 5.10 presents the computations of parameters by CC and from the field tests. The parameters from the field tests are derived from the response of L-equivalents for each mode.

It is observed that the calculated results for L' and C' are sufficiently accurate and consequently the propagation velocity v and characteristic impedance Z_c agree with field tests. The resistance R' found from field tests is always greater than the one found from the CC results. This can be explained by the fact that field test results involve resistances in the measuring system.

5.2.2.2.5 Comparison with other measured results

A typical range of propagation velocity and surge impedance of coaxial and earth-return modes have been collected from field tests carried out in Japan on more than twenty cable systems [25,32,36]. For the coaxial mode, the velocity is within 137–160 m/μs and the surge impedance is 13–30 Ω, independently of the type of cable. This is reasonable because the characteristics of the coaxial mode are determined by the radii of core and sheath conductors and the relative permittivity of the core insulation. On the contrary, the earth-return mode shows a wide range of values for velocity and surge impedance. The tunnel-installed cable can be compared to an overhead line, and the characteristics of the pipe-type cable are quite dependent on the size of the pipe and the permittivity of the insulation inside the pipe [32]. The velocity can be as high as 150–170 m/μs for tunnel installed cables or as low as 30–70 m/μs in the case of pipe-type and directly buried cables. The surge impedance can be as low as 5 Ω for direct burial and pipe-type, or as high as 100 Ω in the case of tunnel-installed cable.

The measured results in Table 5.10 show a satisfactory agreement with those referred above, and thus it can be concluded that the measured results on the RTE 225 kV cable are within a valid range.

5.2.2.3 EMT simulation of field tests on a minor section

5.2.2.3.1 Overall results

The field tests on the minor section are simulated separately for coaxial, inter-sheath, and earth-return modes. A comparison between the field tests and simulation results is given in Table 5.11 with V_{init} and I_{init} referring respectively to the initial voltage and current peaks at the cable sending end. The ratio $V_{\text{init}}/I_{\text{init}}$ allows to calculate Z_c. The propagation speeds are found from waveform oscillating periods in EMT simulation and field tests.

5.2.2.3.2 Coaxial mode

The setup for this test is illustrated in Figure 5.17(a). The IG current and the receiving end voltage are compared in Figure 5.18. The initial voltage peak at the sending end (not shown) is 1769 V in the EMT simulation result. The values at the receiving end are 3,485 V in EMT simulation and 3,480 V in field test, that is, nearly double the sending end voltage. This is expected given that the receiving end is open-circuited

Table 5.11 Comparison of field test and simulation results

Mode	Source	V_{init} (V)	I_{init} (A)	v (m/μs)	Z_c (Ω)
Coaxial	Field	1,752	68.70	157.7	25.5
	EMT	1,769	76.57	159.4	23.1
Inter-sheath	Field	1,221	49.01	114	24.9
	EMT	1,067	46.44	111.9	23.0
Earth-return	Field	1,767	27.04	63.1	65.3
	EMT	1,868	27.30	35.4	68.4

and the coaxial mode attenuation is in general low. The maximum peak current is 76.57 A in EMT simulation and 68.7 A in the field test. The characteristic impedances calculated in Table 5.11 are in agreement with the formula [25]:

$$Z_c = \left(60/\sqrt{\varepsilon_{r_1}}\right)\ln(r_3/r_2) \tag{5.61}$$

which results into 22.9 Ω. The periods of oscillations are 27.1 μs in EMT simulation and 27.4 μs in the field test results and allow calculating the propagation speeds in Table 5.11.

5.2.2.3.3 *Inter-sheath mode*

The setup for this test is given in Figure 5.17(b). The IG current and the receiving end voltage are shown in Figure 5.19.

Though not shown, the voltage peak measured at the sending end of the metallic sheath, that is, after the lead wire, is 1,067 V in the EMT simulation result and 1,221 V in the field test as shown in Table 5.11. The voltage difference between sheath-8 and sheath-4 is 1,221 + 551.6 = 1,772.6 V. The periods of oscillations are estimated from Figure 5.19 as 38.6 μs in the EMT simulation and 38.0 μs in the field test, which result in the velocities given in Table 5.11. The agreement between field test and simulation results is sufficiently good, but there is a small difference in propagation velocities that becomes more evident after the second cycle.

5.2.2.3.4 *Earth-return mode*

The setup for this test is given in Figure 5.17(c). The IG current is shown in Figure 5.20. The earth-return mode shows an over-damped regime, with no oscillations from which we can estimate the propagation velocity and this had to be estimated approximately from $1/\sqrt{L'C'} = 63.1$ m/μs in the field test (see Table 5.11).

5.2.3 *Measured results of switching surges on the major section of the RET cable system*

The total length of the cable is 64 km separated into 17 major sections. The metallic sheaths are grounded at each major section. The first and last major sections are protected by surge arresters against overvoltage of the metallic sheath. These are the underground sections closest to aerial lines and thus more prone to over-voltages.

A series of field tests on cable transients were performed to validate cable modeling in EMT-type simulations. The test on the major section was performed as illustrated in Figure 5.21. The length of a major section is 3,953 m with 2,000 mm² and 2,500 mm² conductors. Complete cable data are given in Table 5.9.

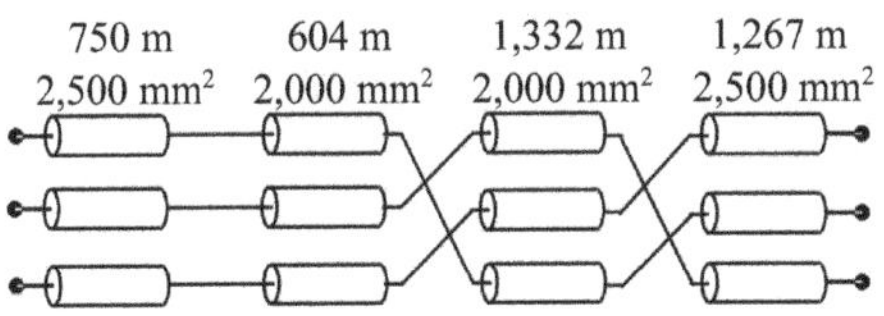

Figure 5.21 Major section of the 225 kV cable used in the field tests

The model for the major section which has two cross-bonding points uses a separate cable model for each minor section as illustrated in Figure 5.21. The cross-bonding of the Boutre-Trans cable uses 300 mm^2 copper wires with an estimated inductance of 20 μH connecting the metallic sheaths of adjacent minor sections [25].

The test circuits are shown in Figure 5.22. A summary of the comparison between field tests and simulations is presented in Table 5.12.

5.2.3.1 Coaxial mode

The setup for this test is given in Figure 5.22(a). The IG current and the sending end core voltage are shown in Figure 5.23 and can be considered as sufficiently close to each other.

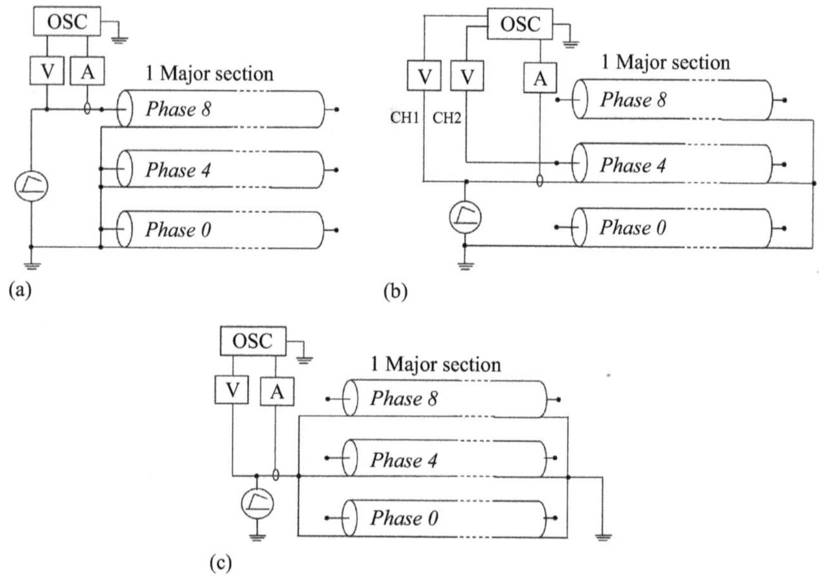

Figure 5.22 Test circuit for a major section. (a) Coaxial mode, (b) inter-sheath mode and (c) earth-return mode

Table 5.12 Field test and EMT-simulation results

Mode	Source	V_{init} (V)	I_{init} (A)	v (m/μs)	Z_c (Ω)
Coaxial	Field	1,697	75.36	158	22.97
	EMTP	1,615	75.97	142	21.3
Inter-sheath	Field	1,817	46	132	39.5
	EMTP	1,622.6	46.29	151	35.0
Earth-return	Field	1,892	69.5	35.1	27.22
	EMTP	1659	55.97	35.3	29.64

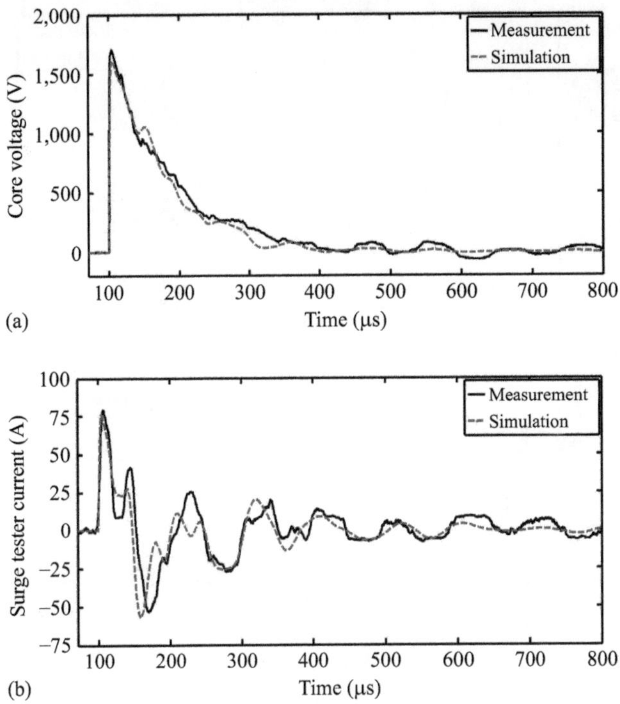

Figure 5.23 Field test and simulation results, coaxial mode excitation, and a major section. (a) Sending end core voltage and (b) IG current

5.2.3.2 Inter-sheath mode

The setup for this test is given in Figure 5.22(b). The sending end core and sheath voltages as well as the IG current are given in Figure 5.24. It is observed that the core voltages are oscillatory whereas the sheath voltages follow early the voltage of the surge tester. The peak voltage between the two sheaths is obtained from the difference of voltages as $788.9 + 833.7 = 1{,}622.6$ V (1,817 V in the field test). The characteristic impedance and velocity found in Table 5.12 are very different from those of the inter-sheath mode on a minor section given in Table 5.11. This is because the propagation modes on a normal-bonded and on a cross-bonded cable are not the same [37]. The agreement between field test and simulation is good, but there is a small difference in propagation velocities.

5.2.3.3 Earth-return mode (sheaths grounded)

The setup for this test is given in Figure 5.22(c). The IG current and the sending end core voltage are given in Figure 5.25.

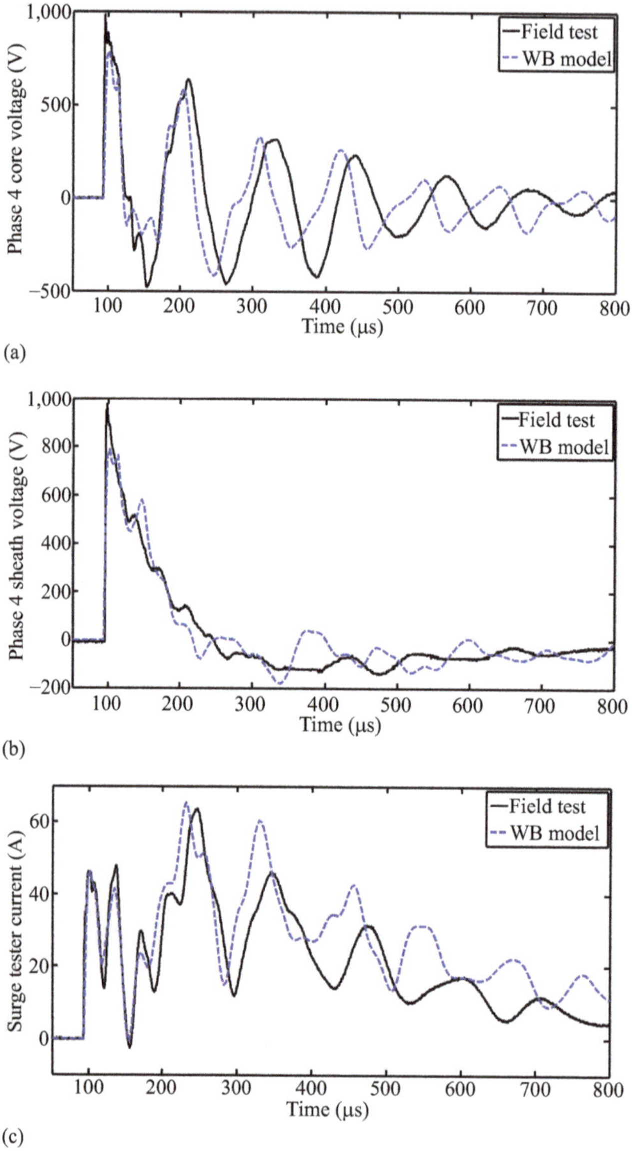

Figure 5.24 Field test and simulation results, inter-sheath mode energization between phases 0 and 4, and a major section. (a) Sending end core voltage, (b) sending end sheath voltage and (c) IG current

5.2.4 Summary

This section describes field measurement of cable transients: (1) equipment and instruments required for the measurement, (2) conditions of a test site, and (3)

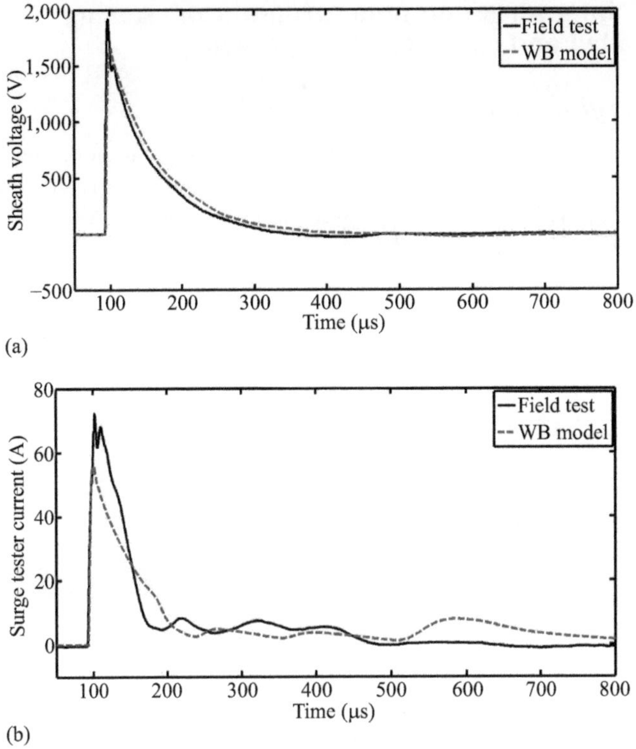

Figure 5.25 Field test and simulation results, earth-return mode excitation, and a major section with sheaths short-circuited. (a) Sending end sheath voltage and (b) surge tester current

time schedule. Then, measured results of switching surges are explained on one minor section, that is, a homogeneous cable, and on a 64 km cross-bonded cable. The field tests were performed on the RTE 225 kV cable system with a total length of 64 km. Because the cable is installed in a hard polyethylene (HDPE) tube which makes the cable configuration eccentric, it is necessary to find the equivalent cylindrical cable with the equivalent radii and permittivity. The details are explained in Section 3.3.4. Based on the equivalent cable, the impedance and admittance are calculated from manufacturer data by using the Cable Constant routine of an EMT-type simulation tools, and calculated results are compared with *RLC* parameters evaluated from field tests. This approach provides a basis for model validation.

The cable is modeled using a frequency-dependent line model and the EMT simulation results are in sufficiently good agreement with field test results. The most important differences are observed for the excitation of the earth-return mode where the actual test setup conditions are more difficult to match.

5.3 Switching surges on underground cables using extended and classical TL approaches

5.3.1 Energization of propagation modes

Figure 5.26 illustrates simulation circuits for cable energization corresponds to the coaxial, inter-sheath, and earth-return modes. The extended and classical TL approaches discussed in Chapter 2 are used in the simulations. Also, Cable 1 shown in Figure 5.27 and Figure 5.28 is adopted in this section.

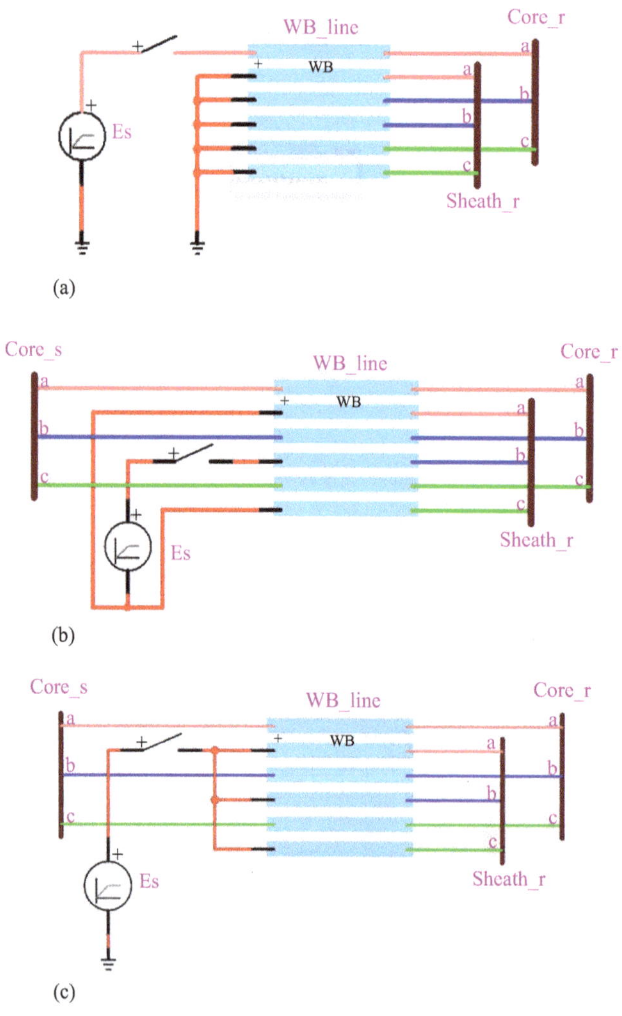

Figure 5.26 Simulation circuits for energization of modes with cable length x. (a) Coaxial mode, (b) inter-sheath mode and (c) earth-return mode

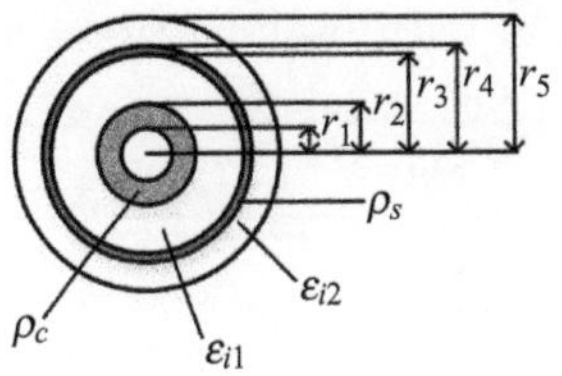

Figure 5.27 Cross-section of a single core cable

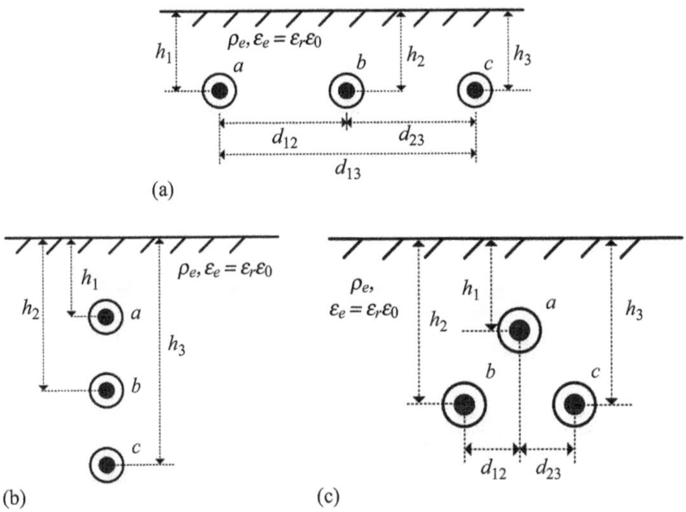

Figure 5.28 Arrangement of three-phase single core cable(a) Flat arrangement of Cable 1, (b) vertical arrangement of Cable 2 and (c) trefoil arrangement of Cable 3

The cross-section of a single core cable is shown in Figure 5.27. Three different cable arrangements are illustrated in Figure 5.28. The parameters adopted into three cables are summarized in Table 5.13 [4].

5.3.1.1 Coaxial mode

Simulation results for coaxial mode energization by extended and classical TL approaches are shown in Figure 5.29 when a step voltage source $E_s = 1$ kV is applied to the circuit in Figure 5.26(a). As expected, no difference is observed between extended and classical TL approaches.

5.3.1.2 Inter-sheath mode

Figure 5.30 shows transient sheath voltages of phase-*b* when a step voltage source $E_s = 1$ kV is applied to the circuit in Figure 5.26(b). Because of the higher attenuation of the inter-sheath mode, the transient sheath voltages by the extended TL approach are damped faster than those by the classical one. Thus, the

Table 5.13 Parameters of cables

	Cable 1	**Cable 2**	**Cable 3**
Voltage (kV)	132	110	150
Arrangement	Flat	Vertical	Trefoil
h_1 (m)	1	1	1.213
h_2	1	1.35	1.3
h_3	1	1.7	1.3
d_{12} (m)	0.35	0	0.05
d_{13}	0.7	0	0.05
d_{23}	0.35	0	0.1
r_1 (cm)	1.03	1.049	0
r_2	1.9	1.78	2.075
r_3	3.45	3.075	4.025
r_4	3.85	3.255	4.276
r_5	4.25	4.075	4.794
ρ_c (Ωm)	1.7×10^{-8}	1.7×10^{-8}	3.19×10^{-8}
ρ_s	2.1×10^{-7}	2.1×10^{-7}	2.228×10^{-7}
ε_{i1}	3.5	3.5	2.68
ε_{i2}	4	4	2.3

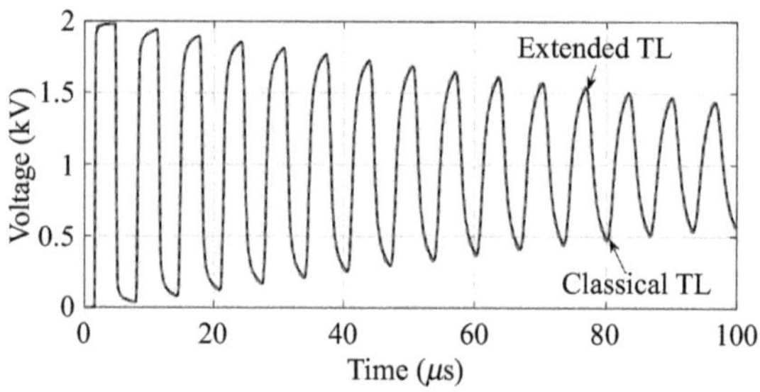

Figure 5.29 Phase-a core voltage at the receiving end: coaxial mode energization with x = 263 m

waveforms by the extended one are smoothly converging to the steady-state voltage 0.667 kV, that is, $2E_s/3$ determined by the inter-sheath mode circuit in Figure 5.26 (b). In the case of $\rho_e = 100$ Ωm, the time delay τ is estimated in Figure 5.30(a) as $\tau = 6.3$ μs and $\tau' = 7.8$ μs, where τ and τ' are calculated by the extended and classical approaches, respectively.

Since the cable length is $x = 263$ m, the propagation velocity c of the inter-sheath mode becomes $c = 41.7$ m/μs and $c' = 33.7$ m/μs, where c and c' are the velocity calculated by the extended and classical TL approaches, respectively. The modal velocity approximately corresponds to the value at about $f = 2$ MHz in the frequency domain.

In the case of $\rho_e = 500$ Ωm in Figure 5.30(b), the higher attenuation and faster propagation velocity are clearly observed in comparison with Figure 5.30(a). Also, no significant differences between $\varepsilon_r = 1$ and 20 are observed.

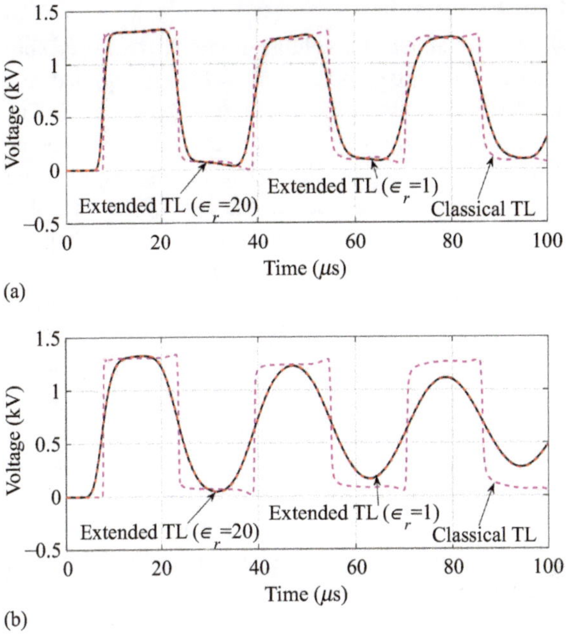

Figure 5.30 Phase-b sheath voltage at the receiving end for inter-sheath mode energization with x = 263 m. (a) ρe = 100 Ωm and (b) ρe = 500 Ωm

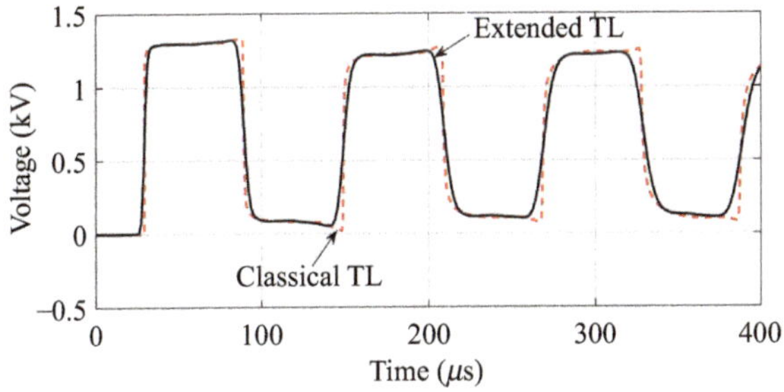

Figure 5.31 Phase-b sheath voltage at the receiving end for inter-sheath mode energization with $\rho_e = 100\ \Omega m$, $\varepsilon_r = 1$ and $x = 1$ km

Figure 5.31 shows the simulation results with the length $x = 1$ km and $\rho_e = 100\ \Omega\text{m}$. A noticeable difference is observed between the results by the extended and classical TL approaches especially at the time when a traveling wave arrives at the receiving end. In this case, $\tau = 27.9\ \mu\text{s}$, $\tau' = 29.6\ \mu\text{s}$, $c = 35.9\ \text{m}/\mu\text{s}$, and $c' = 33.7\ \text{m}/\mu\text{s}$.

5.3.1.3 Earth-return mode

Figure 5.32(a) shows phase-*a* sheath voltage at the receiving end for the observation time 250 μs, when a step voltage source $E_s = 1$ kV is applied to the three-phase sheaths which are short-circuited at the sending end as shown in Figure 5.26(c).

Similarly, the sheath voltage calculated by the extended approach is more attenuated than that by the classical one, and reaches the steady-state voltage *Es*, much faster. A similar observation can be made in the case of the inter-sheath mode application in Section 5.3.1.2. It should be noted that a small voltage less than 100 V appears at around 6 μs, before the earth-return wave arrives at the receiving end after 10 μs. Figure 5.32(b) and (c) shows phase-*a* and phase-*b* sheath voltages at the receiving end for the observation time of 50 μs. It is clear in Figure 5.32(b) that the phase-*a* and phase-*c* voltages are the same with the positive polarity, while that of phase-*b* in Figure 5.32(c) is twice of the phase-*a* voltage with the negative polarity. The velocity corresponds to that given in Section 5.3.1.2, that is, the inter-sheath mode. It is well-known in a field test result of switching surges that the earth-return mode cannot be completed by short-circuiting the three phases at the sending end. A typical field test on a 500 kV transmission line is described in [39]. About 28% of the earth-return mode voltage was measured on phases at the receiving end as the inter-phase mode voltage. In Figure 5.32(b), about 5% of the earth-return mode voltage appears on the phases-*a* and-*c*, and about 10% appears on phase-*b* (middle phase) with the negative polarity in Figure 5.32(c). Because of the more balanced impedances in the cable in Figure 5.28

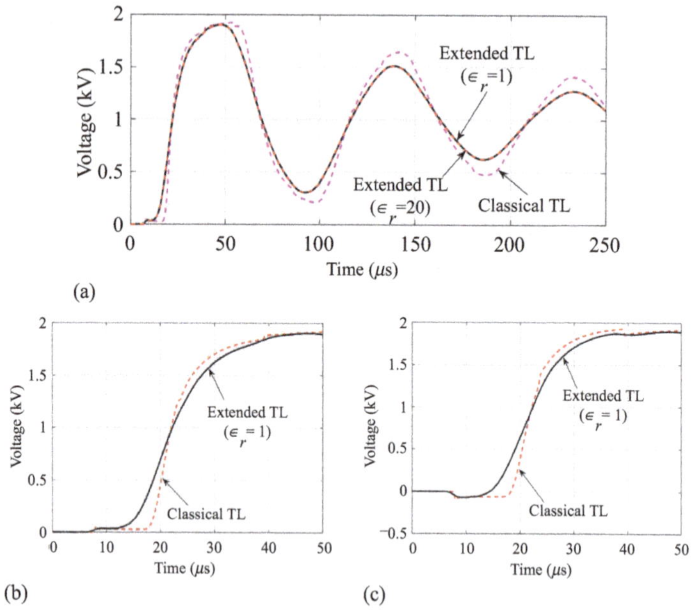

Figure 5.32 Sheath voltages at the receiving end for earth-return mode energization with $\rho_e = 100\ \Omega$m and x = 263 m. (a) Phase-a with t = 250 μs, (b) phase-a = phase-c with t = 50 μs and (c) phase-b with t = 50 μs

(a), the result is reasonable compared with 28% in the 500 kV un-transposed vertical twin circuit line. If a complete earth-return mode is necessary, the three-phase sheaths at the receiving end should be short-circuited. The velocity of the earth-return mode in Figure 5.32 is observed as $c = 19.6$ m/μs and $c' = 14.7$ m/μs.

5.3.2 Energization of a cross-bonded cable

In this section, switching surges on a cross-bonded cable are investigated.

5.3.2.1 Sheath bonding methods

The sheath of a single core cable for AC service behaves like the secondary circuit of a transformer. The current in the conductor induces a voltage to the sheath. When the sheaths of cables are bonded to each other, the induced voltage causes a current to flow through a closed circuit [4]. This current causes losses in the sheath. Various methods of bonding may be used, and must perform the following functions [4]:

- Limit sheath voltages as required by the sheath sectionalizing joints.
- Reduce or eliminate the sheath losses.
- Maintain a continuous sheath circuit to permit fault-current return, and adequate lightning and switching surge protection.

To satisfy these requirements either fully or partially, the cable sheaths are divided into a number of sections by means of sheath sectionalizing joints. The length of these sections is determined by the permissible sheath voltage levels for normal and fault conditions.

5.3.2.1.1 Single-point bonding

Figure 5.33 shows a typical configuration of single-point bonding. It is the simplest form of bonding that consists of arranging for the sheaths of the three cables to be

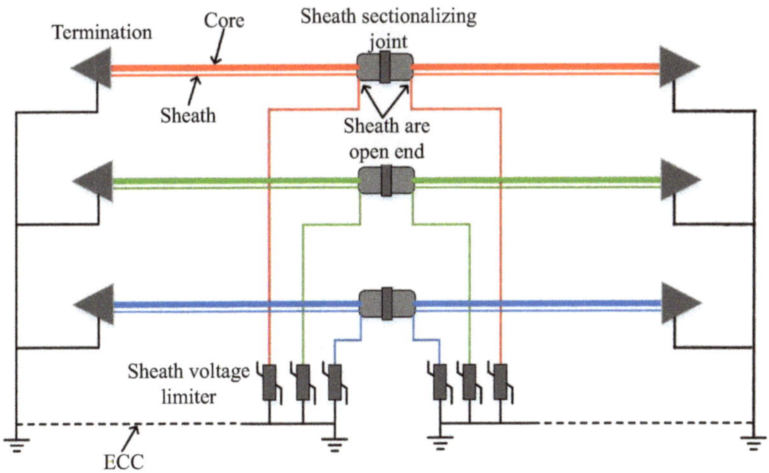

Figure 5.33 Single-point bonding (two sections) with ECC and SVL

connected and grounded at one point only along their length. At all other points, a voltage will appear from the sheath to the ground that will be a maximum at the farthest point from the ground bond. The sheaths must therefore be adequately insulated from the ground. Since there is no closed sheath circuit, except through the sheath voltage limiter (leakage current), current does not normally flow longitudinally along the sheaths and no sheath circulating current loss occurs (sheath eddy loss will still be present). This is typically suitable for cable sections less than 500 m, or one drum length.

The functions of earth continuity cable (ECC) and sheath voltage limiter (SVL) can be also clarified [4]. During a ground fault on the power system, the zero-sequence current carried by the cable conductors returns by whatever external paths are available. Since a single-point bonded cable sheath is grounded at one position only, it cannot, except in the case of a cable fault, carry any of the returning current. This being so, unless some parallel external conductor is available or is provided to serve as an alternative path, the return current can flow only by way of the ground itself. Accordingly, it is recommended that a single-point bonded cable installation be provided with a parallel earth continuity cable that is grounded at both ends of the route. The size of this conductor must be adequate to carry the full expected fault current for the cable system.

Also, the SVLs, which protect sheath insulators and cable jackets, are designed to limit [4]:

- Transient voltages associated with lightning.
- Switching surges.
- Fault initiation.

The use of SVLs is intended to reduce the problems of failures of sheath insulators and cable jackets encountered in the installations. For the single-point bonding, the SVLs are connected between the un-bonded ends of the cable sheaths and ground.

5.3.2.1.2 Cross bonding

Cross bonding consists essentially of sectionalizing the sheaths into minor sections and cross-connecting them so as to approximately neutralize the total induced voltage in three consecutive sections (three equal-length sections), as shown in Figure 5.34.

Unlike the single-point bonding, the sheath circuit on the left circuit is connected to the sheath circuit of a different phase cable on the right side as shown in Figure 5.34. The main advantage of cross bonding is that the induced sheath currents are limited during the normal load operation, and the sheaths form a continuous path from end to end of the cable circuit and are grounded at both ends [4]. Because the sheath current can flow during ground faults, thus ECC can be removed and give an economic advantage. Also, the sheath becomes a more effective screening conductor for ground faults, hence, the voltage induced in parallel cables is less during ground faults in a cross bonded system than a single-point bonded system.

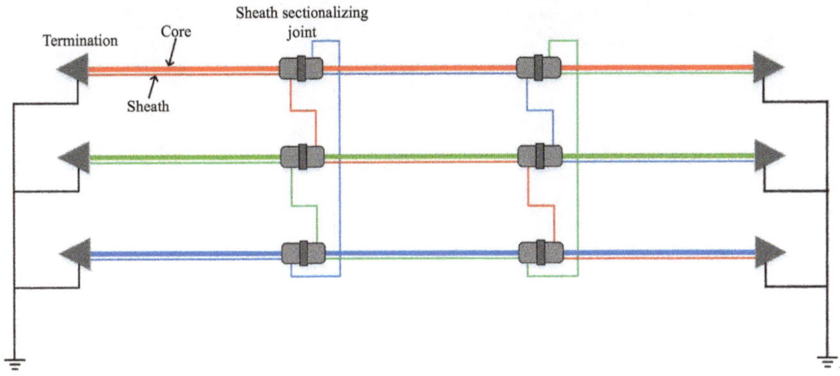

Figure 5.34 Cross bonding without core transposition, a major section

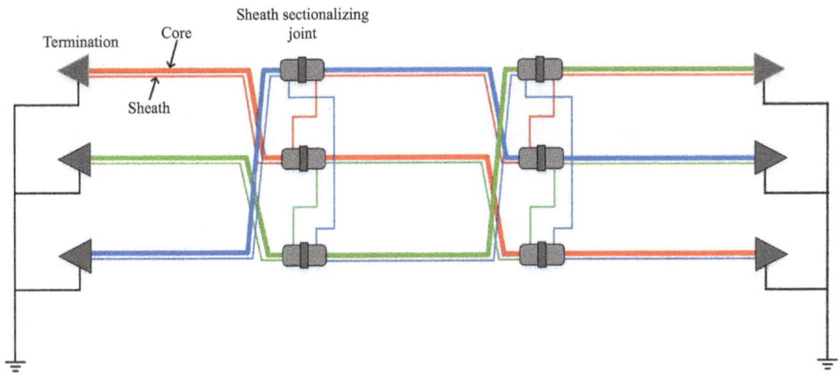

Figure 5.35 Cross bonding with core transposition, a major section

With un-transposed cables, it is impossible to achieve an exact balance of induced sheath voltages unless the cables are laid in trefoil. When the cable cores are transposed at each joint position, as shown in Figure 5.35, the induced sheath voltages will be neutralized irrespective of cable formation. However, the main difficulty of cable with core transposition is that how to do the cable crossing, and still obtain the distance between the cables to ensure proper heat dissipation. In case the cables are closer to each other at the point of crossing there is introduced a bottleneck for the ampacity of the line.

Although the continuous sheath voltage is suppressed by the small sheath current and the balanced sheath voltage, the minor section sheath voltage cannot be suppressed by the cross bonding [4]. Thus, in order to against the minor section sheath voltage, the SVLs are installed at the sheath sectionalizing joint via link box, as shown in Figure 5.36. Figure 5.37 shows a structure of a typical link box, and SVLs are normally arranged by a star connection.

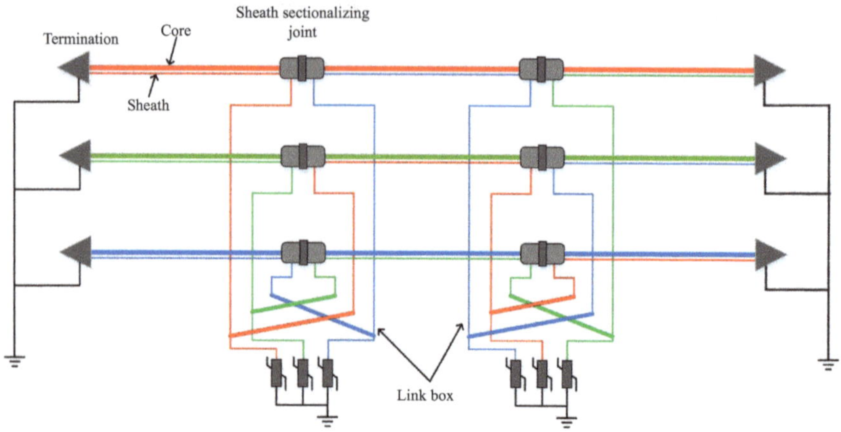

Figure 5.36 Cross bonding without core transposition, a major section with link box and SVLs

Figure 5.37 Link box with cross bonding connection and SVLs. Adapted from [40]

5.3.2.2 Energization of a major section

Figure 5.38 illustrates a field test circuit on a 110 kV cross-bonded cable [4]. Cable 2 is adopted and the geometry is shown in Figure 5.28(b). A 7.3 kV impulse voltage with wavefront duration $T_f \approx 0$ and wave tail $T_t = 40$ μs is applied to the middle phase-core at the sending end. Figure 5.39 shows the measured results (re-drawn from photos [41] recorded by an analog oscilloscope) and Figure 5.40 is EMTP

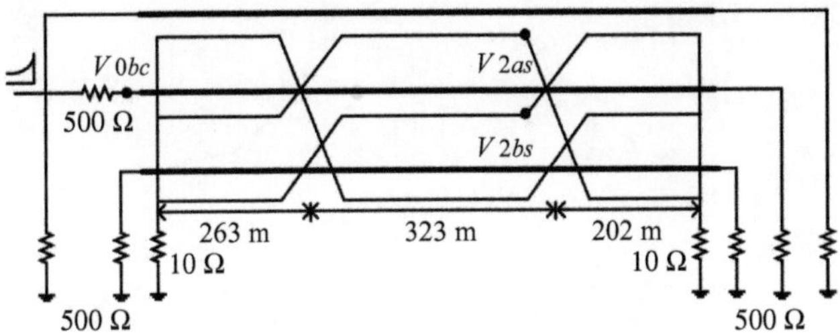

Figure 5.38 Test circuit for Cable 2 with a major section shown in Figure 5.34, $\rho_e = 100\ \Omega m$ and $\varepsilon_r = 1$

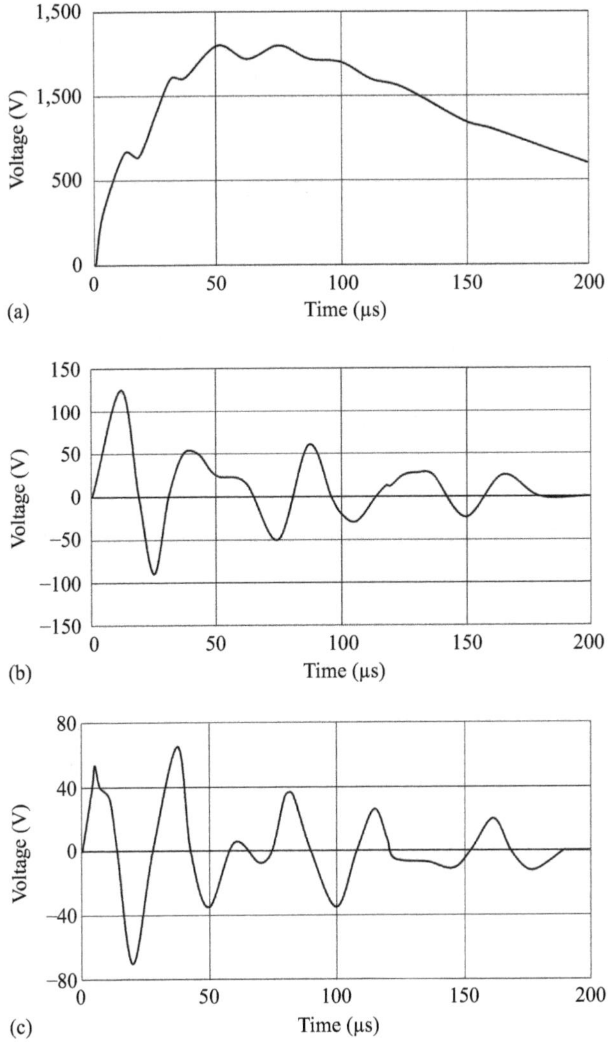

Figure 5.39 Field test results on the cable in Figure 5.38. (a) V_{0bc}, (b) V_{2as} and (c) V_{2bs}

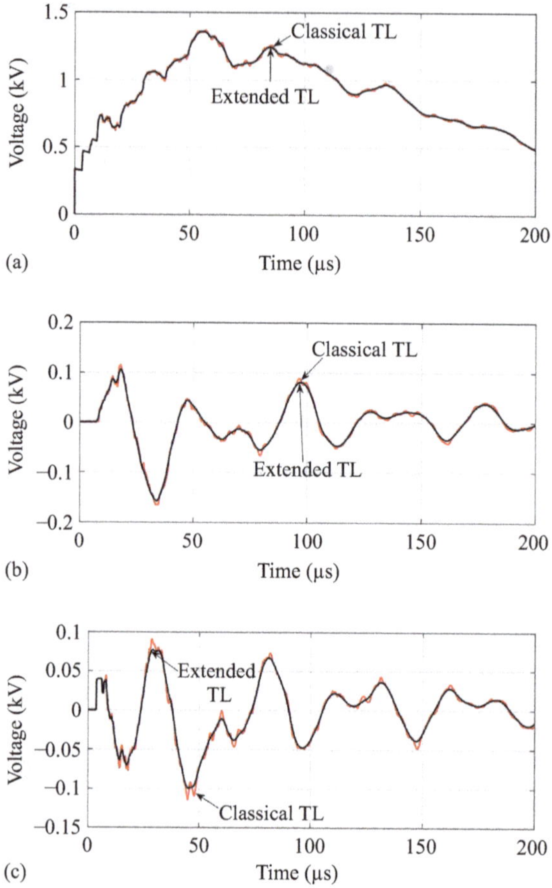

Figure 5.40 EMTP simulation results on the cable in Figure 5.38. (a) V_{0bc}, (b) V_{2as} and (c) V_{2bs}

simulation results of three measured points shown in Figure 5.38 by the extended and classical TL approaches.

The simulation results in Figure 5.40 show reasonable agreement with the field test results. However, higher frequency components are observed in the waveforms calculated by the classical TL approach, which are not observed in the extended TL approach and field test results. Note that it was a common practice to use a damping circuit to eliminate noise when surges were measured by the analog oscilloscope.

Because the shortest length of a minor section is 202 m, the highest transient frequency due to the minor section is estimated to be about 200 kHz. In this frequency region, no significant influence of the extended TL approach compared

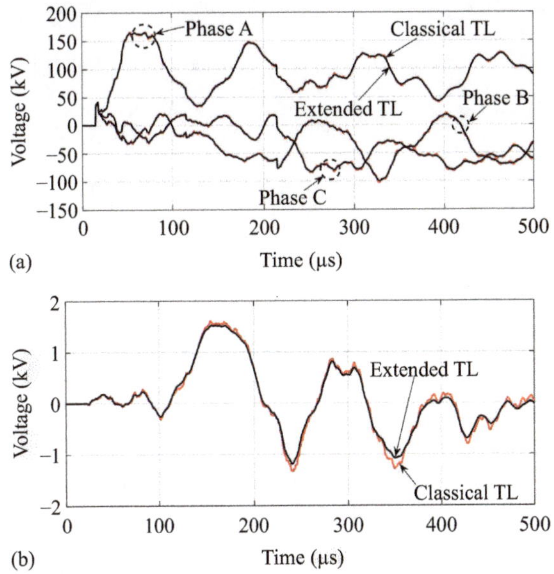

Figure 5.41 *Core and sheath voltages at the receiving end of the cross-bonded cable in Figure 5.38: ρ_e = 100 Ωm and ε_r = 1. (a) Three phase core voltages at the receiving end and (b) sheath voltage at the receiving end*

with the classical approach is expected. Thus, this measured result does not achieve the purpose of demonstrating the advantages of the extended TL approach versus the classical approach. It is required to perform different experiments and corresponding measurements in the future.

5.3.2.3 Energization of three cascaded major sections

Three major sections of Figure 5.38 are cascaded. The three-phase sheaths are short-circuited and grounded at the sending end, and at the end of each major section by R_g = 10 Ω. A 110 kV AC source is sequentially applied to the sending end of cable with closing time: phase-*a*: t = *0*; phase-*b*: t = 0.1; and phase-*c*: t = 0.2 ms. Figure 5.41 shows the three-phase core and sheath voltages at the receiving end of the cable. As expected, no differences in the core voltages are observed between classical and extended TL approaches. Because of sequential energization, the sheath voltage involves various modes. A noticeable difference between the classical and extended TL approaches is observed. The extended TL approach shows a higher damping as explained in Section 5.3.2.2.

Figure 5.42 shows the sheath voltage at the receiving end when Cable 1 shown in Figure 5.28 is adopted. The differences between classical and extended TL approaches are more visible in the flat cable than that in the vertical one as shown in Figure 5.41.

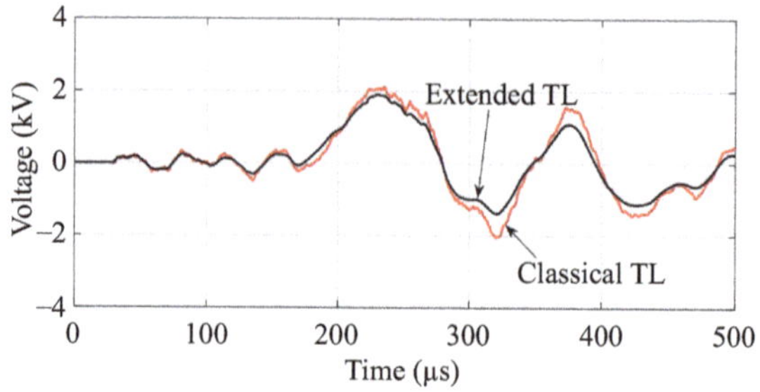

Figure 5.42 Sheath voltage at the receiving end of the cross-bonded cable in Figure 5.28(a): $\rho_e = 100\ \Omega m$ and $\varepsilon_r = 1$

5.3.3 Summary

In this section, the transient simulations on underground cables including a field test are performed and investigated. The following remarks are obtained:

1. A significant difference of waveforms for inter-sheath and earth-return modes is observed between the extended and classical TL approaches.
2. No difference is observed for waveforms of the coaxial mode on a cable between the two approaches.
3. Earth resistivity shows an important effect on the transient waveforms of the inter-sheath and earth-return modes. The effect of earth permittivity is negligible.
4. The cable arrangement shows a minor influence on core and sheath voltages.
5. The extended TL approach is validated with the field test result on a 110 kV cross-bonded cable.
6. It is confirmed that the extended TL approach results in higher damping and smooth convergence to the steady-state compared with those calculated by the classical TL approach.

5.4 Very fast transient (VFT) in gas-insulated substation (GIS, overhead cable)

5.4.1 Very-fast transient (VFT)

It is well-known that lightning strikes to a transmission tower nearby a substation and switching operations in a gas-insulated substation (GIS) produce high-frequency surges. The dominant frequency components involved in the lightning surge are, in general, lower than 1 MHz. The switching surges due to dis-connector (DS) or circuit breaker (CB) operation in the GIS contain the frequency components from some MHz up to more than 100 MHz, and are called “very fast transient (VFT)” or “very fast transient over-voltage (VFTO)” [42–45]. There are a number of publications investigating the VFTs [46–86]. It has been found based on many VFT measurements in laboratories and actual GISs that the VFT frequencies range from some MHz to more than 100 MHz, and the over-voltages range from 1.2 pu to

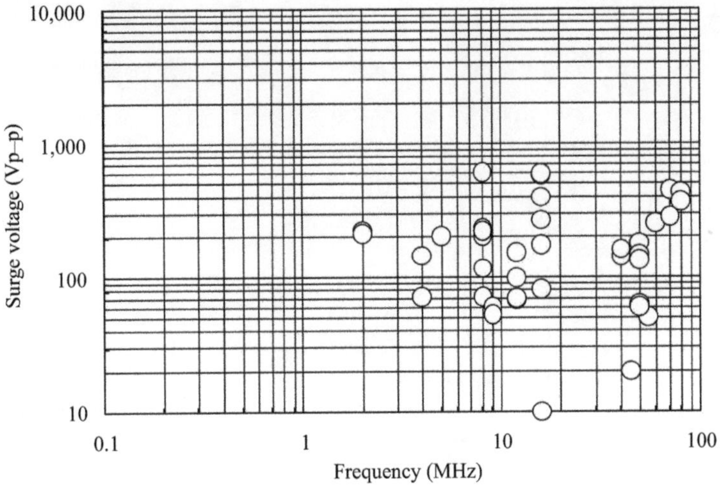

Figure 5.43 Field test results of the oscillating frequency and amplitude of transient voltages in in gas-insulated substations. Adapted from Figure 2 of [60]

3.0 pu on the core conductor of the GISs, and up to 0.7 pu on the metallic enclosure (pipe) [46–60]. Figure 5.43 shows field test results of the peak-to-peak voltage versus frequency of switching surges experienced in many gas-insulated substations in Japan for 10 years [57–60]. It is observed in the figure that the voltage does not decrease as frequency increases. This phenomenon is against the classic TL theory, which tells the attenuation monotonously increases as frequency increases. The phenomena observed in Figure 5.43 cannot be explained by the classic TL approach. The very-high-frequency components observed in the field test results are generated by multiple reflection and refraction of traveling waves along with short gas-insulated buses (GIBs) including DSs and CBs in GISs. Figure 5.44 illustrates the layout of a real 500 kV GIS [69]. The length of each GIB is up to 7 m of which the traveling time is less than $\tau = 2.5$ ns with the propagation velocity being nearly the same as that of the light in free space because of the relative permittivity of SF_6 gas is about 1.0. When a GIB is open-circuited, the dominant transient frequency becomes: $f = 1 / 4\tau = 100$ MHz.

Because of very high-frequency components, the VFTs in the GISs can cause electromagnetic interferences (EMIs) and malfunctions in GIS control circuits [51,55–60,70–73].

Modeling methods for electromagnetic transient (EMT) simulations are proposed in [69,74–77]. It should be noted that these EMT-type simulations are performed by adopting a constant-parameter (CP) line model because the frequency-dependent line model with Carson's earth-return impedance and space admittance in existing EMT-type simulation tools shows numerical instability for VFTs [79–85]. Some papers investigated the electric and magnetic fields associated with VFTs by

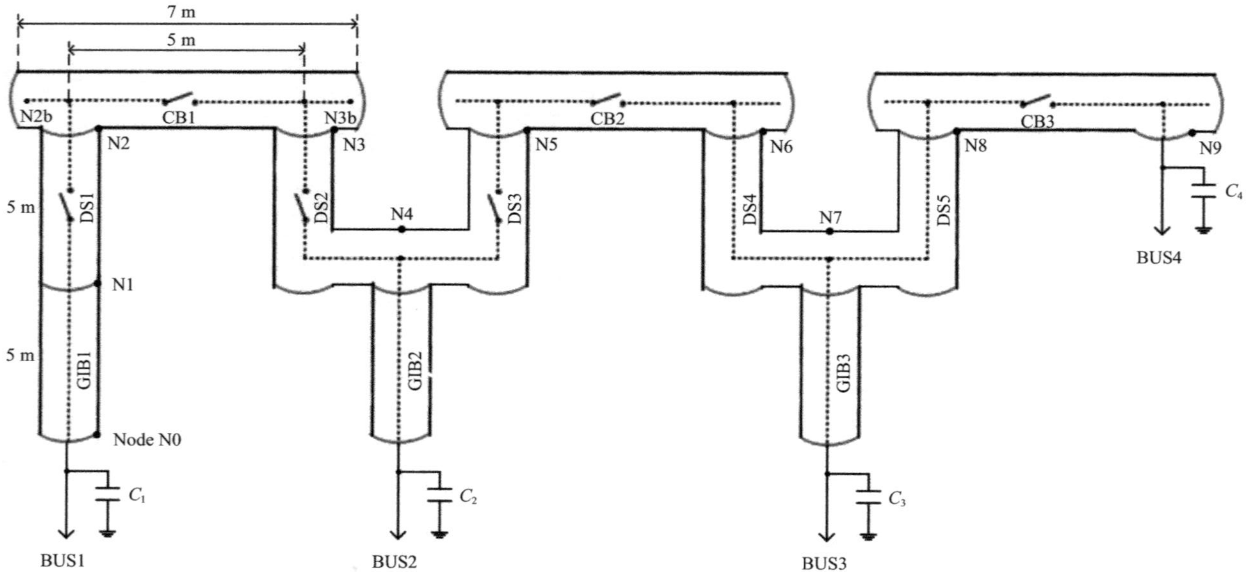

Figure 5.44 Layout of a real 500 kV GIS in Japan. Adapted from [69]

adopting the three-dimensional (3D) full-Maxwell approach with finite-element method (FEM) [61,78]. However, the FEM approach is limited because of its reduced computer performance. Most of the above researches are focused on VFTs on the core of the GISs and gas-insulated circuit breakers (GCBs) because the insulation strength and high-voltage testing of the core is essential for GIS and GCB design and testing. A gas-insulated transmission line (GITL) similar to GIS is investigated in [46] and [86]. The above-mentioned articles are briefly categorized as

1. Measurements of VFTs on prototype GISs and parts of real GISs: [46,54,62,65].
2. Measurements of electromagnetic interferences in GISs and related standards: [51,55,60,66,70,73].
3. Modeling methods of GIB, GCB, dis-connector, etc. for VFT simulations by circuit-theory-based approach, that is, EMT-type, or by numerical electromagnetic analysis (NEA) methods, and the accuracy of the methods: [46,54,61,63,69,74,85].
4. Damping methods of VFTs: [49,67].

5.4.2 Numerical instability of EMT simulation of VFT

This section shows a numerical instability observed in EMT simulation of a very-fast transient (VFT) in a 500 kV gas-insulated substation [79–85]. The instability is associated with the physical phenomenon called as "mode transition" from low-frequency earth-return wave (quasi-TEM mode of wave propagation) to high-frequency surface wave (TM mode = Sommerfeld-Goubau wave) propagation as explained in Chapter 4 [87–95] involved in switching surges in gas-insulated buses (overhead cables). Because Carson/Pollaczek's earth-return impedance [96,97] and space admittance (classical transmission line (TL) approach, see Chapter 2) implemented into existing EMT-type simulation tools cannot deal with the surface wave propagation, the mode transition cannot be handled. Thus, EMT simulation results in numerical instability. The instability is solved by the extended transmission line (TL) approach with accurate formulas for the earth-return impedance and admittance as explained in Chapter 2 [80–83,93], and also by the modified classical TL approach [84,85].

5.4.2.1 Model circuit

Figure 5.45 illustrates a part of a fully gas-insulated 500 kV substation illustrated in Figure 5.44 [69]. The cross-section of gas-insulated buses GIB1 and GIB2 and the dis-connectors DS1 to DS3 are shown in Figure 5.46. Although the cross-section of circuit breakers CB1 and CB2 is a little larger than that in Figure 5.46, the same cross-section is assumed in the EMTP simulation. The length of the GIB, DS, and CB ranges from 5.1 m to 6.8 m. In the EMTP simulation, the GIB and DS length is assumed to be 5 m, and that of the CB is 7 m. There are spacers at the joints of the DSs. It has been shown in [69] that the spacers decrease the VFT frequencies by 5–20% depending on the number of the spacers. VFT over-voltages are not necessarily reduced, because the spacers increase more the reflection/refraction of traveling waves. The spacers are neglected in the simulation. The pipes of the GIBs, DSs, and CBs are grounded to the grounding mesh of the GIS. The grounding resistance (R_g) is assumed to be 2 Ω.

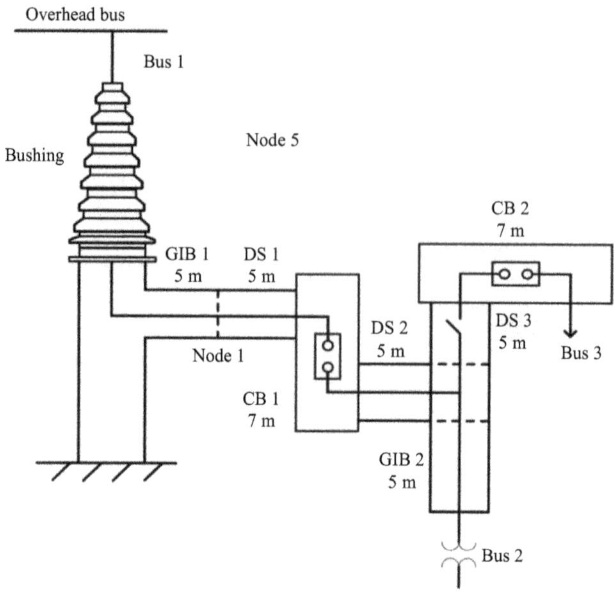

Figure 5.45 A part of a 500 kV GIS shown in Figure 5.44

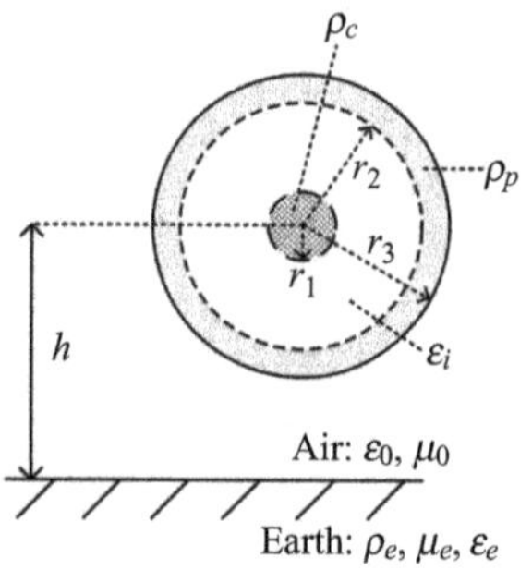

Figure 5.46 Cross-section of a gas-insulated bus: $h = 1.5$ m, core $r_1 = 5$ cm, pipe $r_2 = 18$ cm, and $r_3 = 20$ cm; resistivities: core $\rho_c = 1.68\times10^{-8}$ Ωm, pipe $\rho_p = 2.82\times10^{-8}$ Ωm, earth $\rho_e = 100$, and 2000 Ωm; and relative permittivity: SF_6 gas $\varepsilon_i = 1$, earth $\varepsilon_r = 1$ and 20, and relative permeability of all media: 1.0. Adapted from [69]

In the transient calculation, CB2 is kept open. DS1, CB1, and DS2 are kept closed and charged from a 500 kV AC 60 Hz source (amplitude 408 kV) connected to BUS1. DS3 is closed at $t = 0$. Logarithmic Laplace transform (LLT) [82,83] is adopted to calculate the VFTs. The LLT gives nearly identical results to those calculated by EMTP frequency-dependent wideband model, the details will be explained in Section 5.5.2.3 [84].

5.4.2.2 Propagation function and step response

5.4.2.2.1 Propagation function in the frequency domain

Figure 5.47 shows propagation function $H_1(\omega)$ defined as

$$H_1(\omega) = e^{-\alpha(\omega)x} \tag{5.62}$$

where x is the conductor length.

It is clear in Figure 5.47 that $H_1(\omega)$ monotonously decreases as frequency increases in the classic TL approach following the characteristic of $\alpha(\omega)$ explained in Chapter 4. On the contrary, $H_1(\omega)$ decreases as frequency increases up to f_1. Then it increases until frequency reaches f_2. Because of lower attenuation in a frequency region higher than f_2, $H_1(\omega)$ is much greater than that in the classic TL approach.

The above-explained phenomena have not been investigated/discussed in previous publications.

5.4.2.2.2 Step response in the time domain

Although mode transition has been investigated and discussed in some papers, all the investigations are in the frequency domain. The time responses of the propagation function and transient voltages on a gas-insulated bus are calculated by adopting numerical Laplace transform with logarithmic sampling [55,56] and the effect of the mode transition on the transient responses is discussed in this section.

When a source voltage defined as $E(\omega)$ in frequency domain is applied to the sending end of a conductor, the voltage $V_x(\omega)$ at distance x from the sending end is given in the following equation:

$$V_x(\omega) = e^{-\Gamma(\omega)x}E(\omega) = H(\omega)E(\omega) \tag{5.63}$$

$$\Gamma(\omega) = \sqrt{Z(\omega)Y(\omega)} \tag{5.64}$$

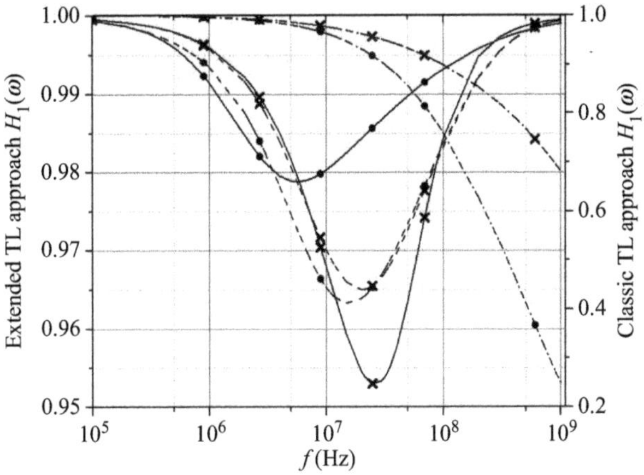

Figure 5.47 Propagation function $H_1(\omega)$ on a gas-insulated bus with $x = 2$ m, ×: $\rho_e = 100\ \Omega$m, •: $\rho_e = 2000\ \Omega$m, –: extended TL approach ($\varepsilon_r = 1$), –: extended TL approach ($\varepsilon_r = 20$), and -•-: classic TL approach ($\varepsilon_r = 1$)

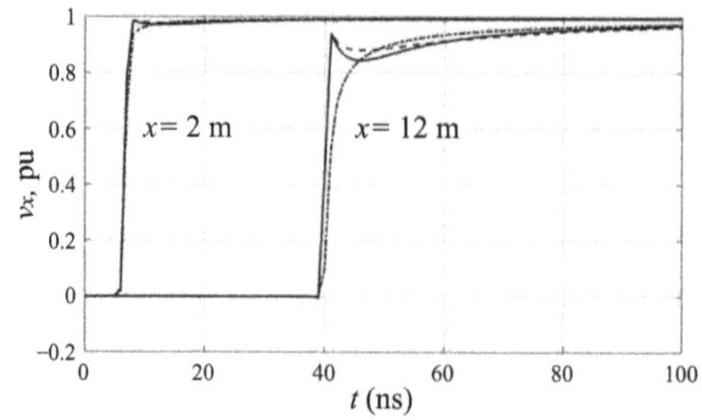

Figure 5.48 Step response on an overhead cable with $h = 1.5$ m and $r = 5$ cm at $x = 2$ m and 12 m, $\rho_e = 100\ \Omega$m: —: extended TL approach ($\varepsilon_r = 1$), –: extended TL approach ($\varepsilon_r = 20$), and -•-: classic TL approach ($\varepsilon_r = 1$)

where $Z(\omega)$ and $Y(\omega)$ are the series impedance and shunt admittance of the conductor. The earth-return admittance is given in [98–101].

The time response $v_x(t)$ is calculated by numerical Laplace to transform as

$$v_x(t) = \mathcal{L}^{-1} V_x(s) \tag{5.65}$$

where s is the Laplace operator.

When $E(\omega)$ is given as $1/j\omega$, that is, a unit step voltage $v_x(t)$ is called a step response. Figure 5.48 shows calculated results of the step responses for a gas-insulated bus in Figure 5.46 at the distance $x = 2$ m and 12 m.

A spike-like voltage is observed at the wavefront of the step responses calculated by the extended TL approach, while no such voltage appears in the step response calculated by the classic TL approach. The spike voltage is a result of the lower attenuation explained in Chapter 4 and Figure 5.47 calculated by the extended TL approach. In the classic TL approach, the attenuation increases monotonously, and the propagation function decreases monotonously as frequency increases as observed in Figure 5.47. Thus, the step responses show a smooth increase to reach 1 pu as time passes. It is estimated that the spike voltage observed in Figure 5.48 results in a sustained high-frequency transient voltage because of a lower attenuation in a high frequency.

5.4.2.3 Simulation results

5.4.2.3.1 Extended and classical TL approaches

Figure 5.49 shows EMT simulation results on Node 1 core voltage using numerical Laplace transform [102,103]: (a) is the result by the extended TL approach and (b) by the classical TL approach. The frequency-dependent line model called the "Wide-band model" prepared in EMTP [28,104–106] gives almost the same results as those by the numerical Laplace transform. Up to 200 ns, the classical TL approach shows a similar voltage wave-form to that obtained by the extended TL approach. Then, it starts to diverge and shows numerical instability as is clear in Figure 5.49(b). This is the reason why the frequency-dependent model in existing EMT-type simulation tools is not used, but the constant parameter (CP or Bergeron) line model is almost always adopted as an alternative approach to avoid the

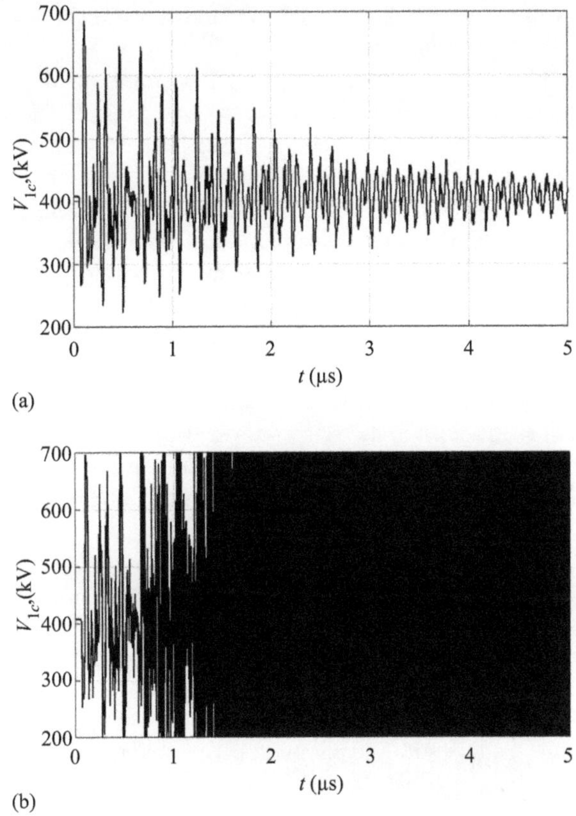

Figure 5.49 Node 1 core voltage. (a) Extended TL approach and (b) classical TL approach

numerical instability in the frequency-dependent model in the VFT simulations in a GIS [47,48,51,54,56–58,67,69,74–76].

As is well-known, Carson/Pollaczek's earth-return impedance cannot be applied to very high-frequency phenomena such as the VFT when the frequency dependence is to be taken into account. The classical TL approach with Carson/Pollaczek's impedance and the space admittance cannot express the mode transition and the surface wave propagation as explained in Section 5.4.2.2 and also as pointed out by Kikuchi *et al.* (see Chapter 4). This fact results in the numerical instability in Figure 5.49(b). It is noteworthy that the earth-return admittance contributes significantly to the high-frequency phenomena on an overhead line including a gas-insulated bus as explained later. If the admittance is considered in an EMT simulation of the high-frequency transients, no numerical instability is observed except that caused by a numerical instability of Pollaczek's impedance itself such as a negative inductance. Also, it should be noted that the numerical instability does not appear in the EMT simulation of a high-frequency transient on an underground cable, because there exists no mode transition and no surface wave propagation on the underground cable.

The extended approach shows a typical VFT waveform and converges to the steady-state voltage 408 kV. The dominant frequency of the VFT in Figure 5.49(a) is observed to be about 10–15 MHz, and a higher frequency component of about 120 MHz is observed.

Figure 5.50 shows Node 1 pipe voltage calculated by the extended TL approach. Since the classical TL approach results in numerical instability as explained above, no result is shown. Similar frequency components to those of the core voltage in Figure 5.49(a) are also observed in the pipe voltage in Figure 5.50.

Figure 5.51 shows frequency spectra of the core voltage given in Figure 5.49 (a) and the pipe voltage in Figure 5.50. The frequencies observed in Figures 5.49(a) and 5.50 are confirmed by the frequency spectra in Figure 5.51. The above-explained numerical instability has been confirmed by numerical Fourier/Laplace transform [102,103], and existing EMT-type simulation tools with the classical TL approach. Further, no instability is observed when the extended TL approach is adopted.

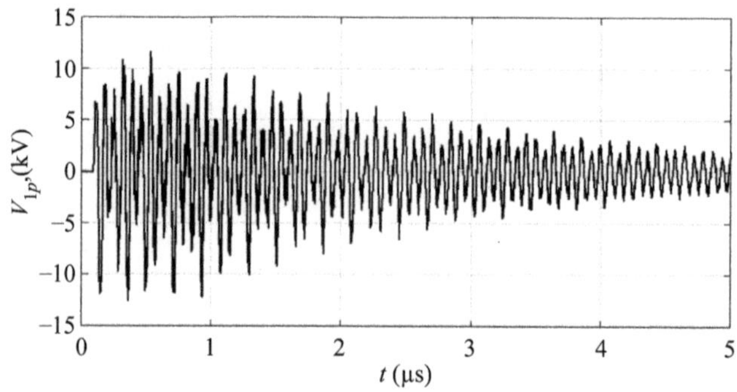

Figure 5.50 Node 1 pipe voltage by the extended TL approach

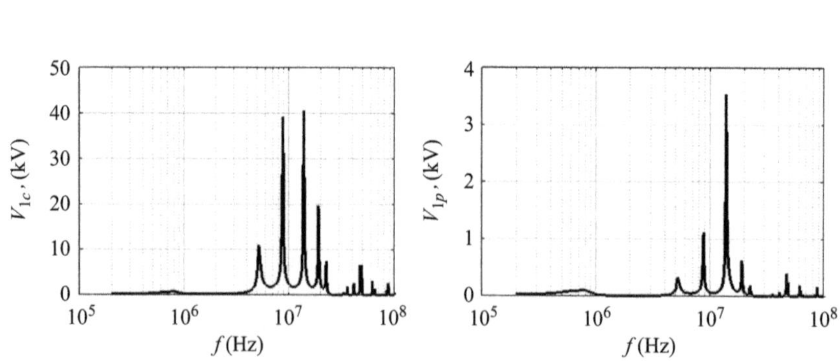

Figure 5.51 Frequency spectra of the core voltage in Figure 5.48(a) and pipe voltage in Figure 5.49

5.4.2.3.2 Modified classical TL approach

As explained in Section 2.3, Carson/Pollaczek's earth-return impedance and Gary/Deri's approximate formulas are modified to consider displacement currents so that these formulas can be used to deal with high-frequency phenomena. In this section, the modified classical TL approach is adopted to investigate the numerical instability of VFT simulation by EMT-type simulation tools [87,88].

1. Model circuit

   Figure 5.52 illustrates a simplified circuit of the 500 kV GIS shown in Figure 5.44 [69]. The pipe is grounded by R_{gs} at the sending end, and by R_{gr} at the receiving end. The core is open-circuited at the receiving end. Bus length x_1 from the sending end (Node 0) to Node 1 is 2 m and length x_2 from Node 1 to the receiving end (Node 2) is 10 m (case 1) and 4 m (case 2). Since the both cases give almost the same results of numerical instabilities in EMTP simulations, only the case of $x_2 = 4$ m (total length $x = 6$ m) is shown in the paper.

   A frequency-dependent line model (Universal Line Model in PSCAD/EMTDC) is used in the EMT simulations.
2. Accuracy of modified classical TL approach

   Figure 5.53 shows simulation results of core voltage V_{2C} at the receiving end in Figure 5.52 with $R_{gs} = R_{gr} = 2\ \Omega$ for $T_{max} = 2.5$ μs by the extended and modified classical TL approaches considering the earth-return admittance. The simulation results in Figure 5.53(b) and (c) by the modified classical approach are nearly the same as those in Figure 5.53(a) by the extended TL approach. Thus, the proposed methodology to modify the impedance formula derived by Carson and an approximate formula by Gary/Deri is satisfactorily accurate. Furthermore, no numerical instability is observed in the EMT simulation results.
3. Numerical instability

   Figure 5.54 shows EMT simulation results with the same circuit conditions as those in Figure 5.53 by the classical TL approach with the space admittance Ys implemented into existing EMT-type simulation tools. Figure 5.54(a) is calculated by Carson's formula and Figure 5.54(b) is calculated by Gary/Deri approximate formula.

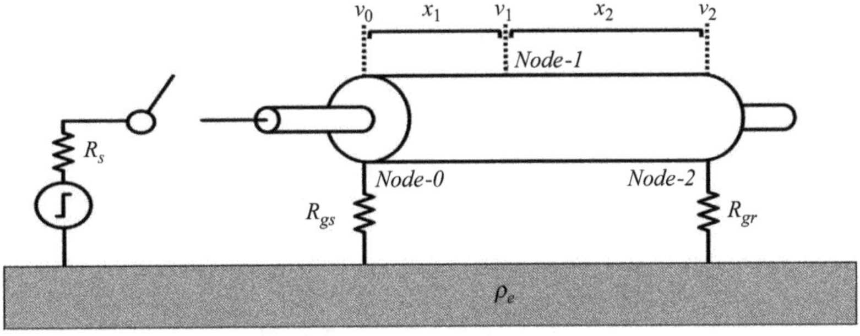

Figure 5.52 Model circuit of a gas-insulated bus. Adapted from [69]

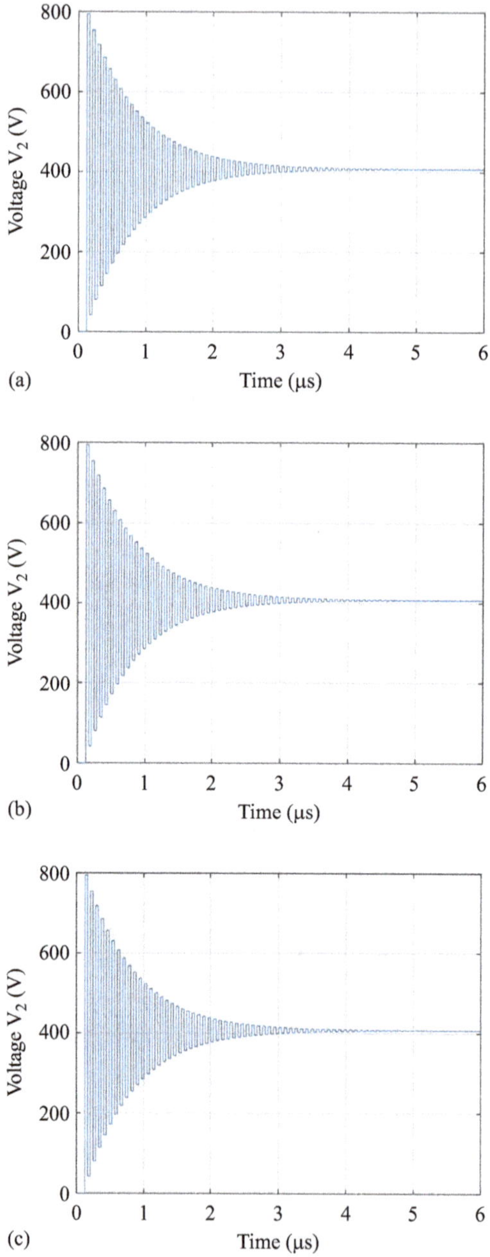

Figure 5.53 EMT simulation results of the core voltage at the receiving end with $R_{gs} = R_{gr} = 2\ \Omega$ for $\varepsilon_r = 1$ by the extended and modified TL approaches. (a) Extended TL approach with accurate Z_e and Y_e, (b) modified classical TL approach with modified Carson Z_c and accurate Y_e and (c) modified classical TL approach with modified approximate Z_c and accurate Y_e

It is observed that the transient voltage in Figure 5.54 is diverging and results in numerical instability. On the contrary, Figure 5.53 calculated by the extended and modified classical TL approaches shows no numerical instability. The transient voltage in Figure 5.53 is gradually attenuated and reaches the steady-state voltage of 408 kV.

If the voltage waveform is carefully observed, a spike voltage appears first at about $t = \tau = 20$ ns for length $x = 6$ m and is amplified at every 2τ in Figure 5.54, while the spike voltage is gradually damped out in Figure 5.53. This voltage amplification in the waveform in Figure 5.54 calculated by the classical TL approach is the cause of the numerical instability in the EMTP simulation. The original spike voltage at $t = \tau$ is generated by the boundary condition at the receiving end because of different propagation characteristics of the coaxial and earth-return modes.

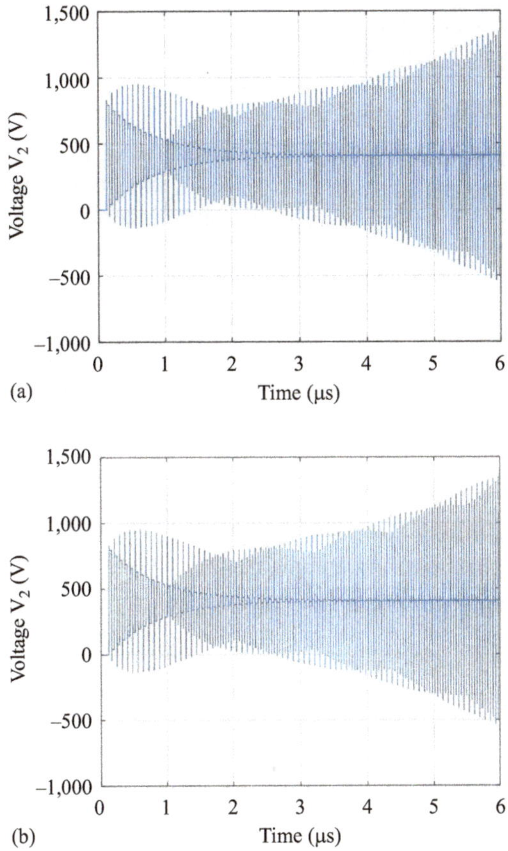

Figure 5.54 EMTP simulation results of the core voltage at the receiving end with $R_{gs} = R_{gr} = 2\ \Omega$ for $\varepsilon_r = 1$ by the classical TL approach. (a) Classical TL approach with Carson's earth-return impedance Z_c and space admittance Y_s (current EMT-type simulation tool) and (b) classical TL approach with approximate Z_e and Y_s

The basic structure of the EMT-type simulation tools and the classical TL approach are within the framework of the TEM mode of wave propagation. Therefore, the EMT-type tools and classical TL approach (Carson's earth-return impedance) cannot be applied physically and theoretically to the mode transition phenomenon which involves a non-TEM mode of wave propagation in a high-frequency region as is well-known. This is the reason for the numerical instability observed in EMT simulation results of VFTs in GISs. The passivity analysis can somehow explain the numerical instability within the framework of the TEM mode.

The numerical instability is related to the mode transition which cannot be dealt with the current EMT-type simulation tools and classical TL approach. Above the critical frequency, the propagation mode becomes a non-TEM. Then, the basic theory of the EMT-type simulation tools collapses, and results in the numerical instability.

(3) Passivity analysis

This section discusses the stability of time-domain simulation using passivity analysis in frequency domain within the framework of a TEM mode of wave propagation. A passive cable model ensures the stability of the time domain simulation provided that the external circuit is also passive; however, a non-passive model may result in unstable simulation. This is a profound problem associated with frequency-dependent transmission line models in EMT-type simulation tools.

The stability (or the passivity) of the transmission line model is ultimately based on the curve-fitted data (i.e., Hermitian matrix derived from fitted propagation and characteristic admittance matrices) [107].

There are two main passivity identification approaches: (1) state space approach and (2) frequency sweep approach [107]. The frequency sweep approach is used as it is easier to implement compared with the former method. The passivity is checked by observing the real eigenvalues of the Hermitian matrix for a wide frequency range (0.5 Hz–1000 MHz). There is a possibility that there can be missing passivity violations in between samples; however, this can be overcome by carefully taking sufficient frequency samples for a given frequency range.

Figure 5.55 shows passivity analysis results in frequency domain corresponding to the time-domain simulation results in Figures 5.53 and 5.54. In the case of the extended TL approach shown in Figure 5.55(a), it can be said that there exists no passivity violation, that is, the negative eigenvalues in the order of 10−6. A similar result is observed in the negative eigenvalues in Figure 5.55(b) and (c).

With the classical TL approach (Carson's earth-return impedance and space admittance implemented into existing EMTPs), the eigenvalues of Hermitian matrix are shown in Figure 5.55(d) and (e). The large passivity violations, that is, large negative eigenvalues are observed in the frequency region of 100 MHz corresponding to the numerical instability in the transient response in Figure 5.54. The largest negative eigenvalue in Figure 5.55(d) is −5.2 at 200 MHz.

5.4.2.3.3 Constant-parameter (CP, Bergeron) line model

The constant parameter (CP or Bergeron) line model has been almost always adopted as an alternative approach [47,48,51,54,56–58,67,69,74–76]. The reason why the frequency-dependent model in existing EMT-type simulation tools is not

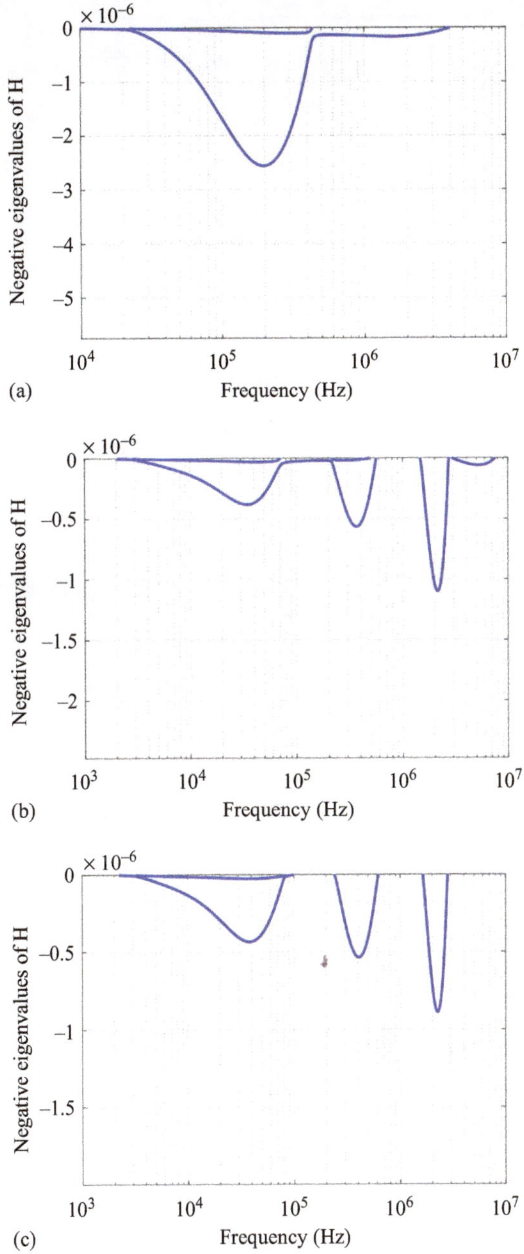

Figure 5.55 Passivity analysis results in frequency domain corresponding to the transient voltage as shown in Figures 5.53 and 5.54. (a) Extended TL approach with accurate Z_e and Y_e, (b) modified classical TL approach with modified Carson Z_c and accurate Y_e, (c) modified classical TL approach with modified approximate Z_c and accurate Y_e, (d) classical TL approach with Carson's earth-return impedance Z_c and space admittance Y_s (current EMT-type simulation tool), and (e) classical TL approach with approximate Z_e and Y_s

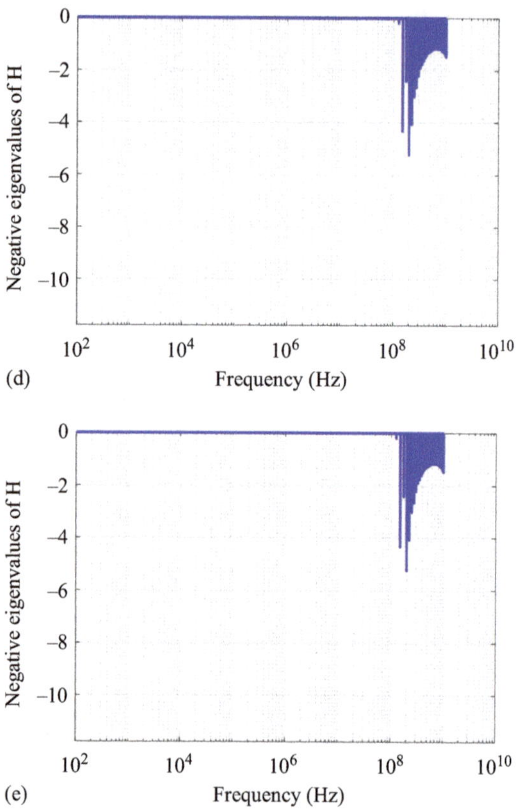

Figure 5.55 *(Continued)*

used is to avoid the numerical instability in the frequency-dependent model in the VFT simulations in a GIS as explained in Sections 5.4.2.2 and 5.4.2.3.

Figure 5.56 shows EMT simulation results by using the CP line model prepared in an EMT-type simulation tool with the extended TL approach, and Figure 5.57 shows EMT simulation results with the classical TL approach. It is clear that no numerical instability is observed in both results. The overall waveshapes are similar to those in Figure 5.53 independently of the extended or classical TL approaches. If carefully observed, there are minor differences in the peak voltage and oscillating frequency. The above results clearly explain the reason for the CP line model being almost always adopted in the VFT simulations.

5.4.3 EMT simulation of VFTs in a 500 kV GIS

This section shows thorough investigations of very-fast transients (VFTs) in a 500 kV GIS based on EMTP simulation results by adopting accurate earth-return impedance and admittance formulas. The effects of the GIS length, pipe grounding, spacers, operating dis-connector (DS), bus charging conditions, etc. on the VFTs are made clear, although no reignition in the operating DS is considered. Analytical formulas

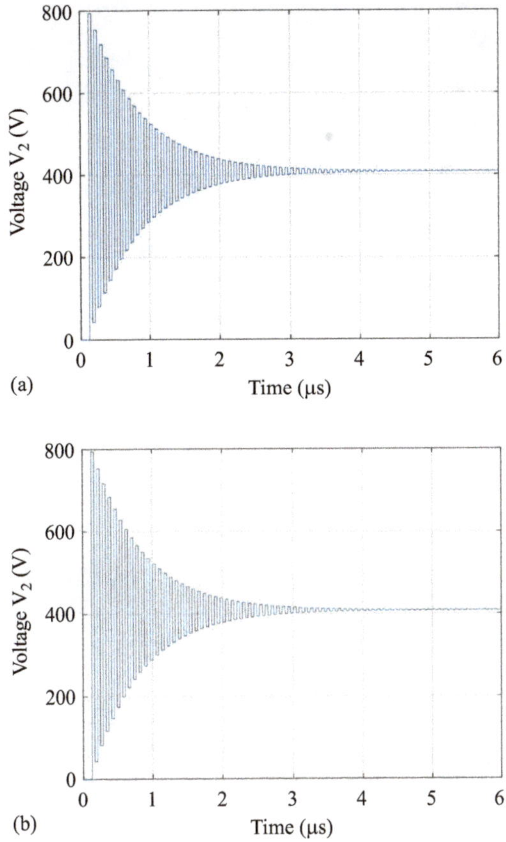

Figure 5.56 EMT simulation results by CP line model with the parameters calculated by the extended TL approach. (a) CP line model with the parameters at f = 10 MHz and (b) CP line model with the parameters at f = 200 MHz

of the lowest (dominant) oscillating frequency of VFTs are derived and show satisfactory agreement with the EMTP simulation results. The formulas are expected to be useful to estimate the VFT oscillating frequency, and also to explain the transient behavior of VFTs in a GIS composed of a number of buses, DSs, and CBs.

5.4.3.1 Model circuit of 500 kV GIS

5.4.3.1.1 500 kV GIS

Figure 5.58 illustrates a single-phase expression of the 500 kV fully gas-insulated substation in Figure 5.44 [69]. In the figure, GIB is a gas-insulated bus, DS is a disconnector, and CB is a circuit breaker. 500 kV transmission lines are connected to BUS1 and BUS4 and lower voltage lines go out from BUS2 and BUS3 through transformers Tr.2 and Tr.3 represented by the stray capacitances C2 and C3, respectively, in Figure 5.44. Spacers are installed at both ends of the DSs. The length of the GIBs, DSs, and CBs ranges from 5.1 m to 6.8 m. The cross-section is shown in

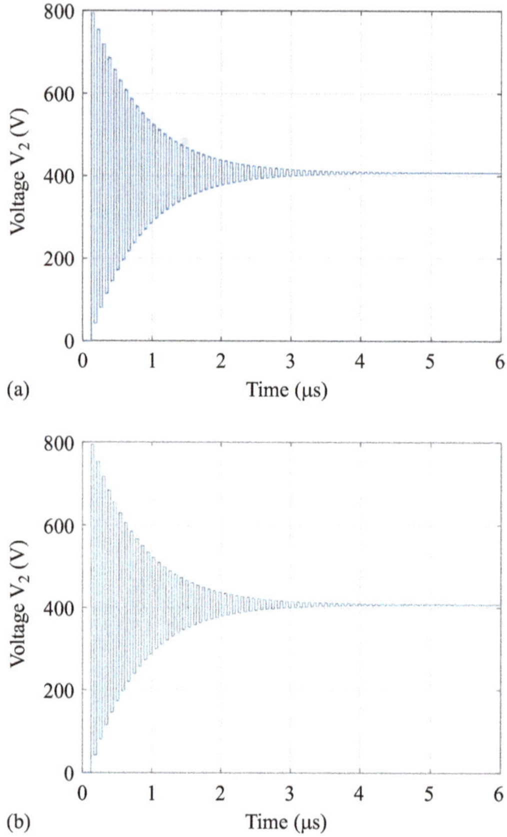

Figure 5.57 EMT simulation results by CP line model with the parameters calculated by the classical TL approach. (a) CP line model with the parameters at $f = 10$ MHz and (b) CP line model with the parameters at $f = 200$ MHz

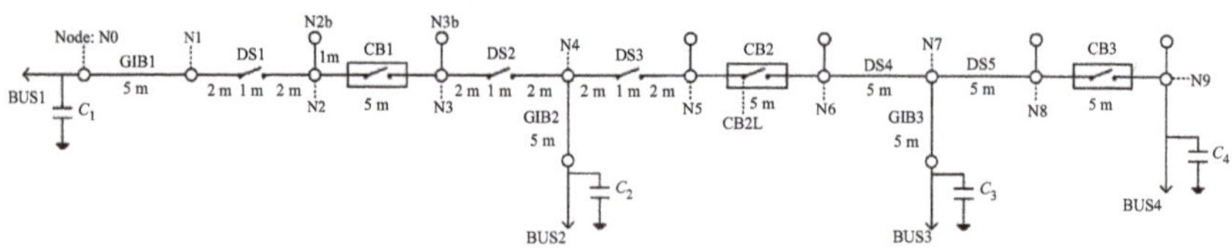

Figure 5.58 Single-phase expression (core circuit) of the 500 kV GIS illustrated in Figure 5.44

Figure 5.59. The cross-section of the GIBs and DSs is about the same, but that of the CBs is a little larger. The pipes of some GIBs, DSs, and CBs are connected to vertical buses of which the pipes are electrically connected to the grounding mesh of the substation, and thus the vertical pipes become equivalent to a grounding resistance.

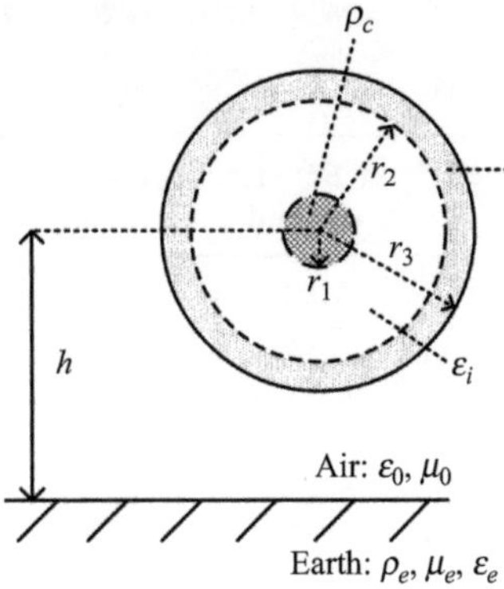

Figure 5.59 Cross-section of GIB: $h = 2.45$ m, core $r_1 = 12.5$ cm, pipe $r_2 = 46$ cm, and $r_3 = 48$ cm; resistivity: core $\rho_c = 1.68\times10^{-8}$ Ωm, pipe $\rho_p = 2.82\times10^{-8}$ Ωm, and earth $\rho_e = 100$ Ωm; and relative permittivity: SF_6 gas $\varepsilon_i = 1$

5.4.3.1.2 EMT simulation model

In the following EMTP simulations, the length of GIBs, DSs, and CBs is assumed to be 5 m, but the CBs have an additional part with a length 1 m on both sides as shown in Figure 5.58. The cross-section of the GIBs, DSs, and CBs is assumed to be the same as that in Figure 5.59. A frequency-dependent Wide-band (WB) model of [28,105,106] is adopted to represent GIBs, DSs, and CBs.

Test circuit

The following three test circuits are considered:

(a) Circuit – A: CB1 open-circuited.
The circuit is composed of BUS1 - GIB1(5 m) - DS1(5 m) - upper part of CB1 (2 m): total length $x_A = 12$ m + 1 m branch of CB1.

(b) Circuit – B: CB1 closed and CB2 open-circuited.
The system composed of Circuit - A and lower part of CB2 + DS2 - left part of CB2 // GIB2: total length $x_B = 27$ m + GIB2 (5 m) + and branches of CB1 and CB2.

(c) Circuit – C: CB1, CB2 closed, and CB3 open-circuited.

The system composed of all the GIBs and DSs, CB1, and CB2 with the lower part of CB3: total length $x_C = 42$ m + GIB2 (5 m), GIB3 (5 m), and CB branches.

The left part of a dis-connector (DS1, DS2, or DS3) is charged by a 500 kV AC source (peak voltage = 408 kV) through source inductance $L_0 = 50$ mH or an infinite bus ($L_0 = 0$) from BUS1 for $t < 0$, and the dis-connector is closed at $t = 0$.

Table 5.14 summarizes test conditions and simulation results.

Table 5.14 Simulation conditions and results: core voltage (kV), pipe voltage (kV), and frequency (MHz)

(a) DS1 closed at $t = 0$ (DS1 left to BUS1 charged by $E = 408.24$ kV for $t < 0$), $R_g = 2\Omega$, C1 and C4 = 10 nF

Case	Circuit	Tr.C	V_{1c}	Max/node	V_{1p}	Max/node	f
A1	A	-	720	880/CB1L	−17.0	(−14.0/BUS1)	6.0
A2	B	-	630	850/CB2L	−13.5	16.0/BUS1	2.3
A3	C	-	540	810/CB3L	14.0	16.0/BUS1	1.7
A4	B	C_2	430	590/CB2L	17.0	26.0/BUS1	1.0
A5	C	C_2, C_3	420	425/CB3L	15.0	23.0/BUS1	0.8

(b) Effect of grounding in Circuit - B: Cases B1 and B2.

(b-1) DS2 operation: $f = 1.0$ MHz in all the cases.

Case	R_g (Ω)	Number of grounding	V_{1c}	Max/node	V_{1p}	Max/node
B11	2	4	455	570/CB2L	13.8	23.5/BUS1
B12	10	4	430	(420/CB2L)	55.5	86.3/BUS1
B13	0.001	4	511	750/CB2L	0	0
B14	2	8	450	575/CB2L	2.4	23.0/BUS1

(b-2) DS3 operation.

Case	R_g (Ω)	Number of grounding	V_{1c}	Max/node	V_{1p}	Max/node	f
B21	2	4	690	(686/CB2L)	12.9	(8.0/BUS1)	5.0
B22	10	4	675	(657/CB2L)	−52.5	(34.0/BUS1)	5.2
B23	0.001	4	710	904/CB2L	0	0	5.0
B24	2	8	683	686/CB2L	0.2	7.9/BUS1	5.0

(c) Effect of spacers corresponding to Cases A1 and A2.

Case	Ref.	Spacer Model	V_{1c}	Max/node	V_{1p}	Max/node	f
C11	A1	CP Line	730	900/CB1L	20.0	(−15.0/BUS1)	5.0
C12	A1	Capacitance	730	900/CB1L	−20.0	(−17.0/BUS1)	5.1
C21	A2	CP Line	660	770/CB1L	−15.5	−16.5/BUS1	2.0
C22	A2	Capacitance	650	800/BUS2	13.0	17.0/BUS1	2.2

(d) Other parameters.

(d-1) Effect of the length of DS3 corresponding to Case B11 (DS3 length 5 m), DS2 closed at $t = 0$: $f = 1.0$ MHz in all the cases.

Case	Circuit	DS3 (m)	V_{1c}	Max/node	V_{1p}	V_p/BUS1
D1	B	3	425	700/CB2L	19.0	24.5
D2	B	1	430	610/CB2L	21.5	24.0

(d-2) Effect of the CB radii (r_1 = 7.5 cm, r_2 = 34.05 cm, and r_3 = 36 cm) corresponding to Cases A2 and A4 (r_1 = 5 cm, r_2 = 18 cm, r_3 = 20 cm), DS1 closing.

Case	Circuit	Tr.C	V_{1c}	Max/node	V_{1p}	Max/node	f
D3	B	-	620	870/CB2L	12.5	15.0/BUS1	2.2
D4	B	C2	430	605/CB2L	15.0	26.0/BUS1	1.0

(e) Source circuit parameters corresponding to Cases A1, DS1 closing.

Case	L1(mH)	C1(nF)	V_{1c}	Max/node	V_{1p}	Max/node	f
E1	50	3	700	840/CB1L	16.0	(−12.5/BUS1)	6.2
E2	50	0.5	590	820/CB1L	16.0	(−12.5/BUS1)	7.5
E3	0	0.5	760	890/CB1L	−17.5	(−14.0/BUS1)	5.8

(f) Operating DS and source conditions in Circuit - C with C1 and C4 = 10 nF and C2 and C3 = 7 nF: f = 0.6 MHz in all the cases.

Case	DS	LeftDS	RightDS	V_{1c}	Max/node	V_{1p}	Max/node
F1	DS1	+E	0	430	(415/Node2)	16.0	24.0/BUS1
F2	DS1	0	−E	−560	(−530/Node2)	16.0	24.0/BUS1
F3	DS1	+E	−E	−700	(−650/Node2)	33.0	48.0/BUS1
G1	DS3	+E	0	500	510/Node2	−15.5	−24.0/BUS1
G2	DS3	0	−E	−470	−500/Node2	−15.0	−24.0/BUS1
G3	DS3	+E	−E	−740	(600/Node2)	−31.0	−48.0/BUS1

Max/node: maximum voltage and the node where the max voltage appears.
( ) indicates the second max voltage.

5.4.3.2 EMT simulation results and discussions

5.4.3.2.1 Effect of GIS length

The effect of total length "x" of a GIS composed of GIBs, GSs, and CBs on VFTs is investigated in Circuit - A (12 m), Circuit - B (27 m), and Circuit - C (42 m). To investigate the effect of x, stray capacitances C_2 and C_3 of transforms connected to BUS2 and BUS3 are not considered.

Figure 5.60 shows Node 1 core voltages (V_{1c}) and Figure 5.61 shows its frequency spectrum. The maximum voltages of the core and pipe decrease as the length increases as observed in Table 5.14(a). The dominant transient frequency, "f", becomes lower as the GIS length increases. The frequency is approximately evaluated by $f = 1/4\tau$, where τ is the travel time, in an open-circuited conductor.

$$\left.\begin{aligned} \text{Circuit} - \text{A}: \tau_A &= \frac{x_A}{c_c} = 40\text{ ns}, \quad f_A = 6.25\text{ MHz} \\ \text{Circuit} - \text{B}: \tau_B &= \frac{x_B}{c_c} = 90.1\text{ ns}, \quad f_B = 2.78\text{ MHz} \\ \text{Circuit} - \text{C}: \tau_C &= \frac{x_C}{c_c} = 140.14\text{ ns}, \quad f_C = 1.78\text{ MHz} \end{aligned}\right\} \tag{5.66}$$

where C_c = propagation velocity of coaxial mode = 0.2997 m/ns at 10 MHz.

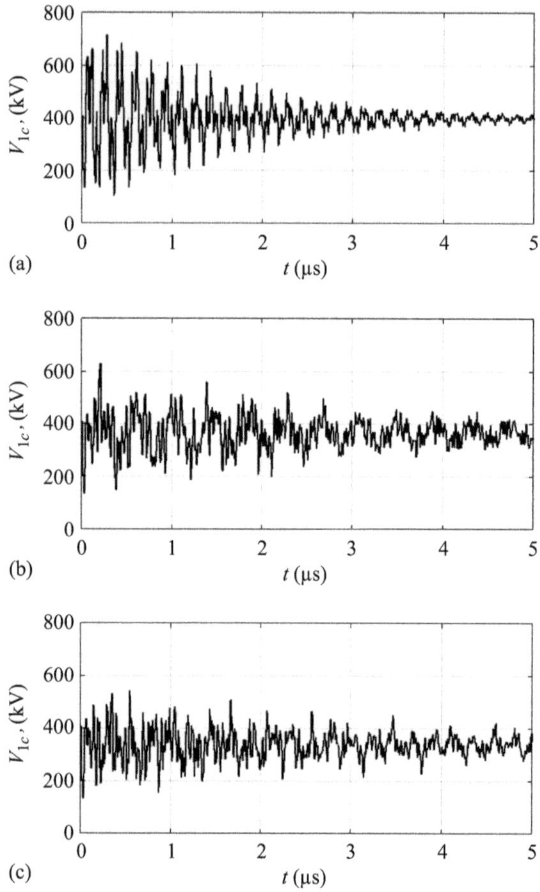

Figure 5.60 Node 1 core voltage in (a) Circuit – A, (b) Circuit – B and (c) Circuit – C

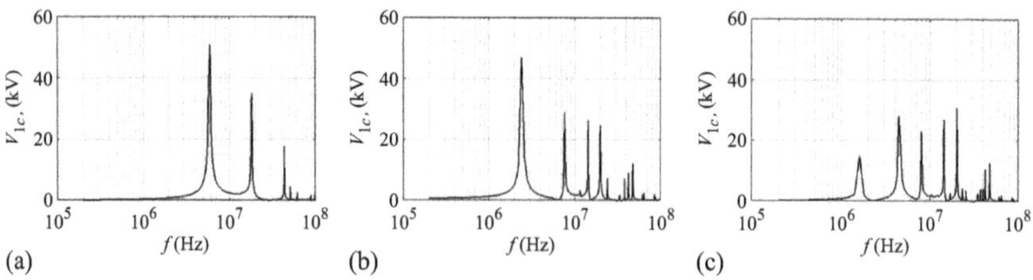

Figure 5.61 Frequency spectrum of Node 1 core voltage. (a) Circuit – A, (b) Circuit – B and (c) Circuit – C

The lowest frequency in Figure 5.61 nearly corresponds to the frequency described in (5.66). The higher frequency components are not dependent on the GIS length, because those are produced by the terminating conditions of circuit components, incoming/outgoing buses, and also pipe grounding. The third, fifth, or sixth harmonics of the lowest frequency components are observed in Circuits - A, -B, and -C. The pipe voltage shows a similar frequency spectrum to that of the core voltage, although the amplitudes are different.

5.4.3.2.2 Effect of transformer stray capacitance

In a real substation, there exist transformers as shown in Figure 5.58. It is a common practice to represent an open-circuited transformer by its stray capacitance. The value of stray capacitances C_2 and C_3 of transformers Tr.1 and Tr.2 is informed to be 7 nF by the utility of the 500 kV GIS.

When a stray capacitance is considered, a VFT waveform, especially the oscillating frequency, differs significantly from that in the cases of no capacitance. The change of the frequency is explained by considering the equivalent length of the capacitance. The per-unit length capacitance, ΔC, of the gas-insulated bus in Figure 5.60 is evaluated by

$$\Delta C = \frac{2\pi\varepsilon_0}{\ln\left(\frac{r_2}{r_1}\right)} = 43.4\ \mathrm{pF/m} \tag{5.67}$$

Then, the equivalent length of the stray capacitance, $C_2 = 7$ nF, is evaluated as

$$x_e = \frac{C_2}{\Delta C} = \frac{7000}{43.4} = 161.3\ \mathrm{m} \tag{5.68}$$

It is easily estimated from the above value that the oscillating frequency becomes much lower. The oscillating frequency f_e can be roughly estimated by rewriting (5.66) considering the fact that no traveling time exists in the capacitance:

$$f_e = 1/4\tau = c_0/4x = 1/4\sqrt{LC} \tag{5.69}$$

where

$$L = \Delta L \cdot x, \quad \Delta L = (\mu_0/2\pi)\ln(r_2/r_1), \quad C = \Delta C \cdot x + C_2 \tag{5.70}$$

From Figure 5.58, $\Delta L = 0.2562$ μH/m. For Circuit-B with $x_B = 27$ m, $L = 6.917$ μH, and $C = 1.172 + 7.0 = 8.172$ nF. Then

$$f_e = 1.052\ \mathrm{MHz} \tag{5.71}$$

In comparison with f in (5.66), the capacitance decreases the frequency by less than a half.

Figure 5.62 shows VFT waveforms of Node 1 core voltage in Cases A4 and A5. The circuit configuration of A4 is the same as that of A2 except for stray capacitance C_2 is considered in Case A4. Similar Case A5 corresponds to Case A3 with stray capacitances C_2 and C_3. Figure 5.63 shows the frequency spectrum. It is clear from a comparison of Figure 5.62 with Figure 5.60, and of Figure 5.63 with Figure 5.61 that the dominant transient frequency becomes much lower when the stray capacitance is considered. The maximum core voltage is decreased by the stray capacitance as observed in Table 5.14(a).

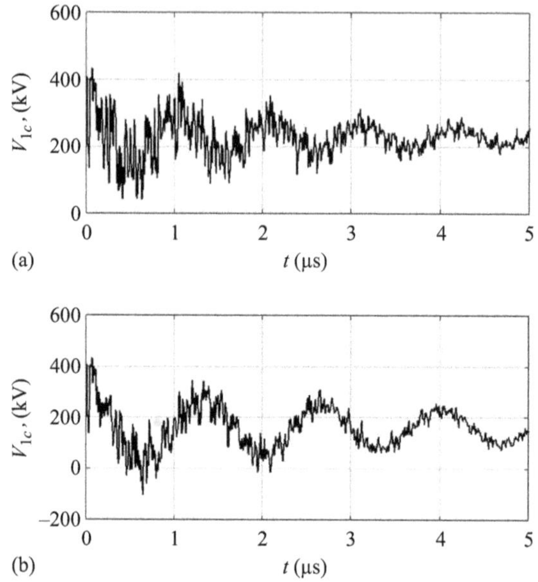

Figure 5.62 Node 1 core voltage in Cases A4 (a) and A5 (b)

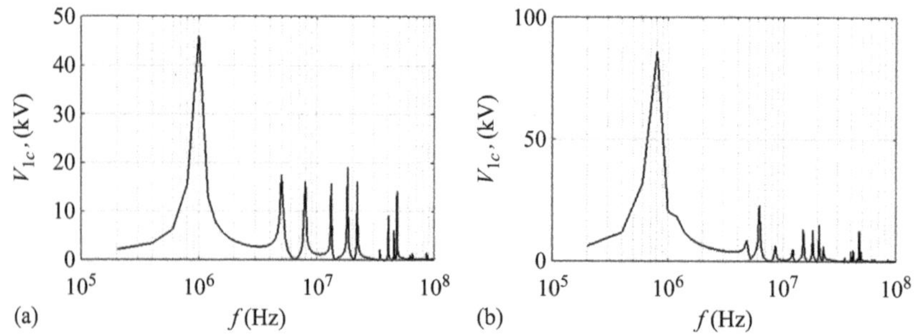

Figure 5.63 Frequency spectrum of Node 1 core voltage in Cases A4 (a) and A5 (b)

5.4.3.2.3 Effect of pipe grounding

Figure 5.64 shows Node 1 voltages in Cases B11 ($R_g = 2\Omega$) and B12 ($R_g = 10\Omega$). Figure 5.65 shows the corresponding frequency spectrum. The greater R_g damps out the oscillating waveform in Figure 5.64(b) much faster than that with lower R_g in Figure 5.64(a). No significant difference of the oscillating frequency is observed between Cases B11 and B12 in Figure 5.65, although the amplitude is much smaller in Case B12 corresponding to Figure 5.64(b). The voltage waveform and the frequency spectrum in Case B14 are quite similar to those in Case B11. In Cases B11–B14, the frequency spectrum of the pipe voltage is nearly the same as that of the core voltage.

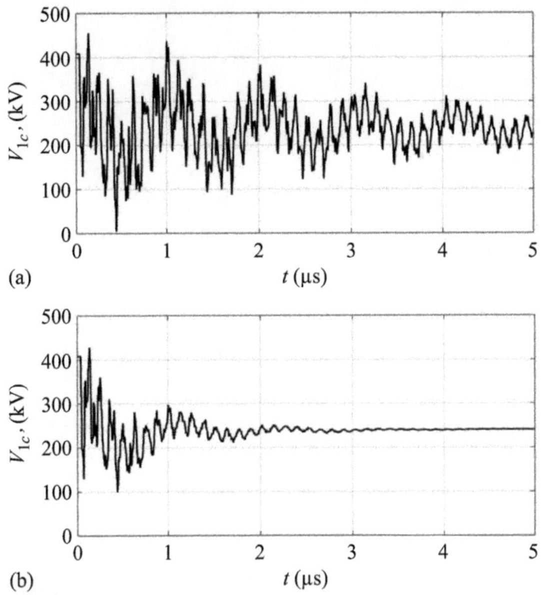

Figure 5.64 Node 1 core voltage in Cases B11 (a) and B12 (b)

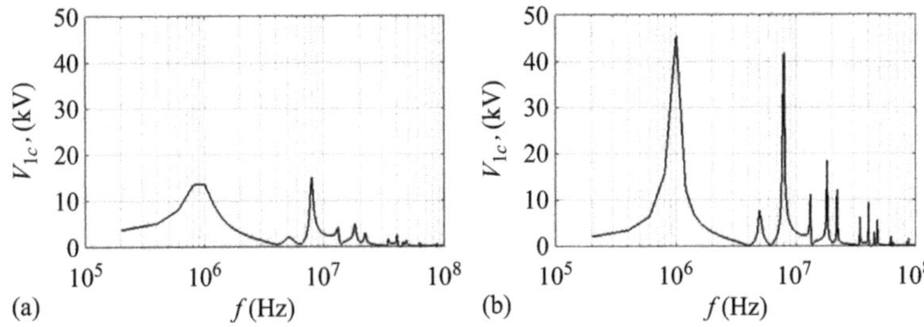

Figure 5.65 Frequency spectrum of Node 1 core voltage in Cases B11 (a) and B12 (b)

Figure 5.66 shows voltage distribution along with the GIS in Case B1: (a) is for the core voltage, and (b) for the pipe voltage. It is clear that the core voltage is the largest either at Node 1 or at the open-circuited remote end, that is, Node CB2L in Circuit - B, and is the lowest at the node of BUS2 where a stray capacitance exists, because the stray capacitance acts like a surge capacitor. The core voltage is the highest in the case of a lower grounding resistance. The pipe voltage tends to decrease toward the remote end and is lower at a grounded node. The pipe voltage is the lowest when the grounding resistance is the smallest.

Although voltage distribution in Cases B21–B24 shows a similar trend to those in Figure 5.66, the core voltages are much greater and the pipe voltage is smaller as

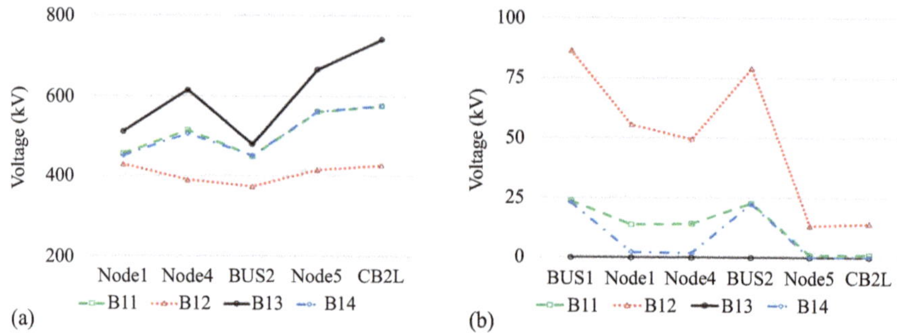

Figure 5.66 Voltage distribution in Case B1. (a) Core and (b) pipe

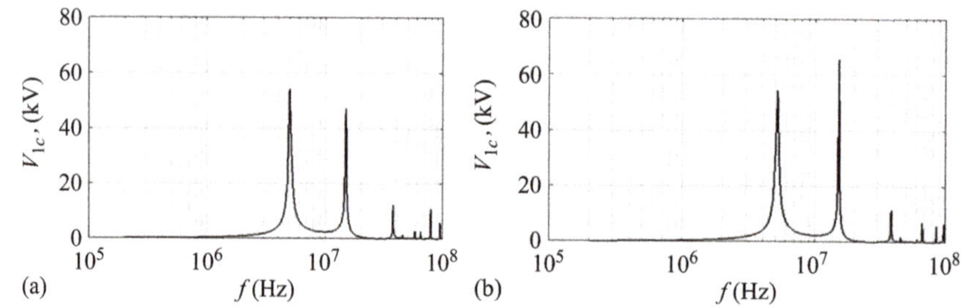

Figure 5.67 Frequency spectrum of Node 1 core voltage in Cases C11 (a) and C12 (b)

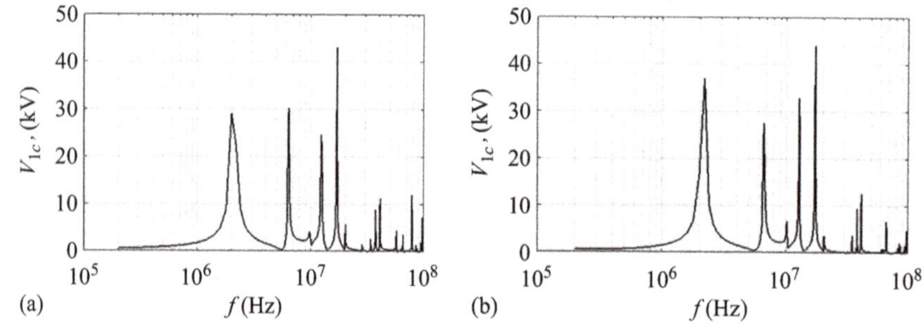

Figure 5.68 Frequency spectrum of Node 1 core voltage in Cases C21 (a) and C22 (b)

in Table 5.14(b-2). The fact indicates that an operating DS influences the core and pipe voltages significantly.

5.4.3.2.4 Effect of spacers

VFT waveforms when considering spacers, either by a CP line model or by a lumped capacitance, are not much different from those neglecting the spacers. However, the oscillating frequency becomes lower as shown in Figure 5.67 for Cases C11 and C12, and in Figure 5.68 for Cases C21 and C22. Figure 5.68(a) is the case of the spacer being represented by a CP line model of which the parameters are given at 10 MHz:

Surge impedance: coaxial mode $Z_{oc} = 31.4\,\Omega$; earth-return mode $Z_{oe} = 367.3\,\Omega$.

Propagation velocity: coaxial mode $c_c = 122.4$ m/μs; earth-return mode $c_e = 287.2$ m/μs.

The lower coaxial mode impedance and velocity are caused by the spacer relative permittivity of 6. Figure 5.67(b) is the case of the spacer being represented by a capacitance which is evaluated from the coaxial mode parameters:

$$\Delta C_{\text{sp}} = \frac{1}{(Z_{oc} \cdot c_c)} = 260.2 \text{ pF/m} \tag{5.72}$$

$$C_{\text{sp}} = \Delta C_{sp} \cdot x_s = 78.1 \text{ pF} \tag{5.73}$$

From a comparison of Figure 5.67 with Figure 5.61(a) and of Figure 5.68 with Figure 5.61(b), it is clear that the oscillating frequency is decreased by the spacers. The decrease is explained similarly to that by a stray capacitance in (5.66) to (5.71).

The core and pipe voltages become greater than those neglecting the spacers, because of multiple reflection and refraction of traveling waves due to the spacers.

5.4.3.2.5 Effect of other bus parameters

1. Length of DS2

   There may be a short-length bus in between longer buses. To represent such a case, the length of DS2 in Circuit-B is varied to 3 m and 1 m corresponding to Case B1 where DS2 length is 5 m. The simulation results are summarized in Table 5.14(d-1). It is observed that the shorter the length, the higher the core and pipe voltages. The voltage waveforms show a minor difference. The frequency spectra in Cases D1 and D2 are nearly the same as that in Figure 5.65 (a) except for the amplitude.
2. CB radii

   In Section 5.4.3.1, the CB radii are assumed to be the same as the DS radii. In this section, the effect of a larger CB size is investigated. Figure 5.69 shows Node 1 core voltage waveforms in Cases D3 and D4, and Figure 5.70 shows the frequency spectrum which is similar to that of the pipe voltage except the amplitude. A larger ratio r_2/r_1 results in a larger surge (characteristics) impedance. The change of the surge impedance from those connected to the CB produces more traveling wave reflection and refraction. However, the overall VFT waveforms and frequency spectrum in Figures 5.69(a) and 5.70(a) are not much different from those in Figure 5.60(b) and Figure 5.61(b). Similarly, Figures 5.69(b) and 5.70(b) show only a minor difference from Figures 5.62(a) and 5.63(a).
3. Effect of source circuit

   Stray capacitance C_1 at BUS1 and C_4 at BUS1 in Figure 5.58 are assumed to be 10 nF considering bushings and transformers in Section 5.4.3.1. If no transformer exists at BUS1 and BUS4, but there are only bushings, C_1 and C_4 should be much smaller than 10 nF. For example, the stray capacitance of a

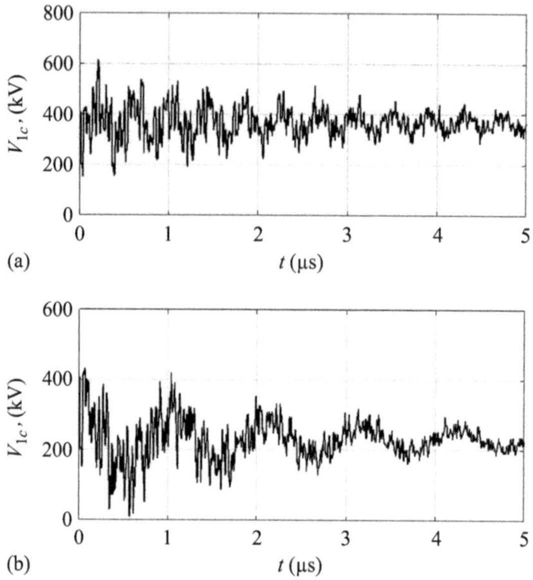

Figure 5.69 Node 1 core voltage in Cases D3 (a) and D4 (b)

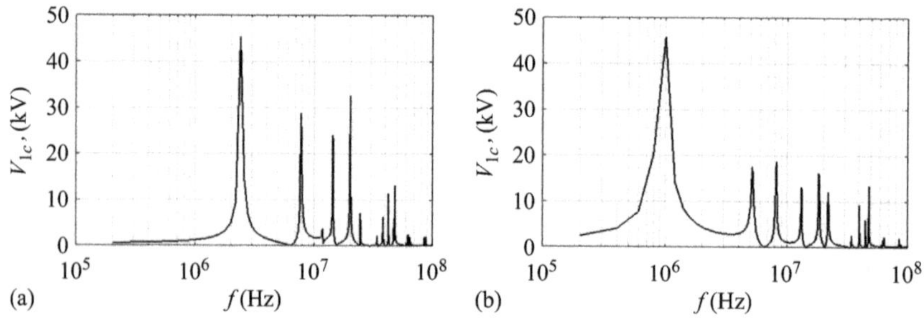

Figure 5.70 Frequency spectrum of Node 1 core voltage in Cases D3 (a) and D4 (b)

500 kV bushing is known to be about 0.5 nF. Considering the fact, the effect of C_1 and C_4 on VFTs is investigated as Cases E1 and E2 in this section. In a large substation, a number of incoming lines are connected to the main bus, and the source impedance of the bus becomes very small. Assuming BUS1 is an infinite bus involving only a bushing stray capacitance, EMT simulations are performed as Case E3 with $L_0 = 0$ in comparison with those in Case E2.

Figure 5.71 shows VFT waveforms at Node 1: (a) for Case E2 and (b) for Case E3. The frequency spectrum is shown in Figure 5.72. It is observed that the core voltage decreases in Figure 5.71(a) and the lowest oscillating frequency becomes higher in Figure 5.72(a) as capacitance C_1 becomes smaller in comparison with

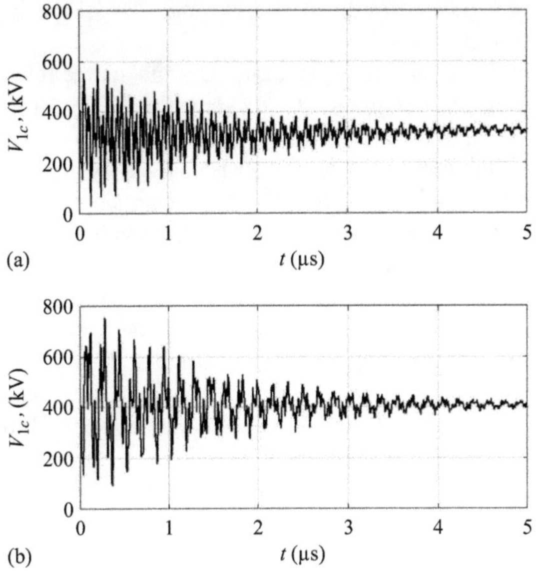

Figure 5.71 Node 1 core voltage in Cases E2 and E3. (a) Case E2: $L_0 = 50$ mH, $C_1 = 0.5$ nF and (b) case E3: $L_0 = 0$, $C_1 = 0.5$ nF

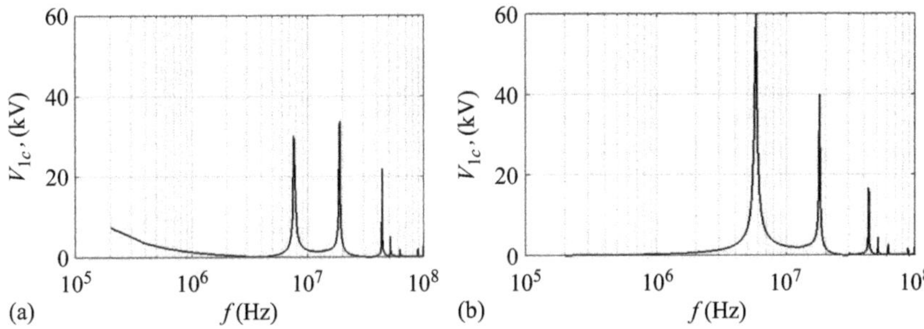

Figure 5.72 Frequency spectrum of Node 1 core voltage in Cases E2 (a) and E3 (b)

those in Case A1 ($C_1 = 10$ nF) and Case E1 ($C_1 = 3$ nF). No significant difference of the pipe voltage is observed.

In Case E3, infinite source, the core, and pipe voltage increase as is well-known. The lowest frequency decreases slightly compared with that in Case E2 ($L_0 = 50$ mH) as observed in Table 5.14(e). It should be noted that C1 does not affect a transient response when an infinite source is used, because the impedance of the infinite source is zero.

(4) Transient condition in circuit-C

It is well-known that the severe VFT overvoltage appears when a part of a GIS is charged by "$+E$" and the remaining part is charged by "$-E$" as many papers investigated the condition. This is similar to the case that a severe overvoltage appears on an overhead line during CB reclosing, that is, the line with residual voltage when the CB is closed.

In this section, the effect of an operating DS and source (charged) conditions on VFTs are investigated in Circuit - C. As shown in Table 5.14(f), the operating DS in Case F is DS1, and the left side of DS1 is charged by positive $E = 408.24$ kV in Case F1. In Case F2, the right side of DS1 is charged by negative E. In Case F3, the DS1 left is charged by $+E$ and the right is charged by $-E$. In Case G, an operating DS is DS3, and the charge (source) conditions in Cases G1 to G3 are the same as those in Cases F1 to F3.

Figure 5.73 shows Node 1 core voltages in Cases F1–F3, and Figure 5.74 shows the frequency spectrum of the voltages in Case F3. It should be clear that the superposition law is satisfied between Cases F1 and F3, that is, F1 + F2 = F3 for both time responses (transient waveforms) in Figure 5.73 and frequency responses. Therefore, only the frequency spectrum in Case F3 are given in Figure 5.74. If the amplitudes of

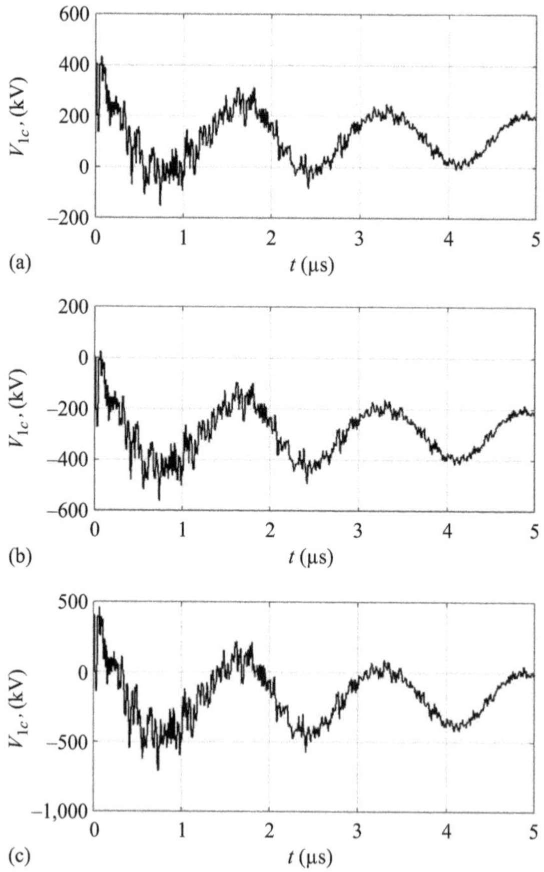

Figure 5.73 Node 1 core voltage in (a) Case F1, (b) Case F2 and (c) Case F3

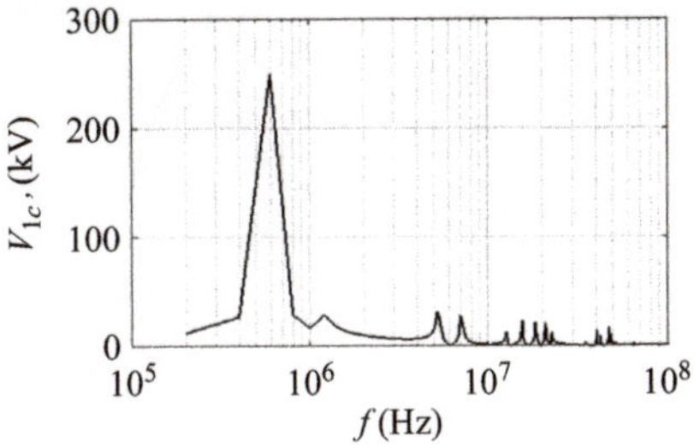

Figure 5.74 Frequency spectrum of Node 1 core voltage in Case F3

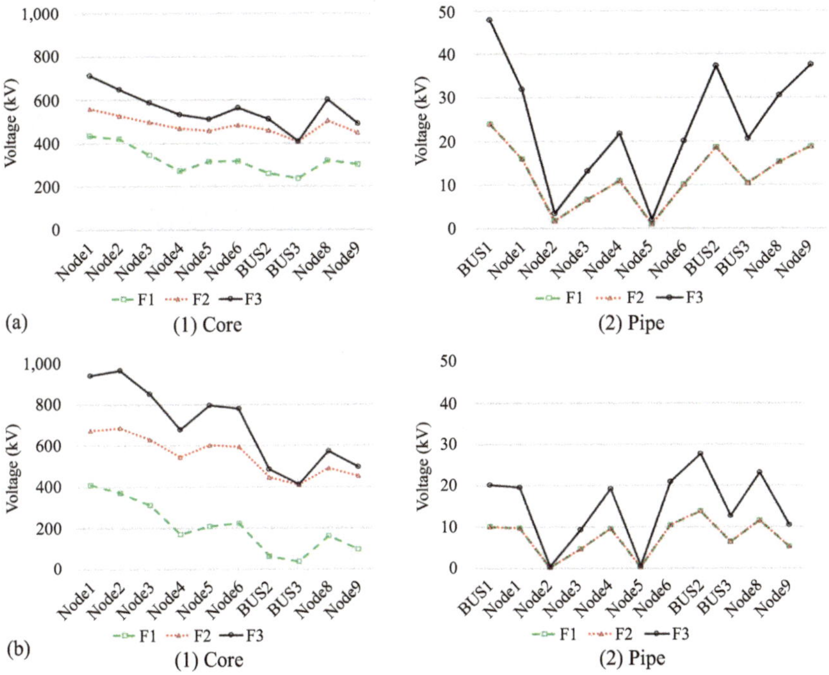

Figure 5.75 Maximum voltage distribution in Cases F1, F2, and F3. (a) $C_1 = C_4 =$ 10 nF and (b) $C_1 = C_4 = 0.5$ nF

each spectrum in Case F1 and F2 are superposed, the correct amplitude in Case F3 is obtained. As already pointed out in previous papers, Case F3 shows higher overvoltages than those in Cases F1 and F2 as is clear in the overvoltage distribution along the GIS in Figure 5.75. It is observed that the pipe overvoltage distribution in Case F2 is nearly the same as that in Case F1, and the sum of the pipe overvoltage in Cases F1 and F2 becomes nearly the same as the pipe overvoltage in Case F3. When $C_1 = C_4 = 0.5$ nF, the core voltage becomes much larger and the pipe voltage becomes smaller than those in the case of $C_1 = C_4 = 10$ nF.

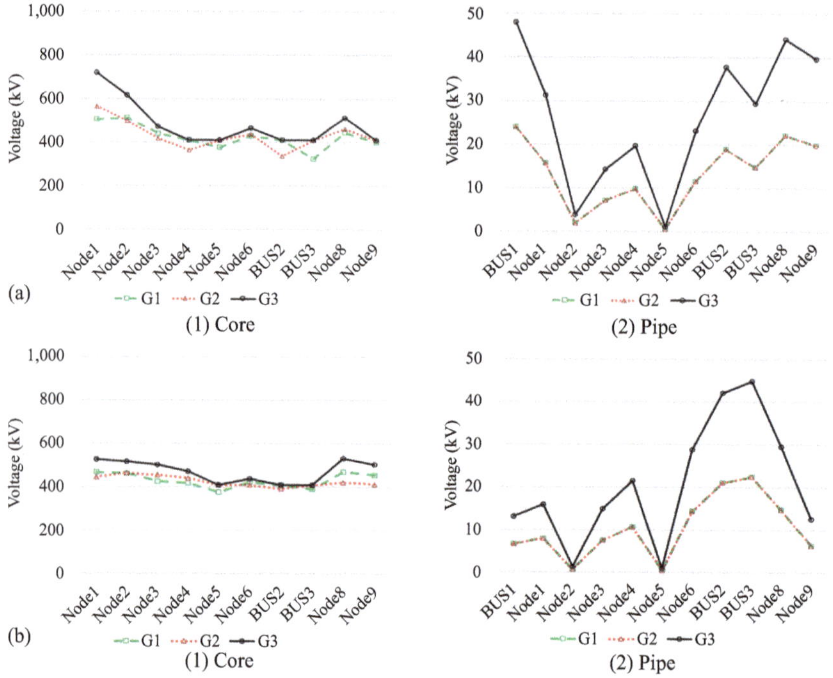

Figure 5.76 Maximum voltage distribution in Cases G1, G2, and G3. (a) $C_1 = C_4 =$ 10 nF and (b) $C_1 = C_4 = 0.5$ nF

Figure 5.76 shows the voltage distribution in Cases G1–G3. No significant difference is observed between Cases F and G for $C_1 = C_4 = 10$ nF. However, the core voltage in Case G is much smaller and the pipe voltage is much higher than those in Case F for $C_1 = C_4 = 0.5$ nF in Figure 5.75(b). In other words, an operating DS influences significantly to the core and pipe voltages when C_1 is small. The voltage waveforms and frequency spectrum in Case G are similar to those in Case F.

5.4.3.3 Theoretical analysis

VFTs are investigated in Section 5.4.3.2 based on EMT simulation results. Although the effects of various circuit parameters/conditions on the VFTs are explained, it is not easy to generalize the investigations. Analytical formulas of estimating the lowest oscillating frequency (f_1) has been shown based on a distributed-parameter circuit theory considering a traveling wave behavior, see (5.66)–(5.71), which approximately explain the EMT simulation results. In this section, analytical formulas for evaluating the frequency and a maximum core voltage are introduced by applying a lumped-parameter equivalent of a gas-insulated bus.

5.4.3.3.1 Lumped-parameter equivalent circuit

A lumped-parameter equivalent circuit of a transmission line is widely applied in the field of power engineering. Although the equivalent cannot take into account traveling wave natures in a short time period of a transient on a distributed-parameter line, it can

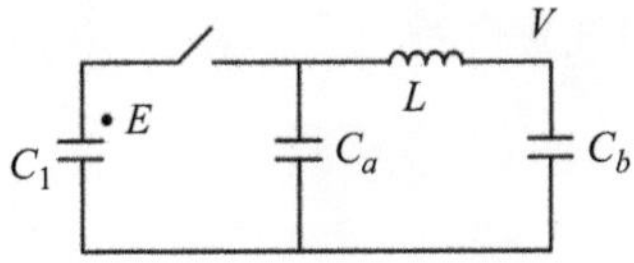

Figure 5.77 Lumped-parameter equivalent circuit of a GIS

explain the transient characteristic in a long time period. Figure 5.77 illustrates a lumped-parameter equivalent circuit of a GIS. In the figure, C_1 represents the source circuit capacitance in Figure 5.58. C_a represents the capacitance of gas-insulated buses (GIBs, DSs, and CBs) defined in (5.66) and also stray capacitance C_2 of BUS2 transformer. Similarly, C_b is to represent the gas-insulated buses and stray capacitances. L in Figure 5.77 represents the inductance of the gas-insulated buses which is estimated by (5.70). C with "$\cdot E$" means that C is charged by voltage $E = 408.24$ kV for $t < 0$.

5.4.3.3.2 Analytical formula and calculated example

Analytical formulas of V in Figure 5.77 are derived as

$$V = A(\omega_3/\omega)^2[1 - \cos(\omega t)]E \tag{5.74}$$

where

$$\left.\begin{aligned} &A = C_1/(C_1 + C_a), C = C_1C_a/(C_1 + C_a) \\ &\omega^2 = \left(C_1\omega_3^2 + C_a\omega_4^2\right)/(C_1 + C_a) \\ &\omega_2^2 = 1/LC_a, \omega_3^2 = 1/LC_b, \quad \omega_4^2 = \omega_2^2 + \omega_3^2 \end{aligned}\right\} \tag{5.75}$$

and DS1/DS2: operation $C_a = C_{g1}$, $C_b = C_{g_2} + C_{\text{stray}}$,

Circuit - A/Cases A1–A3: no C_{stray},

Circuit - B: $C_{\text{stray}} = C_2$, Circuit - C: $C_{\text{stray}} = C_2 + C_3$.

DS3 operation: $C_a = C_{g_1} + C_2$, $C_b = C_{g_2} + C_3$.

When an infinite source E is applied, source-side capacitance C_1 causes no effect on a transient response as already explained. Then, the analytical formula becomes:

$$V = [1 - \cos(\omega t)]E, \quad \omega = 1/\sqrt{LC_b} \tag{5.76}$$

When the damping of a voltage waveform is necessary, multiply $e^{-\alpha t}$ to $\cos(\omega t)$in (5.74) and (5.76) with

$$\alpha = R/2L \tag{5.77}$$

where $R = R_g + \Delta R \cdot x$, R: coaxial mode resistance, L in (5.70).

ΔR can be approximated in the following equation [25]:

$$\Delta R = \sqrt{\rho_c f \times 10^{-7}}/r_1 \ \Omega/\text{m} \tag{5.78}$$

For example at 100 MHz with $r_1 = 5$ cm, $\rho_c = 1.68 \times 10^{-8}$ Ωm,

$$\Delta R = 8.2 \times 10^{-3}\ \Omega/\text{m} \tag{5.79}$$

An analytical waveform for Case A4 in Figure 5.62(a) is shown in Figure 5.78 by a dashed line. The analytical result represents the EMT simulation result qualitatively well.

Table 5.15 shows calculation examples by (5.75) and (5.76) in comparison with EMT results in Table 5.14. The analytical formula of the oscillating frequency "*f*" in Table 5.15 agrees reasonably with the lowest frequency in the EMT

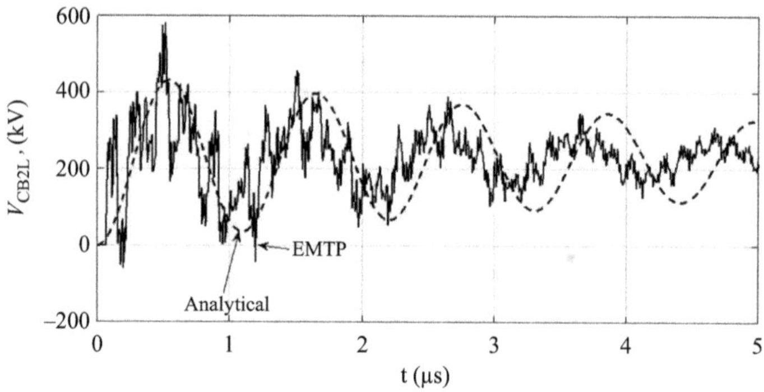

Figure 5.78 Comparison of analytical and EMTP waveforms at the remote end for Case A4 with Rg = 2 Ω

Table 5.15 Lowest oscillating frequency evaluated by (5.75) and (5.76)

(a) By (5.75) for L_0 = 50 mH

Case	**A1**		**A2**		**A3**		**A4**		**A5**	
C_1(nF)	10	0.5	10	0.5	10	0.5	10	0.5	10	0.5
(10)	6.2	7.0	2.1	3.0	1.3	2.1	0.9	2.2	0.7	1.8
EMTP	6.0	7.2	2.3	3.5	1.7	2.5	1.0	2.2	0.8	2.1

Case	**B11**		**B21**		**C12**	**C22**	**E1**	**F3**	**G3**	
C_1 (nF)	10	0.5	10	0.5	10	10	0.3	0.5	10	0.5
(10)	0.9	1.9	4.1	4.2	5.4	1.8	6.4	1.7	0.5	0.7
EMTP	1.0	2.1	5.0	2.1	5.1	2.2	6.2	2.1	0.6	1.0

(b) By (5.76) for infinite source (L_0 = 0)

Case	**A1**	**A2**	**A3**	**A4**	**A5**	**B11**	**C12**	**C22**	**E3**
(10)	6.2	2.1	1.3	0.7	0.5	0.7	5.3	1.7	6.2
EMTP	5.8	2.1	1.3	0.8	0.4	0.8	5.0	2.0	5.8

simulation results except Case B21 with $C_1 = 0.5$ nF. Thus, the analytical formulas might be useful to estimate the frequency. It should be noted that the accuracy of ω in (5.75) cannot be guaranteed when C_1 and C_b are far smaller than C_a, that is,

$$\omega_4^2 \approx \omega_3^2 >> \omega_2^2, \quad \omega \approx \omega_3 \text{ in (5.47) for } \mathrm{C}_1, Cb \ll Ca \tag{5.80}$$

In this case, $\omega^2 = 1/LC_1$ for $C \approx C_1$ can be used.

5.4.4 Summary

Field test results of VFTs are summarized considering electromagnetic disturbances in GIS control circuits which occasionally result in power system operation failures. Numerical instabilities are observed in VFT simulation results by existing EMT-type simulation tools, and the reason for the instability is made clear. Then, the modeling of GIS elements for EMT simulations and the effects of various parameters of the GIS are investigated in detail. Based on the investigations, the following remarks are made clear.

1. The oscillating frequencies of VFTs generated by dis-connector and circuit breaker operations in GISs range from 1 MHz up to 140 MHz in GIS high-voltage circuits, and from 2 MHz to 80 MHz in low-voltage control circuits.
2. Numerical instabilities observed in EMT simulation results are caused by the space admittance of an overhead line implemented into the existing EMT-type simulation tools. No instability appears when the space admittance is replaced by the accurate earth-return admittance.
3. The VFT over-voltages in the test results range from 1.7 pu to 3.0 pu in the GIS main circuits, and from 0.1 pu to 0.7 pu in the metal enclosures (pipe or tank). In the low-voltage control circuits, 20–700 V are observed in the field test results.
   (a) The VFT oscillating frequency is, in principle, proportional to the total length of buses, DSs and CBs in the GIS. However, the frequency is decreased easily less than a half by stray capacitances of transformers and bushings in the GIS, because those are far greater than the bus capacitance.
   (b) Although the pipe grounding reduces the pipe voltage significantly as easily estimated, the core voltage tends to increase as the grounding resistance becomes smaller. The grounding causes only minor influence on the oscillating frequency.
   (c) The position of an operating dis-connector causes a very significant effect on the VFT over-voltages and oscillating frequency. The distance to the open-circuited remote end from the DS is short, the overvoltage and frequency become very high.
   (d) The effect of spacers is similar to that of a stray capacitance, but the effect is less noticeable for the smaller capacitance of the spacers.
   (e) The effect charging conditions is less noticeable than the effect of an operating DS.

4. When the VFT in a control cable is simulated, a VT and a CT are to be modeled in details, and mutual coupling between the cable and grounding mesh of the GIS is to be considered because the mesh voltage cannot be assumed zero.
5. The analytical formulas explained in this section show a satisfactory agreement with the EMT simulation results.

The investigations in this section might contribute a further understanding of VFTs, and the EMT simulation results can be used as a reference because those are calculated by a frequency-dependent line model with the accurate earth-return impedance and admittance. Also the analytical formulas are expected to be useful to a rough estimate of the VFT characteristics.

5.5 EMT simulation in comparison with experimental and FDTD computed results

5.5.1 Modeling of GIS elements/components in EMT simulation

Modeling of various elements and components in a GIS for EMT simulation are briefly explained in Section 5.4.3.1. Reference [74] describes the basic concepts for modeling GIS elements for VFT simulations by EMT-type software. Reference [75] summarizes modeling methods of various elements/components involved in the GIS. Reference [76] shows modeling of GIBs connected to a main bus, pipe grounding with arresters, etc. Reference [108] explains modeling of mutual coupling between an overhead control cable and a grounding mesh in a substation. Reference [69] shows modeling of a VT connected to the control cable and mutual coupling between the control cable and the GIB. Reference [77] describes a detailed model of a CT. Reference [109] gives the values of equivalent capacitances of GIS spacers, bus open-end, CT, VT, etc., and surge impedances of DS, CB, a bushing, and an XLPE cable.

Summarizing the above references and the authors' experiences, the following remarks are made for VFT simulations by EMT-type simulation tools.

1. For VFT frequencies are very high, the earth is assumed as perfectly conducting ($\rho_e = 0$) or an Al plate can be assumed as the earth as often adopted in a laboratory test. In a very high-frequency region, the coaxial mode is completed within the pipe and the pipe conductor becomes a complete shield. Thus, the earth effect is hardly observed in measured and simulation results. The fact has indicated that no frequency-dependent model of a conductor is necessary for an EMT simulation. Further, only a core conductor with the coaxial propagation mode parameters is enough to perform EMT simulation of VFTs, especially when both ends of a pipe are grounded.
2. Pipe grounding has to be carefully considered. The waveform of a pipe voltage is affected by the grounding very much, although the core voltage is less influenced.

3. A branched GIB (branch) causes a high-frequency oscillation and a resultant surge waveform becomes very oscillatory. However, a measured waveform is entirely dependent on the sampling time (highest frequency) of an oscilloscope used for the measurement. If necessary, a branch can be represented by capacitance C_b [109]:

$$C_b = 1/(z_{oc}c_c)\ (\mathrm{F/m}) \tag{5.81}$$

where z_{oc} is the coaxial-mode surge impedance and c_c is the coaxial-mode velocity.
4. A charged GIB involves a DC voltage component which cannot be handled by EMT-type simulation tools. The DC voltage is represented by a low frequency AC voltage.
5. The length of a lead wire connecting a source and a GIB core and grounding a GIB pipe is comparatively large as the GIB length is short. Therefore, the lead wire has to be modeled as a distributed-parameter line, that is, CP (constant parameter) line model prepared in the EMT-type simulation tools. In most cases, the lead wire is vertical, and its surge impedance can be evaluated approximately in the following equation [108]:

$$Z_{ov} = 60[\ln(h/r) - 1] \tag{5.82}$$

where h is the lead wire height = length and r is the radius.
6. If VFTs in a GIS main circuit including metallic enclosures are concerned, a CT and a VT can be represented by an equivalent capacitance [75,109]. When VTFs in a GIS control circuit for an EMD study are concerned, the CT and VT are to be modeled in detail [69,77]. Also, a control cable should be modeled as accurately as possible because wave deformation along the cable influences significantly the calculated VFT current and voltage at the entrance of the control circuit.
7. To analyze a VFT on an overhead control cable due to mutual coupling between grounding meshes, the mesh voltage cannot be assumed zero [108].

5.5.2 Proto-type GIS

5.5.2.1 Test circuit

Figure 5.79 illustrates the cross-section of a GIB which is the same as that of all the GIBs investigated in this paper. The height and radii depend on the voltage class and manufacturers. Figure 5.80 shows the test circuit of a proto-type GIB [46]. The GIB was set at height $h = 1$ m above a Cu plate. Both ends of the pipe (nodes A, B, C, and D) are grounded. Because of a proto-type test, a plastic-made cylinder with the length of about 0.7 m and the relative permittivity of 6 was installed at every 1 to 2 m of the GIB. It has been shown in [46] that an equivalent permittivity of the GIB is 2.3. C_t is an equivalent capacitance of a transformer connected to node D. The left side (node A to node C) of the bus is charged by −45 V and the right (B to D) is charged by +45 V. Node A is short-circuited to node B at $t = 0$.

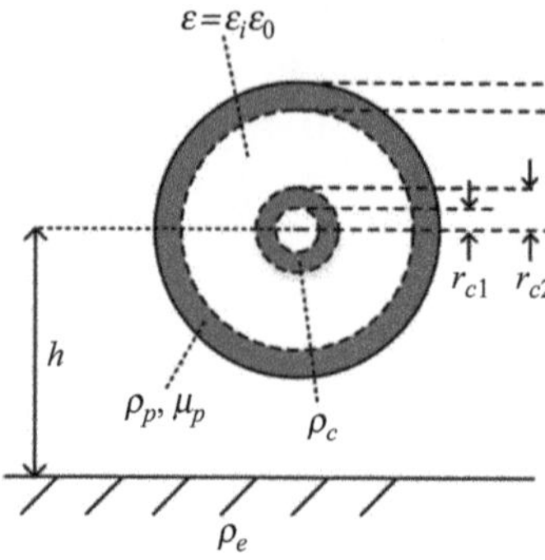

Figure 5.79 GIB cross-section

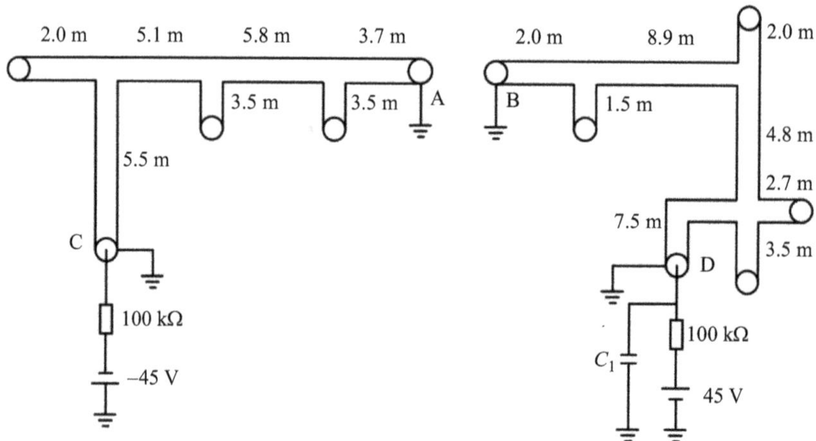

Figure 5.80 Test circuit of a proto-type GIB: $h = 1$ m, $r_{c1} = 5.5$ cm, $r_{c2} = 7.0$ cm, $r_{p1} = 34.05$ cm, $r_{p2} = 34.5$ cm, $\rho_c = \rho_p = 2.8\times10^{-8}$ Ωm, and $\mu_p = \mu_0$

5.5.2.2 Experimental result in the test circuit

Figure 5.81 shows the test results measured by an analog oscilloscope: (a) is for the case of no equivalent capacitance C_t of a transformer, and (b) is for $C_t = 5$ nF.

5.5.2.3 EMT simulation result

5.5.2.3.1 Core voltage Vc

Figure 5.82(a) shows simulation results of core voltage V_c at node C by EMT frequency-dependent line model (Wide-Band = WB model [105–106]) in the case of no C_t for (1) Cu plate, and (2) earth with the resistivity $\rho_e = 100$ Ωm. The voltages in (1) and (2) are almost identical, and thus it is confirmed that the earth resistivity causes only minor effect on the VFT.

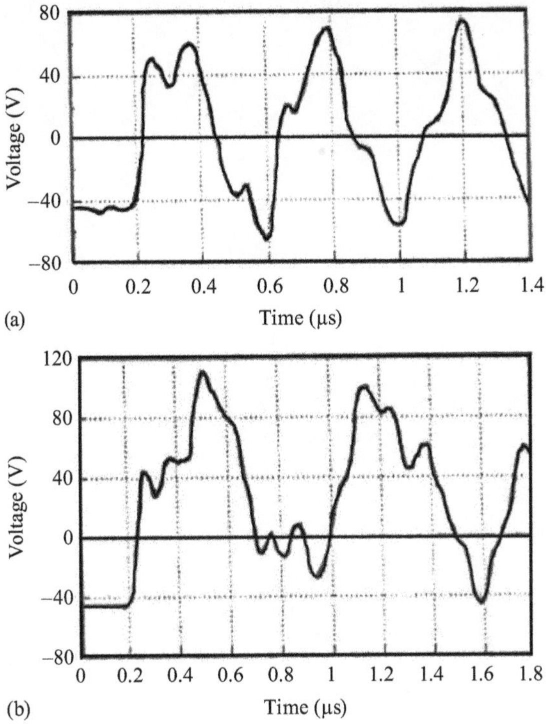

Figure 5.81 Test results of node C voltage. (a) No C_t and (b) $C_t = 5$ nF

Figure 5.82(b) shows V_c calculated by a CP line model (2 MHz) with $\rho_e = 100\ \Omega$m. Figure 5.82(c) is V_c calculated by a single-phase (coaxial mode) CP line model. Again, no significant difference is observed between the voltage V_c wave-forms in Figure 5.82(a)–(c). Thus, note (1) in Section 5.5.1 is confirmed.

Figure 5.83 shows simulation results of core voltage V_c at node C by (a) WB model, (b) CP model, and (c) single-phase CP model in the case of $C_t = 5$ nF for the earth $\rho_e = 100\ \Omega$m. Because of C_t, the oscillating frequency becomes lower than that in the case of no C_t in Figure 5.82. No noticeable difference is observed between the three line models.

Figure 5.84 shows V_c when the branched buses in Figure 5.80 are represented by an equivalent capacitance $C_b = 81$ pF/m evaluated by (5.81) with equivalent relative permittivity $\varepsilon_r = 2.3$ considering spacers in the GIB. Figure 5.84(a) is for no C_t, and (b) for $C_t = 5$ nF by WB and CP models. Because of the almost identical results, only the results by the CP model are shown in the figure. It is observed that the waveforms in Figure 5.84 agree better than those in Figures 5.82 and 5.83 with the test results in Figure 5.81, because of no higher frequency component in Figures 5.81 and 5.84 However, the reason for no high-frequency component in the test results is due to the oscilloscope used in the measurements in 1970s, that is, the

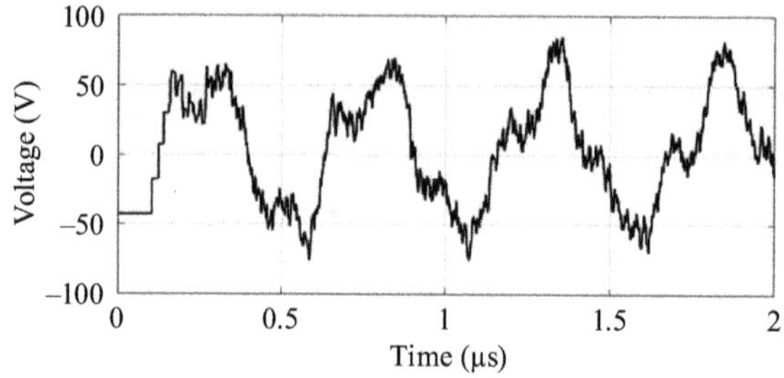

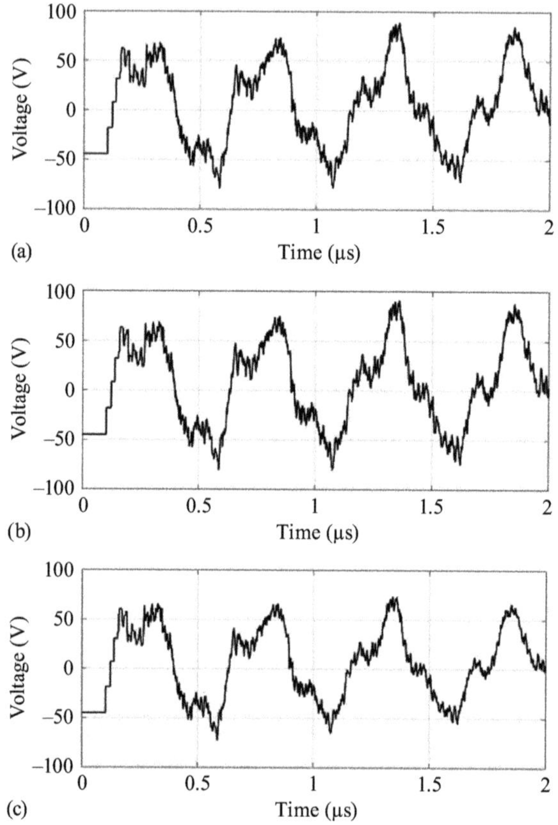

Figure 5.82 Node C voltage of test circuit with branches and no C_t. (a) WB model, (b) CP model and (c) single-phase (core only) CP model

highest frequency which could be measured by the oscilloscope was less than some 10 MHz.

5.5.2.3.2 Pipe voltage Vp

Figure 5.85 shows the pipe voltage at node A in Figure 5.80 when the pipe is open-circuited at nodes A and B. Figure 5.85(a) is by WB model of EMTP and

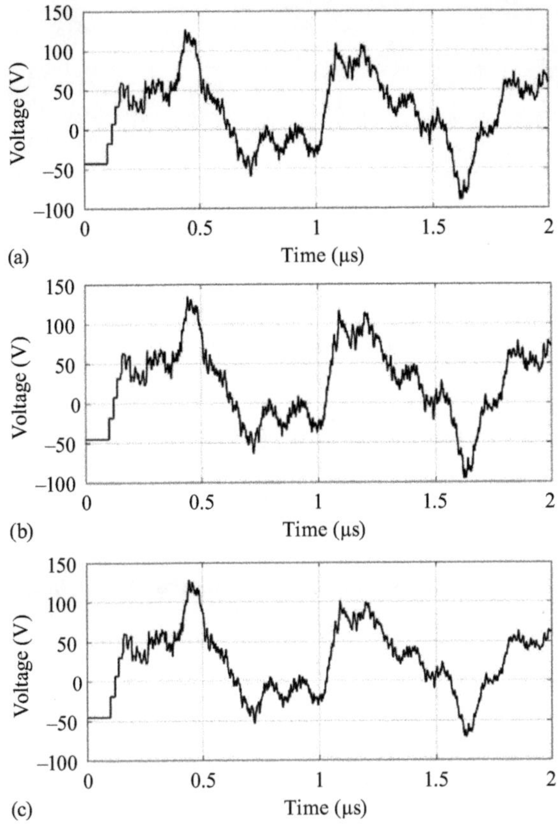

Figure 5.83 Node C voltage of test circuit with branches and $C_t = 5$ nF. (a) WB model, (b) CP model and (c) single-phase (core only) CP model

Figure 5.85(b) is by CP model. No significant difference between (a) and (b) is observed up to $t = 1$ μs. Then the peak value of a spike-like voltage is more damped in (a) due to the frequency-dependent effect of the impedance. It is clear that the pipe voltage is more oscillatory than the core voltage in Figure 5.83.

5.5.3 Ultra-high voltage GIS

A model circuit of an ultra-high voltage (UHV) GIS [49] is illustrated in Figure 5.86. EMT simulation results of DS switching surges at nodes P and Q are shown in Figure 5.87 calculated by CP line model (1 MHz). The EMT results in Figure 5.87 are quite similar to those shown in Figure 5(a) and (b) in [49].

Reference [49] and also [48] and [51] discuss the influence of arc resistance. It is stated that the polarity of the arc affects the VFT overvoltage. In [8], a UHV disconnector with a resistor is developed and tested. In such a disconnector, the influence of the arc becomes not clear.

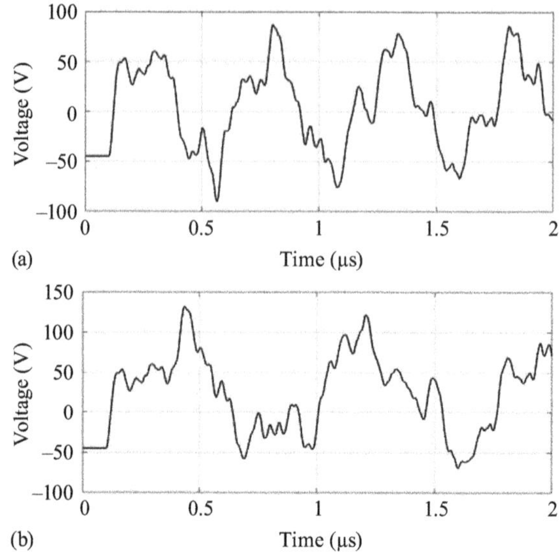

Figure 5.84 Node C voltage with equivalent capacitances of the branches in Figure 5.80. (a) No C_t and (b) $C_t = 5$ nF

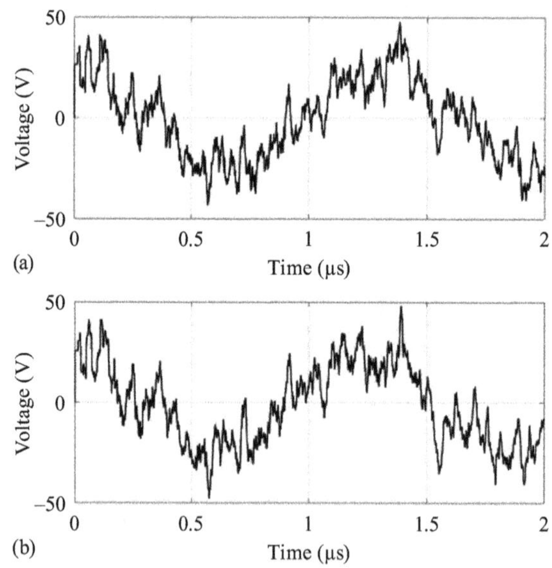

Figure 5.85 Pipe voltage at node A in the open-circuited case with $C_t = 5$ nF. (a) By WB model and (b) by CP model

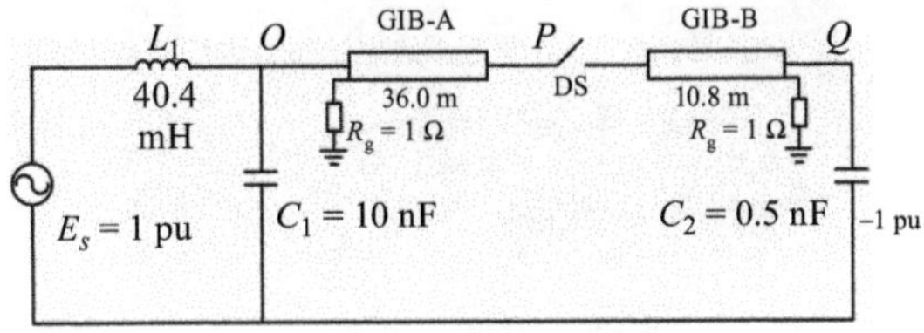

Figure 5.86 Model circuit of an UHV GIS: $h = 1.5$ m, $r_{c1} = 13.5$ cm, $r_{c2} = 15$ cm, $r_{p1} = 76$ cm, $r_{p2} = 77$ cm, $\rho_c = 1.8\times10^{-8}$ Ωm, $\rho_p = 1.5\times10^{-7}$ Ωm, and $\mu_p = 100\mu_0$

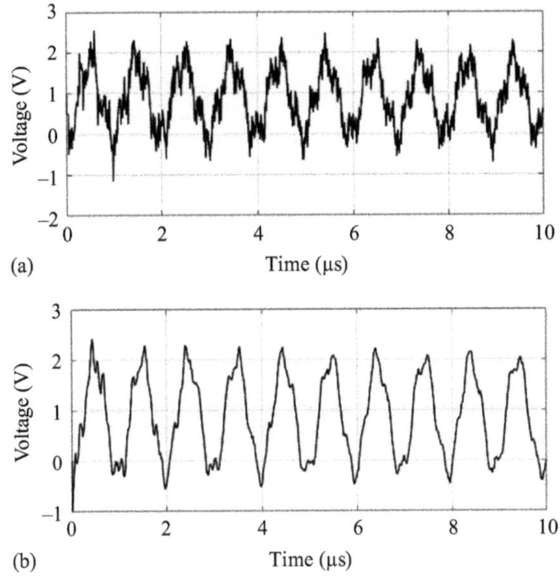

Figure 5.87 EMTP simulation results of core voltages at nodes P and Q. (a) Node P and (b) Node Q

5.5.4 FDTD computation

Numerical electromagnetic analysis (NEA) methods are becoming more and more powerful for analyzing various phenomena associated with both TEM and non-TEM modes of wave propagation and also three-dimensional phenomena [110]. Among various NEA methods, FDTD in time domain is most widely used for transient analysis.

One of the advantages of FDTD is easy computation of a three dimensional transient. For example, a GIB in the 500 kV GIS in Figure 5.88 in Section 5.4.2.3 can be represented three-dimensionally as it is. Then a transient due to DS or CB switching operation in a three-dimensional field can be easily performed, although the required memory and CPU time are very large.

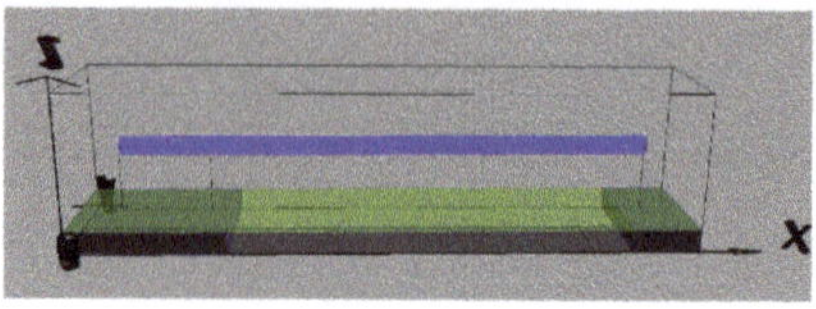

Figure 5.88 FDTD model circuit of a part of GIBs in the GIS

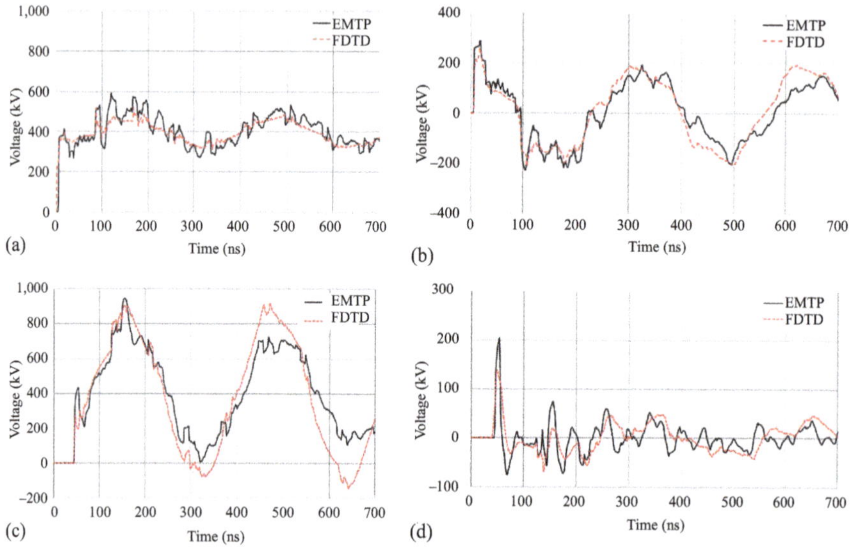

Figure 5.89 Switching surges on a GIB in Figure 5.88 calculated by FDTD [111] and EMTP. (a) Sending-end core, (b) sending-end pipe, (c) receiving-end core and (d) receiving-end pipe

5.5.4.1 Case-A: 500 kV GIB

An example of a GIB model circuit for a transient computation by FDTD is illustrated in Figure 5.52 using VSTL [111]. The model circuit corresponds to Figure 5.88 in Section 5.4.2.3.2 [69]. The total length of the GIB is 12 m, and the pipe is grounded through 2 Ω resistance at the position of 2 m from the sending end and at the receiving end. A source voltage of 408 kV is applied to the sending end, the left-hand side of the model circuit. The earth resistivity is 100 Ωm.

The FDTD computed and EMT simulation results are shown in Figure 5.89. A reasonable agreement between the FDTD and EMT results is observed, although the former is less oscillatory than the latter. The reason for this looks radiation losses in the FDTD computation which cannot be considered in the EMT simulation. Also, the CP line representation of the vertical lead wires for the source and pipe grounding in the EMT simulation produces more oscillation than the FDTD, while the frequency dependence of the vertical lead wires is included automatically in the FDTD computation.

5.5.4.2 Case-B: Partial discharge pulse propagation along GIB with elbow part

To detect a partial discharge along a GIB, a pulse propagation characteristic is investigated using "Maxwell Solver" [112] based on MoM. Figure 5.90 shows a model circuit of the GIB and a computed result. The GIB is composed of an elbow part, of which the left-hand side is a vertical conductor and the left is horizontal. Because the impedance of a vertical conductor differs from that of a horizontal conductor, a reflected traveling wave is generated at the corner of the elbow part when an applied pulse at node-A arrives at the corner. The reflected wave is clearly observed in Figure 5.90 after around 3 ns.

Figure 5.91(a) illustrates an FDTD model circuit corresponding to Figure 5.90. Figure 5.91(b) shows FDTD computed results of the traveling wave responses along the GIB as a function of time. At $t = 0$, an applied pulse starts to propagate to the left and right along the GIB. At $t = 2$ ns, the traveling wave to the left reaches the elbow corner and reflection occurs. The reflected wave travels to the right as observed at $t = 3$ ns, and comes back to the pulse applied node at $t = 4$ ns. The traveling wave behavior corresponds to the reflected current wave in Figure 5.90. Figure 5.91(c) shows the voltage and current waveforms. The current in the figure agrees with the current waveform in Figure 5.90. Note that the polarity of the applied pulse in Figure 5.91(c) is opposite to that in Figure 5.90.

Figure 5.92(a) shows an EMT simulation model of the GIB corresponding to Figures 5.90 and 5.91. Figure 5.92(b) shows EMT simulation results with a pulse voltage applied at $t = 0$.

As discussed in [108,112–116], there exists mutual coupling between the horizontal and vertical parts of the GIB which cannot be handled by a circuit theory. The mutual coupling can be represented in EMT simulation as the surge impedance of the vertical conductor different from that of the horizontal part evaluated from the refraction coefficient at Node 2 based on an FDTD result, or by adopting a nonuniform conductor impedance explained in [115].

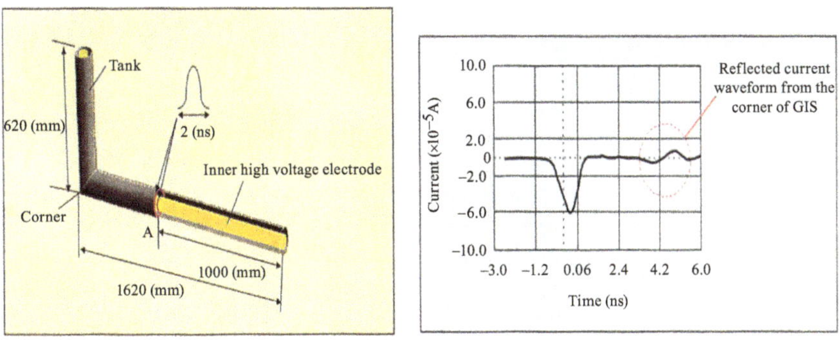

Figure 5.90 Partial discharge pulse propagation along a gas-insulated bus (GIB) by MoM (Maxwell Solver). Adapted from [112]

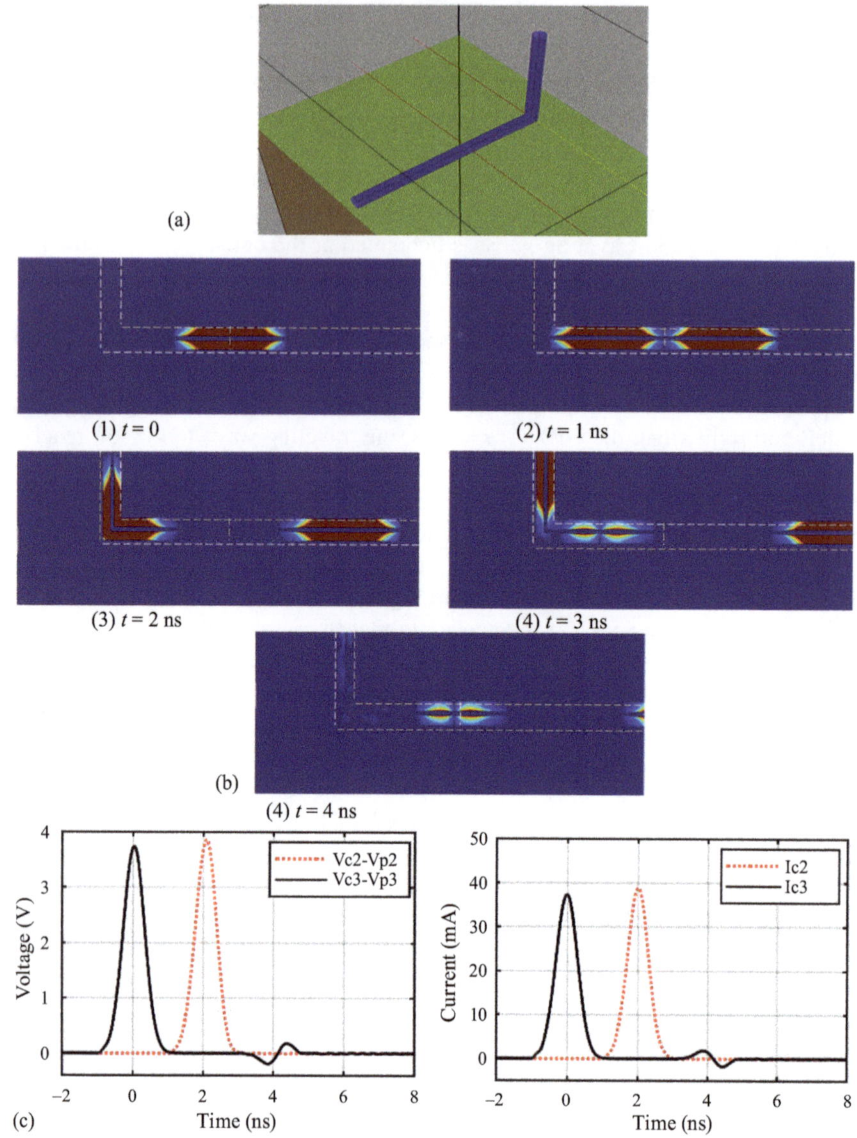

Figure 5.91 FDTD computation of discharge pulse propagation along GIB. Adapted from [113]. (a) FDTD model circuit of GIB with an elbow part, (b) time response of travelling wave along the GIB and (c) time responses of the voltage and current along the GIB

The EMT results qualitatively agree with the FDTD results except the reflected traveling wave. The voltage reflected wave is positive, while it is negative and then becomes positive as observed in Figure 5.91(c). A similar difference is observed in the current waveform. This requires a further investigation. A transient involving

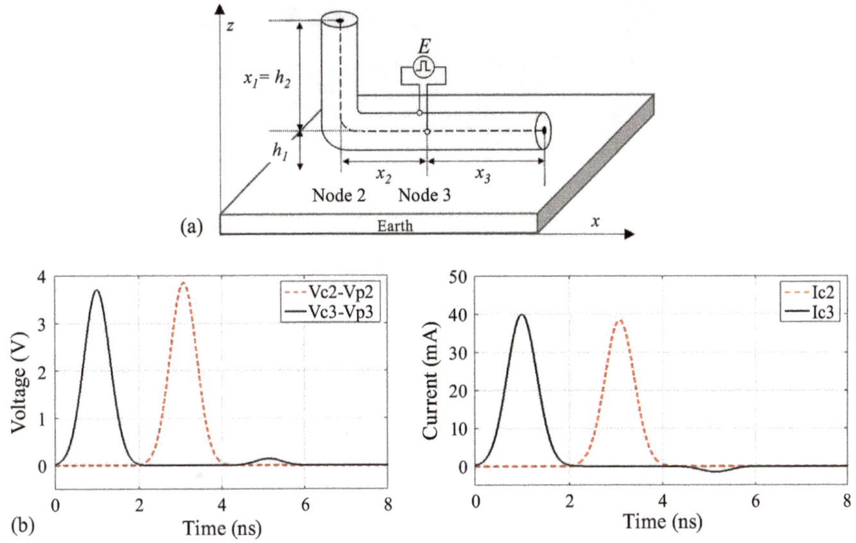

Figure 5.92 EMT model circuit of GIB with an elbow part and simulation results of voltages and currents at Node 3 (bold line) and Node 2 (dotted line). Adapted from [114]. (a) Model circuit: $x_1 = x_3 = 1$ m, $x_2 = 0.62$ m, $h_1 = 0.23$ m, $E = 3.72$ V, $\rho e = 80$ Ωm and (b) EMTP simulation results

mutual coupling between horizontal and vertical conductors can be approximately simulated by EMT-type simulation tools.

5.5.4.3 Case-C: Two-conductor system involving a nonuniform line [117]

Figure 5.93(a) illustrates a model circuit of cascaded homogeneous lines and a lead wire which is nonuniform for separation-distance y_1 at node 2 differs from y_2 at node 3. Figure 5.93(b) and (c) shows FDTD and EMT simulation results of a transient when a step-like voltage ($E = 1$V) is applied to node "a1" to the ground. In the figure, (1) is the energized phase and (2) is the induced phase. The EMT simulation results on phase-*a* in Figure 5.93(c) qualitatively agree with FDTD results in Figure 5.93(b). Thus, transient responses on the energized phase of a nonhomogeneous line can be simulated by EMT-type tools. However, it is rather hard to simulate accurately the responses on the induced phase by the EMT-type simulation tools. The time delay at the sending end of the induced phase cannot be represented by the EMT-type tools, because the delay is due to a direct electromagnetic wave in the air from node "a1" to node "b1."

5.5.5 Summary

EMT simulation results by different line models prepared in the existing EMT-type simulation tools are compared with experimental results in a proto-type GIS.

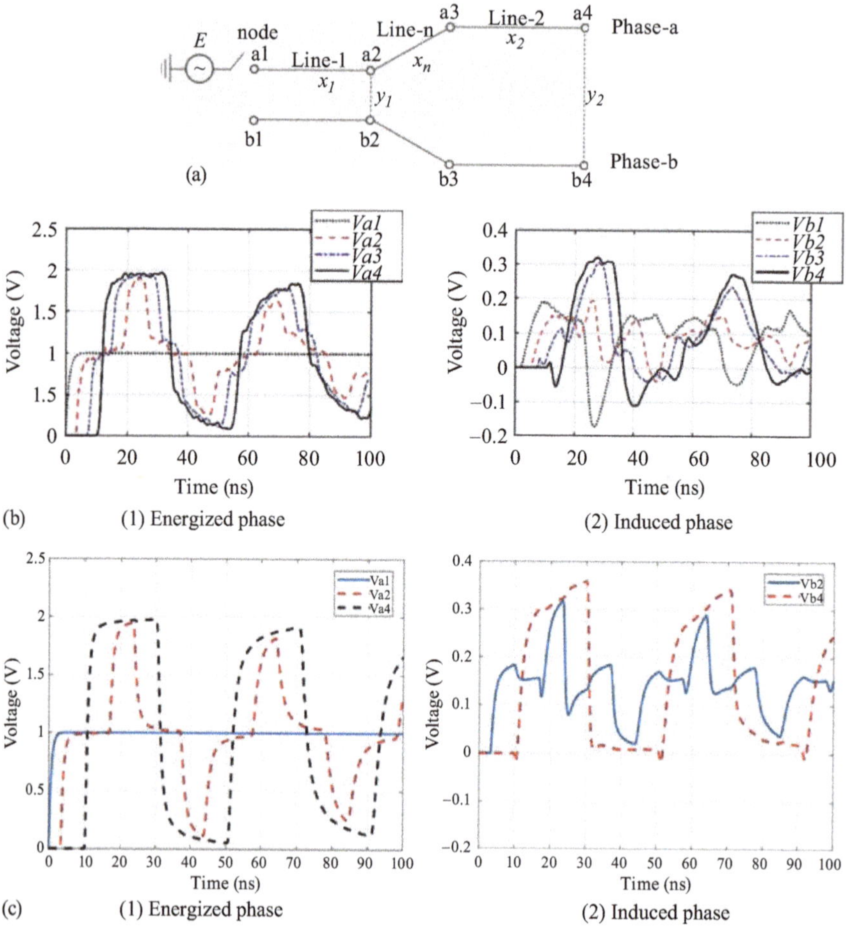

Figure 5.93 Transient on a nonhomogeneous two-conductor system. Adapted from [114]. (a) Non-uniform two-conductors: $x_1 = x_2 = 1$ *m,* $x_n = 1.031$ *m,* $h_1 = h_2 = h_n = 1$ *m,* $\rho e = 100$ *Ωm, (b) FDTD simulation results and (c) EMTP simulation results*

Further, a three-dimensional VFT computation is performed by FDTD in comparison with EMT simulation results.

1. In EMT simulations of the VFT, both frequency-dependent line model and constant-parameter CP model show a reasonable agreement with the test results, because the VFT frequency is very high and the earth resistivity causes only a minor influence. The pipe grounding and lead wires have to be carefully represented.
2. FDTD computed results agree rather well with those by EMT-type simulation tools, although the EMT results are more oscillatory. The reason for this might be the effect of radiation losses which are automatically included in the FDTD

computation. A main disadvantage of the FDTD is a large amount of computer resources.

5.6 Conclusions

In this chapter, the charging and discharging characteristics of underground cables are explained and investigated in Section 5.1. Cable charging and discharging phenomena are becoming more and more important for utilities which are using high voltage (HV) cables for voltage control during periods of light load.

In Section 5.2, the simulations of electromagnetic transients on an RTE 225 kV cable system are thoroughly investigated and studied. Moreover, the simulations are well-validated and compared with the field test results of the cable system.

Section 5.3 deals with transient simulations on homogeneous and cross-bonded cables using extended and classical TL approaches. The influences of the extended TL approach on different mode energizations are also investigated and made clear.

Section 5.4 investigates very-fast transients (VFTs) in gas-insulated substations (GISs). Field test results of VFTs are summarized considering electromagnetic disturbances in GIS control circuits. Also, numerical instabilities are observed in VFT simulation results by using existing EMT-type simulation tools, and the reason for the instability is made clear.

In Section 5.5, a difference of EMT simulation results between frequency-dependent and constant-parameter line models are discussed in comparison with experimental results in a proto-type GIS. Also, FDTD technique is adopted to compare and validate the results performed by EMT-type simulation tool.

Appendix A5.1 Basic formulation for theoretical analysis

Because a single-core coaxial cable is composed of two insulators, that is, one between the core and the metallic sheath and the other between the sheath and the earth. Thus, a model circuit of charging a cable is given as two series capacitors illustrated in Figure A5.1. C1 and G1 represent the core to metallic sheath circuit, and C2 and G2 for the sheath to earth circuit. When switch S is closed at $t = 0$ in the circuit of Figure A5.1, the following solution is obtained:

$$q_1 = (1/(C_1 + C_2)) \cdot [C_1C_2E - C_1 \cdot k \cdot E(1 - \exp(-t/\tau))] \tag{A5.1}$$

$$q_2 = (1/(C_1 + C_2)) \cdot [C_1C_2E + C_2 \cdot k \cdot E(1 - \exp(-t/\tau))] \tag{A5.2}$$

$$\text{where } \tau = (C_1 + C_2)/(G_1 + G_2) \quad \text{and} \quad k = (G_1C_2 - G_2C_1)/(C_1 + C_2) \tag{A5.3}$$

A5.1.1 Charging a cable

When a coaxial cable is charged by a voltage source E, the following relation is obtained from (A5.1). Remind that the metallic sheaths are grounded at both ends, and

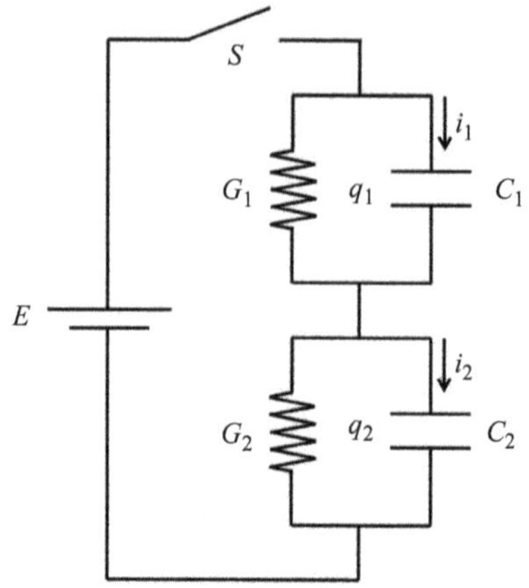

Figure A5.1 Double C–G circuit equivalent to a coaxial cable

thus only the core to sheath insulator with *C1* and *G1* in Figure A5.1 is to be considered:

$$q = Q[1 - \exp(-t/\tau)] \tag{A5.4}$$

$$\text{where } q = CV, Q = CE, \text{ and } \tau = CR = C/G \tag{A5.5}$$

The above equation is rewritten by

$$V = E[1 - \exp(-t/\tau)] \tag{A5.6}$$

In Figure A5.1, source voltage E is given by $R_0\, I_0$. Then

$$V(t) = R_0 I_0[1 - \exp(-t/\tau)] \tag{A5.7}$$

Assume the cable is charged until time T_1 when the voltage reaches V_1. Then the following relation is obtained from (A5.7):

$$V_1 = E[1 - \exp(-T_1/\tau)], \text{ or } (E - V_1)/E = \exp(-T_1/\tau) \tag{A5.8}$$

Taking logarithm of the above equation

$$-T_1/\tau = \ln[(E - V_1)/E] \tag{A5.9}$$

Thus, time *T1* is given in the following form:

$$T_1 = \tau \ln[E/(E - V_1)] \tag{A5.10}$$

A5.1.2 Discharging a cable

When switch S is open-circuited in Figure A5.1, the stored charges in C_1 and C_2 are discharged. In this case, the two closed circuits become independent, that is, a transient solution of the discharge in each circuit can be solved separately. Assume the initial charge is Q. Then, the following solution is obtained:

$$q = Q \cdot \exp(-t/\tau), \quad \tau = C/G = CR \tag{A5.11}$$

Or

$$V = E\exp(-t/\tau) \tag{A5.12}$$

where $E = V_1$ when a discharge starts at $V(\mathrm{t}) = V_1$. Then

$$V(\mathrm{t}) = V_1 \exp(-t/\tau) \tag{A5.13}$$

Assume the cable voltage becomes V_2 at time T_2 by discharge

$$V_2 = E\exp(-T_2/\tau) \tag{A5.14}$$

Thus

$$T_2 = \tau \ln(E/V_2) \tag{A5.15}$$

A5.1.3 Leakage resistance R for cable discharge

The following relation between charge Q, voltage V, and capacitance C of a capacitor is well-known:

$$Q = CV \tag{A5.16}$$

By discharging stored charge Q_1 in a cable from $t = T_1$ to T_2, or from $v = V_1$ to V_2, Q_1 is decreased to Q_2 at $t = T_2$:

$$Q = Q_1 \text{ and } V = V_1 \text{ at } t = T_1, \text{ and } Q = Q_2 \text{ and } V = V_2 \text{ at } t = T_2 \tag{A5.17}$$

The relation between discharging current i and charge Q is defined by

$$i = dQ/dt \text{ or } integral\ idt = Q \tag{A5.18}$$

Assuming current i is constant from $t = T_1$ to T_2,

$$[i_t] = i(T_2 - T_1) = Q_1 - Q_2 \tag{A5.19}$$

Leakage resistance R is obtained by assuming voltage V being given as the average for $t = T_1$ to T_2 from a measured result:

$$\begin{aligned} R = V/I &= (V_1 + V_2)(T_2 - T_1)/\{2(Q_1 - Q_2)\} \\ &= (V_1 + V_2)(T_2 - T_1)/\{2C(V_1 - V_2)\} \end{aligned} \tag{A5.20}$$

Appendix A5.2 Impulse generator (pulse generator)

1. As far as transient voltages within two travel times (2T) of the cable with the length $x = 1$ km are concerned, the charging current of a cable can be neglected. Then, it looks possible to use an ordinary (commercially available) impulse/pulse generator with the maximum voltage of 2–5 kV:

$$T = \mathrm{x}/\mathrm{c}_1 \fallingdotseq 1{,}000[\mathrm{m}]/180\left[\frac{\mathrm{m}}{\mu\mathrm{s}}\right] \fallingdotseq 5.5\ \mu\mathrm{s},\ 2T = 11\ \mu\mathrm{s} \tag{A5.21}$$

where $c_1 = c_0/\sqrt{\varepsilon_i} \fallingdotseq 300/\sqrt{2.8} = 180\left(\frac{m}{\mu s}\right)$ is the coaxial mode propagation velocity and ε_i is the relative permittivity of an insulator between core and sheath.

After some 10 μs, the wavetail of a transient voltage and a current cannot be measured correctly because of the charging current, when a step-function like the voltage is applied.

2. If an impulse wave such as 1/50 μs is to be applied and the observation time is some 10 μs, we have to consider the charging current.
3. It is very much dependent on what one likes to know from the measured results. If only transient voltages at both ends are to be known, an ordinary pulse or impulse voltage generator is enough.
4. If one needs to measure a transient current, we need an impulse current generator (IG) which is quite expensive, but one can produce the IG in a laboratory (see Figures A5.2–A5.5).

In general, a charging current has to be considered in a field test, if the cable length is more than some 100 m. In such a case, we need a large capacity impulse generator, impulse current generator (IG) we call in Japan, which can cover the charging current. The price of the IG is much higher than an ordinary impulse

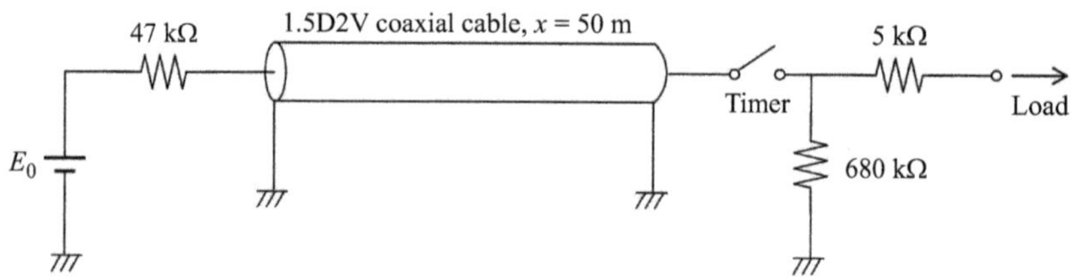

Figure A5.2 PG circuit

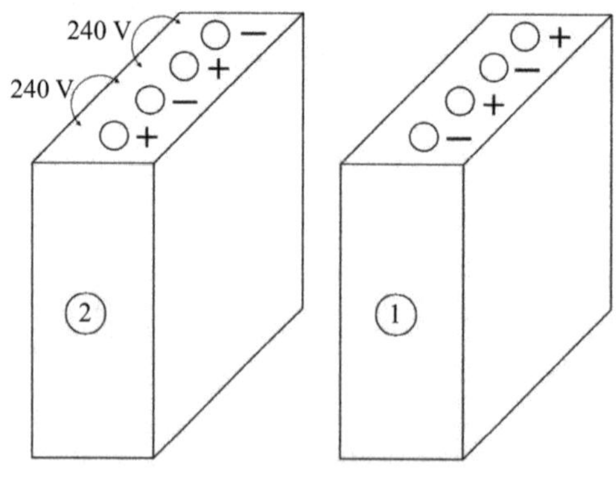

Figure A5.3 Batteries

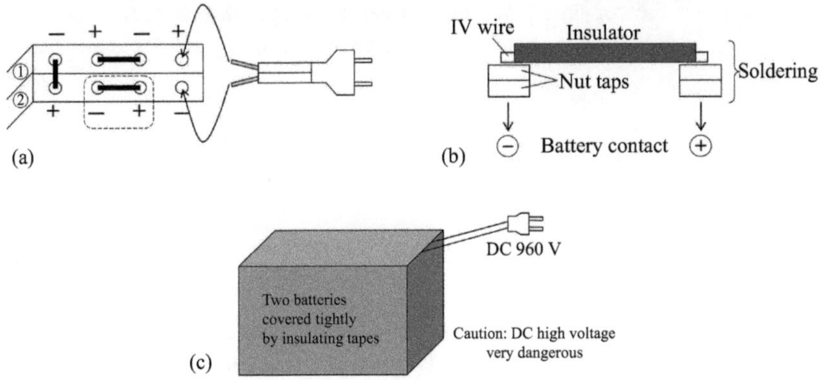

Figure A5.4 Series connection of batteries ① and ②. (a) Connection of 2 batteries, (b) series connection within a battery and (c) finished product of a DC 960 V battery

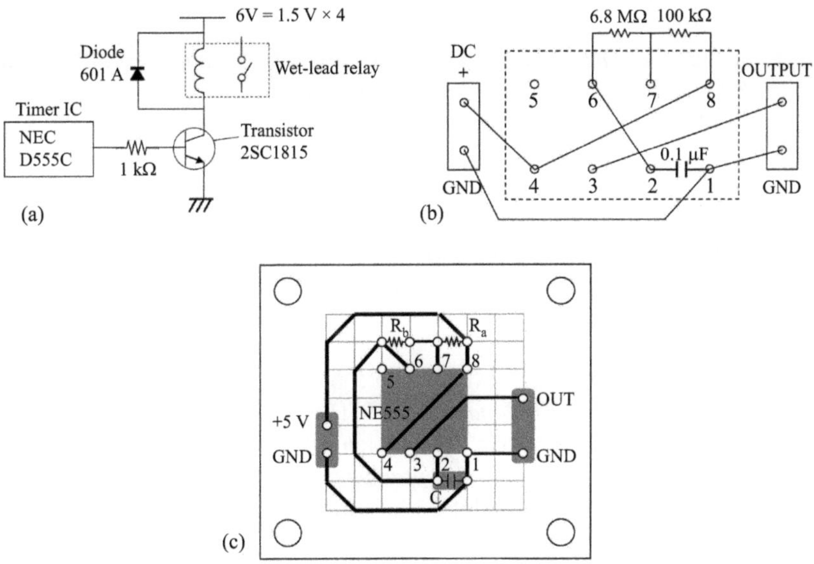

Figure A5.5 Timer circuit. (a) Operating circuit by relay, (b) Timer IC – back side and (c) Figure of timer IC pattern appeared on the back. (a) Operating circuit by relay, (b) Timer IC – back side and (c) Figure of timer IC pattern appeared on the back

generator. Because the surge impedance of a cable is low, it is necessary to prepare a low impedance source (generator). If the source impedance is high, we cannot apply a voltage to the cable. This is easily understood by a theory of a voltage divider. Thus, we need the low impedance source, that is, a capacitor with a DC source to charge the capacitor. We need a gap to trigger the charged capacitor so that an impulse voltage is output. Its capacitance should be greater than that of a tested cable.

1. PG circuit
2. DC source E_0: batteries Panasonic 0160 W(480 V) $\times$ 2 = 960 V
3. Timer

References

[1] IEC. *Tests on Oil-Filled and Gas-Pressure Cables and Their Accessories – Part 1: Oil-Filled, Paper or Polypropylene Paper Laminate Insulated, Metal-Sheathed Cables and Accessories for Alternating Voltages Up to and Including 500 kV*. IEC 60141-1; 1993.

[2] JEC. *Japanese Standard of Cable Testing*. JEC-3401-2006; 2006.

[3] CIGRE WG C4.502. *Power System Technical Performance Issues Related to the Application of Long HVAC Cables*. CIGRE Technical Boucher TB 556; 2013.

[4] Ametani, A., Ohno, T., and Nagaoka, N. *Cable system transients: Theory, modeling and simulation*. Singapore: Wiley; 2015.

[5] Paul, D., Chavadarian, R. and Haddadian, V. "Cable-capacitance discharge time with and without the application of grounding device." *IEEE Transactions on Industry Applications*. 2011; 47(1): 286–91

[6] Ghassemi, F. "Effect of trapped charges on cable SVL failure." *International Conference on Power systems Transients*. Vancouver, Canada, 2013.

[7] Watanabe, Y., and Kosugi, K. Installation of 275 kV POF cable in Tokyo – Part 3 After installation tests. *Fujikura Densen Technical Review*. 1973; 48: 11–26

[8] Takaoka, M., and Siseki, M. "The potential distribution of direct current of cable." *Journal of IEE Japan*. 1971; 91(11): 2126–34

[9] Takaoka, M. "The potential distribution of direct current oil-filled cable." *IEEE Transactions on Power Apparatus and Systems*. 1971; 90(6): 2622–30

[10] Iwata, Z., Shii, H., Kinoshita, M., Hirukawa, H., and Kawai, E. "New modified polyethylene paper proposed for UHV cable insulation." *IEEE Transactions on Power Apparatus and Systems*. 1977; 96(5): 1573–82

[11] Ghassemi, F. *Private Correspondence*. October 2015.

[12] Lafaia, I. *Contributions to Modeling and Simulation of HVAC Cables using Field Test Results*. Polytechnic Montreal Ph.D. Thesis; 2016.

[13] Lafaia, I., Ghassemi, F., Ametani, A., Mahseredjian, J., Haddad, M., and Robson, S. "Experimental and theoretical study of cable discharge phenomena." *IEEE Transactions on Power Delivery*. 2017; 32 (4): 2022–2030

[14] Chmura, L., Jin, H., Cichecki, P., Smit, J. J., Gulski, E., and Vries, F. D. "Use of dissipation factor for life consumption assessment and future life modeling of oil-filled high-voltage power cables." *IEEE Electrical Insulation Magazine*. 2012; 28(1): 27–37

[15] Good Fellow. *Polyimide (PI) – Material Properties*. Available: www.goodfellow.com/E/Polyimide.html (accessed in Jan. 2016).

[16] DuPont. *Summary of Properties for Kapton® Polyimide Films*. Available: www.dupont.com/content/dam/assets/ products-and-services/membranes-films/assets/DEC-Kapton-summary-of-properties.pdf (accessed in Jan. 2016).

[17] Mazzanti, G., and Marzinotto, M. *Extruded cables for high voltage direct current transmission: Advances in research and development*. Singapore: Wiley-IEEE Press; 2013.

[18] Nexans. 60-500 kV High Voltage Underground Power Cables – XLPE Insulated Cables. [pdf file]. Available: www.nexans.co.uk/ eservice/UK-en GB/fileLibrary/Download 540192183/UK/files/Nexans %20High%20Voltage%20Underground.pdf.

[19] Brugg Kable, A.G. High Voltage XLPE Cable Systems – Technical User Guide. Available: nepa-ru.com/brugg_files/02_hv_cable_ xlpe/03_web_xlpe_guide_en.pdf (accessed Jan. 2016).

[20] IEEE Working Group on Insulator Contamination. "Surface resistance measurements on nano-ceramic insulators." *IEEE Transactions on Power Delivery*. 2001; 16(4): 801–805

[21] Brugg Kabel, AG. *Outdoor Terminations*. Available: brugg.nubosys.com/en/products/product-detail-en/productgroup/pg-30/ (accessed in Jan. 2016).

[22] Ametani, A. *Private Communications with RTE*. 2013 to 2014.

[23] Lafaia, I., Mahseredjian, J., and Ametani A. *Report on the Measurements on the Boutre-Trans Cable and Comparison with Simulation Results*. Report submitted to RTE; 2014.

[24] Réseaux de Transport d'Électricité. Filet de Sécurité PACA: Pour Une Sécurisation Electrique Durable de la Région. Available: http://www.rte-france.com/fr/projet/filet-de-securite-paca-pour-une-securisation-electrique-durable-de-la-region, 2015.

[25] Ametani, A. *Distributed-parameter circuit Theory*. Tokyo: Corona Pub. Co.;1990.

[26] Mahseredjian, J., Dennetière, S., Dubé, L., Khodabakhchian, B., and Gérin-Lajoie, L. "On a new approach for the simulation of transients in power systems." *Electric Power Systems Research*. 2007; 77(11): 1514–20

[27] Koçar, I., Mahseredjian, J., and Olivier G. "Weighting method for transient analysis of underground cables." *IEEE Transactions on Power Delivery*. 2008; 23(3): 1629–35

[28] Koçar, I., Mahseredjian, J., and Olivier, G. "Improvement of numerical stability for the computation of transients in lines and cables." *IEEE Transactions on Power Delivery*. 2010; 25(2): 1104–11

[29] Ametani, A. "A general formulation of impedance and admittance of cables." *IEEE Transactions on Power Apparatus and System*. 1980; 99(3): 902–10

[30] Dommel, H. W. *EMTP theory book*. Portland: Bonneville Power Administration; 1986.

[31] Ametani, A. *Cable constants rule book*. Portland: Bonneville Power Administration; 1982.

[32] IEE Japan Working Group. Surge Phenomena and Countermeasures on Cable Systems. *IEE Japan Technical Report No. 366*; 1991.
[33] Gustavsen, B., Martinez, J. A., and Durbak, D. "Parameter determination for modelling system transients—Part II: Insulated cables." *IEEE Transactions on Power Delivery*. 2005; 20(3): 1045–50
[34] Ametani, A., Miyamoto, Y., and Nagaoka, N. "Semiconducting layer impedance and its effect on cable wave-propagation and transient characteristics." *IEEE Transactions on Power Delivery*. 2004; 19(4): 1523–31
[35] Ohtani, T., Ametani, A., and Fujita, I. *Electric Circuit Theory*. Shokohdo Pub. Co.;1992.
[36] Nagaoka, N., Yamamoto, M., and Ametani, A. "Surge propagation characteristics of a POF cable." *Electrical Engineering in Japan*. 1985; 105(5): 67–75
[37] Ametani, A., Miyamoto, Y., and Nagaoka, N. "An investigation of a wave propagation characteristic on a cross-bonded cable." *IEE Japan Transactions on Power and Energy*. 2003; 123(3): 395–01
[38] Lafaia, I., Alatawneh, N., Mahseredjian, J., Ametani, A., Correia de Barros, M. T., Koçar, I., and Naud, A. "Modeling of an underground cable installed in a poly-ethylene tube tfor transient simulations." *IEEJ Power and Energy Society*, Nagoya, Japan, 2015.
[39] Ametani, A., Ohsaki, E., and Honaga, Y. "Surge characteristics on an untransposed vertical line." *IEEJ Transactions on Power and Energy*. 1983; 103(2): 117–24
[40] Gudmundsdottir, U. S. *Modelling of Long High Voltage AC Cables in Transmission Systems*. Aalborg: Aalborg University; 2010.
[41] Nagaoka, N., and Ametani, A. "Transient calculations on crossbonded cables." *IEEE Transactions on Power Apparatus and Systems*. 1983; 102(4): 779–87
[42] IEC. *Insulation Co-ordination Part 1: Definitions, Principles and Rules*. IEC 60071-1; 2006.
[43] CIGRE WG D1.03. *Very Fast Transient Over-Voltages (VFTO) in Gas-Insulated Substations*. CIGRE Technical Brochure 519; 2012.
[44] Boggs, S. A., Chu, F. Y., and Fujimoto, N (Eds.). *Gas-insulated substations, technology and practice*. New York, NY: Pergamon; 1986.
[45] Koch, H. *Gas insulated substations*. United Kingdom: Wiley-IEEE Press; 2014.
[46] Ametani, A. "Surge propagation characteristics on a gas-insulated cables." *IEEJ Transactions on Power and Energy*. 1981; 1(1): 491–97
[47] Murase, H., Oshima, I., Aoyagi, H., and Miwa, I. "Measurement of transient voltages induced by disconnect switch operation." *IEEE Transactions on Power Apparatus and Systems*. 1985; 104(1): 157–65
[48] Ogawa, S., Haginomori, E., Nishiwaki, S., Yoshida, T., and Terasaka, K. "Estimation of restriking transient overvoltage on disconnecting switch for GIS." *IEEE Transactions on Power Delivery*. 1986; 1(2): 95–02
[49] Ozawa, J., Yamagiwa, T., Hosokawa, M., Takeuchi, S., and Kozawa, H. "Suppression of fast transient overvoltage during gas disconnector switching in GIS." *IEEE Transactions on Power Delivery*. 1986; 1(4): 194–01

[50] Nojima, K., Nishiwaki, S., Okubo, H., and Yanabu, S. "Measurement of surge current and voltage waveforms using optical-transmission techniques." *IEE Proceedings C (Generation, Transmission and Distribution)*. 1987; 134(6): 415–22

[51] IEEJ WG for GIS Induced Voltages. Surge Phenomena Induced to Low-Voltage Circuits in Gas-Insulated Substations. *IEEJ Technical Report 273*; 1988.

[52] Fujimoto, N., and Boggs, S. "Characteristics of GIS disconnector-induced short rise time transients incident on externally connected power system components." *IEEE Transactions on Power Delivery*. 1988; 3(3): 961–70

[53] Meppelink, J., Diederich, K. J., Feser, K., and Pfaff, W. R. "Very fast transients in GIS." *IEEE Transactions on Power Delivery*. 1989; 4(1): 223–33

[54] Yanabu, S., Murase, H., Aoyagi, H., Okubo, H., and Kawaguchi, Y. "Estimation of fast transient overvoltage in gas-insulated substation." *IEEE Transactions on Power Delivery*. 1990; 5(4): 1875–22

[55] Wiggins, C. M., Thomas, D. E., Nickel, F. S., Salas, T. M., and Wright, S. E. "Transient electromagnetic interference in substations." *IEEE Transactions on Power Delivery*. 1994; 9(4): 1869–84

[56] Nishiwaki, S., Nojima, K., Tatara, S., Kosakada, M., Tanabe, N., and Yanabu, S. "Electromagnetic interference with electronic apparatus by switching surges in GIS-cable system." *IEEE Transactions on Power Delivery*. 1995; 10(2): 739–46

[57] Working Group of Surges in Protection and Control Systems. *Technology of Countermeasures against Surges in Protection and Control Systems in Japanese Utilities*. Reports of Electric Technology Research Association; 2001.

[58] IEEJ WG for Power System Transient Analysis (Convener Ametani A.). *Power System Transients and EMTP Simulations*. IEE Japan Technical Report 872; 2002.

[59] Matsumoto, T., Kurosawa, Y., Usui, M., Yamashita, K., and Tanaka, T. "Experience of numerical protective relays operating an environment with high-frequency switching surge in Japan." *IEEE Transactions on Power Delivery*. 2006; 21(1): 88–93

[60] Ametani, A., Motoyama, H., Ohkawara, K., Yamakawa, H., and Suga, N. "Electromagnetic disturbances of control circuits in power stations and substations experimented in Japan." *IET Generation Transmission and Distribution*. 2009; 3(9): 801–15

[61] Smajic, J., Holaus, W., Kostovic, J., and Riechert, U. '3D full-Maxwell simulations of very fast transients in GIS." *IEEE Transactions on Magnetics*. 2011; 47(5): 1514–17

[62] Simka, P. "*Dielectric coordination of high voltage gas circuit breakers*", Ph.D. Thesis. Switzerland: ETH Zurich; 2011.

[63] Simka, P., Straumann, U., and Franck, C. M. "SF6 high voltage circuit breaker contact systems under lightning impulse and very fast transient voltage stress." *IEEE Transactions on Dielectrics and Electrical Insulation*. 2012; 19(3): 855–64

[64] Li, C., He, J., Hu, J., Zeng, R., and Yuan, J. “Switching transient of 1000-kV UHV system considering detailed substation structure.” *IEEE Transactions on Power Delivery*. 2012; 27(1): 112–22

[65] Yue, G., Liu, W., Chen, W., Guan, Y., and Li, Z. “Development of full frequency bandwidth measurement of VFTO in UHV GIS.” *IEEE Transactions on Power Delivery*. 2013; 28(4): 2550–57

[66] CIGRE WG C4.208. *EMC within Power Plants and Substations*. CIGRE Technical Brochure 535; 2013.

[67] Burow, S., Straumann, U., Kohler, W., and Tenbohlen, S. “New methods of damping very fast transient overvoltages in gas-insulated switchgear.” *IEEE Transactions on Power Delivery*. 2014; 29(5): 2332–39

[68] Filho, J. G. R., Teixeira, Jr., J. A., Sans, M. R., and Martinez, M. L. B. “Very fast transient overvoltage waveshapes in a 500-kV gas insulated switchgear setup.” *IEEE Electrical Insulation Magazine*. 2016; 32(3): 17–23

[69] Ametani, A., Goto, T., Nagaoka, N., and Omura, H. “Induced surge characteristics on a control cable in a gas-insulated substation due to switching operation.” *IEEJ Transactions on Power and Energy*. 2007; 127(12): 1306–12

[70] IEC. *Electromagnetic Compatibility (EMC) Part 1, "General"*. IEC 61000-6-5; 2001.

[71] IEC. *Electromagnetic Immunity Standard for Testing and Measurement Techniques, Oscillatory Waves Immunity Test*. IEC 61000-4-12; 2001.

[72] ANSI/IEEE. *IEEE Guide for Surge Voltages in Low-Voltage AC Power Circuits*. ANSI/IEEE C62.41; 1980.

[73] Japanese Electro-Technical Commission. *Test Voltage for Low-Voltage Control Circuits in Power Stations and Substations*. IEE Japan JEC-0103; 2004.

[74] Haznadar,Z., Carsimamovic, S., and Mahmutcehajic, R. “More accurate modeling of gas-insulated substation components in digital simulations of very fast electromagnetic transients.” *IEEE Transactions on Power Delivery*. 1992; 7(1): 434–41

[75] Povh, D., Scmitt, H., Volker, O., and Witzmann, R. “Modeling and analysis guidelines for very fast transients.” *IEEE Transactions on Power Delivery*. 1996; 11(4): 2028–35

[76] Rao, M. M., Thomas, M. J., and Singh, B.P. “Frequency characteristics of very-fast transient currents in a 245 kV GIS.” *IEEE Transactions on Power Delivery*. 2005; 20(4): 2450–57

[77] Nishimura, K., Ametani, A., Nagaoka, N., and Baba, Y. “Modeling of a current transformer for electromagnetic transient simulation in a power station.” *CIGRE SC C4 Kushiro Colloquim*, Kushiro, Japan, 2009.

[78] Szewczyk, M., Kutorasinski, K., Wronski, M., Florkowski, M. “Full-Maxwell simulation of very fast transients in GIS: Case study to compare 3-D and 2-D-axisymmetric models of 1100 kV test setup.” *IEEE Transactions on Power Delivery*. 2017; 32(2): 733–39

[79] Ametani, A., Xue, H., Natsui, M., and Mahseredjian, J. “Electromagnetic disturbances in gas-insulated substations and VFT calculations.” *Electric Power System Research*. 2018; 160: 191–98

[80] Ametani, A., Xue, H., Kocar, I., and Mahseredjian, J. "Numerical instabilities of very-fast-transients in EMTP simulation." *IEE Japan High-Voltage Conference*, Okinawa, Japan, 2018.

[81] Xue, H., Ametani, A., Mahseredjian, J., Baba, Y., Rachidi, F., and Kocar, I. "Transient responses of overhead cables due to mode transition in high frequencies." *IEEE Transactions on Electromagnetic Compatibility*. 2018; 60(3): 785–94

[82] Xue, H., Ametani, A., and Mahseredjian, H. "Very-fast transients in a 500 kV gas-insulated substation." *IEEE Transactions on Power Delivery*. 2019; 34(2): 627–37

[83] Xue, H. "General formulation and accurate evaluation of earth-return parameters for overhead/underground cables." École Polytechnique de Montréal, Ph.D. Thesis; 2018.

[84] Liu, J., Ametani, A., Gole, A., and De Silva, J. "Closed-form earth-return impedance formulas for overhead and underground cables considering displacement currents." *IEE Japan High-Voltage Conference*, Tsushima, Japan, 2020.

[85] De Silva, J., Liu, J., Muthumuni, D., and Ametani, A. "EMTP simulation of very-fast transients in gas-insulated substations using modified Carson's formula." *IEEE Transactions on Electromagnetic Compatibility*. 2020; DOI: 10.1109/TEMC.2020.3036211

[86] Benato, R., Di Mario, C., and Koch, H. "High capability applications of long gas insulated lines in structures." *IEEE Transactions on Power Delivery*. 2007; 22(1): 619–26

[87] Kikuchi, A. H. "Wave propagation on the ground return circuit in high frequency regions." *Journal of IEE Japan*. 1955; 75: 1176–87

[88] Kikuchi, H. "Electro-magnetic field on infinite wire at high frequencies above plane-earth." *Journal of IEE Japan*. 1957; 77(825): 721–33

[89] Stratton, J. A. *Electromagnetic theory*. New York, NY: McGraw-Hill Book Co.; 1941.

[90] Sommerfeld, A. *Partial differential equations in physics*. New York, NY: AP; 1949.

[91] Goubau, G. "Surface wave and their application to transmission lines." *Journal of Applied Physics*. 1950; 21(11): 1119–28

[92] Nakagawa, M. "Admittance correction effects of a single overhead line." *IEEE Transactions on Power Apparatus and Systems*. 1981; 100(3): 1154–61

[93] Nakagawa, M. "Further studies on wave propagation along overhead transmission lines: effects of admittance correction." *IEEE Transactions on Power Apparatus and Systems*. 1981; 100(7): 3626–33

[94] Ametani, A., Miyamoto, Y., Baba, Y., and Nagaoka, N. "Wave propagation on an overhead multi-conductor in a high frequency region." *IEEE Transactions on Electromagnetic. Compatibility*. 2014; 56(6): 1638–48

[95] Xue, H., Ametani, A., Mahseredjian, J., Baba,Y. and Rachidi, F. "Frequency response of electric and magnetic fields on overhead conductors with particular reference to axial electric field." *IEEE Transactions on Electromagnetic Compatibility*. 2018; 60(6): 2029–32

[96] Carson, J. R. "Wave propagation in overhead wires with ground return." *The Bell System Technical Journal*. 1926; 5(4): 539–54

[97] Pollaczek, F. "Uber das feld einer unendlich langen wechselstrom durchlossenen einfachleitung." *Elektrische Nachrichtentechnik*. 1926; 3(9): 339–59

[98] Ametani, A., Miyamoto, Y., and Mahseredjian, J. "Derivation of an earth-return impedance of an overhead multi-conductor considering displacement currents." *IEE Japan Transactions on Power and Energy*. 2014; 134(11): 936–40

[99] Nakagawa, M., Ametani, A., and Iwamoto, K. "Further studies on wave propagation in overhead lines with earth return: Impedance of stratified earth." *Proceedings of the Institution of Electrical Engineers*. 1973; 120 (12): 1521–28

[100] Wise, W. H. "Propagation of high-frequency currents in ground return circuits." *Proceedings of the Institute of Radio Engineers*. 1934; 22(4): 522–27

[101] Wise, W. H. "Potential coefficients for ground return circuits." *The Bell System Technical Journal*. 1948; 27(2): 365–71

[102] Ametani, A., and Imanishi, K. "Development of exponential Fourier transform and its application to electrical transients." *Proceedings of the Institution of Electrical Engineers*. 1979; 126(1): 51–56

[103] Nagaoka, N., and Ametani, A. "A development of a generalized frequency-domain transient program-FTP." *IEEE Transactions on Power Delivery*. 1988; 3(4): 1996–04

[104] Mahseredjian, J., Dennetière, S., Dubé, L., Khodabakhchian, B., and Gérin-Lajoie, L. "On a new approach for the simulation of transients in power systems." *Electric Power Systems Research*. 2007; 77(11): 1514–20

[105] Kocar, I., Mahseredjian, J., and Olivier, G. "Weighting method for transient analysis of underground cables." *IEEE Transactions on Power Delivery*. 2008; 23(3): 1629–35

[106] Kocar, I., and Mahseredjian, J. "Accurate frequency dependent cable model for electromagnetic transients." *IEEE Transactions on Power Delivery*. 2016; 31(3): 1281–88

[107] De Silva, J., Gole, A. M., Nordstrom, J., and Wedepohl, L. M. "Robust passivity enforcement scheme for time-domain simulation of multi-conductor transmission lines and cables." *IEEE Transactions on Power Delivery*. 2008; 25(2): 930–38

[108] Ametani, A., Taki, N., Miyazaki, D., Nagaoka, N., and Okabe, S. "Lightning surges on a control cable incoming through a grounding lead." *IEE Japan Transactions on Power and Energy*. 2007; 127(1): 267–75

[109] Ametani, A., Nagaoka, N., Mori, N., and Shimizu, K. "Switching over-voltages on a pipe in a gas-insulated substation." *International Conference on Power Systems Transients*, Seattle, USA, 1997.

[110] CIGRE WG C4.501. *Guideline for Numerical Electromagnetic Analysis Method and Its Application to Surge Phenomena*. CIGRE Technical Brochure 543; 2013.

[111] Central Research Institute of Electric Power Industries. *Virtual Surge Test Lab (VSTL)*. http://criepi.denken.or.jp/; 2013.

[112] Sakaguchi, S., and Oyama, M. "Application of Maxwell solvers to PD propagation: Part III PD propagation in GIS." *IEEE Electrical Insulation Magazine*. 2003; 19(1): 6–12

[113] Ametani, A., Natsui, M., Xue, H., Mahseredjian, J., and Goto, T. "Effect of elbow part of gas-insulated bus on surge propagation in GIS." *IEE Japan High-Voltage Conference*, Montreal, Canada, 2018.

[114] Ametani, A., and Yamamoto, K. "Non-uniform lines – review of the theory and measured / simulation examples." *International Conference on Power Systems Transients*, Perpignan, France, 2019.

[115] Ametani, A. "Wave propagation on a non-uniform line and its impedance and admittance." *Science and Engineering Reviews Doshisha University*. 2002; 43(3): 136–47

[116] Ametani A., Nagaoka N., Baba Y., and Ohno T. *Power system transients: theory and applications*. New York, NY: CRC Press; 2016.

[117] Natsui M., De Silva J., and Ametani A. "Non-homogeneous line – problem associated with a circuit theory and EMTP." *IEE Japan Annual Conference of Power and Energy Society*, 2018.

Chapter 6

Transients in interconnected EHV cable network in Denmark

6.1 Background of EHV cable network in Denmark

In order to mitigate climate change, many countries have set a target to reduce their greenhouse gasses. Due to its abundant natural resources (wind) and green-minded people, Denmark is one of the leading countries in this effort of the world to tackle global warming. As of the year 2011, the wind energy accounted for 29.1% in generation capacity and 28.1% of the electrical production [1].

This change of the generation profile has led to and will continue to lead to the necessity of major upgrades in the Danish transmission grid. Especially, the expansion of Horns Rev, the world's largest offshore wind farm, will require the ability of the Danish transmission grid to transmit increased power from west to east where the electricity is consumed.

In Denmark, receiving public and political pressures to underground its OHLs, Energinet.dk published a report on the future expansion and undergrounding of its transmission grid on the 3rd of April 2008 [2]. The report proposed and compared five principles (A–E in Figure 6.1). From the five principles, the Danish government has selected Principle C in which all new 400 kV lines will basically be undergrounded as in Figure 6.2.

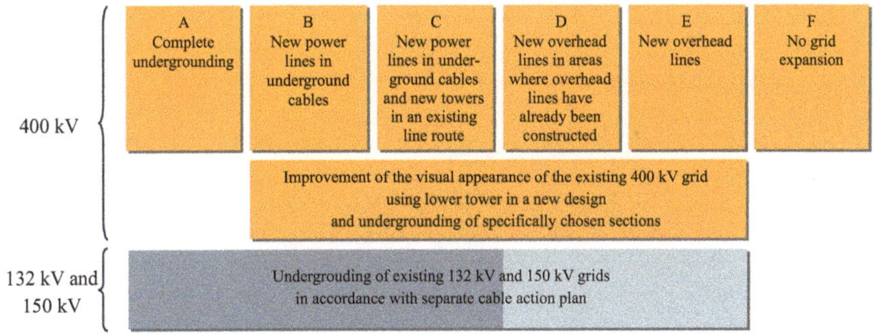

Figure 6.1 Five principles for the future grid expansion. Adapted from [2]

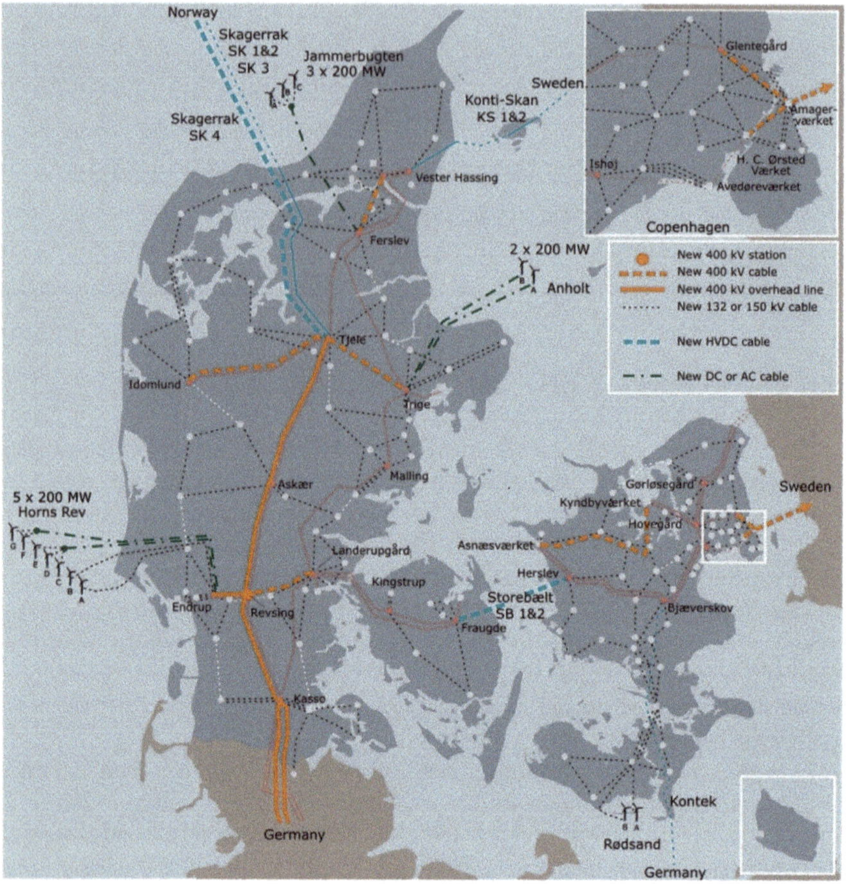

Figure 6.2 Grid expansion plan based on Principle C. Adapted from [2]

6.2 Model setup

6.2.1 Modeled area

The entire 400 kV network in Zealand is included in this model for the temporary overvoltage analysis and the slow-front overvoltage analysis. Figure 6.3 shows the modeled area of the 400 kV network.

Figure 6.4 shows an example of the simulation model created in ATP-Draw. Each component in the figure is explained in this section.

The simulation model was set up also in PSCAD. Figure 6.5 shows an example of the model created in PSCAD.

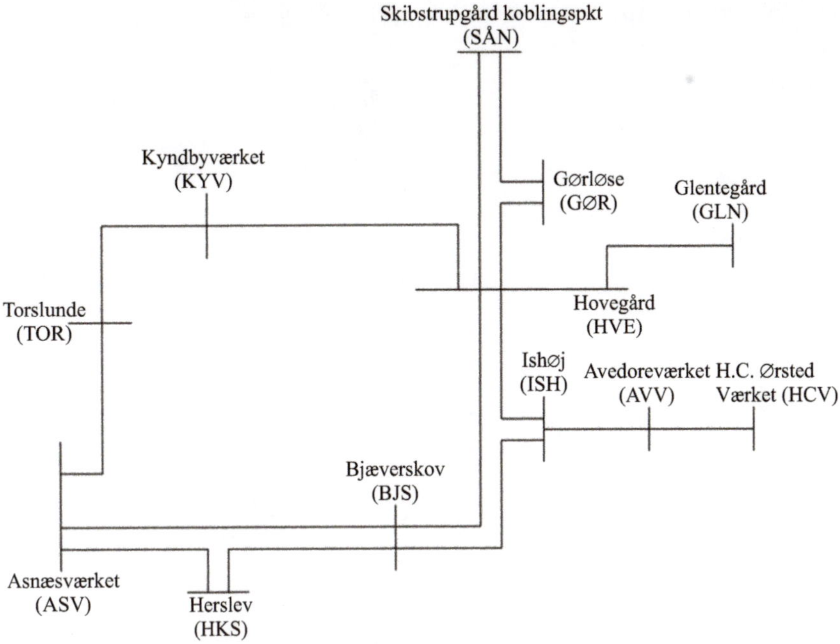

Figure 6.3 Modeled area of the 400 kV network

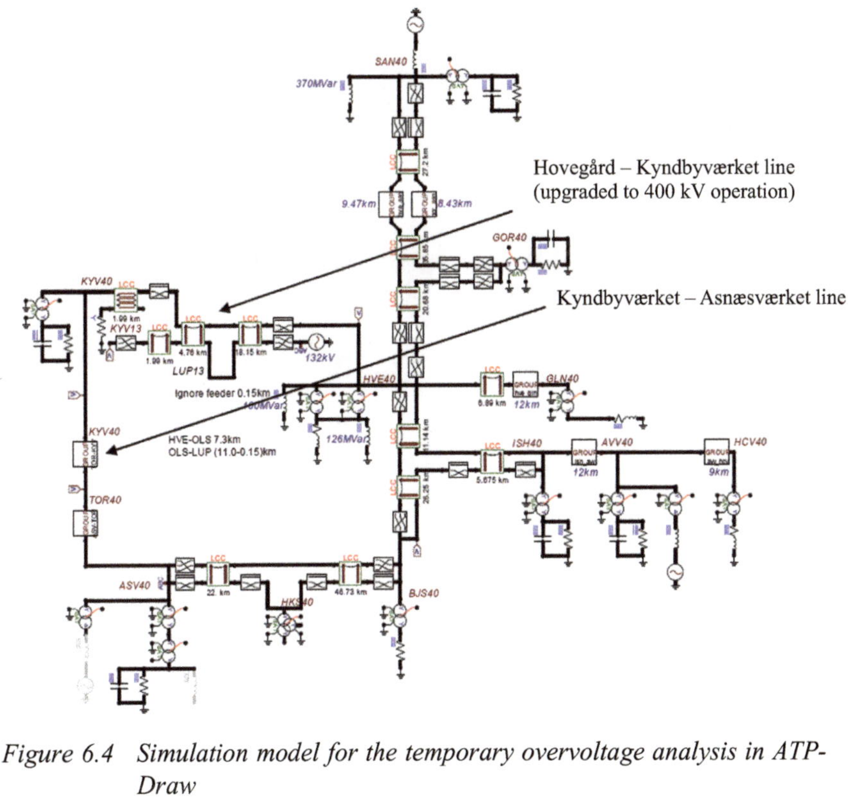

Figure 6.4 Simulation model for the temporary overvoltage analysis in ATP-Draw

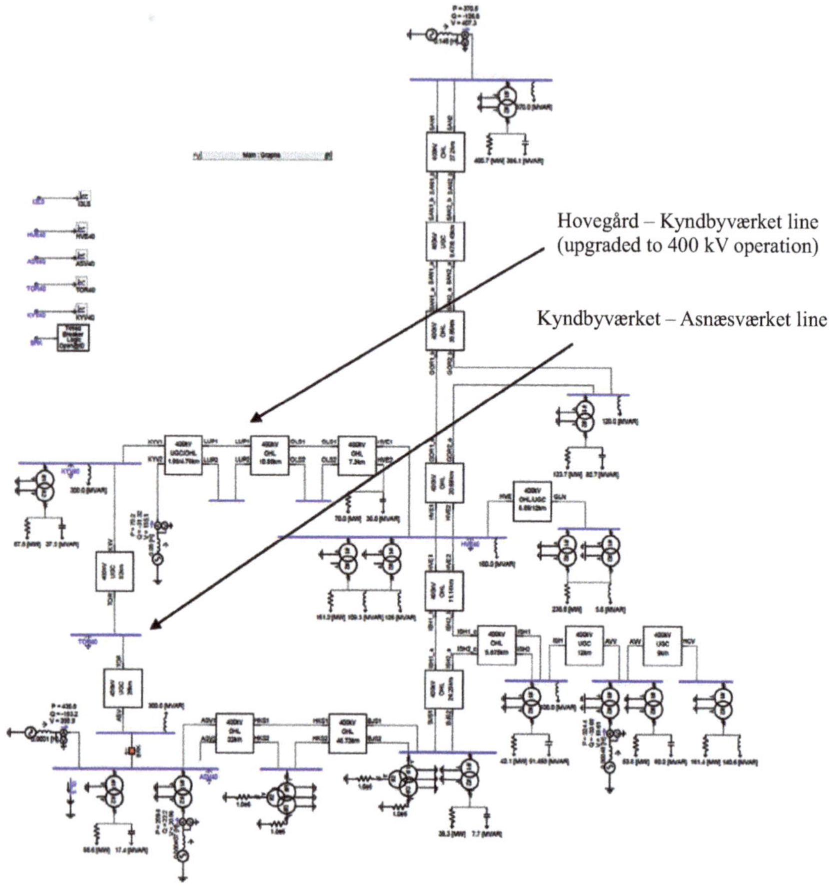

Figure 6.5 Simulation model for the temporary overvoltage and slow-front overvoltage analysis in PSCAD

6.2.2 Underground cable

6.2.2.1 Physical and electrical information

The simulation model includes the following four 400 kV cables:

- Al 1,600 mm^2 XLPE cable (Al sheath)
- Al 400 mm^2 + Cu 1,600 mm^2 XLPE cable (Pb sheath)
- Cu 1,000 mm^2 LPFF cable (Pb sheath + Al armor)
- Cu 1,000 mm^2 LPFF cable (Pb sheath)

Figure 6.6 shows the locations of these 400 kV cables in the network. The first cable type is for the Kyndbyværket–Asnæsværket line. Additionally, this type was applied to the short section of the Hovegård–Kyndbyværket line near the Kyndbyværket substation, where the overhead line is designed as 132 kV. This short section (1.99 km) has to

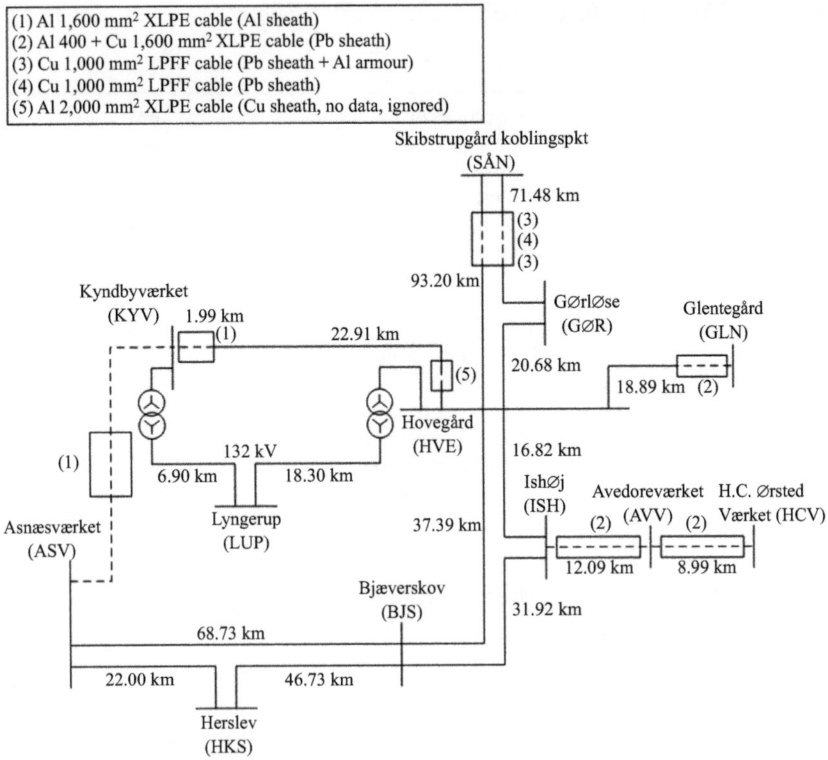

Figure 6.6 Four types of cables in the simulation model

be replaced by 400 kV cables when the Hovegård–Kyndbyværket line is operated at 400 kV.

6.2.2.1.1 Al 1,600 mm² XLPE cable (Al sheath): Kyndbyværket–Asnæsværket line

Physical and electrical information of this type of 400 kV cable is shown in Table 6.1. As the Kyndbyværket–Asnæsværket line has not been constructed yet, it is not possible to obtain exact parameters. Instead, the standard data was assumed based on ABB XLPE Underground Cable Systems – User's Guide (rev. 3) [3].

It is not possible to model semiconducting layers inside and outside the insulation in the subroutine CABLE CONSTANTS [4]. See the details in Section 2.1 of Chapter 2. Hence, semiconductive layers inside and outside the insulation were modeled as a part of the insulation since their resistivity is large enough in comparison to the conductor or metallic sheath [5].

Because of this modification, the insulation appears thicker in the ATP-Draw model than the actual thickness, and the charging capacity of the cable appears smaller than its actual value. In order to compensate for this error, the relative permittivity of the insulation was converted using the following equation [6]:

Table 6.1 Physical and electrical information for the 400 kV cable used at ASV–KYV (1)

Conductor	R1	Inner radius of the tubular core	mm	0
	R2	Outer radius of the tubular core	mm	26.0
	ρc	Resistivity	Ωm	2.84E−08
	μc	Relative permeability		1
Conductor screen	Rsi	Outer radius of conductor screen	mm	28.0
Insulation	μI1	Relative permeability		1
	εI1	Relative permittivity		2.852
	Rso	Outer radius of the insulation	mm	55.0
Metallic sheath	R3	Inner radius of the metallic sheath	mm	58.0
	R4	Outer radius of the metallic sheath	mm	59.2
	ρs1	Resistivity		2.84E−08
	μs1	Relative permeability		1
Outer covering	μI2	Relative permeability		1
	εI2	Relative permittivity		2.4
Completed cable	R5	Average outer radius of cover	mm	63.5

$$\varepsilon I1 = \frac{\ln(R3/R2)}{\ln(Rso/Rsi)}\varepsilon I0 = \frac{\ln(58.0/26.0)}{\ln(55.0/28.0)} \cdot 2.4 = 2.852 \tag{6.1}$$

The Kyndbyværket–Asnæsværket line has both a land part and a submarine part. The submarine part has an armor conductor outside of the metallic sheath. However, the armor was not included in the model mainly because the armor causes only a minor effect on the transient overvoltage, which will be discussed in Section 6.2.2.8.

6.2.2.1.2 *Al 400 mm^2 + Cu 1,600 mm^2 XLPE cable (Pb sheath)*

Physical and electrical information of this type of 400 kV cable is shown in Table 6.2.

Table 6.2 Physical and electrical information for the 400 kV cable used at ISH–AVV–HCV and HVE–GLN (2)

Conductor	R1	Inner radius of the tubular core	mm	0
	R2	Outer radius of the tubular core	mm	25.50
	ρc	Resistivity	Ωm	1.724E−08
	μc	Relative permeability		1
Conductor screen	Rsi	Outer radius of conductor screen	mm	28.80
Insulation	μI1	Relative permeability		1
	εI1	Relative permittivity		2.96
	Rso	Outer radius of the insulation	mm	57.80
Metallic sheath	R3	Inner radius of the metallic sheath	mm	60.25
	R4	Outer radius of the metallic sheath	mm	63.85
	ρs1	Resistivity		2.20E−08
	μs1	Relative permeability		1
Outer covering	μI2	Relative permeability		1
	εI2	Relative permittivity		2.4
Completed cable	R5	Average outer radius of cover	mm	69.60

The core of this cable has copper in the center and aluminum around it. Since CABLE CONSTANTS subroutine and PSCAD cannot deal with two different conductors without insulation between them, the core is assumed to be composed only of copper. Copper has lower resistance than aluminum, and thus this is a conservative assumption.

Physical and electrical information of the cable has been provided from NKT Cables through Energinet.dk. The provided data sheet is shown in Figure 6.7.

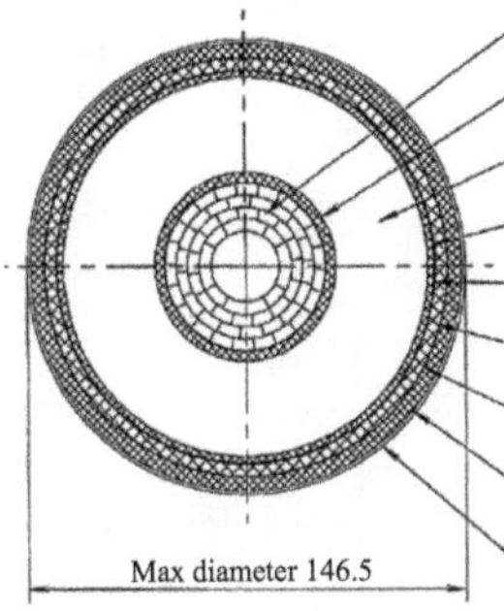

Opbygning:

Leder:	Kobberleder af profiltrade med kvældpulver og en massiv aluminiumkerne.
Leder skærm:	Ekstruderet halvledende materlale.
Isolation:	Ekstruderet PEX-Isolation.
Isolations skærm:	Ekstruderet halvledende materlale.
Båndbevikling:	Bevlkling af halvledende kvældbånd.
skærm:	Blykappe.
Båndbevikling:	Bevlkling af polyesterbånd.
Isolerende kappe:	Ekstruderet isolerende PE, rød.
Halvledende kappe:	Ekstruderet halvledende PE, sort.

Hovedstadsprojektet, 400 kV kabelsystem til GLN-MÅK and HVE

Mekaniske og elektriske data.

Ledertværsnit	mm^2	1600
Lederdiameter	mm	51.0
Diameter over halvledende tape på leder	mm	51.6
Diameter over ekstruderet indre halvleder	mm	57.6
Tykkelse af PEX isolation, nominel	mm	29.0
Diameter over ekstruderet isolation, nominel	mm	115.6
Diameter over ekstruderet ydre halvleder	mm	119.6
Diameter over halvledende kvældbånd, nominel	mm	120.5
Tykkelse af blykappe, nominel	mm	3.6
Diameter over blykappe, nominel	mm	127.7
Diameter over polyesterbånd, nominel	mm	128.2
Tykkelse af PE-kappe med ekstruderet halvledende lag	mm	5.5
Udvendig diameter, nominel	mm	139.2
Udvendig diameter, max	mm	146.5
Kabelvægt (omtrentlig)	kg/m	42
Mintste tilladte bøjeradius	mm	2100
Største tilladte træk i kablet	kN	80 *
Jævnstrømsmodstand af Cu pr. fase ved 20 °C	Ω/km	0.0113
Kapacitet pr. fase	μF/km	0.20
Tilladelig lederstrøm i 1 sekund ved begyndelsestemperatur 90 °C med sluttemperatur 250 °C	kA	228
Tilladelig skærmstrøm i 1 sekund med sluttemperatur 200 °C	kA	40

Figure 6.7 Provided data on Cable (2)

6.2.2.1.3 Cu 1,000 mm² LPFF cable (Pb sheath + Al armor)

Physical and electrical information of this type of 400 kV cable is shown in Table 6.3.

Physical and electrical information of the cable has been provided from Energinet.dk. The provided data sheet is shown in Figure 6.8.

Table 6.3 Physical and electrical information for the 400 kV cable used at HVE–SÅN and GØR–SÅN (3)

Conductor	R1	Inner radius of the tubular core	mm	8.65
	R2	Outer radius of the tubular core	mm	20.65
	ρc	Resistivity	Ωm	1.724E−08
	μc	Relative permeability		1
Conductor screen	Rsi	Outer radius of conductor screen	mm	21.25
Insulation	μI1	Relative permeability		1
	εI1	Relative permittivity		2.56
	Rso	Outer radius of the insulation	mm	45.25
Metallic sheath	R3	Inner radius of the metallic sheath	mm	46.20
	R4	Outer radius of the metallic sheath	mm	49.40
	ρs1	Resistivity		2.20E−07
	μs1	Relative permeability		1
Insulation	μI2	Relative permeability		1
	εI2	Relative permittivity		2.4
Armor	R5	Inner radius of the metallic sheath	mm	53.90
	R6	Outer radius of the metallic sheath	mm	61.90
	ρs1	Resistivity		2.84E−08
	μs1	Relative permeability		1
Outer covering	μI2	Relative permeability		1
	εI2	Relative permittivity		2.4
Completed cable	R7	Average outer radius of cover	mm	66.30

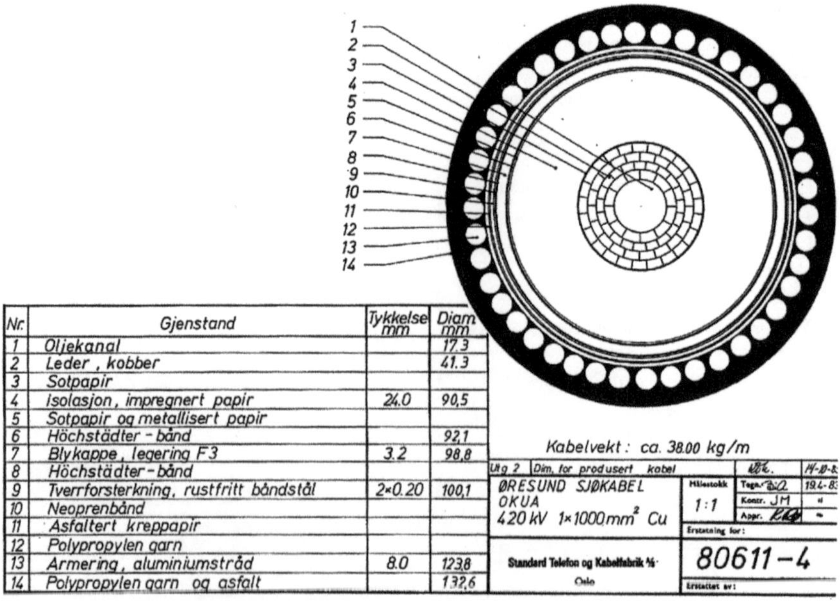

Nr.	Gjenstand	Tykkelse mm	Diam mm
1	Oljekanal		17.3
2	Leder , kobber		41.3
3	Sotpapir		
4	Isolasjon , impregnert papir	24.0	90,5
5	Sotpapir og metallisert papir		
6	Höchstädter - bånd		92,1
7	Blykappe , legering F3	3.2	98,8
8	Höchstädter - bånd		
9	Tverrforsterkning, rustfritt båndstål	2×0.20	100,1
10	Neoprenbånd		
11	Asfaltert kreppapir		
12	Polypropylen garn		
13	Armering , aluminiumstråd	8.0	123,8
14	Polypropylen garn og asfalt		132,6

Kabelvekt : ca. 38.00 kg/m

Utg 2 Dim. for produsert kabel

ØRESUND SJØKABEL
OKUA
420 kV 1×1000 mm² Cu

Målestokk 1:1

Tegn.
Kontr. JM
Appr.

Erstatning for:

Standard Telefon og Kabelfabrik A/S
Oslo

80611–4

Erstattet av:

Figure 6.8 Provided data on Cable (3)

6.2.2.1.4 Cu 1,000 mm² LPFF cable (Pb sheath)

Physical and electrical information of this type of 400 kV cable is shown in Table 6.4.

Physical and electrical information of the cable has been provided from Energinet.dk. The provided data sheet is shown in Figure 6.9.

Table 6.4 Physical and electrical information for the 400 kV cable used at HVE–SÅN and GØR–SÅN (4)

Conductor	R1	Inner radius of the tubular core	mm	8.65
	R2	Outer radius of the tubular core	mm	20.65
	ρc	Resistivity	Ωm	1.724E−08
	μc	Relative permeability		1
Conductor screen	Rsi	Outer radius of conductor screen	mm	21.05
Insulation	μI1	Relative permeability		1
	εI1	Relative permittivity		2.532
	Rso	Outer radius of the insulation	mm	45.25
Metallic sheath	R3	Inner radius of the metallic sheath	mm	46.20
	R4	Outer radius of the metallic sheath	mm	50.60
	ρs1	Resistivity		2.20E−08
	μs1	Relative permeability		1
Outer covering	μI2	Relative permeability		1
	εI2	Relative permittivity		2.4
Completed cable	R5	Average outer radius of cover	mm	55.75

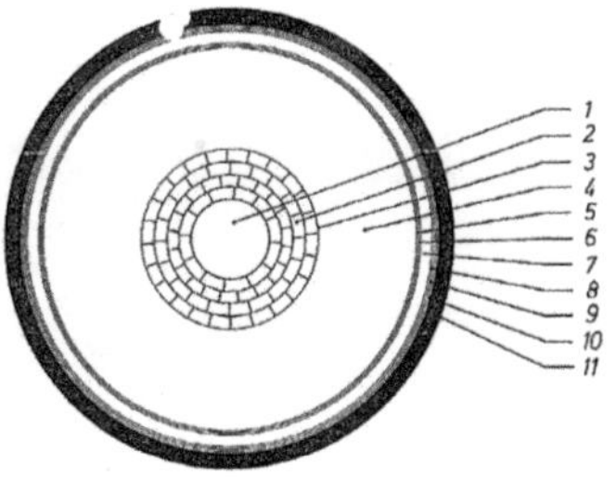

Nr.	Gjenstand	Tykkelse mm	Diam mm
1	Oljekanal		17,3
2	Leder, kobber		41.3
3	Sotpapir		
4	Isolasjon, impregnert papir	24,2	90,5
5	Sotpapir og metallisert papir		
6	Höchstädter bånd		92,1
7	Blykappe, F 3	4,4	101,2
8	Höchstädter bånd		
9	Tverrforsterkning	2×0.20	102,5
10	PP - bånd		102,8
11	Ytre kappe, polyetylen	4.0	111,5

Kabelvekt: ca 34 kg/m

② 10.10-85 Dim. for produsert kabel

OKEE
420 kV 1×1000 mm² Cu
Øresund jordkabel

Målestokk ./.
Kontr. J M
15.4-83

Erstatning for:

Standard Telefon og Kabelfabrik A/S
Oslo

67919−4

Erstattet av:

Figure 6.9 Provided data on Cable (4)

6.2.2.2 Cable layout

All 400 kV cables were assumed to be buried in a flat formation as shown in Figure 6.10. The buried depth of 1.3 m and the phase spacing of 0.3 m are assumed to be common for all four cable models.

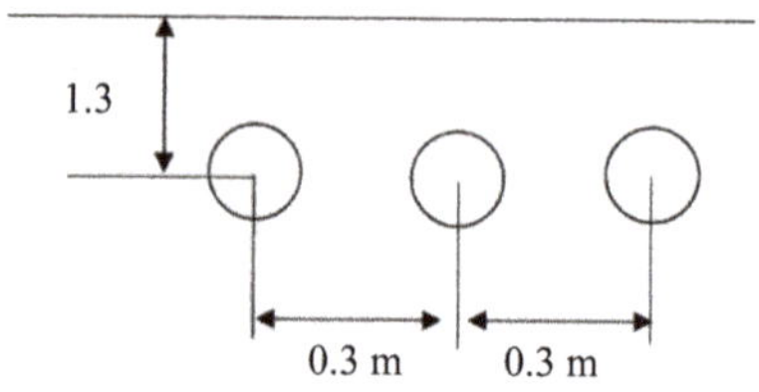

Figure 6.10 Cable layout

6.2.2.3 Cable route

The assumed route of the Kyndbyværket–Asnæsværket line is illustrated in Figure 6.11 by a red dotted line. The cable line is assumed to be composed of land and submarine sections. The length of each section is:

- Asnæsværket to Torslunde: 28 km (only land section)
- Torslunde to Kyndbyværket
    - Land section: 22 km
    - Submarine section: 10 km

Figure 6.11 Cable route

The cable model in the land section is cross-bonded, whereas the cable model in the submarine section is solidly bonded. The length of the minor section and the number of major/minor sections are shown in Table 6.5.

Table 6.5 Cross-bonding of the Kyndbyværket–Asnæsværket line

	Asnæsværket–Torslunde	**Torslunde–Kyndbyværket**
Length of minor section	1,867 m	1,833 m
Number of minor sections	15	12
Number of major sections	5	4

Finally, sheath bonding is modeled as shown in Figures 6.12 and 6.13. Each node between minor sections is named as AT1, AT2, …, AT15 for the Asnæsværket–Torslunde section and KT1, KT2, …, KT14 for the Torslunde–Kyndbyværket section. These node names are later used to specify fault locations or monitoring points for the overvoltage.

6.2.2.4 Modeling of auxiliary components [7]

The setup of cable models requires the modeling of cross-bonding wires, grounding wires, and other grounding conditions. The following settings are commonly applied for all 400 kV cables:

(a) Impedance of cross-bonding wires: 20 μH
  Typical length of the cross-bonding wire from joint to link box is about 10 m. The impedance was assumed to be 1 μH/m.
(b) Impedance of grounding wires: 10 μH
  Typical length of the grounding wire is about 10 m. The impedance was assumed to be 1 μH/m.
(c) Grounding resistance
  Substation: 1 ohm
  Normal Joint: 5 ohm
(d) Ground resistivity: 100 ohm-m

6.2.2.5 Effects of cable models

This section sets up the simulation model both in ATP-EMTP and PSCAD. Different cable models are adopted in ATP-EMTP and PSCAD. The comparison of the cable models is given in Table 6.6.

The resonance overvoltage analysis in Section 6.3 used the Bergeron model in ATP-EMTP since the analysis was done before the release of PSCAD X4.4. PSCAD, after X4.4, has an option to calculate the earth-return impedance by the numerical integration of Pollaczek's equation, which helps to resolve the numerical instability encountered in many cases. All transient analyses, except for the

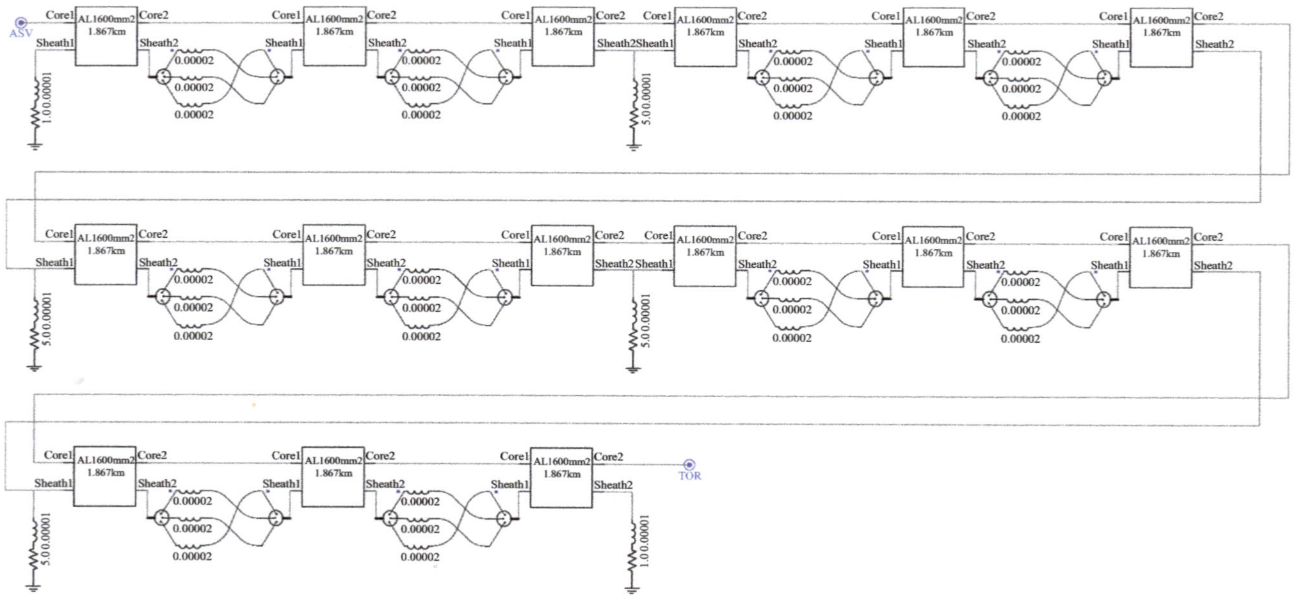

Figure 6.12 Modeling of cross-bonding (Asnæsværket–Torslunde)

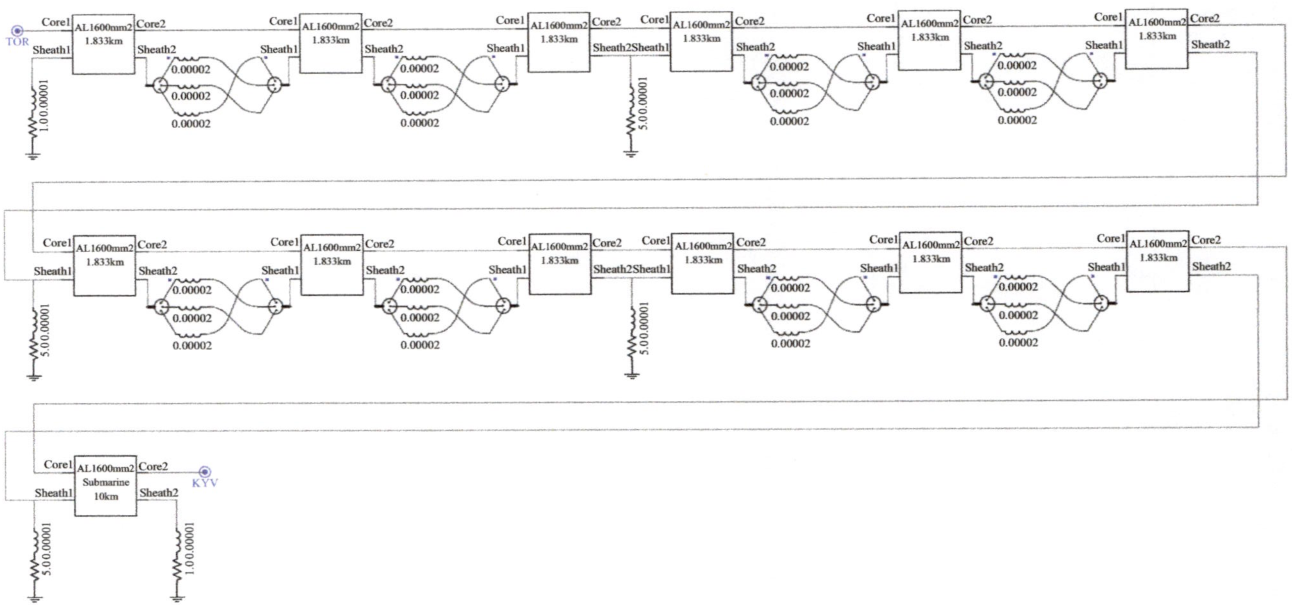

Figure 6.13 Modeling of cross-bonding (Torslunde–Kyndbyværket)

Table 6.6 Comparison of cable models in ATP-EMTP and PSCAD

	ATP-EMTP	PSCAD
Model name	Bergeron	Frequency dependent (phase)
Frequency dependent	Nonfrequency dependent	Frequency dependent
Model domain	Modal domain	Phase domain

resonance overvoltage analysis, were conducted after the release of PSCAD X4.4, and the frequency-dependent (phase) model in PSCAD was used for the analyses.

It is important for the setup of the Bergeron model (constant-parameter (CP) line model) to appropriately select the target frequency, at which the cable impedance and admittance are calculated. It is a common practice, especially for the temporary overvoltage analysis, to set the target frequency at 50 Hz as a severe assumption. The lower target frequency leads to a smaller resistance, which normally causes a higher overvoltage. In this chapter, the target frequency was first set at 50 Hz. After the dominant frequency in the overvoltage is known; however, the target frequency was changed to the dominant frequency in order to obtain more reasonable results.

Since different cable models are adopted, frequency responses of the cable impedances are obtained from both cable models in order to find the difference between these models. Figures 6.14 and 6.15 show the comparison of frequency responses of positive and zero sequence impedances of the Kyndbyværket–Asnæsværket line. The target frequency of the Bergeron model is set to 50 Hz. Considering the main focus of this chapter, frequency responses are obtained in the range 0–2,000 Hz at 10 Hz step.

For both positive and zero sequence impedances, natural frequencies of the cable match between two different models. The results indicate that natural frequencies of the network with the cable, approximately 450 Hz and 1,330 Hz, can be found by both cable models with an equal accuracy.

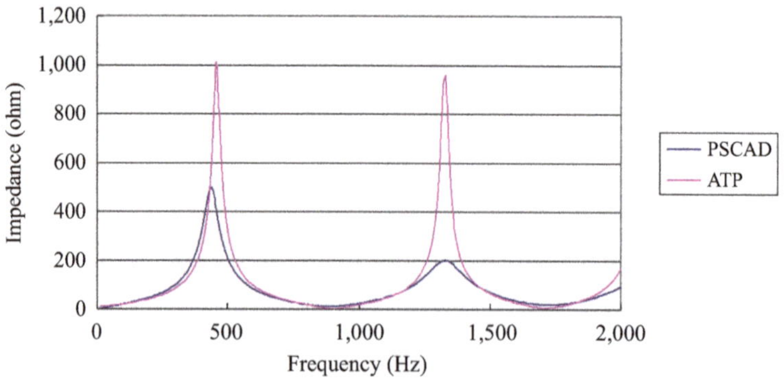

Figure 6.14 Frequency response of positive sequence impedance of the KYV–ASV line

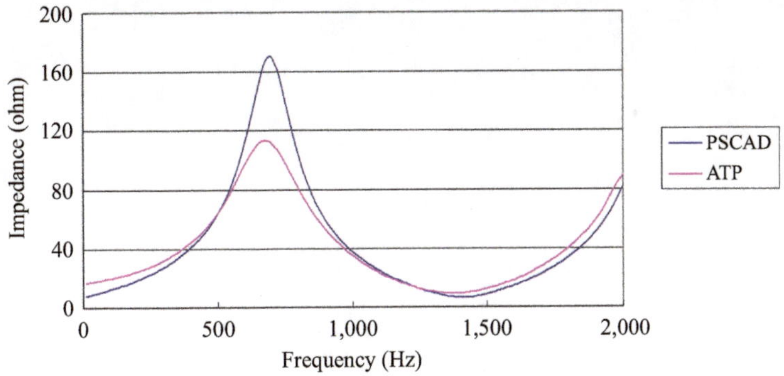

Figure 6.15 Frequency response of zero sequence impedance of the KYV–ASV line

The differences can be more clearly seen in Figures 6.16 and 6.17 where positive sequence resistance and reactance are separately derived. Both resistance and reactance have peaks and zeros at the same frequencies, but the magnitudes of the resistance and reactance at their peaks are approximately two times higher than in the Bergeron model in ATP-EMTP.

As a comparison in the time domain simulation, the Kyndbyværket–Asnæsværket line is energized from the 400 kV voltage source behind the source impedance of 147 mH, which corresponds to the fault current 5 kA. It is expected that the line energization overvoltages with the two cable models become similar because the discrepancies of impedances between the two cable models are observed only around the natural frequencies in Figures 6.14–6.17. Considering the length of the Kyndbyværket–Asnæsværket line, the dominant frequency in the overvoltages will be much lower than the natural frequencies 450 and 1330 Hz.

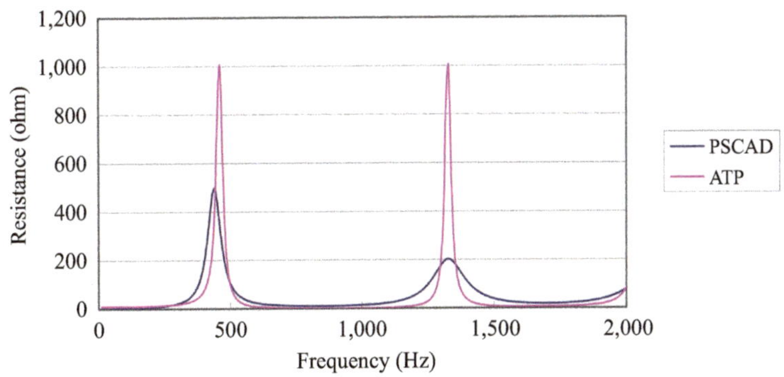

Figure 6.16 Frequency response of positive sequence resistance of the KYV–ASV line

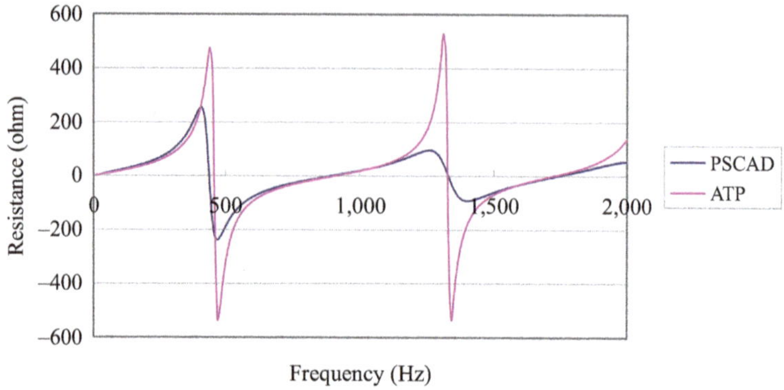

Figure 6.17 Frequency response of positive sequence reactance of the KYV–ASV line

Figures 6.18 and 6.19 respectively show the comparison of the line energization overvoltages at the sending end and the receiving end. The waveforms of the overvoltages are similar in these cases, but the magnitudes of the overvoltages include 5.9% deviation at the sending end and 7.9% deviation at the receiving end. It is estimated that the deviation is caused by the small difference in resistance.

Figure 6.20 shows frequency components contained in the line energization overvoltage. The frequency components are obtained by the Fourier transform of the line energization overvoltages in the time window 0–0.06 s. The figure demonstrates that the results match well between PACAD and ATP-EMTP, and the dominant (transient) frequency is around 130 Hz for both cases.

Since the dominant frequency was found to be around 130 Hz, the target frequency of the Bergeron model was modified to 130 Hz. Figures 6.21 and 6.22 show the comparison of the line energization overvoltages at the sending end and the

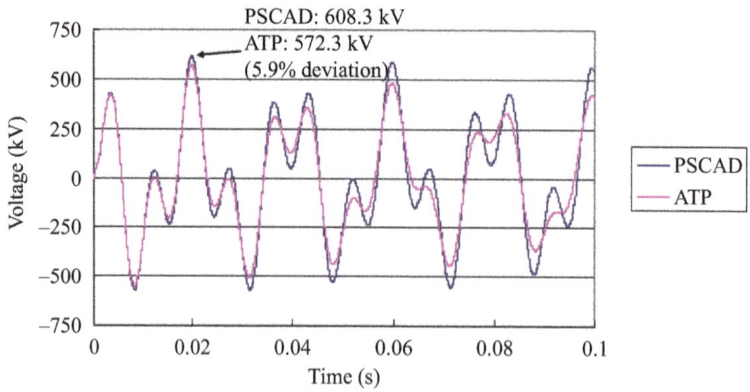

Figure 6.18 Overvoltage caused by the energization of the KYV–ASV line (sending end)

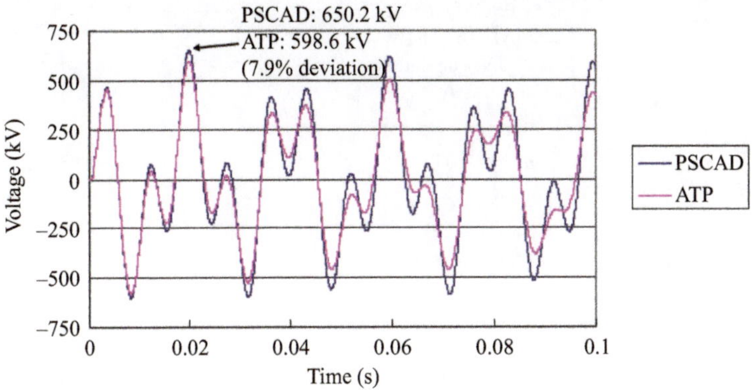

Figure 6.19 Overvoltage caused by the energization of the KYV–ASV line (receiving end)

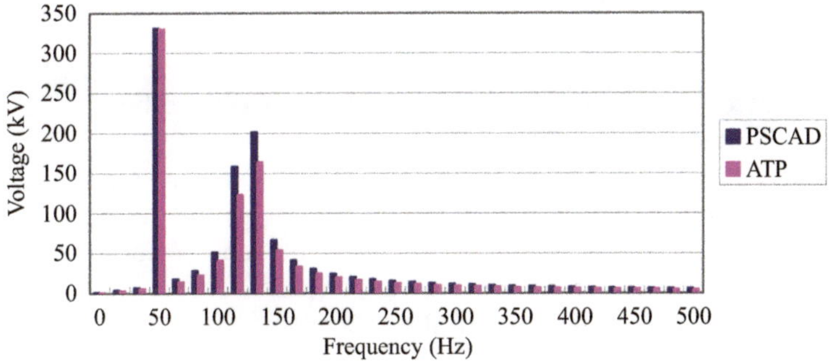

Figure 6.20 Frequency components contained in the overvoltage caused by the energization of the KYV–ASV line (sending end)

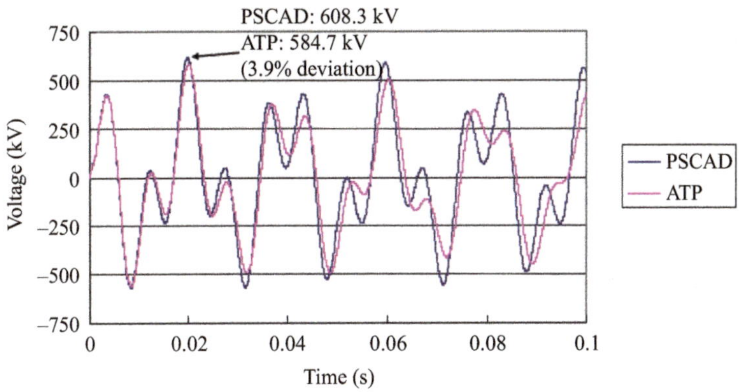

Figure 6.21 Overvoltage caused by the energization of the KYV–ASV line (sending end, target frequency: 130 Hz)

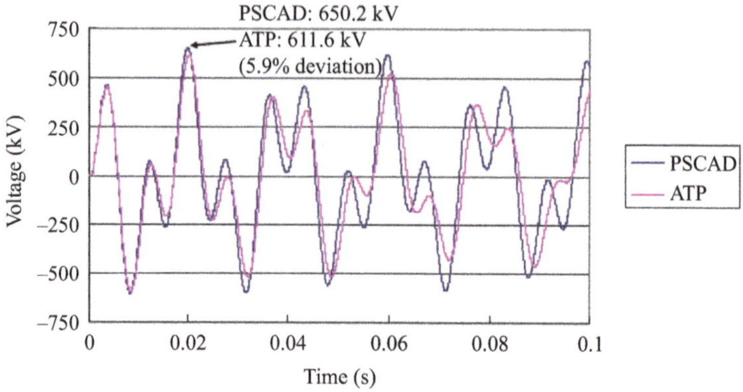

Figure 6.22 Overvoltage caused by the energization of the KYV–ASV line (receiving end, target frequency: 130 Hz)

receiving end, with the modified Bergeron model. With the lower target frequency, the magnitude of the overvoltages becomes higher, and the differences from the results in PSCAD become smaller.

From the results, the Bergeron model produces a similar overvoltage to the frequency-dependent (phase) model during the first 20 ms after the energization. In finding the maximum overvoltage, the choice of cable models between the Bergeron model and the frequency-dependent (phase) model does not have a noticeable impact. On the other hand, the overvoltages after the first 20 ms following energization have notable differences. It may require the use of a frequency dependent (phase) model when matching simulation results to field measurements, although it is not included in the scope of this chapter.

6.2.2.6 Effects of cross-bonding

The sheath cross-bonding has to be modeled as it changes the cable impedances. The analytical formulas and calculated results show that the cable inductances become larger due to the cross-bonding [8]. Because of the larger cable inductance, the propagation velocity is slower for cross-bonded cables [9].

Figure 6.23 shows the energization overvoltage when the Kyndbyværket–Asnæsværket line is energized from the Asnæsværket side. The black and red lines respectively show the overvoltage when the cable line is cross-bonded and solidly bonded. The figure demonstrates that the propagation velocity is slower in the cross-bonded cable case.

6.2.2.7 Effects of span length

It is a common practice to reduce the number of cable joints as much as possible in order to reduce the cost and to shorten the construction duration. However, the cable span length (minor length) is normally limited by restrictions in the transportation or by the sheath voltage. The restrictions in the transportation mainly include:

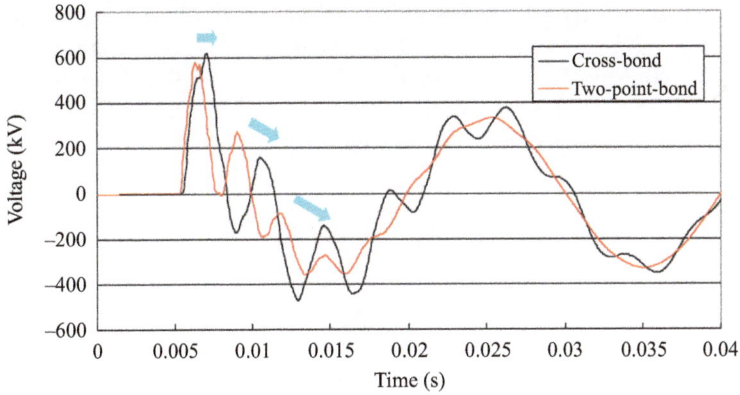

Figure 6.23 Effect of cross-bonding

- Number of turns cables can be wound around a drum
- Weight and height restrictions when a drum car has to go by road or cross bridges

The restriction in the sheath voltage differs from country to country and utility to utility. If the restriction comes from a utility's own safety codes, it may be possible to relax the restriction by applying more strict work codes, for example, to wear specific gloves.

Table 6.5 explains that this chapter assumes the cable span length to be around 1800–1900 m, but this may be a difficult goal to achieve depending on the above-mentioned restrictions. However, when numerous numbers of simulations are performed for statistical switching studies, it is convenient to have larger cable span lengths in order to reduce the computational burden.

This section studies the effects of assuming a longer cable span length. The study was conducted with the Kyndbyværket–Asnæsværket line, but the submarine part was removed from the line. The sensitivity analysis considers the following parameters:

- Cable span length: 2, 4, 6, 8, and 16 km
- Total cable length: 24, 48, 72, 96 km

For Figures 6.24–6.28, the highest overvoltages are compared for cable energization with different cable span lengths. When the cable span length is increased to 4, 6, and 8 km, the errors are below 5%. When the cable span length is further increased to 16 km as in Figure 6.28, the error goes up to 9%.

6.2.2.8 Effects of armor

The Kyndbyværket–Asnæsværket line has both the land part and the submarine part. The submarine part has an armor conductor outside of the metallic sheath, but the armor was not included in the simulation models of this chapter. This section studies the effects of armor when it is included in the simulation model.

As in the land part of the Kyndbyværket–Asnæsværket line, it was not possible to obtain the exact parameters. The standard data were derived from ABB XLPE Submarine Cable Systems, Attachment to XLPE Land Cable Systems – User's Guide (rev. 5) [10] as shown in Table 6.7. R5 and R6 were not available and were determined from a reasonable engineering estimation.

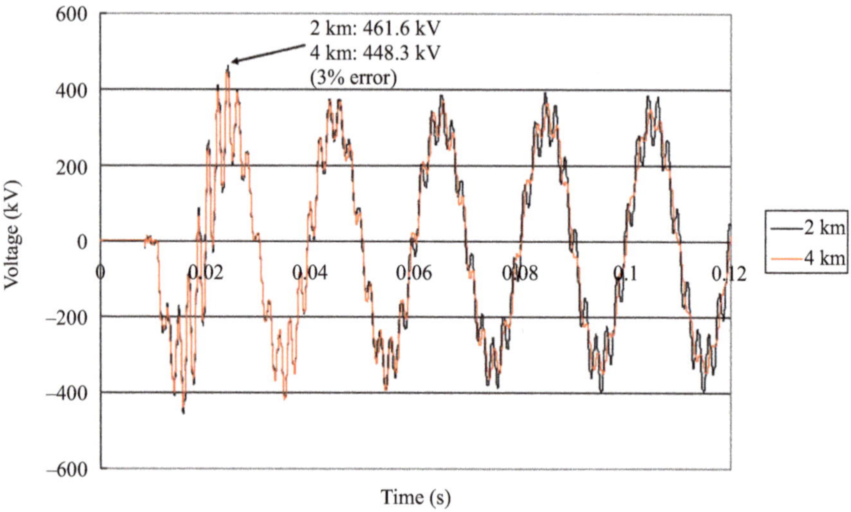

Figure 6.24 Effect of cable span length for 24 km cable line (2 km → 4 km)

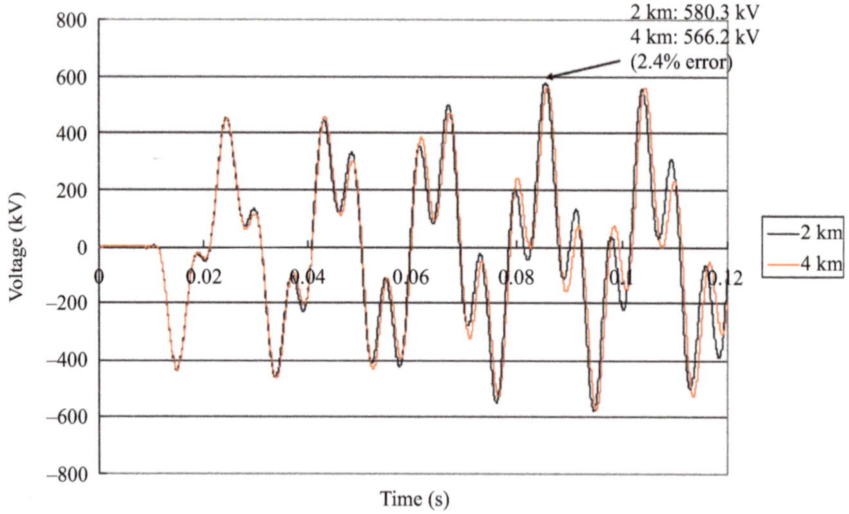

Figure 6.25 Effect of cable span length for 48 km cable line (2 km → 4 km)

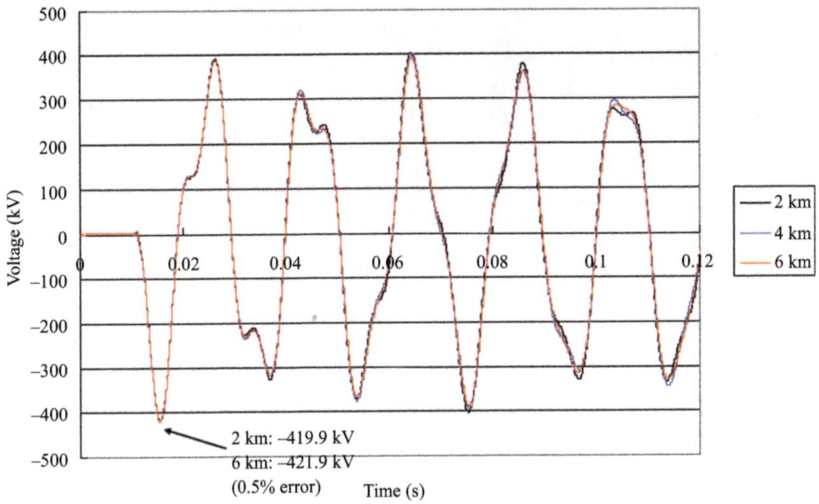

Figure 6.26 Effect of cable span length for 72 km cable line (2 km → 6 km)

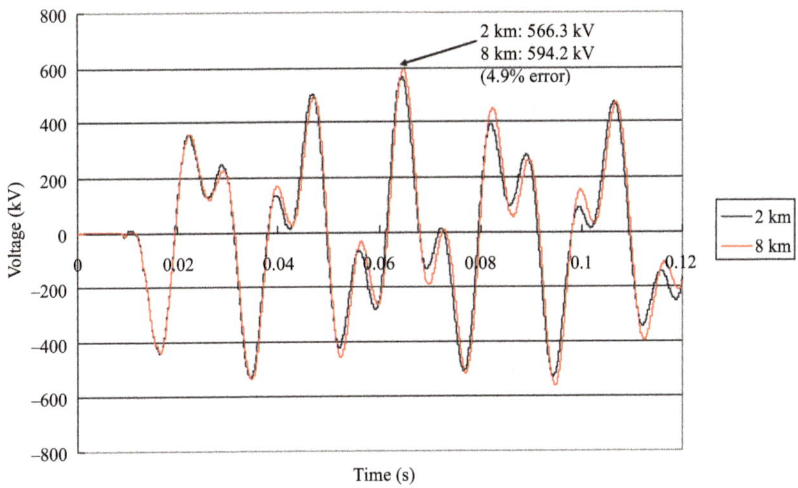

Figure 6.27 Effect of cable span length for 96 km cable line (2 km → 8 km)

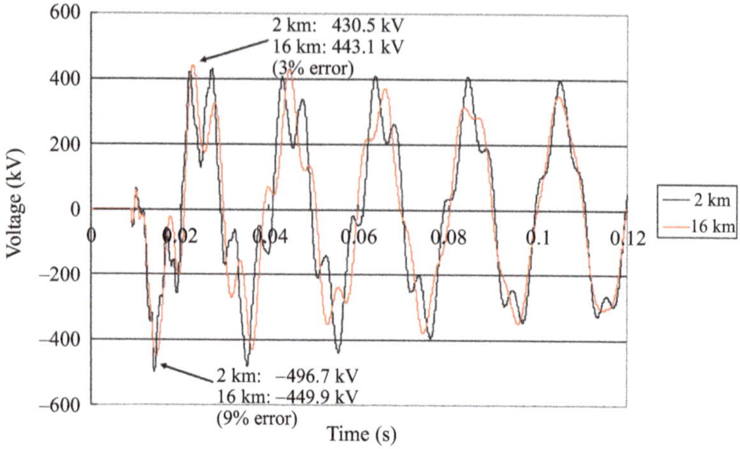

Figure 6.28 Effect of cable span length for 24 km cable line (2 km → 16 km)

Table 6.7 Physical and electrical information of cable (1), submarine part

Conductor	R1	Inner radius of the tubular core	mm	0
	R2	Outer radius of the tubular core	mm	23.7
	ρc	Resistivity	Ωm	2.84E−08
	μc	Relative permeability		1
Conductor screen	Rsi	Outer radius of conductor screen	mm	25.7
Insulation	μI1	Relative permeability		1
	εI1	Relative permittivity		2.853
	Rso	Outer radius of the insulation	mm	52.9
Metallic sheath	R3	Inner radius of the metallic sheath	mm	55.9
	R4	Outer radius of the metallic sheath	mm	59.0
	ρs1	Resistivity		2.20E−07
	μs1	Relative permeability		1
Insulation	μI2	Relative permeability		1
	εI2	Relative permittivity		2.4
Armor	R5	Inner radius of the metallic sheath	mm	63.5
	R6	Outer radius of the metallic sheath	mm	67.5
	ρs1	Resistivity		2.84E−08
	μs1	Relative permeability		1
Outer covering	μI2	Relative permeability		1
	εI2	Relative permittivity		2.4
Completed cable	R7	Average outer radius of cover	mm	70.5

In order to find the effects of armor on the energization overvoltages, the Kyndbyværket–Asnæsværket line was energized from the 400 kV voltage source behind the source impedance of 147 mH, which corresponds to the fault current 5 kA. Figures 6.29 and 6.30 respectively show the comparison of the

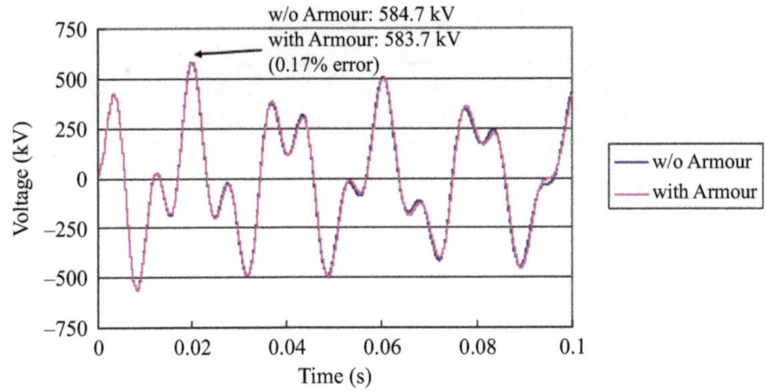

Figure 6.29 Effects of armor on the energization overvoltage (sending end)

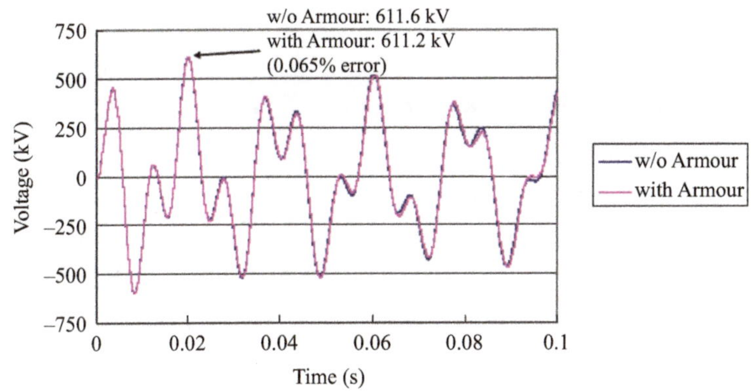

Figure 6.30 Effects of armor on the energization overvoltage (receiving end)

energization overvoltages at the sending end and the receiving end. The waveforms demonstrate that the armor does not have noticeable effects on the energization overvoltages.

6.2.3 Overhead line

6.2.3.1 Conductor and tower configuration

400 kV overhead lines were modeled using a frequency-dependent model (J. Marti model). Input data for ATP-EMTP are shown in Table 6.8, and a cross-section diagram of transmission lines and ground wires are shown in Figure 6.31.

Figure 6.32 shows the typical tower configuration for the 400 kV overhead lines. The layout of transmission lines and ground wires were derived from the typical tower configuration as shown in Table 6.9. This standard layout was applied to all 400 kV overhead lines.

Table 6.8 Overhead line data

Types	Inner radius (mm)	Outer radius (mm)	Resistance per conductor (ohm/km)	Number of conductors in a bundle	Separation between conductors (cm)
Transmission line (Martin)	6.025	18.085	0.04259	2	40
Ground wire	4.65	8.10	0.208	–	–

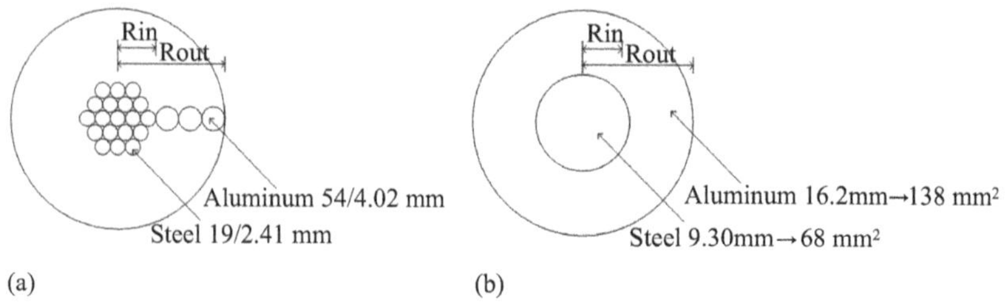

Figure 6.31 Cross-section diagram of transmission line (left) and ground wire (right)

6.2.3.2 Phase configuration

The phase configuration of 400 kV overhead lines is also provided from Energinet.dk. Figure 6.33 shows the phase configuration of the existing 400 kV overhead lines.

As for the Hovegård–Kyndbyværket line, it was assumed that the line has the same phase configuration after the line is upgraded to the 400 kV operation.

6.2.3.3 Comparison between PSCAD and ATP-EMTP [7]

In PSCAD, it is not possible to model a hollow ground wire as shown in Table 6.8 since the inner radius is always considered to be zero. To keep the satisfactory accuracy of the PSCAD model for an overhead line, the sequence impedance of the double circuit tower line between Asnæsværket and Herslev is calculated in PSCAD and ATP-EMTP. It is assumed that the Asnæsværket–Bjæverskov line is cut at Herslev for the convenience in measurement.

Figures 6.34 and 6.35 show a comparison of sequence impedances. A frequency-dependent phase model is used in PSCAD. The comparison shows the error in the model by PSCAD is negligible as it is focused on the frequency component less than 1 kHz.

6.2.4 Models representing apparatuses in the network

6.2.4.1 Transformers

6.2.4.1.1 400/132 kV transformers

Model parameters of all 400/132 kV transformers were derived from the PSS/E data provided by Energinet.dk. Since all the 400/132 kV transformers except the

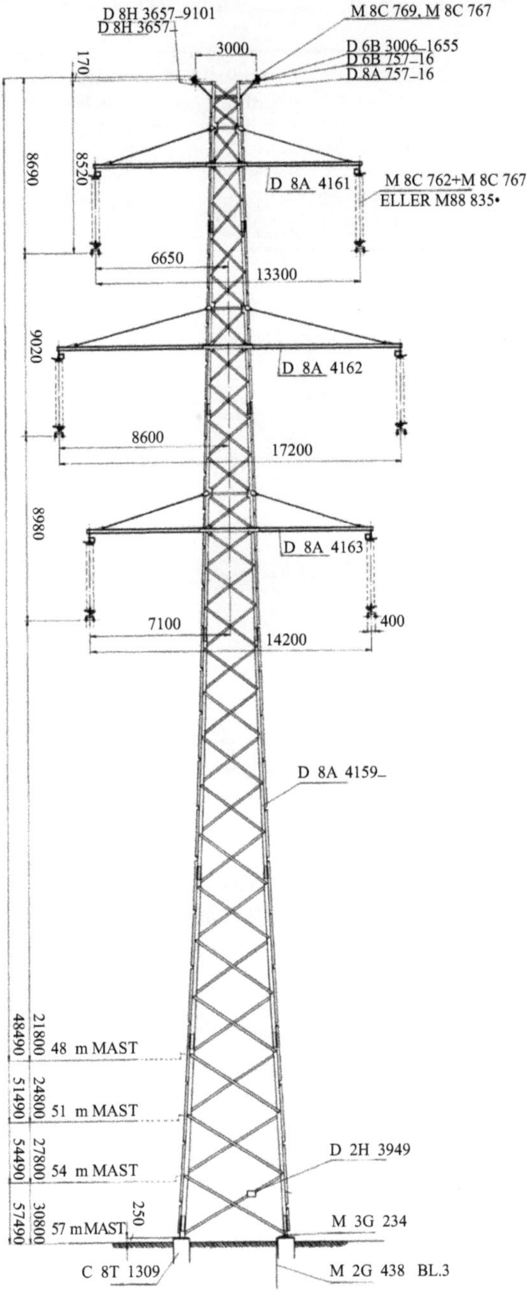

Figure 6.32 Tower configuration of standard 400 kV OHLs

Table 6.9 Layout of standard 400 kV OHLs

Types	Horizontal position (m)	Vertical position (m)
Transmission lines	±6.65	34.0
	±8.60	26.0
	±7.10	20.0
Ground wires	±3.00	41.0

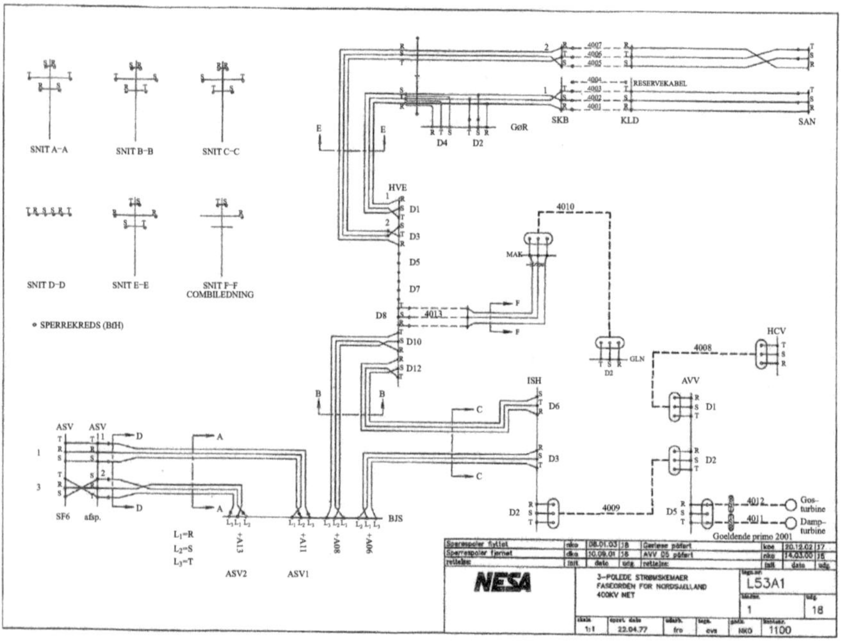

Figure 6.33 Phase configuration of the existing 400 kV OHLs

one in GLN are modeled as two winding transformers in the PSS/E data, only the impedance between the primary and secondary windings is available. In the EMT simulation data, the impedance was split half and half in pu to the primary and to the secondary. The parameters are given in Table 6.10.

The saturation characteristic of 400/132 kV transformers was not available from Energinet.dk. Instead, a severe saturation characteristic within a reasonable extent, which is shown in Table 6.11 and Figure 6.36, has been assumed for the analysis. Here, Phi is the flux linkage. The saturation characteristic was set only in transformer inrush simulations.

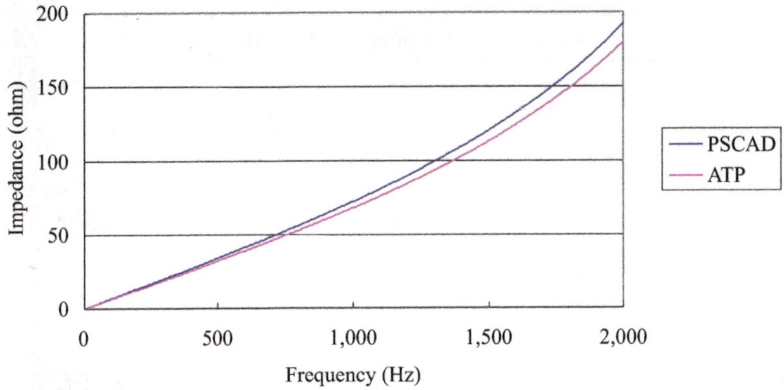

Figure 6.34 Comparison of the positive sequence impedance of 400 kV overhead line

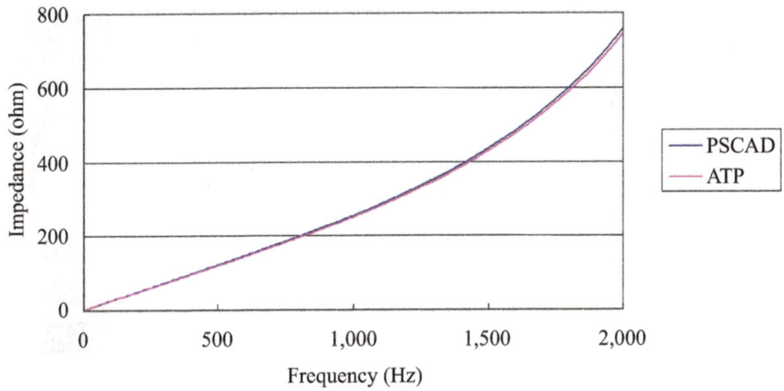

Figure 6.35 Comparison of the zero-sequence impedance of 400 kV overhead line

In the simulations of transformer energizations, it is necessary to include the hysteresis characteristic of a transformer. In ATP-EMTP, the hysteresis characteristic is obtained using the subroutine HYSDAT. Assuming the positive saturation point at 1.4 pu, the hysteresis characteristic in Table 6.12 is obtained. The positive saturation point is, in definition, the point in the first quadrant where the hysteresis loop changes from being multivalued to being single-valued.

The positive saturation point at 1.4 pu could be too severe. In order to perform a sensitivity analysis, another hysteresis characteristic is derived assuming the positive saturation point at 1.3 pu as shown in Table 6.13. The inrush current will

Table 6.10 Model parameters of 400 kV transformers

	R1 (ohm)	R2 (ohm)	X1 (mH)	X2 (mH)
BJS	0.968	0.1054	191.9	20.90
ISH	0.528	0.0575	138.1	15.04
AVV	0.560	0.0610	122.9	13.39
HCV	0.416	0.0453	94.7	10.32
HVE	0.904	0.0984	165.5	18.02
	0.512	0.0558	130.2	14.18
GLN	0.432	0.0470	98.4	10.72
	0.432	0.0470	98.4	10.72
GØR	0.312	0.0340	77.6	8.45
SÅN	0.296	0.0322	65.6	7.14

Table 6.11 Applied saturation characteristic of 400 kV transformers

I (A, peak)	Phi (Wb-T, peak)
0	0
0.05	600.00
1.01	1,200.00
150.00	1,350.00
400.00	1,450.00
650.00	1,530.00
1,250.00	1,700.00

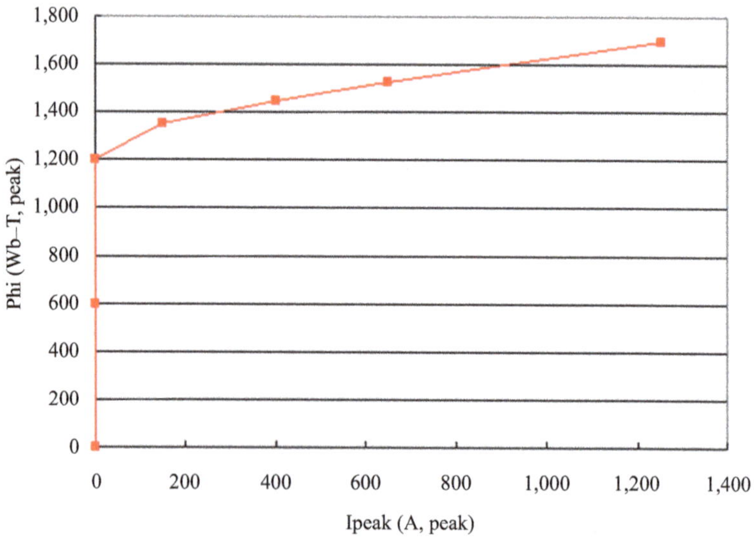

Figure 6.36 Applied saturation characteristic of 400 kV transformers

be reduced when this characteristic is adopted instead of the characteristic in Table 6.12.

6.2.4.1.2 Generator step-up transformers

Model parameters of all the generator step-up transformers were also derived from the PSS/E data provided by Energinet.dk. Table 6.14 summarizes the model parameters of the generator step-up transformers.

6.2.4.1.3 Phase shifting transformer

Model parameters of a phase-shifting transformer in Asnæsværket were obtained from Energinet.dk's document *Modelling of tap changers of the ASNÆSVÆRKETS 400/132 kV TRANSFORMER using the DIgSILENT PowerFactory software* [11].

Table 6.12 Hysteresis characteristic of 400 kV transformers (positive saturation at 1.4 pu)

I (A, peak)	Phi (Wb-T, peak)
−100.000	−1,407.353
−12.500	−1,333.147
7.500	−1,228.235
17.500	−955.294
33.750	827.353
52.500	1,091.765
90.000	1,253.824
166.250	1,364.706
400.000	1,450.000
550.000	1,458.529

Table 6.13 Hysteresis characteristic of 400 kV transformers (positive saturation at 1.3 pu)

I (A, peak)	Phi (Wb-T, peak)
−37.500	−1,310.294
−4.688	−1,241.206
2.813	−1,143.529
6.563	−889.412
12.656	770.294
19.688	1,016.471
33.750	1,167.353
62.344	1,270.588
150.000	1,350.000
206.250	1,357.941

The document contains Tables 6.15 and 6.16 for the phase shifter and the voltage amplitude tap changer.

The tap positions were selected according to the PSS/E power flow data. The tap positions were set at Tap No. 2 for the phase shifter and at Tap No. 15 for the voltage amplitude tap changer, respectively.

Table 6.14 Model parameters of generator step-up transformers

	HV Side	LV Side	R1 (ohm)	R2 (ohm)	X1 (mH)	X2 (mH)
ASV	400 kV	21.0 kV	0.160	0.00044	47.2	0.130
AVV	400 kV	19.5 kV	0.172	0.00041	58.4	0.139
	400 kV	11.5 kV	1.511	0.00125	396.1	0.327
			1.511	0.00125	396.1	0.327
	HV Side	**LV Side**	**R1 (ohm)**	**R2 (ohm)**	**X1 (mH)**	**X2 (mH)**
ASV	132 kV	12.0 kV	0.091	0.0008	25.6	0.211
KYV	132 kV	10.0 kV	2.108	0.0121	134.2	0.770
	132 kV	18.0 kV	0.068	0.0013	13.1	0.244
			0.061	0.0011	12.3	0.229

Table 6.15 Data of phase shifter given at each tap position

Trin Nr.	UHv (kV)	Rt at 500 (%)	Xt at 500 (%)	Loading (%)	Vinkel (°)
1	399.97	0.000	0.000	100	0.00
2	395.73	0.004	0.013	100	1.07
3	391.36	0.009	0.027	100	2.12
4	387.11	0.013	0.040	100	3.15
5	381.91	0.017	0.053	100	4.15
6	378.19	0.022	0.067	100	5.13
7	371.99	0.026	0.080	100	6.08
8	367.70	0.031	0.095	100	7.02
9	371.38	0.031	0.095	100	8.09
10	374.39	0.031	0.095	100	9.18
11	376.53	0.031	0.095	100	10.28
12	378.49	0.031	0.095	100	11.40
13	381.44	0.031	0.095	100	12.53
14	384.12	0.031	0.095	100	13.68
15	386.22	0.031	0.095	100	14.84
16	388.25	0.031	0.095	100	16.02
17	390.05	0.031	0.095	100	17.20
18	391.54	0.031	0.095	100	18.40
19	392.70	0.031	0.095	100	19.61
20	395.34	0.031	0.095	100	20.83
21	396.50	0.031	0.095	100	22.05

Table 6.16 Data of voltage amplitude tap changer given at each tap position

Trin Nr.	UHv (kV)	Rt at 500 (%)	Xt at 500 (%)	Loading(%)	Vinkel (°)
1	447.23	0.197	15,399	100	0
2	443.62	0.197	15,329	100	0
3	440.01	0.196	15,259	100	0
4	436.40	0.196	15,189	100	0
5	432.79	0.196	15,119	100	0
6	429.18	0.195	15,049	100	0
7	425.57	0.195	14,979	100	0
8	421.96	0.194	14,919	100	0
9	418.35	0.194	14,849	100	0
10	414.74	0.194	14,779	100	0
11	411.13	0.193	14,709	100	0
12	407.52	0.193	14,639	100	0
13	403.91	0.193	14,569	100	0
14	399.97	0.192	14,499	100	0
15	396.69	0.193	14,449	100	0
16	393.08	0.194	14,389	100	0
17	389.47	0.195	14,339	100	0
18	385.86	0.196	14,279	100	0
19	382.25	0.197	14,229	100	0
20	378.64	0.198	14,179	100	0
21	375.03	0.199	14,119	100	0
22	371.42	0.200	14,069	100	0
23	367.81	0.201	14,019	100	0
24	364.20	0.202	13,959	100	0
25	360.59	0.203	13,909	100	0
26	356.98	0.204	13,849	100	0
27	353.37	0.205	13,798	100	0

6.2.4.2 Shunt reactors

Power flow analysis concluded that one unit of a 400 kV 300 MVar shunt reactor should be installed at Kyndbyværket and Asnæsværket, respectively, which results in the compensation rate of 100.7%. These shunt reactors should be directly connected to the Kyndbyværket–Asnæsværket line, but not to the buses.

In most analyses, the shunt reactors were modeled as a linear inductance as it yields higher overvoltages. When it is necessary, the shunt reactors are modeled with their saturation characteristics. These saturation characteristics were provided by Energinet.dk, but their capacities were 70 and 140 MVar. Assuming that the saturation characteristic of a 300 MVar shunt reactor was equal to that of a 140 MVar shunt reactor, the saturation characteristic was derived as shown in Table 6.17 and Figure 6.37.

Mutual inductances between phases were not considered since 300 MVar shunt reactors would have to be a single-phase shunt reactor.

Table 6.17 Saturation characteristic of 400 kV 300 MVar shunt reactors

I (A, peak)	Phi (Wb-T, peak)
0	0
116.64	218.32
233.28	436.63
349.93	654.95
466.57	873.26
583.21	1,091.58
699.85	1,298.97
816.50	1,506.37
933.14	1,702.86
1,049.78	1,877.51
1,166.42	2,052.16
1,283.07	2,161.32
1,399.71	2,292.31
1,516.35	2,412.38

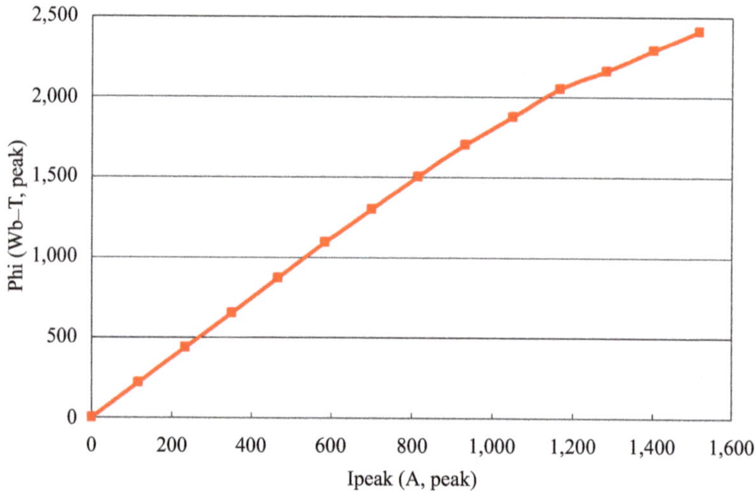

Figure 6.37 Saturation characteristic of 400 kV 300 MVar shunt reactors

6.2.4.3 Surge arresters

The V-I characteristic of existing 400 kV surge arresters was provided by Energinet.dk as shown in Table 6.18 and Figure 6.38. The surge arrester model was built by a nonlinear branch (Type 92 in ATP-EMTP).

Line discharge class 3 was assumed for the existing surge arresters. The energy absorption capability of these surge arresters are given as follows:

Table 6.18 V-I characteristic of surge arresters

I (A)	V (V)
1.0E−03	4.81E+05
1.0E−02	5.09E+05
1.0E−01	5.44E+05
1.0E+00	5.78E+05
1.0E+01	6.13E+05
1.0E+02	6.47E+05
5.0E+02	6.84E+05
1.0E+03	7.04E+05
2.0E+03	7.32E+05
5.0E+03	7.75E+05

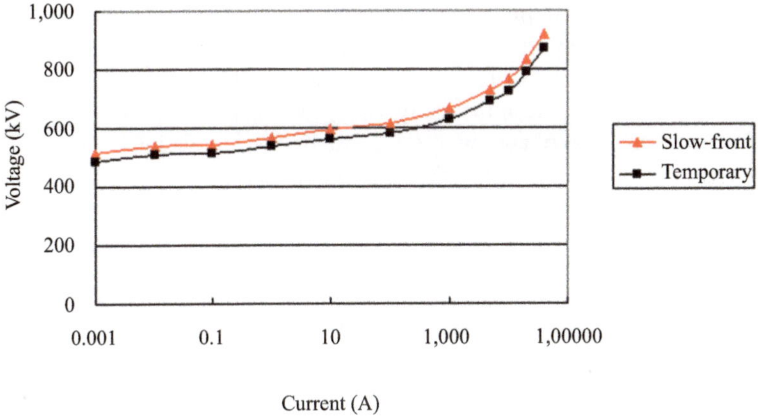

Figure 6.38 V–I characteristic of surge arresters

- 2.6 MJ (= 7.8 kJ/kV × 336 kV) for impulse
- 4.4 MJ (= 13 kJ/kV × 336 kV) for thermal stress

For the new surge arresters installed for the Kyndbyværket–Asnæsværket line, the line discharge class can be upgraded to Class 5 when necessary. Since the energy absorption capability of Class 5 surge arresters is not available for the thermal stress, it was derived by linear approximation.

- 5.7 MJ (= 17 kJ/kV × 336 kV) for impulse
- $9.5\ \text{MJ}\left(= 17\ \text{kJ/kV} \times \dfrac{13\ \text{kJ/kV(Class 3, for thermal stress)}}{7.8\ \text{kJ/kV(Class 3, for impulse)}} \times 336\ \text{kV}\right)$

  for thermal stress

Table 6.19 Withstand voltages of surge arresters

Maximum continuous operating voltage		269 kV (1.16 pu)
Power frequency withstand voltage (1 min)		364 kV (1.58 pu)
Temporary overvoltage	10 sec	370 kV (1.60 pu)
	1 sec	390 kV (1.69 pu)

It is the normal practice in the insulation coordination to ensure that surge arresters are the weakest equipment against overvoltages. They protect other important equipment in a substation, but they themselves are considered sacrificable in unforeseeable events. Since a failure of a surge arrester can lead to the unavailability of equipment which has been protected by it, they should not fail in foreseeable events.

The withstand voltages of surge arresters can be an issue for the temporary overvoltage with low damping. Temporary overvoltages found in Section 6.3 were evaluated against surge arrester ratings, that is, energy absorption capability and withstand voltages. Table 6.19 shows the withstand voltages (phase to ground, rms) of surge arresters assumed in the analysis.

For efficient overvoltage analysis, most of the simulations were performed without the surge arrester model. When a high overvoltage occurs compared with the withstand voltages, the simulation was repeated with the surge arrester model in order to evaluate the effectiveness and energy absorption of the surge arrester.

6.2.4.4 Generators

In EMT simulations, generators are often modeled as an ideal voltage sources without the generator impedance. The generator impedance jX''_d (subtransient direct-axis impedance) of a generator has to be considered outside of the voltage source.

In contrast, in power flow calculations, generators models normally include jX''_d inside themselves. Results of power flow calculations, as a result, do not give us the magnitude and the phase angle of the voltage source behind jX''_d.

In Figure 6.39, the magnitude V_b and phase angle θ_b of the generator terminal voltage is known from the results of the power flow calculation. In order to set up the generator models in EMT simulations, it is necessary to find the magnitude V_g

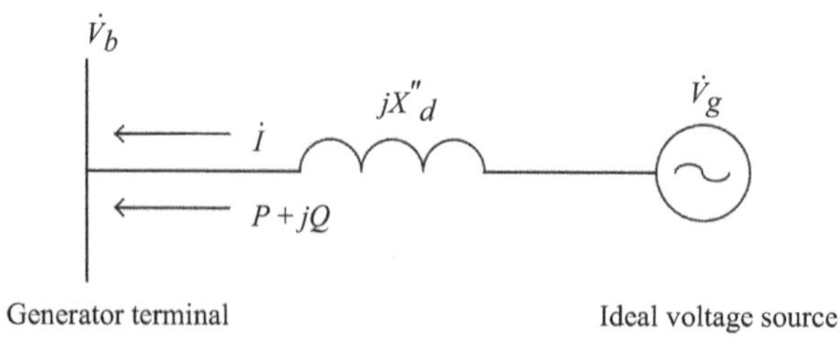

Figure 6.39 Setup of generator models

and phase angle θ_g of the ideal voltage source from the generator terminal voltage and the output of the generator $P+jQ$.

$\dot{V}_b = V_b \cdot e^{j\theta_b}$: generator terminal voltage (phase-to-phase)

$X = X_d$": subtransient direct-axis reactance

$\dot{V}_g = V_g \cdot e^{j\theta_g}$: voltage source behind the subtransient reactance

$$P + jQ = \sqrt{3} \cdot \dot{V}_b \cdot \dot{I}^* = \frac{\dot{V}_b\left(\dot{V}_g^* - \dot{V}_b^*\right)}{-jX}$$

$$= j\frac{\dot{V}_b\left(\dot{V}_g^* - \dot{V}_b^*\right)}{X}$$

$$= j\frac{V_b V_g e^{j(\theta_b - \theta_g)} - V_b^2}{X}$$

$$P = \frac{V_b V_g \sin\left(\theta_g - \theta_b\right)}{X}$$

$$Q = \frac{V_b V_g \cos\left(\theta_g - \theta_b\right) - V_b^2}{X}$$

Modifying the above equations, V_g and θ_g are found by the following equations:

$$\theta_g = \tan^{-1}\frac{P}{Q + \frac{V_b^2}{X}} + \theta_b$$

$$V_g = \frac{XP}{V_b \sin\left(\theta_g - \theta_b\right)}$$

In the off-peak load condition, only two large generators are in-service at Asnæsværket (ASV) and Avedøreværket (AVV). Distributed generations operating in the HV network are not included in the simulation model. It is a common practice since their effects on the EHV network can be ignored and their operating conditions cannot be monitored in the real-time operation.

Table 6.20 shows the input data for generator models obtained from power flow calculation results. Since line parameters in power flow calculations are different from those in EMT simulations, it is sometimes necessary to slightly adjust the values in Table 6.20 in order to have an appropriate initial power flow

Table 6.20 Calculated data for generators

	Rated voltage	P (MW)	Q (MVar)	X (mH)	V_g (kV)	θ_g (°)
ASV	21.0 kV	384	−116.3	0.407	19.24	−109.8
AVV	19.5 kV	331.2	83.8	0.480	20.06	68.4

conditions for the transient overvoltage analysis. Also, in some analyses, network conditions were modified in order to consider severer conditions. In such cases, it is necessary to adjust the generator model setting in order to produce appropriate initial conditions.

6.2.4.5 Loads

Load models were connected to 132 kV buses of 400 kV substations. First, the power flows in the 400/132 kV transformers were obtained from the power flow calculation results. These power flows were then converted to equivalent RL or RC circuits. The equivalent circuits were connected to 132 kV buses as constant impedance load models. When the reactive power was fed from the 400 kV buses, an RL equivalent circuit is adopted. Otherwise, when the reactive power was fed from the 132 kV buses, an RC equivalent circuit is adopted.

The conversion to the equivalent circuits is necessary only in ATP-EMTP. In PSCAD, active and reactive power loads obtained from the power flow in the 400/132 kV transformers can be connected to the 132 kV buses without the conversion. The conversion is then automatically performed inside PSCAD.

Table 6.21 shows the active and reactive loads and the constant impedance load models assumed in the off-peak load condition.

Table 6.21 Load models in off-peak load condition

	Bus #	P (MW)	Q (MVar)	R (ohm)	L (mH)	R (ohm)	C (μF)
ASV	Total	80.5	23.2	-	-	216.4	4.238
BJS	92	38.8	−5.0	-	-	449.1	0.913
ISH	187	42.1	69.7	-	-	413.9	12.733
AVV	82	50.5	−68.6	-	-	345.0	12.532
HCV	156	172.0	167.1	12,497	38,647	-	-
HVE	179	149.9	266.6	8,540	48,344	-	-
KYV	224	77.2	−35.9	-	-	225.7	6.558
GLN	133	265.4	17.0	17,388	3,545	-	-
GOR	144	125.3	−73.0	-	-	139.1	13.336
SAN	52	400.8	−319.0	-	-	43.5	58.276

6.3 Temporary overvoltage

Temporary overvoltages are the greatest concern when studying long EHV AC cables. Because of the large charging capacity and large shunt reactors for the compensation, the natural frequencies of a network with the long EHV AC cables tend to be much lower than those of a conventional network without the EHV cables. As such, in conventional networks, switching operations such as cable energization and fault clearance are considered as causes of slow-front overvoltages, but they need to be considered as causes of temporary overvoltages in a network with long EHV AC cables.

Even though several studies on temporary overvoltages of long EHV AC cables have been published, there is no widely accepted guideline on the temporary overvoltage analysis with the long EHV AC cables [12–15]. When we standardize the temporary overvoltage analysis on long EHV AC cables, there are the following technical difficulties:

- Simulation models need to include a wide area away from the point of interest.
- A severe overvoltage occurs only in particular operating conditions.

As for the first point, it is necessary to model a very wide area away from the point of interest in order to accurately reproduce low-frequency components of a temporary overvoltage. However, it is not always possible to obtain precise information on all equipment. The availability largely depends on a utility's practice. A compromise to some extent is often necessary. In addition, large models can slow down a simulation depending on the adopted models and simulation software, which almost prohibits a daily analysis.

Model reduction is discussed to deal with this problem, but in order to obtain a reduced model, a full model has to be built first. It is not recommended to obtain a reduced model from power flow data since it can match only the fundamental frequency response of the network. As such, the model reduction can contribute to speed up the simulation, but modeling of a very wide area is still necessary.

As for the second point, temporary overvoltages such as the resonance overvoltages and the overvoltages caused by load shedding are low probability phenomena. Only certain and particular network conditions can yield a severe overvoltage that can lead to an equipment failure. Major difficulties exist in finding these certain and particular network conditions. Evaluating a low-probability high-consequence risk is also difficult but must be done when and after the cables are installed. The risks may be avoided via carefully prepared operational countermeasures, but not all utilities can take operational countermeasures as it can be a major burden for system analysts.

Considering these difficulties, this section takes the following steps:

- Start from the full model created in the previous section
- Confirm the full model runs within a reasonable duration which enables analyses of the model on a daily rush basis. Otherwise, consider the reduction of the model.
- Consider the most severe scenario to characterize problems
- Perform sensitivity analysis on source impedances, load levels, and 132 kV networks

6.3.1 Series resonance overvoltage

6.3.1.1 Most severe scenario

We start by identifying the most severe switching scenarios related to the Kyndbyværket–Asnæsværket line.

Figure 6.40 shows the assumed most severe scenario for the Kyndbyværket 132 kV network. The series resonance overvoltage is caused by the energization of the Kyndbyværket–Asnæsværket line from the Kyndbyværket side. The

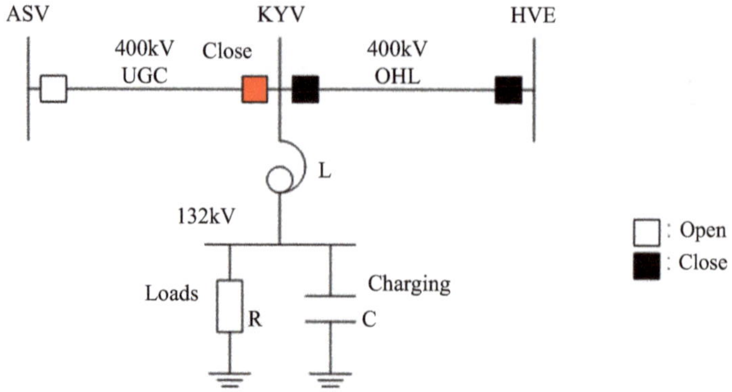

Figure 6.40 Assumed most severe scenario for the KYV 132 kV network

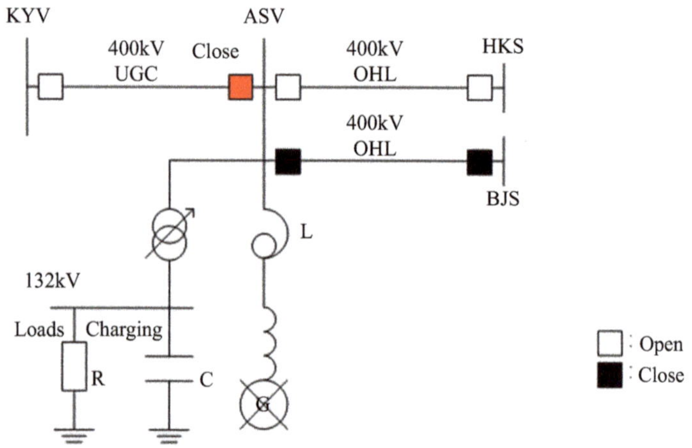

Figure 6.41 Assumed most severe scenario for the ASV 132 kV network

energization overvoltage first travels to the Asnæsværket side (open end) and reflected back to the Kyndbyværket side. A part of the reflected overvoltage travels into the Kyndbyværket–Hovegård line; the other part of the reflected overvoltage travels into the series resonance circuit composed of the 400/132 kV Kyndbyværket transformer and the charging capacity of the Kyndbyværket 132 kV network.

The source impedance and the charging capacity will be adjusted to match the dominant frequency and resonance frequency.

As the loads in the Kyndbyværket 132 kV network will damp the series resonance overvoltage on the secondary side of the 400/132 kV Kyndbyværket transformer, the series resonance overvoltage is more severe under the low load condition.

Figure 6.41 shows the assumed most severe scenario for the Asnæsværket 132 kV network. The series resonance overvoltage is caused by the energization of the Kyndbyværket–Asnæsværket line from the Asnæsværket side. The energization overvoltage first travels to the Asnæsværket side (open end) and reflected back to the Asnæsværket side. A part of the reflected overvoltage travels into the Asnæsværket–Bjæverskov line; the other part of the reflected overvoltage travels into the series resonance circuit composed of the 400/132 kV Asnæsværket transformer and the charging capacity of the Asnæsværket 132 kV network.

The large generator at Asnæsværket is assumed to be out of service since the generator contributes to the damping of the energization overvoltage.

Here, the Kyndbyværket–Asnæsværket line is energized from Bjæverskov since the generator at Asnæsværket does not have a black start capability. If the generator has the black start capability, the energization from the generator at Asnæsværket can be the most severe scenario which needs to be studied.

6.3.1.2 Dominant frequency in energization overvoltage

First, the dominant frequency in an energization overvoltage is studied in the off-peak load condition. When a transmission line, either an overhead line or a cable, is energized from a weaker source, which has a larger source impedance (off-peak load condition), the magnitude of the energization overvoltage becomes lower and the dominant frequency becomes lower. In general, the off-peak load condition is a more severe assumption because of the following reasons:

- Damping from the loads (more important)
  The loads in the Asnæsværket or Kyndbyværket 132 kV network will damp the series resonance overvoltage on the secondary side of the 400/132 kV transformer. The load level has to be low in order to have a severe resonance.
- Damping from the network (transformers, lines, and etc.)

The lower dominant frequency means an extended duration of the overvoltage. The high-frequency components, which raise the magnitude of the overvoltage in a stiff source case, typically die out quickly due to the large damping (losses).

However, this general notion may not be true in the energization of the Kyndbyværket–Asnæsværket line. In Section 6.2.2, the Kyndbyværket–Asnæsværket line was energized from the 400 kV voltage source behind source impedance 147 mH, which corresponds to the fault current 5 kA. In that case, the dominant frequency was already low (100–150 Hz) in the energization, due to the large charging capacity of the cable. If we make the source impedance larger, the dominant frequency goes down toward 50 Hz, and it is hard to observe the energization overvoltage as the magnitude becomes lower. The dominant frequency in the peak load condition is, therefore, studied later in this section.

Here, in the off-peak load condition, the Kyndbyværket–Asnæsværket line is energized from the Kyndbyværket side according to Figure 6.40. Only one large generator in AVV and the equivalent generator in SAN are assumed to be in-service in the simulation. The cable line is energized at the phase-b voltage peak.

The energization overvoltage is shown in Figures 6.42 and 6.43. Figure 6.42 is the voltage waveform at the sending end (Kyndbyværket side), and Figure 6.43 is the voltage waveform at the receiving end (Asnæsværket side cable open terminal). As can be seen from the figure, the waveforms look not like typical energization

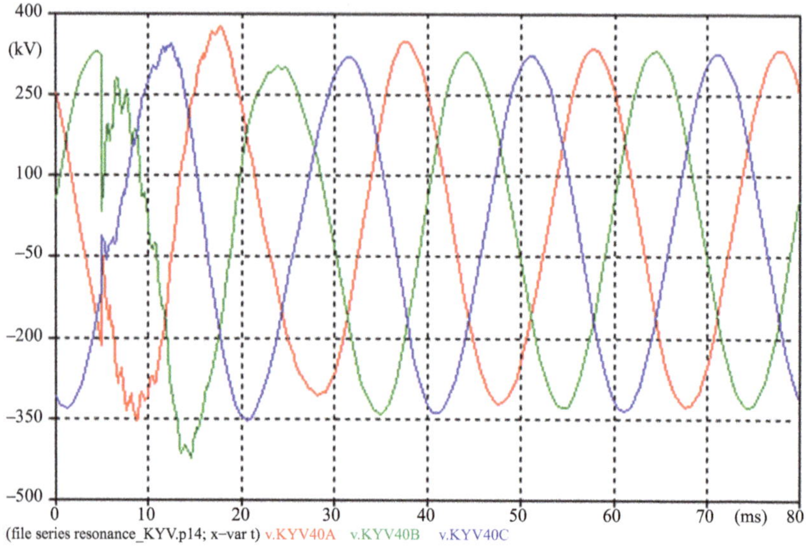

Figure 6.42 Energization overvoltage of the KYV–ASV line from the KYV side (sending end)

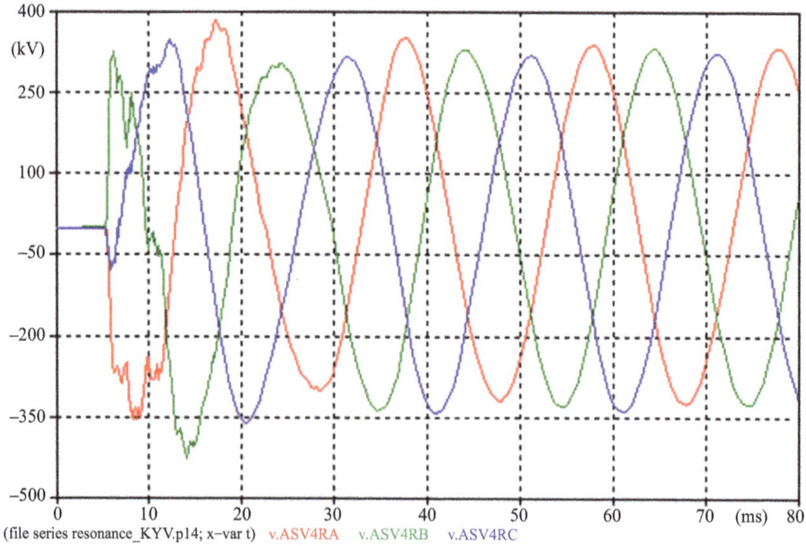

Figure 6.43 Energization overvoltage of the KYV–ASV line from the KYV side (receiving end)

overvoltages due to the length of the cable and the large charging capacity. High-frequency components are contained in the waveforms with very small proportions.

Figure 6.44 shows frequency components contained in the overvoltage caused by the energization of the Kyndbyværket–Asnæsværket line. The Fourier transform is applied to the waveform at the sending end since the series resonance overvoltage is expected at the Kyndbyværket 132 kV bus.

Figure 6.44 demonstrates that frequency components other than the fundamental frequency are not large enough to cause series resonance frequency. The analysis is however continued in order to show that there is no problem related to the series resonance overvoltage.

Figure 6.45 shows the frequency components in 1 Hz step. In order to obtain it, it is necessary, due to the limitation of ATP-EMTP, to run the simulation for 1 s and

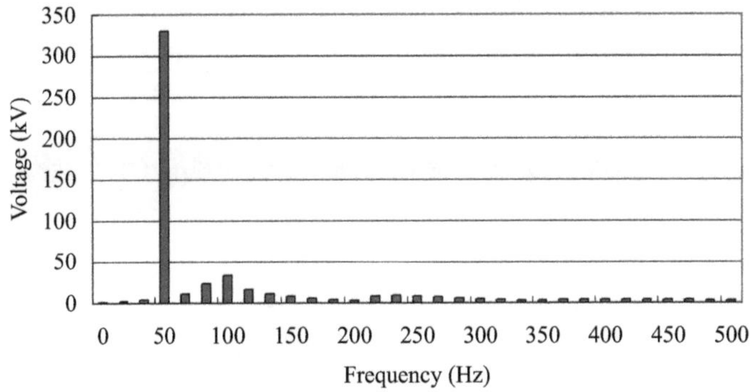

Figure 6.44 Frequency components contained in the overvoltage caused by the energization of the KYV–ASV line from the KYV side

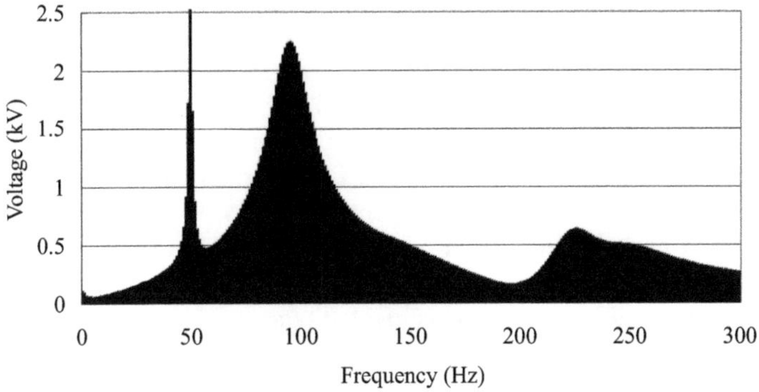

Figure 6.45 Frequency components contained in the overvoltage caused by the energization of the KYV–ASV line from the KYV side (1 Hz step)

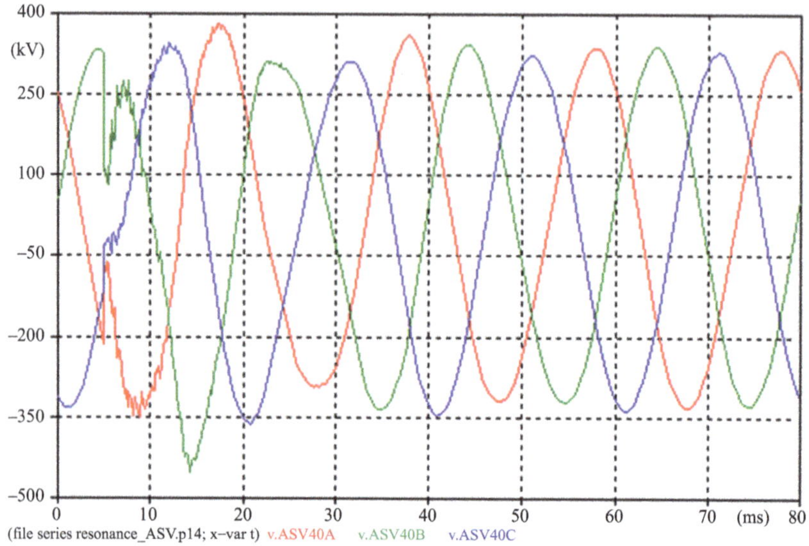

Figure 6.46 Energization overvoltage of the KYV–ASV line from the ASV side (sending end)

apply Fourier transform to the waveform of 1 s. Because of it, the frequency components except for the fundamental frequency are reduced compared to Figure 6.44.

According to Figure 6.45, in order to cause a severe series resonance overvoltage, it is necessary to set the natural frequency of the Kyndbyværket 132 kV network to 90–100 Hz. The amplification from the 400 kV side to the 132 kV side needs to be very high.

Second, the Kyndbyværket–Asnæsværket line is energized from the Asnæsværket side according to Figure 6.41. The cable line is energized at the phase-b voltage peak.

The energization overvoltage is shown in Figures 6.46 and 6.47. Figure 6.46 is the voltage waveform at the sending end (Asnæsværket side), and Figure 6.47 is the voltage waveform at the receiving end (Kyndbyværket side cable open terminal). As in the energization overvoltage from the Kyndbyværket side, high-frequency components are contained in the waveforms with very small proportions.

Figures 6.48 and 6.49 show frequency components contained in the overvoltage caused by the energization of the Kyndbyværket–Asnæsværket line from the Asnæsværket side. The Fourier transform is applied to the waveform at the sending end since the series resonance overvoltage is expected at the Asnæsværket 132 kV bus.

As in the energization from the Kyndbyværket side, the natural frequency of the Asnæsværket 132 kV has to be set to 85–100 Hz in order to cause a severe series resonance overvoltage. The amplification from the 400 kV side to the 132 kV side needs to be very high.

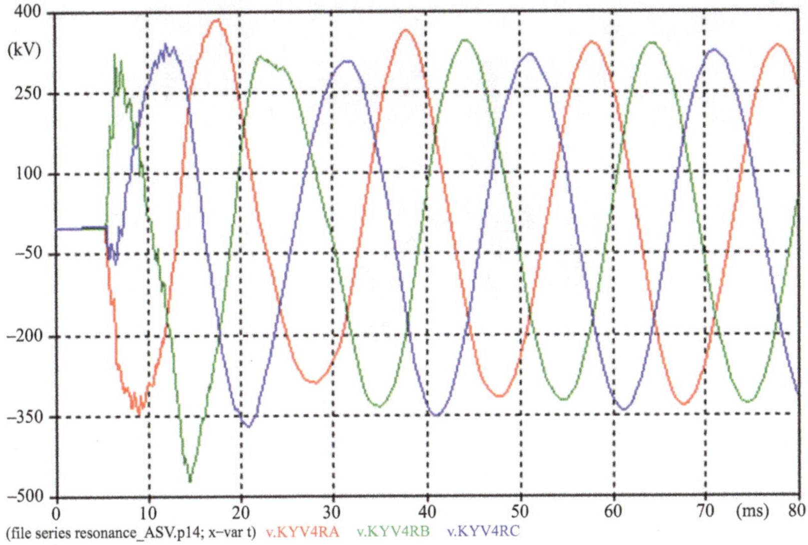

Figure 6.47 Energization overvoltage of the KYV–ASV line from the ASV side (receiving end)

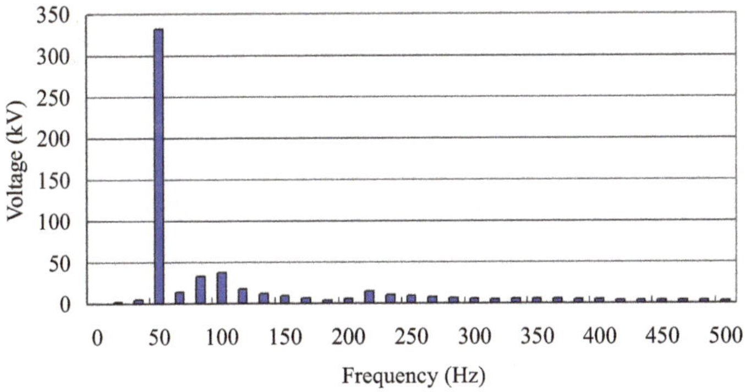

Figure 6.48 Frequency components contained in the overvoltage caused by the energization of the KYV–ASV line from the ASV side

In order to have a larger magnitude of the energization overvoltage, energizations from the stiffer voltage source, namely in the peak load condition, are additionally studied. To make the voltage source stiffer, the large generator at Asnæsværket is assumed to be in-service. In addition, equivalent generators are modeled near the line, that is, at the HVE and BJS 132 kV buses. Adding equivalent

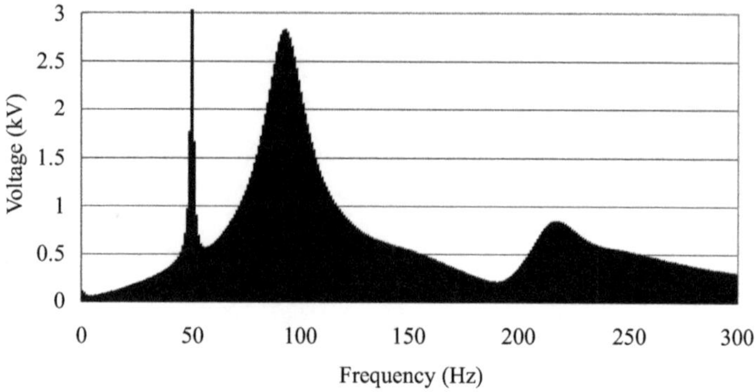

Figure 6.49 Frequency components contained in the overvoltage caused by the energization of the KYV–ASV line from the ASV side (1 Hz step)

Table 6.22 Source impedances of 132 kV equivalent generators

132 kV buses	Short circuit current	Source impedance
Hovegård	19.6 kA	12.4 mH
Bjæverskov	13.3 kA	18.2 mH

generators at the ASV and KYV 132 kV buses would be more efficient to make the voltage source stiffer, but it is not considered since these voltage sources will suppress the series resonance overvoltage at the ASV and KYV 132 kV buses.

The source impedances for these equivalent generators are derived from the results of the short circuit current calculation in the peak load condition as shown in Table 6.22. Only the short circuit current from the 132 kV network is considered

Figure 6.50 shows these additions of generators in the simulation model in the peak load condition.

Figures 6.51 and 6.52 show energization overvoltages of the Kyndbyværket–Asnæsværket line when the line is energized from the Kyndbyværket side. Overall, the waveforms of the overvoltages are similar to the ones observed in the off-peak load condition. The change of the loading condition does not have a significant difference in the energization overvoltage. It is mainly because the fault current level at the Kyndbyværket 400 kV bus is very small, only 11.4 kA, even in the peak load condition.

Only one notable difference is that the energization overvoltages in the peak load condition have more damping. In the off-peak load condition, nonfundamental components can be observed in the waveform even at 80 ms. In contrast, in the peak load condition, we can observe only a fundamental frequency component after 40 ms.

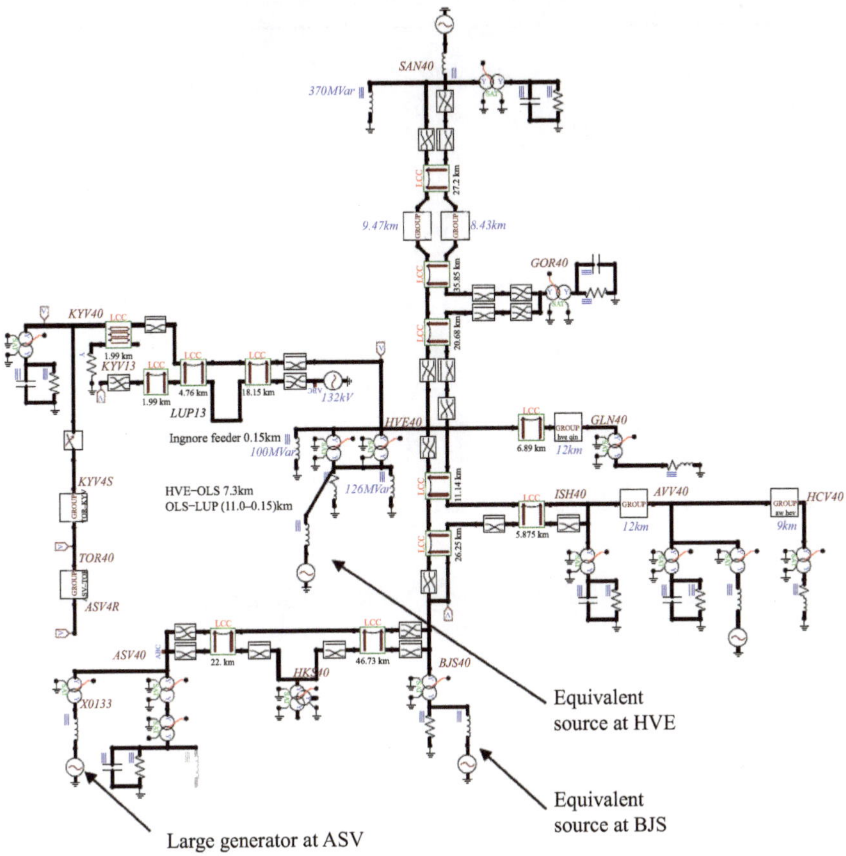

Figure 6.50 Addition of generators to set up the peak load condition

Figures 6.53 and 6.54 show frequency components contained in the overvoltage caused by the energization of the Kyndbyværket–Asnæsværket line. The dominant frequency goes up to 120–125 Hz, which explains the more damping observed in the voltage waveforms. Compared to the off-peak load condition, the distribution of frequency components becomes flatter. It is clear that the off-peak load condition is more severe in terms of the overvoltage which propagates into the Kyndbyværket 132 kV bus.

Second, the Kyndbyværket–Asnæsværket line is energized from the Asnæsværket side according to Figure 6.41. The cable line is energized at the phase-b voltage peak as in the off-peak case.

The energization overvoltage is shown in Figures 6.55 and 6.56. Again, the waveforms of the energization overvoltages are similar to the ones in the off-peak load condition, but with higher damping. It is mainly because the fault current level

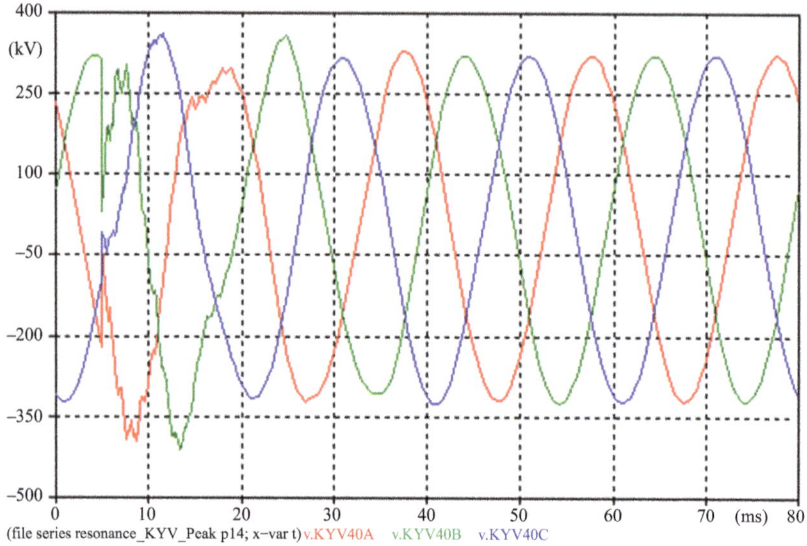

Figure 6.51 Energization overvoltage of the KYV–ASV line from the KYV side (peak load, sending end)

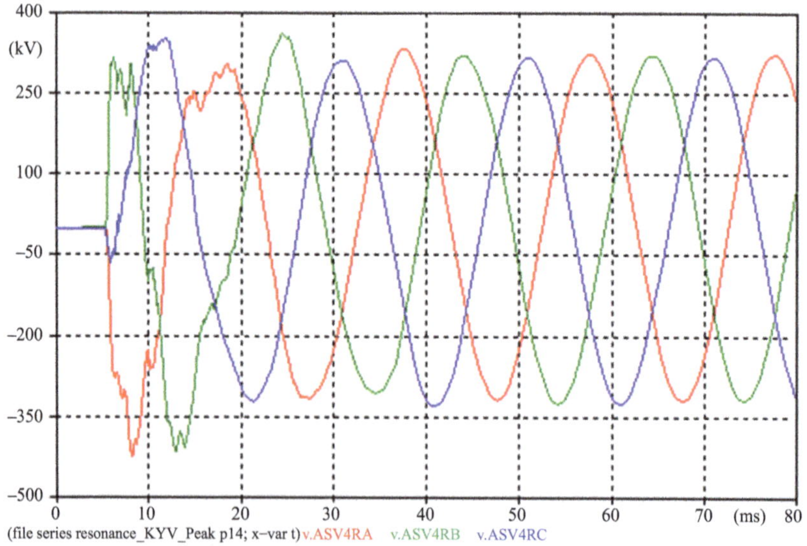

Figure 6.52 Energization overvoltage of the KYV–ASV line from the KYV side (peak load, receiving end)

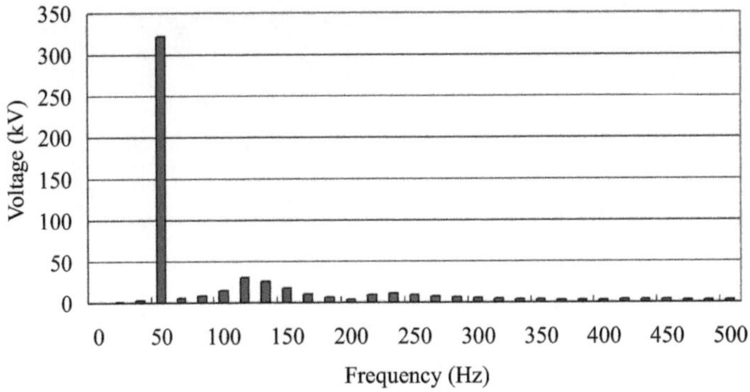

Figure 6.53 Frequency components contained in the overvoltage caused by the energization of the KYV–ASV line from the KYV side (peak load)

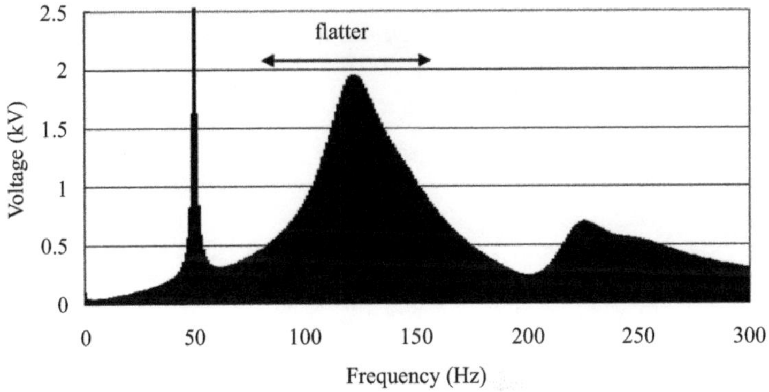

Figure 6.54 Frequency components contained in the overvoltage caused by the energization of the KYV–ASV line from the KYV side (peak load, 1 Hz step)

at the Asnæsværket 400 kV bus is very small, only 11.4 kA, even in the peak load condition.

Figures 6.57 and 6.58 show frequency components contained in the overvoltage caused by the energization of the Kyndbyværket–Asnæsværket line from the Asnæsværket side in the peak load condition. As in the energization from the Kyndbyværket side, the dominant frequency goes up to 120–125 Hz, which explains the better damping observed in the voltage waveforms. Compared to the off-peak load condition, the distribution of frequency components becomes flatter. It is clear that the off-peak load condition is more severe in terms of the overvoltage which propagates into the Kyndbyværket 132 kV bus.

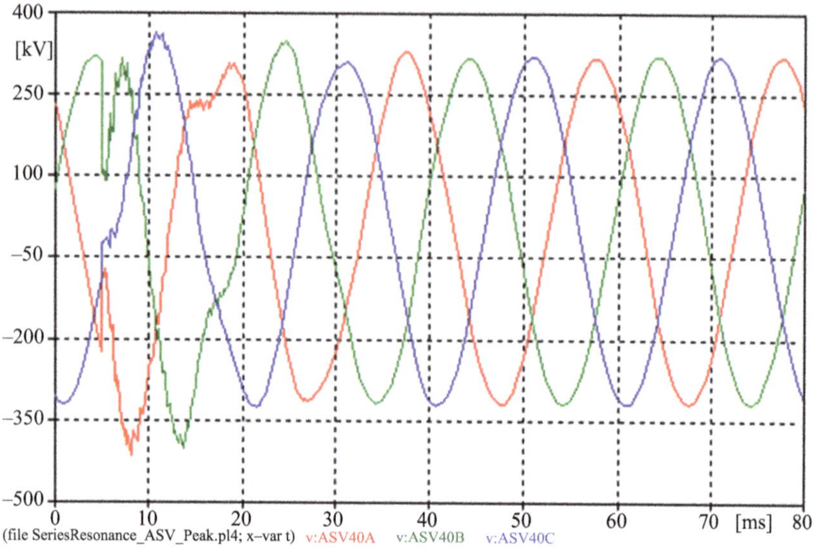

Figure 6.55 Energization overvoltage of the KYV–ASV line from the ASV side (peak load, sending end)

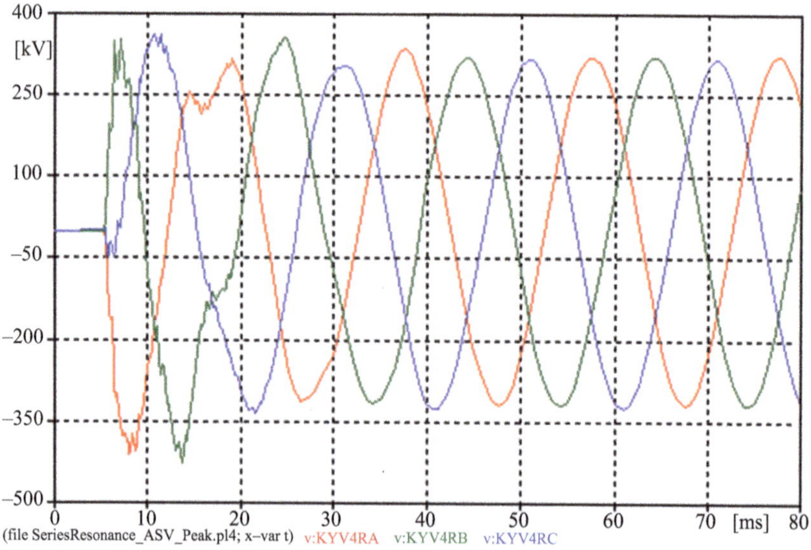

Figure 6.56 Energization overvoltage of the KYV–ASV line from the ASV side (peak load, receiving end)

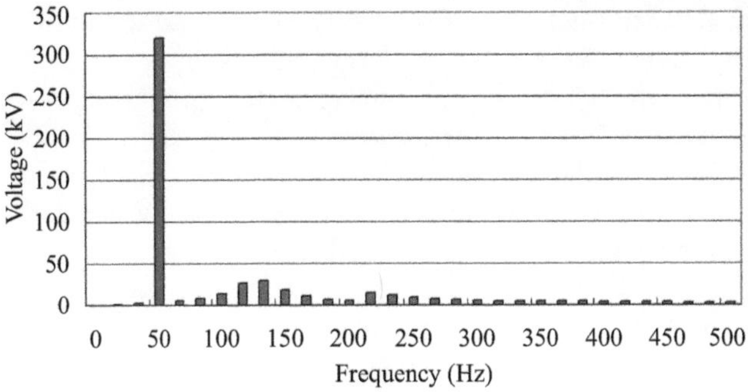

Figure 6.57 Frequency components contained in the overvoltage caused by the energization of the KYV–ASV line from the ASV side (peak load)

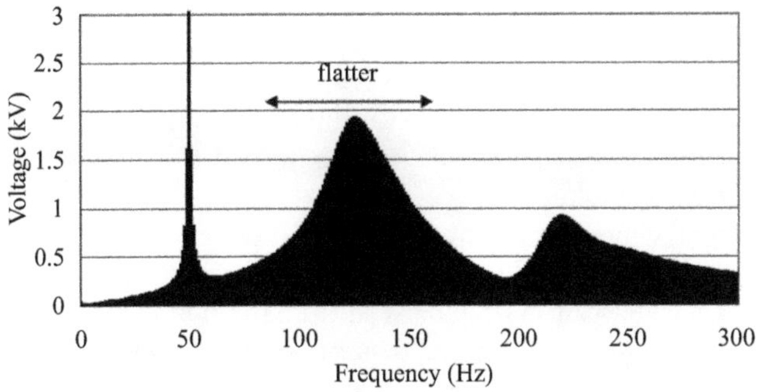

Figure 6.58 Frequency components contained in the overvoltage caused by the energization of the KYV–ASV line from the ASV side (peak load, 1 Hz step)

6.3.1.3 Natural frequency of series resonance circuit

It has been found that the dominant frequency contained in the overvoltage caused by the energization of the Kyndbyværket–Asnæsværket line from the Kyndbyværket side ranges from 85 Hz to 100 Hz. In order to cause the series resonance overvoltage, the natural frequency of the Kyndbyværket 132 kV network has to be between 85 Hz and 100 Hz.

The natural frequency of the Kyndbyværket 132 kV network can be found as $f_n = \frac{1}{2\pi\sqrt{LC}}$ using 400/132 kV transformer inductance L and 132 kV cable capacitance C. In order to set the natural frequency to 90 Hz, the following equation has to be satisfied:

$$\frac{1}{2\pi\sqrt{0.02088 \times C}} = 90$$
$$C = 149.8\ [\mu\text{F}] \tag{6.2}$$

Cable capacitance 149.8 μF corresponds to 819.8 MVar in the 132 kV network. This amount of cable capacitance is too large to assume in the 132 kV network. In terms of the cable length, assuming the typical cable capacitance per length 0.2 μF/km, cable capacitance 149.8 μF corresponds to 132 kV cable of about 750 km in circuit length. From the above, it is reasonable to conclude that the series resonance overvoltage will not occur in the Kyndbyværket 132 kV network due to the installation of the Kyndbyværket–Asnæsværket line.

The series resonance overvoltage is, however, studied from an academic interest. When 132 kV cable capacitance 149.8 μF is connected to the Kyndbyværket 132 kV bus, it results in the natural frequency of the Kyndbyværket 132 kV network to 90 Hz and, at the same time, will raise the steady-state system voltage around the Kyndbyværket substation. A shunt reactor has to be assumed at the Kyndbyværket 132 kV bus, so as to prevent the steady-state overvoltage.

In order to 100% compensate for the large reactive power produced from the cable capacitance 149.8 μF, the shunt reactor $L = 67.65$ mH has to be connected to the Kyndbyværket 132 kV bus. However, this shunt reactor shifts the natural frequency of the Kyndbyværket 132 kV network to 100 Hz, which is not ideal to cause a severe series resonance. In order to shift the natural frequency lower, the cable capacitance is increased to 170.0 μF. It is confirmed that this increase does not cause steady-state overvoltage.

In the above-mentioned condition, the frequency scan is performed to find the natural frequency of the Kyndbyværket 132 kV network. Figure 6.59 shows the result of the frequency scan. As can be seen from the figure, the natural frequency of the Kyndbyværket 132 kV network is successfully set to 95 Hz.

In order to have severe series resonance, it is necessary to have a high-voltage amplification ratio from the Kyndbyværket 400 kV bus to the 132 kV bus. The amplification ratio is shown in Figure 6.60. It is observed in the figure that the amplification ratio is approximately 6 at 88 Hz, which is not very high.

The amplification ratio is highly affected by the loading condition. In terms of the amplification ratio, the assumed condition is already severe since it is derived in the off-peak load condition. If it is necessary to assume the no-load condition in the Kyndbyværket 132 kV network, it significantly increases the amplification ratio. Figure 6.61 shows that the amplification ratio increases to around 25 in the no-load condition.

The same analysis is performed for the Asnæsværket 132 kV network. According to the following equation, 132 kV cable capacitance 77.7 μF is assumed in the Asnæsværket 132 kV network. Again, this cable capacitance corresponds to 425.4 MVar or 388.6 km in circuit length in the Asnæsværket 132 kV network. It is thus concluded that the series resonance overvoltage will not occur in the Asnæsværket 132 kV network due to the installation of the Kyndbyværket–Asnæsværket line:

$$\frac{1}{2\pi\sqrt{(0.0000362 + 0.0402) \times C}} = 90$$
$$C = 77.7\ [\mu\text{F}] \tag{6.3}$$

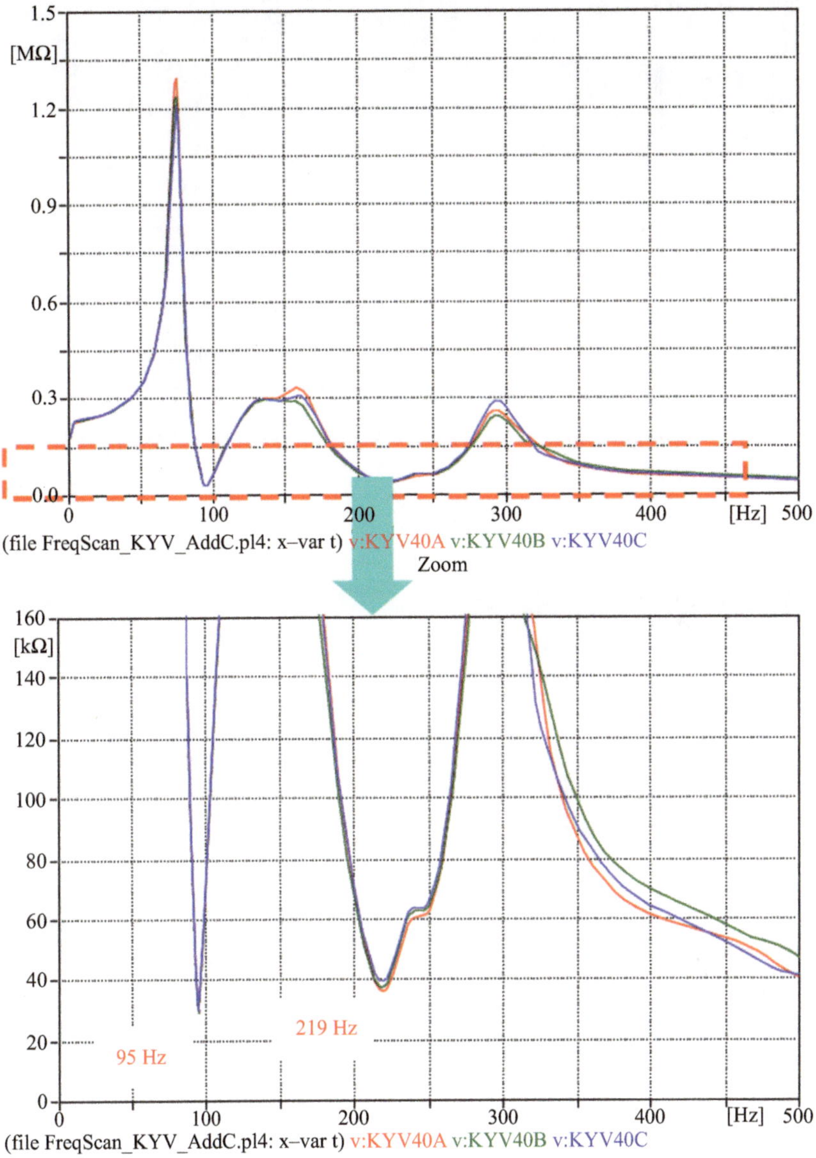

Figure 6.59 Series resonance frequency of the KYV network

Figure 6.62 shows the result of the frequency scan. As can be seen from the figure, the natural frequency of the Asnæsværket 132 kV network is successfully set to 100 Hz. The natural frequency is shifted from 90 Hz to 100 Hz by the inductance $L = 130.4$ mH, which represents the shunt reactor to compensate for the

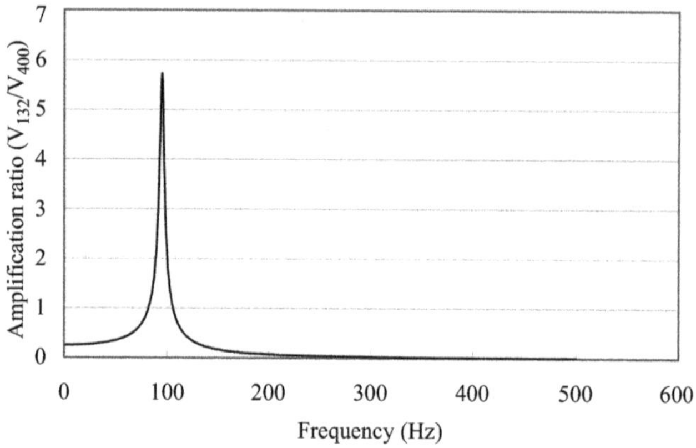

Figure 6.60 Voltage amplification ratio from the KYV 400 kV bus to the 132 kV bus

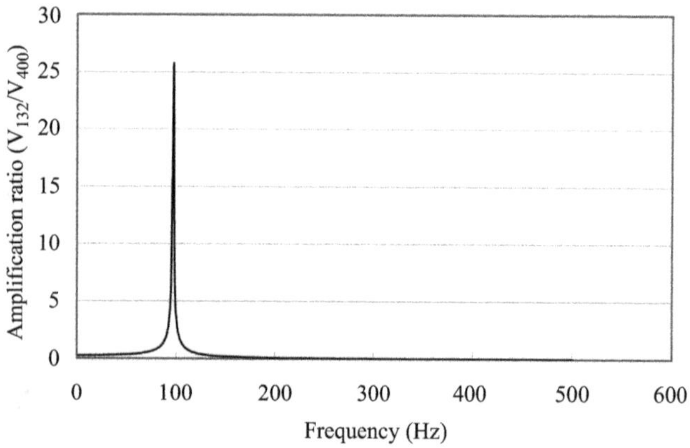

Figure 6.61 Voltage amplification ratio from the KYV 400 kV bus to the 132 kV bus in the no-load condition

large reactive power from the 132 kV cable. It is not possible in this case to lower the natural frequency by increasing 132 kV cable capacitance as it causes steady-state overvoltage.

In order to have severe series resonance, it is necessary to have a high-voltage amplification ratio from the Asnæsværket 400 kV bus to the 132 kV bus. The amplification ratio is shown in Figure 6.63. It is clear in the

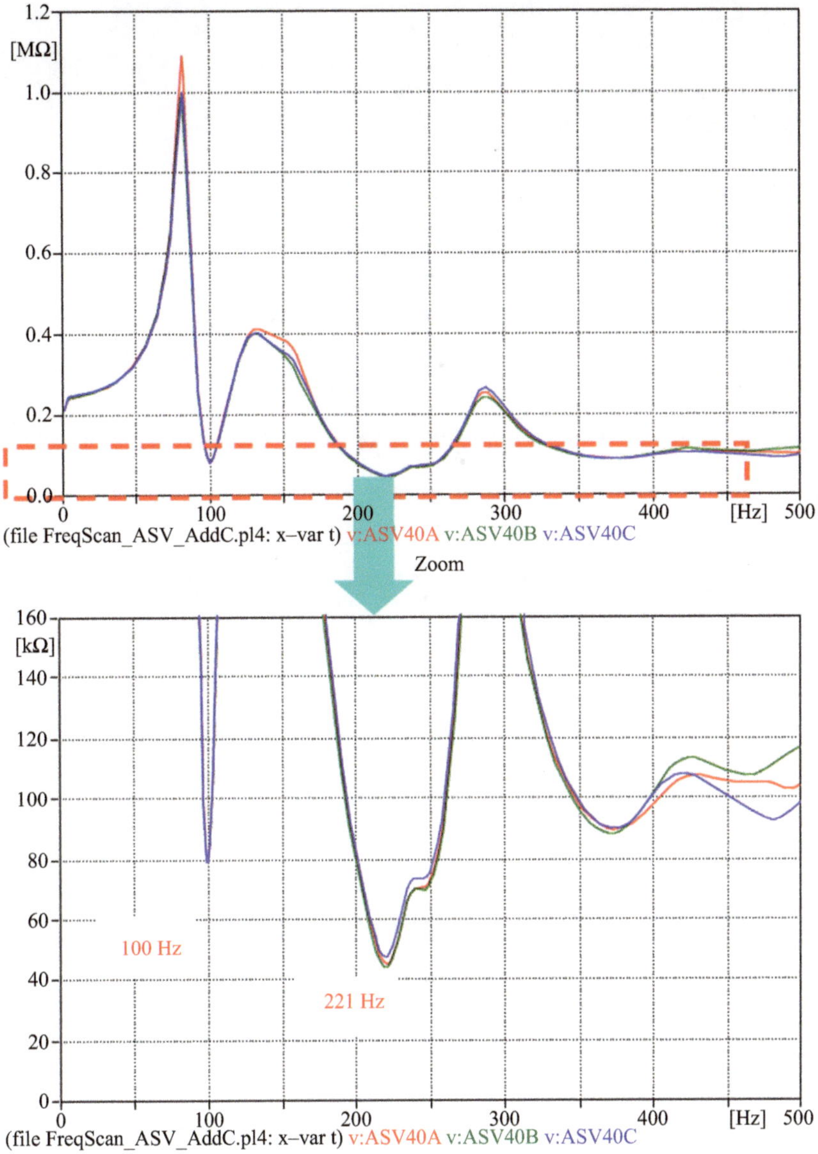

Figure 6.62 Series resonance frequency of the ASV network

figure that the amplification ratio is approximately 3.5 at 100 Hz, which is not very high.

In the no-load condition, the amplification ratio is increased to around 43 as shown in Figure 6.64.

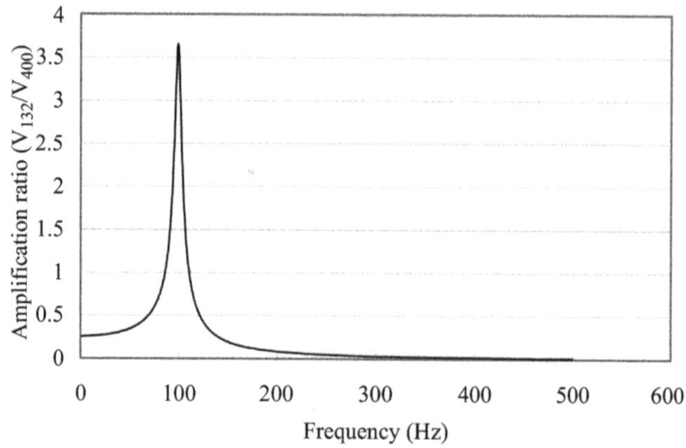

Figure 6.63 Voltage amplification ratio from the ASV 400 kV bus to the 132 kV bus

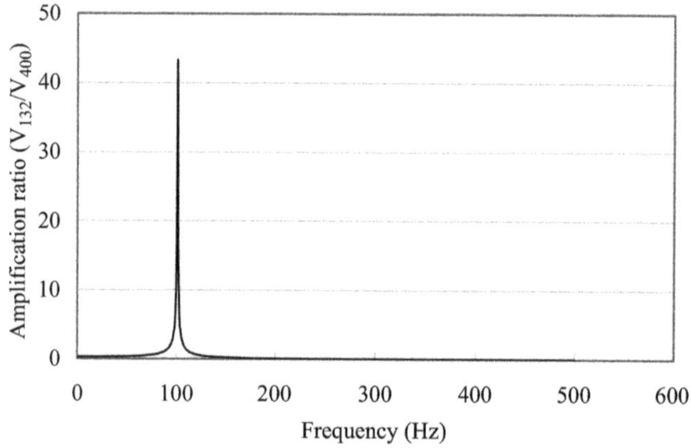

Figure 6.64 Voltage amplification ratio from the ASV 400 kV bus to the 132 kV bus in the no-load condition

6.3.1.4 Simulation results of series resonance overvoltage

The time-domain simulations are performed to find the series resonance overvoltage in the assumed most severe condition. First, the Kyndbyværket–Asnæsværket line is energized from the Kyndbyværket side and the overvoltage in the Kyndbyværket 132 kV bus is monitored as shown in Figure 6.65.

Results of the time-domain simulations are shown in Figure 6.66. In the figure, the upper graph shows the phase to ground overvoltage, and the lower graph shows the phase to phase overvoltage. According to IEC 60071, the standard short-duration power-frequency withstand voltage for the 132 kV equipment is (185), 230, or 275 kV (rms). The same withstand voltages are applied to both the phase to ground overvoltage and the phase to phase overvoltage. Here, the application to the

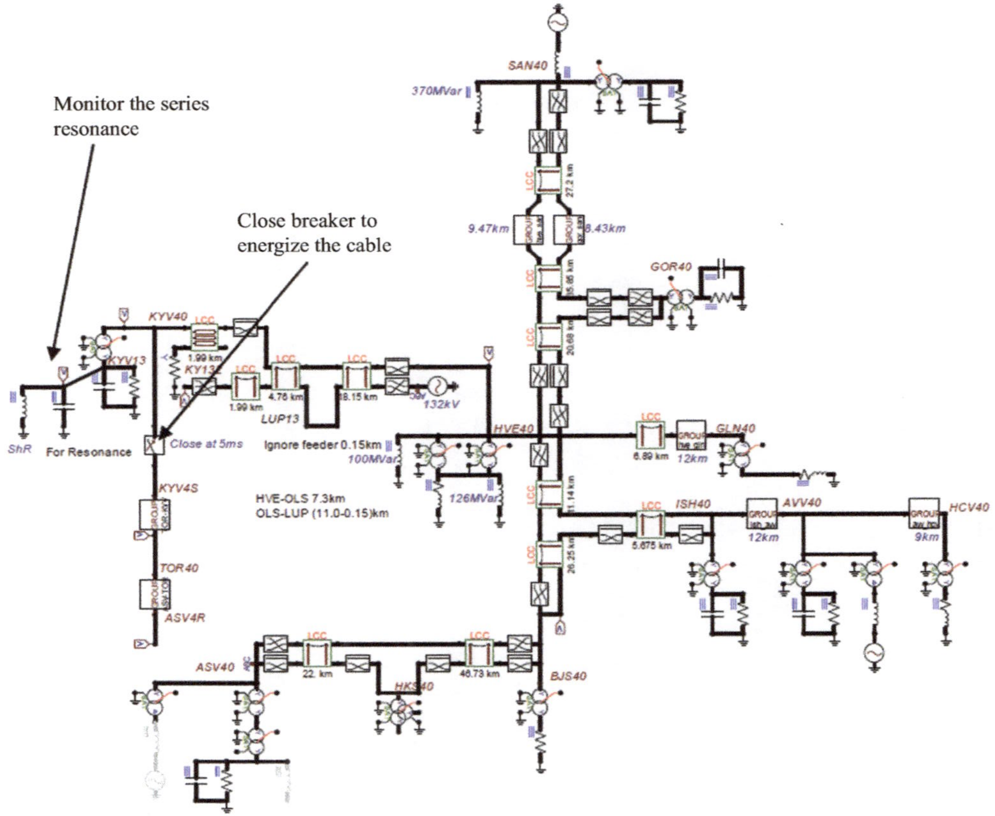

Figure 6.65 Simulation model to find the series resonance overvoltage in the Kyndbyværket 132 kV bus

phase to phase overvoltage depends on the physical construction of the equipment. Figure 6.66 shows that the series resonance overvoltage in the assumed most severe condition does not exceed the withstand voltage. The phase to phase overvoltage decays below 230 kV (peak) after one cycle.

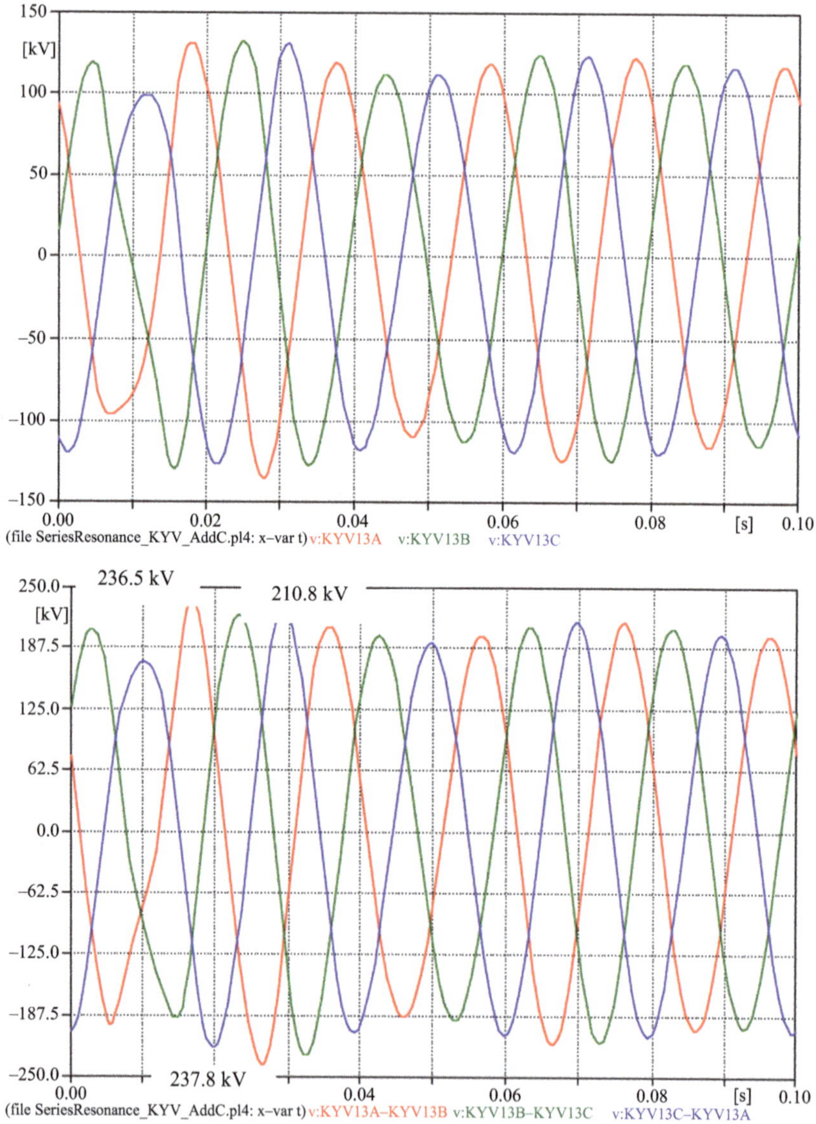

Figure 6.66 Series resonance overvoltage in the KYV 132 kV bus in the off-peak load condition (upper: phase to ground; lower: phase to phase)

The same simulations are repeated with the no-load condition. Figure 6.67 shows the results of the simulations. Only a minor difference is observed compared with the results in the off-peak load condition.

Second, the Kyndbyværket–Asnæsværket line is energized from the Asnæsværket side and the overvoltage in the Asnæsværket 132 kV bus is

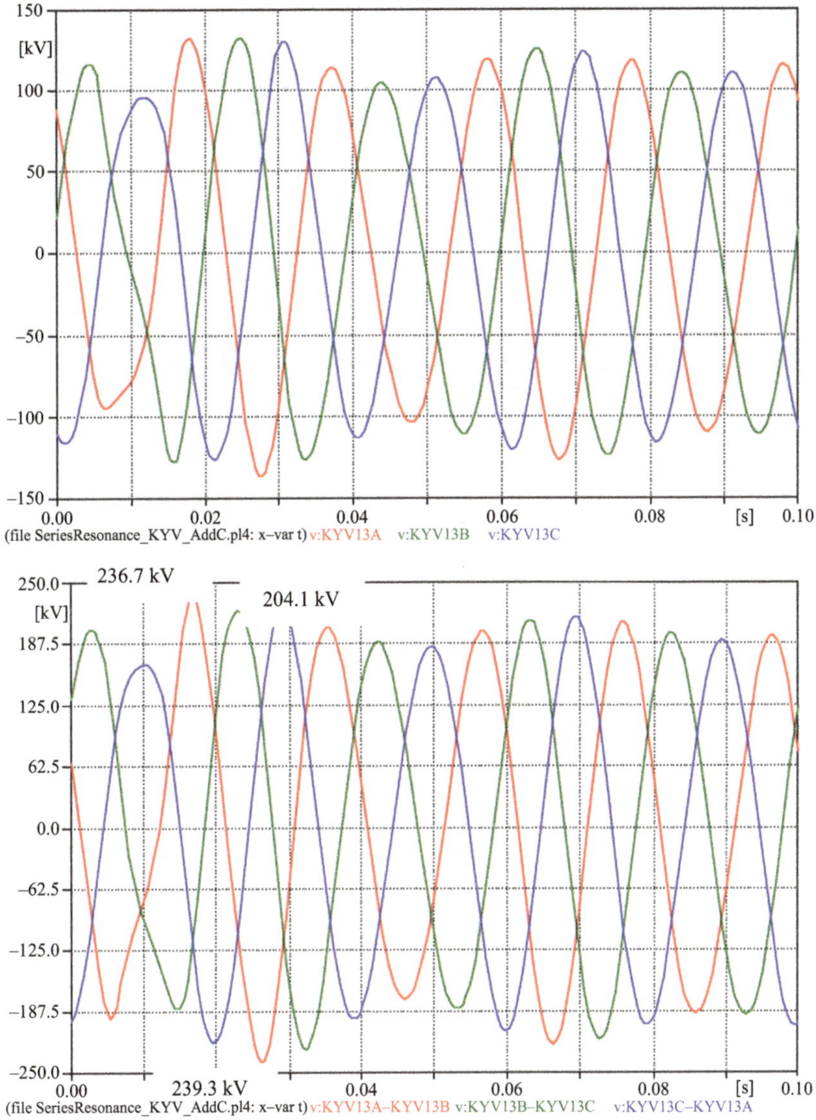

Figure 6.67 Series resonance overvoltage in the KYV 132 kV bus in the no-load condition (upper: phase to ground; lower: phase to phase)

monitored. Figures 6.68 and 6.69 show the series resonance overvoltage found with the off-peak load condition and the no-load condition, respectively.

The overvoltage level is higher compared with the overvoltage in the Kyndbyværket 132 kV bus, but it is still much lower than the withstand voltage as long

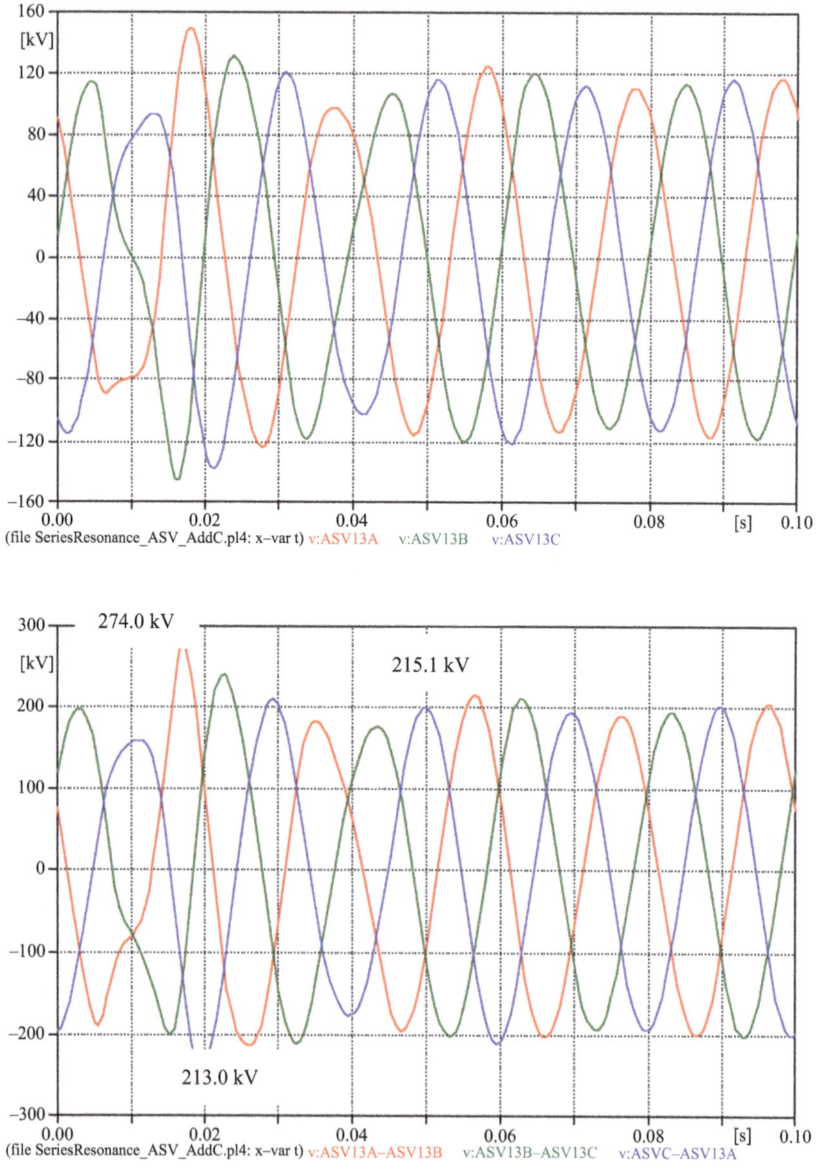

Figure 6.68 Series resonance overvoltage in the ASV 132 kV bus in the off-peak load condition (upper: phase to ground; lower: phase to phase)

as the withstand voltage 230 or 275 kV is selected. The phase to phase overvoltage decays below 230 kV after one cycle. As in the overvoltage in the Kyndbyværket 132 kV bus, the load level has only a minor impact on the overvoltage level. It is because the load level is already low in the off-peak load condition.

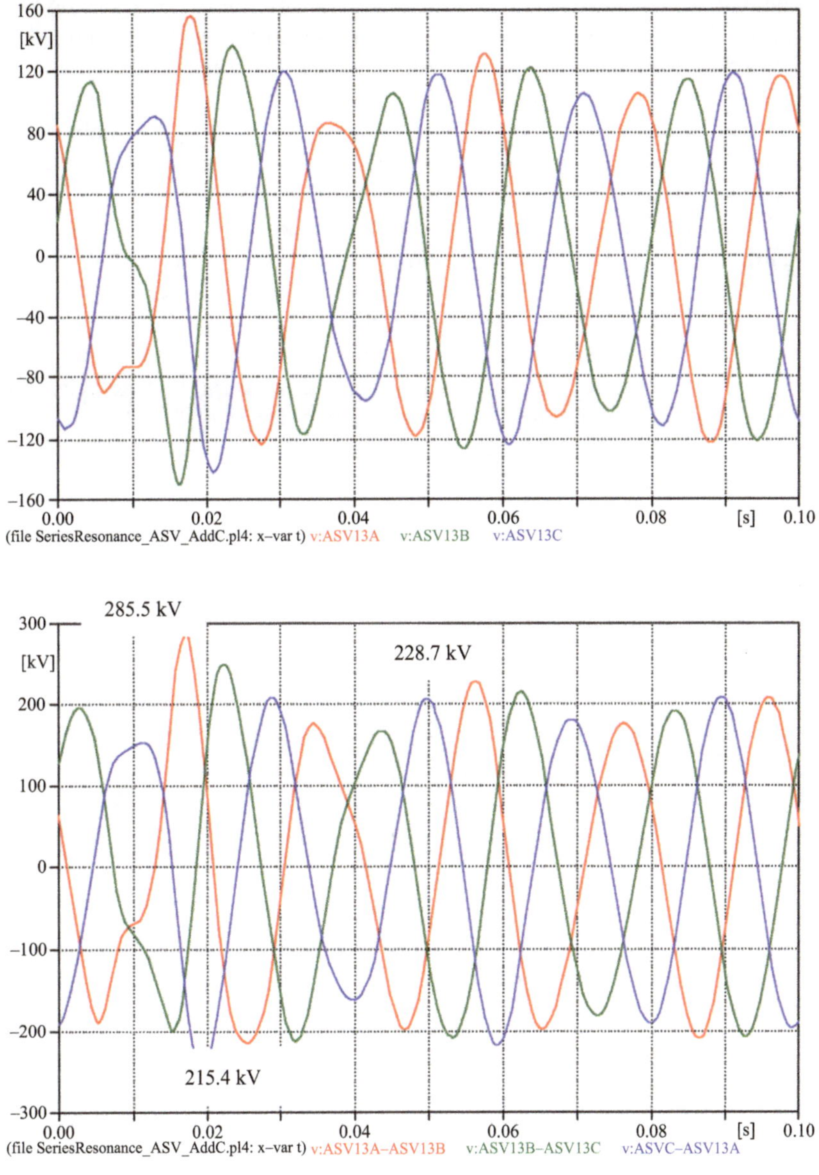

Figure 6.69 Series resonance overvoltage in the ASV 132 kV bus in the no-load condition (upper: phase to ground; lower: phase to phase)

6.3.2 Parallel resonance overvoltage

6.3.2.1 Most severe scenario

The transformer energization will be studied in this section as the most severe switching scenario. In order to cause parallel resonance, a transformer needs to be energized through a cable whose charging current is compensated by shunt reactors.

Figure 6.70 shows the assumed most severe scenario for the energization of the 400/132 kV Kyndbyværket transformer. The transformer is energized through the Kyndbyværket–Asnæsværket line. The voltage source for the energization is assumed on the Bjæverskov side, and the source impedance is adjusted so that the natural frequency of the parallel resonance circuit becomes 100 Hz. The Kyndbyværket–Hovegård line is assumed to be opened in order to have a large source impedance (weak network) and to limit the propagation path of the inrush current.

If the generator at Asnæsværket has black start capability, the Kyndbyværket–Hovegård line can be opened assuming the black start operation from Asnæsværket. It, in theory, causes more severe overvoltages because of limited propagation paths of the inrush current and lower load level.

Figure 6.71 shows the assumed most severe scenario for the energization of the 400/132 kV Asnæsværket transformer and phase-shifting transformer. The transformers are energized through the Kyndbyværket–Asnæsværket line.

As the generator at Kyndbyværket has the black start capability, the most severe scenario assumes the black start operation from Kyndbyværket. As such, the following three lines are assumed not to be in service:

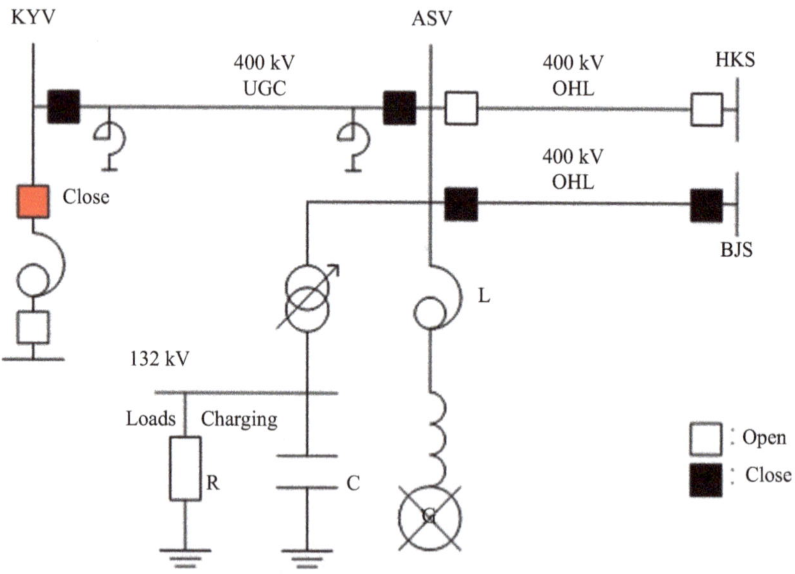

Figure 6.70 Assumed most severe scenario for the energization of KYV transformer

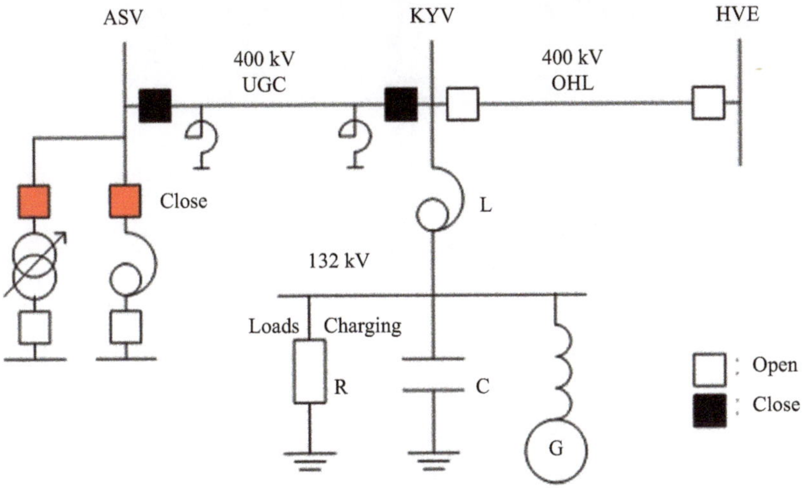

Figure 6.71 Assumed most severe scenario for the energization of ASV transformer

- Asnæsværket–Herslev line
- Asnæsværket–Bjæverskov line
- Kyndbyværket–Hovegård line

It may not be possible to adjust the natural frequency of the parallel resonance frequency exactly at 100 Hz because of the limited number of available generators at Kyndbyværket. Even with that condition, severe overvoltages are often caused due to limited propagation paths of the inrush current and lower load level.

6.3.2.2 Natural frequency of parallel resonance circuit

First, the energization of the 400/132 kV Kyndbyværket transformer is studied. The natural frequencies of the network are found by frequency scan using the simulation model illustrated in Figure 6.72. The operating condition of the network is modified according to the most severe scenario in Figure 6.70.

Since it is expected to have the natural frequency at 100 Hz, the target frequency of all the Bergeron models is set to 100 Hz.

Results of the frequency scan are shown in Figure 6.73. Natural frequencies are found at 92 Hz and 324 Hz in the figure. The magnitude of the impedance is 1306 ohm at 92 Hz and goes down to 878 ohm at 100 Hz.

To have a natural frequency at 92 Hz means that the network is already too weak in the off-peak condition. In order to have a peak at 100 Hz, the dummy source impedance 450 mH is added to the Bjæverskov 400 kV bus as illustrated in Figure 6.72.

The results of the frequency scan after the adjustment are shown in Figure 6.74. The figure shows that the natural frequency is successfully shifted from 92 Hz to 100 Hz. The magnitude of the impedance at 100 Hz is increased to 1376 ohm.

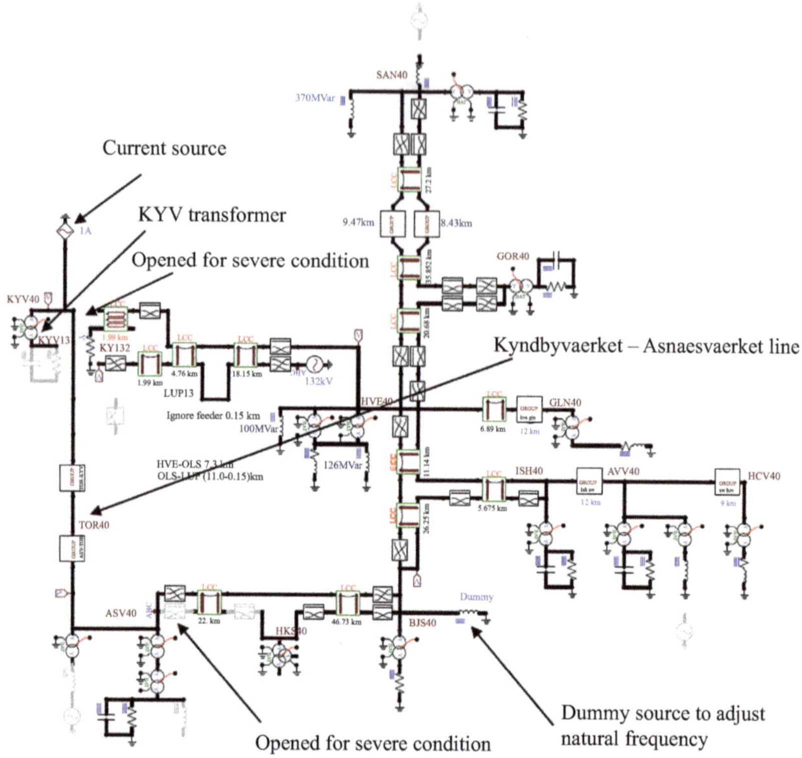

Figure 6.72 Simulation model for frequency scan for the energization of KYV transformer

Second, the energization of the 400/132 kV Asnæsværket transformer and phase-shifting transformer is studied. Assuming the black start restoration from the KYV generator, the operating condition of the network is modified according to the most severe scenario in Figure 6.71.

In the black start restoration scenario, the following steps are assumed:

1. Black start Kyndbyværket 132 kV generator
2. Restore the Kyndbyværket 132 kV load 20 MW
3. Energize the Kyndbyværket 400/132 kV transformer from the secondary side
4. Energize the Kyndbyværket–Asnæsværket line from the Kyndbyværket side
5. Energize the 400/132 kV Asnæsværket transformer or phase shifting transformer

The parallel resonance overvoltage is studied in Step (5).

Figure 6.75 shows the results of the frequency scan. Initially, only one Kyndbyværket 132 kV generator, Machine 224 in the provided PSS/E data, is assumed to be in-service. The source impedance 578.5 mH which corresponds to this generator is connected to the Kyndbyværket 132 kV bus. The figure shows that

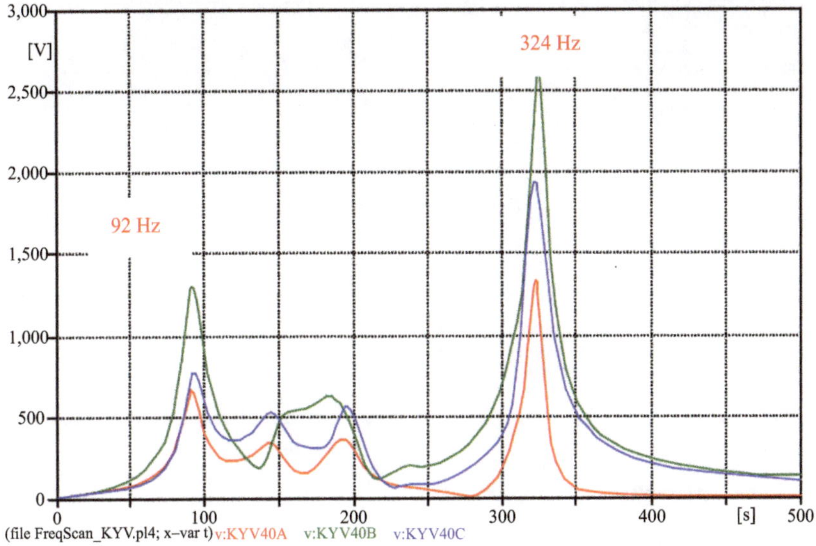

Figure 6.73 Natural frequencies of the network in the energization of KYV transformer in the off-peak load condition

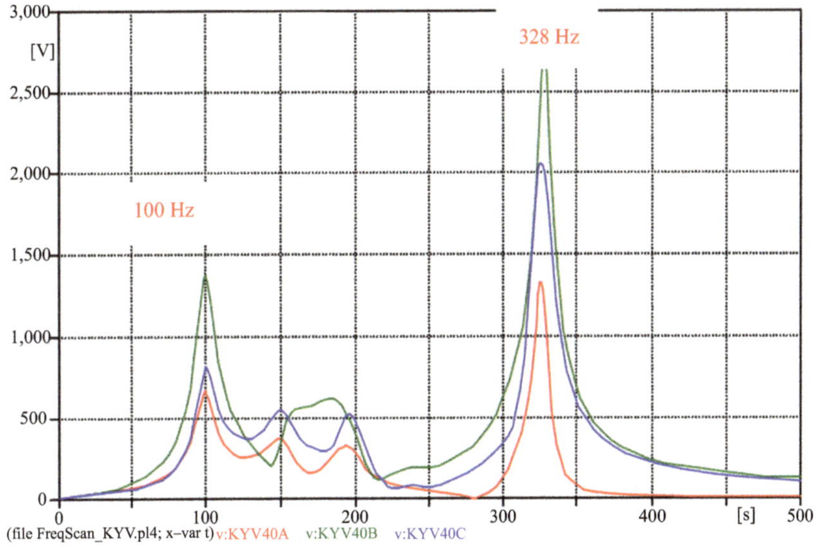

Figure 6.74 Natural frequencies of the network in the energization of KYV transformer after the source impedance adjustment

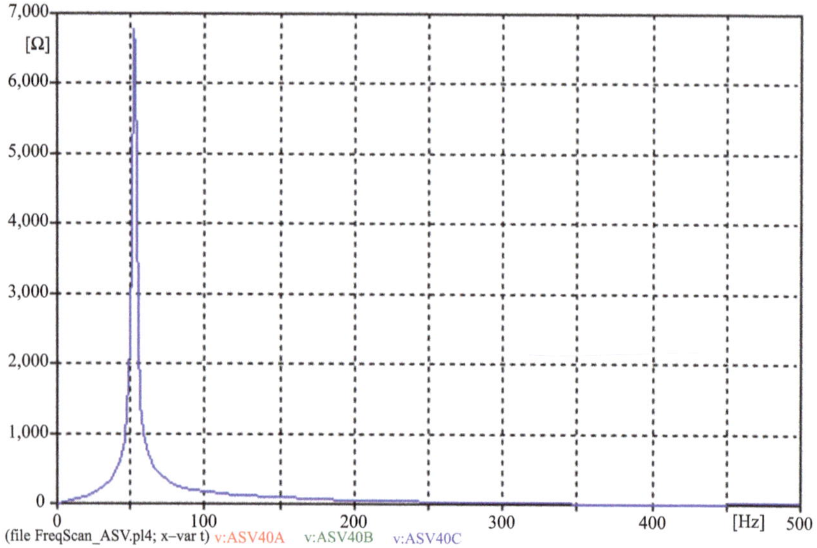

Figure 6.75 Natural frequencies of the network in the energization of ASV transformer in the black start restoration

the network has a very sharp peak at 52 Hz. The magnitude of the impedance is 6773 ohm, but it goes down to 168 ohm at 100 Hz.

Considering the compensation rate of the cable, the natural frequency of the network is located at 50 Hz without a source impedance and loads. The results show that the source impedance and the loads increase the natural frequency only by 2 Hz because of the following reasons:

- The source impedance is very large since the capacity of the assumed black start generator is small (28.57 MVA).
- The source impedance is large since there is only one 400/132 kV Kyndbyværket transformer.
- The load level is very low (20 MW) due to the assumed black start restoration scenario.

In order to shift the natural frequency to 100 Hz, it is necessary to assume a smaller source impedance. Assuming many other generators in the Kyndbyværket 132 kV network are in-service, the source impedance is decreased from 578.5 mH to 8 mH. The source impedance 8 mH corresponds to the fault current level 30.3 kA at the Kyndbyværket 132 kV bus. As shown in Figure 6.76, the natural frequency is successfully shifted to 100 Hz. The magnitude of the impedance is 9,439 ohm at 100 Hz.

As it is assumed that many other generators are in-service, it is not realistic to assume that only 20 MW load is restored at the Kyndbyværket 132 kV. In Figure 6.77, the Kyndbyværket 132 kV load is increased from 20 MW to the

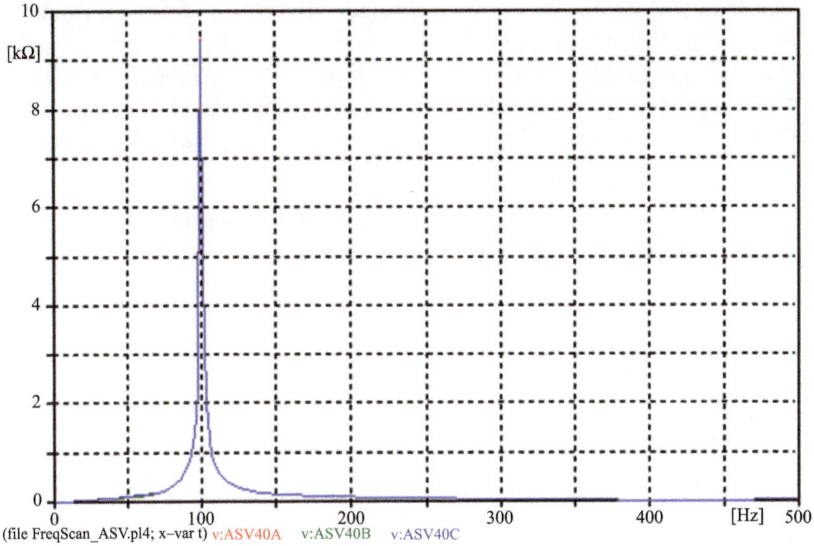

Figure 6.76 Natural frequencies of the network in the energization of ASV transformer after the source impedance adjustment

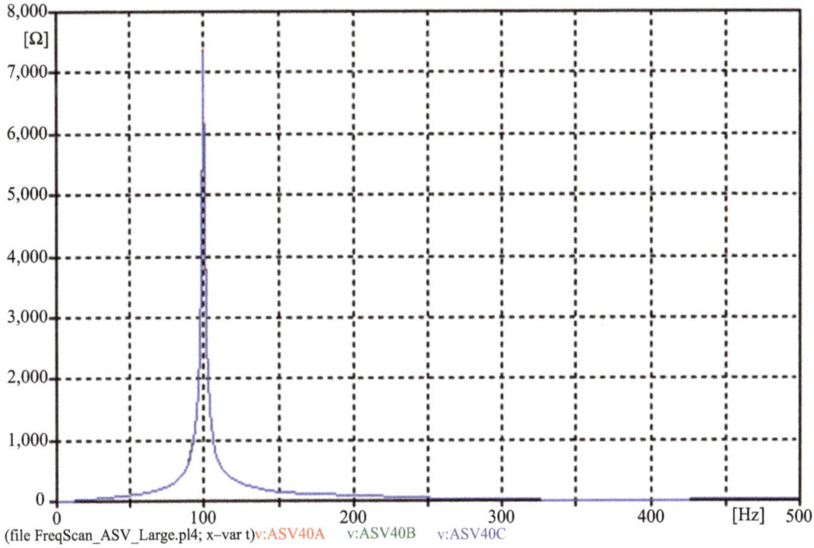

Figure 6.77 Natural frequencies of the network in the energization of ASV transformer after the source impedance adjustment with increased load

off-peak load 67.8 MW. The load increase does not shift the natural frequency, but lowered the magnitude of the impedance to 7,364 ohm.

Even though, the natural frequency is successfully shifted to 100 Hz, it may not be realistic to assume the fault current level 30.3 kA at Kyndbyværket 132 kV bus. In order to make it more realistic the following modified restoration scenario is considered:

1. Black start Kyndbyværket 132 kV generators
2. Restore the Kyndbyværket 132 kV load 67.8 MW (off-peak)
3. Energize the Kyndbyværket 400/132 kV transformer from the secondary side
4. Energize the Kyndbyværket–Hovegård line from the Kyndbyværket side
5. Energize the two Hovegård 400/132 kV transformers
6. Start up generators connected to the Hovegård 132 kV network
7. Restore the Hovegård 132 kV load 151.2 MW (off-peak)
8. Energize the Kyndbyværket–Asnæsværket line from the Kyndbyværket side
9. Energize the 400/132 kV Asnæsværket transformer or phase shifting transformer

The parallel resonance overvoltage is studied in Step (9).

In the modified scenario, Hovegård is restored before Kyndbyværket in Steps (4)–(7). Since Hovegård has two 400/132 kV transformers, it is easier to have a small source impedance. The simulation model for the modified restoration scenario is illustrated in Figure 6.78

In the modified restoration scenario, it is possible to have a source impedance at the Hovegård 132 kV bus as well as the Kyndbyværket 132 kV bus. First, the source impedance at the Kyndbyværket 132 kV bus is increased from 8 mH to 20 mH. The source impedance 20 mH corresponds to the fault current level 12.1 kA. This is a more realistic assumption compared with the original scenario. The source impedance at the Hovegård 132 kV bus is set to 80 mH in order to have the natural frequency at 100 Hz. The source impedance 80 mH corresponds to the fault current level of 3.0 kA.

Figure 6.79 shows the results of the frequency scan. It is observed in the figure that the natural frequency is successfully set at 100 Hz. The magnitude of the impedance at 100 Hz is 18846 ohm, which is extremely high.

6.3.2.3 Simulation results of parallel resonance overvoltage

First, the energization of the Kyndbyværket 400/132 kV transformer is studied. The natural frequency of the network is set to 100 Hz as adjusted in Figure 6.74. The hysteresis characteristic of the transformer is set as Table 6.12, assuming the positive saturation point at 1.4 pu. The remanent flux is assumed in phase-a (85%) and phase-b (−85%).

Results of the time domain simulation are shown in Table 6.23. Since the inrush current highly depends on the switch timing, the switch timing is changed from 0 ms to 19 ms by 1 ms step. The highest overvoltage 469.5 kV is observed when the switch timing is 11 ms. The waveforms of the parallel resonance overvoltage and inrush current with the switch timing 11 ms are shown in Figures 6.80 and 6.81.

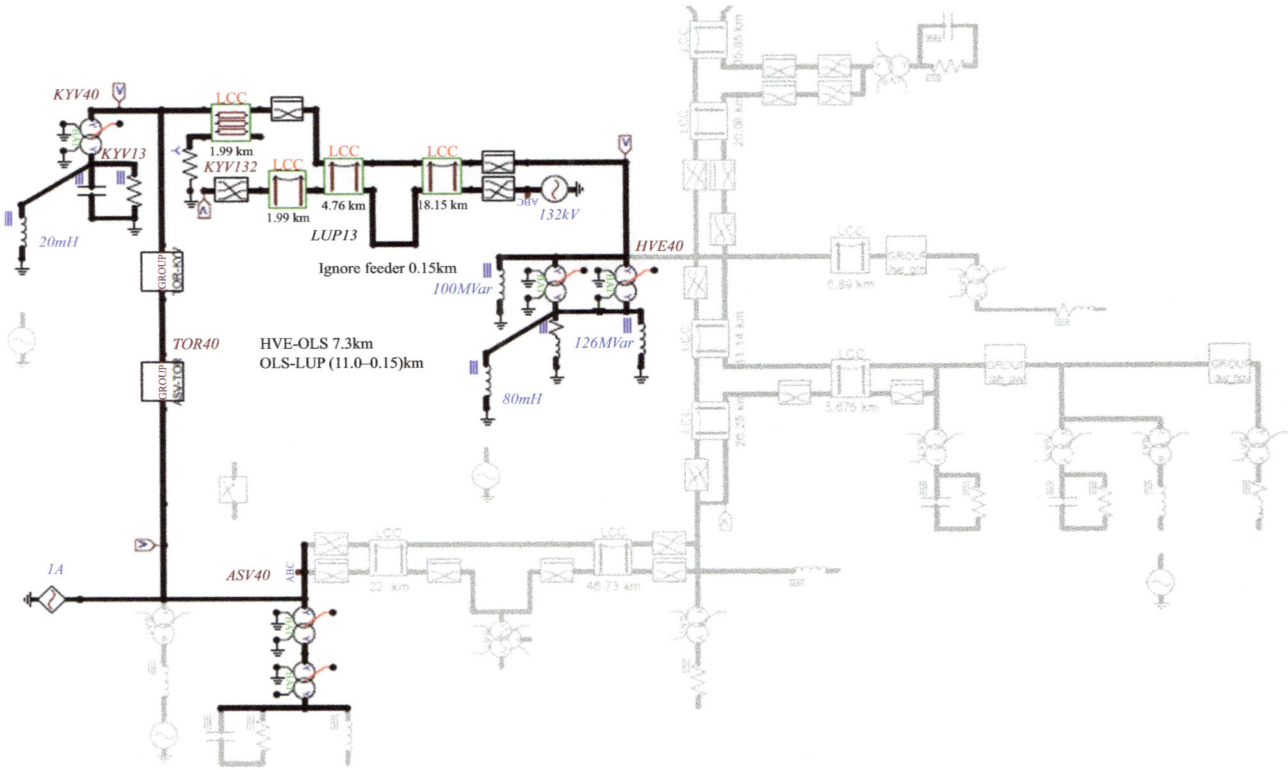

Figure 6.78 Simulation model for frequency scan for the energization of ASV transformer after HVE is restored

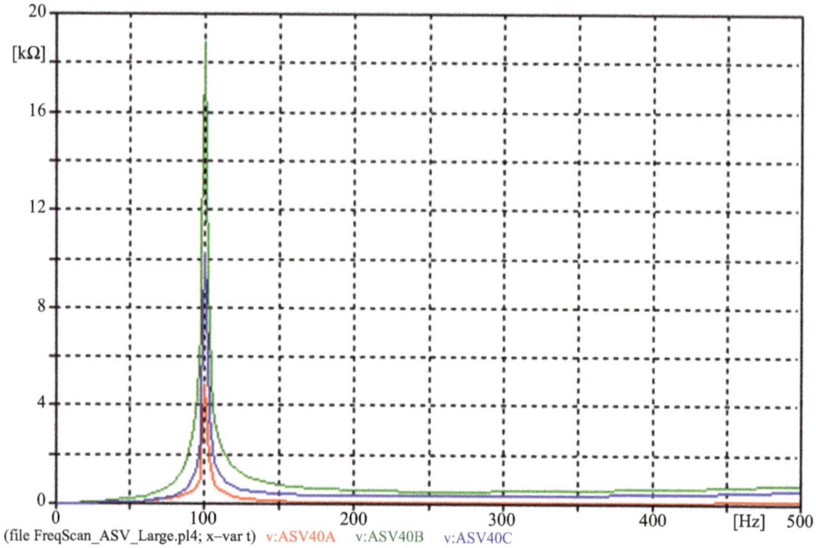

Figure 6.79 Natural frequencies of the network in the energization of ASV transformer after HVE is restored and the source impedance is adjusted

Table 6.23 Parallel resonance overvoltage caused by the KYV transformer energization

Switch time	Overvoltage (kV)		
	Phase-a	Phase-b	Phase-c
0 ms	360.4	411.7	395.0
1 ms	377.9	405.5	398.1
2 ms	394.6	391.1	397.1
3 ms	409.6	393.2	374.2
4 ms	443.8	387.0	385.2
5 ms	435.8	422.5	392.8
6 ms	420.8	427.7	400.4
7 ms	402.7	406.7	410.2
8 ms	436.8	408.4	430.1
9 ms	448.6	409.6	434.7
10 ms	438.8	423.4	459.0
11 ms	433.8	430.5	469.5
12 ms	433.1	422.0	465.8
13 ms	411.5	407.8	435.3
14 ms	401.1	424.5	421.6
15 ms	413.6	428.0	420.2
16 ms	430.9	433.7	398.8
17 ms	424.5	392.9	378.0
18 ms	418.5	400.3	373.6
19 ms	400.6	397.3	381.4

The parallel resonance overvoltage in Figure 6.80 is not very high. Even if the highest overvoltage level is continued for 10 s, the overvoltage is within withstand voltages of related equipment (1.60 pu for the assumed arrester). The magnitude of the inrush current in Figure 6.81 is reasonable as the highest inrush current 4008 A is eight times larger than the rated current of the transformer (350 MVA, 505 A).

Second, the energization of the Asnæsværket 400/132 kV transformer is studied. The natural frequency of the network is set to 100 Hz as adjusted in Figures 6.78 and 6.79. The hysteresis characteristic of the transformer is set as Table 6.12, assuming the positive saturation point at 1.4 pu. The remanent flux is assumed in phase-a (85%) and phase-b (−85%).

Results of the time domain simulation are shown in Table 6.24. Since the inrush current highly depends on the switch timing, the switch timing is changed from 0 ms to 19 ms by 1 ms step. The highest overvoltage 702.6 kV is observed when the switch timing is 14 ms. The waveforms of the parallel resonance overvoltage and inrush current with the switch timing 14 ms are shown in Figures 6.82 and 6.83.

Even though the observed highest overvoltage 702.6 kV is very high as a temporary overvoltage, it is clear from the waveform that the overvoltage exceeding 700 kV lasts only for one impulse. It is thus reasonable to evaluate the overvoltage against the switching impulse withstand overvoltage 1050 kV and conclude the observed overvoltage is within the equipment standards. The highest overvoltage after a half cycle is 518.7 kV, which is also within withstand voltages of related equipment (1.60 pu–10 s for the assumed arrester).

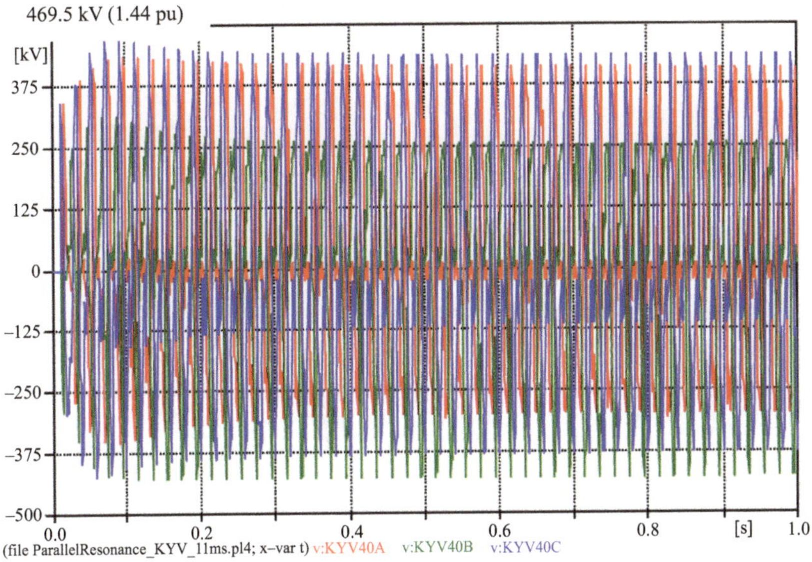

Figure 6.80 Parallel resonance overvoltage caused by the KYV transformer energization

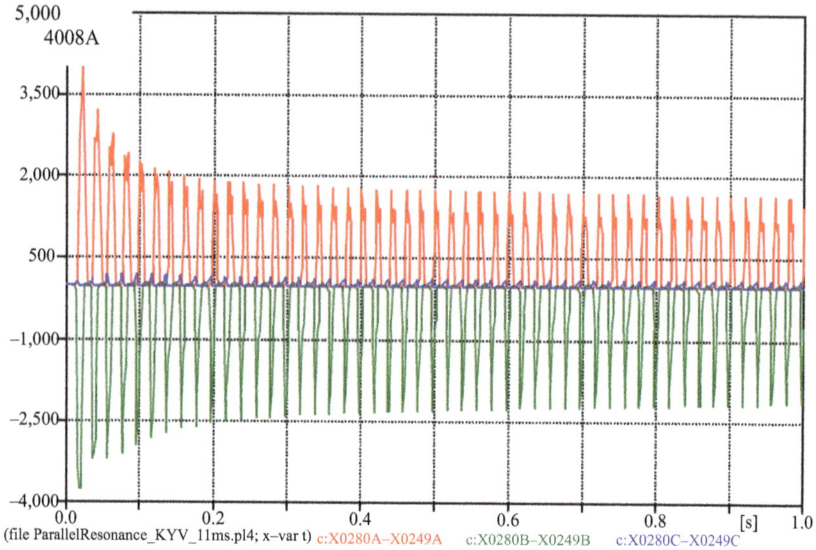

Figure 6.81 Inrush current in the KYV transformer energization

Table 6.24 Parallel resonance overvoltage caused by the ASV transformer energization

Switch time	Overvoltage (kV)		
	Phase-a	**Phase-b**	**Phase-c**
0 ms	567.0	386.6	559.5
1 ms	540.7	357.8	569.2
2 ms	356.8	329.2	387.5
3 ms	326.4	327.8	329.7
4 ms	332.6	330.4	350.3
5 ms	360.5	363.8	492.0
6 ms	399.5	569.2	576.5
7 ms	384.4	457.5	491.5
8 ms	347.0	498.9	435.5
9 ms	553.3	647.6	433.1
10 ms	503.2	647.6	530.4
11 ms	425.6	684.0	617.3
12 ms	632.5	653.1	473.4
13 ms	623.9	676.3	386.9
14 ms	702.6	575.5	344.4
15 ms	664.4	544.8	550.0
16 ms	699.2	419.3	597.1
17 ms	583.7	570.6	436.3
18 ms	594.4	508.1	435.9
19 ms	432.3	409.2	440.1

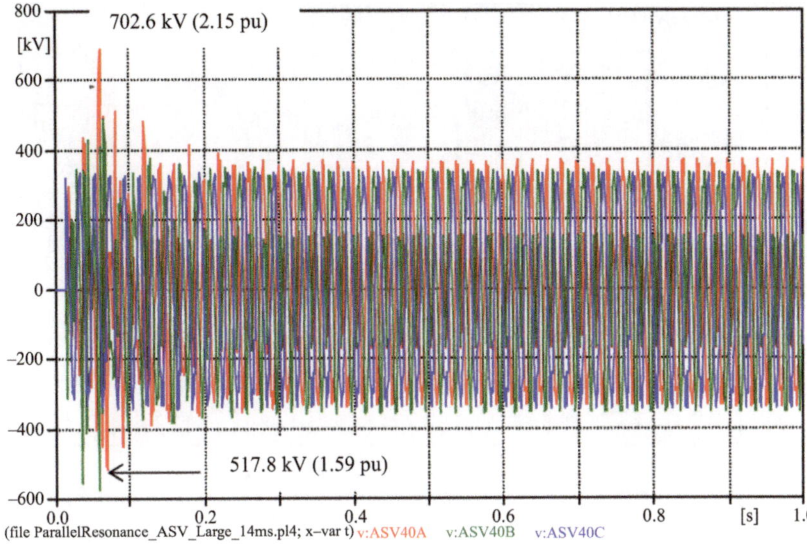

Figure 6.82 Parallel resonance overvoltage caused by the ASV transformer energization

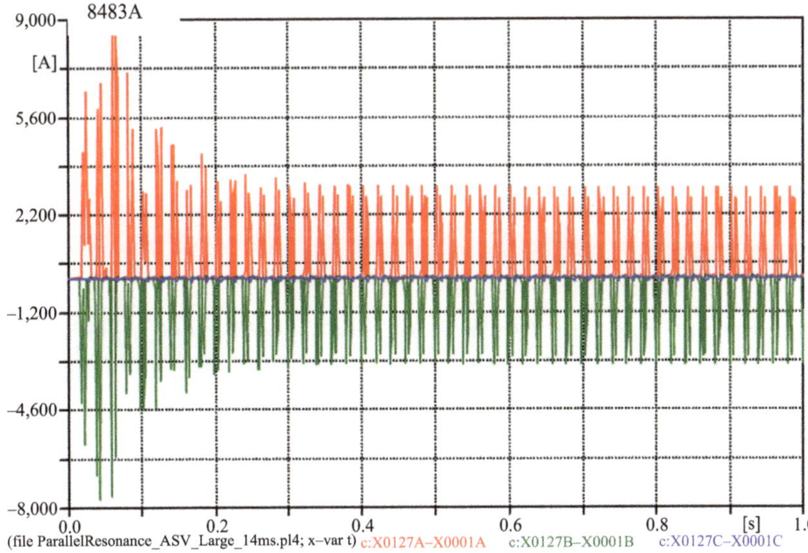

Figure 6.83 Inrush current in the ASV transformer energization

The highest inrush current 8,483 A in Figure 6.83 is too large, compared with the typical inrush current, as it is 12 times larger than the rated current of the transformer (500 MVA, 722 A).

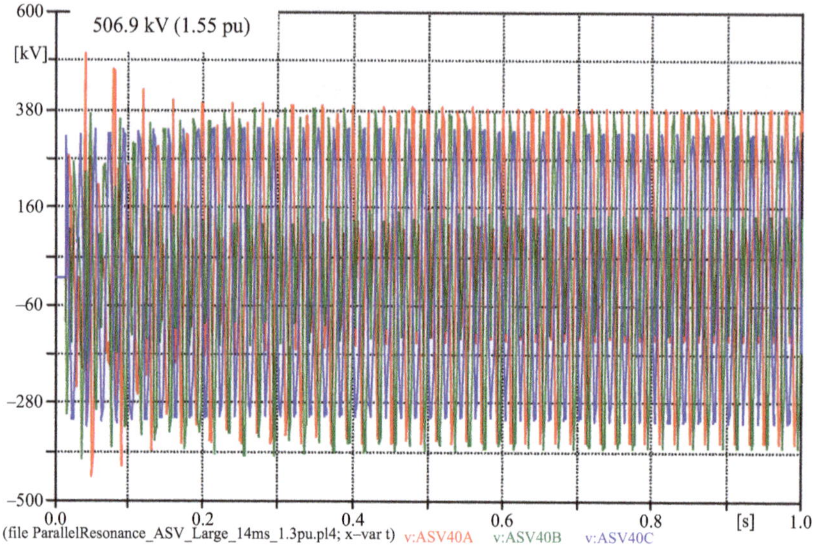

Figure 6.84 Parallel resonance overvoltage caused by the ASV transformer energization (positive saturation point 1.3 pu)

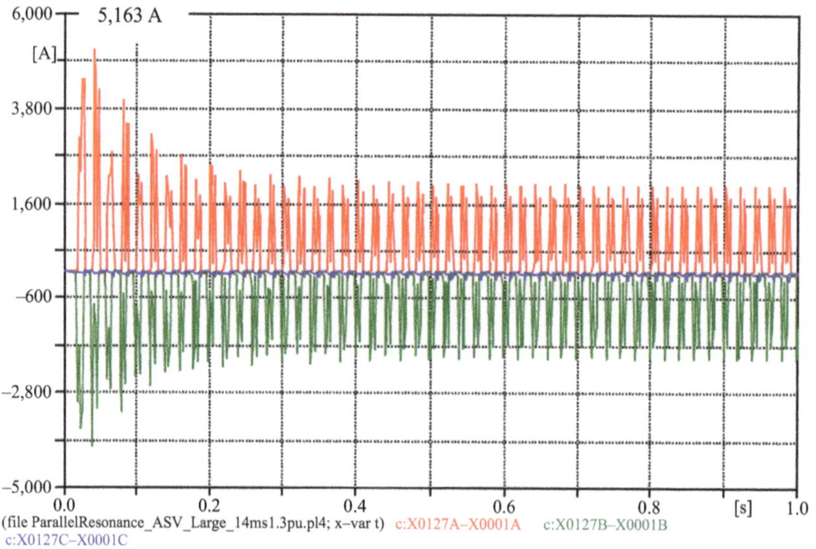

Figure 6.85 Inrush current in the ASV transformer energization (positive saturation point 1.3 pu)

As discussed in Section 6.2.4.1, the positive saturation point 1.4 pu is a severe assumption, which caused the large inrush current 8,483 A. When the positive saturation point 1.3 pu is assumed, the highest overvoltage is lowered to 506.9 kV. The waveforms of the parallel resonance overvoltage and inrush current under this moderate assumption are shown in Figures 6.84 and 6.85. The observed overvoltages are within withstand voltages of related equipment. The highest inrush current 5,163 A is reasonable as it is seven times larger than the rated current of the transformer.

6.3.3 Overvoltage caused by system islanding

6.3.3.1 Study conditions

This system islanding can easily occur when the long cable is installed in a radial network. As the Kyndbyværket–Asnæsværket line is installed as a part of the loop network, the system islanding can only occur in specific conditions. Among these specific conditions, an outage of the Hovegård–Kyndbyværket line (under 400 kV operation) is the most likely precondition in which an Asnæsværket 400 kV bus fault can lead to the system islanding. Figure 6.86 illustrates the network condition that leads to the system islanding. The islanded system is actually connected to the main grid through the 132 kV system, but the connection to the main grid is weak enough to cause the oscillatory overvoltage. The system islanding actually occurs in simulations as the simulation model in PSCAD does not include the connection in the 132 kV system. At this point, the study condition is more severe than the actual system.

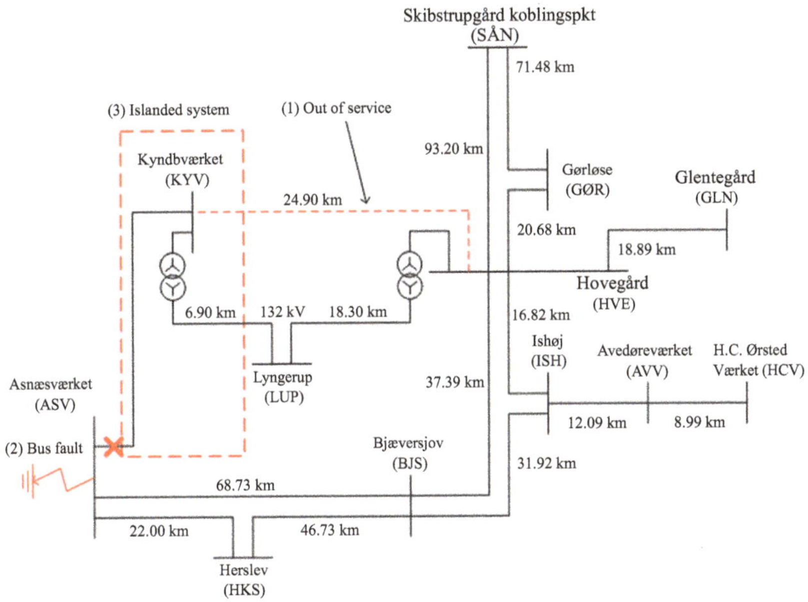

Figure 6.86 ASV 400 kV bus fault and the system islanding

Generators in the Kyndbyværket 132 kV network need to be modeled in order for the islanded system to experience the oscillatory overvoltage. A dummy generator was connected to the Kyndbyværket 132 kV bus through the source impedance whose value was considered as a parameter in order to study different fault current levels. Table 6.25 shows the fault current levels and their associated values of the source impedance considered in the analysis. Since there are no other generators in the islanded system, the calculation of the source impedance is straightforward. The maximum fault current level was set to 40 kA since the typical breaking capability of the 132 kV breakers in the Danish network was 40 kA.

Now, we consider study conditions where the Kyndbyværket 400 kV bus fault results in the system islanding. As the 400 kV transmission lines from Asnæsværket to Hovegård via Bjæverskov are all double circuit lines, the Kyndbyværket 400 kV bus fault will not lead to the system islanding in normal or N-1 preconditions. When we consider an outage of the Asnæsværket–Herslev

Table 6.25 Source impedances of the KYV 132 kV dummy generator

Fault current level at the KYV 132 kV bus (kA)	5	10	20	30	40
Source impedance of the dummy generator (mH)	48.5	24.3	12.1	8.09	6.06

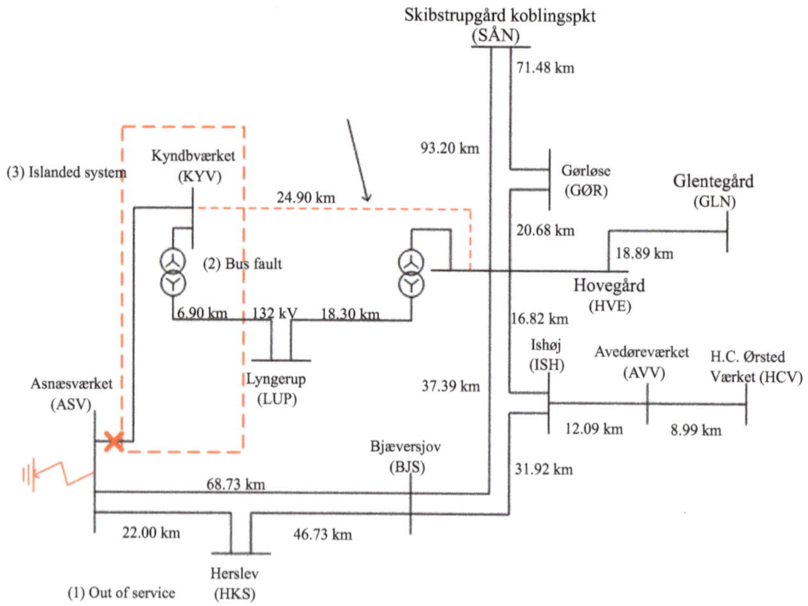

Figure 6.87 KYV 400 kV bus fault and the system islanding

Table 6.26 Source impedances of the ASV 132 kV dummy generator

Fault current level at the ASV 132 kV bus (kA)	10	20	30	40
Fault current fed from the dummy generator (kA)	1.30	11.3	21.3	31.3
Source impedance of the dummy generator (mH)	186	21.5	11.4	7.75

Table 6.27 Study conditions for the overvoltage caused by the system islanding

Fault location	ASV, KYV, TOR 400 kV buses
Fault type	Three-phase fault (no fault resistance)
Fault current level	5, 10, 20, 30, 40 kA at the ASV or KYV 132 kV bus
Shunt compensation	All shunt reactors in service (100.7%). One 150 MVar shunt reactor out of service (75.5%)

line, however, the Kyndbyværket–Asnæsværket line is only weakly connected to the main grid via Bjæverskov in case of the Kyndbyværket 400 kV bus fault. As the Asnæsværket 400 kV bus is approximately 69 km away from Bjæverskov, the oscillatory overvoltage may occur depending on the fault current level at Asnæsværket. Figure 6.87 shows the network condition which may lead to the oscillatory overvoltage.

Different fault current levels were considered also for the Kyndbyværket 400 kV bus fault. Based on the off-peak load condition, the fault current level at the Asnæsværket 132 kV bus was adjusted to 10, 20, 30, and 40 kA by adding a dummy generator to the Asnæsværket 132 kV bus. As the fault current level at the Asnæsværket 132 kV bus is 8.70 kA in the off-peak load condition, the remaining fault current has to be fed from the dummy generator as in Table 6.26.

Table 6.27 summarizes the study conditions for the overvoltage caused by the system islanding. The fault in the Torslunde 400 kV bus was additionally studied assuming there is a switching station at Torslunde. As for the fault type, only the three-phase fault was considered as it was the most severe fault with respect to the overvoltage caused by the system islanding.

While the charging capacity of the Kyndbyværket–Asnæsværket line cannot be changed, two other conditions, the source impedance and the shunt compensation can be changed as the study condition. According to the results of the reactive power compensation study, one unit of 300 MVar shunt reactor needs to be connected to both sides of the Kyndbyværket–Asnæsværket line. In this case, all shunt reactors have to be in service, which results in a compensation rate of 100.7%. When two units of 150 MVar shunt reactors are connected to both sides, one of the four shunt reactors can be taken out of service, which lowers the compensation rate to 75.5%. This condition was studied as a more severe assumption.

6.3.3.2 Simulation results of overvoltage caused by system islanding

6.3.3.3 ASV 400 kV bus fault

6.3.3.3.1 All shunt reactors in service

First, the simulations were performed with all reactors in service. Figure 6.88 shows the overvoltage caused by the system islanding with the fault current level of 5 kA at the Kyndbyværket 132 kV bus. The highest overvoltage 744.6 kV was observed at the open-end of the cable in phase-c.

From the waveform of the overvoltage, it is clear that the resonant frequency is superimposed on the nominal frequency. The resonant frequency can be found by the Fourier transform of the waveform. Figure 6.89 shows frequency components

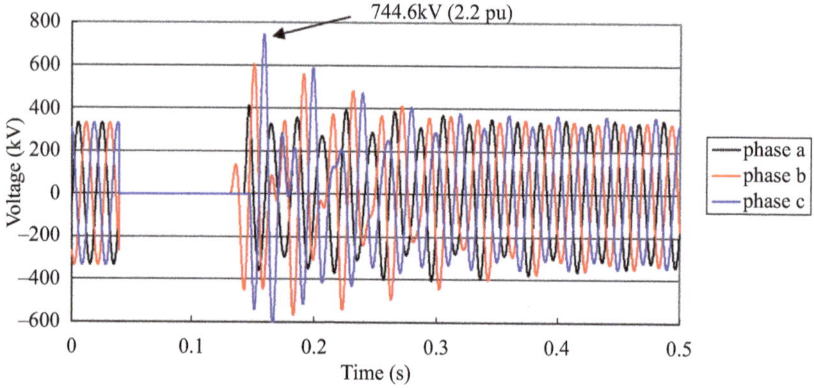

Figure 6.88 Overvoltage caused by the ASV bus fault leading to the system islanding (all shunt reactors in service, fault current level: 5 kA)

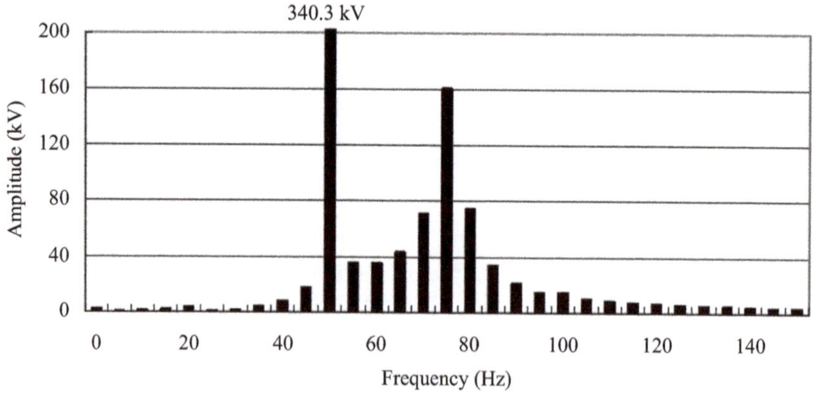

Figure 6.89 Frequency components contained in the overvoltage caused by the ASV bus fault leading to the system islanding (all shunt reactors in service, fault current level: 5 kA)

contained in the overvoltage. The resonant frequency of 75 Hz was found from the Fourier transform of the waveform of the overvoltage.

Here the resonant frequency can also be derived theoretically from Equation (6.4) [16]. Here, L_0 is the source impedance. C and L are the charging capacities of the cable and the inductance of the shunt reactors directly connected to the cable. The theoretical derivation gives the resonant frequency 76.6 Hz, which is close enough to the one derived from simulation results:

$$\omega_0 = \sqrt{\frac{1}{CL_0} + \frac{1}{CL}} \tag{6.4}$$

The overvoltage when the fault current level is increased to 10, 20, 30, and 40 kA is shown in Figures 6.90–6.93.

Figure 6.94 shows the comparison of resonant frequencies derived from the theoretical formula and simulation results. The resonant frequencies derived from

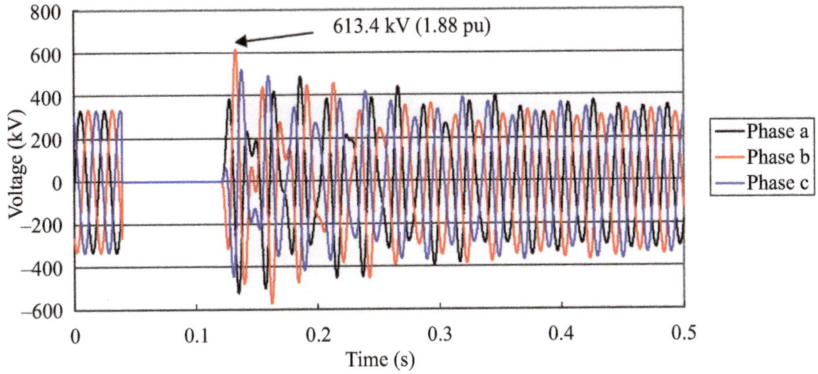

Figure 6.90 Overvoltage caused by the ASV bus fault leading to the system islanding (all shunt reactors in service, fault current level: 10 kA)

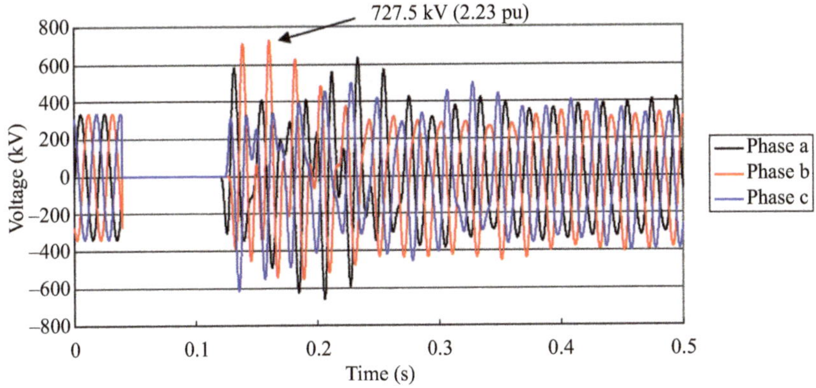

Figure 6.91 Overvoltage caused by the ASV bus fault leading to the system islanding (all shunt reactors in service, fault current level: 20 kA)

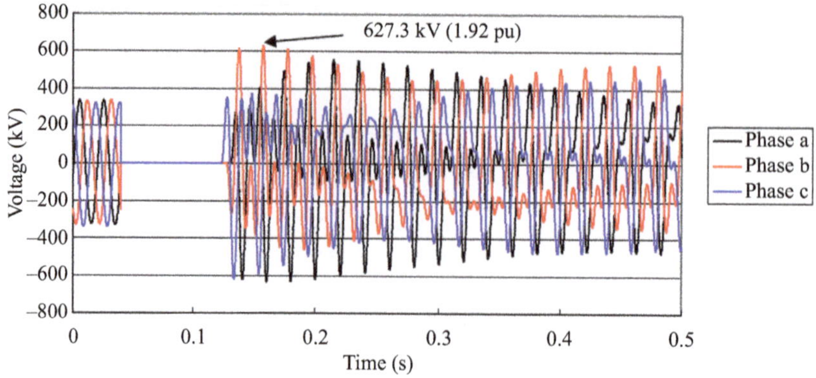

Figure 6.92 *Overvoltage caused by the ASV bus fault leading to the system islanding (all shunt reactors in service, fault current level: 30 kA)*

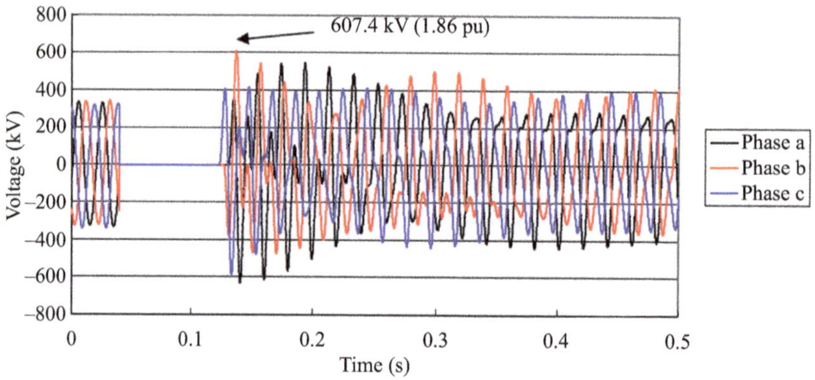

Figure 6.93 *Overvoltage caused by the ASV bus fault leading to the system islanding (all shunt reactors in service, fault current level: 40 kA)*

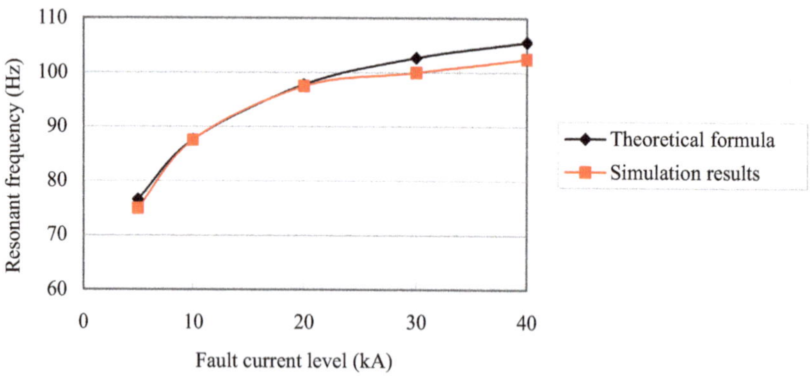

Figure 6.94 *Resonant frequencies derived from theoretical formula and simulation results*

simulation results match closely to those derived from the theoretical formula. The deviation of 2.5 Hz can be considered as a round-off error as the Fourier transform identifies the resonant frequency with a 2.5 Hz step. This demonstrates that overvoltages as expected from Equation (6.4) are produced in the simulations.

Comparing the waveforms of the overvoltages, the most severe overvoltage was caused when the fault current level was set to 5 kA or 20 kA. The initial overvoltage level is slightly higher with the fault current level of 5 kA, but the sustained overvoltage with the frequency around 1 Hz can be observed with the fault current level of 20 kA.

The simulations were, so far, performed without a surge arrester. Here, the overvoltage with the fault current level of 20 kA was evaluated against the ratings of the modeled surge arrester.

Figure 6.95 shows the overvoltage with one surge arrester connected to the open-end of the cable when the fault current level was set to 20 kA. The highest overvoltage was lowered by the surge arrester. The low-frequency oscillation (1 Hz) is not damped within 1 s, but the overvoltage level is lowered to 400 kV (1.22 pu) within 0.4 s. Considering the power frequency withstand voltage (1 min) 1.58 pu of the modeled surge arrester, it can endure the overvoltage shown in Figure 6.95.

Figure 6.96 shows the energy absorbed by the surge arrester in the overvoltage. The energy absorbed by the surge arrester is within the energy absorption capability 9.5 MJ of the modeled surge arrester.

6.3.3.3.2 Effects of one shunt reactor out of service

Here, we discuss the overvoltage with one 150 MVar shunt reactor out of service. As the overvoltage is highest at the open-end of the cable, it is most severe to take one shunt reactor at Asnæsværket out of service.

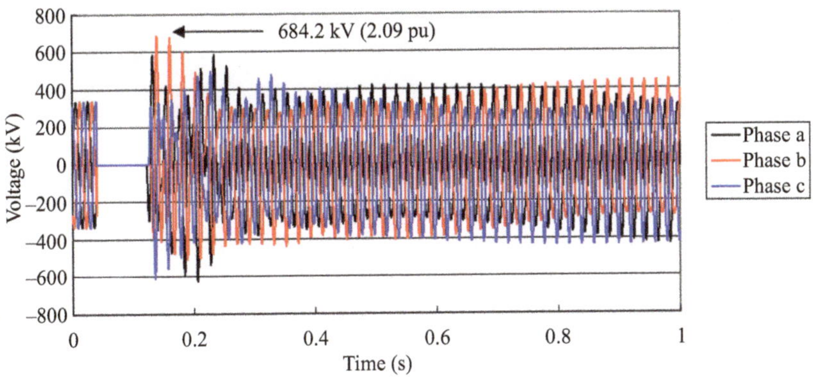

Figure 6.95 Overvoltage caused by the ASV bus fault leading to the system islanding with an arrester (all shunt reactors in service, fault current level: 20 kA)

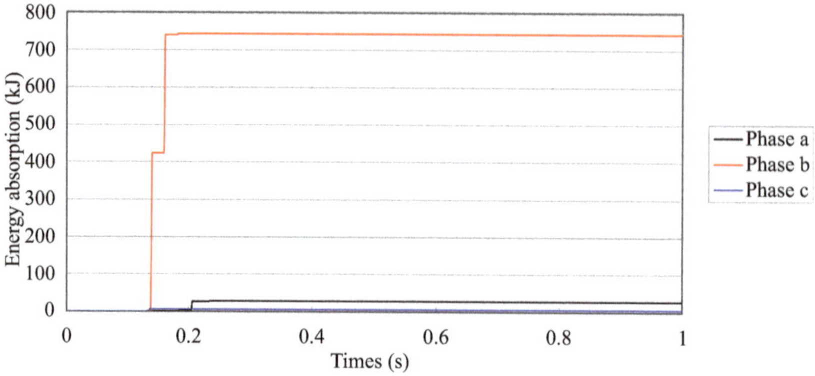

Figure 6.96 Energy absorption by an arrester in the islanding overvoltage (all shunt reactors in service, fault current level: 20 kA)

Figures 6.97–6.101 show the overvoltages caused by the system islanding with the fault current levels of 5–40 kA. The overvoltage level becomes higher due to the lower compensation rate. The sustained overvoltage is observed due to the surplus reactive power supply in the islanded system.

Figure 6.102 shows the comparison of resonant frequencies derived from the theoretical formula and simulation results. The resonant frequencies derived from simulation results match closely to those derived from the theoretical formula, as in the cases with all shunt reactors in service.

Comparing the waveforms of the overvoltages, the most severe overvoltage was caused when the fault current level was set to 5 kA or 40 kA. The initial

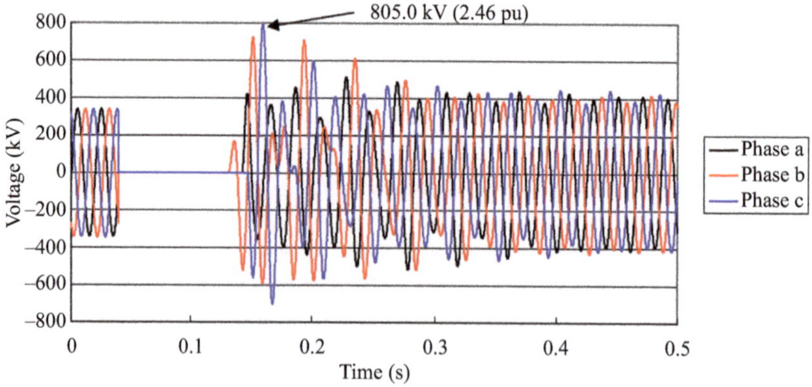

Figure 6.97 Overvoltage caused by the ASV bus fault leading to the system islanding (one shunt reactors out of service, fault current level: 5 kA)

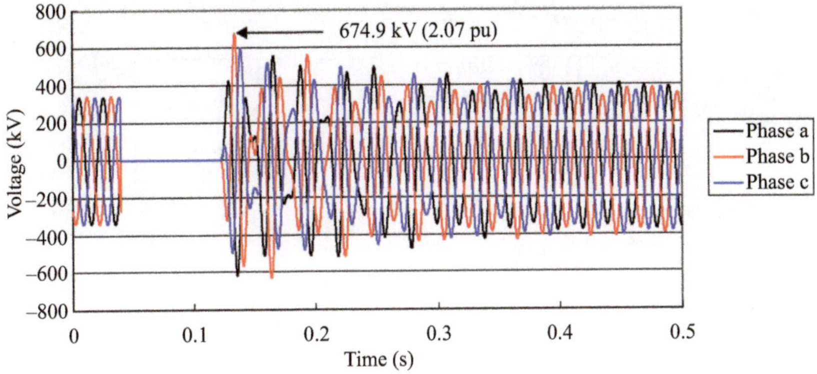

Figure 6.98 Overvoltage caused by the ASV bus fault leading to the system islanding (one shunt reactors out of service, fault current level: 10 kA)

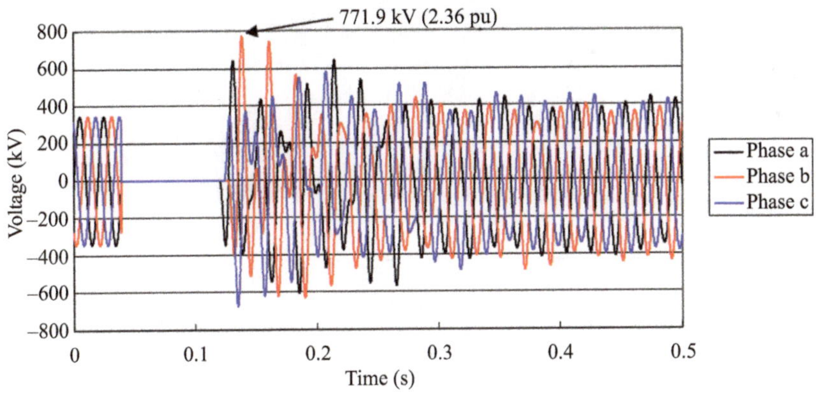

Figure 6.99 Overvoltage caused by the ASV bus fault leading to the system islanding (one shunt reactors out of service, fault current level: 20 kA)

overvoltage level is significantly higher with the fault current level of 5 kA, but the sustained overvoltage with the frequency around 1 Hz can be observed with the fault current level of 40 kA.

Here, the overvoltage with the fault current level of 5 kA was evaluated against the ratings of a surge arrester. Figure 6.103 shows the overvoltage with one surge arrester connected to the open-end of the cable. The overvoltage level is lowered to 400 kV (1.22 pu) within 0.4 s. Considering the power frequency withstand voltage (1 min) 1.58 pu of the modeled surge arrester, it can endure the overvoltage shown in Figure 6.103.

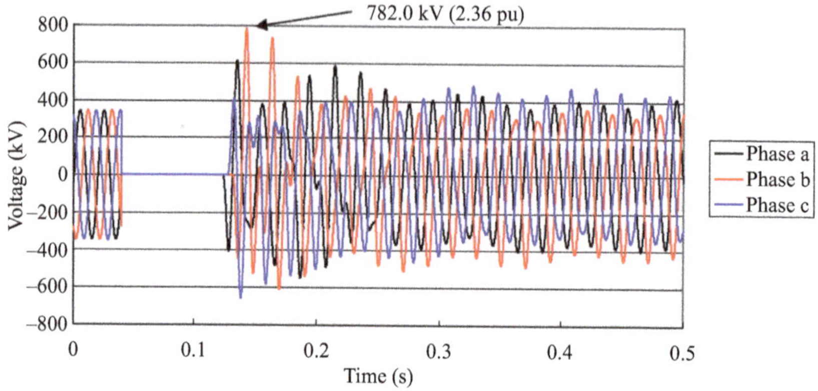

Figure 6.100 Overvoltage caused by the ASV bus fault leading to the system islanding (one shunt reactors out of service, fault current level: 30 kA)

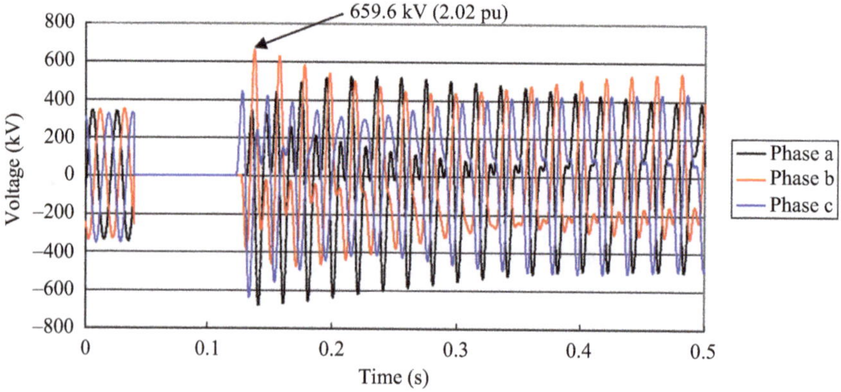

Figure 6.101 Overvoltage caused by the ASV bus fault leading to the system islanding (one shunt reactors out of service, fault current level: 40 kA)

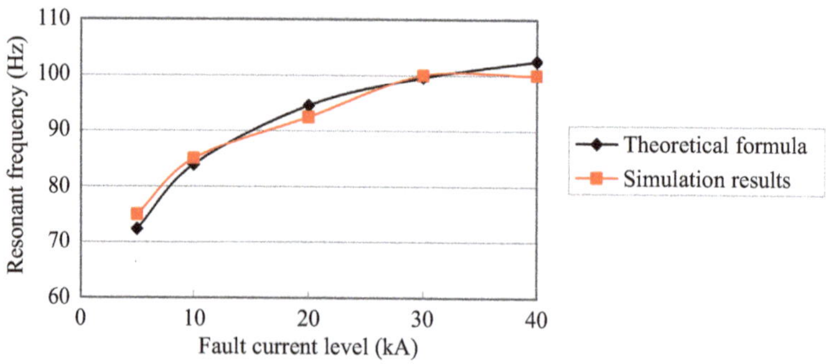

Figure 6.102 Resonant frequencies derived from theoretical formula and simulation results (one shunt reactor out of service)

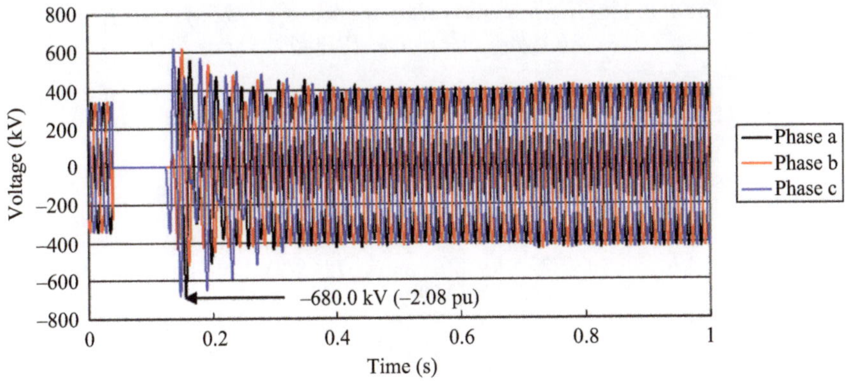

Figure 6.103 Overvoltage caused by the ASV bus fault leading to the system islanding with an arrester (one shunt reactor out of service, fault current level: 5 kA)

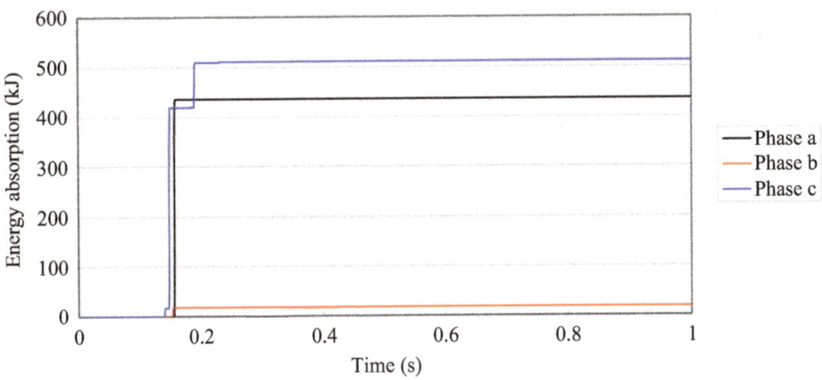

Figure 6.104 Energy absorption by an arrester in the islanding overvoltage (one shunt reactor out of service, fault current level: 5 kA)

Figure 6.104 shows the energy absorbed by the surge arrester in the overvoltage. The energy absorbed by the surge arrester is within the energy absorption capability 9.5 MJ of the modeled surge arrester.

Finally, the highest overvoltages caused by the system islanding due to the Asnæsværket 400 kV bus fault are summarized in Table 6.28. Values in bold show the results with one surge arrester modeled at the open-end of the cable.

6.3.3.3.3 Effects of timing for fault clearing

This part studies the effects of the timing for the fault clearing. The fault clearing always occurs at the current zero, but by changing the timing, it is possible to change the order of the three phases.

Table 6.28 Highest overvoltages caused by the system islanding due to the ASV 400 kV bus fault

Fault current level	All shunt reactors in service	One shunt reactor out of service
5 kA	744.6 kV (2.28 pu)	805.0 kV (2.46 pu) **680.0 kV (2.08 pu)**
10 kA	613.4 kV (1.88 pu)	674.9 kV (2.07 pu)
20 kA	727.5 kV (2.23 pu) **684.2 kV (2.09 pu)**	771.9 kV (2.36 pu)
30 kA	627.3 kV (1.92 pu)	782.0 kV (2.36 pu)
40 kA	607.4 kV (1.86 pu)	659.6 kV (2.02 pu)

Table 6.29 Highest overvoltages caused by the system islanding with different fault clearing timings

Order of the fault clearing	Highest overvoltages	Phase in which highest overvoltages are observed
Phase a→c→b	727.5 kV (2.23 pu)	Phase-b
Phase c→b→a	726.3 kV (2.22 pu)	Phase-a
Phase b→a→c	670.7 kV (2.05 pu)	Phase-c

In this analysis, it was assumed that all shunt reactors were in service. The fault current level at the Kyndbyværket 132 kV bus was set to 20 kA since the most severe overvoltage was observed with this condition.

When the fault current level was 20 kA, the fault clearing occurs in the order of phase-a, -c, and -b. By delaying the fault clearing timing, the order can be changed to phase-c, -b, and -a. Further delay can result in the order of phase-b, -a, and -c. The magnitude of the highest overvoltages is highly affected by the phase angle of the voltage source when the fault is cleared. It is expected that the highest overvoltages are observed in different phases by changing the fault clearing timing.

Table 6.29 summarizes the highest overvoltages and the phases in which the highest overvoltages are observed. As expected, they are affected by the fault clearing timing. The highest overvoltages are observed in the last phase in the order of the fault clearing. The magnitude of the highest overvoltages becomes low for the last order of the fault clearing.

Figure 6.105 shows the waveforms of the overvoltages with different fault clearing timings. The similarity of the waveforms suggests that the frequency components contained in the overvoltages do not change with the fault clearing timing. In addition, it is clear that the fault clearing timing does not have a noticeable effect on the damping characteristic of the overvoltages. These characteristics of the overvoltages are mainly determined by the fault current level, charging capacity, and the compensation rate of the cable line, and are not affected by the fault clearing timing.

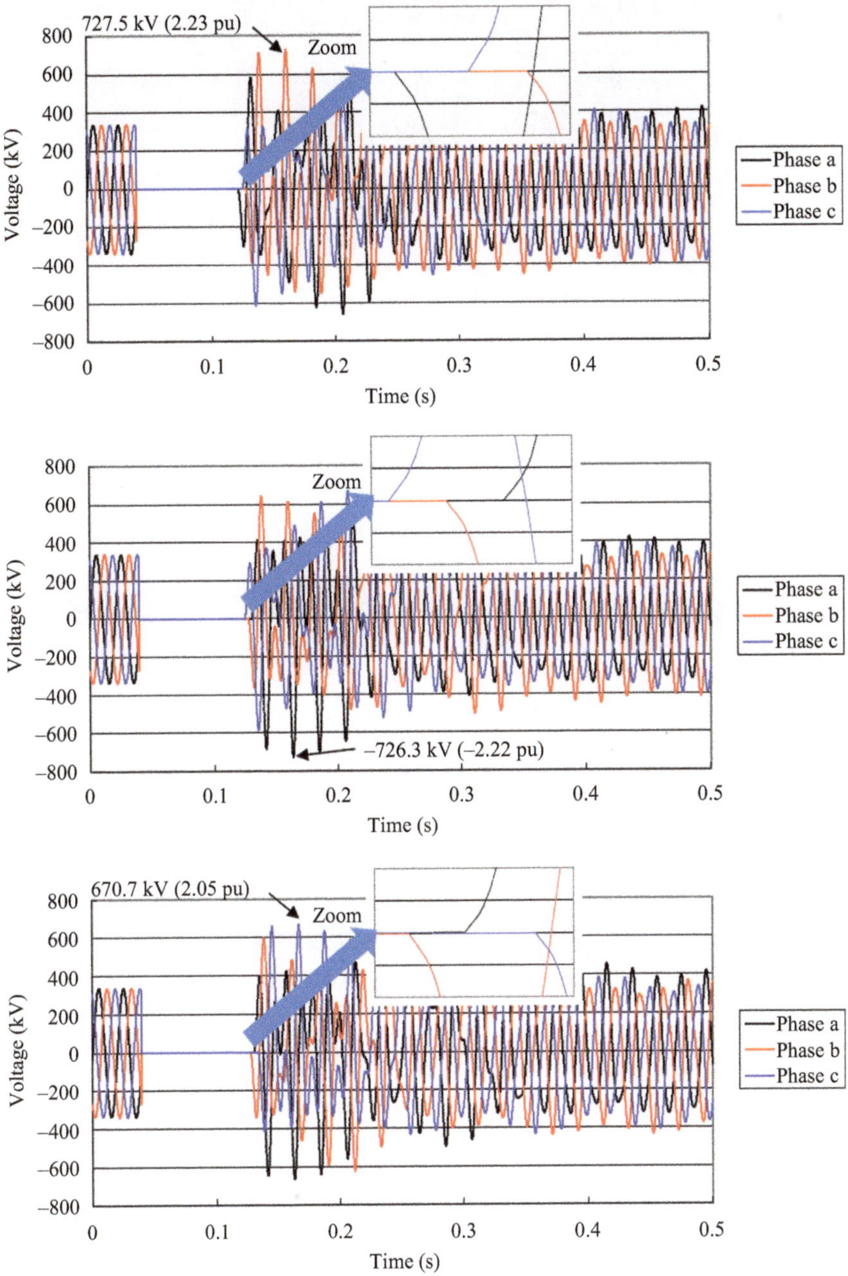

Figure 6.105 Overvoltage caused by the system islanding with different fault clearing timings (top: phase c→b→a; middle: phase b→a→c; and bottom: phase a→c→b)

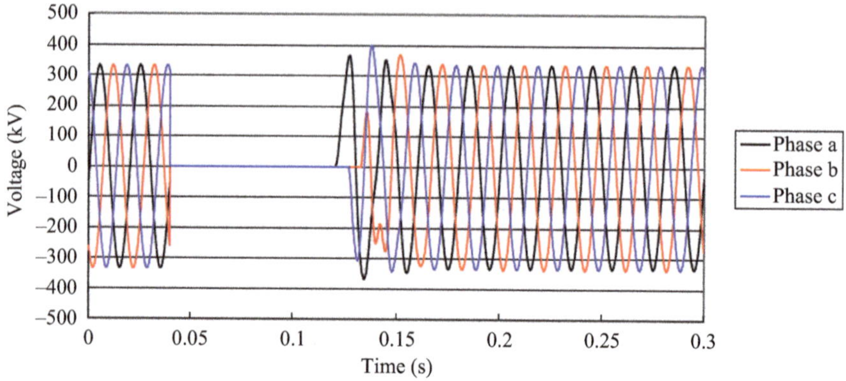

Figure 6.106 Overvoltage caused by the KYV bus fault leading to the system islanding (all shunt reactors in service, fault current level: 10 kA)

6.3.3.4 KYV 400 kV bus fault

As mentioned in Section 6.3.3.1, an outage of the Asnæsværket–Herslev line is considered as the precondition. When the Kyndbyværket 400 kV bus fault occurs, the Asnæsværket 400 kV bus is weakly connected to the main grid via Bjæverskov, which may lead to the oscillatory overvoltage.

Figure 6.106 shows the overvoltages caused by the system islanding with the fault current level of 10 kA. It has been confirmed that the oscillatory overvoltage does not occur with the fault current levels of 10, 20, 30, and 40 kA. The Asnæsværket 400 kV bus is weakly connected to the main grid, but it is enough to damp the oscillatory overvoltage to a harmless level.

6.3.3.5 TOR 400 kV bus fault

The fault in the Torslunde 400 kV bus was additionally studied assuming there is a switching station at the Torslunde as in Figure 6.107. Compared with the Asnæsværket 400 kV bus fault without the switching station, the shunt reactors L in Equation (6.4) becomes twice larger and the charging capacity C becomes half, which affects the overvoltage.

Figures 6.108–6.112 show the overvoltages caused by the system islanding with the fault current levels of 5–40 kA. The overvoltage becomes lower compared with the Asnæsværket 400 kV bus fault without the Torslunde switching station.

6.4 Ground fault and fault clearing over-voltages

6.4.1.1 Study conditions and parameters

The idea of the statistical evaluation is also applied to the evaluation of the ground fault and fault clearing overvoltages since the fault location and the fault timing vary and cannot be controlled. Instead of considering them as random phenomena,

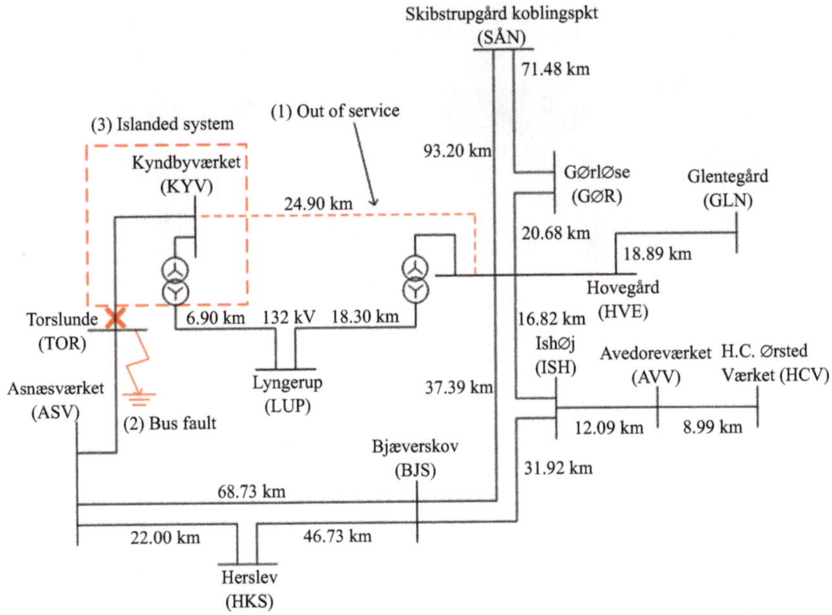

Figure 6.107 TOR 400 kV bus fault and the system islanding

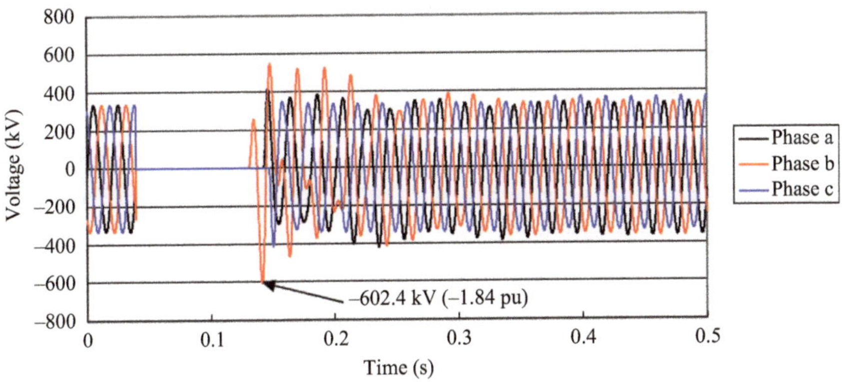

Figure 6.108 Overvoltage caused by the TOR bus fault leading to the system islanding (fault current level: 5 kA)

it is a common practice to cover different fault locations and fault timings with fixed (predefined) steps.

Table 6.30 summarizes the study conditions and parameters for the ground fault and fault clearing overvoltages. The 10 fault locations were selected so that

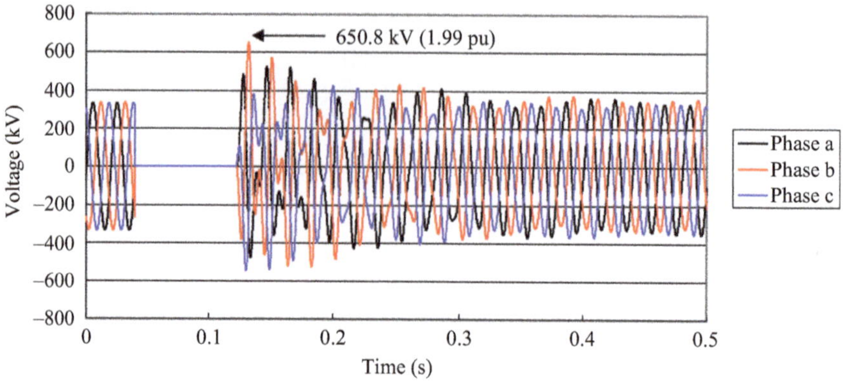

Figure 6.109 Overvoltage caused by the TOR bus fault leading to the system islanding (fault current level: 10 kA)

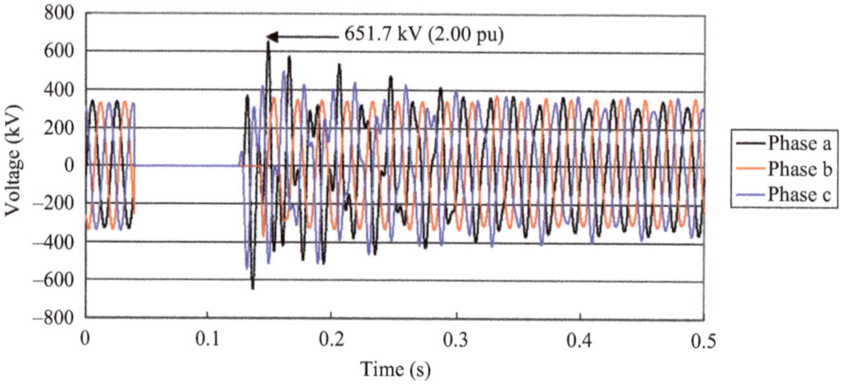

Figure 6.110 Overvoltage caused by the TOR bus fault leading to the system islanding (fault current level: 20 kA)

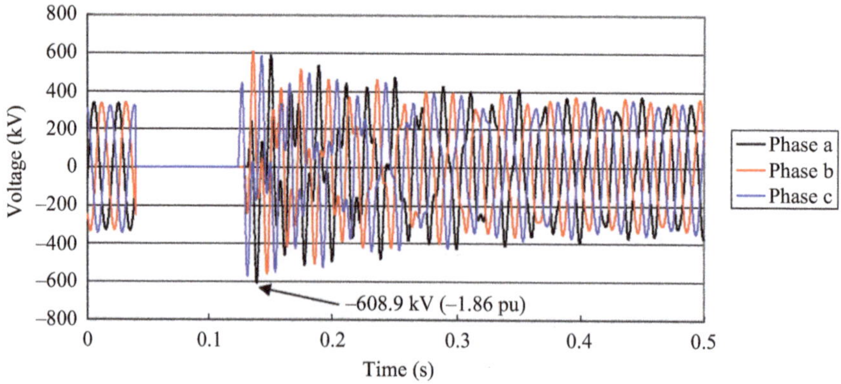

Figure 6.111 Overvoltage caused by the TOR bus fault leading to the system islanding (fault current level: 30 kA)

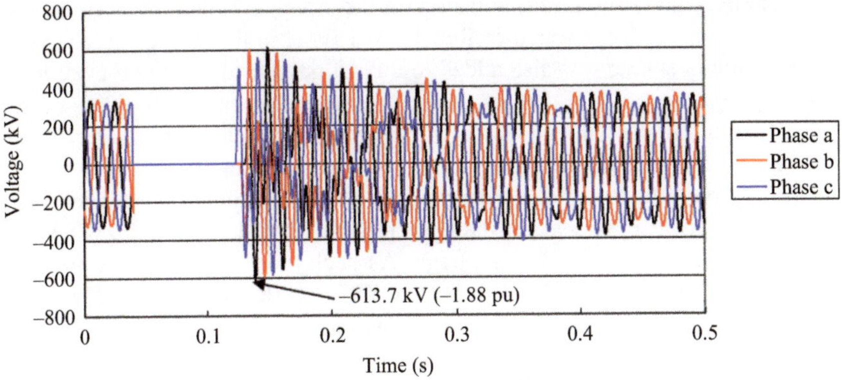

Figure 6.112 Overvoltage caused by the TOR bus fault leading to the system islanding (fault current level: 40 kA)

Table 6.30 Study conditions and parameters for the ground fault and fault clearing overvoltage

Fault type	**Single line to ground fault (core to sheath)**
Fault location	Five points between ASV–TOR and TOR–KYVASV–TOR: AT3, AT6, AT8, AT11, and AT13TOR–KYV: TK4, TK7, TK10, TK13, and TK14
Fault timing	0°–180° (10° step)
Fault clearing time	First current zero at least 40 ms after the fault occurrence
Fault clearing sequence	Line-CBs of all three phases are opened at the above fault clearing time. Sequential switching
Feeding source	Off-peak load condition + dummy source: 15 mH (strong source) and 100 mH (weak source)

the separation between each fault location is limited to 3–6 km. In order to cover ten fault locations and 19 fault timings, 190 simulations were performed. The 2% value is the third-highest overvoltage in these simulations.

As in the energization overvoltage analysis, two feeding networks were considered. One is the strong source case and the other one is the weak source case. Both feeding networks were modeled based on the off-peak load condition and by adding a dummy source to the Asnæsværket 400 kV bus. The source impedance of the dummy source impedance was set to 15 mH and 100 mH in the weak source and strong source case, respectively.

Since the ground fault and fault clearing overvoltages depend on the fault current, the fault current level needs to be set within a reasonable range. In a cable energization overvoltage analysis, the dummy source impedance was set to 1 mH in order to find the effect of a steep rise of an initial overvoltage. However, this setting should be avoided in the ground fault and fault clearing overvoltage analysis as unreasonably large fault current will flow in the simulation with this setting. The dummy source impedance was therefore set to 15 mH in the strong source case.

The dummy source was also added to the off-peak load condition even in the weak source case, but the dummy source impedance was set to a large value (100 mH). Without the dummy source, ground fault and fault clearing overvoltages were almost unnoticeable since the fault current was too small.

In the ground fault and fault clearing overvoltage analysis, the highest overvoltage is not always observed at the open terminal. In addition, there is no open terminal for the ground fault overvoltage. As such, the overvoltages were monitored at AT3, AT6, AT8, AT11, AT13, TK4, TK7, TK10, TK13, and TK14 as well as at Asnæsværket, Torslunde, and Kyndbyværket.

6.4.1.2 Results of the analysis

Figure 6.113 and Table 6.31 show the cumulative probability distributions of the ground fault and fault clearing overvoltages and their characteristics. The overvoltage level is as low as the energization overvoltage of the Kyndbyværket–Asnæsværket

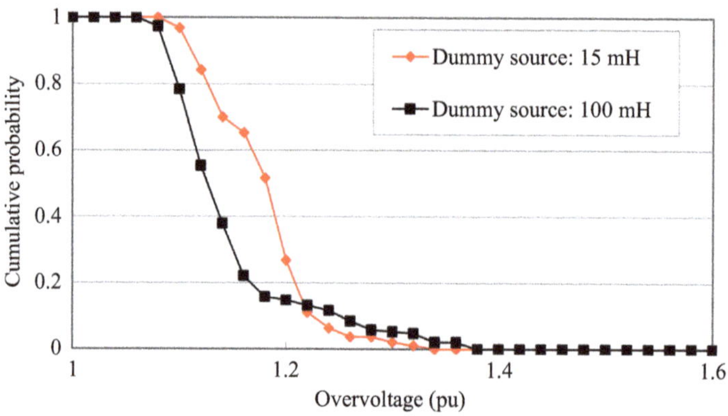

Figure 6.113 Cumulative probability distributions for the ground fault and fault current overvoltage

Table 6.31 Characteristics of probability distributions for the ground fault and fault clearing overvoltage

	Dummy source: 15 mH	**Dummy source: 100 mH**
Max. OV (pu)	1.33	1.37
2% OV (pu)	1.30	1.36
Average OV (pu)	1.17	1.14
Standard dev. (pu)	0.0482	0.0647
Skewness	0.430	1.79
Kurtosis	3.52	5.76

line in the off-peak load condition. The probability distributions are affected by the dummy source impedance, but the overvoltage level is low in both cases.

With the dummy source impedance of 15 mH and 100 mH, the maximum overvoltage was observed when the fault occurs at AT3 and at the phase angle 60° and 70°, respectively. In both cases, the maximum overvoltage was the ground fault overvoltages and was not the fault clearing overvoltages. Figure 6.114 shows the profile of the maximum ground-fault overvoltages along the length of the Kyndbyværket–Asnæsværket line. The figure shows that the maximum overvoltages were observed at the Asnæsværket 400 kV bus.

As the maximum overvoltage was observed at the Asnæsværket 400 kV bus, the voltage waveforms at the Asnæsværket 400 kV bus are given in Figures 6.115–Figure 6.116. The figures show that both ground fault overvoltages and fault clearing overvoltages are highly damped.

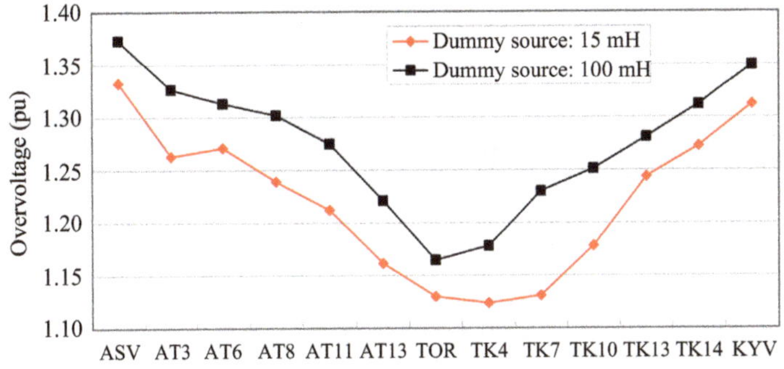

Figure 6.114 Profile of the maximum ground-fault overvoltages

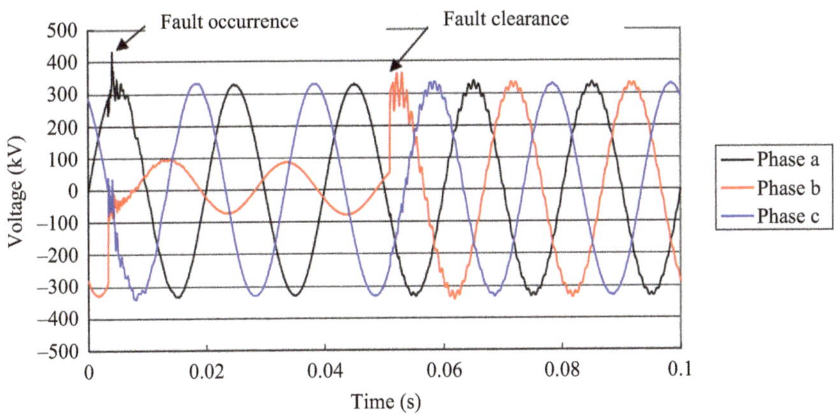

Figure 6.115 Voltage waveforms of the maximum ground-fault overvoltage at the ASV 400 kV bus (dummy source: 15 mH)

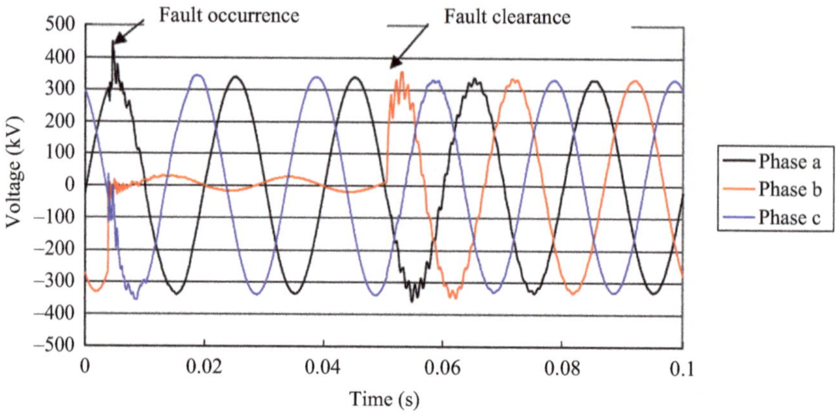

Figure 6.116 Voltage waveforms of the maximum ground-fault overvoltage at the ASV 400 kV bus (dummy source: 100 mH)

6.4.1.3 Results with the sequential switching

The sequential switching has been adopted as a countermeasure to the zero-missing phenomenon. With the sequential switching, the sequence for the fault clearing is different from the ordinal sequence shown in Table 6.30.

According to the fault clearing sequence with the sequential switching, the ground fault and fault clearing overvoltage was analyzed with the following sequence:

Fault occurrence (phase-a):	0.0–0.1 s
Fault clearing (phase-a):	0.05 s
Shunt reactor tripping (phase-b and -c):	0.11 s (only on the Kyndbyværket side)
Line breaker tripping (phase-b and -c):	0.17 s

The time separations between the above steps were set shorter than the actual setting. As long as the overvoltage caused by the previous step is well damped, the shorter time separations have no effect on the overvoltage caused by the next step.

In Sections 6.4.1.1 and 6.4.1.2 with the normal fault clearing sequence, ground faults in different phases were simulated. In this section with the sequential switching, ground faults occur only in phase-a. Since the three phases are well balanced in the Kyndbyværket–Asnæsværket cable model, it is not expected that this has a major impact on the simulation results.

Figure 6.117 and Table 6.32 show the cumulative probability distributions of the ground fault and fault clearing overvoltage and their characteristics. The maximum overvoltage level is lower than the overvoltage with the normal fault clearing sequence approximately by 0.1 pu. The average overvoltage is not affected by the sequential switching. The probability distributions are affected by the dummy source impedance, but the overvoltage level is low in both cases.

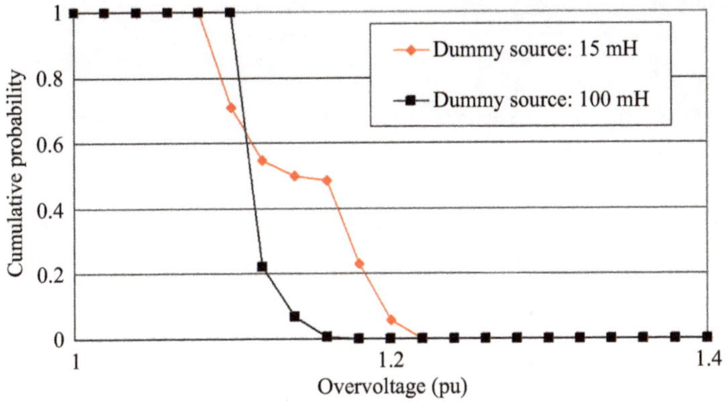

Figure 6.117 Cumulative probability distributions for the ground fault and fault current overvoltage with the sequential switching

Table 6.32 Characteristics of probability distributions for the ground fault and fault clearing overvoltage with the sequential switching

	Dummy source: 15 mH	**Dummy source: 100 mH**
Max. OV (pu)	1.21	1.16
2% OV (pu)	1.21	1.15
Average OV (pu)	1.14	1.11
Standard dev. (pu)	0.0410	0.0145
Skewness	0.167	1.59
Kurtosis	1.42	4.47

With the dummy source impedance 15 mH and 100 mH, the maximum overvoltage was observed when the fault occurs at TK13 and TK17, respectively, and at the phase angle of 130°. In both cases, the maximum overvoltages were the ground fault overvoltages and were not the fault clearing overvoltages. This means that the difference of the maximum overvoltage, compared to the one with the normal fault clearing, was not caused by the difference in the fault clearing sequence. The difference was caused by the faulted phase.

Figure 6.118 shows the profile of the maximum ground-fault overvoltages along the length of the Kyndbyværket–Asnæsværket line. The figure shows that the maximum overvoltages were observed at TK13 with the dummy source impedance 15 mH and at the Asnæsværket 400 kV bus with the dummy source impedance 100 mH.

As the maximum overvoltage was observed at TK13 with the dummy source impedance 15 mH, the voltage waveforms at TK13 are shown in Figure 6.119. The line current at the Asnæsværket side is also shown in the figure for a better understanding of the sequential switching. The figures show that both ground fault overvoltages and fault clearing overvoltages are highly damped.

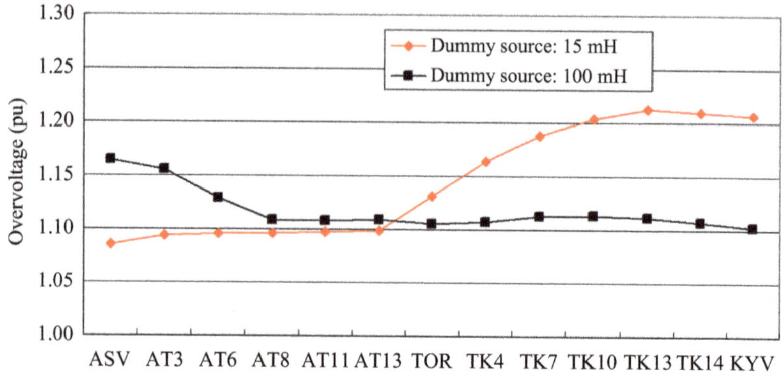

Figure 6.118 Profile of the maximum ground-fault overvoltages with the sequential switching

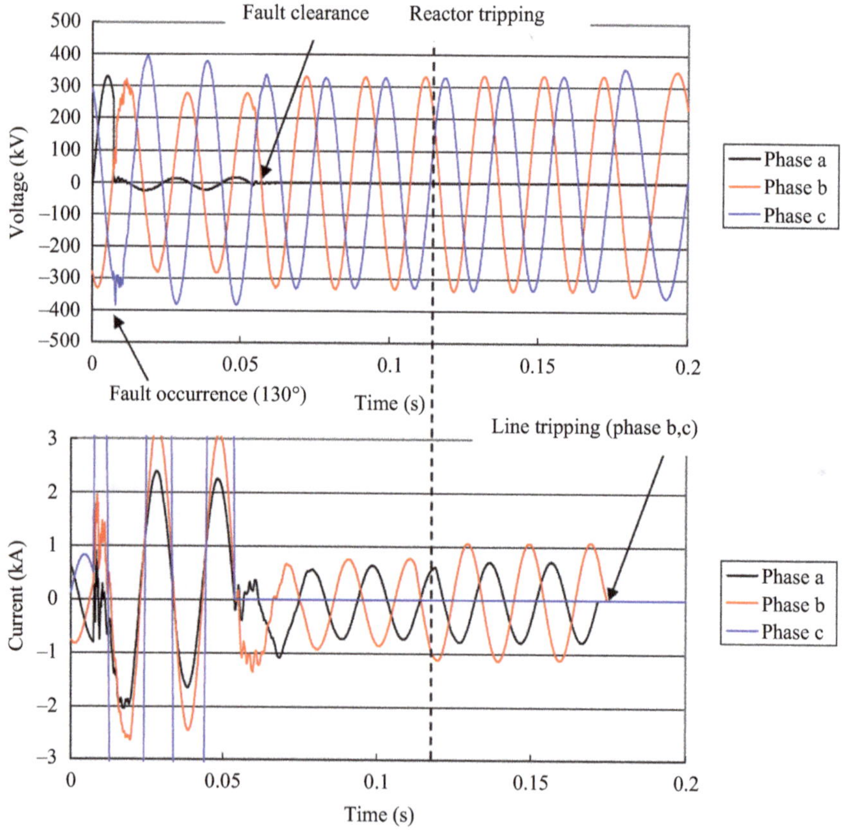

Figure 6.119 Voltage and current waveforms of the maximum ground-fault overvoltage at the ASV 400 kV bus with the sequential switching (dummy source: 15 mH)

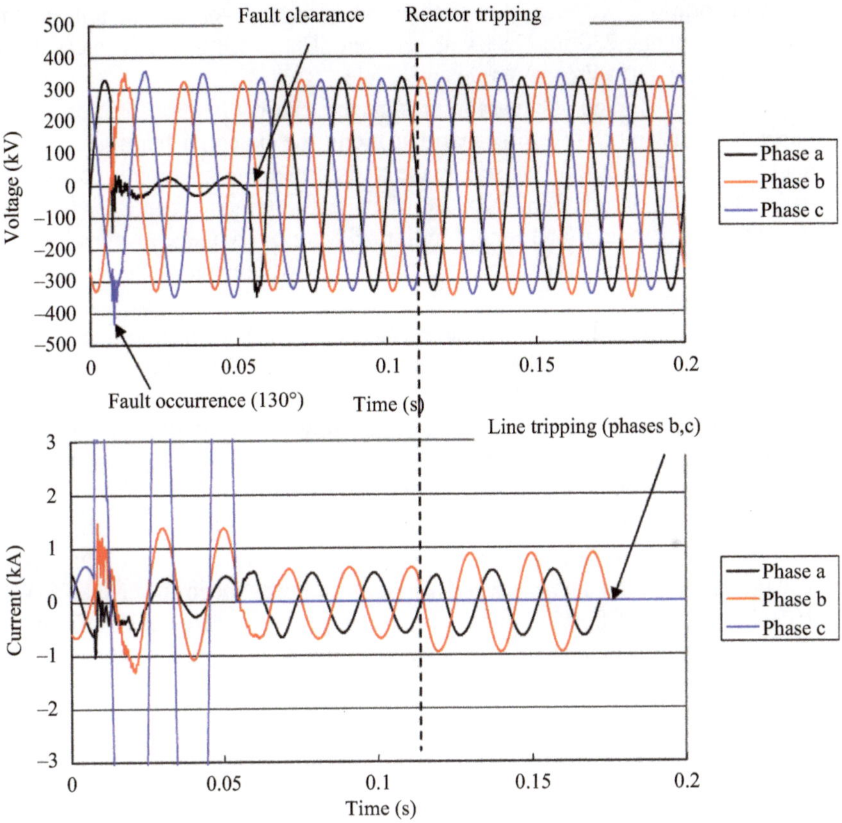

Figure 6.120 Voltage and current waveforms of the maximum ground-fault overvoltage at the ASV 400 kV bus with the sequential switching (dummy source: 100 mH)

As the maximum overvoltage was observed at Kyndbyværket 400 kV bus with the dummy source impedance 100 mH, the voltage waveforms at the bus are shown in Figure 6.120.

6.5 Conclusions

This chapter studied temporary overvoltages and slow-front overvoltages on the Kyndbyværket–Asnæsværket cable line in Denmark.

1. In Section 6.3.1, the series resonance overvoltage caused by the cable energization was studied. The most severe switching scenarios were first identified. The simulations were performed with the most severe assumptions of the Kyndbyværket 132 kV network and the Asnæsværket 132 kV network. The simulations show that the series resonance overvoltages in the assumed most

severe conditions are much lower than the withstand voltages as long as the withstand voltage 230 or 275 kV is selected. The phase to phase overvoltage decays below 230 kV (peak) after one cycle.

2. Section 6.3.2 studied the parallel resonance caused by transformer inrush currents as the most severe scenario. Among the potential harmonic current sources, an inrush current caused by the transformer energization is considered to be the most onerous one because of its high harmonic contents, low frequency, low damping, and long duration. Severe overvoltages were observed with the most severe scenarios, but they were within withstand voltages of related equipment.
3. Section 6.3.3 studied the overvoltage caused by the system islanding. A fault on the Kyndbyværket, Asnæsværket, and Torslunde 400 kV buses was studied with different fault current levels, different shunt compensation degrees, and different timings for the fault clearing. The study showed that the overvoltages were most severe when a fault occurs on the Asnæsværket 400 kV bus, but they were within withstand voltages of related equipment.
4. Section 6.4 studied the statistical distributions of a ground fault and fault clearing overvoltages of the Kyndbyværket–Asnæsværket line. The maximum overvoltage found in the analysis was 1.37 pu, which was much lower than SIWV 1050 kV (3.2 pu). Effects of dummy source impedances were not significant.
5. The sequential switching may be applied to the Kyndbyværket–Asnæsværket line as a countermeasure of the zero-missing phenomenon. The sequential switching does not affect the severity of the overvoltages.

References

[1] *Danish Annual Energy Statistics* 2011. Danish Energy Authority. October 2008. Available: http://www.ens.dk/en-US/Info/FactsAndFigures/Energy_statistics_and_indicators/Annual%20Statistics/Documents/Energy%20in%20Denmark%202010.pdf.

[2] *Technical Report on the Future Expansion and Undergrounding of the Electricity Transmission Grid*. Energinet.dk. April 2008. Available: https://www.stjornarradid.is/library/01–Frettatengt—myndir-og-skrar/ANR/ANR—Raflinur-i-jord/5-Elinfrastrukturudvalget.pdf.

[3] *XLPE Underground Cable Systems – User's Guide (rev. 3)*. ABB.

[4] Ametani, A. "A general formulation of impedance and admittance of cables." *IEEE Transactions on Power Apparatus and Systems*. 1980; 99, 3: 920–910.

[5] Ametani A., Miyamoto Y., and Nagaoka N. "Semiconducting layer impedance and its effect on cable wave-propagation and transient characteristics." *IEEE Transactions on Power Delivery*. 2004; 19(4): 1523–1531.

[6] Gustavsen, B. "Panel session on data for modeling system transients insulated cables", *Proceedings of the IEEE Power Engineering Society Winter Meeting*, 2001.

[7] Ohno, T. "Dynamic Study on the 400 kV 60 km Kyndbyværket – Asnæsværket Line." Ph.D. Thesis, Aalborg University, December 2012.

[8] Ohno, T., Ametani, A., and Bak, C. L. "Derivation of theoretical formulas of sequence currents on underground cable system." *IPST.* 2011. Available: https://www.ipstconf.org/papers/Proc_IPST2011/11IPST014.pdf.

[9] Hall, S. H., and Heck, H. L. *Advanced signal integrity for high-speed digital designs*. Wiley-IEEE Press; 2009.

[10] *XLPE Submarine Cable Systems, Attachment to XLPE Land Cable Systems – User's Guide (rev. 5)*, ABB.

[11] *Modelling of tap changers of the ASNÆSVÆRKETS 400/132 kV TRANSFORMER using the DIgSILENT PowerFactory software*, Energinet.dk (company internal document of Energinet.dk).

[12] Momose, N., Suzuki, H., Tsuchiya, S., and Watanabe, T. "Planning and Development of 500 kV Underground Transmission System in Tokyo Metropolitan Area." *CIGRE Session.* 1998: 37–202.

[13] *Assessment of the Technical Issues relating to Significant Amounts of EHV Underground Cable in the All-island Electricity Transmission System.* Tokyo Electric Power Company, November 2009. Available: http://www.eirgrid.com/media/Tepco%20Report.pdf.

[14] Rebolini, M., Colla, L., and Iliceto, F. "400 kV AC new submarine cable links between Sicily and the Italian mainland: Outline of project and special electrical studies." *CIGRE Session* 2008; C4-116.

[15] Colla, L., Lauria, S., and Gatta, F. M. "Temporary overvoltages due to harmonic resonance in long EHV cables." *IPST* 2007. Available: http://www.ipst.org/techpapers/2007/ipst_2007/papers_IPST2007/Session16/233.pdf.

[16] Karasaki, T., Goto, T., and Ametani, A. "An abnormal overvoltage due to load rejection on ehv underground transmission lines." *International Conference on Power System Transients (IPST)*. September 1995.

Chapter 7

Steady-state and transient behavior of hybrid overhead line-underground cable networks in the Netherlands

7.1 Background of EHV hybrid OHL-cable networks in the Netherlands

Power transmission grids are traditionally based on overhead lines (OHL) and any increase in the grid transfer capacity has been realized by building new OHLs. This technology has been used for decades since it is relatively cheap to construct, easy to maintain, reliable in operation, and well-known by transmission system operators (TSOs) and manufacturers. However, OHLs often cross through (semi) urban or environmentally sensitive areas, causing several problems and adverse consequences.

The main problem with developing new OHLs is the widespread public opposition and environmental impacts. People are reluctant to have OHLs and transmission towers in their surroundings due to their negative impacts; most notably, visual effect, land price decrement, land occupation (right-of-way), effects on forests and territories, health concerns related to the exposure to electromagnetic fields, radio interference, and audible noise [1]. The negative impacts of overhead transmission networks have caused strong political and environmental pressure on system operators and have made the construction of new OHLs more difficult than before. This situation is mostly noticeable in densely populated countries like the Netherlands.

A promising solution for the mentioned obstacles is to extend the grids by means of alternating current (AC) underground cables. There are many factors that make cables attractive alternatives to OHLs. In underground cable systems, the electromagnetic field attenuates much faster with distance from the center of the installation compared to those in OHL systems, although they are higher in cable systems right above the center of the installation [1,2]. In addition, underground cables have limited visual impact and no audible noise. The right-of-way for cables is also narrower compared to that of OHLs. Further, cables are stronger against earthquakes, storms, and freezing temperatures. Therefore, cables can limit or remove the negative impacts of OHLs in locations like densely populated areas and national parks. This enables system operators to develop transmission grids with reduced social and environmental impacts. Grids with full or partial cable-based connections are known as hybrid OHL-cable grids, where transmission connections are composed of OHLs in nonsensitive areas and underground cables in sensitive (and maybe nonsensitive) areas.

A large-scale application of underground cables in transmission grids is not yet a well-practiced technique from the technical point of view. Both system operators and equipment suppliers do not possess much experience with the technical performance of transmission systems with long cables as they are very different than OHLs from electrical, thermal, and mechanical perspectives [2,3]. Cables that need reactive power shunt compensation are generally considered long cables.

From the electrical point of view, major challenges are caused by significant differences between the electrical characteristics of cables and OHLs. The electrical characteristics of cables and OHLs are different since their structure and physical properties are not the same. The conductor in OHLs is surrounded and insulated by air, whereas cables have a multilayer structure consisting of at least four layers, including a core conductor, an insulating material with semiconductor layers at the interfaces, a sheath (screen) conductor, and an outer insulation layer. As a result, per unit length, the series inductance of a cable can be up to five times smaller and the shunt capacitance to earth can be up to thirty times larger than those of an equivalent-rated OHL [2].

The application of cables influences the behavior of a traditionally OHL-based grid in several aspects due to the large cable capacitance and its interaction with the existing inductive grid. Several undesirable operating conditions are expected in hybrid OHL-cable systems, which increase the risk of damage to system components and reduce the reliability of the power supply [2–9]. Therefore, all the technical issues should be investigated before the realization of a cable project.

Years ago, the Dutch government passed a law to limit the total length of OHL networks. This law obliges the system operator, TenneT, to keep the total length of the OHL network constant. This means that for the construction of a new OHL at a given location an existing OHL with the same length at another location has to be undergrounded so that the total OHL length in the grid remains unchanged. As a result, in 2009, TenneT began an extensive research program to study the possibilities of underground cables application in future transmission projects. The 10-year 380 kV cable research program covered major technical issues associated with long EHV cables in order to increase the knowledge and experience of TenneT in the design and operation of hybrid OHL-cable transmission systems.

Currently, the partially undergrounded Randstad 380 kV connection is in operation in the Dutch transmission grid. This is a double-circuit transmission line consisting of a south ring and a north ring. The south ring with 20.1 km length is in the south-west of the Netherlands, out of which 10.8 km is realized by 2,500 mm^2 single-core copper XLPE cables with two cables per phase. This part of the connection is composed of two OHL sections and a cable section in the middle (OHL-cable-OHL) with lengths of 4.4, 10.8, and 6.8 km, respectively. The north ring is with the total transmission length of 58 km, out of which 9.3 km will be realized by single-core copper and aluminum XLPE cables of different diameters. This connection includes five OHL and four cable sections placed in an intermediate sequence, where the shortest cable section will be 1 km long.

Within the Dutch cable research program, several researches were conducted on 380 kV hybrid OHL-cable grids, including studies on steady-state operation, dynamic stability, harmonic resonance behavior, electromagnetic transients, and system reliability. In this chapter, the findings of the steady-state operation, harmonic resonance behavior, and transient behavior will be discussed.

7.2 Approach and modeling

This section elaborates on the study approach and the Dutch grid modeling, where the case study project and the defined cable scenarios for this project are explained. Moreover, this section discusses the developed models of the Dutch transmission grid for simulation of different technical issues. Here it is explained which models are used for steady-state and transient analyses.

7.2.1 Spaak project

The case study is a hypothetical 380 kV double-circuit transmission line, called Spaak project, for which different cable scenarios are defined. The influence of long cables on the system performance can be determined by analyzing the technical issues of cable operation in this transmission line. This transmission line connects the western side of the 380 kV ring to the eastern side. This would be a double-circuit connection between substations Krimpen aan den Ijssel (KIJ) and Dodewaard (DOD) with the transmission length of 80 km. Figure 7.1 shows the single-line diagram of the Dutch 380 kV grid and the Spaak project.

7.2.1.1 Mixed-line configuration

The first important step is to determine the mixed-line configuration, which is specified by the number and the location of OHL and cable sections in the connection. A mixed transmission line is made of series-connected OHL and cable sections. Different configurations can be driven as the result of various possibilities for the number and location of sections.

In this regard, a study case mixed-line configuration was determined in order to investigate all the technical issues for this configuration. In addition, four other configurations were also taken into account to find out the influence of the mixed-line configuration on the system technical performance. In other words, since the number and the location of cable sections are determined in the planning and design stages of a cable project, it is important to compare various mixed-line configurations from the system technical performance perspective.

As a practical approach, cable sections are expected to be utilized mainly at sensitive areas like cities and national parks, where the importance of undergrounding is highlighted. Therefore, it was assumed that each circuit of the Spaak connection includes seven sections as shown in Figure 7.2. There are three cable

Figure 7.1 Map of the Dutch transmission grid. Adapted from TenneT (2018)

sections and four OHL sections per circuit and the two circuits are identical. In the Dutch 380 kV grid, the continuous current ratings (ampacity) of a typical OHL, consisting of a bundle of four sub-conductors, and a typical single-core cable are respectively 4 kA and 1.5 kA (see Table 7.2). Therefore, the cable sections require two parallel cables per phase to have almost the same transmission capacity as the OHL sections. In this research, this is the default mixed-line configuration, used for the analysis of the technical issues, unless a different configuration is mentioned.

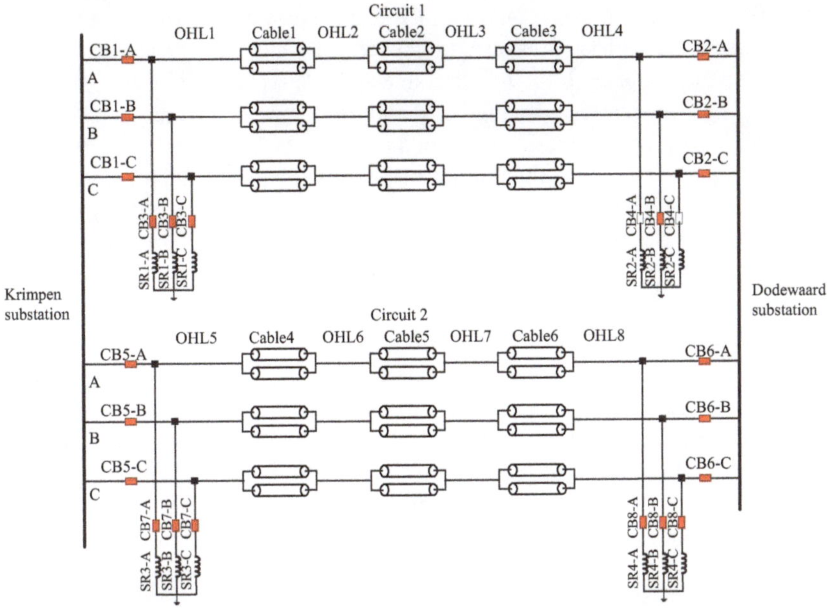

Figure 7.2 Study case mixed-line configuration for the Spaak connection. CB: circuit breaker; SR: shunt reactor

7.2.1.2 Shunt reactors location

At the sending and receiving terminals, the three-phase shunt reactors (SR) are connected through breakers to the circuit right behind the line breakers. This type of compensation is known as line-end compensation (LEC). It should be noted that each SR, that is, SR1 to SR4, is representing a compensation bank. A measure of flexibility will be provided by installing switchable variable reactors or a parallel combination of smaller fixed reactors with the unit size (typical unit sizes are 100 Mvar or 150 Mvar), which can be switched in/out separately.

The connection of shunt reactors to the circuit and behind the line breakers helps to limit post-switching steady-state energization overvoltages in the circuit (especially at the open-end of the circuit due to the Ferranti effect) and to minimize the capacitive current flowing through the line breaker during energization. These benefits are not achieved when the shunt reactors are connected to busbars or to tertiary windings of power transformers at substations. However, for some cable lengths, shunt reactors can be connected to busbars or to tertiary windings if energization overvoltages and capacitive currents are not very large. In this case, the inductive voltage transformer connected to the cable should have enough discharge capability and also the leading current interruption capability of the line breaker should be checked [10].

The installation of separate breakers for the shunt reactors results in an added switching flexibility as well as minimizing the risk of line resonance by decoupling reactors from the disconnected phase(s). Resonance may occur between the reactor inductance, the inter-phase/inter-circuit capacitance, and the cable capacitance if reactors remain connected to the disconnected phase(s) in uneven open-phase conditions (e.g., one phase is disconnected and the other two are energized) [11,12]. In fact, the induced power-frequency voltage in the disconnected phase(s) due to the capacitive coupling with the energized phase(s) can trigger the resonance circuit. The risk of ferroresonance also exists when shunt reactors have a magnetic core. Thus, these risks can be minimized by disconnecting the shunt reactors from the disconnected phase(s) when they are connected via breakers.

7.2.1.3 Alternative mixed-line configurations

It is of importance to investigate the influence of other mixed-line configurations on the system technical performance. Different configurations can be driven by varying the number and the location of OHL and cable sections. In addition to the study case configuration, four mixed-line configurations were selected to be compared with the study case configuration in terms of their influence on the system performance. These configurations are as follows:

1. Configuration 1: OHL-cable
2. Configuration 2: Cable-OHL
3. Configuration 3: OHL-cable-OHL
4. Configuration 4: OHL-cable-OHL-cable-OHL

Figure 7.3 shows the four alternative configurations. Only one circuit is depicted due to the identical configuration of the two circuits.

7.2.1.4 Cable scenarios

Six cable scenarios were defined to determine the impact of the cable length on the system technical performance. In each scenario, it is specified how many kilometers of the total transmission length is realized by the cables. The total transmission length of the Spaak connection from the sending substation to the receiving substation is 80 km. By these scenarios, the cable share varies from 0% (fully-OHL) to 100% (fully-cable) of the transmission length. The cable scenarios are presented in Table 7.1.

The cable share is increasing from zero to 100% in five steps, which means in total six scenarios: 0% cable (fully-OHL), 15% cable, 25% cable, 50% cable, 75% cable, and 100% cable (fully-cable). Different cable shares were chosen to show the influence of the cable length increase on the system performance. When it is mentioned L km cable, it is meant that the total transmission length of the cable sections is L km. Thus, each phase is composed of $2\times L$ km cable (two cables per phase), each circuit $3\times 2\times L$ km cable, and the two circuits $12\times L$ km cable. Table 7.1 presents the total transmission length of the OHL and cable sections.

7.2.2 Grid modeling

This section elaborates on the developed models of the Dutch 380 kV grid for the steady-state and transient studies of the hybrid OHL-cable circuits of the Spaak

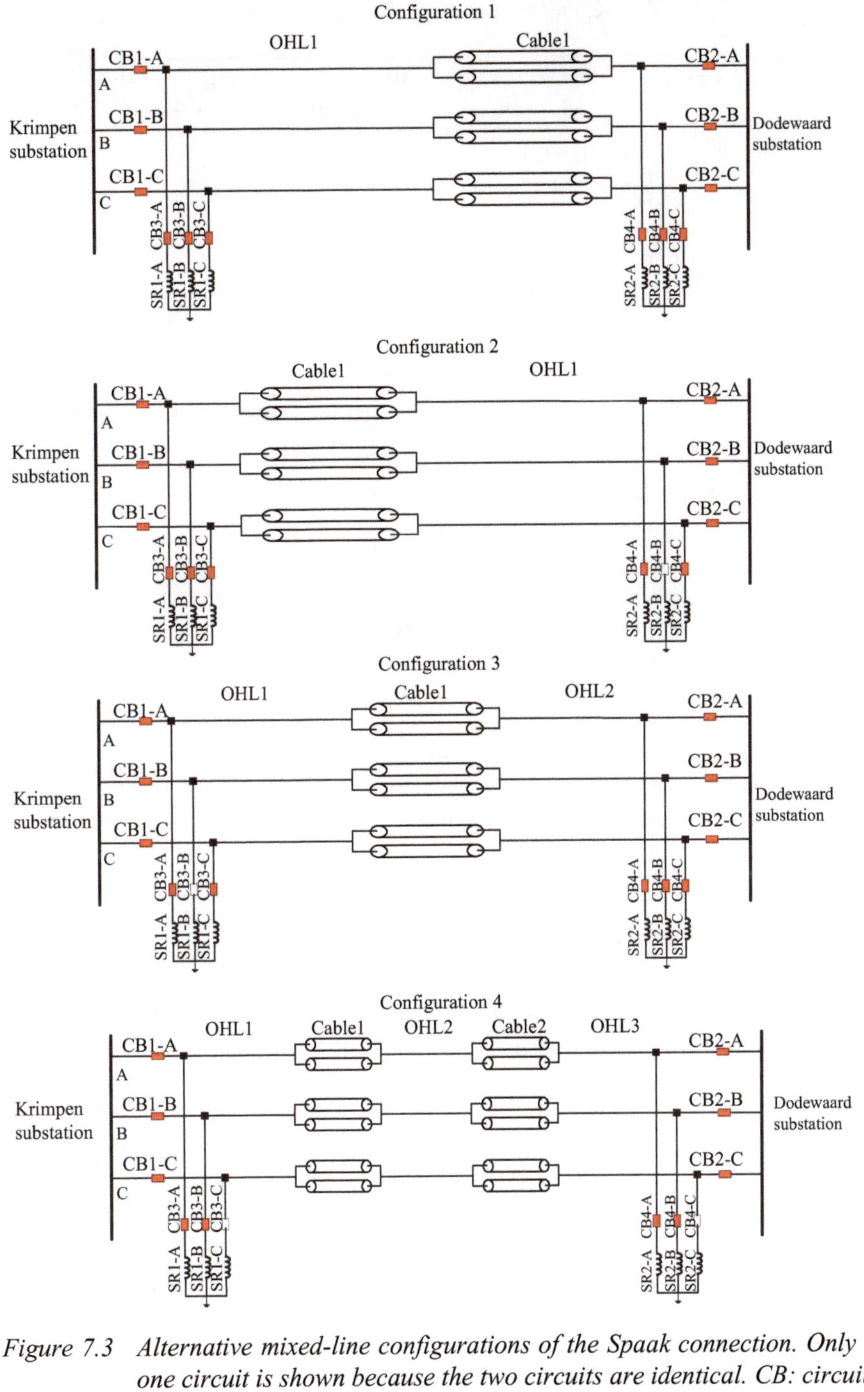

Figure 7.3 Alternative mixed-line configurations of the Spaak connection. Only one circuit is shown because the two circuits are identical. CB: circuit breaker; SR: shunt reactor

Table 7.1 Cable scenarios of the Spaak project

Scenario	OHL length (km)	Cable length (km)
0% Cable (Fully-OHL)	80	0
15% Cable	68	12
25% Cable	60	20
50% Cable	40	40
75% Cable	20	60
100% Cable (Fully-cable)	0	80

connection. It is explained how the grid components are modeled for each of the technical issues and to which extend the grid should be modeled to ensure sufficiently fast and accurate simulations.

7.2.2.1 Grid model for steady-state studies

The steady-state simulations are based on the positive-sequence representation of the whole Dutch transmission grid (110 kV, 150 kV, 220 kV, and 380 kV voltage levels) in the DIgSILENT PowerFactory environment. This model includes representations of substations, transmission lines (OHLs and cables), transformers, shunt reactors, capacitor banks, loads, and all the interconnections with the neighbor grids together with a simplified representation of European transmission grids. By this means, realistic load-flow scenarios and short-circuit powers can be simulated based on the expected future dispatch scenarios.

In the steady-state analysis, the wavelength of current and voltage (6,000 km in a 50 Hz system) is much larger than the network physical dimensions. Therefore, a lumped-parameter model can be used to simulate the steady-state behavior of the transmission lines [13]. Two types of lumped-parameter model are studied here: nominal pi-model and exact (or equivalent) pi-model, where both are two-port models.

The nominal pi-model, as shown in Figure 7.4, is a very simple representation of a line at power-frequency, in which the total resistance (R), inductance (L), capacitance (C), and usually conductance (G) of the line are represented by lumped parameters. Z is the total series impedance and Y is the total shunt admittance of the line:

$$Z = R + j\omega L \tag{7.1}$$

$$Y = G + j\omega C \tag{7.2}$$

This model is only accurate for steady-state simulations and for short- and medium-length lines (lengths up to 240 km). For long transmission lines, the line parameters are distributed uniformly throughout the length of the line and cannot be represented by the nominal pi-model [3,14].

The exact pi-model is an adopted version of the nominal pi-model for simulation of long transmission lines and even transient phenomena (at a single

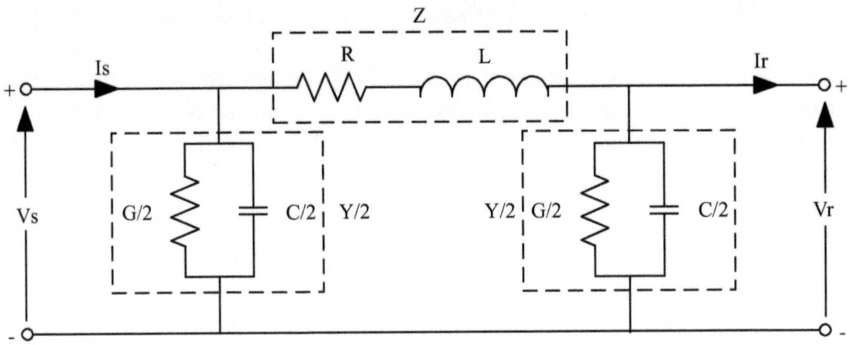

Figure 7.4 Nominal pi-model of a two-conductor transmission line

Table 7.2 Input electrical parameters for 380kV–50Hz OHL and cable sections of the Spaak connection

Type	Cross-section (mm^2)	R+jX (Ω/km)	C (nF/km)	Z_s[(b)] (Ω)	SIL[(c)] (MVA)	I_n[(d)] (A)	S_n[(e)] (MVA)
OHL[(a)]	2,483.6	0.0148+j0.232	15.3	219.7	657.3	4,000	2,635
XLPE cable	2,500	0.0227+j0.1712	231.5	48.5	2976.2	1,500[(f)]	987.27

(a) Bundle of four subconductors.
(b) Surge impedance: $Z_s = \sqrt{\frac{L}{C}}$.
(c) Surge impedance loading for a three-phase system calculated at V_n=380 kV.
(d) Rated current (ampacity).
(e) Apparent power at ampacity I_n (thermal power limit): $S_n = \sqrt{3}V_nI_n$.
(f) This value can be increased to 2 kA for only 24 h continuous operation.

frequency) as it takes the distributed nature of the line impedance and admittance into account by applying correcting factors in the Z and Y [15–17].

Table 7.2 presents the applied electrical parameters for 380 kV-50 Hz OHL and cable sections of the Spaak connection. The OHL is composed of a bundle of four sub-conductors with a continuous current rating (ampacity) of 4 kA in total. The continuous current rating of a single cable is 1.5 kA, but this value can be increased to 2 kA for 24 h.

7.2.2.2 Grid model for transient studies

Power system harmonic and transient simulations require models in which the frequency-dependency of system components is taken into account. Such models can be developed in specialized programs for electromagnetic transient (EMT) simulations.

The EMT simulation results are only accurate when three requirements are satisfied: firstly, the input data used for the modeling of the components should be accurate; secondly, the level of details put into the model of each component should be sufficient; and thirdly, the extent to which the area around the point of interest is modeled should be enough. Each of these requirements has a substantial influence on the simulation results. In the literature, the second and third requirements are referred to as the modeling depth.

7.2.2.2.1 Modeling depth

An important issue in the simulation of electromagnetic transients is the modeling depth or the depth of complexity in the representation of the grid. The size and the detail level of the modeled area around the event location (disturbance location) can significantly affect both the simulation accuracy and the simulation time. The simulation results are inaccurate when the modeled area is too small or with insufficient details. On the other hand, the required simulation time can substantially increase when the size of the modeled area or the level of details are increased. Therefore, there is a trade-off between the simulation accuracy and the simulation time, which can be properly handled by determining a sufficient modeling depth.

It is recommended in the literature to represent the grid up to two busbars from the event location [18], but this is not always sufficient. In [2,19,20], it is proposed to divide the grid into three zones, as illustrated in Figure 7.5, where each zone is represented by a different level of details:

1. *Study zone*: This is the part of the grid near the event location, which has to be represented by detailed frequency-dependent models. OHLs and cables should

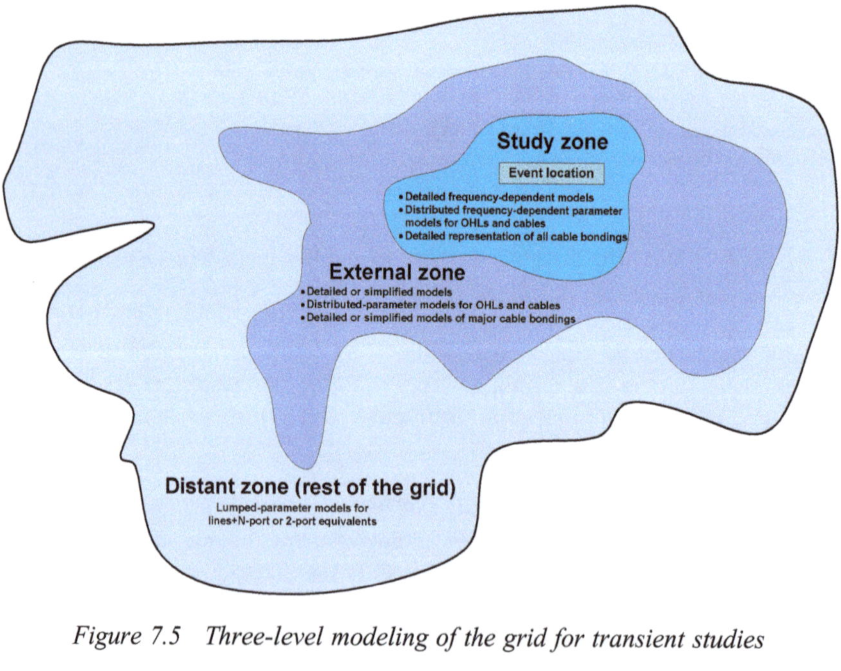

Figure 7.5 Three-level modeling of the grid for transient studies

be represented by distributed frequency-dependent parameter models and all cable bondings should also be included with details.

2. *External zone*: This is the part of the grid farther than the studied zone from the event location; therefore, network components can be represented by detailed or simpler models. OHLs and cables can be represented by distributed-parameter models like the Bergeron model, which uses values of the line parameters at a single frequency. In this zone, only major cable bondings are represented by detailed or simplified models, for example, the use of ideal cross-bondings when cables are cross-bonded.
3. *Distant zone (rest of the grid)*: This zone is the distant part of the grid, which is very far from the event location and has no contribution to the transient voltages and currents. This zone can be represented by one or two busbars depth where lumped-parameter line models are used. This will reduce the simulation inaccuracy (by decreasing the chance of erroneous high voltage) but at the expense of a slightly longer simulation time. Finally, an N-port or 2-port equivalent can be used to represent the remaining parts of the grid.

It should be pointed out that finding the border of these tree zones is mainly based on a try and error approach, although some methods are proposed in the literature to find the extend of the detailed study zone. For example, in [19], the authors proposed a method that can be used to estimate the depth of the modeling area by using the grid layout. Do note that the level of details is also strongly dependent on the type of the investigated phenomenon at the event location. To address this challenge, frequency-dependent network equivalents (FDNE) have been developed that can be used to represent the external and distant zones [21–24]. The FDNE modules are already included in many EMT programs to accelerate simulations of large power systems.

7.2.2.2.2 Overview of the Dutch 380 kV grid model

A thorough frequency-dependent model of the whole Dutch 380 kV transmission system was developed in PSCAD/EMTDC for the harmonic and transient studies [25,26]. The whole 380 kV grid was modeled in order to ensure the maximum accuracy of the results despite of long simulation time. However, the sensitivity analyses revealed that the major part of the simulation time is spent for the cables of the Spaak connection, especially when the cables are long. This is because of the complexity of cable models and the use of a small numerical time step for the simulation of transients in the cable circuits.

The Dutch 380 kV grid model includes detailed representations of substations, transmission lines, three-phase three-limb power transformers, and shunt reactors. Each generation unit was modeled by a voltage source, a series impedance, and a step-up transformer with values calculated based on a specific load-flow snapshot.

The lower voltage levels, that is, 220 kV and below, were modeled by equivalent parallel RL or RC loads. A common approach in the harmonic and transient studies of EHV systems is to either disregard the downstream network or to model them by parallel RC or RL circuits on the secondary side of the transmission transformers, where they are sized based on the load-flow dispatch. These simplifications can decrease the required time for the modeling and simulation with the expense of some inaccuracies in simulation results.

In [27], the authors studied the impact of the downstream network modeling on the harmonic frequency impedance (frequency scan) and switching overvoltages of the Dutch 380 kV grid, where it has been concluded that the type of the downstream network (OHL or cable) has a great influence while the parameters of the end customer loads have only minor impact. In addition, it has been concluded that the lower voltage lines can be modeled by nominal pi-model without losing the accuracy of the simulated harmonic impedance and maximum overvoltage. However, the use of the parallel RC equivalent for the downstream network results in some disagreements in the harmonic frequency impedance seen from the EHV level with that of the detailed model. On the other hand, regarding the transient overvoltages, a good agreement between the parallel RC equivalent and the detailed model was observed.

Sections 7.2.2.2.3–7.2.2.2.6 elaborate on the component models used for the representation of the Dutch 380 kV grid in PSCAD/EMTDC. The modeling was performed according to the guidelines and recommendations in the literature [18,28].

7.2.2.2.3 OHL and cable modeling in the Dutch 380 kV grid model

In the developed EMT model for the Dutch 380 kV grid, OHLs and XLPE cables have been represented by the frequency-dependent phase model of PSCAD to ensure maximum simulation accuracy. This model is based on the Universal Line Model theory [29]. The input data of the model is based on the actual geometry and material properties of the represented transmission lines. It should be noted that the EMT programs widely use the Cable Constants routine developed in the 1970s based on the study presented in [30]. EMTP-RV and ATP-EMTP use this routine while PSCAD uses an approximation of this routing according to the formulas in [31].

Tables 7.3 and 7.4 show the geometry and material properties used as input parameters to model the OHLs and cables of the Spaak connection in PSCAD. The cable is a single-core 2500 mm^2 copper conductor cable with the XLPE insulation. The semiconductive layers and the additional copper wire screen[1] are included in the main insulation, where the thickness, permittivity, and permeability of the insulation are modified according to the methods proposed in [32] and [33]. The proposed methods in [32] were verified in [34] too. It has been shown in [35] and [36] that the semiconductive layers should be included in the cable model when simulating transients.

The cable metallic sheath should be connected to the earth to provide a path for the cable charging current and fault currents and to nullify the electric field outside the cable. A proper sheath grounding is also necessary to improve the cable performance by reducing the sheath current (thus reduction of the sheath losses and improve the cable ampacity due to the lower heat dissipation) and by reducing the

[1]The copper wires of the screen are laid helical around the cable and not straight along the cable.

Table 7.3 Geometry and material properties used to model the OHL sections of the Spaak connection

Conductor properties	
Conductor style	Solid
Outer radius	16.2 (mm)
DC resistance	0.0488 (Ω/km)
Relative permeability	1
Sag	11.1 (m)
Bundling	
Number of bundled conductors	4
Bundle configuration	Symmetrical
Bundle conductors spacing	0.5 (m)
Configuration	
Ideal transposition of circuits	Disabled
Shunt conductance	1×10^{-11} (S/m)

Table 7.4 Geometry and material properties used to model cable sections of the Spaak connection (single-core 2,500 mm² copper conductor cable with XLPE insulation)

	Cable layer			
	Core conductor	**First insulation**[a]	**Screen conductor (metallic sheath)**	**Second insulation (outer sheath)**
Outer radius (mm)	30.65	63.1	63.7	71.5
Resistivity (Ωm)	1.98×10^{-8}	–	1.68×10^{-8}	–
Relative permeability	1	1.17	1	0.76
Relative permittivity	–	2.79	–	3.05

[a]The thickness, permittivity, and permeability of the insulation are modified to include the semi-conductive layers and the copper wire screen.
*Earth return approximation/ resistivity/permeability: Saad/100 Ωm/1.

sheath voltage (thus decreasing the breakdown probability of the cable insulation) [37]. In a normal operating condition, the sheath carries the cable charging current plus the current induced in it by the current of the main conductor, which results in power losses and (over)voltages in the sheath.

An effective bonding of the cable sheath to the earth at sufficient points is a very important factor in operation of long cables. By an effective bonding, the sheath current is reduced and limited to flow over a short length before going into the earth; moreover, the sheath voltage is also reduced and the electric field is completely confined within the cable. There are different methods of sheath bonding to the earth, among those the common types are single-point bonding, two-points bonding, multiple-points bonding, and cross-bonding. The cross-bonding is the most effective and common bonding method for long cables [37–39].

When a cable circuit is cross-bonded, the cable phases are sectionalized and divided to equal lengths over the route, called minor sections. At the end of each minor section, the sheath continuities are broken and they are cross-bonded. Sheath Voltage Limiters (SVLs) are connected to the earth at these intermediate cross bonding positions (i.e., between two minor sections) to dissipate sheath voltage surges. Finally, sheaths are bonded and grounded at the ends of each major section, which is made of three minor sections. In some cases, the cable core conductors may also be transposed for further reduction of sheath currents. By using the cross-bonding technique and grounding in between, the induced sheath currents are largely canceled out and the sheath (over)voltages are reduced. The cross-bonding affects the cable impedance and admittance matrices since the system parameters are changed by transposition and grounding of the cable sheaths [40,41].

Figure 7.6(a) shows an illustration of the applied sheath cross-bonding for the Spaak connection. This bonding method has been already used in the commissioned

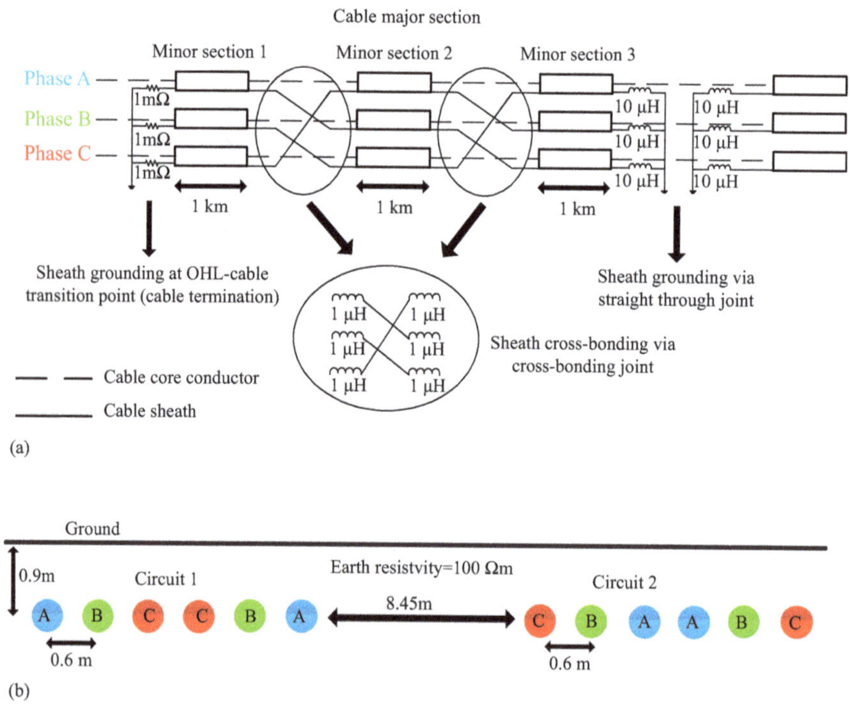

Figure 7.6 (a) Schematic of a cable major section and sheath cross-bonding and (b) cable cross-sectional layout with the flat formation in an open cable trench. The depth and distances are referred to the center of the cable

cable projects in the Dutch 380 kV grid (Randstad South-Ring and North-Ring projects). Therefore, the same bonding approach is considered in this research. It is assumed that each minor section is 1 km long. So, when for instance a cable section in Figure 7.2 is 30 km long, it is made of 10 major sections and 30 minor sections. In practice, cable cross-bonding joints are used to connect cable minor sections. This allows cross-bonding or termination of sheath conductors and also the direct connection of core conductors without transposing.

It is very important to model the cable cross-bondings with sufficient details in order to have an accurate transient simulation. In the developed EMT model of the Dutch 380 kV grid in PSCAD, all cable cross-bondings are represented in full detail as shown in Figure 7.6(a). A detailed model of each cross-bonding was applied as the discrete representation of each minor section is the most accurate way of modeling [42]. The bonding wires are represented by 1 μH inductance at the cross-bonding joints. The grounding wires are represented by 10 μH inductances at the straight-through joints (connection of two major sections) and 1 mΩ resistances at the cable terminations [43].

Figure 7.6(b) shows a cross-sectional layout of the cable circuits, which are laid in a flat formation in an open cable trench. The mutual electric and magnetic couplings between all 12 cables are taken into account in the developed EMT model of the Dutch 380 kV grid in PSCAD. A flat formation improves heat dissipation and thereby increases the ampacity of the cable. Other cable formations are also possible like a trefoil formation, which is advantageous in terms of space and electromagnetic field issues.

7.2.2.2.4 Transformer and shunt reactor modeling

A detailed model of the three-winding three-limb power transformers was used in the EMT model of the Dutch 380 kV grid. This model includes the core saturation and parasitic capacitances of the transformer. The air-core shunt reactors are represented by a multilayer model; however, the mutual couplings (mutual inductances) between windings of reactors are not taken into account since this a typical simplification used when performing transient studies [44]. More elaboration on the modeling of transformers and shunt reactors is available in [33].

7.2.2.2.5 Capacitor banks modeling

In the Dutch 380 kV grid, capacitor banks exist at some substations for voltage control and they are switched in or out depending on the load-flow and the voltage level. Equivalent RLC circuits were used to represent the 380 kV capacitor banks in the EMT model of the Dutch 380 kV grid.

7.2.2.2.6 Lower voltage grids modeling

As it was previously mentioned in Section 7.2.2.2.2, the lower voltage levels, that is, 220 kV and below, are represented by equivalent parallel RL or RC loads. The fixed load component in PSCAD is used for this purpose. This component models the load characteristics as a function of the voltage magnitude and frequency, where the load active and reactive powers are considered separately using the following expressions [26,45]:

$$P = P_0 \left(\frac{V_l}{V_{l_0}} \right)^{NP} (1 + K_{PF} dF) \tag{7.3}$$

$$Q = Q_0 \left(\frac{V_l}{V_{l_0}} \right)^{NQ} (1 + K_{QF} dF) \tag{7.4}$$

where P is the equivalent load active power (Watt); P_0 is the rated active power per phase (Watt); V_l is the load voltage (Volt); V_{l_0} is the rated load voltage (Volt); NP is the dP/dV voltage index for active power; K_{PF} is the dP/dF frequency index for active power; Q is the equivalent load reactive power (VAR); Q_0 is the rated reactive power per phase (VAR); NQ is the dQ/dV voltage index for reactive power; and K_{QF} is the dQ/dF frequency index for reactive power.

7.3 Shunt compensation allocation in hybrid OHL-cable systems

Cables have different electrical characteristics compared to OHLs, which results in excessive reactive power production and in some cases cable overloading when they are operated in parallel with OHLs. Several operation problems are associated with the reactive power of long cables, which should be handled by an optimal reactive power shunt compensation. In addition, necessary countermeasures should be applied to control the loading of overloaded cables which are in parallel operation with OHLs.

It is crucial to have a sufficient size of shunt compensation since both under- and over-compensation increase the risk of undesirable conditions like over-voltages and the zero-missing phenomenon [2,46]. The size of shunt compensation has to be determined in the first step of system impact studies as it has a direct influence on the rest of technical issues like harmonic and transient operations.

This section describes the shunt compensation sizing in hybrid OHL-cable transmission systems and addresses the most important parameters for an optimal compensation allocation. In addition, the active power flow and the cable over-loading are analyzed. Simulation results show that various parameters including the cable length, shunt reactor location, mixed-line configuration, and system voltage level can substantially affect the required size of shunt compensation. The out-comes of this study are important in the design process of every hybrid OHL-cable circuit as they can be used to choose the best compensation approach. The com-pensation sizes obtained in this section will be used in the Sections 7.4–7.7 for the analysis of other technical issues.

7.3.1 Cable capacitive behavior

A useful parameter in the study of electrical power systems is the surge impedance loading (SIL) or natural loading. SIL of a transmission line is the megawatt (MW) loading of the line at which a natural reactive power balance occurs. In other words,

SIL is the power loading at which reactive power is neither produced nor absorbed by the line and can be expressed as below [14]:

$$\mathrm{SIL} = \frac{V^2}{Z_S}, \quad \text{where } Z_S = \sqrt{\frac{L}{C}} \tag{7.5}$$

where V is the system voltage and Z_s is the line surge impedance. L is the line series inductance and C is the line shunt capacitance, both per unit length.

SIL is a useful parameter to reveal the reactive power behavior of a transmission line. When a line is loaded above its SIL, it acts as a reactor and absorbs reactive power and when it is loaded below its SIL, it acts as a capacitor and generates reactive power. Compared to an OHL, a cable has a larger shunt capacitance and a smaller series inductance, thus the surge impedance (Z_s) of the cable is several times smaller than that of the OHL and consequently, the SIL of the cable is several times larger than that of the OHL.

The thermal rating of a line is the power transmission capacity at its ampere capacity (ampacity). During normal operation, cables are always loaded below their SIL because the cable SIL is several times larger than its thermal rating. Hence, cables behave as capacitors and produce reactive power. On the contrary, SIL of an OHL is several times smaller than its thermal rating; therefore, OHLs are normally operated above their SIL and behave as inductors, that is, consuming reactive power [14,47].

7.3.2 Negative effects of cable reactive power

The large surplus reactive power produced by cables results in steady-state voltage rise and under-excitation of synchronous generators (in severe cases self-excitation), which can impose severe stresses on system equipment and increase the risk of system transient instability [2,47,48]. Another risk of long cables operation is the large capacitive current in the line circuit breakers, which can exceed the type-tested breaking capability of the breaker [3,47,49,50].

Always a part of the cable ampacity is occupied by its capacitive charging current whilst this part becomes larger by increasing the cable length and consequently the remaining transfer capacity for the active power gets smaller. At a critical length, L_{critical}, the capacitive current equals the cable ampacity so that no transfer capacity is left for the active power. This is one of the main limitations in the application of long AC cables, which is elaborated and analyzed in references [47,51–53].

Figure 7.7 illustrates the voltage profile and the flows of current, reactive power, active power, and apparent power in a lossless cable [53]. The optimal operating condition in a lossless cable happens when the voltage of the sending (V_s) and the receiving (V_r) terminals are equal. This leads to an equal distribution of the reactive power between the two terminals. In this situation, the maximum cable length, $L_{\max}$, is twice of L_{critical} and the charging current equals to the ampacity at each end.

In Figure 7.8, the maximum transferable active power through a lossless cable is plotted versus the cable length for three voltage levels [47]. The calculated

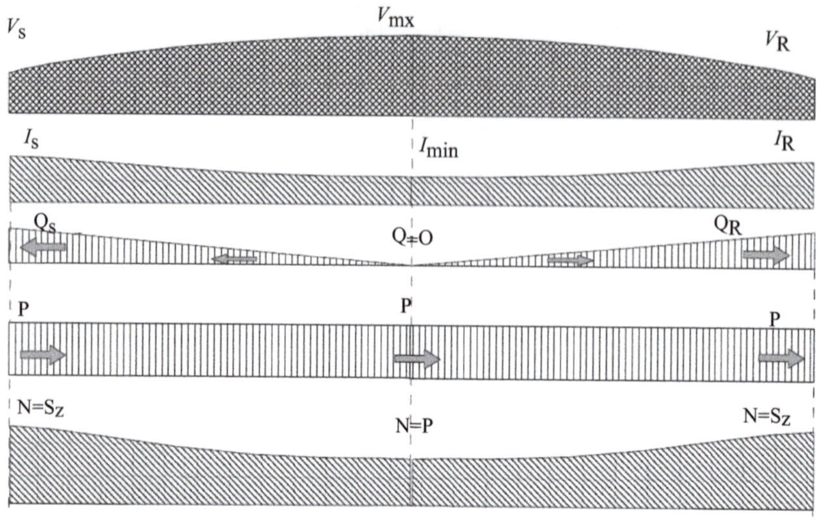

Figure 7.7 Voltage profile (V) and flows of current (I), reactive power (Q), active power (P), and apparent power (S) in a lossless EHV cable operated with equal terminal voltages (optimal operation condition). Adapted from [53]. S and R denote the sending and receiving terminals, respectively

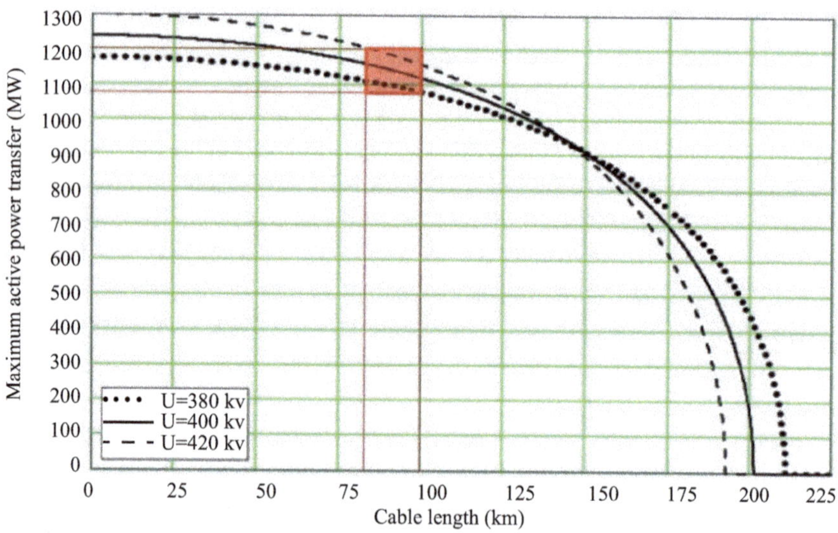

Figure 7.8 Maximum active power transfer, P_{max}, at the thermal limit versus the cable length, calculated for different terminal voltages. The cable losses are neglected and terminals have equal voltages (optimal operation condition). Adapted from Reference [47]. The red box shows an example case

powers are for a three-phase system with 2,500 mm^2 copper conductor single-core cables. For 380 kV cables, the maximum length is around 210 km, implying that at this length the active power that can be transferred through the cable is zero. According to this figure, it is beneficial from the transfer capacity aspect to operate short cables under higher voltages and long cables under lower voltages.

The operation of a lossless cable with different terminal voltages causes an asymmetrical reactive power profile (nonoptimal operation condition). This results in the flow of more reactive power toward the terminal with the lower voltage. In this case, the maximum transferable active power can be calculated as a function of the sending and the receiving terminal voltages by using nonlinear equations [47].

It should be noted that for a lossy cable, the optimal operating condition and the maximum active power transfer are achieved when there is a particular voltage deviation (depending on the system parameters) between the sending and receiving terminals.

7.3.3 Reactive power compensation

Taking into account the negative impacts of the cable surplus reactive power, it is necessary to allocate reactive power shunt compensation with an optimal size at proper locations. The shunt compensation is realized by using fixed- or variable-size shunt reactors and it reduces the reactive power being injected to the rest of the system.

As it was previously mentioned in Section 7.2.1.2, shunt reactors should be connected directly to the hybrid circuit at the two ends of the circuit or be distributed at multiple locations along the circuit. Alternatively, for some cable lengths, shunt reactors can be installed at substation busbars or at tertiary windings of power transformers if energization overvoltages and capacitive currents are not large. When the increase of the active power transfer capacity is required, shunt reactors should be distributed along the cable especially for cables longer than the critical length. A distributed compensation can also reduce the required size of compensation and boost the system transient stability [48,49]; however, the distributed compensation may not be possible in some situations due to the land occupation and visual impact of reactor banks as well as reliability issues [47].

The shunt compensation can be realized by fixed-size reactors or by controllable shunt compensation devices like tapped-winding reactors, Thyristor Controlled Reactors (TCR), and shunt-connected flexible AC transmission systems (FACTS) based on voltage-source converters, like the static synchronous compensator (STATCOM). Power electronics-based apparatus like TCRs and STATCOMs are more costly, but they give a continuous regulation capability and a fast dynamic response despite the fast response is not strictly required for the steady-state reactive power control. Continuous regulation can also be provided if a sufficient number of discrete regulation steps in the desired range are available [52]. In this aspect, tapped-winding reactors or reactor banks with a sufficient number of switchable fixed-size reactors are favorable and cost-effective solutions.

In the literature, several arrangements for the location of shunt reactors are proposed. In [54], the authors proposed different arrangements for shunt reactors as

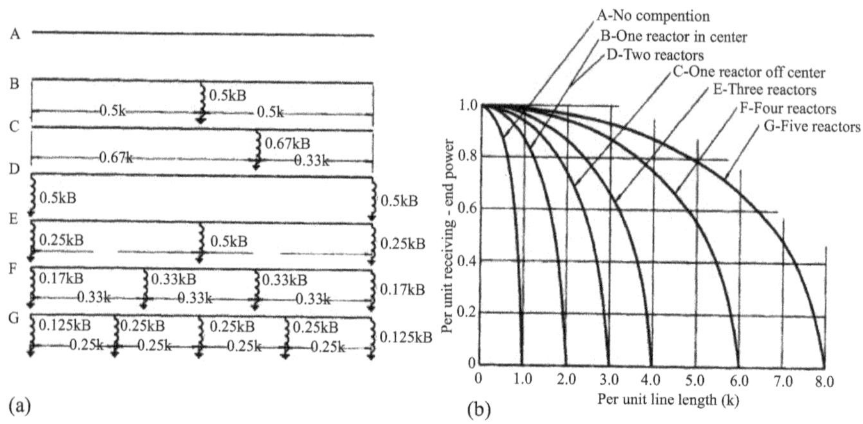

Figure 7.9 (a) Various arrangements of shunt compensating reactors and (b) line performance with a unity power factor load, pu values. Adapted from [54]

shown in Figure 7.9(a), where it is assumed that a load with a unity power factor is supplied. The curves in Figure 7.9(b) show the effectiveness of the proposed arrangements by the extent to which the thermal rating of the cable is available for the active power transmission as a function of the cable length. One per unit (pu) receiving-end power means the active power equal to the thermal rating of the cable. One pu line length is the cable length at which the total charging current equals the current rating of the cable (i.e., critical length). In general, it can be concluded that the best efficiency in the active power transfer can be archived by a distributed compensation in which shunt reactors are placed at several points along the circuit. However, in practice it is usually preferred to place the shunt reactors at the ends of the circuit to minimize the land occupation and visual impacts.

7.3.4 Simulation results

The steady-state study was performed for the Spaak connection with the six cable scenarios presented in Section 7.2.1.4. The mixed-line configuration under study is as shown in Figure 7.2, unless a different configuration is mentioned. The line-end compensation (LEC) with direct connection of reactors to the circuit was applied due to the reasons explained in Section 7.2.1.2. Several sensitivity analyses were performed to determine the impacts of different parameters on the steady-state design and operation of cable circuits. The steady-state simulations were carried out in the PowerFactory model of the Dutch transmission grid.

7.3.4.1 Sizing criteria

In this study, a shunt compensation sizing approach based on the criteria addressed in [2,47,51,55,56] was used. This approach is according to the technical concerns related to the operation of long EHV cables. In this regard, the following four criteria were used for the sizing of shunt reactors, which are mainly based on the no-load open-end operation as the situation with the most severe reactive power balance:

1. *Sudden voltage step (or rapid voltage change (RVC))*: This is a quick transition in RMS voltage between two steady-state conditions during which the voltage does not exceed the dip/swell thresholds [57]. Based on this sizing criterion, the sudden voltage jump at the sending terminal after the no-load energization should be less than 3% (according to the Dutch grid code) to reduce the disturbance to customers. In addition, the new steady-state voltage at the sending terminal should be less than the swell threshold of 1.1 pu.
2. *Voltage along the line*: this constraint is related to the voltage on the hybrid circuit following the no-load energization. The voltage must be less than 1.1 pu to allow line reclosure and avoid damages to line-connected devices. Protection equipment like surge arrestors must be checked in terms of temporary overvoltage (TOV) capability. The voltage at all OHL-cable transition points (cable terminations) and open-end has to be smaller than 1.1 pu.

   It is worth noting that due to the Ferranti effect, the voltage at the open-end is the highest voltage during the no-load energization when the shunt compensation is not utilized. However, the highest voltage along a shunt-compensated hybrid circuit may be at a point different than the open-end. Therefore, in this study the attention is given to the voltage of all points along the circuit and not only to the open-end voltage.

   The Ferranti effect states that in case of no-load operation of a transmission line, the steady-state voltage at the receiving open-end (V_r) increases beyond the sending-end voltage (V_s), which can be expressed as below:

$$V_r = \frac{V_s}{1 - \dfrac{\omega^2 LCl^2}{2}} \tag{7.6}$$

   where L is the line series inductance per unit length, C is the line shunt capacitance per unit length, and l is the line length. ω is the system angular frequency ($\omega = 2\pi f_0$, where f_0 is the system power frequency). The source inductance (i.e., the equivalent short-circuit inductance of the system) is included in L. This means that the steady-state voltage rise at the open-end is higher when the system short-circuit power is lower (i.e., larger L).
3. *Capacitive breaking current of circuit breakers*: when a large temporary overvoltage occurs, the fast line switching-off at the energizing terminal is an option under the condition that the capacitive breaking current of the line circuit breaker be sufficient. When shunt compensation is applied, it must be checked that the no-load current of the hybrid circuit is not larger than the type-tested capacitive breaking current of the circuit breaker. This is an important requirement in order to avoid circuit breaker failures [50,58]. For the sake of this study, 400 A is selected as the most strict limit as well as the preferred value according to [58], although higher limits can also be used.

4. *Cable charging current*: in situations where the cable length is longer than the critical length, shunt compensation must ensure that the cable charging current stays below the thermal limit (ampacity) of the cable. This can be realized by installing the shunt reactors at appropriate intermediate locations along the cable. In this study, 2 kA limit is chosen for this criterion since each individual cable of the Spaak connection can withstand 2 kA for 24 h, whereas the energization only lasts for tens of seconds or maximum a few minutes.

7.3.4.2 Load-flow scenarios

It is important to do system studies for different load-flow scenarios such as high wind power generation or low conventional power generation. The shunt reactor sizing under different load-flow scenarios is crucial to take the impact of different power dispatches and short-circuit powers into the account. Two load-flow scenarios were chosen for this study:

1. Load-flow A: In this scenario, a large power transport is expected from the west to the east of the country. The short-circuit powers of the Krimpen and Dodewaard substations are 29.7 and 20.4 GVA, respectively. The steady-state voltages at the Krimpen and Dodewaard substations are around 1.03 pu.
2. Load-flow B: In this scenario, there is a large transport of power from the north-east to the west and south-west of the country. The short-circuit powers of the Krimpen and Dodewaard substations are 20.1 and 21 GVA, respectively. The steady-state voltages at the Krimpen and Dodewaard substations are around 1.08 pu. It should be pointed that the steady-state voltage profiles are high in this load-flow scenario.

7.3.4.3 Global compensation

The steady-state operation of the Spaak connection was studied under two assumptions. Firstly, according to the best-end switching approach, the hybrid OHL-cable circuit is energized from the substation with a higher short-circuit power (the Krimpen substation in load-flow A and the Dodewaard substation in the load-flow B). The best-end switching results in a lower sudden voltage jump at the sending-end and also lower steady-state overvoltage at the open-end (due to the lower Ferranti effect: (7.6)), which leads to further saving on the compensation size.[2]

Secondly, when both circuits should be connected, the second circuit is energized after the complete connection of the first circuit to the system at both ends. It was also assumed that the required compensation for each circuit is equally distributed at the two ends (SR1=SR2 and SR3=SR4 in Figure 7.2). In addition, the required compensation sizes for the first energized circuit and the second energized circuit can be different.

Table 7.5 presents the minimum required size of shunt compensation for different cable lengths in the circuit. The results are categorized for the two load-flow

[2]In the case of disconnection, it is better to begin the switching from the weak-end in order to minimize sudden voltage changes.

Table 7.5 Minimum required shunt compensation size (Q_{sh}) and degree (K_{sh}) for the single-circuit and double-circuit operations of the Spaak connection

Cable length (km)	**Q_{sh} (Mvar) (K_{sh} [%])**			
	Load-flow A		**Load-flow B**	
	Single-circuit	**Double-circuit**	**Single-circuit**	**Double-circuit**
12 (15%)	50 (19.8%)	102 (20.2%)	176 (69.8%)	366 (72.6%)
20 (25%)	214 (50.9%)	430 (51.2%)	356 (84.8%)	722 (86%)
40 (50%)	620 (73.8%)	1,245 (74.1%)	750 (89.3%)	1,560 (92.9%)
60 (75%)	1,030 (81.7%)	2,062 (81.8%)	1,160 (92.1%)	2,380 (94.4%)
80 (Fully-cable)	1,440 (85.7%)	2,883 (85.8%)	1,600 (95.2%)	3,226 (96%)

scenarios and two modes of operation, namely single-circuit and double-circuit operations. The reported sizes for the double-circuit operation are the total size of both circuits (i.e., SR1+SR2+SR3+SR4). The values between parenthesizes are the compensation degrees, calculated as below:

$$K_{sh}(\%) = \frac{Q_{sh}}{N \times \omega\, ClV_n^2} \times 100 \qquad (7.7)$$

where Q_{Sh} is the total shunt compensation size in Mvar, C is the cable capacitance in F/km, l is the cable length in km, and V_n is the system nominal voltage, which is 380 kV in this study. The factor N denotes the number of cables, which is 6 and 12 respectively for the single-circuit and the double-circuit operations. In this equation, the reactive power produced by the OHL sections of the hybrid circuit is ignored since it is negligible compared to that of the cable sections.

Figure 7.10 shows the minimum required compensation size and degree for different cable lengths and the two load-flow scenarios when both circuits are in operation. It can be noticed that the minimum required compensation size becomes higher by increasing the cable length albeit a full compensation ($K_{sh} = 100\%$) is not required. When the compensation degree gets closer to 100%, the resonance frequency of the circuit formed by the shunt compensation inductance and the cable capacitance gets closer to the fundamental frequency. In addition, the probability and the severity of the zero-missing phenomenon is higher when the degree of compensation is higher [46].

According to Table 7.5, the required compensation size for the double-circuit operation is slightly higher than double of that for the single-circuit operation. There is also a substantial difference between the results of the two load-flows with higher compensation degrees for load-flow B. The main reason is the high voltage profile in the region of the Spaak connection in the load-flow B, which is already around 1.08 pu before the cable energization. Therefore, a larger compensation is required to keep the voltages within the permissible limits.

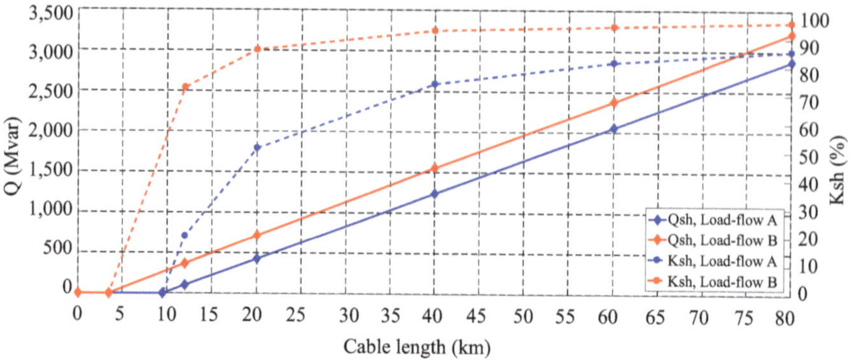

Figure 7.10 Minimum required shunt compensation size (Q_{sh}) and degree (K_{sh}) versus the cable length for the double-circuit operation of the Spaak connection

It has been shown that the load-flow scenario has a considerable impact on the shunt compensation size. The steady-state voltage prior to the cable energization and the short-circuit power of the substation from which the circuit is energized can significantly affect the size of shunt compensation. Hence, when operation in several load-flow scenarios is expected, a global compensation capable of satisfying the sizing criteria in all the expected scenarios is necessary. This means that the installed capacity for shunt compensation should be based on the highest calculated size among different load-flow scenarios. In this regard, in the case of the Dutch transmission system, future operation in both the load-flows A and B together with other load-flow scenarios could be expected. With the assumption that the most extreme scenarios are the load-flows A and B, the global compensation sizes are those obtained for the load-flow B as reshown in Table 7.6.

7.3.4.4 Most decisive sizing criterion

It is of high interest to determine the sizing criterion that requires the largest compensation size in order to be satisfied. Among the four sizing criteria, the cable charging current during the no-load energization is found to be the less strict criterion. The cable charging current variation by increasing the cable length prior to the shunt compensating is shown in Figure 7.11. None of the cable currents exceeds the limit (2 kA) even for longer lengths up to around 103 km in the load-flow A and around 92 km in the load-flow B. Comparing the first circuit and the second circuit energizations, the cable currents are identical for lengths up to 40 km. However, the currents get slightly different for lengths longer than 40 km with higher values for the second circuit energization.

Figure 7.12 shows the minimum required shunt compensation degree for the double-circuit operation versus the cable length calculated based on the other three sizing criteria. It can be seen that the capacitive breaking current of the

Table 7.6 Minimum required shunt compensation size (Q_{sh}) and degree (K_{sh}) for the global compensation of the Spaak connection

Cable length (km)	Q_{sh} (Mvar) (K_{sh} [%])	
	Single-circuit operation	**Double-circuit operation**
12 (15%)	176 (69.8%)	366 (72.6%)
20 (25%)	356 (84.8%)	722 (86%)
40 (50%)	750 (89.3%)	1560 (92.9%)
60 (75%)	1160 (92.1%)	2380 (94.4%)
80 (Fully-cable)	1600 (95.2%)	3226 (96%)

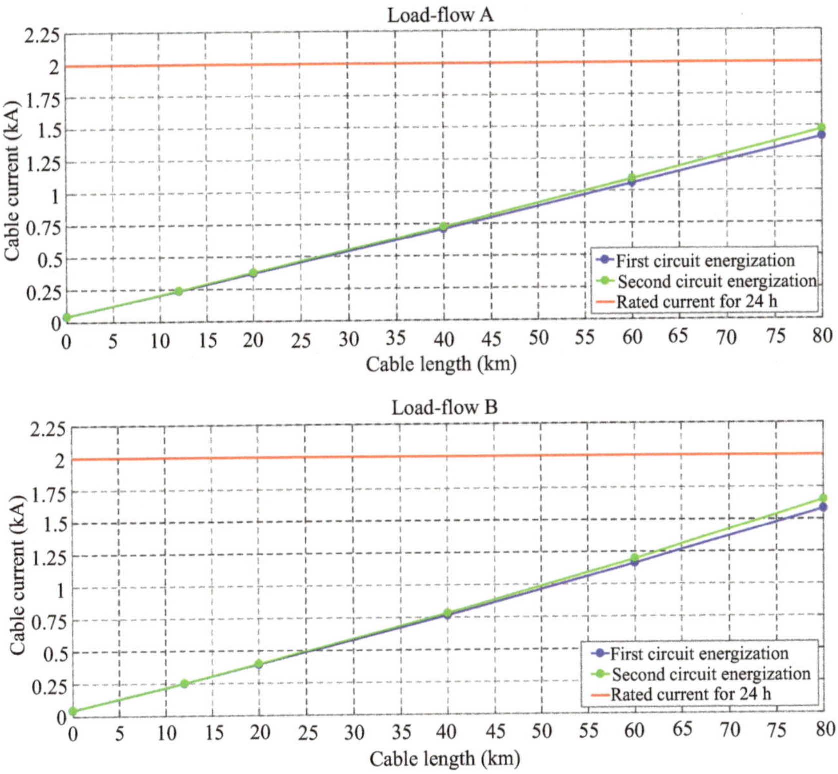

Figure 7.11 Cable charging current versus the cable length after the no-load energization of the Spaak connection without shunt compensation in the load-flows A and B. The plotted currents are for the first cable section from the sending-end, which is: Cable 1 in the load-flow A and first circuit energization; Cable 4 in the load-flow A and second circuit energization; Cable 3 in the load-flow B and first circuit energization; Cable 6 in the load-flow B and second circuit energization (see Figure 7.2)

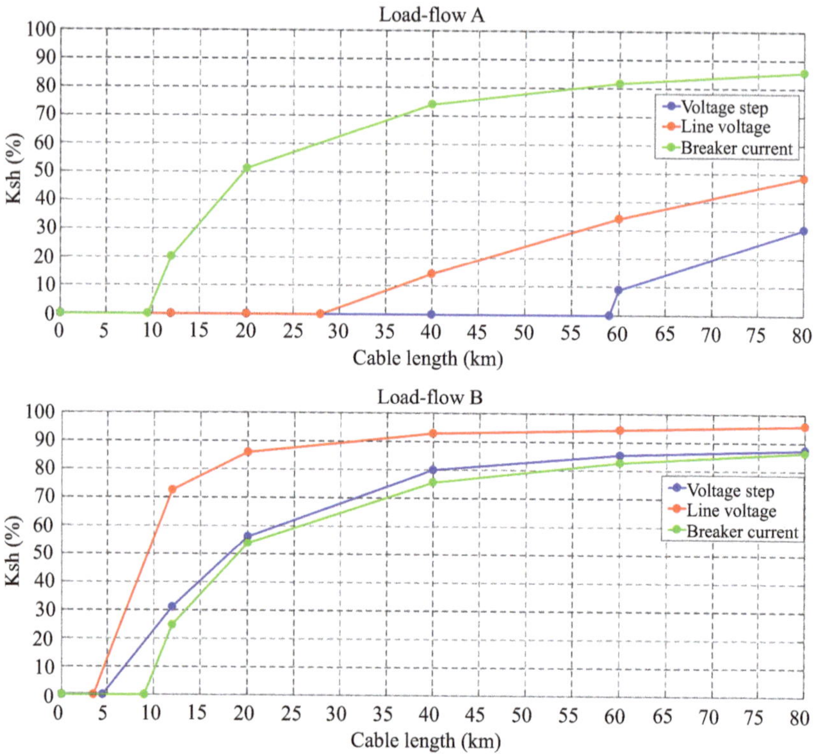

Figure 7.12 Minimum required shunt compensation degree versus the cable length calculated based on different sizing criteria for the double-circuit operation of the Spaak connection in the load-flows A and B

circuit breaker is decisively the most determinant criterion in the load-flow A. This means that the size of compensation is dictated by this criterion and any further savings on shunt reactors can be achieved with the use of higher rated circuit breakers. However, in contrast to the load-flow A, the voltage along the hybrid circuit is the most decisive sizing criterion in the load-flow B despite the decreasing difference between the three sizing criteria by increasing the cable length. It should be noted that for the Spaak connection with 80 km transmission length, the most decisive sizing criterion is not dependent on the mixed-line configuration, compensation location, and operation mode (single-circuit or double-circuit operation), whereas this dependency exists when the transmission length is longer.

It is necessary to note that the computed compensation sizes related to the capacitive breaking current criterion are based on the 400 A target value as the recommended value in the standard [58]. However, in some cases, the alternative solution to reduce the compensation size is to type test the circuit breakers for a higher capacitive breaking current and perform the compensation sizing based on this current as the target value. This results in further saving on the shunt compensation in scenarios like the load-flow A. For instance, the capacitive current in the line circuit breaker before shunt compensating in the load-flow A is reported in Table 7.7. The circuit breaker should be either type tested for interrupting the total capacitive current of the circuit or, if it cannot interrupt the total current, shunt reactors should be sized to decrease the capacitive current down to the interrupting level of the breaker.

Figure 7.13 shows the results of the sensitivity analysis on the circuit breaker capacitive breaking current rating. The minimum required compensation degree is plotted versus the cable length for two rated values, 400 A and 500 A, when operating in the load-flow A. The utilization of a 500 A-rated circuit breaker, compared to a 400 A, results in up to 20% saving in the compensation size, which can be a considerable value for long cable lengths.

Table 7.7 Capacitive current flowing through the line circuit breaker before shunt compensating in the load-flow A

Cable length (km)	Current (A)
12	484
20	756
40	1437
60	2140
80	2862

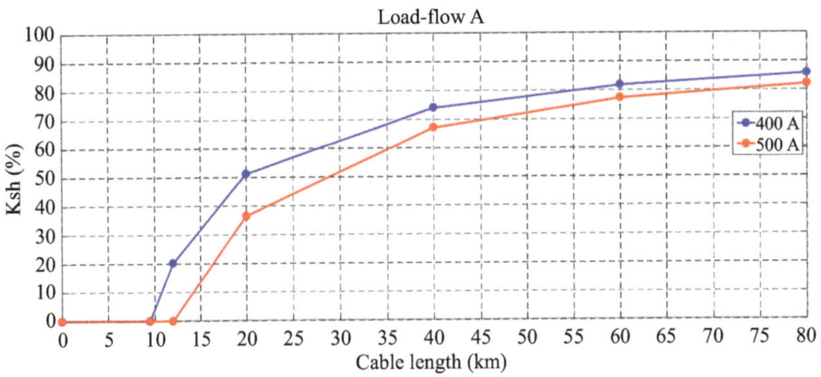

Figure 7.13 Minimum required shunt compensation degree versus the cable length for 400 A and 500 A ratings for the capacitive breaking current of the line circuit breaker

7.3.4.5 Distributed compensation

In addition to the shunt compensation size, the location of shunt compensation impacts the steady-state and transient operations of the system. The line-end compensation (LEC) was investigated in the Sections 7.3.4.3 and 7.3.4.4. This type of compensation is easy to be realized in practice and well-known for system operators. There are also less-practiced compensation arrangements in which shunt reactors are distributed along the circuit. As it was previously mentioned in Section 7.3.3, the realization of these arrangements in practice is certainly more difficult than the line-end compensation due to the higher cost and the need for the construction of more intermediate facilities along the line route. However, the distributed compensation requires a smaller compensation size compared to the line-end compensation and it improves the system operation by increasing the active power transfer capacity and better voltage control along the circuit. In addition, it is more reliable in case of failure in one of the reactor banks because the rest of the reactors will remain in operation and only a percentage of the total compensation will be lost.

In this section, the performance of different distributed compensation arrangements is investigated in terms of the required compensation size. Three arrangements, which differ in the number and location of shunt reactors, are compared with the line-end compensation:

1. *Arrangement A3*: In this arrangement, two reactors are connected to the circuit at the sending and receiving ends right behind the line circuit breakers (same as LEC), plus an additional shunt reactor at the fourth OHL-cable transition point (cable termination) of each circuit. In the case of 80 km cable (fully-cable scenario), since there is no OHL-cable transition point, the additional reactor is located at the middle of the cable.
2. *Arrangement A6*: This arrangement is called cable-end compensation (CEC). In CEC, reactors are located at each OHL-cable transition point of each circuit (i.e., one reactor per transition point, which is in total six reactors per circuit of the Spaak project). In the case of 80 km cable (fully-cable scenario), reactors are placed at six equally distributed intermediate points along the cable with a distance of 11.43 km from each other.
3. *Arrangement A8*: This arrangement has eight reactors per circuit. It is similar to the arrangement A6, but with two more reactors at the sending and receiving ends of each circuit right behind the line circuit breakers.

Figure 7.14 shows the minimum required shunt compensation degree versus the cable length for different distributed compensation arrangements when both circuits are in operation (double-circuit operation) in the load-flows A and B. The sizes were calculated according to the most decisive sizing criterion, which is the breaker capacitive breaking current in the load-flow A and the voltage along the line in the load-flow B.

For the load-flow A, the application of the distributed compensation can result in up to 1.6% decrease in the minimum required compensation size compared to

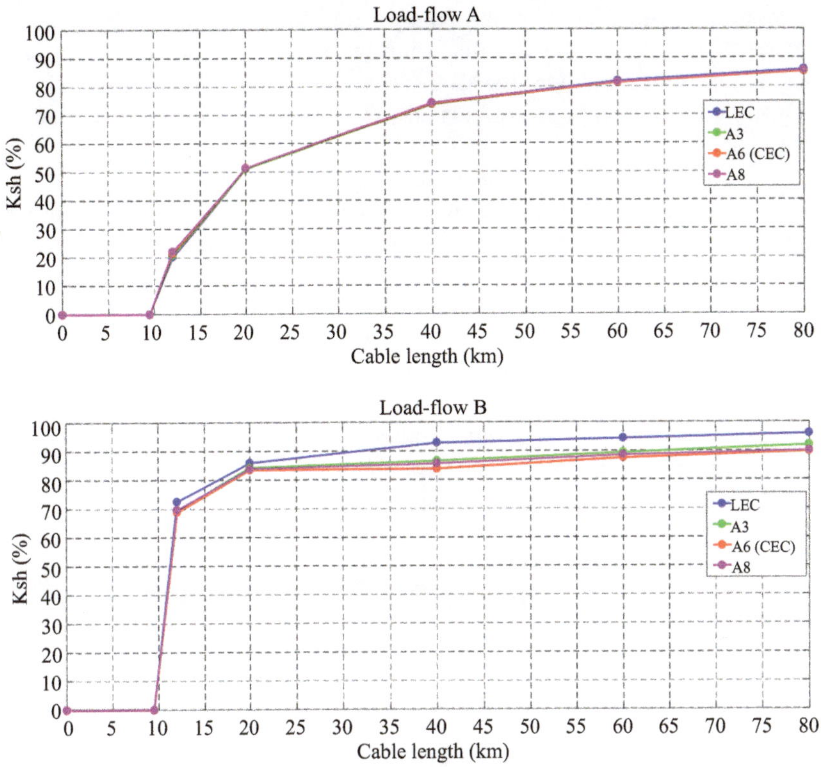

Figure 7.14 Minimum required shunt compensation degree versus the cable length for the double-circuit operation of the Spaak connection in the load-flows A and B with the line-end compensation (LEC) and three distributed compensation arrangements

that of the line-end compensation. In addition, for load-flow B, the application of the distributed compensation can decrease the minimum required compensation size up to 8.8%. This is due to the fact that the distributed compensation provides very effective control on overvoltages along the hybrid circuit (especially for circuits with long cables). Such an effective control on the voltage leads to the need for a smaller compensation size compared to the line-end compensation for the cases like the load-flow B in which the voltage is the most decisive sizing criterion. For very long cables, it can be even more crucial to utilize the distributed compensation for an effective voltage control and to increase the active power transfer capacity of the cable.

7.3.4.6 Impact of mixed-line configuration

Since the number and the location of cable sections are determined in the planning and design stages of cable projects, it is important to compare various mixed-line configurations (i.e., different number and location of cable sections) from the shunt compensation perspective. For this purpose, the four mixed-line configurations presented in Section 7.2.1.3 and the study case configuration (Figure 7.2) are compared under the condition that all these configurations are compensated at the line-ends (i.e., LEC) with sizes calculated based on the sizing criteria presented in Section 7.3.4.1.

The minimum required compensation degree for the five mixed-line configurations versus the cable length is shown in Figure 7.15. For load-flow A, there is a negligible difference between the configurations, whereas the difference is considerable for load-flow B as it can be up to 60% in some cases. This means hundreds of Mvar differences in the total size of compensation. The difference between configurations is however decreasing by increasing the cable length in the circuit due to the decreasing dissimilarity between the configurations.

Moreover, as shown for the load-flow B, Configuration 1 requires the highest compensation degree (even overcompensation for cable lengths between 15 km and 60 km), whereas Configuration 2 requires the lowest compensation degree. These

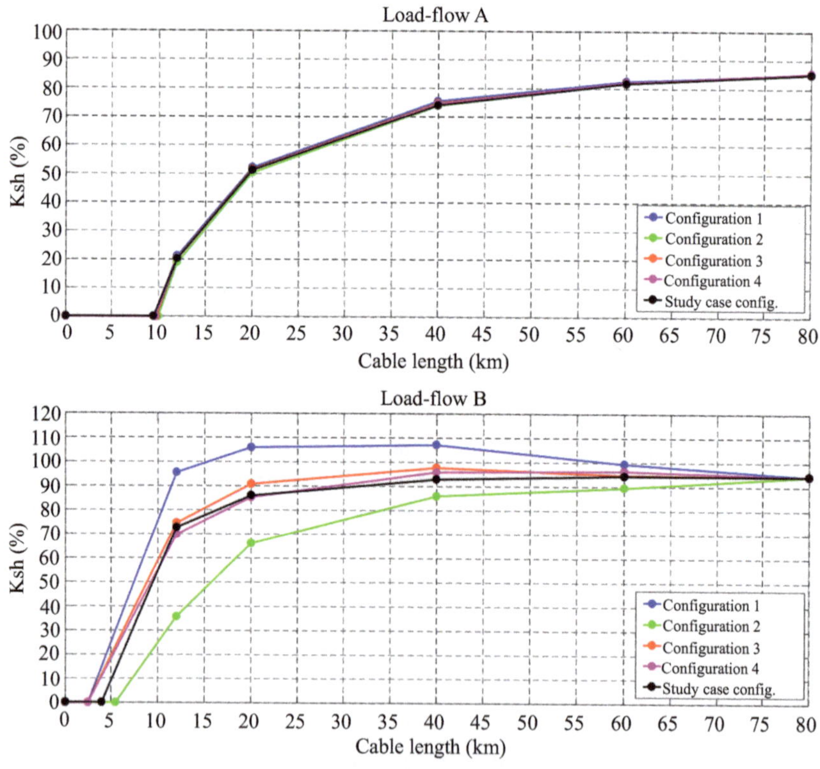

Figure 7.15 Minimum required shunt compensation degree versus the cable length for different mixed-line configurations for the double-circuit operation of the Spaak connection in the load-flows A and B

two configurations will obviously represent an opposite behavior when the circuit is being energized from the other end (i.e., the Krimpen substation). Thus, it can be concluded that the location of the cable sections with respect to the substation from which the circuit is being energized can significantly affect the overvoltages on the circuit and consequently the required compensation size. This point should be always taken into account in the determination of the switching sequence for the energization of cable circuits.

7.3.5 Cable overloading

In addition to the determination of the optimal size and location of shunt reactors, it is necessary to investigate the power flow and cable loadings when the hybrid OHL-cable circuit is under operation. The power transferred through a hybrid circuit increases by increasing the cable length due to the decrease of the total series impedance of the circuit. It can be expected that the cable rated power be exceeded for cables longer than a particular length depending on the grid topology and the load-flow scenario [2,55].

To investigate this issue for the Spaak connection, it is assumed that the circuits are compensated with the sizes reported in Table 7.6 and the currents at both ends of each cable section, as shown in Figure 7.16, are observed during the single-circuit and double-circuit operations. For the case of 80 km cable (fully-cable scenario), since there is only one cable section, the currents are measured at the sending terminal, the receiving terminal, and four points along the circuit with the distances of 11.43 km, 34.29 km, 45.72 km, and 68.57 km from the sending terminal.

Figure 7.17 shows the cable currents for different cable lengths during the single-circuit and double-circuit operations of the Spaak connection in the load-flow A. The cable currents are exceeding the rated current when the connection is operated with a single circuit and the cable length is longer than approximately 37 km, whereas the loading stays below the rated current when the connection is operated with double circuits. This means that in the double-circuit operation when one of the circuits is switched out of service, special attention should be given to the loading of

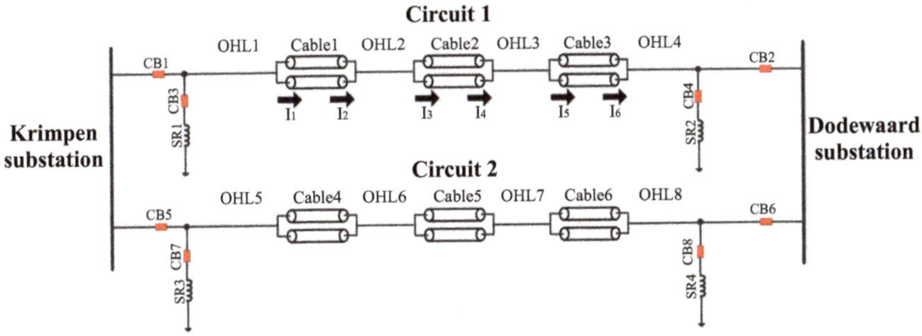

Figure 7.16 Simulated cable currents shown on the single-line diagram of the Spaak connection

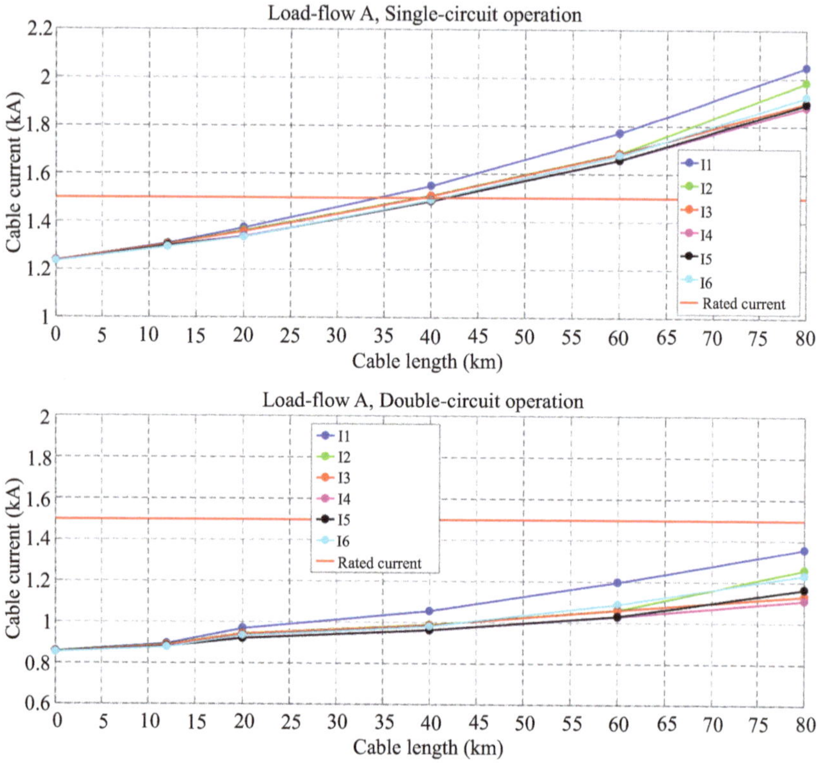

Figure 7.17 Cable currents versus the cable length for the single-circuit and double-circuit operations of the Spaak connection in the load-flow A

the remained connected circuit (i.e., the n-1 criterion should be checked). The control of power flow can be realized by using series reactors, phase-shifting transformers or changing the dispatch scenario.

Figure 7.18 depicts the injected active power to the circuit at the sending terminal. It can be seen that the injected active power increases for longer cables as the result of the lower series impedance of the cable compared to that of OHL.

7.4 Resonance behavior of hybrid OHL-cable systems

The application of long EHV cables in existing transmission systems raises serious concerns regarding the risk of low-order resonance frequencies. The large cable capacitance together with the inductance of inductive components, like transformers and shunt reactors, may result in series and parallel resonance circuits leading to temporary overvoltages when excited. Resonance frequencies in systems with long cables can be considerably lower than those in systems without cables, especially during periods of low short-circuit power. As a result, low resonance frequencies are

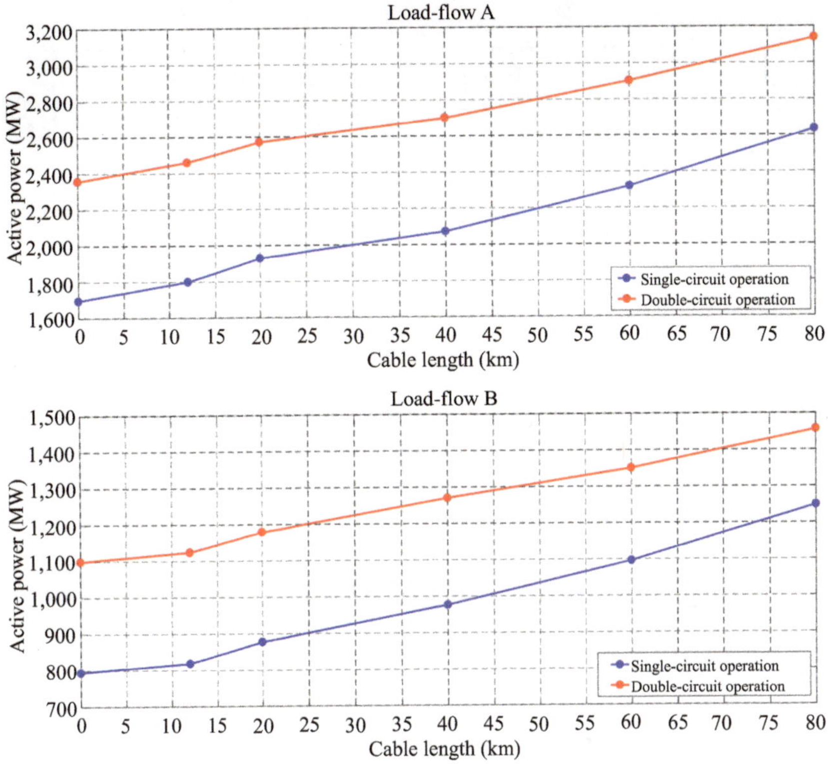

Figure 7.18 Injected active power at the sending terminal versus the cable length for the single-circuit and double-circuit operations of the Spaak connection in the load-flows A and B

considered among the most limiting factors for the large-scale application of cables in transmission systems.

This section investigates the impact of the Spaak connection on the resonance behavior of the Dutch transmission system by performing the frequency-scan analysis in the PSCAD model of the grid, as described in Section 7.2.2.2. The analysis is carried out for the six cable scenarios (introduced in Section 7.2.1.4) with sensitivity analyses on the influence of the shunt compensation size, shunt compensation location, and mixed-line configuration on the resonance behavior of the grid. The simulation results are analyzed and compared based on two defined criteria: the order of first resonance frequency and the number of resonance frequencies.

7.4.1 Resonance in electrical circuits

Resonance is a phenomenon in electrical circuits consisting of both capacitors and inductors during which the energy stored in the electric field, that is, in capacitors,

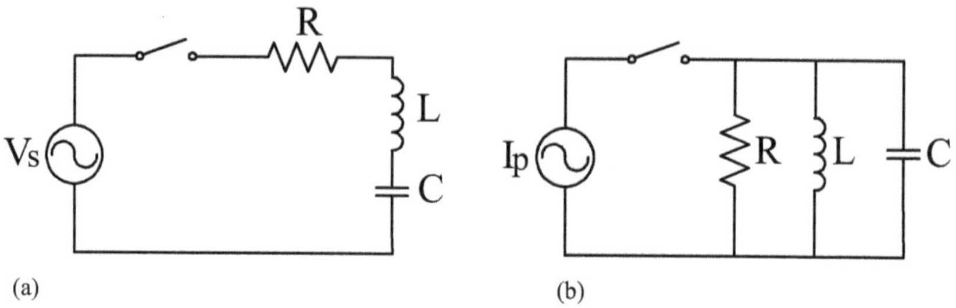

Figure 7.19 Series (a) and parallel (b) resonance circuits

is periodically transformed into the energy stored in the magnetic field, that is, in inductors, and vice versa. In other words, when a resonance occurs in an electrical system, the electromagnetic energy periodically oscillates between capacitors and inductors.

Resonance in electrical circuits has two types: series and parallel. This can be simply explained by simple series and parallel RLC circuits shown in Figure 7.19. The total impedance of the circuits is

$$Z_{\text{series}} = R + j\left(\omega L - \frac{1}{\omega C}\right) \tag{7.8}$$

$$Z_{\text{parallel}} = \frac{R}{1 + jR\left(\omega C - \frac{1}{\omega L}\right)} \tag{7.9}$$

The condition for resonance occurrence is that the stored electric and magnetic energies be equal. This condition is satisfied when the capacitive reactance equals the inductive reactance ($X_C = -X_L$), which only happens when the driving frequency is

$$f_n = \frac{1}{2\pi\sqrt{LC}} \tag{7.10}$$

This frequency is called natural frequency or resonance frequency. When a series or parallel resonance circuit is fed by respectively a voltage or current source with a frequency equal to its natural frequency, the capacitive and inductive elements of the circuit are canceling out each other; thus, the imaginary part of Z becomes zero and the total impedance of the circuit becomes a pure resistance.

In reality, electrical circuits are not as simple as shown in Figure 7.19 forming a perfect series or parallel resonance circuit. Usually, circuits are more complex and involve mixed parallel and series combinations of inductors and capacitors. These circuits are known as series–parallel resonance circuits and have the ability to resonance in the series mode or the parallel mode depending on the type and frequency of the excitation.

The magnitude and the phase of the driving impedance of an example series-parallel resonance circuit are shown in Figure 7.20, where they are plotted versus the frequency [11]. When the series-parallel resonance circuit is fed by an

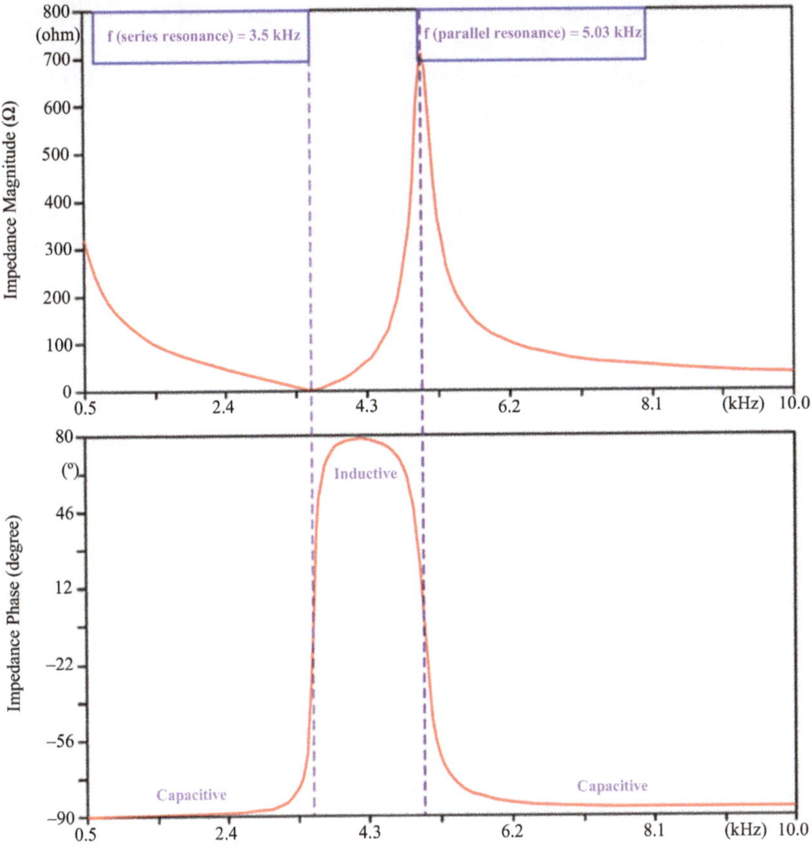

Figure 7.20 The magnitude and the angle of the driving impedance of a series-parallel resonance circuit as a function of frequency. Adapted from [11]

alternative voltage source, the series resonance circuit gets excited. In this case, at low frequencies, the total circuit impedance given by (7.8) is dominated by the reactance of the capacitor and the phase angle approaches $-90°$ (since the current is leading the voltage). This can be seen in Figure 7.20. As the frequency increases, the reactance of the inductor becomes larger so that it becomes equal to the capacitive reactance at the resonance frequency given by (7.10). In this situation (series resonance), the circuit series impedance is purely resistive with phase angle equal zero and the voltages across the inductor and capacitor reach high values (called resonance overvoltages) with opposing phase angles. For frequencies above the resonance frequency, the inductive reactance dominates the circuit characteristics and the phase angle approaches $+90°$ (since the current is lagging the voltage).

In series resonance circuits, the circuit impedance is minimum at the resonance frequency, while it is larger at lower and higher frequencies. The small impedance at the resonance frequency (dip in the plot) is the indication of a series resonance. The resistor R damps the oscillations and acts as a current limiter in the circuit, where the current can only reach up to $I_s = \frac{V_S}{R}$. Without the resistor, the lossless circuit is an ideal (undamped) series resonance circuit, where the voltages and currents increase without any restriction and reach infinity in theory.

When the series–parallel resonance circuit is fed by an alternative current source, the parallel resonance circuit gets excited. In this case, at low frequencies, the total impedance given by (7.9) is dominated by the inductive reactance and the phase angle approaches +90°. As the frequency increases, the reactance of the capacitor becomes larger until it becomes equal to the inductive reactance at the resonance frequency given by (7.10). In this case, as shown in Figure 7.20, the impedance is purely resistive with a phase angle equal to zero. The currents in the inductor and capacitor branches reach high values (and produce high overvoltages) with opposing phase angles during the parallel resonance. As the frequency increases above the resonance frequency, the capacitive reactance dominates the circuit characteristic and the phase angle approaches −90°.

In parallel resonance circuits, the circuit impedance is maximum at the resonance frequency, while it is smaller at lower and higher frequencies. The high impedance at the resonance frequency (peak in the plot) is the indication of parallel resonance. The resistor R damps the oscillations and acts as a voltage limiter, where the voltage can only reach up to $V_p = I_p R$. Without the resistor, the parallel resonance circuit is undamped and voltage and currents can theoretically reach infinity.

7.4.2 Typical grid topologies leading to resonance in cable systems

Generally, actual power systems are much more complex than simple RLC circuits as they consist of numerous capacitors and inductors that form various paths for energy exchange between them. In other words, different paths lead to various series and parallel resonance circuits with several resonance frequencies.

Resonance overvoltages are categorized as temporary overvoltages (TOVs). In power systems with long cables, the resonance frequencies are low because of the large cable capacitance. On the other hand, the system resistance is frequency-dependent and will be smaller at the low-order resonance frequencies. As a result, the resulting TOVs are weakly damped and last long when the resonance occurs in a cable system. This sustained nature of resonance TOVs in cable systems can impose a high-risk situation for system equipment, where surge arresters are one of the most vulnerable components as their withstand voltages and energy absorption capabilities can be exceeded by severe TOVs.

The voltages and currents that excite the resonance circuits in an actual power system can be originated from switching actions or short-circuits in the system (damped oscillations) and/or from the existing background harmonic sources (driven oscillation) [59,60]. In the latter case, the existing background harmonics produced by sources like frequency controllers of motors, HVDC connections, and convertors can be amplified due to the resonance. The background harmonic amplification can cause power quality issues and is unequal among the three phases when the cable system is electrically asymmetrical [61].

In the Sections 7.4.2.1 and 7.4.2.2, some of the most common topologies in transmission cable systems that can result in resonance are described.

7.4.2.1 Series resonance topologies

An example of a series resonance circuit in an actual power transmission system is shown in Figure 7.21, where the cable capacitance and the transformer leakage inductance form a series resonance circuit [3,50]. When the parallel transmission line is energized, the generated transient overvoltage wave due to the energization propagates into the series resonance circuit as well as to the surrounding grid. If the part of the wave that travels into the series resonance circuit contains the natural frequency (f_n) of the circuit, the circuit gets excited and produces resonance overvoltages at the secondary side of the transformer.

To determine the possibility and the severity of the series resonance occurrence, two parameters should be identified: (1) the natural frequency of the series resonance circuit, (2) the dominant frequencies contained in the transient energization overvoltage. The dominant frequencies of the transient overvoltage could be very low when the transmission line being energized is a long cable (depending on the source impedance and the cable length). Low-frequency overvoltages can travel over long distances due to the weak damping of the grid in low frequencies and therefore cause resonance overvoltages at distant locations [8,50].

Another series resonance circuit in power systems is created when a transformer is in a series connection with a cable and they are energized together by a

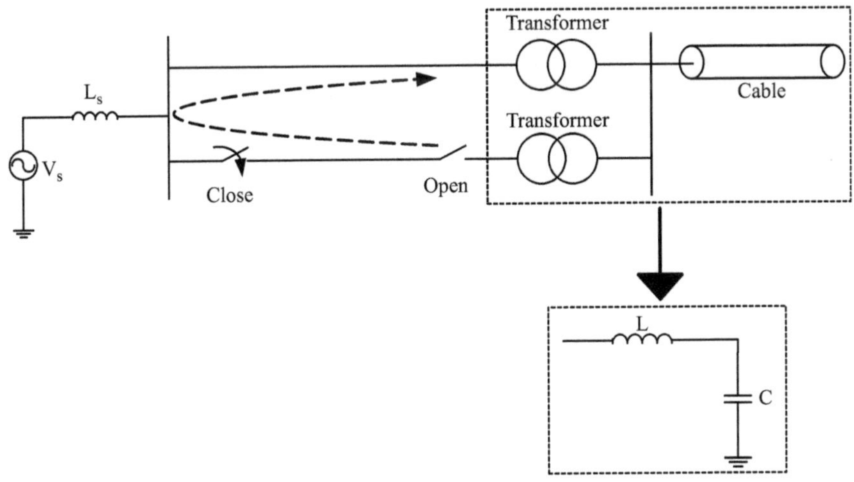

Figure 7.21 Series resonance circuit created by the transformer and cable, excited when the parallel transmission line is being energized. Adapted from [50]

voltage source, as shown in Figure 7.22. This situation is not common in meshed grids as it is very unlikely to energize the cable and transformer together. However, it can occur in offshore grids where transformers and subsea cables may be energized together.

7.4.2.2 Parallel resonance topologies

Parallel resonance circuits can be created in various ways in an actual power system and can be excited by a harmonic current. Harmonic currents are generated by cable energization, transformer energization, power electronics, HVDC links, etc. One of the most common cases is a parallel resonance circuit excited by the inrush current of a transformer being energized. The transformer energization inrush current has high contents of harmonic frequencies, which are low order, low damped, and with a long duration [50,59].

An example case of a parallel resonance circuit is shown in Figure 7.23(a), where a transformer is energized through a cable. The cable reactive power is

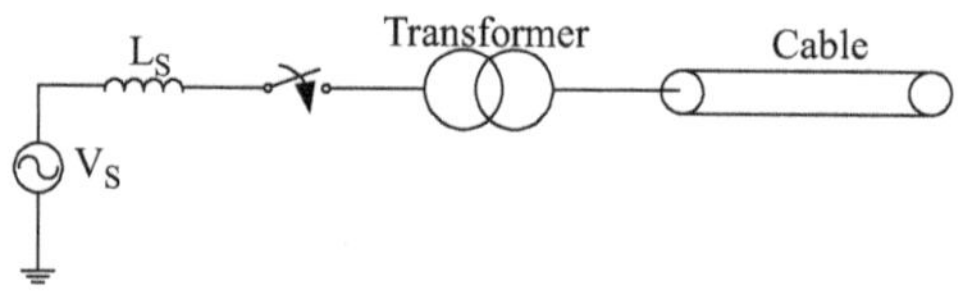

Figure 7.22 Series resonance circuit created by the transformer and cable, excited when they are energized together

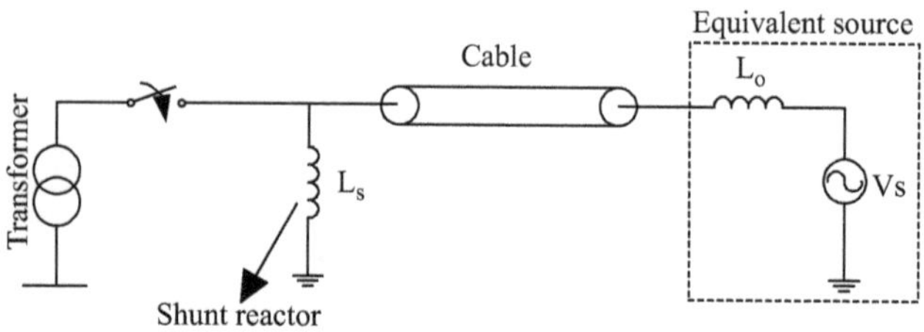

(a)

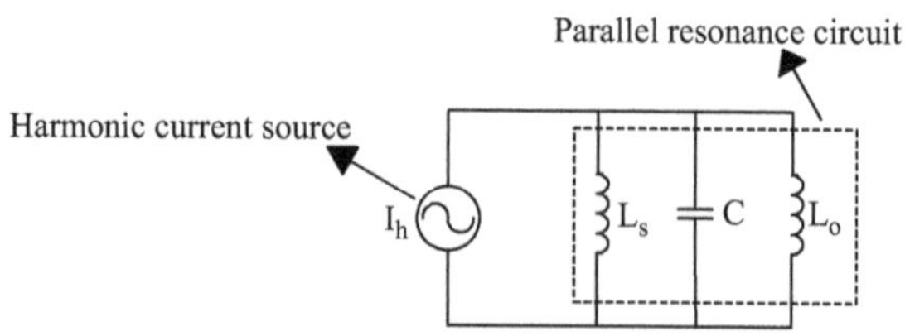

(b)

Figure 7.23 Parallel resonance circuit created by the cable and shunt reactor. The resonance circuit is excited by the transformer inrush current compensated by a shunt reactor. When the transformer is being energized, harmonic currents contained in the inrush current flow through the parallel resonance circuit, composed of the cable capacitance (C), the shunt reactor inductance (L_s), and the equivalent source inductance (L_0). This is illustrated by the harmonic equivalent circuit depicted in Figure 7.23(b), in which a harmonic current source is used to express the transformer inrush current. Therefore, the voltage source is removed since it only generates the fundamental frequency voltage.

The natural frequency of the parallel resonance circuit in Figure 7.23(b) can be expressed as

$$f_n = \frac{1}{2\pi}\sqrt{\frac{1}{L_S C} + \frac{1}{L_0 C}} \tag{7.11}$$

The most dominant harmonic content in a transformer inrush current is the second-order harmonic (100 Hz in 50 Hz systems), which has a higher proportion compared to the other harmonic contents. The natural resonance frequency of the circuit is 50 Hz when the degree of the reactive power compensation is 100%, ignoring the equivalent source inductance (L_0). To obtain a circuit with natural resonance frequency of 100 Hz, the equivalent source impedance should be $L_0 = \frac{L_s}{3}$ [50]. However, during normal operating conditions, the equivalent source impedance is usually much smaller than $\frac{L_s}{3}$ since the typical meshed power grids are strong. Thus, the natural frequency of 100 Hz can occur in weak networks such as in the black start operations. Severe parallel resonance overvoltages are expected in this situation since the load level is generally low in weak networks. Therefore, this type of parallel resonance can be mainly expected in weak grids [4,50].

Another example of the parallel resonance in cable systems is the resonance during the fault clearing and system islanding when one end of a long cable is open and a part of the grid together with the long cable is separated from the rest of the grid. However, system islanding is unlikely in today's interconnected meshed grids, but it can be expected in connections to offshore wind farms. More information regarding resonance overvoltages caused by system islanding can be found in references [50,62,63].

7.4.3 Resonance behavior of the Dutch 380 kV grid with the Spaak connection

It is of importance to investigate how the resonance behavior of a grid would change when a long cable connection is realized in the grid. This includes determination of the resonance points, resonance frequencies, and their series or parallel nature by means of the frequency-scan analysis (also called sweep frequency response analysis) of the grid. In a frequency-scan analysis, one per-unit current with the frequency of interest is injected at a point of interest and afterwards the resulting response voltage is calculated. This voltage is equal to the per-unit impedance in the frequency domain seen from the point of interest. The scan is performed over a range of frequencies and it leads to identify resonance frequencies, sensitive locations, and the impact of different system parameters on the resonance frequencies.

In cable systems, the first resonance frequency at the point of interest can be estimated by (7.10), where L and C can be substituted by the following simplified equations:

$$L = \frac{V^2}{2\pi f_0 S_{\text{system}}} \tag{7.12}$$

$$C = \frac{S_{\text{cable}}}{2\pi f_0 V^2} \tag{7.13}$$

where V is the system voltage, f_0 is the system power frequency, S_{system} is the system short-circuit power at the point of interest including the shunt compensation power, and S_{cable} is the cable reactive power. By using (7.12) and (7.13) in (7.10), (7.14) can be driven to estimate the first resonance frequency at the point of interest:

$$f_1 = f_0 \sqrt{\frac{S_{\text{system}}}{S_{\text{cable}}}} \tag{7.14}$$

where f_1 is the first resonance frequency. It can be deduced from this equation that the first resonance frequency is directly proportional to the short-circuit power of the grid at the point of interest. In other words, the first resonance frequency decreases when the grid becomes weaker and vice versa.

In the Sections 7.4.3.1–7.4.3.6, the frequency response of the Dutch 380 kV grid is evaluated to determine the impact of long cables in the Spaak connection on the resonance frequencies of the grid. Moreover, the influence of the shunt compensation size, shunt compensation location, and mixed-line configuration on the frequency scan of the grid is investigated too.

7.4.3.1 Parameter selection

The frequency scan was performed in the frequency range of 1 Hz–2.5 kHz for the positive sequence impedance. Resonance frequencies below 2.5 kHz (50th harmonic in a 50 Hz system) are of interest as the dominant harmonic frequencies are within this range. All simulations were carried out when both circuits of the Spaak connection were in operation. The mixed-line configuration is as shown in Figure 7.2 with the cable scenarios defined in Table 7.1. Shunt compensation sizes are according to the values reported in Table 7.6.

7.4.3.2 Evaluation criteria

A set of criteria is needed to evaluate the frequency-scans of different cases with and without cables and with different system parameters. Two criteria are used here with respect to the applied analysis approaches in literature [2,8,20,64,65]:

1. *Order of the first resonance frequency*: This is the most important criterion for the analysis of the frequency scan results. The system is more vulnerable when the first resonance frequency is low as the harmonic emissions in transmission systems are mainly at the low-order harmonic frequency range. In addition, the

resulting resonance TOVs are larger and lower damped when the resonance frequency is lower.

2. *Number of resonance frequencies below 2.5 kHz*: It is very crucial to check the number of resonance frequencies in an specific frequency range. The interested frequency range for the analysis should cover the low-order harmonic frequencies since the dominant harmonic frequency emissions at transmission levels are mainly low in the order. This criterion shows how the potential risk of resonance can increase when cables are installed.

7.4.3.3 Impact of the Spaak connection

The frequency scan was carried out at four substations: Krimpen, Dodewaard, Maasvlakte, and Eemshaven with short-circuit powers of about 30, 20, 28, and 11 GVA in the selected snapshot, respectively. One scenario in addition to the six cable scenarios presented in Table 7.1 is the grid without the Spaak connection. The comparison of this scenario with the other scenarios reveals the influence of realizing this project as well as the influence of the increasing cable length on the frequency response of the grid.

Figure 7.24 shows the frequency scan results when the grid is scanned from the Krimpen substation for three scenarios: without the Spaak connection, with the fully-OHL Spaak connection, and with the fully-cable Spaak connection (with LEC). The magnitude and the phase of the driven impedance are plotted versus frequency. As it was previously explained, at the resonance frequencies the phase angle becomes zero (i.e., imaginary part of the impedance is zero). It can be noticed in this figure that one of the substantial influences of the increasing cable length is the decrease of the first resonance frequency and a considerable increase in the number of resonance frequencies. The first resonance frequency is 125 Hz (close to the second harmonic frequency) when the Spaak connection is realized only with cables, while it is 158 Hz without the Spaak connection and 161 Hz with the fully-OHL Spaak connection. The low first resonance frequency of the grid without the Spaak connection is because of the cables in the Randstad South-Ring and North-Ring connections, which are nearby of the Krimpen substation.

A detailed analysis of the frequency scan results for all the cable scenarios is depicted in Figure 7.25, where the first resonance frequency and the number of resonance frequencies seen from the four substations are plotted for the six cable scenarios.

It can be seen that the first resonance frequency decreases with the increasing cable length in the circuit. This decrease was expected since, according to Table 7.6, the percentual increase of the compensation degree is less than the total percentual increase of the cable length, which results in a lower first resonance frequency according to (7.14). In the worst case, the first resonance frequency has decreased to 125 Hz, which is for the fully-cable scenario scanned from the Krimpen substation. Moreover, compared to the other substations, the first resonance frequency seen from the Eemshaven substation is almost unchanged when the cable length is increased from 15% to 100%. This is because of the relatively far distance of the Spaak connection from the Eemshaven substation. Finally, the

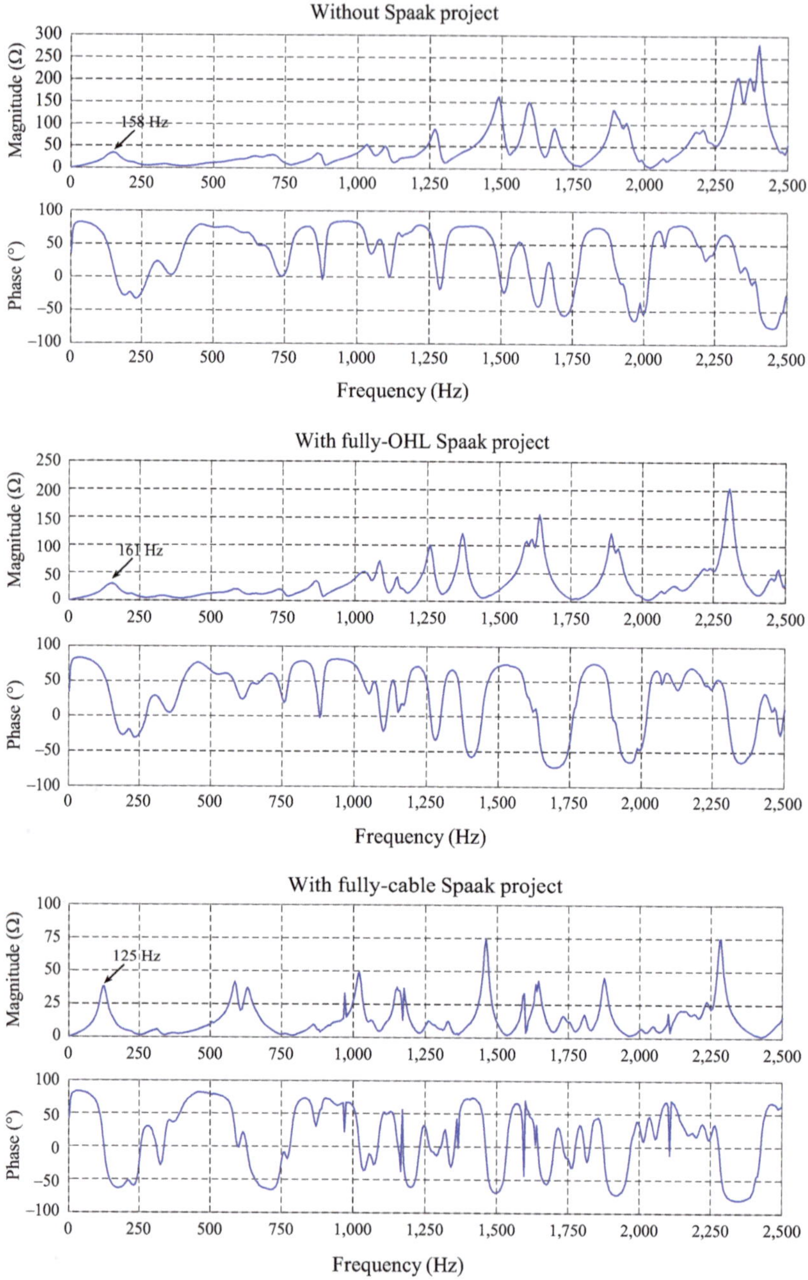

Figure 7.24 Frequency scan of the Dutch 380 kV grid from the Krimpen substation without the Spaak connection, with the fully-OHL Spaak connection, and with the fully-cable Spaak connection

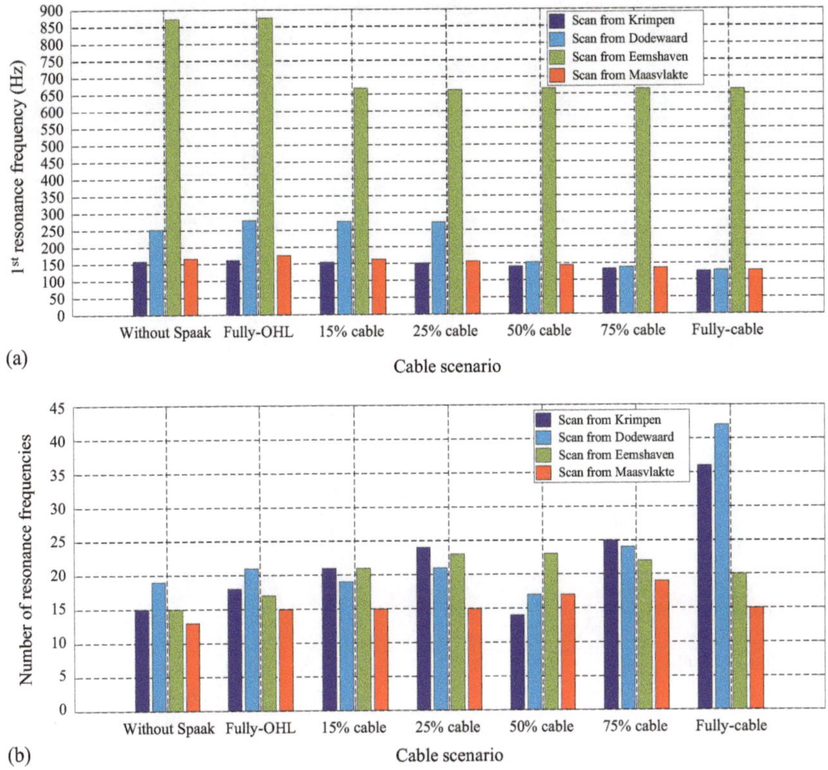

Figure 7.25 Resonance behavior of the Dutch 380 kV grid scanned from four substations for the six cable scenarios of the Spaak connection: (a) first resonance frequency (f_1) and (b) number of resonance frequencies below 2.5 kHz

number of resonance frequencies below 2.5 kHz shows a strong dependency on the cable length and the point of interest (i.e., the substation at which the scan is performed).

7.4.3.4 Impact of shunt compensation

The shunt compensation allocation, that is, locating and sizing shunt reactors, is one of the most important issues regarding to the operation of long cables in transmission systems. The allocation should be carried out according to a detailed steady-state study, as it was explained in Section 7.3.

From the harmonic resonance perspective, it is of interest to investigate the changes in the frequency scan of the system with and without shunt reactors. Figure 7.26 shows the frequency scan of the Dutch 380 kV grid for the 50% cable and the fully-cable scenarios of the Spaak connection, both with and without shunt compensation. It can be noticed that the first resonance frequency and the number of

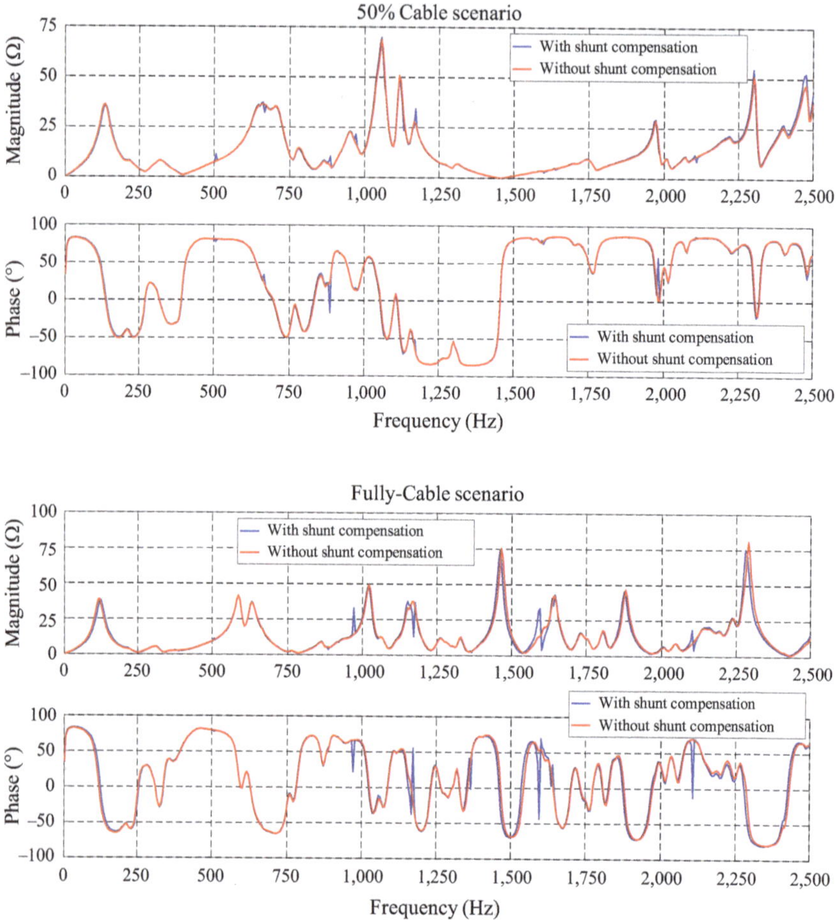

Figure 7.26 Frequency scan of the Dutch 380 kV grid for the 50% cable and the fully-cable scenarios of the Spaak connection with and without shunt compensation. The scan was performed at the Krimpen substation

resonance frequencies increase in both cable scenarios when the shunt reactors are added, although the increase in the first resonance frequency is very small. In addition, the difference between with and without compensation cases is larger for the fully-cable scenario compared to the 50% cable scenario.

Figure 7.27 shows the first resonance frequency and the number of resonance frequencies for the six cable scenarios with and without shunt compensation scanned from the Krimpen substation. It can be seen that the existence of shunt compensation has a small impact on the first resonance frequency, while the impact becomes larger when the cable length is increased. This is because of the bigger shunt compensation size for the longer cables, which has a bigger

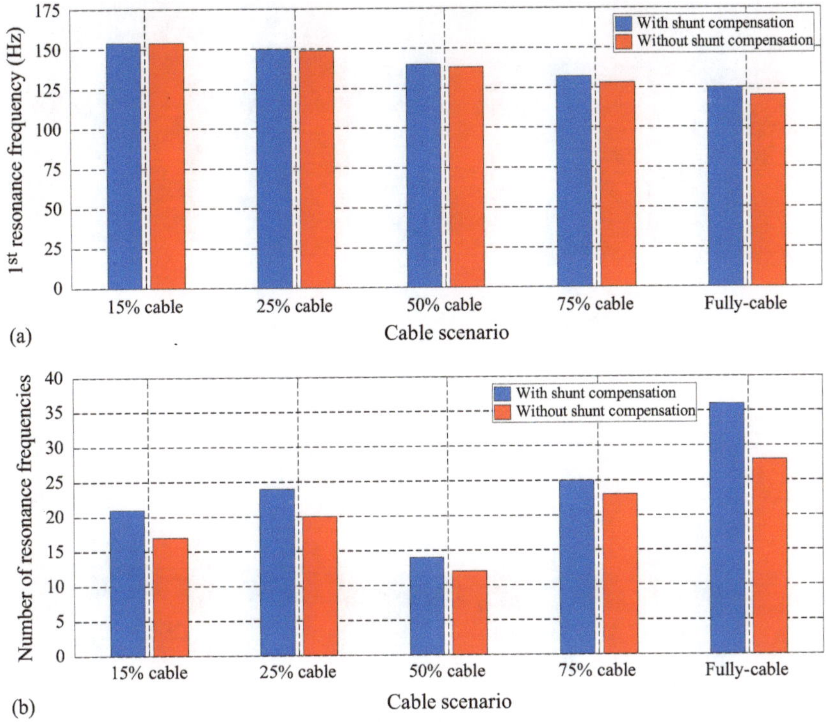

Figure 7.27 Resonance behavior of the Dutch 380 kV grid for the six cable scenarios of the Spaak connection with and without shunt compensation: (a) first resonance frequency (f_1) and (b) the number of resonance frequencies below 2.5 kHz. The scan was performed at the Krimpen substation

contribution to the system short-circuit power in (7.14) (S_{system}) and therefore causing a higher first resonance frequency compared to the case without compensation. Furthermore, the number of resonance frequencies increases when shunt reactors are operated in the grid. This means a higher probability of resonance occurrence when cables are compensated by shunt reactors compared to the case without shunt compensation.

7.4.3.5 Impact of shunt compensation location

The influence of shunt reactor locations on the frequency response of the grid is shown in Figure 7.28. Two cases are compared: (1) line-end compensation (LEC), as previously described in Section 7.2.1.2 and shown in Figure 7.2, and (2) cable-end compensation (CEC), in which, as a distributed compensation, shunt reactors are located at each OHL-cable transition point (i.e., six reactors per circuit in the case of the Spaak project). The sizes for CEC are according to the reported values in Figure 7.14 for the load-flow B.

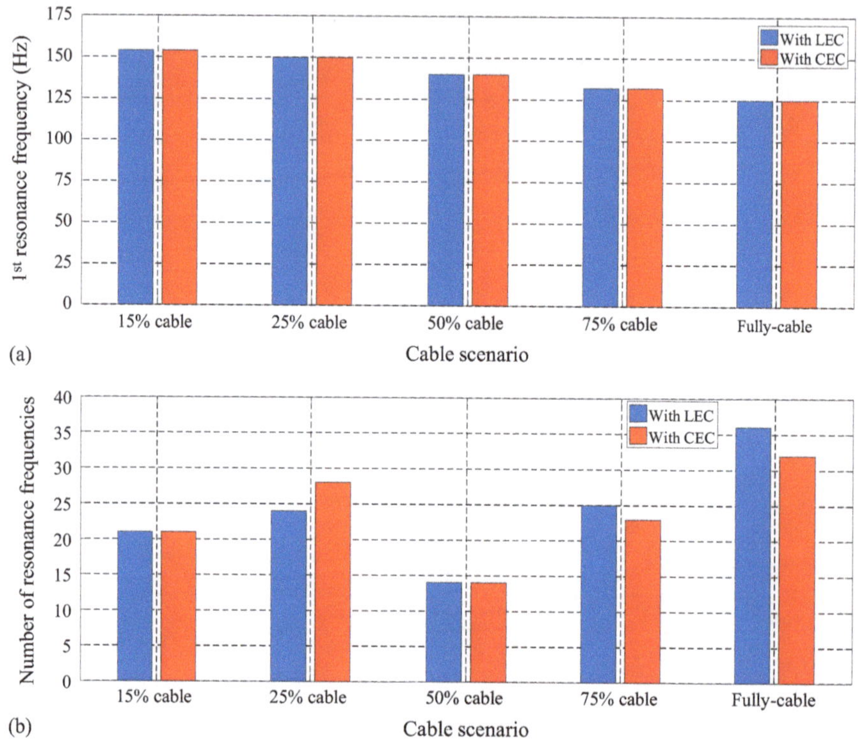

Figure 7.28 Resonance behavior of the Dutch 380 kV grid for the six cable scenarios of the Spaak connection with LEC and CEC: (a) first resonance frequency (f_1) and (b) the number of resonance frequencies below 2.5 kHz. The scan was performed at the Krimpen substation

According to the simulation results, the location of shunt compensation has no impact on the first resonance frequency, whereas it can increase or decrease the number of resonance frequencies depending on the cable length.

7.4.3.6 Impact of mixed-line configuration

Similar to the steady-state analysis, it is of importance to investigate the influence of the mixed-line configuration on the resonance behavior of the system. For this purpose, the four mixed-line configurations presented in Section 7.2.1.3 and the study case configuration (Figure 7.2) are compared in terms of the first resonance frequency and the number of resonance frequencies. All the configurations are compensated at line-ends (LEC) with the compensation degrees reported in Figure 7.15 for the load-flow B.

Figure 7.29 shows the impact of the five mixed-line configurations on the resonance behavior of the Dutch 380 kV grid when the scan is performed at the

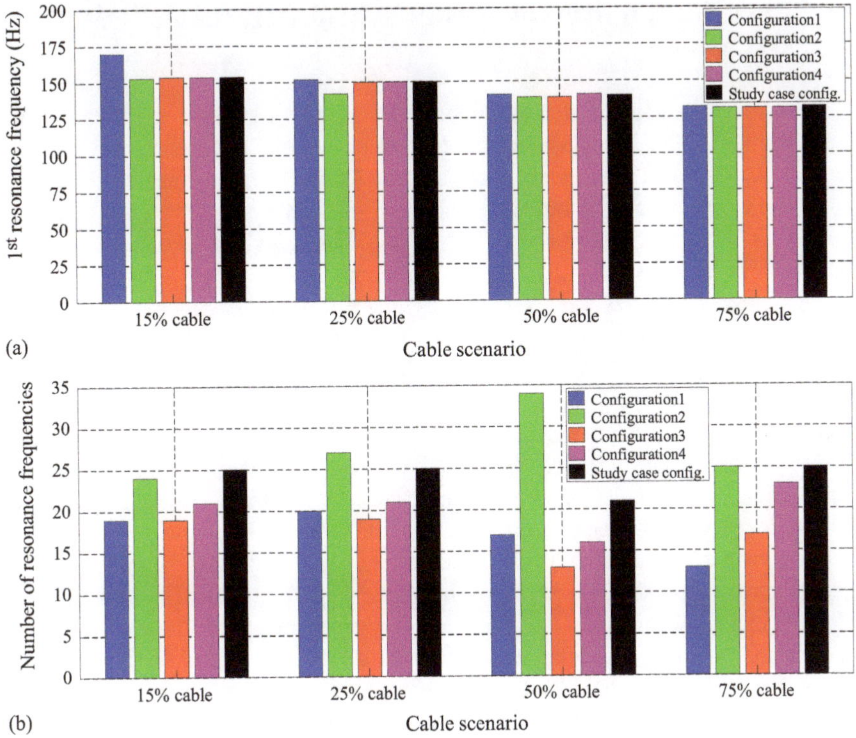

Figure 7.29 Resonance behavior of the Dutch 380 kV grid for the six cable scenarios of the Spaak connection with different mixed-line configurations: (a) first resonance frequency (f_1) and (b) the number of resonance frequencies below 2.5 kHz. The scan was performed at the Krimpen substation

Krimpen substation. As the results show, the first resonance frequency decreases with the increasing cable length in all the five mixed-line configurations, whilst, for a given cable scenario, the difference between the resonance frequencies of the configurations is not very significant. However, the difference is larger from the number of resonance frequencies perspective, for example, the number of resonance frequencies for the 50% cable scenario in Configuration 2 is almost three times those in Configuration 3.

These results can be used in the planning and design stages of cable projects for the selection of the best mixed-line configuration in terms of resonance behavior. For example, let's assume that it is decided to build the project with 60 km cable (i.e., 75% cable scenario); according to the simulation results, Configuration 1 is the best option as it provides a higher first resonance frequency and a lower number of resonance frequencies compared to the other configurations.

7.5 Energization overvoltages in hybrid OHL-cable systems

The transient behavior of an EHV hybrid OHL-cable system is significantly different from an OHL-based system because of substantial differences between the electrical characteristics of OHLs and cables. As a result, several questions have been raised concerning the transient operation of such systems. A very important question is how energization overvoltages would be in hybrid OHL-cable systems and which parameters can affect their significance.

In general, energization overvoltages in EHV systems can reach very high values and are considered among the severest overvoltages stressing the insulation of system components. These overvoltages have a statistical behavior caused by the random behavior of the circuit breaker closing time. Thus, the required insulation level should be specified by means of a statistical approach by which the distribution of energization overvoltages is determined. The statistical distribution of energization overvoltages has been used for decades for the insulation coordination studies, as it has been recommended by the IEC 60071-2 standard too [66].

The distribution of energization overvoltages in purely OHL or cable systems are well investigated in the literature; however, overvoltage distribution can considerably change when both OHLs and cables are used in a transmission line. In this section, energization overvoltages in the hybrid OHL-cable circuits of the Spaak connection are studied. The study includes various sensitivity analyses to identify the impact of system parameters and topology on overvoltages.

7.5.1 Simulation considerations

The transient studies were performed by the use of the distributed frequency-dependent parameter model of the Dutch 380 kV grid in PSCAD, described in Section 7.2.2.2. Surge arresters are not included in this model in order to simulate the worst overvoltages. The magnitude, frequency, and duration of switching overvoltages are dependent on several parameters, mainly switching instant, short-circuit power, cable length, shunt compensation size and location, and mixed-line configuration.

The mixed-line configuration is shown in Figure 7.30 (which is the study case configuration defined in Section 7.2.1.1) with the cable scenarios defined in Table 7.1. Shunt compensation sizes are according to the values reported in Table 7.6. Overvoltages due to the no-load energization of circuit 1 of the Spaak connection were studied with the assumption that the parallel circuit (circuit 2) is out of service (i.e., single-circuit energization). The switching was performed from the Krimpen substation while the circuit was open at the other side, as illustrated in Figure 7.30.

When overvoltages are expressed in per unit, the base value (1 pu) is the peak value of the phase-to-earth nominal voltage (310.27 kV). The power-frequency and the transient overvoltages are the two components of the total switching overvoltage, so the obtained maximum overvoltages depend on the power-frequency

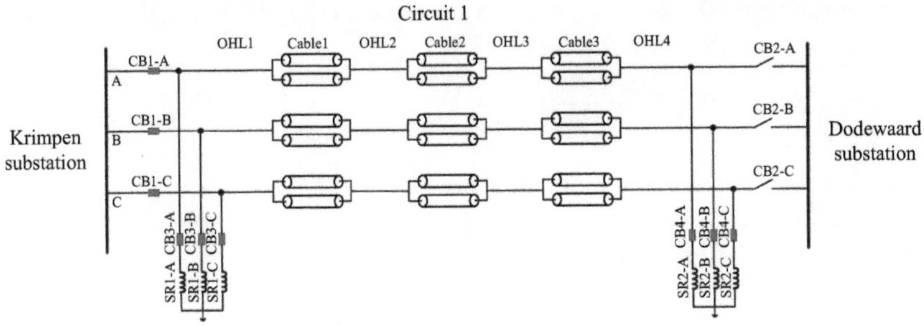

Figure 7.30 Single-circuit energization of the Spaak connection from the Krimpen substation, while the second circuit (Circuit 2) is out of service. CB: circuit breaker; SR: shunt reactor

voltage, which in this study is 1.076 pu phase-to-earth (334 kV). The short-circuit power at the Krimpen substation is 26.3 GVA (corresponding to 37.14 kA fault current or 20 mH equivalent source inductance calculated at 408.88 kV (334 kV $\times \frac{\sqrt{3}}{\sqrt{2}}$)).

In the transient studies of cable systems, the determination of the right numerical integration time step is a critical factor to obtain an accurate simulation of transient voltages and currents. As a rule of thumb, the time step should be smaller than the traveling time of the wave in the shortest piece of the line. In this way, the wave propagation in all lines can be accurately simulated.

To calculate the traveling time of the shortest line piece, the wave propagation velocity should be calculated by means of (7.15):

$$v = \sqrt{\frac{1}{LC}}\,(\mathrm{m/s}) \tag{7.15}$$

The inductance and capacitance of the conductor are required for the calculation of the wave velocity. For a cylindrical conductor with isolation, the inductance and capacitance can be calculated by making use of the following well-known equations:

$$C = \frac{2\pi\varepsilon}{\ln\left(\frac{b}{a}\right)}(\mathrm{F/m}) \tag{7.16}$$

$$L = \frac{\mu}{2\pi}\ln\left(\frac{b}{a}\right)(\mathrm{H/m}) \tag{7.17}$$

where a and b are respectively the radius of the conductor and the insulation, ε is the permittivity of the insulation consisting of the relative permittivity and the vacuum permittivity ($\varepsilon = \varepsilon_0\varepsilon_r$), and μ is the permeability of the conductor consisting of the relative permeability and the vacuum permeability ($\mu = \mu_0\mu_r$).

By substituting these values in (7.15), wave velocity can be obtained as below:

$$v = \sqrt{\frac{1}{LC}} = \sqrt{\frac{1}{\mu\varepsilon}} = \frac{C_0}{\sqrt{\varepsilon_r \mu_r}}\ (\mathrm{m/s}) \tag{7.18}$$

where C_0 is the light velocity in the free space (3×10^8 m/s). The wave propagation velocity in the OHL and the cable can be calculated by using the relative permittivity and permeability values in Table 7.4:

$$v_{\mathrm{OHL}} \approx C_0 = 3 \times 10^8\ (\mathrm{m/s}) \tag{7.19}$$

$$v_{\mathrm{cable}} = \frac{C_0}{\sqrt{2.79 \times 1.17}} = 1.6605 \times 10^8\ (\mathrm{m/s}) \tag{7.20}$$

The shortest line piece in the Dutch 380 kV grid model is 900 m. This is the length of the minor cable sections in the Randstad South-Ring connection nearby the Spaak connection. Therefore, the wave traveling time in these pieces is:

$$t = \frac{\mathrm{length}}{V_{\mathrm{cable}}} = \frac{900}{1.6605 \times 10^8} = 5.42\ \mu\mathrm{s} \tag{7.21}$$

Taking this traveling time into account, the selected simulation time step for the transient studies was 4.8 μs in order to keep the balance between the simulation accuracy and the simulation time.

7.5.2 Energization overvoltages

It was assumed that in all simulations the three poles of CB1 (circuit breaker 1 in Figure 7.30) are closed simultaneously at the voltage peak of phase C ($t = 0.071$ s). In addition, shunt reactors were energized together with the line at the same time.

Figure 7.31 shows the voltage at the Krimpen substation after the no-load energization of the six cable scenarios. In Figure 7.31a, which shows three-phase voltages for the 50% cable scenario, the overvoltage peaks are 367 kV (1.18 pu), 377 kV (1.22 pu), and 356 kV (1.15 pu) respectively for phase A, phase B, and phase C.

As depicted in Figure 7.31(b), the overvoltage peak in phase C is larger for other scenarios and reaches up to 389 kV (1.25 pu) for the fully-OHL, 375 kV (1.21 pu) for the fully-cable, 373 kV (1.2 pu) for the 50% cable, and 371 kV (1.2 pu) for 15% cable scenarios. The fully-OHL scenario has the highest initial overvoltage peak due to the absence of shunt compensation and faster wave traveling time in it. However, the overvoltage damps out much faster in the fully-OHL scenario than in the scenarios with cable.

Furthermore, the voltage peaks happen at different instants depending on OHL and cable section lengths, which are determining the wave propagation pattern. However, overvoltages are damped in the first few milliseconds with high-frequency low-amplitude oscillations for all the cable scenarios. The difference between final steady-state voltages in Figure 7.31(b) is negligible, as it was expected due to the shunt compensation sizing by the steady-state analysis. Without shunt compensation,

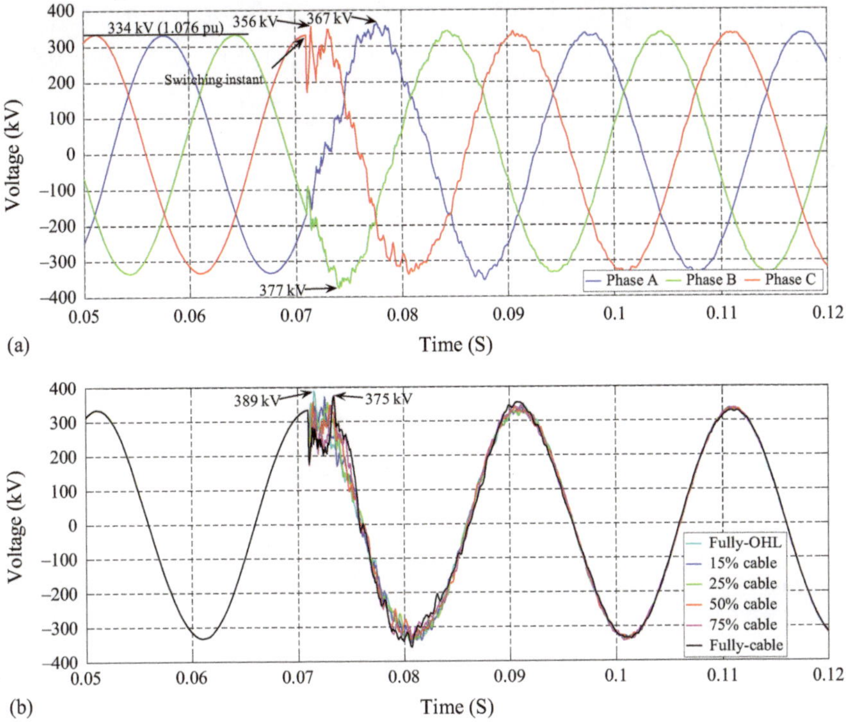

Figure 7.31 Voltage at the Krimpen substation (sending-end) after the no-load energization of the Spaak connection: (a) phase voltages for the 50% cable scenario and (b) phase C voltage for the six cable scenarios

the difference between the final steady-state voltages of different cable scenarios can be up to tens of kilovolts, whereas it is just in the range of a few kilovolts with the use of an optimum shunt compensation. Since the overvoltages last for a few milliseconds, the switching transient overvoltages can be classified as slow-front overvoltages. All overvoltage peaks are far below the phase-to-earth switching impulse withstand voltage (SIWV) proposed in [67], which is 850, 950, or 1,050 kV (peak value) for an equipment with 420 kV highest RMS voltage.

Immediately after switching the circuit breaker, the energizing wave propagates through the circuit resulting in reflection and refraction waveforms at the locations where the conducting medium is changing, that is, cable joints and transition points (TP) between OHL and cable sections. Therefore, during this transient period, voltage variations along the circuit can be significant, depending on the location as it is illustrated in Figure 7.32. In this figure, the voltage of phase C at the six transition points is plotted for the 50% cable scenario.

It can be seen that all the overvoltage peaks for the studied cable scenarios are below the SIWV. In addition, each of these six transition points experiences different overvoltage peaks at different moments, depending on the distance of the

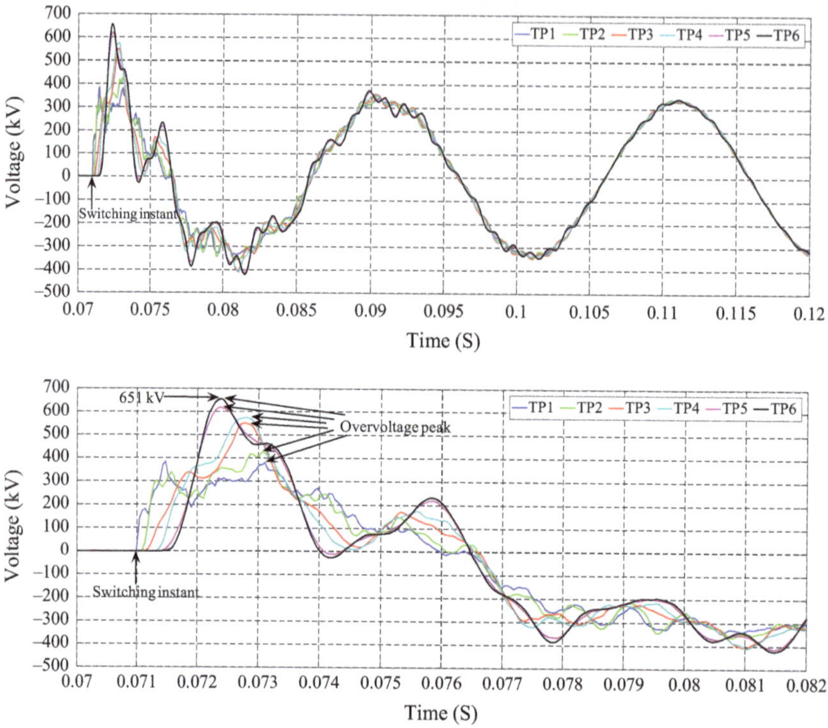

Figure 7.32 Phase C voltage at OHL-cable transition points after the no-load energization of the 50% cable scenario. The lower figure zooms in the first 0.082 s of the upper figure

transition point from the sending-end. The overvoltage peak occurs sooner and with a higher value when the transition point is further from the sending terminal. In other words, the overvoltage peak becomes higher when moving toward the open-end (as was expected) and therefore the highest transition-point overvoltage is at TP6.

Figure 7.33 shows the energization voltage at the open-end for the 50% cable scenario, where the highest overvoltage with the amplitude of 650 kV (2.09 pu) is in phase C since the switching is performed at its peak voltage.

Figure 7.34 depicts the voltage of phase C at the open-end for the six cable scenarios. The highest overvoltage peak with an amplitude of 676 kV (2.18 pu) is for the 15% cable scenario. The amplitude and the frequency of voltage oscillations become smaller for the scenarios with higher cable share. Moreover, with the increasing cable share, it takes a longer time for the first traveling surge to reach the open-end of the circuit since the wave traveling speed in the cable is almost half of its speed in the OHL (1.66×10^8 versus 3×10^8 m/s). This results in a longer time for the wave to reach the open-end when the cable is longer.

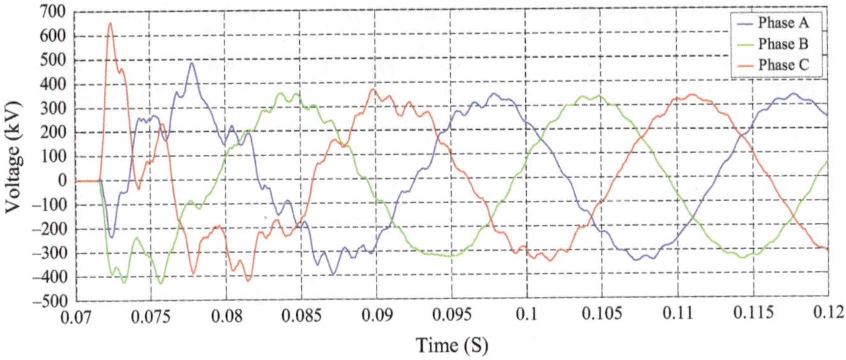

Figure 7.33 Open-end voltage after the no-load energization of the 50% cable scenario

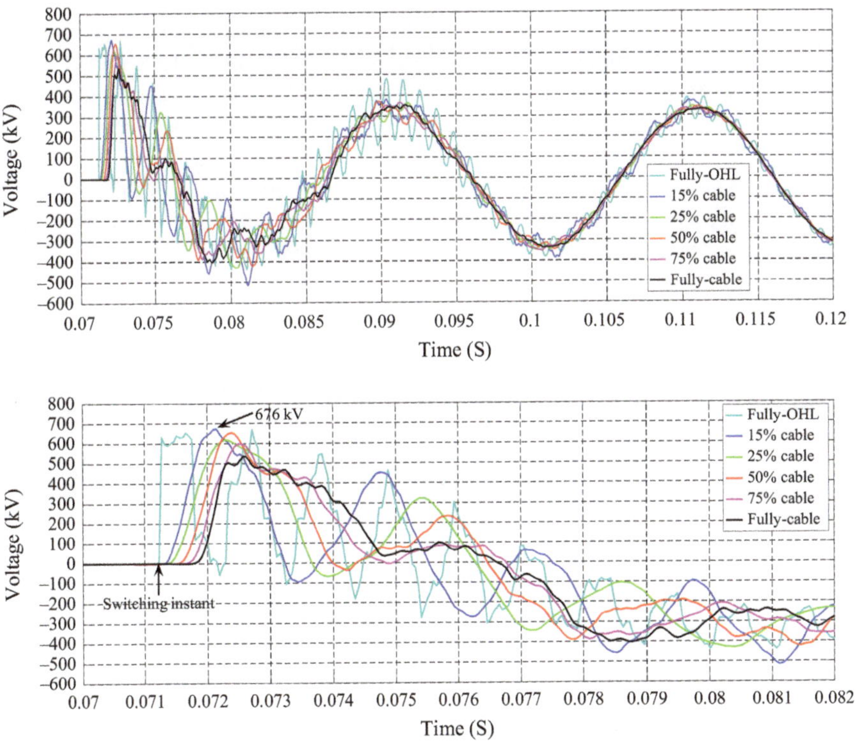

Figure 7.34 Phase C voltage at the open-end after the no-load energization of the six cable scenarios. The lower figure zooms in the first 0.082 s of the upper figure

7.5.3 Statistical analysis of overvoltages

The determination of the statistical distribution of energization overvoltages is highly recommended for the insulation coordination studies due to the statistical nature of switching actions and the insulation strength. The risk of an insulation failure can be calculated by the comparison of the overvoltage probability distribution and the insulation breakdown probability distribution [66,68,69].

The statistical nature of energization overvoltages is a result of the random behavior of the circuit breaker closing time. The breaker contacts can be switched at any point on the voltage waveform with a pole closing span. The pole closing span is the time difference between the first and the last pole to close due to different stochastic variations in the operating time of poles [3,70].

The detailed study of the statistical distribution of line energization overvoltages (OHL or cable) has been addressed in [71–80]. In [75] and [76], the statistical distribution of energization overvoltages in EHV cables and OHLs were compared, where it has been concluded that overvoltages in a cable are lower than those in an OHL. It is has been reported in [81] that energization overvoltages in a cable circuit are in general lower than those in an OHL-cable circuit. In [82], authors concluded that energization overvoltages of a hybrid OHL-cable circuit are less than those in a purely OHL circuit and higher than those in a purely cable circuit. However, these conclusions cannot be generalized as the studies neglected the breaker pole span and are only for one circuit topology and set of system parameters.

In this regard, a more comprehensive study is required for hybrid OHL-cable systems where the impacts of various system topologies and parameters are investigated, notably the cable length, configuration of the hybrid OHL-cable circuit, short cable sections, and system short-circuit power. In practice, the distribution of overvoltages can be significantly different from one system to another and this can be identified by a sensitivity analysis.

The aim of this section is to answer the mentioned questions and perform sensitivity analyses on the system parameters to find out how the statistical distribution of transient energization overvoltages changes in hybrid OHL-cable circuits. In addition, overvoltages due to the presence of trapped charges (residual charges) in the cable are investigated too.

7.5.3.1 Statistical behavior of circuit breaker

The breaker closing time (T_{close}) is the instant at which the breaker connects a phase to the voltage source and it can be expressed as follows [3,77]:

$$T_{\text{close}} = T_{\text{command}} + \underbrace{\Delta T_{\text{known}} + \Delta T_{\text{stochastic}}}_{\Delta T_{\text{operating}}} \tag{7.22}$$

where $T_{command}$ is the instant at which the breaker receives the closing command, and $\Delta T_{operating}$ is the breaker operating time representing the required time for the control and mechanical parts of the breaker to operate. ΔT_{known} is the known and predictable part of the operating time (from measurements and/or adaptive control) and $\Delta T_{stochastic}$ is the stochastic variation (unpredicted variation) in the operating time, which indicates the randomness of the closing time.

$\Delta T_{stochastic}$ can be different between the poles due to the separate mechanical systems and it has a random behavior according to the normal distribution [3,70–80,82]. The different stochastic variations of the operating time can result in energization of phases at different instants, which is known as the pole closing span. In other words, the pole closing span is the time difference between the first and the last pole to close.

With regard to (7.22), the statistical behavior of the breaker closing time can be represented if two parameters are determined: (1) mean closing time, T_{mean}, where $T_{mean} = T_{command} + \Delta T_{known}$, and (2) pole closing span or distribution of pole closing times around the mean closing time, $\Delta T_{stochastical}$. A sufficient number of these two parameters should be simulated in order to obtain accurate energization overvoltage distributions [77,80,83].

7.5.3.2 Statistical simulation approach

The statistical analysis of energization overvoltages is more accurate when a larger number of switching times are simulated, but this requires a long computation time especially when a big and complex grid is modeled. For instance, each single switching simulation in the PSCAD model of the Dutch 380 kV grid can take up to a few hours to be completed, depending on the simulated cable length. Such a long simulation time is mainly caused by the small numerical integration time step ($\Delta t = 4.8\ \mu s$) and a large number of cables.

In [72–74], 100 random line energizations were performed to produce the overvoltage distributions, whilst the analysis was conducted in [75] by 200 random line energizations. The literature has recommended a minimum of 100 simulations to obtain a sufficiently accurate overvoltage distribution [77,80,82,83]. In this study, 400 energizations were carried out for each study case to maximize the accuracy of the obtained statistical distributions. The applied approach for the selection of T_{mean} and $\Delta T_{stochastic}$ (pole span) is illustrated in Figure 7.35:

1. T_{mean}: It was assumed that the line and shunt reactor breakers receive the closing command simultaneously and they have the same T_{mean}. The mean closing time can be at any point of the power-frequency voltage cycle. 100 random points (with uniform distribution) were chosen by using the multiple-run feature of PSCAD.
2. $\Delta T_{stochastic}$ *(pole span)*: For each T_{mean}, four sets of pole closing span were simulated for each breaker. The pole span was determined by a random number generator using the normal (Gaussian) distribution. The mean value of the normal distribution was T_{mean} and the standard deviation was 1 ms ($\sigma = 1$ ms). The normal distribution curve was truncated at -3σ and $+3\sigma$.

The phase-peak method was applied to obtain the overvoltage probability distribution for each study case [66]. This means that from each energization simulation, the peak value of the phase-to-earth overvoltage in each phase at the point of interest was included in the overvoltage probability distribution. Therefore,

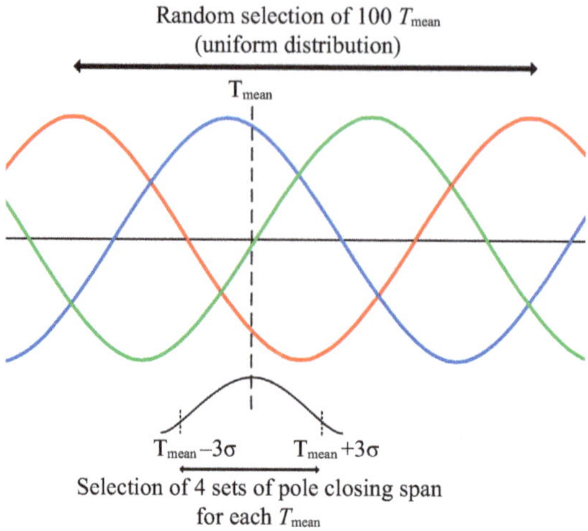

Figure 7.35 Statistical variation of the breaker closing time over a cycle of the power-frequency voltage

the representative probability distribution for each case has 1,200 overvoltage peak values (3 peak values per simulation × 400 simulations).

7.5.3.3 Simulation results and analysis

The effects of cable length, shunt compensation size/location, short cable sections, short-circuit power, trapped charges (residual charges), and circuit configuration on the energization overvoltages are studied in this section. **The point of interest is always the line open-end**, unless a different location is mentioned.

7.5.3.3.1 Impact of cable length

The statistical switching analysis was performed on the six cable scenarios (presented in Table 7.1) to find the influence of cable length on the distribution of energization overvoltages. Figure 7.36 shows the obtained maximum overvoltages at the sending-end, open-end, and all the six OHL-cable transition points (TP). The fully-OHL and the fully-cable scenarios are excluded since the transition points do not exist for these scenarios. It can be seen that for each cable scenario the maximum overvoltage becomes higher at transition points closer to the open-end as the highest overvoltage is at the open-end.

Table 7.8 presents the maximum, mean, standard deviation, and 2% values, which are the four key values of the statistical distributions. The 2% value is the overvoltage value that the probability of this value being exceeded is 2% and is used for the insulation coordination studies [66].

The complementary cumulative distribution functions (here called cumulative probabilities) of energization overvoltages of the six cable scenarios are given in Figure 7.37. For a given overvoltage on the horizontal axis, the corresponding value on the vertical axis shows the probability that overvoltages exceed the given overvoltage.

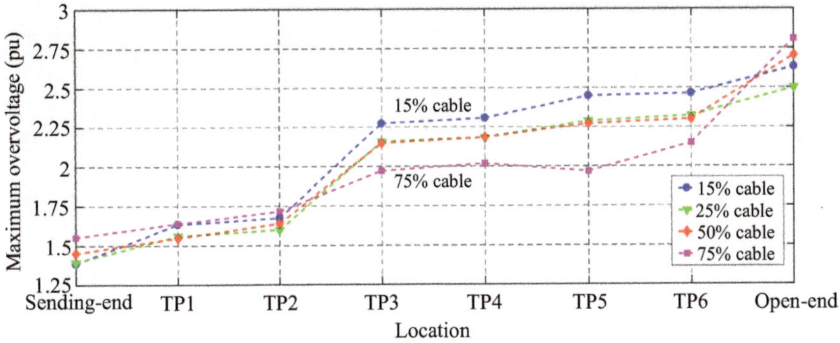

Figure 7.36 Maximum energization overvoltage at different locations along the hybrid OHL-cable circuit

Table 7.8 Key values of the overvoltage distributions

Scenario	Overvoltage (pu)			
	Max.	**Mean**	**Stand. dev.**	**2% value**
Fully-OHL	2.826	1.821	0.358	2.509
15% cable	2.631	1.756	0.262	2.346
25% cable	2.499	1.694	0.245	2.269
50% cable	2.705	1.689	0.259	2.319
75% cable	2.810	1.626	0.256	2.329
Fully-cable	1.908	1.501	0.163	1.820

As it is shown in Table 7.8, for the fully-OHL scenario, the maximum overvoltage (2.826 pu), mean overvoltage (1.821 pu), and 2% value (2.509 pu) are the highest among all the scenarios. In contrast, the lowest values are for the fully-cable scenario with the maximum overvoltage, mean overvoltage, and 2% value of 1.908 pu, 1.501 pu, and 1.820 pu, respectively. The standard deviation is also largest for the fully-OHL scenario and smallest for the fully-cable scenario. In addition, in Figure 7.37, the probability of overvoltages is the highest for the fully-OHL scenario (for overvoltages above 1.65 pu), whereas it is always the lowest for the fully-cable scenario. These findings coincide with the conclusions of [75] and [76] that line energization overvoltages in cables are lower than those in OHLs.

According to Table 7.8, the hybrid scenarios produce overvoltages lower than those of the fully-OHL scenario and higher than those of the fully-cable scenario. In addition, in Figure 7.37 and for overvoltages above 1.65 pu, the overvoltage probabilities of the hybrid scenarios are between those of the fully-OHL and the fully-cable scenarios.

Based on the simulation results, it can be concluded that a partial or full replacement of an overhead transmission line with cables results in lower energization

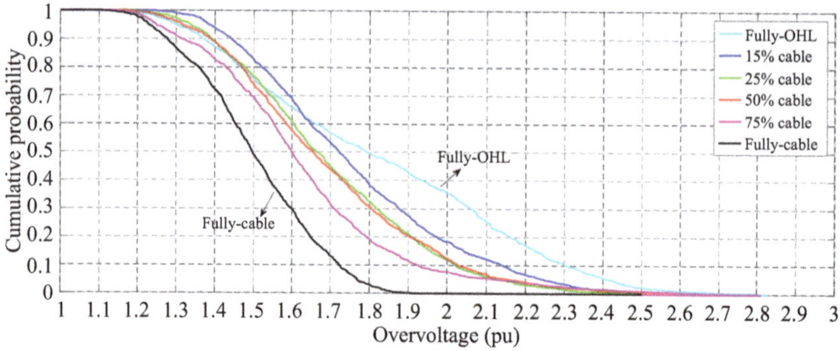

Figure 7.37 Cumulative probability distributions of energization overvoltages of the six cable scenarios

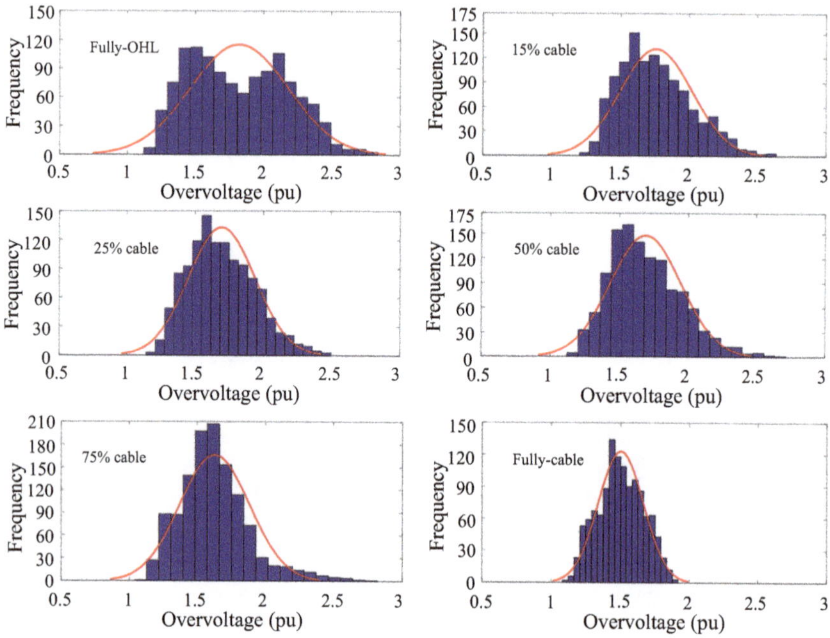

Figure 7.38 Probability distributions of energization overvoltages of the six cable scenarios with fitted normal distribution curves

overvoltages. It should be noted that for the partial cabling, this conclusion is a rule of thumb and may not be always valid due to its dependency on the system topology and parameters (e.g., see Sections 7.5.3.3.4, 7.5.3.3.5, and 7.5.3.3.7 for three exceptions).

The probability distributions of energization overvoltages of the six cable scenarios with fitted normal distribution curves are shown in Figure 7.38. It is a usual approach to compare the probability distributions of random energization

Table 7.9 Skewness and kurtosis of the energization overvoltages distributions

Scenario	Skewness	Kurtosis
Fully-OHL	0.239	2.084
15% cable	0.581	2.867
25% cable	0.453	2.802
50% cable	0.640	3.262
75% cable	1.012	4.886
Fully-cable	0.016	2.321

cases with the normal distribution. The *y*-axis (height of rectangle) is the frequency with which the overvoltages in the specified range (width of rectangle) have occurred. Table 7.9 presents the skewness (S) and kurtosis (K) values of the overvoltage distributions, calculated as expressed below:

$$S = \frac{1}{1200}\sum_{i=1}^{1200}\left(\frac{x_i - \mu_v}{\sigma}\right)^3 \tag{7.23}$$

$$K = \frac{1}{1200}\sum_{i=1}^{1200}\left(\frac{x_i - \mu_v}{\sigma}\right)^4 \tag{7.24}$$

where x_i is the overvoltage value (i.e., sample value), μ_v is the mean value of the overvoltages distribution, and σ is the standard deviation of the overvoltages distribution. There are 1,200 overvoltage values in each probability distribution.

The skewness is a measure describing how symmetrically the data are distributed around the mean value. The skewness of any symmetric distribution like the normal distribution is zero. A negative skewness indicates that the data has a longer tail to the left side of the mean value and a positive skewness indicates that the data has a longer tail to the right side. The overvoltage probability distributions of all the scenarios have positive skewness and are spread out more to the right side of the mean values. The fully-cable scenario has the smallest skewness indicating that its overvoltages probability distribution is the most symmetrical one.

The kurtosis is a measure of how outlier-prone (heavy- or light-tailed) a distribution is. Higher kurtosis indicates that the distribution is more outlier-prone (heavier-tailed) and lower kurtosis indicates that the distribution is less outlier-prone (lighter-tailed). The kurtosis of the normal distribution is 3. According to Table 7.9, the fully-OHL scenario with the kurtosis of 2.084 is the less outlier prone scenario.

It has been observed that the energization overvoltages can substantially change when the share of cable in the circuit is varied. To interpret the simulation results, the significant physical differences between cables and OHLs, which influence the energization overvoltages, should be discussed.

The first difference between cables and OHLs is the propagation velocity and the behavior of the modal waves. According to the modal decomposition theory, a cross-bonded three-phase single-core cable system can be decomposed into six modes: one coaxial mode, two inter-phase modes, two inter-sheath modes, and one ground mode (in contrast to only one inter-phase mode and one earth-return mode for an OHL). Each mode has its own attenuation constant and propagation velocity [50,84–86]. The wave propagation velocity in cables is much lower than that of OHLs as the cable coaxial mode is the fastest with the propagation velocity of $\frac{C_0}{\sqrt{\varepsilon_r \mu_r}}$ m/s, where ε_r and μ_r are respectively the relative permittivity and the relative permeability (compared to $C_0 = 3 \times 10^8$ m/s for OHLs). For a typical cable, the propagation velocity is approximately between 1.6×10^8 m/s and 1.8×10^8 m/s. If a cable is cross-bonded, the propagation velocity is slightly lower than these values [50].

During transients, when an energizing wave propagates into the cable core conductor, all the six modal waves are excited and propagate toward the open-end while they reflect and refract at each impedance discontinuity point (impedance mismatch point), where the line surge impedance changes. Therefore, the resulting overvoltage at a given location on the circuit is the superposition of all the modal waves at that location. As it is discussed in [76], the maximum energization overvoltage at the cable open-end is determined by the superposition of the inter-phase mode (as the dominant mode) and the coaxial mode (as the superimposed mode), where the cycle time of the inter-phase mode (dominant mode) is dependent on the short-circuit power and the cable length. The overvoltages of the cable coaxial mode are low (compared with those of the inter-phase mode in OHLs) and also the cycle time of the cable inter-phase mode is generally long. These two factors result in lower energization overvoltages for cables compared to OHLs.

The second difference between cables and OHLs is the existence of cross-bonding points for cables, which makes it more difficult for the waves to propagate to the open-end [75]. Furthermore, the third difference is the smaller surge impedance of cables compared to that of OHLs, which leads to lower overvoltages in cables for the same switching surge current [75,87].

In addition to the mentioned differences between cables and OHLs, an explanation about the reflection and refraction of the propagating waves in hybrid OHL-cable circuits is also useful for the interpretation of the simulation results. When an electromagnetic wave propagating along a hybrid circuit arrives at an impedance discontinuity point, as shown in Figure 7.39, a part of the incident wave is reflected

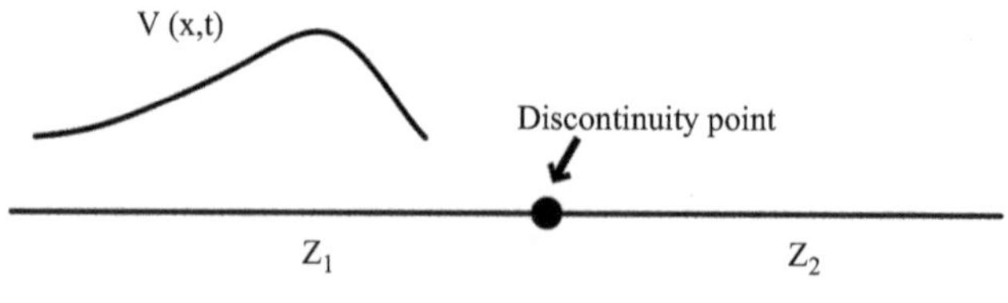

Figure 7.39 Traveling voltage wave arriving at an impedance discontinuity point at the junction of two transmission lines

back and a part of the wave is transmitted beyond the point. The impedance discontinuity point can be a sheath cross-bonding point or an OHL-cable transition point. The reflected voltage wave (V_R) and the transmitted voltage wave (V_T) can be calculated by the following equations [88]:

$$V_R = \frac{Z_2 - Z_1}{Z_2 + Z_1}.V \tag{7.25}$$

$$V_T = \frac{2Z_2}{Z_2 + Z_1}.V \tag{7.26}$$

where V is the incident voltage wave, Z_1 is the surge impedance of the first transmission line through which the incident wave is traveling, and Z_2 is the surge impendence of the second transmission line.

In the Spaak connection, the surge impedances of the OHLs and cables are around 220 Ω and 50 Ω, respectively. So, according to (7.25) and (7.26), when a voltage wave traveling through an OHL section reaches a transition point to a cable section, about 63% of the wave is reflected back with the negative polarity (with respect to the incident wave) and about 37% of the wave is transmitted to the cable. On the contrary, when a wave traveling through a cable section reaches a transition point to an OHL section, about 63% of the wave is reflected back into the cable with the positive polarity (with respect to the incident wave) and 163% of the wave is transmitted to the OHL.

When a circuit consists of a cascade connection of several OHL and cable sections, like the circuits of the Spaak connection, the propagating switching surge produces reflected and transmitted waveforms at all the impedance discontinuity points. This leads to a very complex propagation pattern that can be analyzed with the help of a lattice diagram [13]. In this situation, the voltage at a given point on the circuit is determined by the superposition of all the waves at every instant of time. The highest overvoltages occur when the specific condition in terms of the system topology and parameters (e.g., the number and location of sections and the short-circuit power) are present for the superposition of the waves.

7.5.3.3.2 Impact of compensation size

The shunt reactors are generally sized and located based on a steady-state analysis to control the power-frequency voltage and the capacitive current in the line breakers [49]. To investigate the impact of shunt reactors on the statistical distribution of energization overvoltages, the statistical analysis was conducted without shunt compensation and the results were compared with those of the shunt-compensated circuit.

Table 7.10 presents the key values of the overvoltage distributions and Figure 7.40 shows the cumulative probabilities of energization overvoltages with and without shunt compensation. In all the cable scenarios, except the fully-cable scenario, the maximum overvoltage and the 2% value are lower when the circuit is energized without shunt compensation. In addition, for these scenarios, the probability of high overvoltages (above 1.85 pu) is lower when the circuit is energized without shunt compensation, although the post-switching steady-state overvoltages

Table 7.10 Key values of the overvoltage distributions without shunt compensation (values between parentheses are with shunt compensation)

Scenario	Overvoltage (pu)			
	Max.	**Mean**	**Stand. dev.**	**2% value**
15% cable	2.418 (2.631)	1.748 (1.756)	0.240 (0.262)	2.267 (2.346)
25% cable	2.393 (2.499)	1.676 (1.694)	0.217 (0.245)	2.144 (2.269)
50% cable	2.355 (2.705)	1.699 (1.689)	0.215 (0.259)	2.153 (2.319)
75% cable	2.134 (2.810)	1.653 (1.626)	0.185 (0.256)	2.024 (2.329)
Fully-cable	2.083 (1.908)	1.613 (1.501)	0.154 (0.163)	1.920 (1.820)

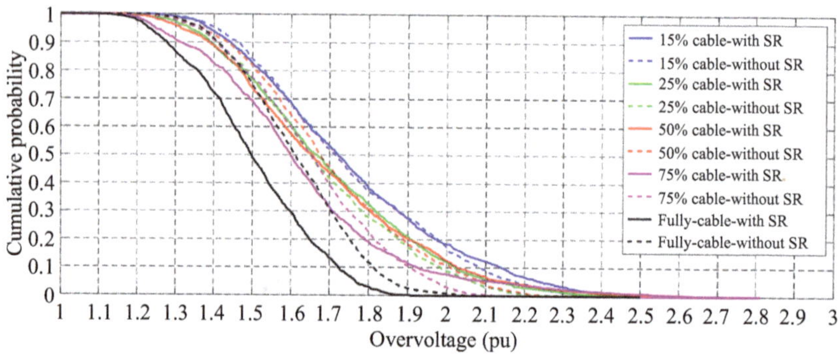

Figure 7.40 Cumulative probability distributions of energization overvoltages with shunt compensation (solid lines) and without shunt compensation (dashed lines)

will be higher due to the Ferranti effect. This could be due to the superposition of the energization transients of the reactors and the line when they are energized together.

7.5.3.3.3 Impact of compensation location

As it was discussed in Section 7.3.4.5, the location of shunt reactors can be at the line ends (line-end compensation (LEC)) or distributed at multiple points along the route of the circuit. The technical convenience of the former or the latter has been assessed in [56,89], where the shunt compensation allocation for different study cases has been analyzed. The LEC is the most common scheme while the distributed techniques have additional advantages, most notably the increase of the available transfer capacity for the active power and the need for a lower degree of compensation [49].

To deduce the influence of compensation location on the statistical distribution of energization overvoltages, overvoltages of the cable-end compensation (CEC) were compared with those of the LEC. Therefore, instead of two reactor banks at two substations, six reactors banks (one at each cable termination) were considered with the sizes reported in Figure 7.14 for the load-flow B.

Table 7.11 presents the key values of the overvoltage distributions and Figure 7.41 shows the cumulative probabilities of energization overvoltages with CEC and LEC. The key values and cumulative probabilities of overvoltages in all CEC scenarios are lower than those of the LEC scenarios, indicating another advantage of the distributed compensation over the line-end compensation.

7.5.3.3.4 Impact of short cable sections

In practice, short cable sections may be utilized in a hybrid circuit (e.g., for water crossing). The short traveling time of the wave in short cable sections results in consecutive reflections and refractions at termination points leading to different overvoltages. To investigate this issue, six cases with short cable sections were defined for the 15% cable scenario. As illustrated in Figure 7.42, the lengths of the OHL sections are unchanged in all the cases and only the lengths of the cable sections are varied while the total cable length is constant (12 km). Case 1 is the base case. Cases 2, 3, and 4 have a 1 km-long cable section at the sending side,

Table 7.11 Key values of the overvoltage distributions with CEC (values between parentheses are with LEC)

Scenario	**Overvoltage (pu)**			
	Max.	**Mean**	**Stand. dev.**	**2% value**
15% cable	2.413 (2.631)	1.708 (1.756)	0.245 (0.262)	2.245 (2.346)
25% cable	2.255 (2.499)	1.648 (1.694)	0.214 (0.245)	2.117 (2.269)
50% cable	2.288 (2.705)	1.639 (1.689)	0.219 (0.259)	2.095 (2.319)
75% cable	2.046 (2.810)	1.568 (1.626)	0.194 (0.256)	1.952 (2.329)
Fully-cable	1.876 (1.908)	1.493 (1.501)	0.155 (0.163)	1.783(1.820)

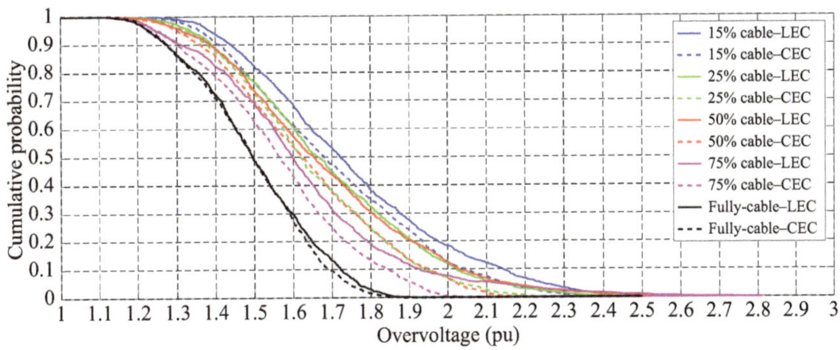

Figure 7.41 Cumulative probability distributions of energization overvoltages with LEC (solid lines) and CEC (dashed lines)

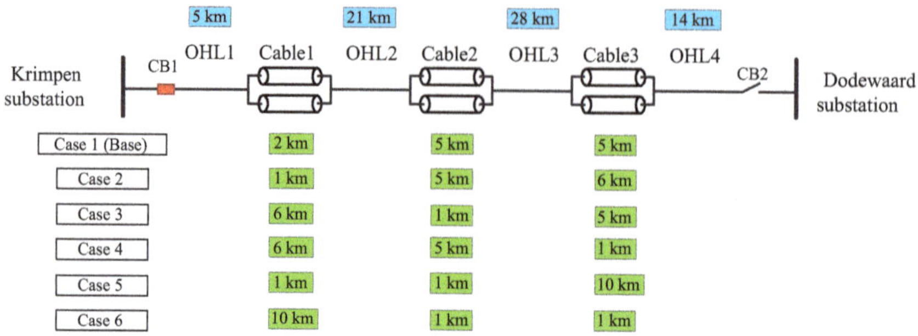

Figure 7.42 Cases of short cable sections for the 15% cable scenario (lengths of the OHL sections are unchanged)

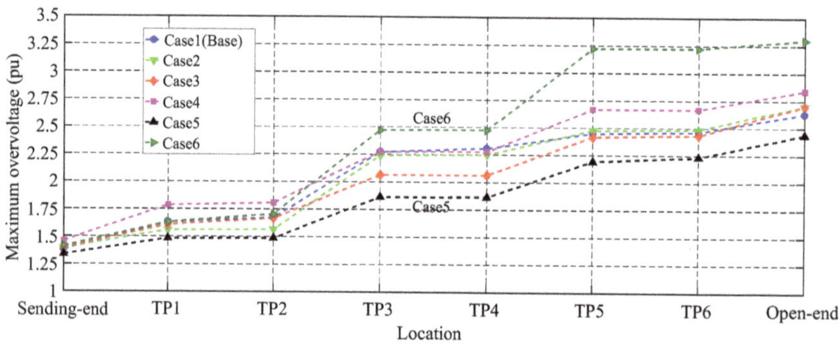

Figure 7.43 Maximum energization overvoltage at different locations along the hybrid OHL-cable circuit for the cases in Figure 7.42

middle, and the receiving side of the circuit, respectively. Cases 5 and 6 have 1 km-long cable sections at the middle plus the sending or receiving sides.

The maximum overvoltages and the cumulative probabilities of energization overvoltages are shown in Figures 7.43 and 7.44, respectively. It can be observed that Case 6 has the highest maximum overvoltage (3.3 pu) and overvoltage probability among all the cases, even higher than those of the fully-OHL scenario in Section 7.5.3.3.1. On the contrary, Case 5 has the lowest maximum (2.44 pu) and probability of overvoltages. In addition, after Case 6, the highest maximum and probability of overvoltages are produced by Case 4.

It can be concluded that for a given cable scenario, higher energization overvoltages are expected when very short cable sections are at the open-end of the circuit. For the studied cases, overvoltages are up to 0.86 pu (equal to 267 kV)

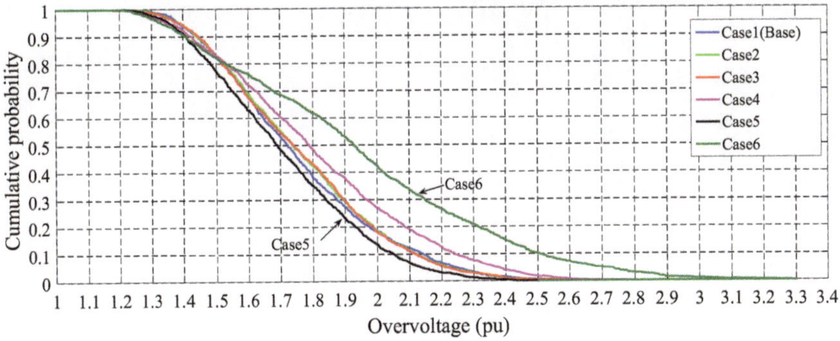

Figure 7.44 Cumulative probability distributions of energization overvoltages for the cases in Figure 7.42

higher than those of the cases without short cable sections. In contrast, the existence of longer cable sections at the open-end results in lower overvoltages. This is due to the short traveling time of the propagating wave in the short cable sections and the successive reflections and refractions.

7.5.3.3.5 Impact of short-circuit power

The impact of the short-circuit power of the supply network on the statistical distribution of energization overvoltages was studied by the use of an equivalent grid model consisting of the hybrid circuit and a lumped parameter inductive source. In fact, the Thevenin equivalent was used instead of the whole grid model with distributed generation because it is more flexible and convenient to adjust the short-circuit power by the variation of the equivalent source impedance. A frequency dependent network equivalent (FDNE) can be also an option to represent the entire grid if it can properly represent the generation and short-circuit power of the grid [21].

In the equivalent model of the grid, the power-frequency voltage at the sending-end was adjusted to the same voltage in the whole grid model (1.076 pu). The statistical analysis was carried out for five source inductances: 15, 20, 25, 50, and 100 mH. The 20 mH source inductance is equal to the equivalent short-circuit inductance calculated at the sending substation in the whole grid model.

Figure 7.45 shows the maximum energization overvoltage and 2% value for different supply networks. For a given source inductance, the lowest and highest values are for the fully-cable and the fully-OHL scenarios, respectively. The only exception is the 50% cable scenario with 25 mH source inductance, for which the maximum overvoltage (3.36 pu) is higher than that of the fully-OHL scenario although the corresponding 2% value is still lower. In addition, for all source impedances, the hybrid scenarios produce overvoltages higher than those of the fully-cable scenario and most likely lower than those of the fully-OHL scenario. This conclusion is in line with the previous conclusions.

It can be also noticed that each cable scenario has a different behavior than the other scenarios when the short-circuit power is varied. For the fully-OHL scenario,

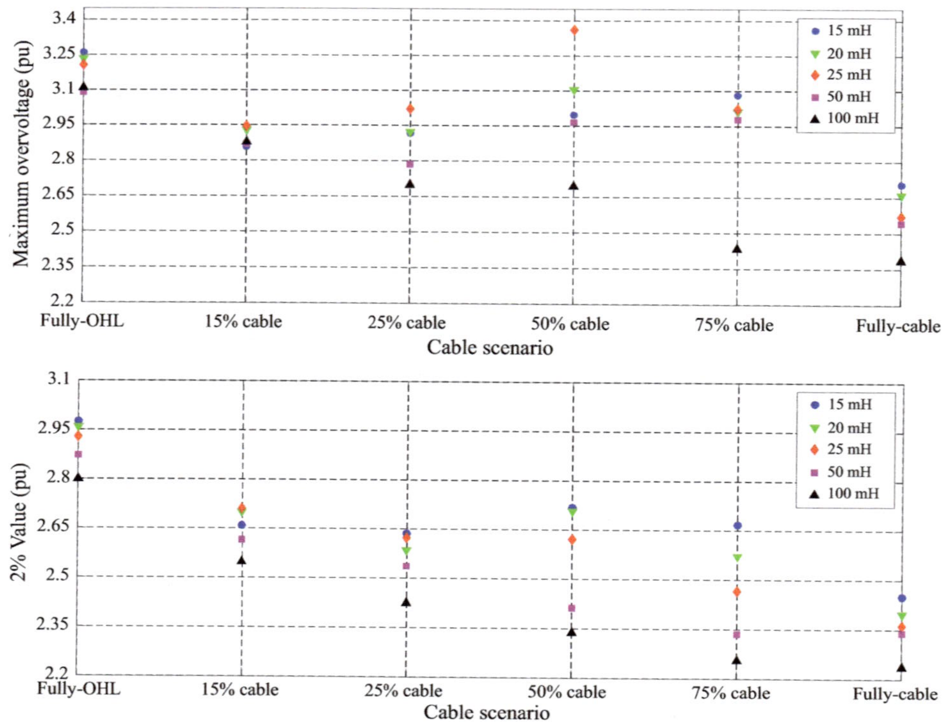

Figure 7.45 Maximum energization overvoltage and 2% value for different supply network short-circuit powers

the maximum overvoltage decreases when the source inductance increases from 15 mH to 50 mH and thereafter slightly increases for 100 mH; but, the 2% value always decreases with increasing the source impedance (weaker supply networks). For the fully-cable scenario, both the maximum overvoltage and 2% value decrease by the increase of the source impedance. For the hybrid scenarios, the dependency of overvoltages on the supply network strength changes case-by-case.

As it was previously mentioned, the maximum energization overvoltage at the cable open-end is determined by the superposition of the inter-phase mode (as the dominant mode) and the coaxial mode (as the superimposed mode), where the cycle time of the inter-phase mode is dependent on the short-circuit power and the cable length [76]. For larger source impedances (weaker supply networks), the rate of rise of voltage is lower and the cycle time of the inter-phase mode (dominant mode) is longer. Thus, by the time the inter-phase mode reaches its maximum at the open-end, the coaxial mode has traveled back and forth to the open-end several times and has become highly damped. As a result, the transient energization overvoltages will be likely lower when the source impedance is larger (weaker supply network).

The comparison of the simulated overvoltages of the whole grid model in Table 7.8 with those of the equivalent model with 20 mH source inductance (as the corresponding inductance with the short-circuit power of the whole grid model) shows that the simulated overvoltages in the equivalent grid are higher. This is due to the fact that the actual damping and wave propagation to the rest of the grid are not represented by the equivalent grid.

7.5.3.3.6 Impact of trapped charges (residual charges)

Trapped charges may exist on disconnected cables and result in very high overvoltages when the voltage of the trapped charges has opposite polarity to that of the switching surge [90]. The study was conducted by initial charging of the cables with three single-phase 30 kV DC voltage sources and then the statistical switching of the breakers whilst the DC sources were removed.

Table 7.12 and Figure 7.46 show the key values of the overvoltage distributions and the cumulative probability distributions, respectively. The maximum

Table 7.12 Key values of the overvoltage distributions with trapped charges (values between parentheses are without trapped charges)

Scenario	Overvoltage (pu)			
	Max.	**Mean**	**Stand. dev.**	**2% value**
15% cable	3.319 (2.631)	1.763 (1.756)	0.275 (0.262)	2.380 (2.346)
25% cable	2.788 (2.499)	1.702 (1.694)	0.257 (0.245)	2.289 (2.269)
50% cable	2.849 (2.705)	1.695 (1.689)	0.273 (0.259)	2.388 (2.319)
75% cable	2.780 (2.810)	1.614 (1.626)	0.254 (0.256)	2.330 (2.329)
Fully-cable	2.036 (1.908)	1.498 (1.501)	0.167 (0.163)	1.828 (1.820)

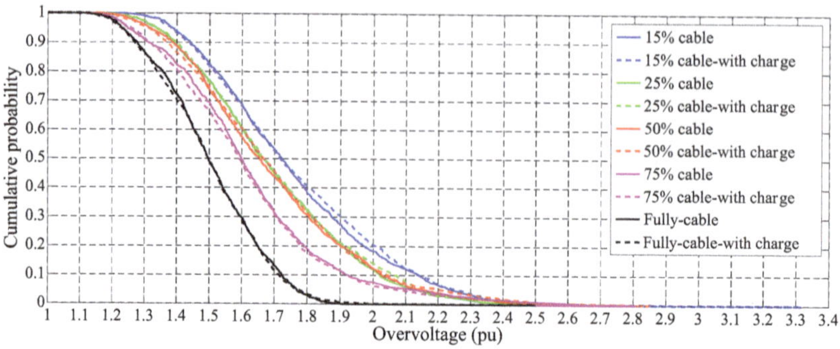

Figure 7.46 Cumulative probability distributions of energization overvoltages without trapped charges (solid lines) and with trapped charges (dashed lines)

overvoltages are up to 0.68 pu higher than those of the cases without trapped charges, whereas the difference in 2% values is much less and only up to 0.07 pu. Moreover, the differences between the cumulative probabilities are negligible, indicating that the overvoltages are generally not higher except for particular switching instants at which the resulting switching surge and the voltage of the trapped charges cause a very high overvoltage together. It should be noted that higher overvoltages can be expected if the residual voltage on the cables is higher than 30 kV.

7.5.3.3.7 Impact of mixed-line configuration

It is also important to investigate the influence of the mixed-line configuration on the statistical distribution of energization overvoltages. For this purpose, the four mixed-line configurations presented in Section 7.2.1.3 and the study case configuration (Figure 7.2) are compared. As concluded in Section 7.3.4.6, the required compensation size for each configuration is different, so the reported values in Figure 7.15 for the load-flow B were used.

The cumulative probabilities of energization overvoltages are shown in Figure 7.47. It can be noticed that for a given cable scenario, the difference between the cumulative probabilities of different configurations can be up to 35%. Therefore, the number and the distribution of the OHL and cable sections can substantially affect overvoltages due to the consequent changes in the wave propagation pattern. In addition, Configuration 1, in which a long cable section is at the open-end of the circuit, produces the lowest overvoltages for all the scenarios. This finding is in line with the conclusions of Section 7.5.3.3.4. The reason that Configuration 1 produces lower overvoltages than the other configurations is as follows: when the traveling voltage wave through the OHL section reaches the transition point to the cable section, about 63% of the incident wave is reflected back (see (7.25) and (7.26)) and only 37% of the incident wave is transmitted to the

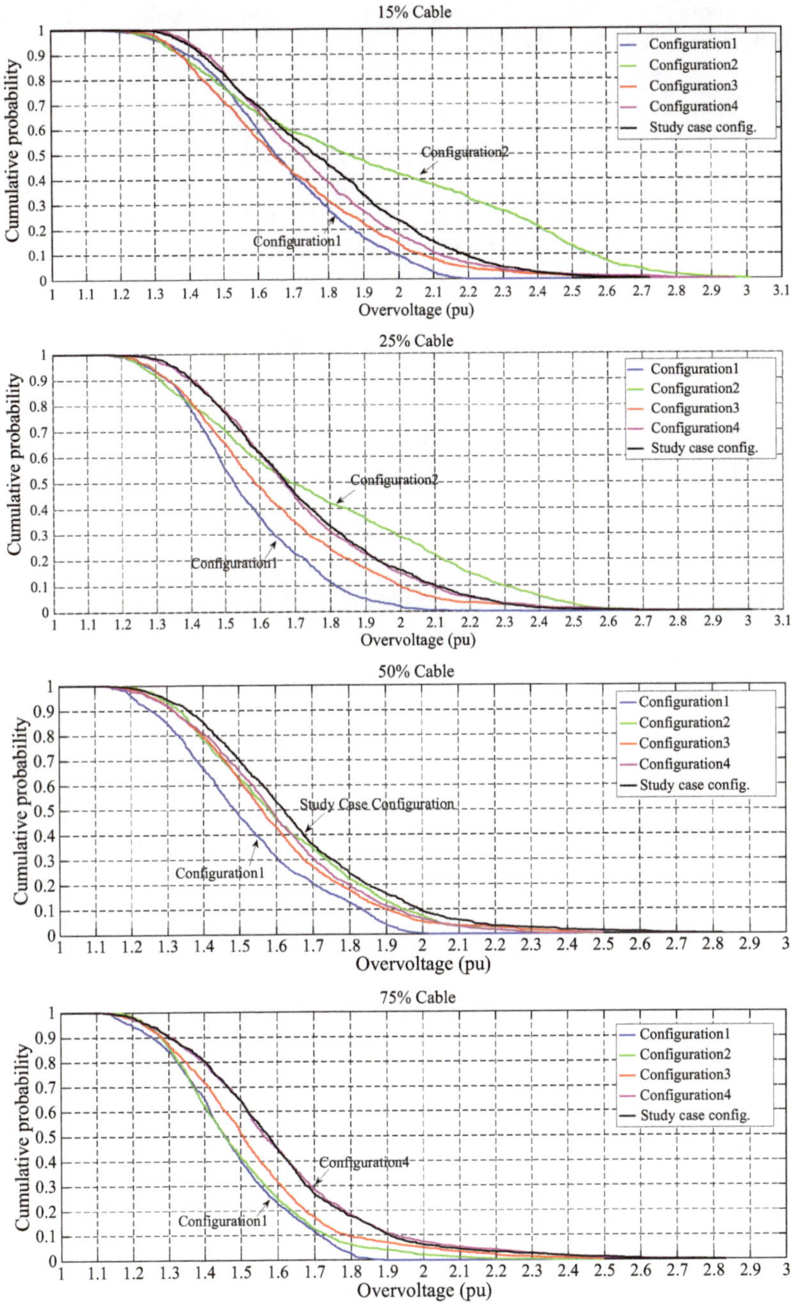

Figure 7.47 Cumulative probability distributions of energization overvoltages for different mixed-line configurations

cable and moving toward the open-end. So, the resulting overvoltage at the open-end of the circuit will be reduced.

7.5.4 Discussions

The statistical distribution of energization overvoltages plays an important role in the determination of the required insulation levels for EHV components. In Section 7.5, the distribution of energization overvoltages has been thoroughly studied for hybrid OHL-cable circuits by means of numerous statistical switching simulations.

It has been found that a hybrid OHL-cable circuit produces energization overvoltages higher than those of a fully-cable circuit and most likely lower than those of a fully-OHL circuit with the same transmission lengths. This means that the probability of experiencing high energization overvoltages and stressing system components is higher in a fully-OHL circuit than in a partial or fully undergrounded circuit.

Energization overvoltages are dependent on the number, locations, and lengths of the OHL and cable sections because of the consequent changes in the energizing wave propagation. For a given cable scenario, the existence of long cable sections at the open-end of the circuit results in lower energization overvoltages compared to the case that short cable sections are at the open-end. In addition, the difference between the overvoltage cumulative probabilities of different mixed-line configurations can be also substantial, where, as an example, it was up to 35% for the studied configurations in this section. This shows the high impact of the number and the distribution of the OHL and cable sections on energization overvoltages.

Furthermore, energization overvoltages can be substantially affected by the system parameters, most notably the short-circuit power of the supply network and the shunt compensation size and location. The transient energization overvoltages are likely lower when the network short-circuit power is lower (i.e., weaker supply network), whereas the post-switching steady-state overvoltages will be higher due to the Ferranti effect at the open-end of the circuit. Moreover, the distributed reactive power compensation along the circuit, like CEC, results in lower overvoltages compared to the line-end compensation.

Finally, in hybrid OHL-cable circuits, extremely high overvoltages (higher than 3 pu) may occur in specific cases such as the presence of trapped charges on the cable, the presence of very short cable sections at the open-end of the circuit, and certain short-circuit powers of the supply network. Special attention should be paid to these situations as they can lead to a component failure and/or exceeding the energy absorption capacity of surge arresters and sheath voltage limiters.

7.6 De-energization transients of hybrid OHL-cable systems

In Section 7.5, energization transients of hybrid OHL-cable circuits were studied. Transient voltages and currents are also initiated when a circuit breaker is being opened to de-energize a circuit or a component. De-energization transients are associated with several transient phenomena, which may become more severe when the circuit is fully or partially undergrounded with cables.

This section aims to analyze de-energization transients in the hybrid OHL-cable circuits of the Spaak connection. The voltages on the disconnected phases and the transient recovery voltages (TRV) across the contacts of line circuit breakers are assessed, as the analysis is performed for two cases: with and without reactors connected to the disconnected phases. The TRV is a very important parameter determining how significant the risk of breaker re-ignition or restrike after the current interruption is.

7.6.1 Capacitive current interruption

The interruption of capacitive currents can be challenging for circuit breakers. Figure 7.48 shows a simple circuit with a capacitive load. When the capacitive current is interrupted at current zero, the voltage is at its peak if a 90° phase shift between the voltage and current is assumed. As it is illustrated in Figure 7.49, at the

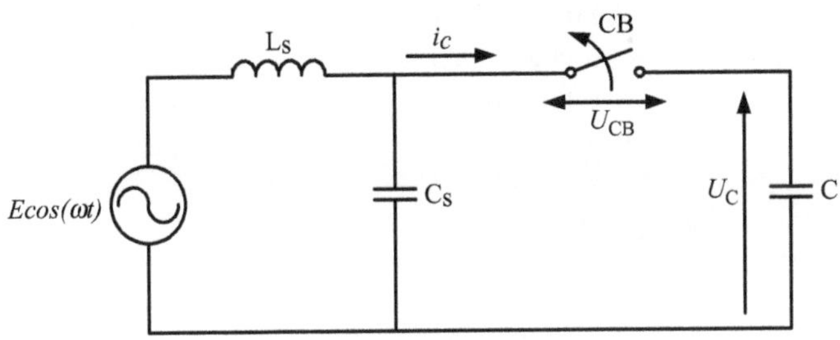

Figure 7.48 A simple circuit for a single-phase capacitive load switching

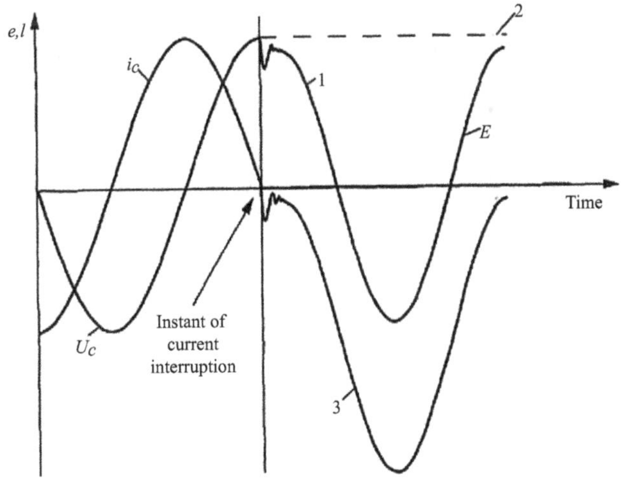

Figure 7.49 Current and voltages after the capacitive current interruption—1: source voltage; 2: capacitor voltage; 3: voltage across the breaker terminals, and E is the peak value of the source voltage. Adapted from [13].

instant of current interruption, the capacitor is fully charged with the voltage $U_c = E + \Delta U$, where E is the peak of the source voltage. This means that the load voltage is higher than 1 pu of the source voltage because of the Ferranti rise at the load side prior to the breaker opening. The voltage difference, ΔU, can be calculated as follows [3]:

$$\Delta U = \frac{1}{1 - \omega^2 L_S C} - 1 \approx \frac{Q}{P} = \frac{I_{cap}}{I_{sc,loc}} \text{(pu)} \tag{7.27}$$

where L_s is the source inductance and C is the load capacitance, as shown in Figure 7.48. ω is the system angular frequency ($\omega = 2\pi f_0$, where f_0 is the system power frequency). P is the local short-circuit active power, Q is the reactive power of the capacitor, $I_{\text{sc,loc}}$ is the local short-circuit current, and I_{cap} is the capacitive current. It can be deduced from this equation that a larger capacitor (i.e., a larger capacitive current) results in a larger load voltage.

As it can be seen in Figure 7.49, after the current interruption, the capacitor voltage remains constant while the source-side voltage makes a transient excursion from $E + \Delta U$ to E and then continues its power-frequency course. A half cycle later, the source voltage reaches the next peak with a reversed polarity, so the voltage across the breaker terminals (U_{CB}) becomes approximately twice of the source-side peak voltage ($U_{CB} = 2E + \Delta U \approx 2pu$) [13].

The recovery voltage across the breaker terminals after the current interruption has two successive parts: the TRV part where high-frequency oscillations occur followed by the steady-state voltage where transients are decayed and power-frequency oscillations are present. The TRV is dependent on the parameters of the circuit being interrupted, mainly the RLC circuit behavior and the traveling waves in the circuit. A re-ignition or restrike occurs when the TRV exceeds the breakdown voltage of the dielectric between the contacts of the circuit breaker. The re-ignition is a breakdown occurring within a quarter of a power-frequency cycle (i.e., 5 ms in a 50 Hz system) after the current interruption and the restrike is a breakdown occurring later. Re-ignition breakdowns involve a lower amount of energy than restrike breakdowns since they occur at relatively low recovery voltage levels, so they are in general harmless and considered as a part of the interruption process. However, restrike breakdowns are associated with a considerable amount of energy in the breakdown and are the main hazards (with severe consequences) of the capacitive current interruption [3].

Figure 7.50 shows the current and voltages after a capacitive current interruption followed by multiple restrikes. If a restrike occurs near the peak of the recovery voltage, the capacitive load suddenly discharges itself via the arc channel, leading to a restrike current with a peak value of $2E\sqrt{\frac{C}{L_s}}$ and a frequency of $\frac{1}{2\pi\sqrt{L_s C}}$. At the first zero crossing of the transient current, the capacitor voltage is $+3E$ (while it was $-E$ at the instant of restrike). If the circuit breaker interrupts the transient current at the first zero crossing of the current (in general at an odd number of restrike-current half-cycles), the TRV increases to $4E$ in a half-cycle later. A second restrike may occur at this instant if this high recovery voltage

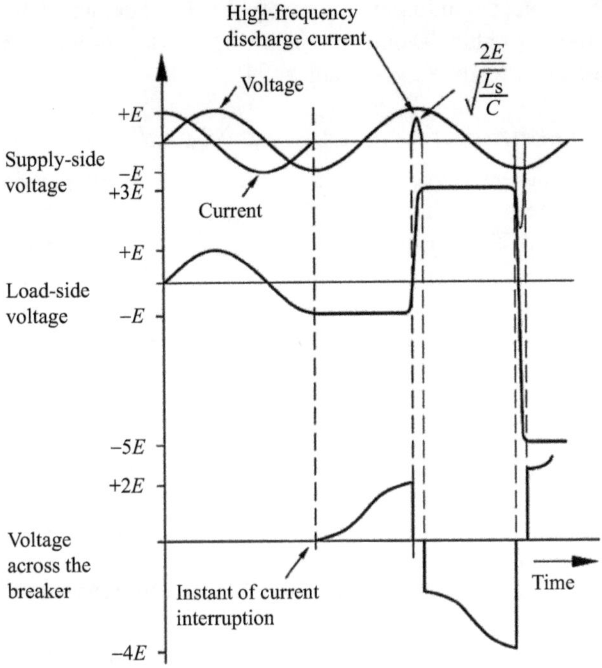

Figure 7.50 Current and voltages after a capacitive current interruption followed by multiple restrikes. Adapted from [13]

exceeds the breakdown voltage of the dielectric between the breaker contacts. If that occurs and the breaker interrupts the transient current again at the first current zero crossing (or at odd number of restrike-current half-cycles), the voltage at the load side of the breaker becomes $5E$ and the peak recovery voltage becomes $6E$. The next restrike is very likely to take place at this moment as the TRV would be much higher than the withstand capability of the circuit breaker. Multiple restrikes are very dangerous as they lead to capacitive voltage escalation with a high risk of severe damages to the cable and surrounding system equipment [3,13,91].

The above-described phenomenon should be studied when cables are disconnected from the grid since a cable has almost a similar behavior as a large capacitor. In fact, the natural discharge of a cable can be explained using the simple equations of a parallel RC-circuit [92]. This applies to the situation where the capacitive current is being interrupted at one end while the other end of the cable is already open. The capacitive current interruption becomes more challenging for breakers when the charging current of a long cable is not sufficiently compensated by shunt reactors directly connected to the cable [50]. This is due to the fact that the difference between the cable voltage and the source voltage will be large (i.e., ΔU in (7.27)).

In Reference [58], the preferred values for the capacitive current interruption capability of circuit breakers have been defined. However, manufacturers can build and supply breakers with higher capacitive current interruption capabilities based on the customer's need. An important point is that breakers which have to perform capacitive current switching should have a low or very low probability of restrike to minimize the risk of restrike overvoltages [93].

7.6.2 De-energization transients of the Spaak connection

In this section, transient voltages and currents after the de-energization of Circuit 1 of the Spaak connection, while Circuit 2 is assumed to be out-of-service, are presented and analyzed. Two cases can be distinguished: (1) disconnection of both the circuit and shunt reactors and (2) disconnection of the circuit while the shunt reactors are left connected to the circuit. Transient voltages and currents are substantially different between these two cases due to the consequent changes in the electrical behavior of the circuit caused by the connection or disconnection of the shunt reactors.

The cable scenarios and shunt compensation sizes are according to the presented values in Tables 7.1 and 7.6, respectively. For all simulations, at $t = 0.15$ s, the open command was given to the breakers at both sides, where the breakers were assumed to be ideal and interrupt the current at the first current zero crossing.

7.6.2.1 Switching off both the circuit and shunt reactors

Figure 7.51 shows the voltage at the Krimpen substation after the de-energization of the shunt reactors and the hybrid circuit with the 50% cable scenario (i.e., all the line and shunt reactor breakers are open). It can be seen that the source voltage does not experience a change and continues its power-frequency course since around 90% of the cable reactive power is compensated in this scenario.

When the line and shunt reactor breakers are opened, the voltage on the disconnected circuit is a decaying DC-voltage, as shown in Figure 7.52. The initial voltage caused by the voltage jump is different between the phases, which is due to the sequence of current interruption in the phases (normally about 3.33 ms time difference between the phase disconnections) and the mutual capacitive coupling between the OHL phases. The unloaded circuit has a dominant capacitive behavior and the magnetic coupling between the phases can be ignored.

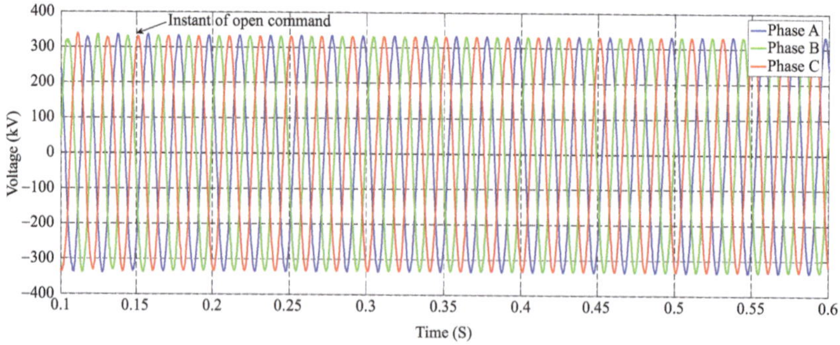

Figure 7.51 Voltage at the Krimpen substation after the de-energization of the 50% cable scenario and opening the shunt reactor breakers

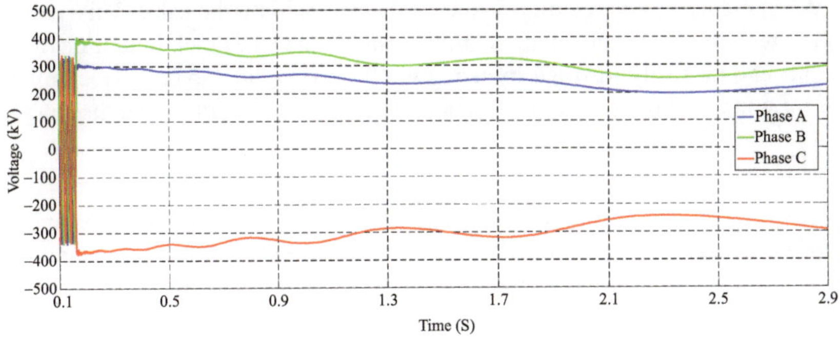

Figure 7.52 Voltage at the end of the disconnected circuit (Krimpen side) after the de-energization of the 50% cable scenario and opening the shunt reactor breakers

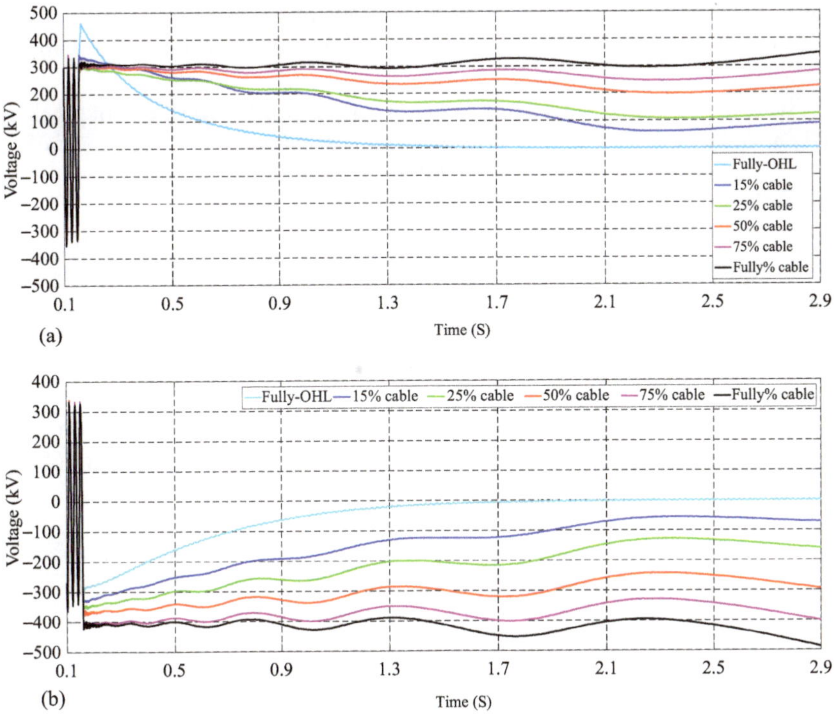

Figure 7.53 Voltage at the end of the disconnected circuit (Krimpen side) after the de-energization of the six cable scenarios and opening the shunt reactor breakers: (a) phase A and (b) phase C

Figure 7.53 shows the voltages of phases A and C after the de-energization of the six cable scenarios. The required time for the complete discharge of the trapped voltage increases by increasing the cable length (up to tens of seconds or minutes depending on the cable length).

When inductive voltage transformers (VT) are present, the cables' discharge is much faster, around several hundred milliseconds, due to the quick saturation of VTs [2]. This can be easily justified by using the following equations representing the transient voltage of a capacitor (V_c) in a parallel RC-circuit during the discharge:

$$V_C(t) = V_{\max} e^{-(t/\tau)}, \quad \text{where } \tau = RC \tag{7.28}$$

where $V_{\max}$ is the initial voltage of the capacitor at the de-energization instant, τ is the time constant of the circuit, R is the resistance, and C is the capacitance. By means of this equation, the required time to discharge the cable from $V(t_1)$ to $V(t_2)$ can be found:

$$\Delta t = t_2 - t_1 = \tau \ln\left(\frac{v(t_1)}{v(t_2)}\right) \tag{7.29}$$

The time constant of a cable is much larger than that of an OHL due to the very large cable capacitance. Thus, the time constant of the hybrid circuit increases when the cable share is increased, resulting in a much longer time for a complete discharge according to (7.29).

Figure 7.54 depicts the recovery voltages across the contacts of circuit breaker 1 (CB1) after the de-energization of the 50% cable scenario. The highest peak of the recovery voltage is in phase B and about 2.33 pu (724 kV), which is the result of the higher initial voltage of this phase compared to that of the other phases in the disconnected circuit (see Figure 7.52). This is a rather high value, so the withstand capability of the dielectric between the breaker contacts and the risk of restrike should be analyzed. Figures 7.55 and 7.56 show the recovery voltage across phases A and C of circuit breaker 1 after the de-energization of the six cable scenarios, respectively.

7.6.2.2 Switching-off the circuit

When a hybrid OHL-cable circuit is de-energized while shunt reactors are left connected to the phases (i.e., only line breakers are opened), the residual voltages

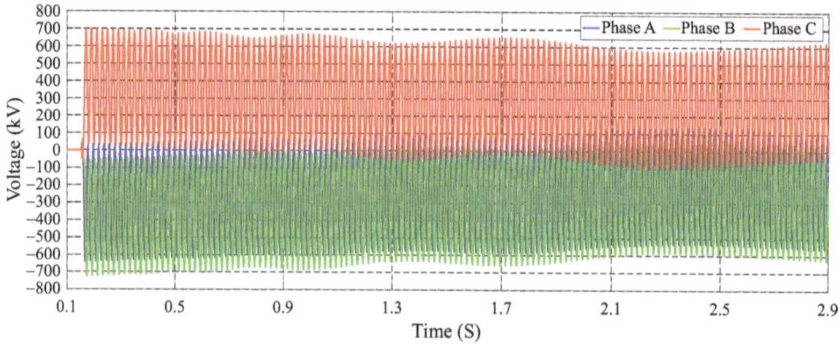

Figure 7.54 Recovery voltage across circuit breaker 1 (CB1) after the de-energization of the 50% cable scenario and opening the shunt reactor breakers

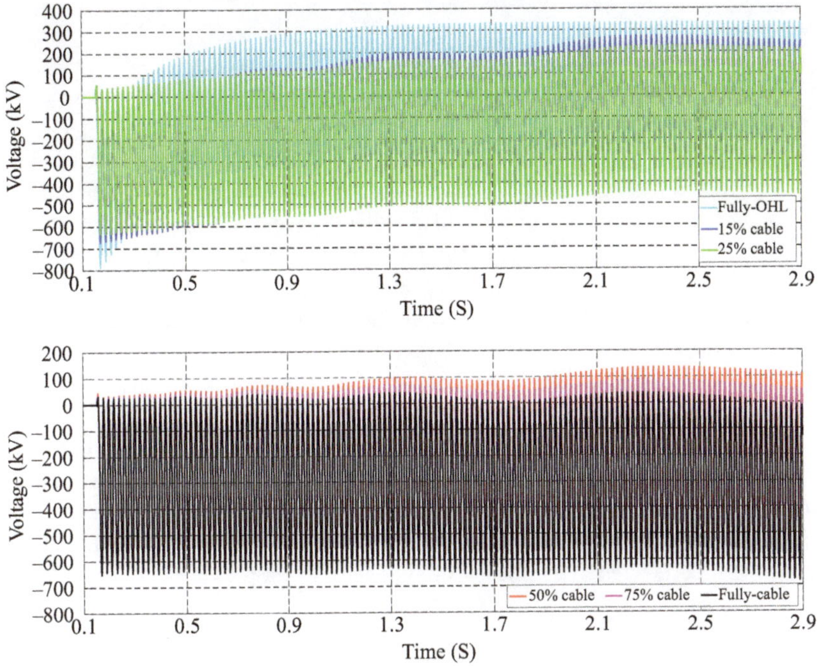

Figure 7.55 Recovery voltage across phase A of circuit breaker 1 (CB1) after the de-energization of the six cable scenarios and opening the shunt reactor breakers

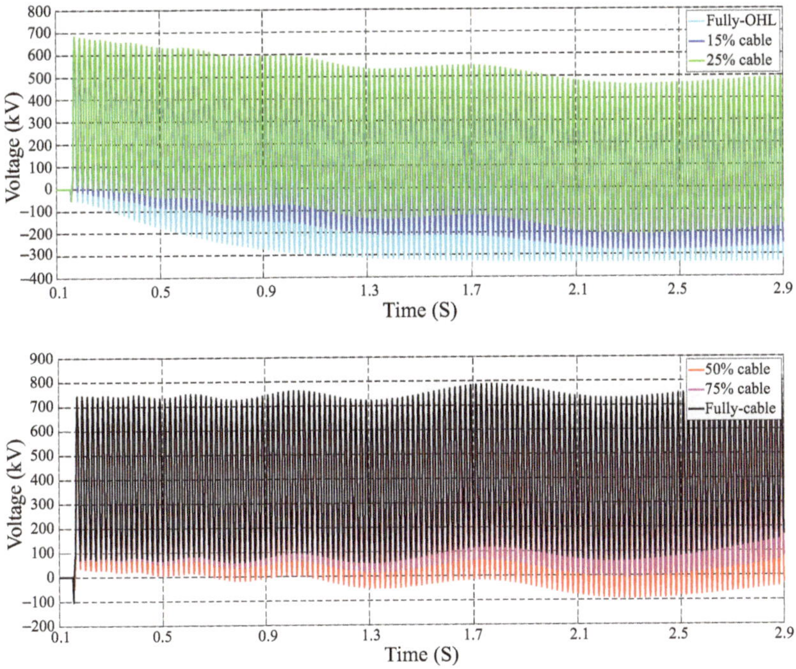

Figure 7.56 Recovery voltage across phase C of circuit breaker 1 (CB1) after the de-energization of the six cable scenarios and opening the shunt reactor breakers

on the disconnected phases are no longer decaying DC-voltages and are instead decaying oscillatory AC-voltages with superimposed frequencies.

A de-energized reactor-connected cable phase oscillates with its natural frequency, determined by the cable capacitance and the shunt reactor inductance. The natural resonance frequency is almost 50 Hz if the degree of compensation is close to 100%, which is the typical case for circuits with long cables, and it will be lower than 50 Hz if the shunt compensation degree is lower. When the circuit is even slightly unsymmetrical, each phase has a slightly different resonance frequency. Due to the mutual inductive coupling between shunt reactor phases and capacitive coupling between circuit phases, the resulting voltage in each phase is superimposed as the sum of its own voltage and the two induced voltages by the other phases. In this situation, adding up three different voltages with different frequencies and phase shifts results in voltage amplification or attenuation [94,95].

Figure 7.57(a) shows the voltage at the end of the disconnected circuit (Krimpen side) after the de-energization of the 15% cable scenario while the shunt reactors are still connected. The phase voltages will all drop to almost zero after a

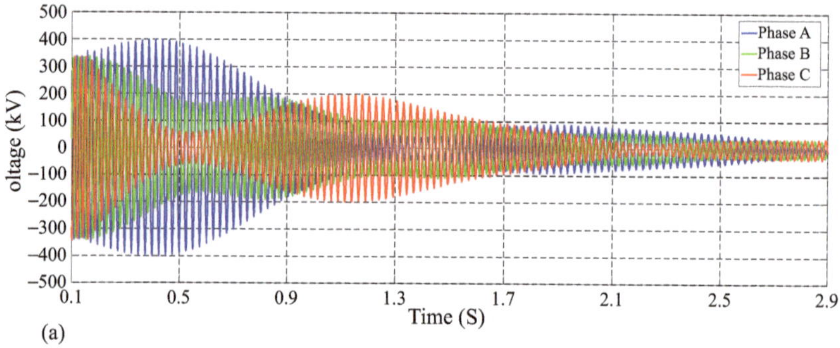

(a)

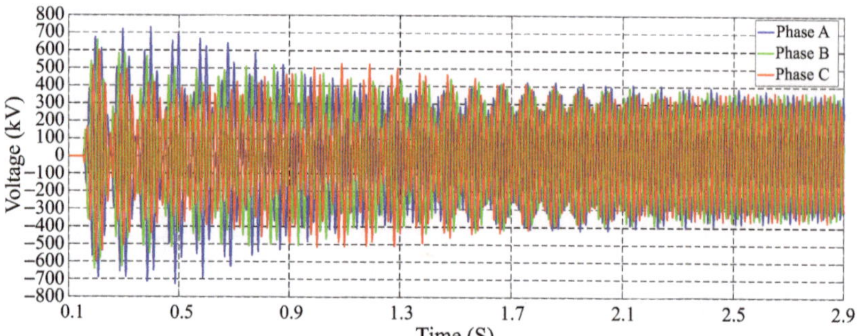

(b)

Figure 7.57 Voltage after the de-energization of the 15% cable scenario while the shunt reactors are still connected: (a) at the end of the disconnected circuit (Krimpen side) and (b) across circuit breaker 1 (CB1)

few seconds, where each phase experiences a different voltage transition pattern. The maximum voltage peak is 1.29 pu (400 kV) in phase A. As it was mentioned in Section 7.2.2.2, the mutual inductances between the shunt reactor phases were not considered in the EMT model of the Spaak connection. These mutual inductances may lead to a higher voltage increase in one or more of the phases, as addressed in references [5,96].

The cable discharge is much faster when reactors remain connected to the disconnected circuit as it only takes a few seconds for a complete discharge [2,50,92]. In fact, shunt reactors provide a lower resistive path to the ground and accelerate the cable discharge process. Therefore, the required time for the complete discharge of the cable depends on the system configuration and its characteristics, mainly the connection of shunt reactors and VTs.

Figure 7.57(b) shows the recovery voltage across the contacts of circuit breaker 1, where the peak voltage is 2.35 pu (728 kV) in phase A. This is a rather high value. The recovery voltage reaches the nominal power-frequency value

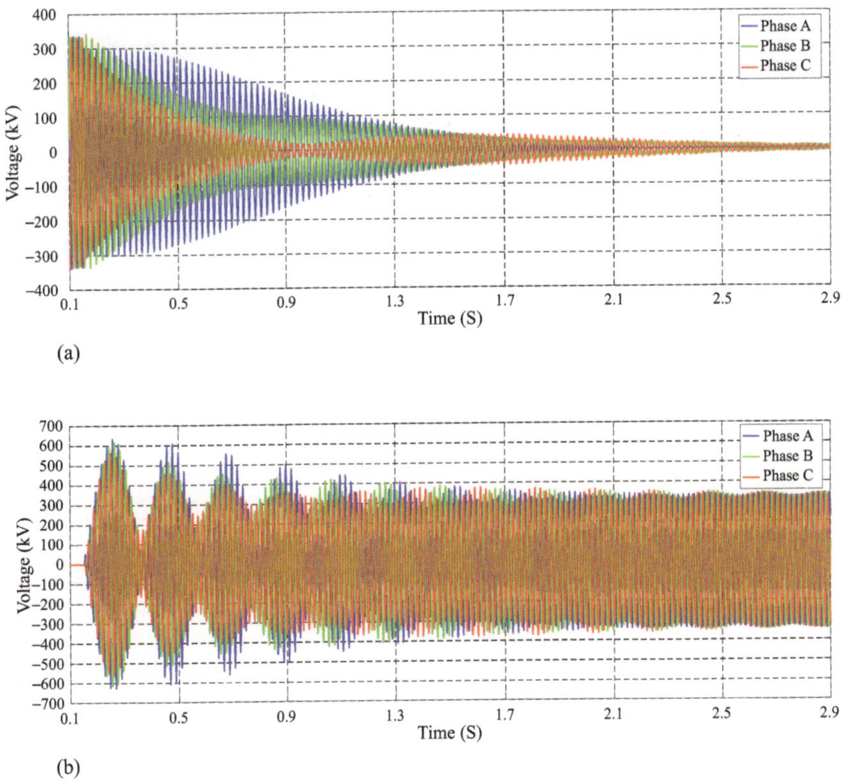

Figure 7.58 Voltage after the de-energization of the 25% cable scenario while the shunt reactors are still connected: (a) at the end of the disconnected circuit (Krimpen side) and (b) across circuit breaker 1 (CB1)

within 1.5 s after the current interruption as the voltage on the disconnected circuit has being damped out.

Resonance overvoltages are also expected in disconnected shunt-compensated hybrid OHL-cable circuits. Resonance may occur between the reactor inductance, inter-phase/inter-circuit capacitance, and cable capacitance if reactors remain connected to the disconnected phase(s) in uneven open-phase conditions (e.g., one phase is open and the other two are closed) or to a fully disconnected three-phase circuit. In other words, the induced power-frequency voltage in the disconnected phase by the capacitive coupling with energized phases can trigger the resonance circuit formed by the cable capacitance, the shunt reactor inductance, and the capacitive coupling between the de-energized phase and the energized phases. Ferroresonance may also occur when magnetic-core shunt reactors are used for the reactive power compensation. These risks can be minimized by connecting the shunt reactors via breakers to the circuit so that they can be disconnected from the de-energized phases [11,12].

The voltage on the disconnected circuit and the recovery voltage across the breaker contacts for the rest of the cable scenarios are shown in Figures 7.58–7.61.

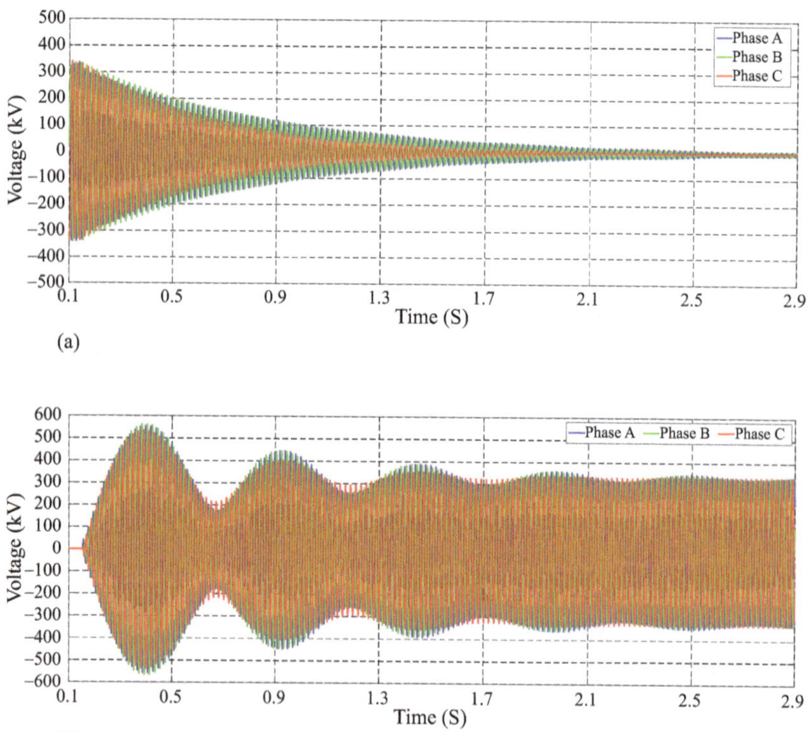

Figure 7.59 Voltage after the de-energization of the 50% cable scenario while the shunt reactors are still connected: (a) at the end of the disconnected circuit (Krimpen side) and (b) across circuit breaker 1 (CB1)

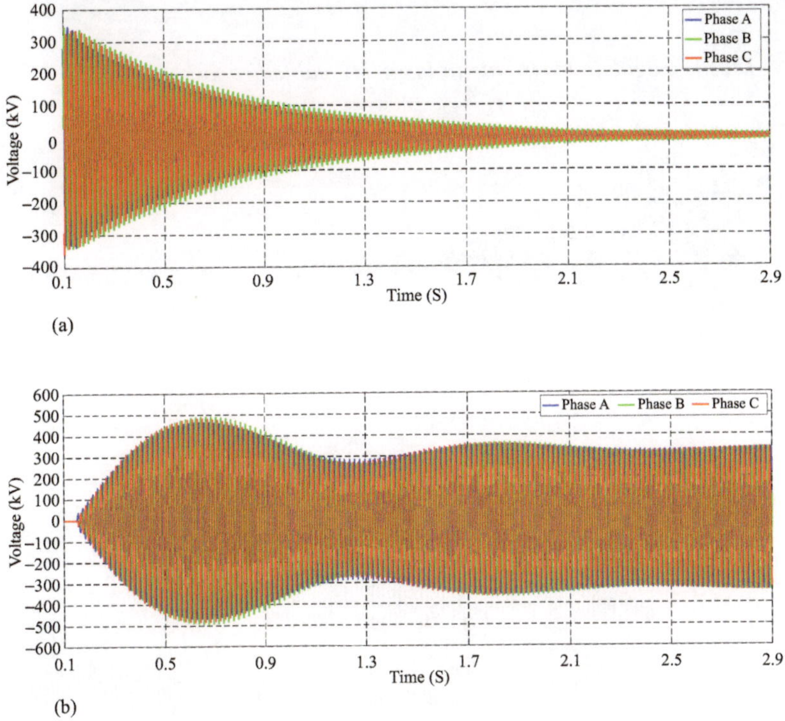

Figure 7.60 Voltage after the de-energization of the 75% cable scenario while the shunt reactors are still connected, (a) at the end of the disconnected circuit (Krimpen side) and (b) across circuit breaker 1 (CB1)

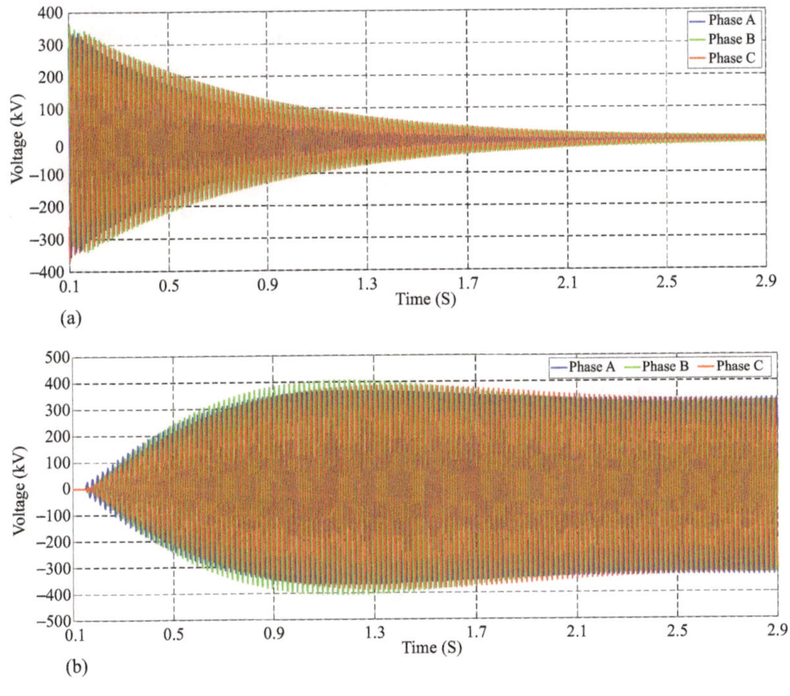

Figure 7.61 Voltage after the de-energization of the fully-cable scenario while the shunt reactors are still connected: (a) at the end of the disconnected circuit (Krimpen side) and (b) across circuit breaker 1 (CB1)

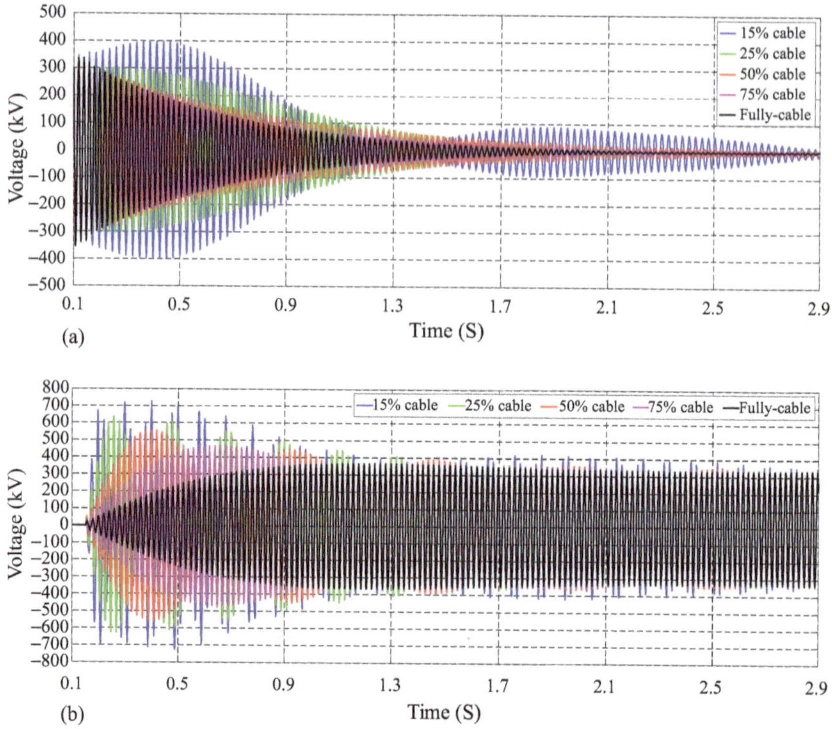

Figure 7.62 Voltage of phase A after the de-energization of the six cable scenario while the shunt reactors are still connected: (a) at the end of the disconnected circuit (Krimpen side) and (b) across circuit breaker 1 (CB1)

Figure 7.62 compares the phase A voltage for all the six cable scenarios. It can be seen that with the increasing cable share in the circuit, the voltages of the disconnected phases become less modulated and more symmetrical between the phases. This is due to the smaller capacitive coupling between the phases when the OHL share in the circuit has been decreased. In fact, the capacitive coupling only exists between the OHL sections, whereas it is almost zero between the cable sections as the cable sheath is grounded.

In addition, it can be also seen in Figure 7.62(b) that the maximum recovery voltage becomes lower with the increasing cable share in the circuit, as it has decreased from 2.35 pu (728 kV) for the 15% cable scenario to 1.3 pu (404 kV) for the fully-cable scenario. The rate of rise of the recovery voltage is also lower when the cable length in the circuit is longer. This is due to the reason that for the longer cable shares, the voltage of the disconnected phase becomes a decaying AC-voltage with less superimposed frequencies, because of smaller coupling with the other phases, and with a lower oscillation frequency between the cable capacitance and the reactor inductance.

7.7 Zero-missing phenomenon in cable systems

As it was mentioned in Section 7.3, the operation of long cables requires reactive power compensation by means of shunt reactors. In most situations, shunt reactors should be directly connected to the hybrid circuit and energized together with the cable to minimize post-switching steady-state overvoltages and to control the capacitive current in the line breaker [2,48].

The simultaneous energization of the cable and reactors may cause the zero-missing phenomenon, which is the situation that the current through the line breaker does not cross the zero value for several cycles (even up to several seconds). In this situation, it is difficult or even impossible to safely open the healthy phases if a fault occurs in the circuit during energization when the zero-missing current is still present. Therefore, the system is more vulnerable and unprotected against faults when proper countermeasures are not devised at the design stage of the cable project.

Several countermeasures are available for the zero-missing phenomenon, while each one has advantages and disadvantages depending on the grid topology and specifications. Synchronized switching at the voltage peak, sequential switching, and the use of circuit breakers equipped with pre-insertion resistors are the mostly studied countermeasures in the literature [50,97–99].

In Reference [97], switching at the voltage peak was applied to avoid the zero-missing phenomenon based on an insulation coordination study. In Reference [98], the authors studied the use of pre-insertion resistors to minimize the zero-missing current, where they proposed a simple formula to approximate the resistor value. The expanded version of [98] has been presented in reference [99], where extra countermeasures of the zero-missing phenomenon have been also discussed. In addition, an iterative process for more accurate determination of the pre-insertion resistor size has been provided in reference [99]. In Reference [50], a general overview of the countermeasures and a more detailed study of the sequential switching countermeasure can be found.

There are three important aspects regarding countermeasures of the zero-missing phenomenon that should be always investigated: (a) the resulting transient overvoltages and inrush currents after applying a countermeasure, (b) the impact of the circuit breaker mechanical delay on the effectiveness of an applied countermeasure, and (c) the influence of hybrid OHL-cable circuits and the cable length on the effectiveness of an applied countermeasure.

This section performs an in-depth investigation on the countermeasures of the zero-missing phenomenon and determines the requirements, benefits, and deficiencies of each method. Countermeasures are compared in terms of resulting currents and voltages after their implementation. High transient overvoltages, a voltage dip/swell, and high inrush currents may occur when a particular countermeasure is applied. Applying a countermeasure should result in neither switching overvoltages nor inrush currents over the permissible limits.

Countermeasures were applied for the Spaak connection, for which six cable scenarios were defined in Section 7.2.1.4 and shunt compensation sizes were calculated in Section 7.3. It is of high importance to take the impact of the cable length into account since a countermeasure may be the optimal solution only for a certain range of the cable length, depending on system specifications. Moreover, the switching sequence of breakers is determined for each countermeasure. This can be implemented in practice by applying the controlled switching (also known as synchronized or point-on-wave switching) in combination with single-pole operated circuit breakers.

7.7.1 Zero-missing phenomenon

The zero-missing phenomenon can be simply illustrated by the cable system (i.e., the cable and the shunt reactor) shown in Figure 7.63. The cable system is energized by a voltage source.

When the reactor is being energized, the charging current has two components: an AC-component and a decaying DC-component:

$$I_{SR}(t) = \underbrace{e^{-\left(\frac{R}{L}\right)t}\left[\frac{-V_{\max}}{|Z|}\sin(\varphi-\theta)\right]}_{\text{DC component}(I_{\text{SR,dc}})} + \underbrace{\frac{V_{\max}}{|Z|}\sin(\omega t+\varphi-\theta)}_{\text{AC component}(I_{\text{SR,ac}})} \tag{7.30}$$

where R, L, and Z are respectively resistance, inductance, and impedance of the reactor. $V_{\max}$ is the voltage amplitude, φ is the initial phase angle of the voltage (i.e., the point on the voltage waveform at which the breaker is closed), and θ is the load angle. As it can be seen, the initial value of the DC-component depends on the initial phase angle and its decaying time depends on the time constant, expressed by $\tau = \frac{L}{R}$. The DC-component is maximum (equal to the amplitude of the AC-component at $t = 0$) when $\varphi - \theta$ is 90°, and is zero when $\varphi - \theta$ is zero.

The current through the circuit breaker (CB1) when the circuit is being energized is as below:

$$\begin{aligned} I_{\text{CB1}}(t) &= I_{\text{SR}}(t) + I_{\text{cable}}(t) \\ &= \underbrace{I_{\text{SR,dc}}(t)}_{\text{DC comp.}} + \underbrace{\left(I_{\text{SR,ac}}(t) + I_{\text{cable}}(t)\right)}_{\text{AC component}} \end{aligned} \tag{7.31}$$

$$I_{\text{SR,ac}}(t) = -K_{sh} \times I_{\text{cable}}(t) \tag{7.32}$$

where I_{CB1} is the current through the circuit breaker, I_{SR} is the energization current of the shunt reactor, and I_{cable} is the cable charging current. K_{Sh} is the shunt compensation degree and shows the percentage of the cable reactive power which is compensated by the shunt reactor (see (7.7)). Due to simplicity reasons, it is assumed that in (7.31) and

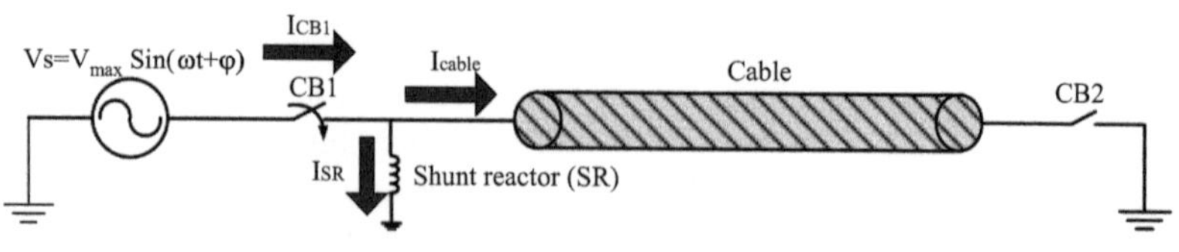

Figure 7.63 Single-line diagram of a simple shunt-compensated cable system. CB: circuit breaker

(7.32), the cable is a pure capacitor and the resistance of the shunt reactor is zero in the AC-component (these simplifications have a negligible influence).

As it is shown in (7.32), the AC-component of the reactor current has an opposite phase angle with the cable charging current. Therefore, in (7.31), the AC-component of the reactor current is partially (or completely if the compensation degree is 100%) canceling out the cable charging current. When the compensation degree is higher than 50%, the AC-component in (7.31) may become smaller than the DC-component, so that the total current (I_{CB1}) does not cross the zero point until the DC-component is sufficiently damped. Thus, the zero-missing phenomenon can occur if two conditions are satisfied: (1) energization of the shunt reactor together with the cable and (2) shunt compensation degree larger than 50% ($K_{sh} > 50\%$).

Assume that the length of the cable in Figure 7.63 is 60 km and the shunt compensation degree is 92.1% (according to the compensation sizing results presented in Table 7.6). The system is energized in two cases: (1) closing each phase at the zero voltage and (2) closing each phase at the voltage peak. Note that this can be realized in practice by means of the controlled switching and single-pole operated circuit breakers. The resulting currents and voltages after energization at the zero voltage and at the voltage peak (obtained by PSCAD simulations) are respectively shown in Figures 7.64 and 7.65.

Figures 7.64(a) and 6.64(b) respectively show the cable charging current (I_{Cable}) and the shunt reactor current (I_{SR}) when each phase is energized at the zero voltage. The

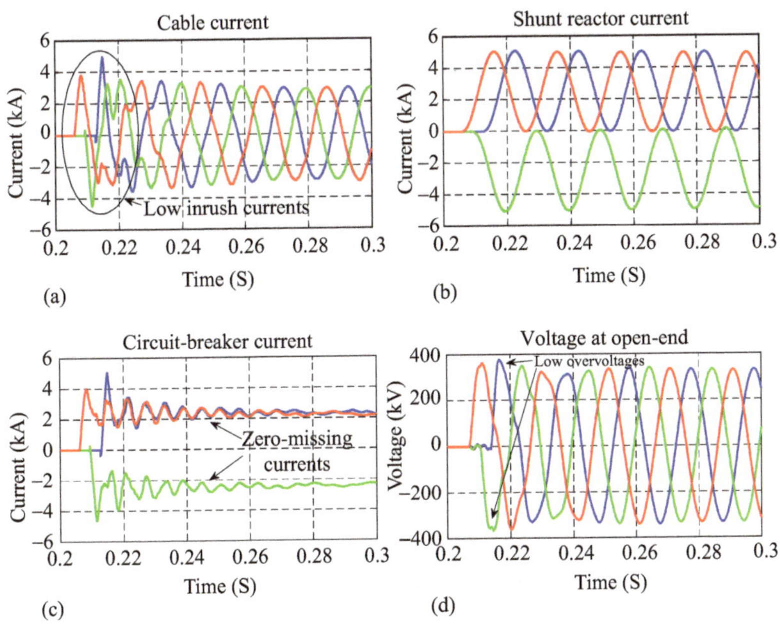

Figure 7.64 Currents and open-end voltages after the energization of the circuit in Figure 7.63 at the zero voltage (60 km cable, K_{sh} = 92.1%)

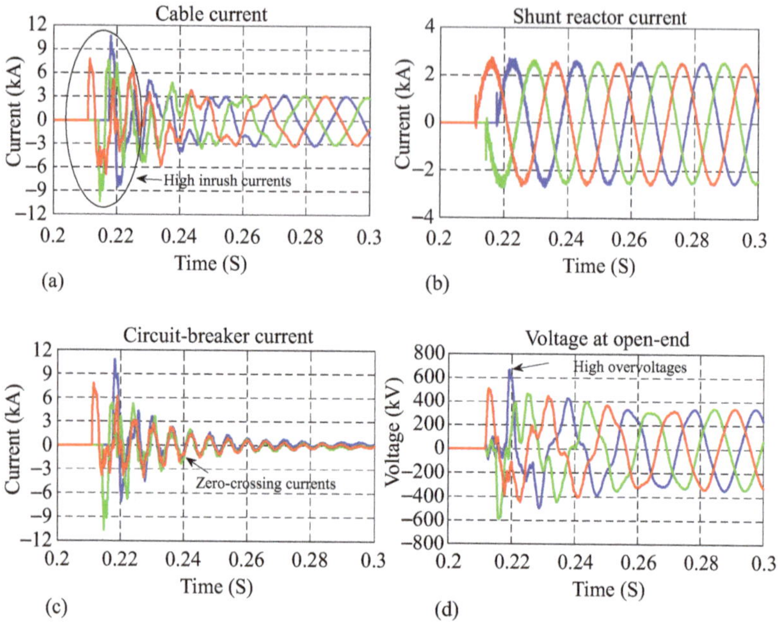

Figure 7.65 Currents and open-end voltages after the energization of the circuit in Figure 7.63 at the voltage peak (60 km cable, K_{sh} = 92.1%)

DC-component of the shunt reactor current is maximum since φ=0 and $\theta \approx 90^{\circ}$ (see (7.30)). However, the reactor current always touches the zero point because its maximum DC-offset is equal to the amplitude of its AC-component. Figure 7.64(c) shows the current through the circuit breaker ($I_{CB1} = I_{Cable} + I_{SR}$). The total AC-component is very small due to the cancelation of 92.1% of the cable charging current by the AC-component of the reactor current. Thus, the total current flowing through the circuit breaker is mainly DC and is not crossing the zero point. The main benefit of switching at voltage zero is that the cable charging inrush currents (Figure 7.64(a)) and the transient switching overvoltages (Figure 7.64(d)) are minimized.

In contrast, the DC-component is zero if each phase is closed at the voltage peak ($\varphi = 90^{\circ}$ and $\theta \approx 90^{\circ}$ in (7.30)), as shown in Figure 7.65(b). This means that the zero-missing phenomenon does not occur (Figure 7.65(c)). However, in comparison to the switching at the zero voltage, the cable charging inrush currents (Figure 7.65(a)) and the transient switching overvoltages (Figure 7.65(d)) are higher.

7.7.2 Operation criteria

The implementation of a countermeasure can cause high and very steep inrush currents and/or transient switching overvoltages. High inrush currents and voltage surges increase dielectric and mechanical stresses on circuit breakers and other system components. These stresses can cause immediate or gradual damages of equipment and affect the system reliability [3].

It is important to define a set of criteria based on the system operation limits and standards to evaluate voltages and currents after implementation of a countermeasure. Three criteria are defined here based on the specified requirements for circuit breakers, insulation coordination, and power quality which are addressed by manufacturers, IEC standards, and the grid code. These criteria are:

1. *Amplitude and frequency of inrush current:* The countermeasure should not cause high inrush currents recognized to be harmful for breakers and other components [3,58]. For capacitor banks, standardized peak value and frequency of capacitive inrush currents are 20 kA and 4.25 kHz, respectively [58,70]. These values are used in study since the cable energization inrush current is almost similar to that of an equivalent capacitor bank. However, it should be noted that the amplitude and the rate-of-rise of the cable inrush current are normally lower than those of an equivalent capacitor bank due to the relatively higher surge impedance of cables [3].
2. *Transient switching overvoltages:* The peak values of the phase-to-earth transient energization overvoltages after implementation of a countermeasure should remain below the switching impulse withstand voltage (SIWV) of 3.38 pu (1,050 kV), as the specified value in Reference [67] for 380 kV equipment[3].
3. *RMS voltage step:* The IEC standard [57] and the Dutch grid code impose limits for the quick changes in the RMS voltage between two steady-states. This criterion is mainly related to the power quality and minimization of the disturbance to consumers. After switching, RMS voltages should be within the voltage dip and swell thresholds of 0.9 pu (342 kV) and 1.1 pu (418 kV), respectively. In addition, the rapid voltage change (RVC) should be smaller than 3% (11.4 kV). The time frame for the calculation of the RMS value is half of the power-frequency cycle (10 ms in 50 Hz systems).

7.7.3 Countermeasures

The countermeasures of the zero-missing phenomenon fall into one of the following three categories:

1. *Prevention methods*: The zero-missing current will be avoided by the application of these countermeasures.
2. *Mitigation methods*: These measures do not prevent the occurrence of the zero-missing current, but they minimize the decaying time of the DC-component by increasing the system damping (i.e., smaller time constant), so that the current through the breaker crosses the zero point faster.
3. *Handling methods*: These measures are neither helping to avoid nor to mitigate the zero-missing current. However, they help system operators to safely open the breakers when a fault occurs in the hybrid circuit at the instant of, or after, energization when the zero-missing phenomenon is still present.

[3]The IEC 60071-1 standard has specified 850, 950, or 1,050 kV (peak value) for the phase-to-earth switching impulse withstand voltage of equipment with 420 kV highest RMS voltage.

In this study, the following six countermeasures are analyzed and compared, where detailed elaborations on them can be found in Section 7.7.4:

1. Simultaneous cable and reactors energization at the voltage peak
2. In advance cable energization
3. In advance reactor energization
4. Sequential switching
5. Opening of the faulty phase(s)
6. Increasing the DC-offset damping

7.7.4 Simulation results

This section presents the simulation results when the countermeasures are tested on the Spaak connection. The mixed-line configuration is as shown in Figure 7.2 with the cable scenarios defined in Table 7.1. Shunt compensation sizes are reported in Table 7.6, which are determined by the steady-state analysis. The energization of one of the circuits from the Krimpen substation was simulated and analyzed by the use of the developed grid model in PSCAD, as the model was described in Section 7.2.2.2. Transient overvoltages are expressed in per unit, where the base value (1 pu) is the peak value of the phase-to-earth nominal voltage (310.27 kV). Overvoltages were recorded at the open-end (unless differently specified) since the highest overvoltages are occurring there.

In the first step, it is important to investigate the influence of the cable length on the severity of the zero-missing phenomenon. Figure 7.66 shows the current through the line breaker for the six cable scenarios when each phase is energized by the controlled switching at the zero voltage. With the increasing cable share, the zero-missing phenomenon becomes more severe in terms of the amplitude of the DC-component and the required time for the first zero crossing. This is due to a larger DC-component (because of bigger compensation size, that is, smaller $|Z|$, in (7.30)) and a smaller AC-component (because of a higher degree of compensation) when the cable length is increased. As the result, the current through the line breaker is dominantly a high-amplitude and low-damped direct current.

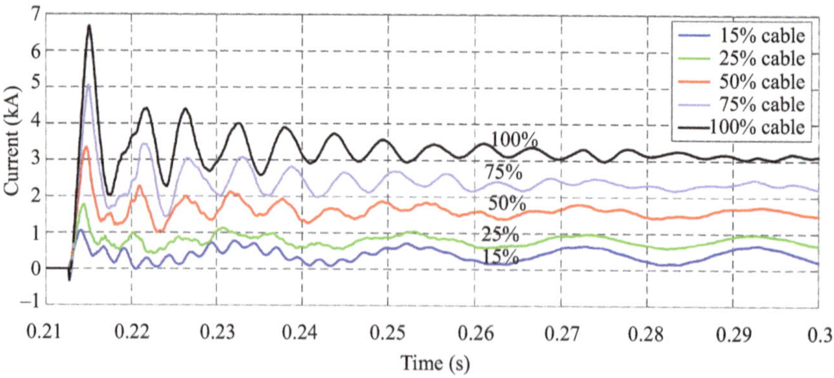

Figure 7.66 Line breaker current (I_{CB1}, phase A) for the six cable scenarios when each phase is energized at the zero voltage

7.7.4.1 Simultaneous cable and reactors energization at the voltage peak

This technique proposes to energize each cable phase together with shunt reactors of that phase at its voltage peak (requires controlled switching and single-pole operated breakers). This is a prevention countermeasure, but higher switching overvoltages and higher inrush currents are expected due to the energization at the voltage peak. Figure 7.67 shows the resulting current when this measure is applied. With increasing the cable length, the inrush current peak value becomes higher (around 13 kA for the 100% cable (fully-cable) scenario) due to the larger circuit capacitance; however, the peak values and the frequencies of the inrush currents are still within the specified limits.

Table 7.13 shows the maximum overvoltage at the open-end of the circuit when energization is performed at the voltage peak (i.e., when this countermeasure is applied) and at the zero voltage. It can be seen that in some cases the resulting

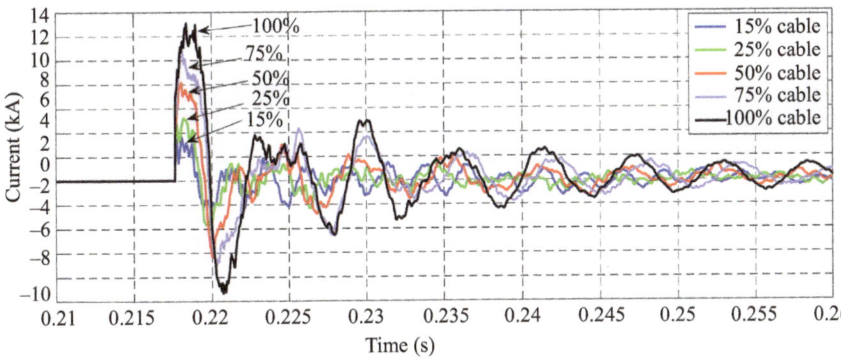

Figure 7.67 Line breaker current (I_{CB1}, phase A) for the six cable scenarios when each phase is energized at the voltage peak (Countermeasure 7.7.4.1)

Table 7.13 Overvoltages at the circuit open-end after energizing each phase at the voltage peak and the zero voltage

Scenario	Maximum overvoltage (pu)					
	Synchronized switching at zero voltage			Synchronized switching at voltage peak		
	Phase A	Phase B	Phase C	Phase A	Phase B	Phase C
15% cable	1.128	1.186	1.199	2.159	1.650	2.182
25% cable	1.125	1.202	1.141	1.973	2.056	1.724
50% cable	1.128	1.193	1.147	2.224	2.111	1.682
75% cable	1.215	1.180	1.170	2.153	1.892	1.647
100% cable (fully-cable)	1.209	1.135	1.151	2.063	1.763	1.615

overvoltages by switching at the voltage peak are almost double of those resulted by switching at the zero voltage. Although the overvoltages are higher, they are still below the specified limit of 3.38 pu for the studied cable scenarios.

It should be pointed out that the resulting overvoltages are dependent on the system topology and parameters such as the number/location/length of OHL and cable sections, the steady-state voltage prior to the switching, and the system short-circuit power. The distribution and length of OHL and cable sections can significantly affect the resulting overvoltages due to the changes in reflections and refractions of the propagating wave at the transition points. These remarks were proved in Section 7.5 by the statistical analysis of energization overvoltages.

This countermeasure requires an accurately controlled switching to close each pole of the breaker at its voltage peak. As it was previously mentioned in Section 7.5, breakers are in practice not ideal and they show variations in their operating times [3,70]. Variations in the operating time are either predictable or stochastic. Predictable variations (e.g., long-term aging) do not affect the effectiveness of the controlled switching and can be compensated by an adaptive control. However, stochastic variations (inherent scatter) of the operating time can result in the energization of the cable and reactors at a point different than the voltage peak. This limitation of the controlled switching can be described by the statistical distribution function of the scatter [3].

A statistical analysis was carried out to understand the influence of stochastic variations in the breaker closing time on the probability distribution of energization overvoltages. For each breaker pole, the closing time was varied by a normal (Gaussian) distribution. The mean value was the instant at which the voltage peak of that pole occurs, the standard deviation (σ) was 1 ms, and the distribution curve was truncated at -3σ and $+3\sigma$. Fifty closing times were simulated for each pole to obtain an accurate set of statistical data [3,75,97]. The statistical simulations were implemented in PSCAD by the use of the multiple-run feature and random number generators based on the normal distribution.

Table 7.14 shows the maximum, mean, standard deviation, and 2% values of the obtained overvoltage distributions for the six cable scenarios. The 2% value is

Table 7.14 Key values of the energization overvoltage distributions when countermeasure 7.7.4.1 is applied

Scenario	**Overvoltage (pu)**			
	Max.	**Mean**	**Stand. dev. (σ)**	**2% value**
15% cable	2.406	1.879	0.279	2.393
25% cable	2.378	1.755	0.271	2.283
50% cable	2.355	1.861	0.216	2.134
75% cable	2.188	1.850	0.248	2.181
100% cable (Fully-cable)	2.151	1.818	0.218	2.132

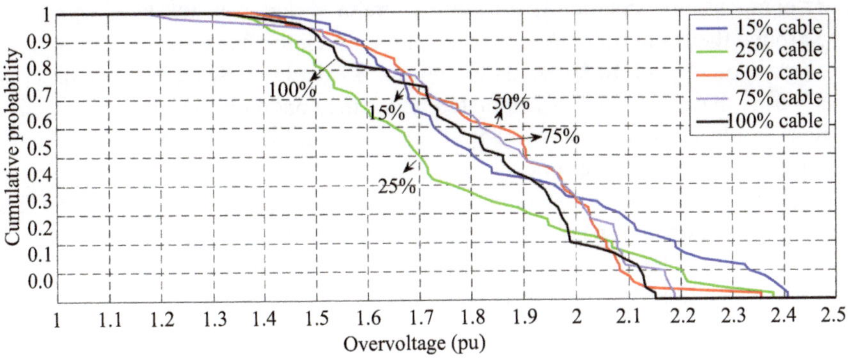

Figure 7.68 Cumulative probability distributions of overvoltages at the circuit open-end due to stochastic variations in the breaker closing time when Countermeasure 7.7.4.1 is applied

the overvoltage value that the probability of this value being exceeded is 2% [66]. From each energization simulation, the peak value of the phase-to-earth overvoltage on each phase at the circuit open-end was included in the overvoltage probability distribution (known as the phase-peak method [75]). It can be seen that for the studied scenarios, the maximum recorded overvoltage is 2.406 pu, which is below the specified limit of 3.38 pu.

Figure 7.68 shows the cumulative probability of overvoltages obtained by the statistical variation of the closing time. For a given voltage level, the vertical axis shows the cumulative probability of overvoltages exceeding that voltage level.

When the stochastic variation in the pole closing time exceeds a threshold, the cable and reactors are energized at a point on the voltage waveform at which the DC-component of the line breaker current is larger than the AC-component. This means that the zero-missing current occurs although it was supposed to be avoided by energizing at the voltage peak as the aimed closing point. Therefore, this countermeasure will be successful only if the stochastic variation in the breaker closing time is not exceeding the threshold. This threshold is called here breaker closing variation threshold (BCVT).

The BCVT shows the maximum variation in the breaker closing time (ΔT_{BCVT}) around the instant of the voltage peak (T_{peak}), for which the breaker current crosses the zero point before a given time limit. If the pole is not closed at $T_{\text{peak}} - \Delta T_{\text{BCVT}} < T < T_{\text{peak}} + \Delta T_{\text{BCVT}}$, the absolute value of the DC-component of the current through the line breaker will be larger than the absolute value of the AC-component, thus the total current does not cross the zero point before the given time limit. The time limit can be defined based on the preference of the system operator.

To calculate the BCVT, first, the line breaker current should be calculated as a function of the breaker closing time. This can be done by the use of system parameters including the OHL and cable impedances, shunt reactor impedance, and shunt compensation degree. Then, the maximum variation in the breaker closing time around the instant of the voltage peak, for which the line breaker current has a first zero-crossing before the given time limit, is calculated by an iterative process. This means that the BCVT is calculated under the condition that the zero-missing current should not occur, or when it occurs, it should disappear within the given time limit after the closing time.

Table 7.15 shows the calculated BCVT for different time limits and the five cable scenarios of the Spaak connection. The values between parentheses are the corresponding angles in degrees. As it is shown, for a given cable scenario, BCVT increases when the time limit increases.

For a given time limit, the BCVT is smaller for longer cables since the higher degree of compensation, required for longer cables, results in a line breaker current with a smaller AC-component but a larger DC-component (see Figure 7.66). The dominant DC-component is very low-damped (due to the large time constant of shunt reactors) and sensitive to variations of the closing time. In other words, in (7.30), any change in $\sin(\varphi - \theta)$ is magnified by $\frac{V_{max}}{|Z|}$, which is larger for longer cables due to bigger compensation size (i.e., smaller $|Z|$). Thus, for longer cables, any variation in the closing time has a more noticeable impact on the DC-component of the line breaker current.

Knowing the BCVT is crucial because it specifies the required accuracy of breakers used for this countermeasure. In some cases, this countermeasure is not a reliable solution when the breaker delay can exceed the BCVT.

7.7.4.2 Energization in sequence

The main drawbacks of the previous countermeasure, in which both the cable and reactors were energized together at the voltage peak, were high transient overvoltages and inrush currents. However, the zero-missing phenomenon can be also avoided when the shunt reactors are energized with an intentional delay before or after the instant of cable energization. In this countermeasure, with a controlled time delay between energizations, the transient overvoltages are minimized by energizing the cable at the zero voltage and the zero-missing current is avoided by energizing the reactors at the voltage peak.

The key parameter for the success of this countermeasure is to keep the time delay between the cable energization and the shunt reactors energization as short as possible. Otherwise, a long delay may lead to a voltage dip/swell, RVC larger than 3%, and under/over excitation of nearby generators [48].

This prevention countermeasure can be realized in two ways, depending on which of cable or reactor is energized first. They are called here *in advance cable energization* and *in advance reactor energization* countermeasures. The application of these methods requires the controlled switching and single-pole operated breakers. Breakers planned to close at voltage peaks (i.e., reactor breakers) should be tested to comply with the BCVT to ensure that the zero-missing current is avoided. However, breakers switching at the zero voltage should not necessarily comply with the BCVT requirement since they do not affect the DC-offset of the current.

Table 7.15 BCVT for different time limits and the five cable scenarios of the Spaak connection

Scenario	Time limit								
	10 ms	**20 ms**	**40 ms**	**60 ms**	**80 ms**	**100 ms**	**500 ms**	**1 s**	**2 s**
15% cable	±0.71 ms (±12.8°)	±1.42 ms (±25.6°)	±1.43 ms (±25.7°)	±1.43 ms (±25.8°)	±1.44 ms (±25.9°)	±1.45 ms (±26°)	±1.54 ms (±27.7°)	±1.67 ms (±30°)	±1.67 ms (±30°)
25% cable	±0.21 ms (±3.7°)	±0.57 ms (±10.3°)	±0.57 ms (±10.3°)	±0.58 ms (±10.4°)	±0.58 ms (±10.4°)	±0.58 ms (±10.5°)	±0.64 ms (±11.6°)	±0.72 ms (±13°)	±0.87 ms (±15.7°)
50% cable	±0.13 ms (±2.3°)	±0.38 ms (±6.8°)	±0.38 ms (±6.9°)	±0.38 ms (±6.9°)	±0.39 ms (±7°)	±0.39 ms (±7.1°)	±0.46 ms (±8.3°)	±0.55 ms (±9.9°)	±0.69 ms (±12.5°)
75% cable	±0.08 ms (±1.5°)	±0.27 ms (±4.8°)	±0.27 ms (±4.9°)	±0.27 ms (±4.9°)	±0.28 ms (±5°)	±0.28 ms (±5.1°)	±0.35 ms (±6.3°)	±0.43 ms (±7.7°)	±0.53 ms (±9.6°)
100% cable (Fully-cable)	±0.04 ms (±0.8°)	±0.15 ms (±2.7°)	±0.16 ms (±2.8°)	±0.16 ms (±2.8°)	±0.16 ms (±2.9°)	±0.16 ms (±2.9°)	±0.21 ms (±3.8°)	±0.27 ms (±4.8°)	±0.32 ms (±5.8°)

7.7.4.2.1 In advance cable energization

In this countermeasure, each cable phase is energized at the zero voltage and a quarter of a cycle later (5 ms in 50 Hz systems), at the voltage peak, the reactors of that phase are energized. The voltage at the open-end and the line breaker current after applying this countermeasure for the 75% cable scenario (60 km cable) are shown in Figure 7.69. The results are compared with those of Countermeasure 7.7.4.1 (i.e., simultaneous cable and reactor energization at the voltage peak). The transient overvoltage and the inrush current are considerably lower with this countermeasure since the cable is energized at the zero voltage. Thus, the main advantage of this measure is the minimization of the stress on the line breaker and system components.

The results in Figure 7.69 are for the case in which the time delay between cable energization and reactors energization is 5 ms (optimal case). However, the delay may be longer in practice. The RVC and voltage swell can exceed the limits when the time delay becomes long.

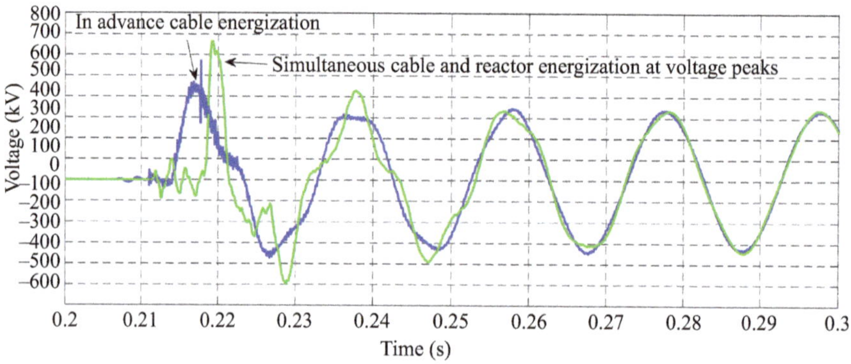

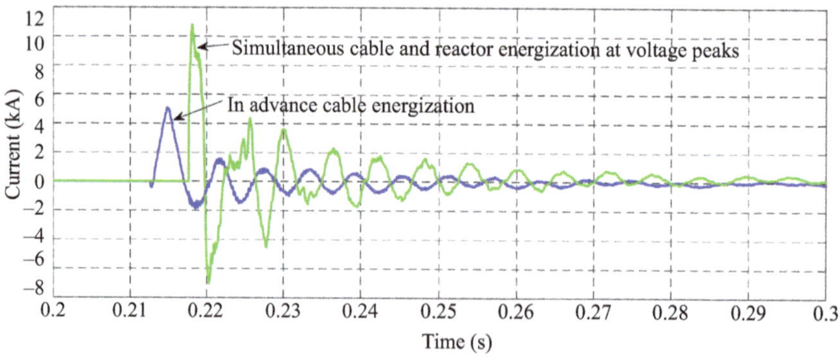

Figure 7.69 Open-end voltage (phase A) and line breaker current (I_{CB1}, phase A) after the implementation of Countermeasure 7.7.4.2.1 (with 5 ms delay) and Countermeasure 7.7.4.1 (75% cable scenario, K_{sh} = 92.1%)

Figure 7.70 shows the RMS voltage step at the Krimpen substation after energizing the 75% cable scenario (60 km cable) with three-time delays between the cable and reactors energizations: 5 ms, 65 ms (3+1/4 of a cycle), and 1.05 s (5+1/4 of a cycle). It can be seen that the voltage step exceeds the 3% limit and goes beyond the swell threshold of 1.1 pu (1.1×380 kV=418 kV) with both 65 ms and 1.05 s delays. Such a high sensitivity to delay is the result of the high steady-state voltage of 1.076 pu before the energization, leaving a margin of only 0.024 pu for the voltage step.

In addition to the initial voltage before the energization, the cable length and the short-circuit strength of the substation from which the circuit is energized affect the amplitude of the voltage step. The effect of cable length on voltage step, when the time delay is 1.05 s, is shown in Figure 7.71. Only voltage steps in the 15% and 25% cable scenarios are below the 3% and 1.1 pu limits, whereas the step exceeds

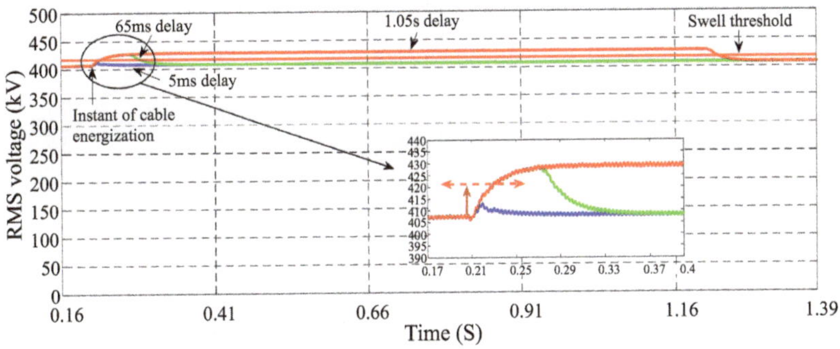

Figure 7.70 RMS voltage step at the Krimpen substation after the implementation of Countermeasure 7.7.4.2.1 with three different time delays (75% cable scenario, K_{sh} = 92.1%)

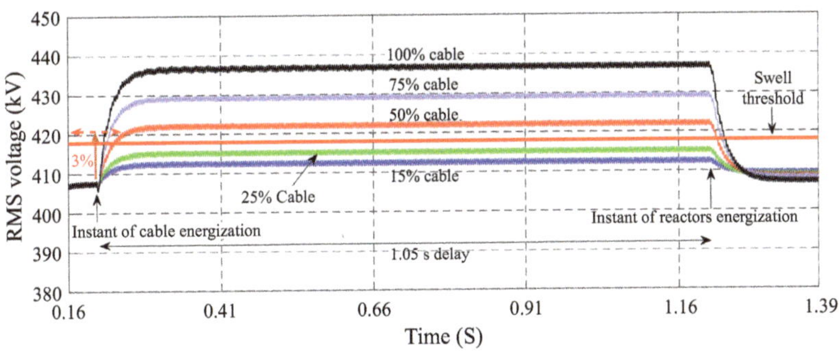

Figure 7.71 RMS voltage step at the Krimpen substation after energizing the six cable scenarios with Countermeasure 7.7.4.2.1. The time delay is 1.05 s

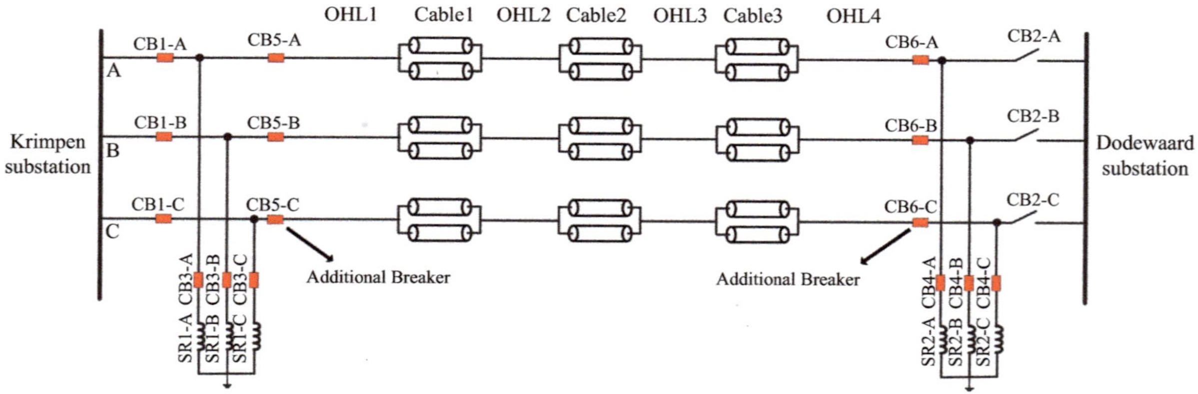

Figure 7.72 Structure of the shunt-compensated hybrid OHL-cable circuit with Countermeasure 7.7.4.2.2. CB: circuit breaker; SR: shunt reactor

both limits with increasing the cable length. Therefore, the main limitation of this countermeasure is the expected large voltage step due to a potential long delay between the cable energization and reactors energization.

7.7.4.2.2 In advance reactor energization

The implementation of this countermeasure requires two additional single-pole operated breakers to be installed at the both sides of the hybrid circuit, as shown in Figure 7.72. This allows that breakers 1 and 3 be closed at the voltage peak to energize reactor 1 (SR1) before the cable is energized. After 5 ms delay, the additional breaker, which is installed at the beginning of OHL1, will be closed at the zero voltage to energize the cable. At the next voltage peak after the cable energization, reactor 2 (SR2) is energized by closing breaker 4 and the other additional breaker, which is installed at the end of OHL4. Since both reactors are energized at the voltage peak, the zero-missing phenomenon is prevented.

Figure 7.73 shows the open-end voltage and the line breaker current after the energization of the 75% cable scenario (60 km cable) when this countermeasure is

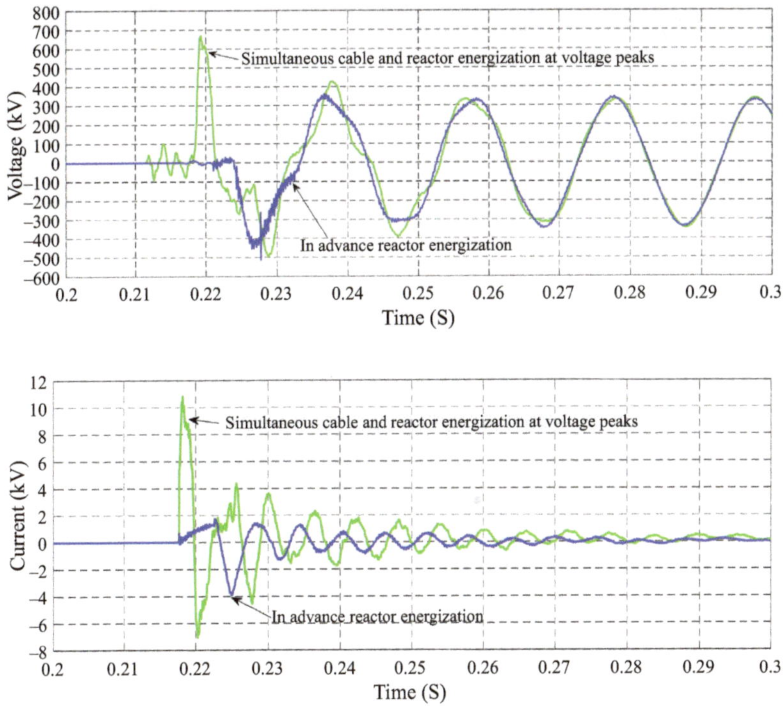

Figure 7.73 Open-end voltage (phase A) and line breaker current (I_{CB1}, phase A) after the implementation of Countermeasure 7.7.4.2.2 (with 5 ms delay) and Countermeasure 7.7.4.1 (75% cable scenario, K_{sh} = 92.1%)

applied. It is evident that the transient overvoltage and the inrush current are lower compared to those of the previous countermeasures.

Figure 7.74 shows the voltage step at the Krimpen substation after the energization of the 75% cable scenario with time delays of 5 ms, 65 ms, and 1.05 s between the cable and reactors energizations. Unlike Countermeasure 7.7.4.2.1 (in advance cable energization), the voltage step is not exceeding the 3% limit and is also far above the dip threshold (0.9 pu). The reason for not exceeding the 3% limit is the size of reactor 1, which according to Table 7.6 is between 34.9% (for the 15% cable scenario) to 47.6% (for the 100% cable scenario) of the total cable reactive power. Thus, in advance, energization of SR1 has a much smaller influence on the voltage than the case of cable energization in advance. The reason for not exceeding the dip threshold, in addition to the comparatively small size of SR1, is the high steady-state voltage before the energization (1.076 pu). As Figure 7.75

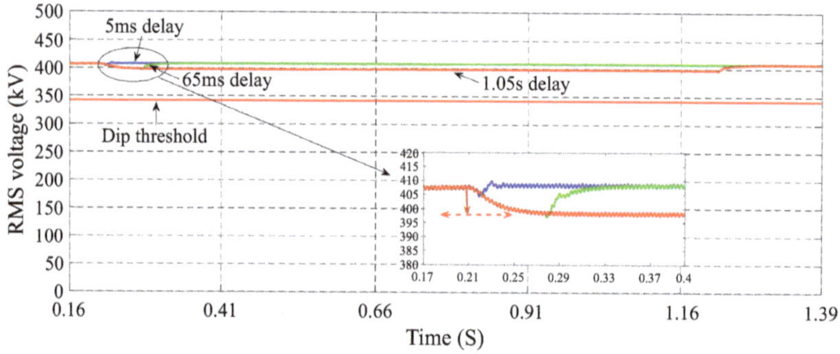

Figure 7.74 RMS voltage step at the Krimpen substation after the implementation of Countermeasure 7.7.4.2.2 with three different time delays (75% cable scenario, K_{sh} = 92.1%)

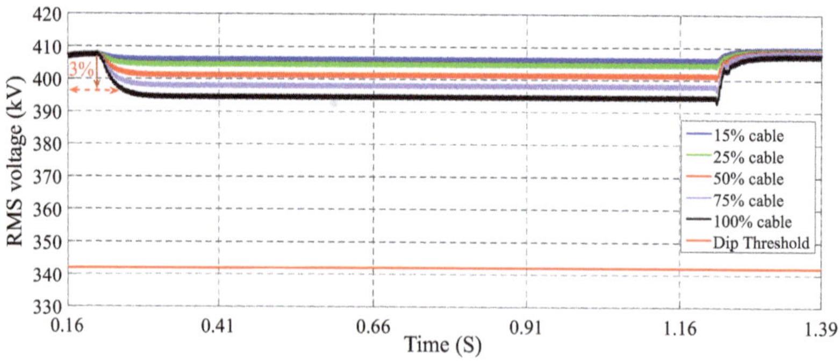

Figure 7.75 RMS voltage step at the Krimpen substation after energizing the six cable scenarios with Countermeasure 7.7.4.2.2. The time delay is 1.05 s

shows, the 3% limit is only exceeded when the cable share is 100% (80 km). It can be concluded that this countermeasure has less negative consequences than the previous measures in the circumstances that the pre-switching steady-state voltage is high.

7.7.4.3 Sequential switching

Sequential switching is a handling countermeasure, which applies remedial switching actions to protect the system in the case that a fault occurs in the circuit during the zero-missing phenomenon [50]. By applying this measure, the cable and reactors are energized together at the zero voltage (by the use of controlled switching and single-pole operated breakers) to minimize the transient overvoltages and the inrush currents.

When a fault occurs in the circuit during energization, the line breaker(s) of the faulty phase(s) can interrupt the current since it crosses the zero point. In fact, the fault current is superimposed on the AC-component of the current through the line breaker and therefore the total AC-component becomes larger than the DC-component. However, the line breaker(s) of healthy phase(s) cannot interrupt the current due to the zero-missing phenomenon.

In order to open the line breaker(s) of the healthy phase(s), the degree of compensation should be decreased to 50% or less. When the compensation degree is dropped to 50% or less, the AC-component of the line breaker current becomes larger than the DC-component, so the line current crosses the zero point. The decrease of the compensation degree to 50% is the preferred case; in fact, the leading current through the line breaker and overvoltages along the circuit are more reduced compared to the case in which the degree of compensation is lower than 50%. Moreover, to limit the overvoltages, it is preferable not to disconnect the reactors at the open-end.

The line breakers should be type-tested for higher capacitive (leading) current interruption capability since the capacitive current flowing through the line breaker can be large after the decrease of the compensation degree to 50% or less (e.g., for 60 km cable with 50% compensation, the capacitive current through the line breaker is about 1 kA). A drawback of this countermeasure is that the zero-missing problem is transferred to the next set of breakers when the line breaker of a healthy phase fails. Current differential relays are also required for realization of this countermeasure [50].

As an example of this countermeasure, assume that the Spaak connection with 75% cable (60 km cable) is energized together with the shunt reactors at the zero voltage. A single-line-to-ground (SLG) fault is applied at $t = 0.32$ s in phase B of OHL3 (see Figure 7.2). Figure 7.76 shows the current through the line breaker when the sequential switching countermeasure is applied. The breaker of phase B (CB1-B) is opened by the protection scheme 120 ms after the fault occurrence. The application of this countermeasure does not cause high transient overvoltages and inrush currents since the cable and reactors are energized at voltage zero.

It is worth mentioning that the currents through the reactor breakers of the faulty phase(s) are not crossing the zero point since reactors are discharging into the

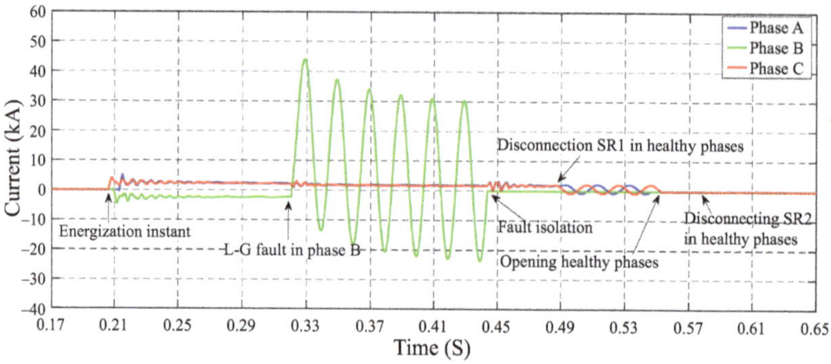

Figure 7.76 Line breaker current (I_{CB1}) when the sequential switching countermeasure is applied (75% cable scenario, $K_{sh} = 92.1\%$)

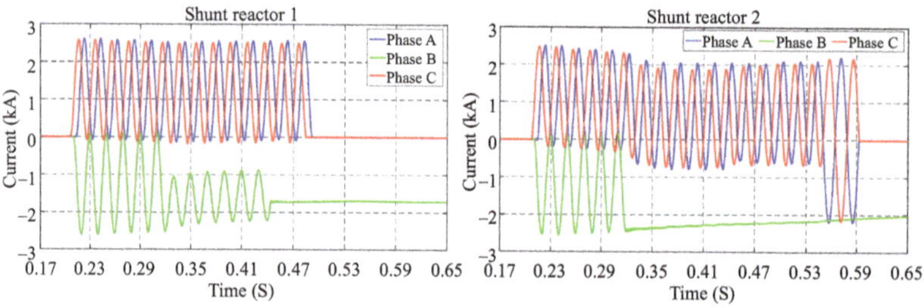

Figure 7.77 Current through shunt reactors 1 and 2 (I_{SR1} and I_{SR2}) when the sequential switching countermeasure is applied (75% cable scenario, $K_{sh} = 92.1\%$)

fault. This is shown in Figure 7.77. So, as a common practice, reactor breakers do not operate until the reactor currents cross the zero point [100].

7.7.4.4 Opening faulty phase(s)

In this handling countermeasure, only the line breaker(s) of the faulty phase(s) is/are opened when a fault occurs in the circuit. The healthy phase(s) is/are opened after the DC-offset is sufficiently damped and the line breaker current crosses the zero point.

The main benefit of this countermeasure is that it is cheap and only needs single-pole operated breakers, which are already commonly used in transmission systems. However, the waiting time for damping of the DC-component in healthy phase(s) can be up to several seconds due to the large time constant of shunt reactors. Leaving healthy phase(s) connected to the system may cause resonance in the disconnected phase(s) after the fault is removed [11]. In addition, applying this measure on cable projects near a generator may not be possible as this measure causes a prolonged unbalanced operation [50].

7.7.4.5 Increasing DC-offset damping

In the cable systems, the total X/R ratio is typically large, which is resulting in a low damped DC-offset. By this mitigation countermeasure, the X/R ratio and as a result the DC-offset damping time are decreased by using an additional resistor. In addition, the cable and reactors are energized together at the zero voltage to minimize the overvoltages and the inrush currents.

One way of adding the resistor is to use a breaker equipped with the pre-insertion resistor (closing resistor) for the line or shunt reactors [3,70]. The resistor (which its size can be in the order of the line surge impedance [100]) is connected through an auxiliary contact in parallel with the main contact (breaking chamber), as it is shown in Figure 7.78(a). The circuit is energized through the resistor for a few milliseconds (e.g., approximately 8–12 ms for ABB breakers [70]) before the closure of the arcing contact. These breakers are more expensive than ordinary breakers and controlled switching. The cost can be even much higher if a new breaker is needed to be developed.

As an alternative case, a separate resistor can be connected in series with an ordinary breaker. The resistor is bypassed by a parallel disconnector or breaker as it is illustrated in Figure 7.78(b). The breaker can be a lower voltage breaker with 380 kV insulation since it is not supposed to interrupt a fault current. The main advantage of this topology is the controllability of the resistor bypass time. However, an additional cost for the purchase of an extra breaker/disconnector and the resistor is imposed.

Figure 7.79 shows the influence of the pre-insertion resistor size on the DC-component of the line breaker current when each phase of the Spaak connection with 75% cable (60 km cable) is energized at zero voltage. The DC-component

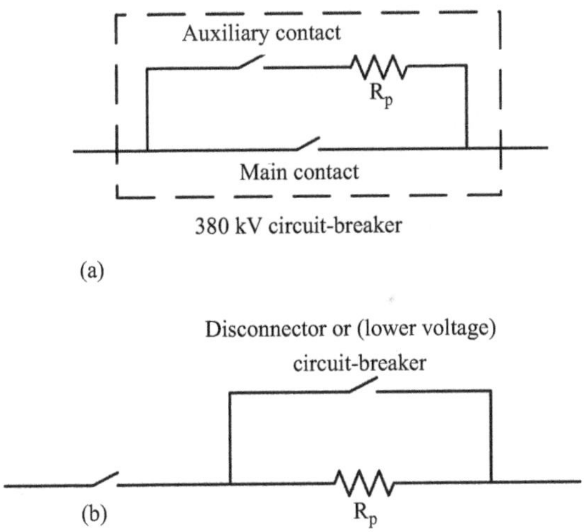

Figure 7.78 Use of resistors to mitigate the zero-missing phenomenon: (a) a circuit breaker equipped with the pre-insertion resistor and (b) a resistor in series with an ordinary circuit breaker

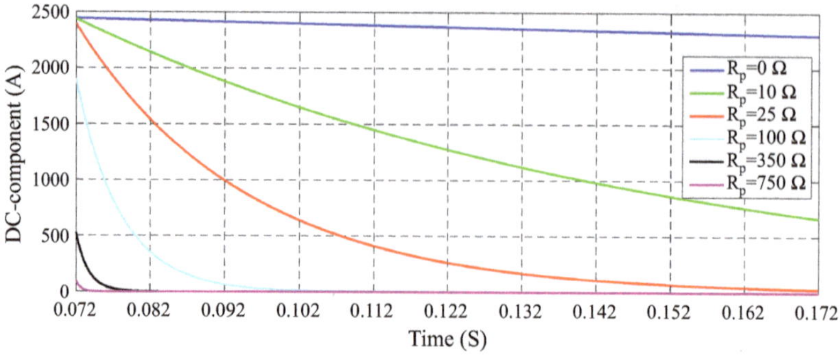

Figure 7.79 DC-component of the line breaker current (I_{CB1}) for different pre-insertion resistor sizes when each phase is energized at the zero voltage (75% cable scenario, K_{sh} = 92.1%)

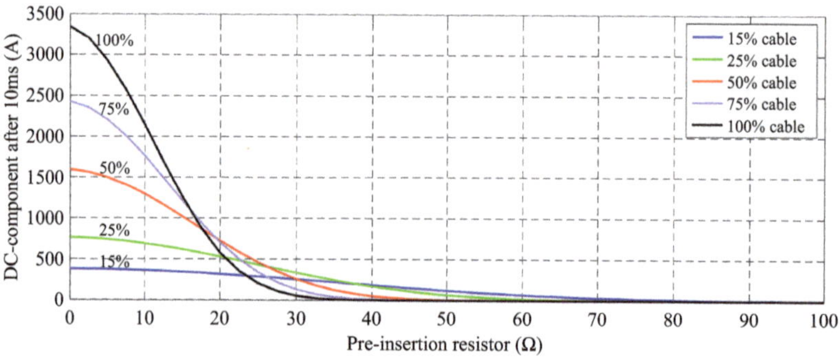

Figure 7.80 DC-component of the line breaker current (I_{CB1}) after 10 ms versus the size of the pre-insertion resistor for the six cable scenarios. Each phase is energized at the zero voltage

decreases faster with increasing the resistor size; however, it is crucial to find the optimum size of the resistor to increase the effectiveness and to minimize the cost. Large resistors act as an open-circuit and small resistors do not cause any improvement in the DC-offset damping.

Reference [99] proposed an iterative process to solve differential equations for the calculation of the pre-insertion resistor size. A formula based on energy equations (energy that the pre-insertion resistor should dissipate) is also proposed in References [98] and [99]. This formula gives an approximation of the resistor value without being required to perform the iterative process, although is not always accurate due to simplifications.

If a breaker equipped with the pre-insertion resistor (Figure 7.78(a)) is used, the resistor should be sized so that the zero-missing phenomenon disappears by the moment that the resistor is bypassed. Figure 7.80 shows the DC-component

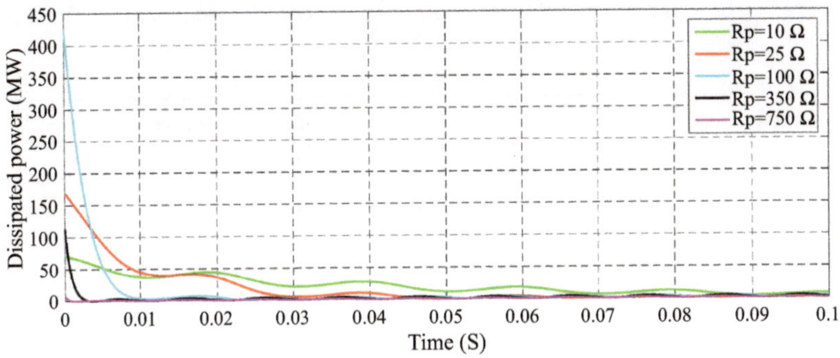

Figure 7.81 Dissipated power in different pre-insertion resistor sizes when each phase is energized at the zero voltage (75% cable scenario, K_{sh} = 92.1%)

after 10 ms (average of 8–12 ms [70,99]) versus the size of pre-insertion resistor for the six cable scenarios when each phase is closed at the zero voltage. When the resistor size is increased, the DC-component experiences a steeper drop for the scenarios with a higher cable share. This is due to the smaller X/R ratio of the hybrid circuits with higher cable share (as a result of the smaller X/R ratio of cables compared to OHLs), in which adding an additional resistance has a more remarkable impact.

The main limitation of this countermeasure is the thermal design of the resistor. The dissipated power in the resistor is shown in Figure 7.81. This countermeasure requires specially designed resistors to tolerate huge power dissipation for a couple of milliseconds. This imposes extra costs and complexity to the countermeasure.

7.7.5 Discussions

Several countermeasures of the zero-missing phenomenon were analyzed and compared. The applied methodology and findings are applicable to any network since the study is based on practical assumptions like the use of a hybrid OHL-cable circuit and the three system operation criteria. Moreover, simulation results were obtained by electromagnetic transient (EMT) studies on the model of an actual power transmission grid. It should be stressed that the influence of the system topology and parameters (like the configuration of hybrid OHL-cable circuit, power-frequency voltage, short-circuit power, etc.) on the effectiveness of a countermeasure should always be considered.

Table 7.16 presents a summarized comparison between the technical advantages and disadvantages of the investigated countermeasures. Some of the countermeasures need specially designed circuit breakers, control strategies, and resistances, which can impose extra costs and complexity on the system design and operation. The implementation of Countermeasures 7.7.4.2.1 (in advance cable energization) and 7.7.4.2.2 (in advance reactor energization) is more sophisticated than commonly

Table 7.16 Advantages and disadvantages of the countermeasures of the zero-missing phenomenon

Countermeasure	Advantages	Disadvantages
Simultaneous cable and reactors energization at the voltage peak (Countermeasure 7.7.4.1)	• Prevention countermeasure • Well-known by system operators • Easy to implement	• Requires controlled switching • Very high overvoltages and inrush currents for long cables • Breaker operation delay should not exceed the BCVT
In advance cable energization (Countermeasure 7.7.4.2.1)	• Prevention countermeasure • Minimized overvoltages and inrush currents for long cables	• Requires controlled switching • Requires a control mechanism to handle the switching sequence of the cable and reactors • Breaker operation delay should not exceed the BCVT[1] • Risk of voltage dip/swell for long cables
In advance reactor energization (Countermeasure 7.7.4.2.2)	• Prevention countermeasure • Minimized overvoltages and inrush currents for long cables	• Requires controlled switching • Requires a control mechanism to handle the switching sequence of the cable and reactors • Breaker operation delay should not exceed the BCVT[1] • Risk of voltage dip/swell for long cables • Requires two additional three-phase breakers
Sequential switching (Countermeasure 7.7.4.3)	• Minimized overvoltages and inrush currents for long cables • Implemented by some system operators	• Requires controlled switching • Handling countermeasure (no prevention) • Requires breakers with higher capacitive current interruption capability • Risk of transferring the zero-missing problem to the next set of breakers if the line breaker of a healthy phase fails
Opening faulty phases(s) (Countermeasure 7.7.4.4)	• Minimized overvoltages and inrush currents • Easy to implement	• Requires controlled switching • Handling countermeasure (no prevention) • Leaves the system unprotected for a very long time when the cable is long • Risk of resonance in the disconnected phase(s) • May not be possible to be utilized near a generator due to causing a prolonged unbalance operation

(Continues)

Table 7.16 (Continued)

Countermeasure	Advantages	Disadvantages
Increase DC-offset damping (Countermeasure 7.7.4.5)	• Lower overvoltages and inrush currents (minimized if controlled switching at zero voltage is applied) • Known by system operators	• Mitigation countermeasure (no prevention) • Requires pre-insertion resistors with a very high thermal design when the cable is long

(1) Only the breakers operating at the voltage peak should comply with the BCVT.

applied techniques like Countermeasure 7.7.4.1 (simultaneous cable and reactors energization at the voltage peak) since they require multiple control systems, which are coordinated together with a communication system. However, the advantages of these countermeasures over a traditionally used measure, in terms of resulting overvoltages and inrush currents, were validated in Figures 7.69 and 7.73, where the results were compared with those of Countermeasure 7.7.4.1.

7.8 Conclusions

During recent years, the utilization of long EHV underground cables in transmission grids has gained lots of attention among transmission system operators. This tendency is due to the widespread public and governmental support to limit the use of overhead transmission lines as they have visual effects, land occupation, possible environmental impacts, and property price decrement when they cross through (semi)urban or environmentally sensitive areas.

Building full or partial cable-based grid connections with long sections of underground cables, at least in sensitive locations like populated areas and national parks, can be a solution to remove or limit the disadvantages of OHLs. Grids with such connections are known as hybrid OHL-cable grids, where transmission connections are composed of OHLs in nonsensitive areas and underground cables in sensitive (and maybe nonsensitive) areas.

Although the utilization of long EHV cables[4] in transmission grids is very encouraging from societal and environmental perspectives, new challenges can arise mainly from the technical point of view. Cables are very different from OHLs from electrical, thermal, and mechanical perspectives and these differences may lead to several undesirable operating conditions, which increase the chance of damage to system components and reduce the reliability of the power supply.

[4]Cables that need reactive power shunt compensation are generally considered as long cables. For a three-phase 380 kV cable circuit with a single cable per phase, a long cable is associated to cables roughly longer than 10 km transmission length.

Therefore, it is required to have an in-depth knowledge and a set of guidelines for planning, designing, and operating of transmission systems with a large share of cables.

This chapter addressed the most important technical issues related to the operation of power systems composed of both OHLs and cables (hybrid OHL-cable systems). The research was conducted for a hypothetical 80 km double-circuit connection in the future Dutch 380 kV grid, called the Spaak project. For this purpose, accurate steady-state and EMT models of the entire Dutch transmission grid were used to ensure the maximum simulation accuracy. These models were thoroughly discussed in Section 7.2.

1. Section 7.3 studied the steady-state operation and the shunt compensation allocation of hybrid OHL-cable systems. This is a crucial issue in the design and impact study of cable projects as it leads to find the optimal size and location of the shunt reactors. According to the simulation results, it is necessary to investigate the influence of different load-flow scenarios on the allocation of shunt reactors. It was also discovered that both the minimum required compensation size and degree increase by increasing the cable length while a 100% compensation degree can be only needed for very long cables in systems operated at voltages higher than approximately 1.05 pu. For long cables, shunt reactors should be connected to the circuit and behind the line breakers as this limits post-switching steady-state energization overvoltages in the circuit and minimizes the capacitive current through the line breaker during energization. Furthermore, a distributed compensation along the circuit can decrease the minimum required size of compensation in comparison to the traditional type of compensation at the two remote ends of the circuit.
2. In Section 7.4, the technical issues related to the resonance behavior of hybrid OHL-cable systems were investigated. This is a very important issue as the higher risk of low-order resonance overvoltages raises serious concerns regarding the large-scale application of cables. In fact, the application of long cables can significantly decrease the resonance frequencies to low-order ranges. Based on the simulation results, the risk of resonance frequencies close to the second and third harmonics is substantially higher when long cables are operated in the grid. In addition, the resonance impedance magnitudes and the number of resonance frequencies increase when cables are applied.
3. Section 7.5 studied the energization overvoltages phenomenon in hybrid OHL-cable systems by a statistical analysis approach. Energization overvoltages in HV systems can reach very high values and are considered among the severest overvoltages stressing the insulation of system components. Therefore, the significance of these overvoltages should be precisely determined for the insulation coordination studies. By means of extensive statistical simulations, it was found that a hybrid OHL-cable circuit with a given transmission length produces energization overvoltages higher than those of a fully cable circuit and most likely lower than those of a fully-OHL circuit with the same transmission lengths. It should also be noted that energization overvoltages are very dependent on the number, locations, and lengths of the OHL and cable

sections. Other parameters like short-circuit power and presence of trapped charges on the cables have also considerable impact on the overvoltages.

4. Section 7.6 elaborated on the technical issues related to the de-energization of hybrid OHL-Cable systems as they are associated with several transient phenomena. These issues may become more severe when the circuit is fully or partially undergrounded with cables. It was found that when both the cable and shunt reactors are de-energized, the cable behavior is very similar to that of a parallel RC circuit. This means that the trapped voltage on the disconnected cable is a decaying DC-voltage, which discharges through the insulation resistance. In contrast, when a hybrid OHL-cable circuit is de-energized while shunt reactors have been left connected to the phases, the residual voltages on the disconnected phases are decaying oscillatory AC-voltages with superimposed frequencies instead of decaying DC-voltages.
5. Section 7.7 studied the zero-missing phenomenon in cable systems. The simultaneous energization of the cable and reactors may cause the zero-missing phenomenon, which means that the current through the line breaker does not cross the zero value for several cycles (even up to several seconds). In this case, it is difficult or even impossible to safely open the healthy phases if a fault occurs in the circuit during the energization when the zero-missing current is still present. Six countermeasures of the zero-missing phenomenon were introduced and analyzed. The simultaneous energization of the cable and reactors at voltage peak is a common prevention countermeasure; however, the resulting transient overvoltages and inrush currents should be carefully analyzed. Other countermeasures are sequential switching, opening of the faulty phase(s), and increasing the DC-offset damping by resistors. These methods result in lower transient overvoltages and inrush currents, however they require specific technical design and components as elaborated in Section 7.7.

References

[1] Burges, K., Bomer, J., Nabe, C. and Papaefthymiou, G. “Study on the comparative merits of overhead transmission lines versus underground cables.” Ecofys Germany GmbH, May 2008.

[2] CIGRE Working Group C4.502. “Power system technical performance issues related to the application of long HVAC cables.” CIGRE Technical Brochure no. 556, October 2013.

[3] Smeets, R., van der Sluis, L., Kapetanovic, M., Peelo, D., and Janssen, A. *Switching in electrical transmission and distribution systems*. Hoboken, NJ: Wiley; 2015.

[4] Bak, C. L., and da Silva, F. F. “High voltage AC underground cable systems for power transmission: A review of the Danish experience, Part 1.” *Electric Power Systems Research*. 2016;140: 987–994.

[5] da Silva, F. M. F. “Analysis and simulation of electromagnetic transients in HVAC cable transmission grids.” *Ph.D. Dissertation*, Department of Energy Technology, Aalborg University, Aalborg, Denmark, 2011.

[6] Gudmundsdottir, U.S. "Modelling of long high voltage AC cables in transmission systems." *Ph.D. Dissertation*, Department of Energy Technology, Aalborg University, Aalborg, Denmark, 2010.

[7] Benato, R. and Napolitano, D. "Overall cost comparison between cable and overhead line including the costs for repair after random failures." *IEEE Transactions on Power Delivery*. 2012;27(3): 1213–1222.

[8] Tokyo Electric Power Company, "Assessment of the technical issues relating to significant amounts of EHV underground cable in the all-island electricity transmission system." November 2011.

[9] Bak, C. L. and da Silva, F. F. "High voltage AC underground cable systems for power transmission: a review of the Danish experience, part 2." *Electric Power Systems Research*. 2016;140: 995–1004.

[10] Ametani,A., Nagaoka, N., Baba, Y., and Ohno, T. Power System Transients: Theory and Application, CRC Press, 2014.

[11] CIGRE Working Group C4.307. "Resonance and ferroresonance in power networks." CIGRE Technical Brochure no. 569, February 2014.

[12] Iliceto, F., Cinieri, E., and Di Vita, A. "Overvoltages due to open-phase occurrence in reactor compensated EHV lines." *IEEE Transactions on Power Apparatus and Systems*. 1984;PAS-103(3): 474–482.

[13] van der Sluis, L. *Transients in power systems*. Chichester, UK: Wiley; 2001.

[14] Grainger, J. J., and Stevenson Jr.,W. D. *Power system analysis*. Singapore: McGraw-Hill, Inc., 1994.

[15] Dommel, H. W. *Electromagnetic transients program manual (EMTP theory book)*. Portland, OR: Bonneville Power Administration; 1986.

[16] Schavemaker, P., and van der Sluis, L. *Power system essentials*. Chichester, UK: Wiley, 2008.

[17] Tleis, N. *Power systems modelling and fault analysis*. Oxford, UK: Newnes; 2008.

[18] CIGRE Working Group 02, Study Committee 33, "Guidelines for representation of network elements when calculating transients." CIGRE Technical Brochure no. 39, 2000.

[19] da Silva,F. F., Bak, C. L., and Holst, P. B. "Estimation of the required modeling depth for the simulation of cable switching in a cable-based network." *IEEE Transactions on Power Delivery*. 2012;27(4): 1902–1908.

[20] da Silva, F. F., Bak, C. L., and Holst, P. B. "Study of harmonics in cable-based transmission networks." in *44th* CIGRE Session. Paris, France, 2012.

[21] Morales, J., Mahseredjian, J., Sheshyekani, K., Ramirez, A., Medina, E., and Kocar, I. "Pole-selective residue perturbation technique for passivity enforcement of FDNEs." IEEE Transactions on Power Delivery. DOI:10.1109/TPWRD.2018.2810706, March 2018.

[22] Gustavsen,B., and Jeewantha De Silva, H. M. "Inclusion of rational models in an electromagnetic transients program: Y-parameters, Z-parameters, S-parameters, transfer functions." *IEEE Transactions on Power Delivery*. 2013;28(2): 1164–1174.

[23] Ubolli, A. and Gustavsen, B. "Multiport frequency-dependent network equivalencing based on Simulated time-domain responses." *IEEE Transactions on Power Delivery*. 2012;27(2): 648–657.

[24] Sheshyekani, K., R. Karami, H., Dehkhoda, P., Paolone, M., and Rachidi, F. "Application of the matrix pencil method to rational fitting of frequency-domain responses." *IEEE Transactions on Power Delivery*. 2012;27(4): 2399–2408.

[25] Ametani, A. *Numerical analysis of power system transients and dynamics*. UK: IET, Power and Energy Series 78; 2015.

[26] Manitoba HVDC Research Centre. "User's guide on the EMTDC analysis for PSCAD power system simulation." Winnipeg, Manitoba, Canada, 2010.

[27] Barakou, F., Bollen, M. H. J., Mousavi-Gargari, S., Wouters, P. A. A. F., and Steennis, E. F. "Downstream network modeling for switching transients in EHV networks containing cables." in *IEEE* Power Technology Conference. Manchester, UK, June 2017.

[28] IEEE Modeling and Analysis of System Transients Working Group, "Modeling guidelines for switching transients." Special Publication, Modeling and Simulation Working Group 15.08.

[29] Morched, A., Gustavsen, B., and Tartibi, M. "A universal model for accurate calculation of electromagnetic transients on overhead lines and underground cables." *IEEE Transactions on Power Delivery*. 1999;14(3): 1032–1038.

[30] Ametani, A. "A general formulation of impedance and admittance of cables." *IEEE Transactions on Power Apparatus and Systems*. 1980;99(3): 902–910.

[31] Wedepohl, L. M., and Wilcox, D. J. "Transient analysis of underground power transmission systems." *Proceedings of the Institution of Electrical Engineers (IEE)*. 1973;120(2): 253–260.

[32] Gustavsen, B. "Panel session on data for modeling system transients: Insulated cables." in IEEE Power Engineering Society (PES) Winter Meeting, 2001.

[33] Wu, L. "Impact of EHV/HV underground power cables on resonant grid behavior." Ph.D. Dissertation, Department of Electrical Engineering, Eindhoven University of Technology, Eindhoven, The Netherlands, October 2014.

[34] Steinbrich, K. "Influence of semiconducting layers on the attenuation behaviour of single-core power cables." *IEE Generation, Transmission and Distribution*. 2005;152(2): 271–276.

[35] Gustavsen, B., Martinez, J. A., and Durbak, D. "Parameter determination for modeling system transients—Part II: Insulated cables." *IEEE Transactions on Power Delivery*. 2005;20(3): 2045–2050.

[36] Ametani, A., Miyamoto, Y., and Nagaoka, N. "Semiconducting layer impedance and its effect on cable wave-propagation and transient characteristics." *IEEE Transactions on Power Delivery*. 2004;19(4): 1523–1531.

[37] "IEEE Guide for Bonding Shields and Sheaths of Single-Conductor Power Cables Rated 5 kV through 500 kV." IEEE Standard 575, 2014.

[38] CIGRE Working Group B1.19. "General guidelines for the integration of a new underground cable system in the network." CIGRE Technical Brochure no. 250, 2004.

[39] CIGRE Working Group B1.18. "Special bonding of high voltage power cables." CIGRE Technical Brochure no. 283, 2005.

[40] Nagaoka, N., and Ametani, A. "Transient calculations on crossbonded cables." *IEEE Transactions on Power Apparatus and Systems*. 1983;102 (4): 779–787.

[41] Ametani, A., Miyamoto, Y., and Nagaoka, N. "An investigation on a wave propagation characteristic on a crossbonded cable." *IEEJ Transactions on Power and Energy*. 2003;123-B(3): 395–401.

[42] Lafaia, I., Mahseredjian, J., Ametani, A., Correia de Barros, M. T., Kocar, I., and Fillion, Y. "Frequency and time domain response of cross-bonded cables." *IEEE Transactions on Power Delivery*. 2018;33(2): 640–648.

[43] Gudmundsdottir, U. S., Gustavsen, B., Bak, C. L., and Wiechowski, W. "Field test and simulation of a 400-kV cross-bonded cable system." *IEEE Transactions on Power Delivery*. 2011;26(3): 1403–1410.

[44] "Insulation Co-Ordination-Part 4: Computational Guide to Insulation Co-Ordination and Modelling of Electrical Networks." IEC 60071-4, 2004.

[45] Manitoba HVDC Research Centre. User's guide on the use of PSCAD. Winnipeg, Manitonba, Canada, 2010.

[46] Khalilnezhad, H., Popov, M., van der Sluis, L., Bos, J. A., de Jong, J. P. W., and Ametani, A. "Countermeasures of zero-missing phenomenon in (E)HV cable systems." *IEEE Transactions on Power Delivery*. 2018;33(4): 1657–1667.

[47] Gatta, F. M., Geri, A., Lauria, S., and Maccioni, M. "Steady-state operation conditions of very long EHVAC cable lines." *Electric Power Systems Research*. 2011;81: 1525–1533.

[48] Khalilnezhad, H., Popov, M., Bos, J. A., and Jansen, K. P. J. "Influence of partial undergrounding on the transient stability of EHV power transmission systems." *Electric Power Systems Research*. 2016;131: 126–138.

[49] Khalilnezhad, H., Chen, S., Popov, M., Bos, J. A., de Jong, J. P. W., and van der Sluis, L. "Shunt compensation design of EHV double-circuit mixed OHL-cable connections." in *IET* International Conference on Resilience of Transmission and Distribution Networks. Birmingham, UK, 2015.

[50] Ametani, A., Ohno, T., and Nagaoka, N. *Cable system transients: Theory, modeling and simulation*. Hoboken, NJ: Wiley, July 2015.

[51] Colla, L., Gatta, F. M., Geri, A., Lauria, S., and Maccioni, M., "Steady-state operation of very long EHV AC cable lines." in *IEEE* Power Technology Conference. Bucharest, Romania, June-July 2009.

[52] Gatta, F. M., and Lauria, S. "Very long EHV cables and mixed overhead-cable lines. Steady-state operation." in *IEEE* Power Technology Conference. St. Petersburg, Russia, June 2005.

[53] Colla, L., Gatta, F. M., Iliceto, F., and Lauria, S. "Design and operation of EHV transmission lines including long insulated cable and overhead section." in *IEEE* Power Engineering Conference. November-December2005.

[54] Dougherty, J. J., and Schifreen, C. S. "Long cable lines-Alternating Current with reactor compensation or direct current." *Transactions of the American Institute of Electrical Engineers. Part III: Power Apparatus and Systems*. 1962;81(3): 169–178.

[55] Gatta, F. M., Geri, A., Lauria, S., and Maccioni, M. "Steady-state operation conditions of very long EHVAC cable lines: Two case studies." *Electric Power Systems Research*. 2012;83: 160–169.

[56] Lauria, S., Gatta, F. M., and Colla, L. "Shunt compensation of EHV cables and mixed overhead-cable lines." in *IEEE* Power Technology Conference. Lausanne, Switzerland, July 2007.

[57] "Electromagnetic Compatibility (EMC)-Part 4-30: Testing and Measurement Techniques-Power Quality Measurement Methods." IEC 61000-4-30, 2014.

[58] "High-voltage Switchgear and Controlgear-Part 100: Alternating-Current Circuit-Breakers." IEC 62271–100, 2008.

[59] CIGRE Working Group B1.47, "Implementation of long AC HV and EHV cable systems." CIGRE Technical Brochure no. 680, March 2017.

[60] Colla, L., Lauria, S., and Gatta, F. M. "Temporary overvoltages due to harmonic resonance in long EHV cables." in International Conference on Power System Transients (IPST). Lyon, France, June 2007.

[61] Jensen, C. F. "Harmonic background amplification in long asymmetrical high voltage cable systems." in International Conference on Power System Transients (IPST). Seoul, Republic of Korea, June 2017.

[62] Akhmatov, V. "Excessive over-voltage in long cables of large offshore windfarms." *Wind Engineering*. 2006;30(5): 375–383.

[63] Karasaki, T., Goto, T., and Ametani, A. "An abnormal overvoltage due to load rejection on EHV underground transmission lines." in International Conference on Power System Transients (IPST). Lisbon, Portugal, 1995.

[64] Khalilnezhad, H., Popov, M., van der Sluis, L., Bos, J. A., and de Jong, J. P. W. "Influence of long EHV AC underground cables on the resonance behavior of the Dutch transmission system." in *IEEE* Power and Energy Society (PES) General Meeting. Boston, MA, USA, July 2016.

[65] Bollen, M., Ackeby, S., Jansson, H., and Jonsson, M. "Using transfer impedance to study harmonic resonance due to AC cables in a transmission system." in *CIGRE*. Lund, Sweden, 2015.

[66] "Insulation Co-Ordination-Part 2: Application Guide." IEC 60071-2, 1996.

[67] "Insulation Co-Ordination-Part 1: Definitions, Principles and Rules." IEC 60071-1, 2006.

[68] CIGRE Working Group 21/33, "Insulation co-ordination for HV AC underground cable systems." CIGRE Technical Brochure no. 189, June 2001.

[69] Hileman, A. R. *Insulation coordination for power systems*. New York, NY: Marcel Dekker:1999.

[70] *Live tank circuit breakers: Buyer's guide*. 6th ed. Ludvica, Sweden; ABB; 2014.

[71] CIGRE Working Group 13.02, "Switching overvoltages in EHV and UHV systems with special reference to closing and reclosing transmission lines." *Electra*. 1973;30: 70–122.

[72] CIGRE Working Group 13.05, "The calculation of switching surges." Electra. 1971;19: 67–78.

[73] CIGRE Working Group 13.05, "The calculation of switching surges-II. Network representation for energization and re-energization studies on lines fed by an inductive source." Electra. 1974;32: 17–42.

[74] CIGRE Working Group 13.05, "The calculation of switching surges-III. Transmission line representation for energization and re-energization studies with complex feeding networks." Electra. 1979;62: 45–78.

[75] Ohno, T., Bak, C. L., Ametani, A., Wiechowski, W. and Sorensen, T. K. "Statistical distribution of energization overvoltages of EHV cables." *IEEE Transactions on Power Delivery*. 2013;28(3): 1423–1432.

[76] Ohno, T., Ametani, A., Bak, C. L., Wiechowski, W., and Sørensen, T. K. "Analysis of the statistical distribution of energization overvoltages of EHV cables and OHLs." in International Conference on Power System Transients (IPST). Vancouver, Canada, July 2013.

[77] Mestas, P., and Tavares, M. C. "Relevant parameters in a statistical analysis-application to transmission-line energization." *IEEE Transactions on Power Delivery*. 2014;29(6): 2605–2613.

[78] Hamza, A. H., Ghania, S. M., Emam, A. M., and Shafy, A. S. "Statistical analysis of switching overvoltages and insulation coordination for a 500 kV transmission line." in 18th International Middle East Power Systems Conference. Cairo, Egypt, December 2016.

[79] Daud, M. Z., Ciufo, P., and Perera, S. "Statistical analysis of overvoltages due to the energization of a 132 kV underground cable." in 6th ECTI Conference, Pattaya. Thailand, May 2009, pp. 54–57.

[80] Martinez, J. A., Natarajan, R., and Camm, E. "Comparison of statistical switching results using Gaussian, uniform and systematic switching approaches." in IEEE Power Engineering Society Summer Meeting. Seattle, USA, July 2000, pp. 884–889.

[81] Itoh, Y., Nagaoka, N., and Ametani, A. "Transient analysis of a cross-bonded cable system underneath a bridge." *IEEE Transactions on Power Delivery*. 1990;5(2): 527–532.

[82] Khalilnezhad, H., Popov, M., Bos, J. A., de Jong, J. P. W., and van der Sluis, L. "Investigation of statistical distribution of energization overvoltages in 380 kV hybrid OHL-Cable systems." in International Conference on Power System Transients (IPST). Seoul, Republic of Korea, June 2017.

[83] Ibrahim, A. I., and Dommel, H. W. "A knowledge base for switching surge transients." in International Conference on Power System Transients (IPST). Montreal, Canada, 2005.

[84] Lafaia, I., Mahseredjian, J., Ametani, A., Correia de Barros, M. T., Kocar, I., and Fillion, Y. "Frequency and time domain response of cross-bonded cables." *IEEE Transactions on Power Delivery*. 2018;33(2): 640–648.

[85] da Silva, F. F., and Bak, C. L. *Electromagnetic transients in power cables*. London, UK: Springer, 2013.

[86] Jensen, C. F., Gudmundsdottir, U. S., Bak, C. L., and Abur, A. "Field test and theoretical analysis of electromagnetic pulse propagation velocity on crossbonded cable systems." *IEEE Transactions on Power Delivery*. 2014;29(3): 1028–1034.

[87] CIGRE Working Group B1.30, "Cable systems electrical characteristics." CIGRE Technical Brochure no. 531, April 2013.

[88] Greenwood, A. *Electrical transients in power systems*. Hoboken, NJ: Wiley, 1971.

[89] Benato, R., and Paolucci, A. "Operating capability of long AC EHV transmission cables." *Electric Power System Research*. 2005;75: 17–27.

[90] Ghassemi, F. "Effect of trapped charges on cable SVL failure." *Electric Power Systems Research*. 2014;115: 18–25.

[91] da Silva, F. F., Bak, C. L., and Holst, P. B. "Switching restrikes in HVAC cable lines and hybrid HVAC cable/OHL lines." in International Conference on Power System Transients (IPST). Delft, The Netherlands, June 2011.

[92] Lafaia, I., Ghassemi, F., Ametani, A., et al. "Experimental and theoretical analysis of cable discharge." *IEEE Transactions on Power Delivery*. 2017;32(4): 2022–2030.

[93] "IEEE Guide for the Application of Capacitance Current Switching for AC High-Voltage Circuit Breakers Above 1000 V." IEEE Standard C37.012, 2014.

[94] Bak, C. L., Wiechowski, W., Søgaard, K., and Mikkelsen, S. D. "Analysis and simulation of switching surge generation when disconnecting a combined 400 kV cable/overhead line with shunt reactor." in International Conference on Power System Transients (IPST), Lyon, France, June 2007.

[95] Bak, C. L., Baldursson, H., and Oumarou, A. M. "Switching overvoltage in 60 kV reactor compensated cable grid due to resonance after disconnection." in 8th International Conference on Electric Power Systems, High Voltages, Electric Machines (POWER '08), 2008.

[96] da Silva, F. F., and Bak, C. L. *Electromagnetic Transients in Power Cables*. London, UK: Springer, 2013.

[97] Gudmundsdottir, U. S., and Holst, P. B. "Solving zero-missing with cable energization at voltage peak, based on insulation coordination study results." in International Conference on Power System Transients (IPST), Vancouver, Canada, July 2013.

[98] da Silva, F. F., Bak, C. L., Gudmundsdottir, U. S., Wiechowski, W., and Knardrupgard, M. R. "Use of a pre-insertion resistor to minimize zero-missing phenomenon and switching overvoltages." in IEEE Power & Energy Society (PES) General Meeting, Calgary, Canada, October 2009.

[99] da Silva, F. F., Bak, C. L., Gudmundsdottir,U. S., Wiechowski, W., and Knardrupgard, M. R. "Methods to minimize zero-missing phenomenon." *IEEE Transactions on Power Delivery*. 2010;25(4): 2923–2930.

[100] *Live tank circuit breakers: Application guide.* 1.2 ed. Ludvica, Sweden: ABB; 2013.

Chapter 8
Conclusions

8.1 Chapter 2: Series impedance and shunt admittance

8.1.1 Section 2.1: Formulation of series impedance and shunt admittance

Section 2.1 describes the formulas of the series impedance and shunt admittance of a distributed-parameter line such as overhead cables (lines) and underground cables.

8.1.2 Section 2.2: Review of existing formulas of earth-return impedance and admittance

Section 2.2 reviews existing formulas of the earth-return impedance and admittance of overhead cables (lines) and underground cables which are derived under the assumption of quasi-TEM or TEM mode of wave propagation.

8.1.3 Section 2.3: Accurate and approximate earth-return impedance formulas for overhead cable (line) and Section 2.4: Accurate and approximate earth-return admittance formulas for overhead cable (line)

Section 2.3 describes accurate and approximate formulas of the earth-return impedance for overhead cables (lines) and Section 2.4 for an accurate earth-return admittance.

8.1.4 Section 2.5: Accurate and approximate earth-return impedance formulas for underground cable and Section 2.6: Accurate and approximate earth-return admittance formulas for underground cable

Section 2.5 explains accurate and approximate formulas of the earth-return impedance for underground cables and Section 2.6 for an accurate earth-return admittance. Also, the derivation of the formulas is described in detail.

8.1.5 Section 2.7: Derivation of the modified earth-return Green function for MoM-SO technique

Section 2.7 describes the derivation of a modified earth-return Green function which is essential in the Method of Moments – Surface Admittance Operator

(MoM-SO) technique to remove possible numerical instabilities for overhead cables (lines).

8.1.6 Section 2.8: Comparison of calculated impedance and admittance by different methods

Section 2.8 shows a comparison of the earth-return impedance and admittance calculated by extended and classical TL approaches. Based on the comparison, the following remarks are made clear:

(a) Overhead line

1. The earth resistivity influences the resistance of the earth-return impedance and shunt conductance in a high-frequency region, while it influences the inductance in a low-frequency region. The shunt capacitance noticeably depends on the earth's resistivity in a middle-frequency region. Note that the classical TL approach cannot consider the conductance and the frequency-dependent capacitance.
2. The effect of the earth's permittivity is noticeable in a high-frequency region.
3. The original MoM-SO shows a numerical instability in a high-frequency region because of assuming uniform current distribution in the Green function.

(b) Underground cable

1. The influence of the earth resistivity and permittivity on the earth-return resistance is less noticeable in underground cables than that in overhead lines.
2. The earth's resistivity influences the inductance in a wide range of frequencies.
3. The earth resistivity and permittivity influence the conductance in a rather high-frequency region, while these influence the capacitance in a rather wide range of frequencies.

(c) The influence of the integral reference and various formulas is made clear.

8.1.7 Section 2.9: Impedance and admittance at f = 0 for HVDC line

Section 2.9 proposes a distortion-less line concept to evaluate the impedance and admittance of an HVDC line, and investigates the basic electrical parameters of the line. The following remarks are made clear.

1. Carson's earth-return impedance cannot be used to study extremely low-frequency phenomena in an HVDC line as well as very high-frequency phenomena.
2. A conventional approach even with accurate earth-return impedance and admittance formulae shows that the propagation velocity converges to zero and the characteristic impedance tends infinity.

3. A distortion-less line model is proposed to evaluate the conductance from known values of the DC resistance, space capacitance, and inductance which is given as a converged value of the accurate earth-return inductance.
4. From a monitored voltage in an HVDC transmission line, the attenuation constant of the line is calculated. Then, the conductance and inductance of the line are easily estimated at various locations along the line.
5. The frequency responses of the propagation-related parameters in a low-frequency region are calculated using the proposed approach. The results are physically more reasonable than those calculated by the conventional approach.

Section 2.10 explains theoretical formulations of external electromagnetic field components generated from a multi-phase overhead line and underground cable.

8.2 Chapter 3: Modeling of cables

Chapter 3 explains the modeling of cables by the well-known simulation techniques which are widely used and implemented into various EMT-type simulation tools. The modeling methods based on numerical electromagnetic analysis (NEA) approaches are also discussed.

8.2.1 Section 3.1: Transmission line models in EMT-type simulation tools

In Section 3.1, the typical models of overhead lines and underground cables prepared in existing EMT-type simulation tools are reviewed and described.

8.2.2 Section 3.2: Modeling of frequency-dependent soil parameters

Section 3.2 reviews existing frequency-dependent (FD) soil models. Also, the calculated results of FD soil resistivity and permittivity are investigated and compared with independently measured data. The following characteristics are observed and summarized.

(a) Characteristics of FD soil models
  1. The values of FD soil resistivity and FD soil relative permittivity decrease significantly as frequency increases.
  2. The FD soil relative permittivity show much larger values at a low frequency, that is, $f < 10$ kHz than that at a high frequency.
  3. As the resistivity at 100 Hz (ρ_0) increases, the FD soil relative permittivity decrease.
  4. The FD soil relative permittivity calculated by Longmire/Smith (LS) model and Scott (SC) model have minor differences.
  5. As frequency increases, the differences between the constant soil (CS) resistivity ρ_{cs} and the FD soil resistivity increase.
  6. The differences between the CS resistivity and the FD soil resistivity are more significant for the large values of ρ_0 and ρ_{cs}.

(b) Comparison with measured data

The FD resistivity and relative permittivity calculated by various FD soil models are compared with the independently measured data. The FD electrical parameters calculated by the LS, Alipio/Visacro (AV), and SC models match qualitatively with the independent wide-band measured data of soil, and the results of LS model show a better agreement than the other two models.

8.2.3 Section 3.3: Various cable installation

Section 3.3 explains the modeling of a cable installed underneath a steel bridge, a tower-installed cable, a tunnel-installed cable, a cable installed in an HDPE tube, and a submarine cable. Based on the investigations, the following remarks are made clear.

(a) Tower-installed cable

Based on experimental and theoretical observations together with EMT simulation results, the following remarks have been obtained when lightning strikes a tower where a power cable is installed.

1. Because of the short distance between a tower structure and a power cable, the self-surge impedance of the tower is nearly the same as the mutual impedance between the tower and the cable and thus no significant voltage difference appears between the tower and the cable. Also, the cable core voltage is nearly the same as the cable sheath voltage due to the electrostatic shield effect of the cable metallic sheath.
2. The cable voltage is determined by electromagnetic coupling between a tower and the cable when the cable sheath at the tower top is grounded through a grounding lead. When the sheath is grounded to a tower arm at the top, the cable voltage is determined by conductive coupling, that is, the sheath voltage becomes nearly the same as the arm voltage. In general, the conductive coupling being greater than the electromagnetic coupling, the cable voltage to earth becomes higher in the case of the sheath is grounded to the arm.
3. The voltage difference between the tower and the cable sheath becomes the largest at the tower bottom, and is found to be greater in the case of the sheath being grounded through a grounding lead. The phenomenon may be a result of a lower voltage of the cable at the tower top.
4. It is suggested to adopt an impedance of a vertical conductor to simulate a lightning surge on a cable installed along with a tower, although a conventional approach of adopting a horizontal conductor equivalence is very convenient.

(b) Cable installed in an HDPE tube

This section also has dealt with a transient simulation on a cable installed in high-density polyethylene (HDPE) tube which is eccentric to the cable with an air gap. The impedance and the capacitance of the HDPE tube installed cable have been evaluated by COMSOL considering the real geometry of the system and by EMT-type simulation tools neglecting the HDPE tube and the air gap.

1. No significant difference in the impedance is found between the calculated results by COMSOL and the EMT-type simulation tools.
2. On the contrary, the capacitance of the cable is found to be far smaller when the air gap and the tube are taken into account.
3. An equivalent insulator to the HDPE tube, the air gap, and the cable outer insulator has been presented. The outer radius of the equivalent insulator is determined, and the equivalent insulator becomes concentric to the cable. The equivalent permittivity is calculated from the total capacitance of the HDPE tube, the air gap, and the cable sheath insulator. By adopting this equivalent insulator, the cable capacitance calculated by the EMT-type simulation tools becomes nearly the same as that calculated by the COMSOL.
4. A transient on an underground cable installed in the HDPE tube is simulated by the EMT-type simulation tools with this model and the results show reasonable agreement with a measured result on the cable. A further improvement is necessary, but it requires information on the unknown parameters such as the earth resistivity, grounding resistances of the impulse generator, and so on.
5. Finally, the effect of the HDPE tube and the air gap on a cable transient is examined and it has been made clear that the air gap affects the transient significantly because its permittivity is far smaller than that of the HDPE tube and the cable outer insulator.

8.2.4 Section 3.4: Cable bonding

Section 3.4 explains the details and methods of cable bonding. Various bonding methods, such as a normal-bonded cable, a normal cross-bonded cable, and a transposed cable, are made clear.

8.2.5 Section 3.5: Numerical electromagnetic analysis

Numerical electromagnetic analysis (NEA) methods such as Finite Element Method (COMSOL), Method of Moments (NEC), and Finite-Difference Time-Domain (FDTD) are explained and reviewed in Section 3.5. The NEA methods are very powerful and useful when the impedance and admittance of a cable are hardly calculated by existing EMT-type simulation tools, for example, a phenomenon involving a non-TEM mode of wave propagation and a three-dimensional phenomenon.

8.3 Chapter 4: Wave propagation characteristics of overhead and underground cables

In this chapter, the wave propagation characteristics of overhead cables (lines) and underground cables are calculated based on extended and classical TL approaches. In general, the following characteristics of propagation constants and external electromagnetic field components are observed.

(a) Overhead cables (lines) and underground cables

1. Sommerfeld-Gaubau wave propagation can be observed at high frequencies for an overhead line and an overhead cable. This characteristic shows mode transition from a low-frequency earth-return wave to a high-frequency surface wave.
2. A similar mode transition of the axial electric field component can be observed at high frequencies on an overhead conductor.
3. In the surface wave propagation, the earth resistivity and permittivity have no influence on attenuation constants and axial electric field components.
4. A significant difference of modal propagation constants on underground cables is observed between the extended and classical TL approaches. The attenuation constant and phase velocity of the earth-return and inter-sheath modes increase as frequency increases. Such phenomena cannot be discussed in calculated results by the classical TL approach.
5. The cable configuration, such as flat, vertical, or trefoil arrangement, shows a minor influence on the modal propagation constants of underground cables.
6. The propagation constant of the coaxial mode on overhead and underground cables is independent of earth-return parameters.

(b) External electromagnetic field components generated by a multiphase underground cable

The external electromagnetic field components generated by a multiphase underground cable are investigated and studied based on excitation currents of uni- and bi-directional propagation. The formulation using bi-directional current propagation allows the considerations of terminating conditions and length of underground cables on the calculations of external electromagnetic field components. Also, the proposed formulation can be applied to arbitrary cable arrangements, such as flat, vertical, and trefoil configurations.

Considering a typical three-phase underground cable, the lateral and total external electromagnetic field components at the various field observation points using the two-phases and single-phase energizations are investigated and analyzed. The following general conclusions are summarized.

1. The Phases *a*–*b* and Phase *b* energizations give nonsymmetrical and symmetrical distributions of external electromagnetic field components in respect of the center of the flat cable arrangement. The external electromagnetic field components evaluated by NEC agree well with the results calculated by the analytical solutions.
2. If the receiving end of the cable is loaded by the non-characteristic impedance of the cable, the resonant phenomena are observed in external electromagnetic field components due to the mismatch of termination. The load resistance at the receiving end of the cable has major influences on the amplitude of the resonance. The source inductance generally shows negligible influences on the components of external electromagnetic field components.

3. The cable length has significant influences on the resonance and the resonant frequency of the external electromagnetic field components. The first resonant peak shifts to a lower frequency as the cable length increases. Also, the earth's resistivity shows important influences on external electromagnetic field components.
4. In general, the influences of FD soil models on external electromagnetic field components mainly occur at the high-frequency region together with the large earth resistivity. Also, the FD soil model produces more resonances on external electromagnetic field components with different load conditions in a high-frequency region than the results evaluated by the CS model.

(c) Cross-bonded cable

A comprehensive study of frequency domain responses for HVAC cross-bonded cables has been performed. The results explained in this section suggest the reader for making appropriate choices for modeling cables within computing time and accuracy constraints.

1. It is shown that cable inter-sheath modes induced because of cross-bonding, affect the propagation characteristics of inter-core modes.
2. Three alternatives for modeling cross-bonded cables with major sections have been described and analyzed. Modeling cross-bonded systems through homogeneous equivalents become inaccurate in the frequency domain when the frequency increases. This is because it cannot represent the inter-sheath modes.
3. In time-domain studies, the models based on homogeneous equivalents are inaccurate in sheath voltage results.
4. Detailed modeling of the first and last major sections does not improve accuracy substantially and limits the numerical integration time step.
5. Accuracy becomes very important when designing cable protections (i.e., sheath voltages limiters) and in such cases, full discrete modeling of each major section should be used.
6. The full discrete model is the most accurate, but computing times can become an issue, due to large surrounding networks and/or small numerical integration time-step requirements for short section lengths of such a model. In that case, it is acceptable to use the homogenous equivalents when considering only the core voltage results.

8.4 Chapter 5: Steady-state and transient characteristics on three-phase cables

In this chapter, the charging and discharging characteristics of cables are explained in Section 5.1. In Section 5.2, the EMT simulations on an RTE 225 kV cable system are described in comparison with field test results. Section 5.3 deals with switching surges on underground cables using extended and classical TL approaches. Section 5.4 investigates very-fast transients (VFTs) in gas-insulated substations (GISs). In Section 5.5, a difference of EMT simulation results between FD and CP line models is discussed in comparison with experimental results in a proto-type GIS. Also, the FDTD computations and EMT simulation results are compared and discussed.

8.4.1 Section 5.1: Cable discharge

The cable charging and discharging phenomena have received new consideration from utilities that are using high-voltage (HV) cables for voltage control during periods of light load. This section has investigated discharge characteristics of 275 kV POF cables based on laboratory and field tests, theoretical analysis, and EMT simulations. The following remarks are made clear:

1. The natural discharge of cable systems can be explained using simple formulas of parallel *RC* circuits. This has been verified from the simulation of field test results for two cable systems.
2. The factors affecting the discharge time are the cable capacitance, which is easily found in manufacturer catalogs, the leakage resistance of cable insulation, and external components, such as insulator poles, bushing, and cable sealing ends. Leakage in insulators exposed to air is highly affected by temperature, humidity, and contamination.
3. Leakage in the cable insulation and external components can be estimated based only on the voltage level, types of cable insulation (EPR, XLPE, and paper), insulator material (composite and ceramic), and the number of parallel insulator poles connected to the cable terminations. Estimation of leakage and knowledge of cable capacitance directly provide a sufficiently accurate estimate of the discharge time.
4. The discharge time constants observed in the two field tests on 275 kV POF cables of 20 km range from 1.5 h to 3.5 h. The time to discharge the cable to a safe voltage level (25 V according to IEC 60479-1) is in theory ten times the discharge time constant.
5. The leakage resistance of POF cables is about 74.5 GΩ-km (3.7 GΩ for 20 km). For XLPE insulated cables, the value is 1/10 of that. The leakage resistance of 275 kV insulators is about 100 (porcelain) to 1600 MΩ (EPDM) per element. For silicone insulators, this value can rise to 26.5 GΩ.

8.4.2 Section 5.2: Field measurement of cable transients and EMT simulation

This section describes field measurement of cable transients: (1) equipment and instruments required for the measurement, (2) conditions of a test site, and (3) time schedule. Then, measured results of switching surges are explained on one minor section, that is, a homogeneous cable, and on a 64 km cross-bonded cable. The field tests were performed on the RTE 225 kV cable system with a total length of 64 km. Because the cable is installed in a hard polyethylene (HDPE) tube which makes the cable configuration eccentric, it is necessary to find the equivalent cylindrical cable with the equivalent radii and permittivity. The details are explained in Section 3.3.4.

Based on the equivalent cable, the impedance and admittance are calculated from manufacturer data by using the Cable Constant routine of an EMT-type simulation tool, and calculated results are compared with *RLC* parameters evaluated from field tests. This approach provides a basis for model validation.

The cable is modeled using a frequency-dependent line model and the EMT simulation results are in sufficiently good agreement with field test results. The most important differences are observed for the excitation of the earth-return mode where the actual test setup conditions are more difficult to match.

8.4.3 Section 5.3: Switching surges on underground cables using extended and classical TL approaches

Transient simulations on underground cables including a field test are performed and investigated in Section 5.3. A significant difference of waveforms for inter-sheath and earth-return modes is observed between the extended and classical TL approaches. No difference is observed for waveforms of the coaxial mode on a cable between the two approaches. Moreover, earth resistivity shows an important effect on the transient waveforms of inter-sheath and earth-return modes. The effect of earth permittivity is negligible. The cable arrangement shows a minor influence on core and sheath voltages. Furthermore, the extended TL approach is validated with a 110 kV cross-bonded cable. It is confirmed that the extended TL approach results in higher damping and smooth convergence to the steady-state compared with those calculated by the classical TL approach.

8.4.4 Section 5.4: Very fast transient (VFT) in gas-insulated substation (GIS, overhead cable)

In Section 5.4, the field test results of VFTs are summarized considering electromagnetic disturbances in GIS control circuits which occasionally result in power system operation failures. Numerical instabilities are observed in VFT simulation results by using existing EMT-type simulation tools, and the reason for the instability is made clear. Then, the modeling of GIS elements for EMT simulations and the effects of various parameters of the GIS are investigated in detail. Based on the investigations, the following remarks are made clear.

1. The oscillating frequencies of VFTs generated by disconnector and circuit breaker operations in GISs range from 1 MHz up to 140 MHz in GIS high voltage circuits, and from 2 MHz to 80 MHz in low voltage control circuits.
2. Numerical instabilities observed in EMT simulation results are caused by the space admittance of an overhead line implemented into the existing EMT-type simulation tools. No instability appears when the space admittance is replaced by the accurate earth-return admittance.
3. The VFT overvoltages in the test results range from 1.7 pu to 3.0 pu in the GIS main circuits, and from 0.1 pu to 0.7 pu in the metal enclosures (pipe or tank). In the low voltage control circuits, 20 V to 700 V are observed in the field test results.

(a) The VFT oscillating frequency is, in principle, proportional to the total length of buses, DSs, and CBs in the GIS. However, the frequency is decreased easily less than a half by stray capacitances of transformers and bushings in the GIS, because those are far greater than the bus capacitance.
(b) Although the pipe grounding reduces the pipe voltage significantly as easily estimated, the core voltage tends to increase as the grounding resistance becomes smaller. The grounding causes only a minor influence on the oscillating frequency.
(c) The position of an operating dis-connector causes a very significant effect on the VFT over-voltages and oscillating frequency. The distance to the open-circuited remote end from the DS is short, and the overvoltage and frequency become very high.
(d) The effect of spacers is similar to that of a stray capacitance, but the effect is less noticeable for the smaller capacitance of the spacers.
(e) The effect of charging conditions is less noticeable than the effect of an operating DS.

4. When the VFT in a control cable is simulated, a VT and a CT are to be modeled in detail, and mutual coupling between the cable and grounding mesh of the GIS is to be considered because the mesh voltage cannot be assumed zero.
5. The analytical formulas explained in this section show a satisfactory agreement with the EMT simulation results.

The investigations in this section might contribute a further understanding of VFTs, and the EMT simulation results can be used as a reference because those are calculated by a frequency-dependent line model with the accurate earth-return impedance and admittance. Also, the analytical formulas are expected to be useful for a rough estimate of the VFT characteristics.

8.4.5 Section 5.5: EMT simulation in comparison with experimental and FDTD computed results

In Section 5.5, the EMT simulation results by different line models prepared in the existing EMT-type simulation tools are compared with experimental results in a proto-type GIS. Furthermore, a three-dimensional VFT computation is performed by FDTD together with comparisons of EMT simulation results.

1. In EMT simulations of the VFT, both FD and CP line models show a reasonable agreement with the test results, because the VFT frequency is very high and the earth resistivity causes only a minor influence. The pipe grounding and lead wires have to be carefully represented.
2. FDTD computed results agree very well with those by EMT-type simulation tools, although the EMT results are more oscillatory. The reason for this might be the effect of radiation losses which are automatically included in the FDTD computation. A main disadvantage of the FDTD is a large amount of computer resources.

8.5 Chapter 6: Transients in the interconnected EHV cable network in Denmark

This chapter studied temporary overvoltages and slow-front overvoltages on the Kyndbyværket–Asnæsværket cable line in Denmark.

8.5.1 Section 6.1: Background of EHV cable network in Denmark

In order to mitigate climate change, many countries have set a target to reduce their greenhouse gasses. Due to its abundant natural resources (wind) and green-minded people, Denmark is one of the leading countries in this effort of the world to tackle global warming. As of the year 2011, the wind energy accounted for 29.1 % in generation capacity and 28.1 % of the electrical production.

This change of the generation profile has led to and will continue to lead to the necessity of major upgrades in the Danish transmission grid. Especially, the expansion of Horns Rev, the world's largest offshore wind farm, will require the ability of the Danish transmission grid to transmit increased power from west to east where the electricity is consumed.

In Denmark, receiving public and political pressures to underground its OHLs, Energinet.dk published a report on the future expansion and undergrounding of its transmission grid on the 3rd of April 2008. The report proposed and compared five principles. From the five principles, the Danish government has selected Principle C in which all new 400 kV lines will basically be undergrounded.

8.5.2 Section 6.2: Model setup

The entire 400 kV network in Zealand is included in this model for the temporary overvoltage analysis and the slow-front overvoltage analysis. An example of the simulation model is created in ATP-Draw, and also is set up in PSCAD.

The simulation model includes the following four 400 kV cables:

- Al 1,600 mm^2 XLPE cable (Al sheath)
- Al 400 mm^2 + Cu 1,600 mm^2 XLPE cable (Pb sheath)
- Cu 1,000 mm^2 LPFF cable (Pb sheath + Al armor)
- Cu 1,000 mm^2 LPFF cable (Pb sheath)

The first cable type is for the Kyndbyværket–Asnæsværket line. Additionally, this type was applied to the short section of the Hovegård–Kyndbyværket line near the Kyndbyværket substation, where the overhead line is designed as 132 kV. This short section (1.99 km) has to be replaced by 400 kV cables when the Hovegård–Kyndbyværket line is operated at 400 kV.

The setup of cable models requires the modeling of cross-bonding wires, grounding wires, grounding conditions, and other auxiliary components which are explained in this section.

8.5.3 Section 6.3: Temporary overvoltage

Temporary overvoltages are the greatest concern when studying long EHV AC cables. Because of the large charging capacity and large shunt reactors for the compensation, the natural frequencies of a network with the long EHV AC cables tend to be much lower than those of a conventional network without the EHV cables. As such, in conventional networks, switching operations such as cable energization and fault clearance are considered as causes of slow-front overvoltages, but they need to be considered as causes of temporary overvoltages in a network with long EHV AC cables.

In this section, the series resonance overvoltage caused by the cable energization was studied. The most severe switching scenarios were first identified. The simulations were performed with the most severe assumptions of the Kyndbyværket 132 kV network and the Asnæsværket 132 kV network. The simulations show that the series resonance overvoltages in the assumed most severe conditions are much lower than the withstand voltages as long as the withstand voltage 230 or 275 kV is selected. The phase to phase overvoltage decays below 230 kV (peak) after one cycle.

Also, Section 6.3 studied the parallel resonance caused by transformer inrush currents as the most severe scenario. Among the potential harmonic current sources, an inrush current caused by the transformer energization is considered to be the most onerous one because of its high harmonic contents, low frequency, low damping, and long duration. Severe overvoltages were observed with the most severe scenarios, but they were within withstand voltages of related equipment.

Section 6.3 also studied the overvoltage caused by the system islanding. The faults on the Kyndbyværket, Asnæsværket, and Torslunde 400 kV buses were studied with different fault current levels, different shunt compensation degrees, and different timings for the fault clearing. The study showed that the overvoltages were most severe when a fault occurs on the Asnæsværket 400 kV bus, but they were within withstand voltages of related equipment.

8.5.4 Section 6.4: Ground fault and fault clearing over-voltages

The idea of the statistical evaluation is also applied to the evaluation of the ground fault and fault clearing overvoltages since the fault location and the fault timing vary and cannot be controlled. Instead of considering them as random phenomena, it is a common practice to cover different fault locations and fault timings with fixed (predefined) steps.

This section studied the statistical distributions of ground fault and fault clearing overvoltages of the Kyndbyværket–Asnæsværket line. The maximum overvoltage found in the analysis was 1.37 pu, which was much lower than SIWV 1,050 kV (3.2 pu). Effects of dummy source impedances were not significant.

8.6 Chapter 7: Steady-state and transient behavior of hybrid overhead line-underground cable networks in the Netherlands

8.6.1 Section 7.1: Background of EHV hybrid OHL-cable networks in the Netherlands

Years ago, the Dutch government passed a law to limit the total length of OHL networks. This law obliges the system operator, TenneT, to keep the total length of the OHL network constant. This means that for the construction of a new OHL at a given location, an existing OHL with the same length at another location of the grid has to be undergrounded so that the total OHL length in the grid remains unchanged. As a result, in 2009, TenneT began an extensive research program to study the possibilities of underground cables application in future transmission projects. The 10-year 380 kV cable research program covered major technical issues associated with long EHV cables in order to increase the knowledge and experience of TenneT in the design and operation of hybrid OHL-cable transmission systems.

Within the Dutch cable research program, several researches have been conducted on 380 kV hybrid OHL-cable grids, including studies on steady-state operation, dynamic stability, harmonic resonance behavior, electromagnetic transients, and system reliability. In this chapter, the findings of the steady-state operation, harmonic resonance behavior, and transient behavior were discussed.

8.6.2 Section 7.2: Grid modeling

This section elaborated on the research approach and the Dutch grid modeling, where the case study project and the defined cable scenarios for this project were explained. Moreover, this section discussed the developed models of the Dutch transmission grid for simulation and analysis of steady-state and transient phenomena.

8.6.3 Section 7.3: Shunt compensation allocation in Hybrid OHL-cable systems

The steady-state operation and the shunt compensation allocation of hybrid OHL-cable systems is a crucial issue in the design and impact study of cable projects as it leads to find the optimal size and location of the shunt reactors. Various technical problems are associated with the reactive power of long cables, where these problems should be prevented or mitigated by an optimal shunt compensation of the reactive power. The most significant problems are the decrease of the cable transfer capacity, steady-state voltage rise, under-excitation of synchronous generators (in severe cases self-excitation), and large capacitive currents in the line circuit breakers.

1. According to the simulation results, it is crucial to investigate the influence of different load-flow scenarios on the allocation of shunt reactors. The results could significantly vary from one load-flow to another as the short-circuit

power and voltage levels change. Moreover, the most decisive sizing criterion is dependent on the load-flow scenario, the cable length, and the shunt reactor's location. It was also discovered that both the minimum required compensation size and degree increase by increasing the cable length while a 100% compensation degree can be only needed for very long cables in systems operated at voltages higher than approximately 1.05 pu. Moreover, it is necessary to investigate the risk of cable overloading in case of parallel operation with OHLs or operation in a meshed network. A larger amount of power tends to flow through the cable connections due to the lower series impedance of cables compared to OHLs. The cable overloading can be avoided or mitigated by countermeasures like the use of phase-shifting transformers or series reactors.

2. For long cables, shunt reactors should be connected to the circuit and behind the line breakers as this limits post-switching steady-state energization overvoltages in the circuit (especially at the open-end of the circuit due to the Ferranti effect) and minimizes the capacitive current through the line breaker during energization. These benefits are not achieved when shunt reactors are connected to busbars or to tertiary windings of power transformers at substations. However, for some cable lengths, shunt reactors can be connected to busbars or to tertiary windings if energization overvoltages and capacitive currents are not very large. It was also concluded that shunt reactors should be energized at the same time or with a short delay (depending on the cable length) after the cable energization to limit the capacitive current, reactive power injection to the system, and steady-state overvoltages.
3. Furthermore, a distributed compensation along the circuit can decrease the minimum required size of compensation in comparison to the traditional type of compensation at the two remote ends of the circuit. This is due to the fact that the distributed compensation provides a very effective voltage control along the circuit (especially in circuits with long cables), which leads to the need for a smaller compensation size. Another advantage of the distributed compensation is the increase of the active power transfer capacity of the cable, which is needed for very long cables when the cable has almost reached its critical length. In addition, the distributed compensation is more reliable in the case of failure in one of the reactor banks because the rest of the reactors remain in operation. To summarize, it is necessary to apply a distributed compensation for very long cables (and recommended for long cables) to achieve the mentioned advantages.

8.6.4 Section 7.4: Resonance behavior of hybrid OHL-cable systems

The resonance behavior of hybrid OHL-cable systems is a very important technical challenge due to the higher risk of low-order resonance overvoltages. In fact, the application of long cables can significantly decrease the resonance frequencies to low-order ranges.

1. According to the simulation results, the risk of resonance frequencies close to the second and third harmonics is substantially higher when long cables are operated in the grid. In addition, the resonance impedance magnitudes and the number of resonance frequencies increase when cables are applied. Another concern is the amplification of the existing background harmonics above the accepted limits, which can cause overvoltages and power quality issues.
2. Temporary overvoltages produced by a resonance in a cable system are weakly damped and last long, where such sustained resonance overvoltages can impose hazardous operating conditions for system components with the risk of failure. Surge arresters are one of the most vulnerable components since they are not intended to mitigate TOVs and therefore their energy absorption capabilities can be exceeded.
3. From the resonance point of view and for future cable projects, it is advised to investigate all available possibilities for the number, locations, and lengths of the cable and OHL sections as well as the size and location of the shunt reactors. It was shown that these parameters have impact on the resonance behavior of the grid. The short-circuit power has also a direct impact on the order of the resonance frequency, where a higher short-circuit power leads to a higher first resonance frequency. In near future, a short-circuit power decrease is expected due to the increasing share of renewable power generation and the decreasing share of conventional generation, which will result in lower resonance frequencies.

8.6.5 Section 7.5: Energization overvoltages in hybrid OHL-cable systems

Energization overvoltages in HV systems can reach very high values and are considered among the severest overvoltages stressing the insulation of system components. Therefore, the significance of these overvoltages should be precisely determined for the insulation coordination studies. It has been strongly advised in the literature and by IEC standards that the insulation coordination study should be based on a statistical simulation to derive and assess the statistical distribution of energization overvoltages and their significance.

1. By means of extensive statistical simulations, it was found that a hybrid OHL-cable circuit with a given transmission length produces energization overvoltages higher than those of a fully cable circuit and most likely lower than those of a fully-OHL circuit with the same transmission lengths. This means that the probability of experiencing high energization overvoltages and stressing system components is higher in a fully-OHL circuit than in a partial or fully undergrounded circuit.
2. Energization overvoltages are also very dependent on the number, locations, and lengths of the OHL and cable sections because of the consequent changes

in the energizing wave propagation. For a given cable scenario, the existence of long cable sections at the open-end of the circuit results in lower energization overvoltages compared to the case that short cable sections are at the open-end. The cable energization with trapped charges on the cable can generate even higher overvoltages on both the core and sheath conductors and increase the risk of equipment failure.

3. Furthermore, energization overvoltages can be substantially affected by the system parameters, most notably the short-circuit power of the feeding network and the shunt compensation size and location. The transient energization overvoltages will be likely lower when the network short-circuit power is lower (i.e., weaker supply network), whereas the post-switching steady-state overvoltages will be higher due to the Ferranti effect at the open-end of the circuit. Moreover, a distributed reactive power compensation along the circuit, like cable-end compensation, results in lower transient overvoltages compared to the line-end compensation.
4. It should be again noted that extremely high overvoltages (higher than 3 pu) may occur in hybrid OHL-cable circuits for specific cases like the presence of trapped charges on the cable, the existence of very short-cable sections at the open-end of the circuit, and certain short–circuit powers. A special attention should be paid to these cases as they can lead to a component failure and/or exceeding the energy absorption capacity of surge arresters and sheath voltage limiters.
5. Regarding the energization transients, it is worth noting that a particular attention should be given to high inrush currents due to the back-to-back energization of cable circuits (parallel cables energization). This is a similar phenomenon to the back-to-back capacitor banks switching, although the amplitude, frequency, and the rate-of-rise of the cable inrush current are normally lower than those of an equivalent capacitor bank due to the relatively higher surge impedance of cables. However, in some cases, the cable inrush current can obtain very large values, where mitigation measures should be considered when the circuit breaker ratings are exceeded.

8.6.6 Section 7.6: De-energization transients of hybrid OHL-cable systems

Transient voltages and currents are also initiated when a circuit breaker is being opened to de-energize a circuit or a component. De-energization transients are associated with several transient phenomena, which may become more severe when the circuit is fully or partially undergrounded with cables. In this regard, the voltages on the disconnected phases and the transient recovery voltages (TRV) across the contacts of circuit breakers should be assessed.

1. It was found that when both the cable and shunt reactors are de-energized, the cable behavior is very similar to that of a parallel RC circuit. This means that the trapped voltage on the disconnected cable is a decaying DC-voltage, which discharges through the insulation resistance. The required time for a complete discharge of a long cable is very long (up to tens of seconds or minutes

depending on the cable length) and it becomes even longer with the increasing cable length since the time constant of the circuit increases. When inductive voltage transformers (VTs) are present, the cables discharge is much faster due to the quick saturation of VTs.

2. When a hybrid OHL-cable circuit is de-energized while shunt reactors have been left connected to the phases, the residual voltages on the disconnected phases are decaying oscillatory AC-voltages with superimposed frequencies instead of decaying DC-voltages. In this situation, the cable discharge is much faster compared to the discharge of a stand-alone cable. However, resonance overvoltages (or ferroresonance when reactors have a magnetic core) may occur in the disconnected phase(s) when a capacitive coupling exists with the other energized phases. In fact, the induced power-frequency voltage in the disconnected phase(s) by the capacitive coupling with the energized phase(s) can trigger the resonance circuit formed by the reactor inductance, the inter-phase/inter-circuit capacitance, and the cable capacitance. It is always advised to install separate breakers for the shunt reactors that are directly connected to the circuit. This results in an added switching flexibility as well as minimizing the risk of line resonance/ferroresonance by decoupling the reactors from the disconnected phase(s).
3. In the simulated scenarios, high transient recovery voltages across the terminals of the line circuit breaker were observed, especially for the scenarios with a shorter cable. In fact, the peak transient recovery voltage is lower due to the smaller capacitive coupling between the phases when the cable share in the circuit is higher. High transient recovery voltages increase the probability of the circuit breaker restrike due to the breakdown of the dielectric between breaker contacts. Restrike overvoltages can be very dangerous and harmful for power system components and therefore breakers which have to perform capacitive current switching should have a low or very low probability of restrike.

8.6.7 Section 7.7: Zero-missing phenomenon in cable systems

The simultaneous energization of the cable and reactors may cause the zero-missing phenomenon, which means that the current through the line breaker does not cross the zero value for several cycles (even up to several seconds). In this case, it is difficult or even impossible to safely open the healthy phases if a fault occurs in the circuit during the energization when the zero-missing current is still present. Therefore, the system is more vulnerable and unprotected against faults when proper countermeasures are not devised at the design stage of the cable project. Several countermeasures are available to prevent, mitigate, or handle the zero-missing phenomenon, where the effectivity of each countermeasure is strongly dependent on the system parameters and topology like the configuration of the hybrid OHL-cable circuit, the power-frequency voltage, and the short-circuit power.

1. Six countermeasures of the zero-missing phenomenon were introduced and analyzed, where the applied methodology and findings are applicable on any

grid since the study is based on realistic assumptions for the system parameters and topology. It should be stressed that the influence of system topology and parameters (like the configuration of hybrid OHL-cable circuit, power-frequency voltage, short-circuit power, etc.) on the effectivity of a countermeasure should always be considered. Some of the countermeasures need specially designed circuit breakers, control strategies, and resistances, which can impose extra costs and complexity on the system design and operation.

2. The simultaneous energization of the cable and reactors at voltage peak is a common prevention countermeasure; however, the resulting transient overvoltages and inrush currents should be carefully analyzed. The impact of stochastic variations in the breaker closing time as well as the breaker compliance with the breaker closing variation threshold (BCVT) should be investigated. To decrease the transient overvoltages and inrush currents, energization of the cable and reactors in sequence, as a prevention countermeasure, can be applied. A voltage dip/swell is the main risk of applying this countermeasure for long cable lengths.
3. Other countermeasures are sequential switching (handling countermeasure), opening of the faulty phase(s) (handling countermeasure), and increasing the DC-offset damping by resistors (mitigation countermeasure). These methods result in lower transient overvoltages and inrush currents. The DC-offset damping by resistors is a costly but relatively robust measure for which the thermal design of the resistor is a challenge.

Index

www.ingramcontent.com/pod-product-compliance
Lightning Source LLC
La Vergne TN
LVHW051047210926
848486LV00001B/43